Optoelektronik in der Technik
Optoelectronics in Engineering

Vorträge des 6. Internationalen Kongresses
Proceedings of the 6th International Congress

Laser 83 Optoelektronik

Herausgegeben von / Edited by W. Waidelich

Mit 537 Abbildungen

Springer-Verlag
Berlin Heidelberg New York Tokyo 1984

Dr. rer. nat. Wilhelm Waidelich

o. Professor, Vorstand des Instituts für Medizinische Optik der Universität München
Leiter der Abteilung Angewandte Optik der Gesellschaft für Strahlen- und
Umweltforschung, Neuherberg

CIP-Kurztitelaufnahme der Deutschen Bibliothek.
Optoelektronik in der Technik: Vorträge d. 6. Internat. Kongresses Laser 83 =
Optoelectronics in engineering / Hrsg.: W. Waidelich. –
Berlin ; Heidelberg ; New York ; Tokyo : Springer. Teilw. mit d. Verlagsorten Berlin, Heidelberg,
New York. NE: Laser ; PT. 6. 1983 (1984).

ISBN-13: 978-3-540-12779-6   e-ISBN-13: 978-3-642-82121-9
DOI: 10.1007/978-3-642-82121-9

2362/3020—543210

Vorwort

"LASER OPTOELEKTRONIK", der von der Münchener Messe und Ausstellungsgesellschaft mbH seit 1973 in zweijährigem Turnus abgehaltene Kongress mit Fachmesse, hat sich zur bedeutendsten internationalen Veranstaltung seiner Art entwickelt. Fachkonferenz und Fachmesse ermöglichen einen umfassenden Überblick über den neuesten Stand dieses aufstrebenden Gebietes.

Die auf dem Kongress "LASER OPTOELEKTRONIK" im Sommer 1983 gehaltenen Vorträge werden wieder in zwei Bänden ("OPTOELEKTRONIK IN DER TECHNIK" und "OPTOELEKTRONIK IN DER MEDIZIN") publiziert und so einem breiteren Interessentenkreis zugänglich gemacht.

Der hiermit vorgelegte Band "OPTOELEKTRONIK IN DER TECHNIK" enthält Übersichts- und Fachreferate aus den Themenbereichen Lasersysteme für Forschung, optoelektronische Komponenten, optronisches und lasertechnisches Messen und Prüfen, Laser in der Materialbearbeitung, optoelektronische Signalübertragung, Laser und Optoelektronik in der Weltraumtechnik, Laser in der Umweltmesstechnik, optoelektronische Displays, Laser-Chemie, photovoltaische Solartechnik, industrielle Umsetzung der Lasertechnologie.

Die Verbindung optischer und elektronischer Komponenten hat sich bei der Lösung vieler Aufgaben als äusserst fruchtbar erwiesen und technische Möglichkeiten eröffnet, die es ohne diese Kombination nicht geben würde. Der Kongressband enthält aktuelle Informationen und Ergebnisse aus diesem Fachgebiet und soll zu Technologietransfer, zu neuen Anwendungen sowie zu weiteren Entwicklungen anregen.

Für das Zustandekommen sei den Autoren und erneut dem Springer Verlag herzlich gedankt.

München, im Dezember 1983   W. Waidelich

Preface

LASER OPTOELECTRONICS, the Congress and Trade Fair held by the Münchener Messe-
und Ausstellungsgesellschaft mbH every two years since 1973, has developed into
the leading international event of its kind. Both the Congress itself and the
Trade Fair provide a detailed and comprehensive insight into the latest state of
the art in this up-and-coming field of technology.

The papers presented at the LASER OPTOELECTRONICS Congress in summer 1983 will
again be published in two volumes (OPTOELECTRONICS IN TECHNOLOGY and OPTOELECTRONICS
IN MEDICINE) and thus made available to a large group of readers.

The present volume OPTOELECTRONICS IN TECHNOLOGY contains general and technical
papers covering subjects such as laser systems in research, optoelectronic
components, optronic and laser-based measuring in testing, laser technologies in
material processing, optoelectronic signal transmission, laser and optoelectronics
in space technologies, lasers in environmental metrology, optoelectronic displays,
laser chemistry, photovoltaic solar technologies, and the use of laser technologies
in industry.

The combination of optical and electronic components has proved to be extremely
useful for solving numerous problems and tasks. This combination therefore provides
unprecedented technical options and opportunities. In presenting the latest
information and results in this area of research, the present volume is intended
to promote a transfer of technology, new applications and further developments
in the years to come.

We would like to express our sincere gratitude to the authors and the Springer
Publishing House for making this publication possible.

Munich, December 1983   W. Waidelich

Inhaltsverzeichnis – Contents

Sitzungsleiter – Session Chairmen XIX

Referenten – Contributors XXI

LASER SYSTEME FÜR FORSCHUNG
LASER SYSTEMS FOR RESEARCH WORK

C. K. Rhodes
Vacuum Ultraviolet and Extreme Ultraviolet Generation with Excimer Lasers 3

A. Luches, V. Nassisi, M. R. Perrone
Experimental Studies of a KrCl Discharge Laser 9

H. J. Eichler, H. Koch, R. Tornow
Extending the Lifetime of CuII Hollow Cathode Lasers 14

E. Endruschat, J. Hamisch, H. Koch, N. Niedrig, W. Tornow
UV-CuII-Laser Induced Electron Beam Generation with High Specific Brightness . 19

S. Donati, G. Martini
Frequency Stabilization of He-Ne Commercial Tubes:
A Comparison of Various Techniques 24

P. Lokai, K. Hohla, D. Basting
Raman Conversion of Dye Lasers: A Coherent High Power
Light Source in the VUV and the IR 31

B. J. Stocker
Q-switching and Cavity Dumping of CO_2 Waveguide Lasers 36

H. J. A. Bluyssen, A. F. van Etteger
EQ-switched and Cavity Dumped Tunable CO_2 Laser and the Generation
of Very Short Optically Pumped Far Infrared Laser Pulses 40

J. Badziak, M. Borzecki, A. Chojnacka, Z. Dźwigalski,
A. Kalbarczyk, Z. Kurzyński, L. Perliński, J. Teter
Dual-Beam Large Aperture CO_2 Amplifier 45

K. Kitagawa
A Study of Gas-Throat Mixing Gasdynamic Laser Driven by Combustion of
C_6H_6 and N_2O 50

VIII

O P T O E L E K T R O N I S C H E   K O M P O N E N T E N
U N D   S E N S O R E N

O P T O E L E C T R O N I C   C O M P O N E N T S
A N D   S E N S O R S

K. Berchtold, J. Angerstein
Optoelektronische Komponenten und Sensoren
Optoelectronic Components and Sensors 55

B. Leskovar
Recent Advances in High-Speed Photon Detectors 68

K. D. Stock, K. Möstl
Ge-Photoelemente für radiometrische Messungen im nahen Infrarot
Ge-Photoelements for Radiometric Measurements in the Near Infrared 75

H. Harjunmaa
Si:In Infrared Detectors at Low Temperatures: a New Mode of Operation 80

B. Aulbach, W. Pfeiffer
Digitale Restlichtkamera im Tieftemperaturbetrieb
Digital Low-Light-Level Camera for Low-Temperature Operation 85

H. Gündel, W. Bohmeyer, W. Kabel, H. Volkmann
*Pyroelektrische Polymerfolien als Sensoren zur Leistungsmessung
von Impulsgaslasern*
Pyroelectric Polymerized Foils Used as Sensors for Power
Measuring of Pulsed Gas Lasers 90

O P T R O N I S C H E S   U N D   L A S E R T E C H N I S C H E S
M E S S E N   U N D   P R Ü F E N

L A S E R S   A N D   O P T O E L E C T R O N I C S
I N   M E A S U R I N G   A N D   T E S T I N G

H. Rottenkolber
Optische Prüfverfahren auf dem Vormarsch
Optical Test Techniques Moving Forward 97

OPTRONISCHES MESSEN UND PRÜFEN

OPTRONIC MEASURING AND TESTING

M. Prammer, R. Horak
Echtzeit-Bildverarbeitung im industriellen Einsatz
Real Time Image Processing for Industrial Applications 108

W. Eikhorst, S. Boseck
*Erkennen metallischer Gefügestrukturen durch Korrelation mit
artifiziellen Mustern an einer lichtoptischen Diffraktionsanlage*
Evaluating Metal Surface Properties by Correlation with
Artificial Pattern Using Light Optical Defraction 113

Th. Kreis, H. Kreitlow, W. Jüptner, G. Sepold
Sensorsystem auf der Basis eines Laserscanners
zur Kantenfindung in der Materialbearbeitung
Sensor System Based on a Laser Scanner for
Edge Detection in Material Processing 118

St. P. Buchanan
Synthetic Horizontal Interlace, a Technique for Doubling
the Horizontal Resolution of Imaging Automatic Trackers 124

K. J. Schell, J. Tijhof
Ink Layer Thickness Monitor (I.T.M.O.) 127

Ch. Buxbaum
Berührungslose und zerstörungsfreie Messung der thermischen Abstrahlung/
Absorption und der Rauheit der technischen Oberflächen mittels Lichtleiter
Contactless and Non-destructive Measurement of Thermal Radiation (or Absorption)
as well as Surface Roughness Using Optical Fiber Probes 133

D. Wittmer, W. Pfeiffer
Örtlich und zeitlich aufgelöste Aufnahme und Auswertung
schwacher transienter Leuchterscheinungen
Spatially and Temporally Resolved Measurement and
Evaluation of Weak Transient Luminous Events 137

LASERTECHNISCHES MESSEN UND PRÜFEN

LASERS IN MEASURING AND TESTING

A. F. Fercher, H. Z. Hu
Two-Wavelength Heterodyne Interferometry 142

H. Zorc
Laser Densitometry of High Optical Densities 147

R. E. Kunz
Self-calibrating Automatic Laser Focus Control System 150

R. Leßing
Farbige Weißlichtholographie
Coloured White Light Holography 155

H. Kreitlow, Th. Kreis, W. Jüptner
Optimierung der automatisierten Auswertung von Specklegrammen
beim Einsatz eines schnellen Fouriertransformators
Optimization of Automated Specklegram Evaluation
Using a Fast Fourier Transformer 159

B. Ruth, D. Haina, W. Waidelich
Ein Speckle-Verfahren zur Schwingungsanalyse
Analysis of In-Plane Vibrations Using a Speckle Method 165

A.-A. Beeck, B. Garbe, W. Geier, U. Hunsche, J. Plischke
Erfassung von Deformationen an einem untertägigen Salzpfeiler
mit Hilfe kohärentoptischer Meßverfahren
In-situ Deformation Measurements of a Rock-Salt Pillar
Using Methods of Coherent Optics 170

X

HOLOGRAPHISCHE INTERFEROMETRIE

HOLOGRAPHIC INTERFEROMETRY

B. Breuckmann, M. Schiller, W. Thieme
Vergleichende Holographie
Comparative Holography 182

A. Krüger, H.-A. Crostack
*Abbildung von akustischen Platten- und Volumenwellen mit Hilfe
der holographischen Interferometrie zur verbesserten Fehlerbeschreibung
bei der Ultraschallprüfung*
Imaging of Acoustic Lamb and Volume Waves by Holographical Interferometry
in order to Improve Defect Description in Ultrasonic Testing 186

E. Müller, V. Hrdliczka
*Kompensation von Bildmodifikationen und dreidimensionale
Deformationsberechnung aus Doppelpulshologrammen*
Determination of Image Modifications and Three-Dimensional
Displacements by Double Pulsed Holograms 193

G. Schönebeck
*Die Holographie, ein Hilfsmittel zur genaueren Bestimmung
von Verformungen und Dehnungen an Bauteilen im Vergleich zu
Berechnungsmethoden der Festigkeits- und Schwingungslehre*
The Holography, a Mean to Determine the Exact Deformation and
Strain of Structures in Comparison with the Methods of the
Stress and Vibrating Calculation 198

J. Geldmacher, H. Kreitlow, E. Vogt, B. Dirr
*Holographische Schwingungsanalyse an rotierenden Bauteilen mit Hilfe
eines objektlagen- und schwingungsphasenbezogenen Lasertriggerverfahrens*
Holographic Vibration Analysis of Rotating Objects
by Means of Double-Pulse-Laser Triggering Related
to Object Position and Vibration Phase 203

J. J. Timkó
*Angewandte Holographie bei den Zweiphasenströmungen
der chemischen Verfahrenstechnik*
Applied Holography by Two Phase Flow in Chemical Engineering 208

R. Feiertag, M. Trundt
*Echtzeitholographie mit dem Dauerstrichlaser
bei Schwingungsuntersuchungen*
Real-time Holography of Vibrations by CW-Laser 212

W. Jüptner, J. Geldmacher, G. Sepold, D. Plathner
*Messung der Verformungen durch Schwerkrafteinfluß an Großbauteilen
mit einem speziellen holographischen Interferometer*
Measurement of Object Deformation Caused by Gravity
Using a Specially Designed Holographic Interferometer 217

LASER IN DER MATERIALBEARBEITUNG
LASERS IN MATERIAL PROCESSING

J. P. Carstens
The Future of High-Power Laser Welding 225

CO$_2$-LASER: SYSTEME UND DIAGNOSTIK
CO$_2$-LASERS: SYSTEMS AND DIAGNOSTICS

L. Bakowsky, P. Loosen, E. Beyer, G. Herziger
Criteria to Match Future CO$_2$-Laser Systems to Material Processing 237

M. Sharp, P. Henry, W. M. Steen, G. C. Lim
An Analysis of the Effect of Mode Structure
on Laser Material Processing 243

P. Loosen, L. Bakowsky, G. Herziger, F. Rühl
Diagnostic of High Power CO$_2$-Lasers 247

R. Rothe, P. Steinlein, G. Sepold
Meßgerät zur On-Line-Kontrolle von Laserstrahlen
bei der Werkstoffbearbeitung
On-Line Control Measuring of Laser Beams
during Material Processing 254

E. Beyer, A. Donges, P. Loosen, G. Herziger
Optical Feedback during Laser Material Processing 259

P. Arnold
Ein Gerät zur Analyse der Intensitätsverteilung im
Laserstrahl während der Materialbearbeitung
An Apparatus for Measuring the Intensity Profile of
a Laser Beam during Material Processing 264

K. Kakizaki, N. Nishida, T. Takahashi
Multi-Kilowatt CO$_2$ Lasers for Material Processing 270

A. S. Kaye
Catalytic Recombination in a Multi-Kilowatt CO$_2$ Laser 275

Li Zaiguang, Li Shimin, Liu Donghua, Han Nianseng, Li Feng,
Li Jiarong, Wang Hansheng, Chen Zhuhai, Qiu Junlin
Untersuchung der Entladungsstabilität von 2KW CW CO$_2$ Lasern
mit transverser elektrischer Anregung
Investigations on Discharge Stabilities of Transversally
Excited 2 kW CW CO$_2$-Lasers 280

W. Abel, D. Schuöcker, B. Walter
1,2 KW Gastransportlaser mit ungefaltetem Resonator
1,2 kW Gas Flow Laser with Unfolded Resonator 285

XII

CO_2-LASER: ANWENDUNG

CO_2-LASERS: APPLICATION

A. Gukelberger
Einige praktische Beipiele über Einsatzmöglichkeiten von
CO_2-Hochleistungslasern beim Schweissen und Härten
Several Typical Applications of High Power CO_2-Lasers
for Welding and Hardening 289

G. Sepold
Beitrag zum Schneiden mit CO_2-Lasern
Cutting with High Power CO_2-Lasers, State of the Art Report 294

P. Wirth, I. Sarady, K. Nilsson
Neueste Trends beim Schneiden und Schweissen mit CO_2-Lasern
in der metallverarbeitenden Industrie
Latest Trends in Cutting and Welding with CO_2-Lasers
in Metal Manufacturing 300

J. Sellner
Das Schneiden von metallischen und nichtmetallischen
Werkstoffen mit einem 1,2 KW CW Laser
Cutting of Metallic and Non-Metallic Materials
with a 1,2 kW CW-Laser 305

K. Wissenbach, L. Bakowsky, H.-G. Treusch, G. Herziger
Laser Hardening Process Parameters 312

R. Becker, G. Sepold, R. Chatterjee-Fischer
Aspekte des Laserstrahlhärtens
Aspects of Laser-Hardening 317

A. Walker, D.R.F. West, W. M. Steen
Laser Surface Alloying of Ferrous Materials with Carbon 322

V. M. Weerasinghe, W. M. Steen
Laser Cladding 327

A. K. Al-Sufi, H. J. Eichler, I. Ostwaldt, J. Salk
Metallabscheidung aus Elektrolyten mit Lasern
Laser Induced Metal Deposition from Electrolytic Solutions 332

H. W. Bergmann, B. L. Mordike, T. Bell
Laser Melting of Cast Irons 335

H. U. Fritsch, G. J. Barton
Zusammenhänge zwischen Eigenschaften und Versuchsparametern
beim Laseroberflächenumschmelzen von Eisenwerkstoffen
Relationship between Properties and Process Parameters in
Laser Surface Melting of Iron Base Alloys 350

ABSORPTION, GASDURCHBRUCH, SCHOCKWELLEN

ABSORPTION, GAS BREAKDOWN, SHOCK WAVES

E. W. Kreutz, E. Beyer, H. G. Treusch, W. Zimmer
Conductivity Measurements on Absorption of Semiconductors
under Intense Laser Excitation 355

R. Poprawe, E. Beyer, L. Bakowsky, G. Brumme, G. Herziger
Limitation of Laser Processing Intensity by Laser Induced Gas Breakdown 361

E. Beyer, L. Bakowsky, R. Poprawe, G. Herziger
*Formation and Influence of Laser Induced Plasma
during CO_2-Laser Welding* 367

J. Munschau
Erzeugung und technische Anwendungen laserinduzierter Stoßwellen
Generation and Technical Applications of Laser Induced Shock Waves 373

Nd-YAG-LASER: SYSTEME UND ANWENDUNG

Nd-YAG-LASERS: SYSTEMS AND APPLICATION

E. Heumann, J. Kleinschmidt, K. Rühle, K. Vogler, G. Wiederhold, W. Zschocke
*Verfahren zur Strukturierung von Materialoberflächen
mittels Laserstrahlung*
Methods for the Structuring of Material Surfaces by Laser Radiation 379

H.-G. Treusch, L. Bakowsky, K. Wissenbach, G. Herziger
Metal Drilling with Lasers 383

J. Junghans, E. Huber
*Variable Dämpfung von Laserstrahlung hoher Intensität
bei Präzisionsbearbeitungen*
Variable Attenuation of High Intensity Laser Radiation
Used for Precision Materials Processing 389

J. Junghans
*Industrielle Einsatzmöglichkeiten von Lasermaterial-
bearbeitungsmaschinen mit periodisch gepulsten Lasern*
Industrial Applications of Laser Materials Processing
Machines with Periodically Pulsed Lasers 394

H. P. Schwob, U. Gruetzner
YAG-Laser-Bearbeitungszentrum
YAG-Laser Machine Center 399

E. Heumann, S. I. Schastak
*Festkörper-Lasersysteme mit annähernd rechteckförmigem
Intensitätsprofil durch kombinierte Anwendung von nichtlinearen
Absorptionsprozessen und Phasenkonjugation*
Solid State Laser Systems with Nearly Rectangular Intensity
Profile Formed by Combination of Nonlinear Absorption
Processes and Phase Conjugation 404

Z. Drozd, L. Boruc
Qualitätsaspekte der Laserbearbeitung von Dünnschichten
Quality Aspects of Thinfilm Laser Processing 410

J. Buchholz
*Laserfeinbearbeitung in der Mikroelektronik zur Bearbeitung
von Masken und integrierten Schaltungen*
Lasers in Microelectronics for Mask Repair and Cutting on IC and LSI 415

XIV

OPTOELEKTRONISCHE SIGNALÜBERTRAGUNG
OPTOELECTRONIC SIGNAL TRANSMISSION

M. Börner
Die zukünftige Entwicklung der optischen Nachrichtentechnik
Future Developments of Optical Communications 425

A. Naab
Einsatz der Glasfasertechnik im Netz der deutschen Bundespost
mit Schwerpunkt im Ortsnetz
Application of Fibre Optics in the Telecommunication Network
of the Deutsche Bundespost, especially in the Local Area Network 435

A.-E. Schurr, R. Deufel
Kapazitätsarme PIN-Photodioden für 1,1 bis 1,55 μm
PIN-Photodiodes with Low Capacitance for 1.1 to 1.55 μm 439

G. Arnold, P. Marschall, E. Schlosser
1,3 μm V-Nutlaser als Sender für optische Nachrichtenübertragungssysteme
1.3 μm V-Groove-Laser, a Source for Optical Communication Systems 443

J. W. M. Biesterbos, G. A. Acket
Optical Feedback Noise Characteristics of Semiconductor Lasers 447

B. Hillerich, M. Rode
Wellenlängenmultiplexer und -demultiplexer mit Beugungsgittern
Wavelength Multiplexer Using Preferentially Etched Silicon Gratings 449

C. Le Sergent
Mid Infrared Waveguides 453

LASER UND OPTOELEKTRONIK
IN DER WELTRAUMTECHNIK
LASERS AND OPTOELECTRONICS
IN SPACE TECHNOLOGY

E. D. Hinkley
Lasers and Advanced Optics in Space 457

W. Englisch, M. Endemann, W. Diehl, W. Wiesemann
CO_2-Laser-Sende-Empfänger-System mit Gigabit/s-übertragungskapazität
Stand der Technik
CO_2-Laser Transceiver System with Gbit/s Capability for
Space-to-Space Communications – State of Technology 470

W. Wiesemann, W. Diehl, M. Rother, M. Endemann, W. Englisch
Raumflugtauglicher CO_2-Laser
Space Qualifiable CO_2-Laser 476

B. Furch, E. Bratengeyer
Optimierung eines Mach-Zehnder-Wellenleitermodulators in $LiNbO_3$
Optimization of a $LiNbO_3$ Interferometric Waveguide Modulator 481

W. Englisch, M. Endemann
Conical Scan-Tracking-Technik mit kohärentem optischen Empfang bei 10 μm
Conical Scan Spatial Tracking Technique with Coherent Optical Detection at 10 μm 486

H. K. Philipp, E. Bonek, A. L. Scholtz, W. R. Leeb
Homodynempfang von amplituden- und phasenmodulierten Infrarotsignalen
Homodyne Reception of Amplitude Modulated and Phase Modulated Infrared Signals . 490

F. Malota, H. Herrmann, K. Kain
Testergebnisse eines CO_2-Laser-Heterodyn-Entfernungsmessers
Test Results of a CO_2-Laser-Heterodyne Rangefinder 495

A. Popescu, W. R. Leeb, B. Furch, A. L. Scholtz
Experimentelle Untersuchungen an Halbleiterlasern für
optische Nachrichtensysteme im Weltraum
Investigation of Semiconductor Lasers for Space Data Links 500

M. Endemann, W. Englisch
Spaceborne Lidars for Global Meteorological Measurements 505

LASER IN DER UMWELTMESSTECHNIK
LASERS IN ENVIRONMENTAL MEASURING TECHNIQUES

J. Werner, K. W. Rothe, H. Walther
Monitoring of the Stratospheric Ozone Layer Using a XeCl Excimer Laser 513

G. Cecchi, P. Mazzinghi, L. Pantani, C. Susini
LIDAR Investigation of Oil Films on Natural Waters 517

P. Bruscaglioni, G. Del Fante, A. Ismaelli, L. Lo Porto, G. Zaccanti
A Laser Transmissometer with Variable Receiver's Field of View
for Measurements of Forward Scattered Power in Dense Media 523

Ch. Werner, F. Bachstein, E. Gelbke
Erprobung eines augensicheren Laser-Schrägsichtmeßgerätes
auf dem Flughafen München-Riem
Test of an Eye-Safe Slant-Visual-Range Measuring Device
at Munich-Riem Airport 528

L. Pantani, I. Pippi, B. Radicati
Analysis of Two Different Techniques for Lidar Visibility Measurements 534

F. Köpp, R. L. Schwiesow, Ch. Werner
Remote Measurement of Wind Profiles Using the DFVLR Laser Doppler Anemometer ... 539

OPTOELEKTRONISCHE DISPLAYS
OPTOELECTRONIC DISPLAYS

B. Kazan
Opto-electronic Display Technologies, State-of-the-Art and Trends 547

LASER - CHEMIE
LASER CHEMISTRY

F. P. Schäfer
Preparative Organic Photochemistry with Lasers 561

H. J. Neusser
Multi-Photon Ionization and Laser Mass Spectrometry 571

W. Becker, S. Dähne, K. Junge, E. Klose
*Fluoreszenzspektroskopie hoher Zeitauflösung
und hoher Empfindlichkeit*
Fluorescence Spectroscopy of High Time Resolution
and High Sensitivity 582

F. Spieweck
Einfluß der Jodzellenreinheit auf die Frequenz I_2-stabilisierter Laser
Influence of the Iodine Cell Purity on the Frequency of I_2-Stabilized Lasers . 587

Cao Wen Zhuang, Wu Bao Ye
A New Method for Designing a Paraboloidal Reflector 593

M. Schneider, J. Wolfrum
UV-Laserinduzierte Chlorwasserstoffabspaltung aus Alkylhalogeniden
UV-Laserinduced Hydrogen Chloride Elimination from Alkylhalides 598

LASER SICHERHEIT
LASER SAFETY

R. Weiner
Status of Laser Safety Requirements 605

L. M. Cook, K.-H. Mader, N. J. Brown, G. R. Wirtenson
*Leached AR Surfaces for High Energy Optics:
Scale-Up and Performance* 611

PHOTOVOLTAISCHE SOLARTECHNIK
PHOTOVOLTAIC SOLAR TECHNOLOGY

O. Hintringer
Perspektiven der Photovoltaik
Future and Perspectives of Photovoltaics 619

F. Pfisterer, H. W. Schock
Hocheffiziente Dünnschichtsolarzellen-Systeme
High Efficient Thin Film Solar Cell Systems 631

A. Wagner
Lichtblitzgerät für Solarzellen-Kennlinien-Messungen
Flasher for Solar Cell Measurements 636

H. Kaase, J. Metzdorf
*Messung der spektralen und integralen Empfindlichkeit
von photovoltaischen Solarelementen*
Measurement of Spectral and Integral Responsivity
of Photovoltaic Solar Elements 641

VDI - Technologiezentrum :

INDUSTRIELLE UMSETZUNG

DER LASERTECHNOLOGIE

INDUSTRIAL TRANSFER

OF LASER TECHNOLOGY

W. R. F. Gosling
The Commercial Laser Market in Western Europe 649

G. K. Klauminzer
State of Laser Markets and Technology in the U. S. 654

K. Kakizaki
Laser Applications in Japanese Industry 658

A. Sona
R & D Programs and Industrial Application of Lasers in Italy 663

I. J. Spalding, A. C. Walker
Laser Technology in Great Britain 667

G. Rauscher
Lasertechnologie in der Bundesrepublik Deutschland
Laser Technology in the German Federal Republic 673

Sitzungsleiter – Session Chairmen

Dr. D. Basting
Lambda Physik GmbH & Co. KG, Göttingen

Laser-Chemie / Laser Sicherheit

Prof. Dr. K. Gürs
Battelle-Institut e.V., Frankfurt/Main

Laser und Optoelektronik in der Weltraumtechnik

Prof. Dr.-Ing. H. Herziger
Institut für Angewandte Physik
Technische Hochschule Darmstadt

Laser in der Materialbearbeitung

Dr. B. Hill
Philips GmbH, Forschungslab. Hamburg

Optoelektronische Bildaufnahme, Bild- und Datenaufzeichnung

Dr. O. Hintringer
Siemens AG, München

Optoelektronische Displays

Prof. Dr. K. L. Kompa
Max-Planck-Institut für Quantenoptik
Garching

Laser-Systeme für Forschung

Prof. Dr. F. Lanzl
Institut für Optoelektronik
DFVLR, Oberpfaffenhofen/Wessling

Optronisches und Lasertechnisches Messen und Prüfen

Dr. S. Maslowski
AEG-Telefunken, Kommunikationstechnik
Ulm

Optoelektronische Signalübertragung

Ing. E. Sommer
ELBEG Elektronik GmbH, München

Optoelektronische Komponenten und Sensoren

Photovoltaische Solartechnik

Dr. R. Stemme
VDI-Technologiezentrum, Berlin

Industrielle Umsetzung der Lasertechnologie

Prof. Dr. H. Tiziani
Institut für Technische Optik
Universität Stuttgart

Optronisches und Lasertechnisches Messen und Prüfen

Dr. J. Wagner
Bayerwerk Leverkusen

Laser-Chemie

Prof. Dr. W. Waidelich
Institut für Medizinische Optik
München

Wissenschaftliche Leitung

Prof. Dr. H. Weber
Fachbereich Physik
Universität Kaiserslautern

Laser in der Materialbearbeitung

Dipl.-Phys. Dr. Ch. Werner
Institut für Optoelektronik
DFVLR, Oberpfaffenhofen/Wessling

Laser in der Umweltmeßtechnik

Mader K.-H.
Marschall P.
Martini G.
Mazzinghi P.
Metzdorf J.
Möstl K.
Mordike B.L.
Müller E.
Munschau J.

Naab A.
Nassisi V.
Neusser H.J.
Niedrig H.
Nilsson K.
Nishida N.

Ostwaldt I.

Paddock J.P.
Pantani L.
Perlinski L.
Perrone M.R.
Pfeiffer W.
Pfisterer F.
Pippi I.
Plathner D.
Plischke J.
Poprawe R.
Prammer M.

Qui Junlin

Radicati B.
Rauscher G.
Rhodes C.K.
Rode M.
Rosen H.-G.
Rothe K.W.
Rothe R.
Rottenkolber h.
Rühl F.
Rühle K.
Russel Ph.B.
Ruth B.

Salk J.
Sarady I.
Sellner J.
Sepold G.
Sharp M.
Sona A.
Spalding I.J.
Spieweck F.
Susini C.
Schäfer F.P.
Schastak S.I.
Schell K.J.
Schiller M.
Schlosser E.
Schneider M.

Schock H.W.
Schönebeck G.
Schuöcker D.
Schwiesow R.L.
Schwob H.P.
Schurr A.-E.

Steen W.M.
Steinlein P.
Stock K.D.
Stocker B.J.

Takahashi T.
Teter J.
Thieme W.
Tijhof J.
Timkó J.J.
Tornow R.
Tornow W.
Treusch H.-G.
Trundt M.

Van Etteger A.F.
Vogler K.
Vogt E.
Volkmann H.

Wagner A.
Waidelich W.
Walker A.

Walker A.C.
Walter B.
Walther H.

Wang Hansheng
Weerasinghe V.M.
Wei Kuang-hui
Weiner R.Welge K.H.
Werner Ch.
Werner J.
West D.R.F.
Wiederhold G.
Wirtenson G.R.
Wirth P.
Wissenbach K.
Wittmer D.
Wolfrum J.
Wu Bao Ye

Zaccanti G.
Zimmer W.
Zorc H.
Zschocke W.

Laser Systeme für Forschung

Laser Systems for Research Work

Vacuum Ultraviolet and Extreme Ultraviolet Generation with Excimer Lasers

C. K. RHODES
Department of Physics, University of Illinois at Chicago
P.O. Box 4348, Chicago, Illinois 60680, U.S.A.

Abstract

High spectral brightness rare gas halogen (RGH) sources can be used to generate coherent extreme ultraviolet radiation by either harmonic generation mechanisms or direct multiquantum excitation of appropriate gain media. In order to demonstrate the basic characteristics of these two approaches, recent comparative measurements have been made. With the use of a 4 GW 193 nm (ArF*) system operating at a pulse duration of $\sim$ 10 ps, harmonic generation has been studied in several atomic and molecular media and used to generate $\sim$ 20 kW at 64.3 nm and $\sim$ 200 W at 38.6 nm. In addition, stimulated emission in molecular hydrogen, on both the Lyman and Werner bands excited by two quantum absorption at 193 nm, has resulted in the generation of radiation as short as 117.6 nm at an efficiency on conversion approaching one percent. It has been concluded that the latter method is superior for the generation of short wavelength radiation. Extension of these results to both shorter wavelengths and higher power levels requires an extended study of the basic character of high order nonlinear processes in the ultraviolet. Recent studies of collision-free multiply-charged ion production with irradiation at 193 nm point to an anomalously strong coupling to high Z materials, with processes involving as many as 99 quanta being observed. These findings strongly suggest that the direct excitation of inversions by appropriate multiquantum processes in the region below 100 nm in certain atomic systems can be generated with existing laser instrumentation. This report discusses (1) the properties of stimulated emission in H_2, (2) the findings arising from the studies of multiply-charged ion production, and (3) the observation of intense, tunable stimulated emission at 93 nm with 193 nm radiation from inner-shell excited $4s4p^6n\ell$ configurations of krypton excited by a four quantum process.

Discussion

A. Stimulated emission in hydrogen

Intense vacuum ultraviolet stimulated emission in molecular hydrogen, on both the Lyman and Werner bands, following excitation by two quantum absorption at 193 nm on the $X\ ^1\Sigma_g^+ \rightarrow E,F\ ^1\Sigma_g^+$ transition, has been observed. The shortest wavelength seen in the stimulated spectrum was 117.6 nm corresponding to the $C\ ^1\Pi_u \rightarrow X\ ^1\Sigma_g^+$ (2-5) $Q(1)$ transition which appears to be inverted with a mechanism involving electron collisions.

4

The maximum energy observed in the strongest stimulated line was $\sim$ 100 μJ, a value corresponding to an energy conversion efficiency of $\sim$ 0.5%. The pulse duration of the stimulated emission is estimated from collisional data to be $\sim$ 10 ps, a figure indicating a maximum converted vacuum ultraviolet power of $\sim$ 10 MW.

Table I contains the transitions and corresponding wavelengths observed [1] and Fig. 1 illustrates the general pattern of emission observed.

B. <u>Collisionless multiquantum ionization of atoms</u>

Recent progress in the generation of coherent high intensity vacuum ultraviolet and extreme ultraviolet radiation has stimulated interest in the study of the production of highly stripped and excited ions by multiquantum processes [2,3]. We report herein results of experiments examining processes of the type

$$N\gamma + X \rightarrow X^{q+} + qe^{-} \tag{1}$$

for which observed values of N and q range as high as 99 and 10, respectively.

The experimental arrangement used to detect the production of highly ionized species consists of a double focussing electrostatic energy analyzer (Comstock) which, in the present experiment, is operated as a time-of-flight mass spectrometer (Fig. 2). The energy analyzer is positioned in a vacuum vessel which is continuously evacuated to a background pressure of $\sim 10^{-7}$ Torr. Materials to be investigated are introduced into the vacuum container at pressures of typically $\sim 10^{-6}$ Torr. The 193 nm ArF* laser [4] beam ($\sim$ 10 ps, $\sim$ 4 GW) is focussed by a f = 50 cm lens in front of the energy analyzer's entrance iris resulting in intensities of $\sim 10^{14}$ W/cm^2. Ions formed in the focal region are collected into the analyzer with an extraction field

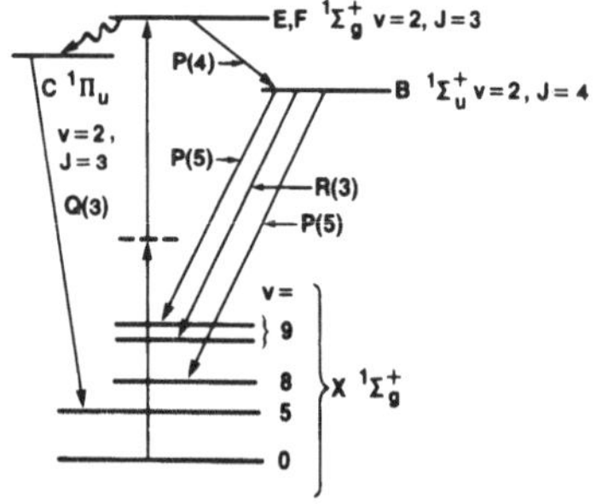

Wavelengths in nm:

E,F → B	B → X			C → X
2-2 P (4)	2-8 P(5)	2-9 P (5)	2-9 R (3)	2-5 Q (3)
922.2	153.2	158.5	157.1	117.8

Fig. 1. Stimulated emission from H$_2$ following E,F←X (2-0) Q(3) two-photon excitation with 2 ArF* (193 nm) quanta. For the E,F→C transition, see text.

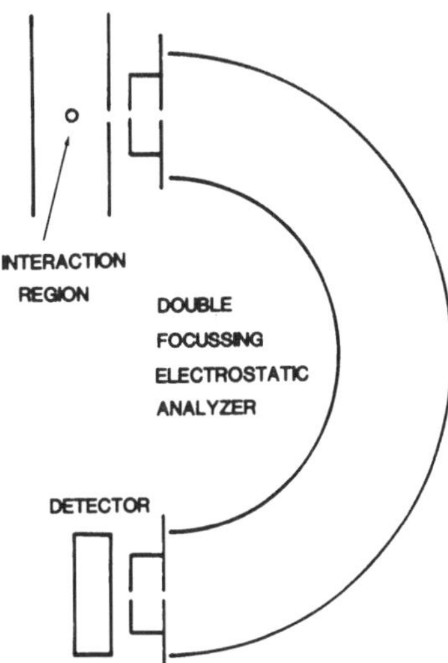

Fig. 2. Experimental system used in ion detection

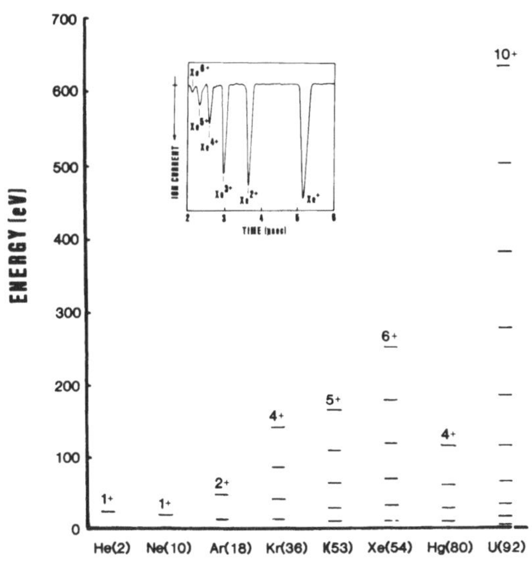

Fig. 3. Data concerning multiple ionization of atoms with 193 nm irradiation at $\sim 10^{14}$ W/cm^2. Plot of total ionization energies of the observed charge states as a function of atomic number (Z). Inset: Typical time-of-flight ion current signal for xenon

Table I. Transitions and corresponding wavelengths of the observed stimulated emission

Transition	Wavelength Å Present work	Wavelength Å Previous work [1,2]	Excited X→E,F transition	Transition	Wavelength Å Present work	Wavelength Å Previous work [1,2]	Excited X→E,F transition
E → B				E → B			
0–0 P(1)	11210	11159.1		2–1 P(3)	8370	8369.23	
				2–0 P(3)	7544.1	7544.06	
B → X							
0–3 R(0)	1275.2	1274.6		B → X			
0–3 P(2)	1280	1279.5	0→0	1–6 P(4)	1440.9	1440.7	0→2
0–4 R(0)	1333.6	1333.5	Q(0)	1–6 R(2)	1428.8	1428.9	Q(2)
0–4 P(2)	1339	1338.6		1–7 P(4)	1499.6	1499.6	
0–5 R(0)	1394.4	1393.7		1–7 R(2)	1487.6	1487.7	
0–5 P(2)	1399.8	1399		1–8 P(4)	1557.4	1557.6	
0–6 R(0)	1455.5	1454.9		1–8 R(2)	1545.4	1545.7	
0–6 P(2)	1460.9	1460.2					
				C → X			
				2–5 Q(2)	1176.3	1176.8	
E → B	Not measured			E → B			
				2–2 P(4)	9222	9222.04	
B → X							
1–6 P(3)	1435.7	1436.2		B → X			
1–7 R(1)	1486.1	1486.2	0→2	2–8 P(5)	1531.5	1532.1	0→2
1–7 P(3)	1494.7	1495.2	Q(1)	2–9 R(3)	1570.8	1571.4	Q(3)
1–8 R(1)	1544.4	1544.9		2–9 P(5)	1584.9	1585.5	
1–8 P(3)	1552.9	1553.4					
C → X				C → X			
2–5 Q(1)	1175.4	1175.8		2–5 Q(3)	1177.7	1178.3	

of 1000 V/cm and detected with a microchannel plate at the exit of the energy
analyzer. The observed ionic charge states together with a typical time-of-flight
spectrum for Xe are given in Fig. 3.

The experimental data display two unusual features. First, the magnitude of the
total energy which can be communicated to the atomic system is unexpectedly large,
especially for high Z materials. The total energy investment of $\sim$ 633 eV, a value
equivalent to 99 quanta, needed to generate U^{10+} from the neutral atom represents
the highest energy value reported for a collision-free nonlinear process. The
removal of the tenth electron from uranium, which requires $\sim$ 133 eV if viewed as an
independent process, requires a minimum of 21 quanta. Second, the ionic distributions do not fall off rapidly towards higher ionic states as one would expect if
stepwise multiphoton ionization were to dominate the coupling.

It is concluded that the conventional treatments of multiquantum ionization do not
correspond to our experimental findings for high-Z materials. Moreover, the data
strongly indicate that the shell structure of the atom is an important physical
property governing the strength of the coupling. In order to consolidate the observations into a single physical picture, a mode of interaction involving radiative
coupling to a collective motion of an atomic shell is proposed [5].

C. <u>Stimulated emission at 93 nm in multiquantum excited krypton</u>

Strong, stimulated emission in the extreme ultraviolet following multiphoton excitation of Kr using a 193-nm ArF* laser has been observed. The ArF* laser pulse [4],
with an output power of 1 GW, 10 ps duration, and 5 cm^{-1} bandwidth, was focussed
with a f = 50 cm lens into a differentially pumped cell similar to one used for
harmonic generation in the extreme ultraviolet [6]. The cell was attached to the
entrance slit of a 1 m VUV monochromator (McPherson 225) and the generated XUV
radiation was detected by an optical multichannel analyzer (OMA PAR).

With Kr pressures between 100 Torr and 1000 Torr, stimulated emission in krypton at
93 nm is observed. Significantly, this result experimentally establishes the selectivity of multiquantum processes for the excitation of atomic inner-shell states,
since the excited level $4s4p^6n\ell$ is populated by a four quantum process at 193 nm
from the ground state $4s^24p^6$ configuration [7]. Overall, the observations can be
understood by the reactions

$$4\gamma(193 \text{ nm}) + Kr(4s^24p^6) \rightarrow Kr(4s4p^6n\ell) \tag{2}$$

$$\gamma'(93 \text{ nm}) + Kr(4s4p^6n\ell) \rightarrow 2\gamma'(93 \text{ nm}) + Kr(4s^24p^5n\ell) \tag{3}$$

which illustrate the direct excitation step and the subsequent stimulated emission.
It appears that $(n\ell)$ is (4d) and (6s) in these experiments. The 93 nm radiation is
moderately tunable, since the autoionization rate of the upper $4s4p^6n\ell$ level confers
a substantial width ($\sim$ 100 cm^{-1}) on the bandwidth of the system [8]. The tunability

has been experimentally demonstrated in this case over a region of ~ 600 cm^{-1}, a fact which can be explained by the presence of a number of closely spaced lines. Incidentally, the selective promotion of an inner-shell electron in this example reinforces the conclusion derived from earlier studies concerning the influence of the atomic shell structure on the multiquantum coupling strength [5]. The maximum efficiency observed in the initial experiments for conversion to 93 nm from the excitation at 193 nm corresponds to $\sim 10^{-4}$. Latest results indicate that it may be possible to increase the efficiency by one to two orders of magnitude. To our knowledge, this system represents the first inner-shell transition laser and the shortest wavelength reported for stimulated emission.

Conclusions

On account of the unusual spectral brightness available from RGH media, particularly picosecond systems, it seems clear that major implications for short wavelength production are present. Given the scaling characteristics of RGH systems, at least the 1 TW range, the feasibility of a laboratory scale coherent source operating at kilovolt quantum energies is strongly implied.

Acknowledgements

The author wishes to acknowledge the expert technical assistance of M. J. Scaggs and J. R. Wright in addition to numerous fruitful discussions with K. Boyer, H. Egger, T. S. Luk, H. Pummer, M. Shahidi, and T. Srinivasan. This work was supported by the Office of Naval Research, the Air Force Office of Scientific Research under grant no. AFOSR-79-0130, the National Science Foundation under grant no. PHY81-16626, and the Avionics Laboratory, Air Force Wright Aeronautical Laboratories, Wright Patterson Air Force Base, Ohio.

References

1. H. Pummer, H. Egger, T. S. Luk, T. Srinivasan, C. K. Rhodes: "Vacuum ultraviolet stimulated emission from two-photon excited molecular hydrogen", Phys. Rev. A (in press)
2. A. L'Huillier, L. A. Lompre, G. Mainfray, C. Manus: Phys. Rev. Lett. **48**, 1814 (1982)
3. T. S. Luk, H. Pummer, K. Boyer, M. Shahidi, H. Egger, C. K. Rhodes: In AIP Conference Proceedings, Vol. 100, Excimer Lasers - 1983, ed. by H. Egger, H. Pummer, C. K. Rhodes (American Institute of Physics, New York, 1983)
4. H. Egger, T. S. Luk, K. Boyer, D. F. Muller, H. Pummer, T. Srinivasan, C. K. Rhodes: Appl. Phys. Lett. **41**, 1032 (1982)
5. T. S. Luk, H. Pummer, K. Boyer, M. Shahidi, H. Egger, C. K. Rhodes: "Anomalous collision-free multiple ionization of atoms with intense picosecond ultraviolet radiation", Phys. Rev. Lett. (in press)
6. H. Egger, R. T. Hawkins, J. Bokor, H. Pummer, M. Rothschild, C. K. Rhodes: Opt. Lett. **5**, 282 (1980)
7. K. Codling, R. P. Madden: J. Res. Natl. Bur. Std. **76A**, 1 (1972); K. Codling, R. P. Madden: Phys. Rev. A **4**, 2261 (1971)
8. D. L. Ederer: Phys. Rev. A **4**, 2263 (1971)

Experimental Studies of a KrCl Discharge Laser

A.Luches, V.Nassisi and M.R.Perrone

Università degli Studi, Dipartimento di Fisica, Lecce, Italy

INTRODUCTION

Rare-gas halide lasers are recognized as the most effective sources of coherent UV radiation. Over the past few years substantial efforts have been done in order to improve efficiency and scalability of these devices. At present the highest energies and efficiencies have been obtained from KrF and XeCl lasers. Few experimental data are available about KrCl discharge-pumped lasers (1). In fact, the low energy and efficiency obtained in early demonstrations of stimulated emission in KrCl determined the neglect of this laser. However, radiation at 222 nm would be of considerable interest for photochemical applications and for the annealing of semiconductor devices, due to the strong coupling of UV radiation with materials.

In a recent paper (2) we demonstrated that output energies comparable with those of KrF and XeCl can be obtained also with KrCl, provided that specific electrical powers about one order of magnitude higher than those required for KrF and XeCl are supplied to the lasing medium.

In the present paper we report the results of a careful optimization of the KrCl laser parameters. The output energy was investigated as a function of pump intensity, gas mixture composition and pressure. Till now, a maximum laser energy of 220 mJ (>3 J/1), with a power conversion efficiency of 0.9%, has been obtained from a 0.09% HCl/ 10.1% Kr/ 1.5% He/ 88.3% Ne mixture at a total pressure of 2,600 torr and at a specific power loading of about 23 MW/cc.

APPARATUS AND RESULTS

The laser used in these experiments employs an UV preionized transverse discharge (3). The electrodes are 35 cm long, 3.5 cm wide with a separation of 2 cm. The active discharge width is 1 cm, then the active volume is about 70 cc. The laser windows are made of UV grade fused silica. The cavity mirrors are mounted externally to the windows, to prevent contact with the corrosive gas mixtures. They consist of a flat aluminized silica reflector (R=80%) and different dielectric-coated output couplers with reflectivity between 10 and 80%. The distance between the mirrors is 60 cm.

The first purpose of our measurements was the optimization of the laser energy and efficiency as a function of gas mixture composition, total fill pressure and optical feedback.Research grade purity He, Ne and Kr were used. HCl was diluted in He at a concentration of 5%. The best results were obtained from an HCl/Kr/He/Ne mixture at the total pressure and concentration stated above. It was observed that the laser output peaks at a coupling-mirror reflectivity of about 20%.

Figure 1 shows the dependence of the output energy and peak power on HCl concentration. The laser output energy and pulse length increase with HCl partial pressure up to 2.4 torr.

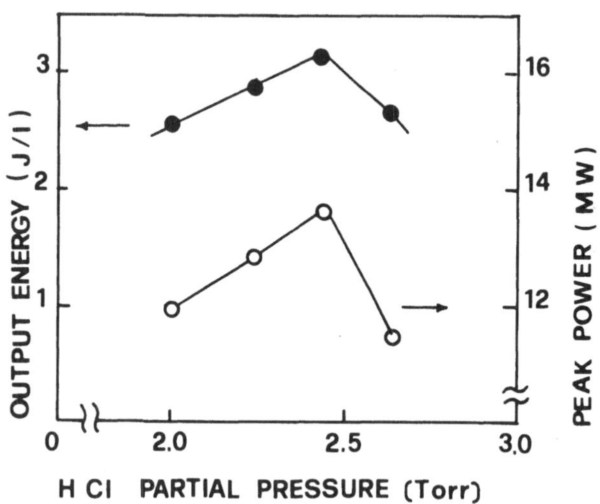

Fig. 1. Laser output energy vs HCl concentration at fixed partial pressures of Kr (270 torr) and Ne (2,300 torr).

The roll over in output energy at higher HCl concentrations could be due to absorption resulting in photodissociation of HCl and photodetachement of Cl^-. It would be possible to minimize the last effect by increasing the breakdown voltage, in order to reduce the dissociative attachment processes.

Figure 2 shows the effect of Kr concentration on laser output. The laser energy and pulselength increase with Kr pressure up to 270 torr. When He is used as buffer gas instead of Ne, the output energy decreases more than 30%. This behaviour, observed also by Sze (1), results from hotter electron energy distribution in Ne, which favours larger excitation rates to metastable states .

In all the investigated mixtures a linear variation of the output energy with total pressure was observed, up to the safe pressure limit of our device (about 2,600 torr).

For a given mixture and a fixed charging voltage, the laser pulse length (FWHM) Δt does not vary with total pressure.

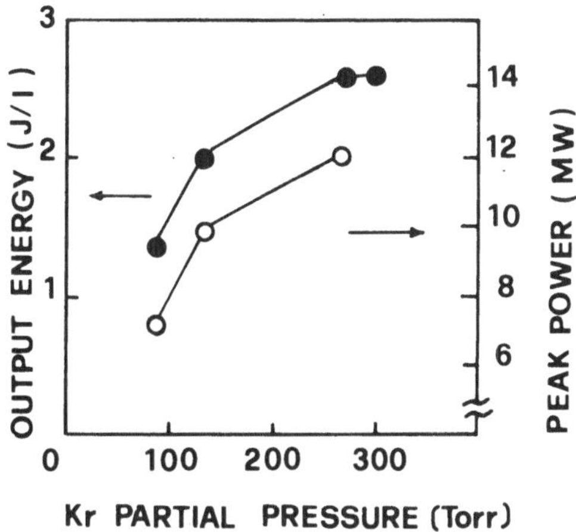

Fig. 2. Laser output vs Kr concentration, at fixed partial pressures of HCl (2 torr), He (40 torr) and Ne (2,300 torr).

To avoid inherent problems in handling corrosive and toxic HCl gas, the less aggressive BCl_3 and CCl_4 were tested as chlorine carriers. Using not completely optimized mixtures (230 torr Kr, 2,400 torr He) we obtained an output energy of 55 mJ (Δt = 10 ns) with BCl_3 at a pressure of 0.06 torr, and of 1 mJ (Δt = 10 ns) with CCl_4 at a pressure of 0.18 torr. Using the same Kr/He mixture we obtained 105 mJ (Δt = 13 ns) with HCl at a partial pressure of 2.0 torr.

Further work is in progress to improve these results, even if better results than from HCl-containing mixtures are not expected. In fact, polyatomic molecules absorb strongly UV radiation, making less effective the preionization radiation. Furthermore, electronegative compounds like BCl_3 and CCl_4 prevent uniform and stable glow discharges (4).
Figure 3 shows the variation of the output energy and laser pulse duration Δt with the ratio of the applied electric field at breakdown E_b to the gas density N, at the optimum gas mixture composition and maximum total pressure. The linear growth of output energy and Δt with E_b/N was observed in all the investigated mixtures. It was also found that E_b/N increases linearly with charging voltage, at least up to the highest voltage possible in our device, which is 45 kV (2). In fact, higher charging voltages decrease the risetime of the gap voltage. This in turn leads to a higher voltage at breakdown.

The strong dependence of the laser output energy on E_b/N can be ex-

plained by the fact that the percentage of discharge power going into
the formation of metastables and resonance-trapped Kr states increases
with the applied electric field. Then it comes out that the laser
action starts early with increasing E_b values, as experimentally observed.

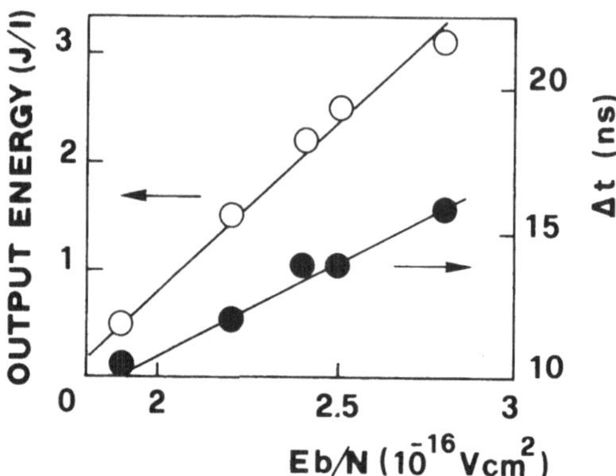

Fig. 3. KrCl lasing energy and pulselength (FWHM) vs E_b/N at
2,600 torr fill pressure. The gas mixture is 0.09% HCl/ 10.1% Kr/
1.5% He/ 88.3% Ne.

Figure 4 shows the time evolution of the discharge voltage, fluorescence and lasing for the best gas mixture at a charging voltage of
32 kV. It is evident that high optical losses, occurring at the gas
breakdown, inhibit laser action for an appreciable fraction of the
power deposition time. The persistence of fluorescence after the
lasing blow out proves the formation of absorbing species rather
than the loss of excited state population. This point is supported
also by the fact that the KrCl laser pulse is characterized by an
exponentially decaying tail, observed also by Sze (1). This fact could
perhaps also explain the feature that the laser pulse length does
not vary with the total pressure.

Finally, it has been found that all the investigated mixtures have a
quite long static fill lifetime. At a pulse repetition rate of 0.5 Hz,
the output energy drops by about 15% over the first 50 pulses, and
the KrCl lifetime to half energy is about 300 shots. Gas circulation
and purification should greatly improve the fill lifetime.

As a conclusion, we can affirm that significantly enhanced output
energy at 222 nm has been obtained from a KrCl laser. Higher energies
could possibly be obtained at higher pressures and specific power
loadings than those achievable with our device. However, since the
working parameters of our device are not too far from those of

commercial excimer lasers, the results of our work are expected to be applicable to those lasers.

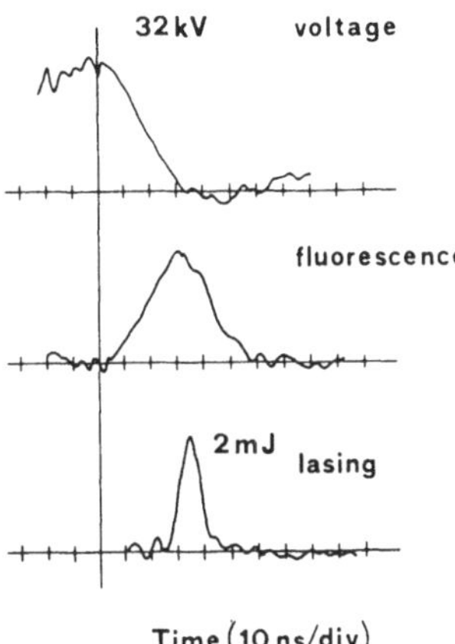

Fig. 4. Time evolution of the discharge voltage, fluorescence and lasing near threshold.

AKNOWLEDGMENT

Work supported in part by ENEA under contract C/452.

REFERENCES

(1) SZE,R.C.: IEEE J. Ouantum Electron. QE-15 (1979) 1338

(2) ARMANDILLO,E., LUCHES,A., NASSISI,V. and PERRONE,M.R.: Appl. Phys. Lett. 42 (1983) 860

(3) ARMANDILLO,E., BONANNI,F. and GRASSO,G.: Optics Commun. 42 (1982) 63

(4) BURNHAM,R.: Optics Commun. 24 (1978) 161

Extending the Lifetime of CuII Hollow Cathode Lasers

H. J. EICHLER, H. KOCH and R. TORNOW
Optisches Institut   TU Berlin
Straße des 17. Juni 135
1000 Berlin 12

1. Introduction

Since 1964 (1) hollow cathode lasers represent an alternative to the
commonly used positive column lasers. Hollow cathode lasers exhibit a
lot of considerable advantages: lower noise, less cataphoretic demix-
ing of gas components, metal vapour production due to cathode sputtering, often lower laser thresholds and a large number of new laser lines
not known in positive column lasers (2). Still a commercial hollow cathode laser is not yet available due to the short lifetime of the prototypes presented up to now (3).

Due to its strong UV-lines (248 - 270 nm, up to 0.9 W output power (4))
the CuII-laser is of special interest. Therefore the two factors limiting its lifetime are investigated: the laser mirror degradation and the
alteration of the hollow cathode geometry due to sputtering.

2. Laser mirror degradation

Operating the UV-CuII-laser for a few hours causes already a considerable reduction of the reflectivity of the internally mounted dielectric
laser mirrors. Fig. 1 shows an example of the spatial distribution of
the reflectivity of a used mirror with formerly 98.7 % reflectivity.
A considerable overall loss and an additional loss at the position of
the laser spot can be recognized. Due to the fact that the front and
the rear side of the mirror substrate exhibit such a contamination pattern and that a mirror cleaning with meta-cresole (5) was possible (upper curve Fig. 1) indicate that the contaminations develop most probably by UV-induced deposition of polymerized hydrocarbons. Similar observations have been reported recently from storage ring free electron laser experiments where vacua of 10^{-10} Torr are common (6). Transverse
flow protections (7), cataphoretic confinement discharges (8), liquid
nitrogen traps, mirror heating and oil free ultra high vacuum technology have been suggested to reduce mirror degradation, but these methods
have not yet been proved in lifetime tests (9).

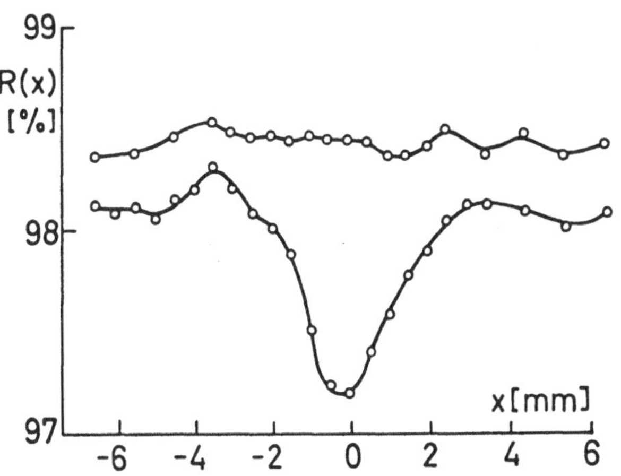

Fig. 1 Spatial distribution of the laser mirror reflectivity after use (lower curve) and after cleaning with m-cresole (upper curve).

3. Alteration of the hollow cathode geometry due to sputtering

Cathode sputtering produces metal vapour in a very convenient manner but on the other hand it causes substantial changes in the cathode geometry. The result of a lifetime test with a cylindrical copper hollow cathode of formerly 40 mm length and 4 mm diameter is displayed in Fig. 2: after maintaining a Ne-discharge with 0.3 A discharge current and 8 mbar fillgas pressure for 400 hours spherical hollows have been separated and the optical axis is completly blocked. The change in geometry occurs since at some areas of the cathode surface more erosion due to sputtering than condensation of rediffusing metal vapour takes place and vice versy at other surface elements. This is obvious from the scanning electron microscope picture in Fig. 3.

Therefore other authors have tried to avoid sputtering at all, by a careful cathode heating that creates liquid metal films at the cathode surface (10), (11). In contrast to these very sophisticated concepts another method is presented here that exploits rather than avoids sputtering. Already White (12) has shown that under sputtering action special hollow cathode geometries can develop that are nearly unalterable for more than 7000 hrs. The shape of the outermost spherical hollow with conical aperture in Fig. 2 closely resembles the hollow cathodes introduced by White. When a different initial hollow cathode geometry was chosen: a 15 mm long cylindrical bore of 4 mm diameter where only one spherical hollow could develop a successful 1000 hour lifetime could be dmonstrated for 0.15 A/cm^2 current density (at cathode surface) and 8 mbar Ne-prssure. Fig. 4 shows a cross section of the shape that developed after 200 hrs and did not alter for the remaining 800 hrs (controlled by x-ray photography).

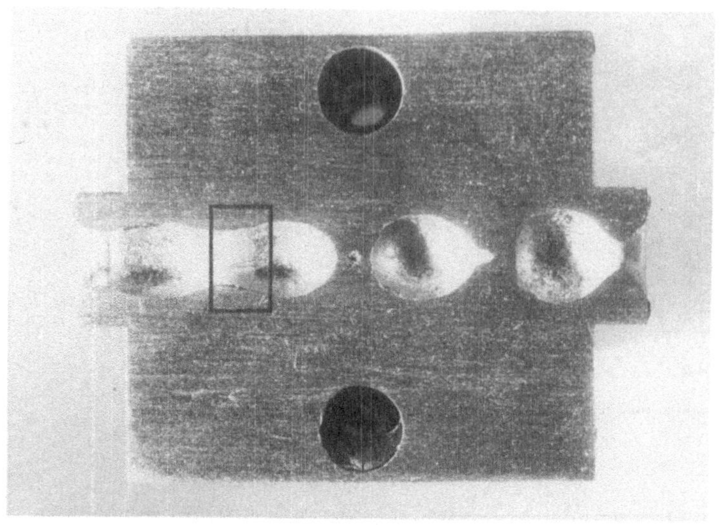

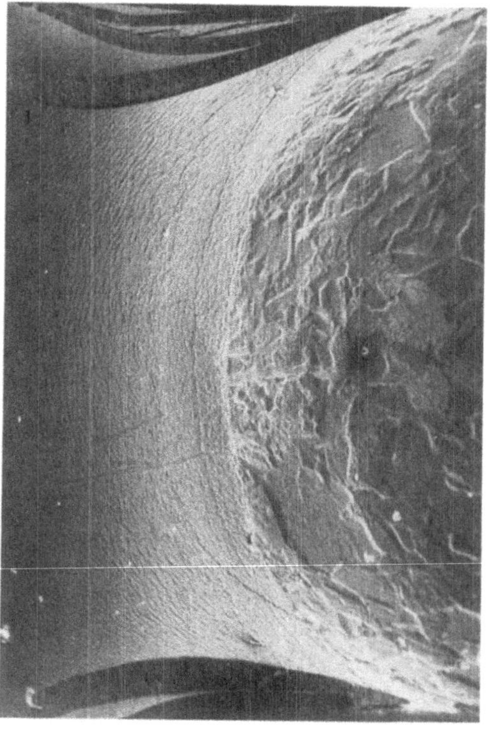

Fig. 2 A formerly cylindrical hollow cathode after 400 hrs life time test.

Fig. 3 Scanning electron microscope picture of the area indicated in Fig 2.

4. Conclusion

Contrary to the hollow cathodes introduced by White (12) the stable geo
metry presented here can be applied to laser- and absorption spectroscopy devices because two aperture are available. A laser tube should
be constructed by a sequence of these cathode elements and anodes as
proposed in the schematic drawing in Fig. 5.

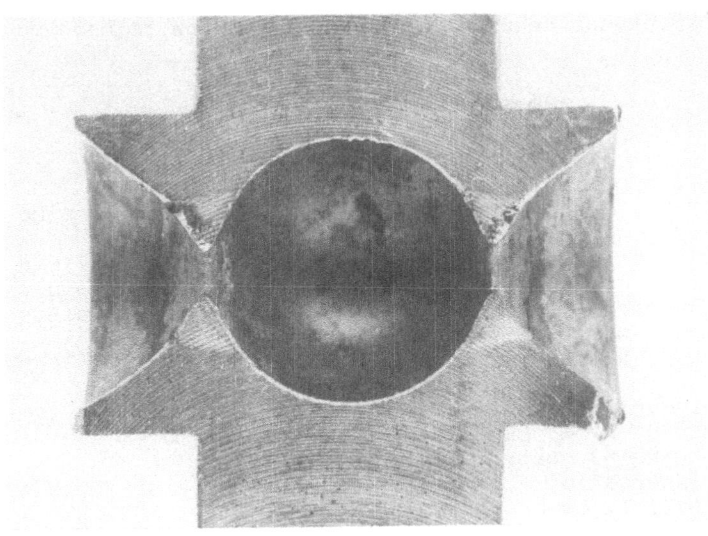

Fig. 4 Cross section of a hollow cathode with stable geometry after 1000 hrs lifetime test.

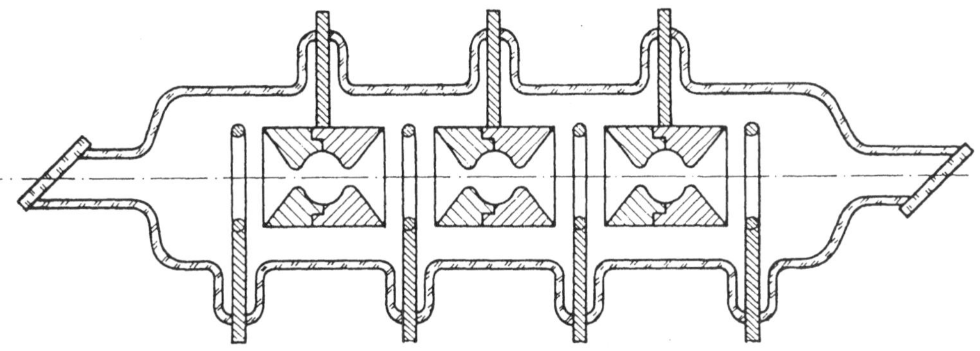

Fig. 5 Laser tube design with spherical hollow cathodes

This design can be applied to other types of lasers too where hollow cathode discharges are advantageous to positive columm discharges, like He-Ne, He-Kr$^+$, He-Cd$^+$-lasers etc. (2). These lasers require considerable lower discharge currents so that in this case a much longer life time of the cathodes can be expected. Other advantages of the spherical hollow cathodes are the much lower gas clean up rate (12) and the lower metal vapour loss compared to conventional hollow cathodes. Thus the established pressure controlled gas reservoirs like in commercial He Cd$^+$ lasers (Liconix, Kimmon) can be employed and expensive materials like gold or isotopes can be used as laser species economically by covering the inner cathode surface with a thin film of the respective

material. A patent has been applied for the hollow cathode concept presented here.

References

1. J. SMITH, J. Appl. Phys. $\underline{35}$, 723 (1964)
2. W. K. SCHUEBEL, Proc. Int. Conf. Lasers '79 (STS Press, McLean, USA 1980), p. 431
3. Panel discussion at Lasers '79 "Hollow Cathode Lasers: Can they be-come practical and compact cw-sources in the visible and the ultra-violet?" Chairman: W. T. SILFVAST, Proc. Int. Conf. Lasers '79 (STS Press, McLean, USA 1980), p. 903
4. H. J. EICHLER, R. MOLT, J. L. QIU and W. MARTIN, Appl. Phys. $\underline{B26}$, 49 (1982)
5. Merck-catalogue "Reagenzien, Diagnostika, Chemikalien" 241 (1981)
6. P. ELLEAUME, D. A. G. DEACON, M. BILLARDON and J. M. ORTEGA, Advan-ce Program CLEO'83, paper ThJ3, p. 69 (1983)
7. M. YANG, Dissertation, Ludwig-Maximilians-Universität, München (1981)
8. J. R. FENDLEY, I. GOROG, K. G. HERNQUIST and C. SUN, RCA Rev. $\underline{30}$, 422 (1969)
9. H. J. EICHLER, H. KOCH, R. MOLT and R. P. TORNOW, DGaO/SGOEM Tagung paper A4 (1982) unpublished
10. K. G. HERNQVIST, IEEE J. Quantum Electr. $\underline{QE-14}$, 129 (1978)
11. R. D. Reid, M. S. thesis, Colorado State University (1978)
12. A. D. WHITE, J. Appl. Phys. $\underline{30}$, 711 (1959).

UV-CuII-Laser Induced Electron Beam Generation with High Specific Brightness

E. ENDRUSCHAT, J. HAMISCH, H. KOCH, H. NIEDRIG, W. TORNOW
Optical Institute TU-Berlin
Straße des 17. Juni 135
1000 Berlin 12

In electron microscopes the beam sources are usually based on two principles: thermionic- and fieldemission. A figure of merit for the electron gun is given by the "brightness" R, an analogon to the luminance
in light optics. The brightness is a measure of the electron current
delivered from the source per emitting area into the unit solid angle:

$$R := \frac{\partial^2 I}{\partial A \partial \omega} \qquad (1)$$

I := emitted current
A := emitting area
ω := solid angle into which I is emitted

With thermionic cathodes brightness values of approximately $R = 10^6$
$Acm^{-2}sr^{-1}$ (corresponding current 0,1..1 mA) are achievable /1/. The
higher values up to $R = 10^8 \ Acm^{-2}sr^{-1}$ can be obtained with fieldemission guns /2/. Here the maximum currents are in the order of some microamperes. While most thermionic cathodes can be operated in conventional high vacuum ($p \approx 10^{-4}..10^{-7}$ mbar), field emission guns need an ultra high vacuum ($p \leq 10^{-9}$ mbar) for proper operation.

Another principle of electron emission is provided by the photoelectric
effect. But due to the low photoelectric yields of $10^{-6}..10^{-4} \ AW^{-1}$ in
the proximity of the photoelectric threshold, light intensities produced by conventional uv-sources are not sufficient to deliver a desirable photocurrent density of $i \geq 0.7 \ Acm^{-2}$ from a photocathode. A
cw-uv-laser beam aimed at a photocathode, however, should yield favour-
able results due to the superiour focusing properties of lasers.

Thus a simple laser photoelectron gun (LPEG) has been developed /3/ to
be operated with a uv-cuII-laser designed at our institute /4/. The
principle of such a LPEG is shown in figure 1. The uv light ($\lambda = 248.6$ nm)
from the laser is focused onto a flat disk-type photocathode PC. The
emitted photoelectrons are accelerated in the approximately homogeneous electric field between cathode PC and anode A. After passing the

aperture hole in the anode the electron beam can be further analyzed.

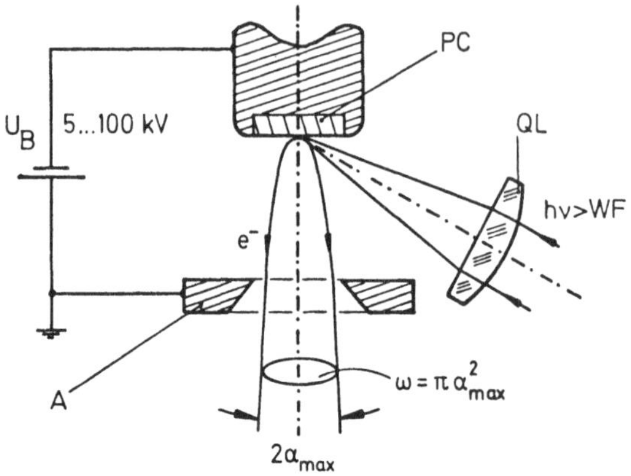

Fig. 1
Principle of an laser photoelectron gun

PC : photocathode
 A : anode
QL : quartzlens

For such a system it is easy to give a rough approximation for the
mean brightness $\bar{R}$ (homogeneous field assumed) /3/ :

$$\bar{R} := I/A \cdot \omega \quad (2) \quad \text{with} \quad I = \eta P \quad (3) \quad \text{and} \quad \omega = \Pi\Delta E/e_o U_B \quad (4)$$

follows:

$$\bar{R} = \frac{\eta P\, e_o U_B}{\Pi A_F \Delta E} \quad (5)$$

P : laserpower on cathode
η : photoelectric yield
ΔE : energy spread of the photoelectrons
U_B : accelerating voltage
A_F : emission area = area of laserfocus

To measure the brightness of the LPEG a simple electron optical device
has been constructed. The whole experimental setup is shown schematically in figure 2. To determine the angular distribution of the photoelectrons the beam was deflected with the coils C1 over the pinhole MB
of the detection system DS. The magnetic electron lens ML was not excited and the beam aperture was not limited by any of the changeable pinholes AB. Plotting the detection signal versus the deflection current
with the scope SC gave the angular distribution. Measurement of the current distribution on the cathode has been performed by imaging the emission area into the plane of the detection system. In order to obtain
a good resolution with the magnetic lens the aperture was limited by
means of an adequate pinhole AB. For the same reason the detection pinhole was changed from 100..300 μm to 5..20 μm. Deflecting the image of
the emission area with the coils C2 delivered the current distribution
on the cathode. The total emitted current was measured with the revolve
able faraday cup FC. The electron optical chamber was evacuated by mean

of a turbopump system to a pressure of about $2 \cdot 10^{-6}$ mbar. The setup
has been tested with a conventional thermionic cathode with a tungsten
emitter.

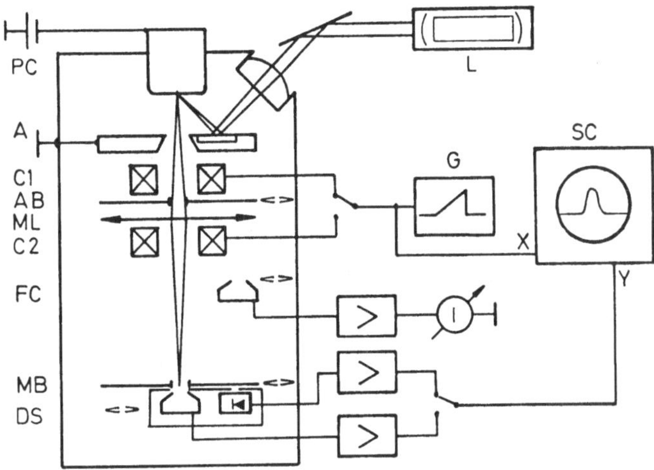

Fig. 2 Scheme of the experimental setup

Experimental results

First preliminary measurements with a Si-photocathode yielded a mean
brightness of 2 Acm^{-2}sr^{-1}. The emitted current was 0.5 nA, the laser
power was about 0.3 mW and the accelerating voltage was set to 30 kV.
The emission area has been determined to $5 \cdot 10^{-6}$ cm^{-2} (17 x 37 μm, ellip-
tic). Figure 3 shows scans through the maxima of the spatial and the angular distribution of the emitted photoelectrons.

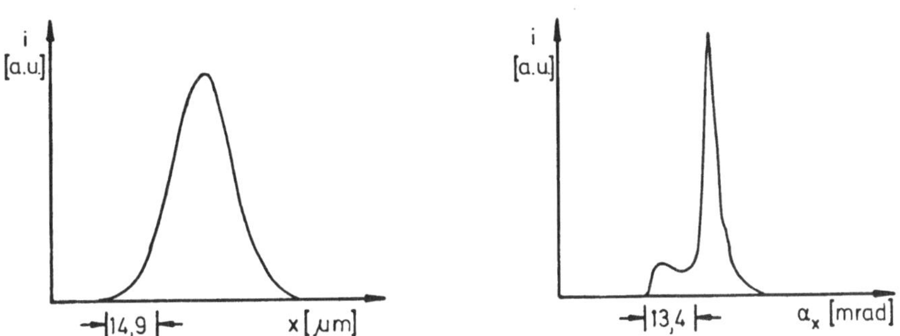

Fig. 3 Scans through the maxima of the distributions of the photoelectrons
a) Spatial current distribution on the photocathode
b) Angular distribution of the emitted current

Discussion
The measured brightness seems to be inferior when compared with the data
of modern thermionic guns. But a more realistic comparison is provided

22

when the brightness is related to the same emission current and the same energy. A suitable quantity is the "specific brightness $r := R/I$ (6). The measured value of $\bar{r} = 5 \cdot 10^9$ $cm^{-2}sr^{-1}$ is of the same order of magnitude as that of conventional thermionic guns with W-cathode when operating with some microamperes. When testing the experimental unit with a commercial thermionic gun (Siemens) a brightness of 570 $Acm^{-2}sr^{-1}$ has been achieved emitting a current of 4,7 µA at a voltage of 30 kV. The resulting specific brightness was only $1.2 \cdot 10^8$ $cm^{-2}sr^{-1}$. The cross-over area had a magnitude $1.1 \cdot 10^{-4} cm^{-2}$.

A considerable increase of the brightness can be expected by improving the light optics, laserpower and photoelectric yield. A system consisting of components with the following parameters seems very realistic: Stable uv-cuII-laser with 100 mW output in the TEM_{oo}-mode, a uv-condensor which is able to focus the light into a spot of 1 $µm^2$, an advanced photocathode with a yield of 10^{-4} AW^{-1}, a maximum energy spread of 1 eV and a voltage of 30 kV. Putting these data into expression (5) yields a <u>mean brightness $\bar{R}$ of 10^7 $Acm^{-2}sr^{-1}$</u> corresponding to a current of 10 µA. That would be sufficient for most applications of electron optical devices.

The main problem to achieve this extrapolated brightness will be to employ a cathode material with sufficient quantum yield which is resistent to the extreme light intensities (up to 10^7 Wcm^{-2}) in the focus. Another problem is that the effective workfunction of the photocathode has to remain stable when it has been exposed to air. Because of these reasons the application of modern high-yield photoemissive materials commonly used in multipliers and camera tubes /5/ might not be appropriate.

Despite these problems chances are good to be able to create a photo-electron gun with high brightness. These guns will not replace conventional cathodes but they posess potential capabilities of some interest for scientific and industrial research. By employing pulsed uv-lasers these photoelectron guns would generate intensive ultrashort electron pulses with promising applications in time resolved electron microscopy /6/. Other important applications of uv-laser induced electron emission are the use in photoelectron emission microscopes /7/ and eventually the development of highly spin polarized electron gun with high brightness /8/.

References

/1/ H. HOFFMANN
    Optik $\underline{36}$, Heft 4, 494 (1972)
    F. J. HOHN, T. H. P. CHANG, A. N. Broers, G. S. FRANKEL, E. T. PETERS
    D. W. LEE
    Appl. Phys. 53(3), 1283, (1982)
/2/ K. H. GAUKLER, R. SPEIDEL, F. FORSTER
    Optik $\underline{42}$, 391 (1975)
/3/ F. E. ENDRUSCHAT
    Diplomarbeit, Optisches Institut TU Berlin (1982)
/4/ H. J. EICHLER, H. KOCH, R. MOLT, J. L. QIU, W. MARTIN
    Appl. Phys. $\underline{B26}$, 29 (1981)
/5/ A. H. SOMMER
    "Photoemissive Materials", J. Wiley & Sons Inc., New York, London
    Sidney, Toronto (1968)
/6/ G. MOUROU, S. WILLIAMSON
    Appl. Phys. Lett. $\underline{41}$, 44 (1982)
/7/ H. BETHGE, G. GERTH, D. MATERN
    Proc. of the 10 th Int. Congr. on Electron Microscopy,
    Hamburg, W.-Germany, Aug. 17-24 1982, Vol. 1, p. 69
/8/ J. KESSLER
    Texts and Monographs in Physics, Springer Verlag Berlin, Heidelberg
    New York, 155 (1976)

Frequency Stabilization of He-Ne Commercial Tubes: A Comparison of Various Techniques

S. Donati   Università   Pavia/Italia

G. Martini   Università   Pavia/Italia

Abstract

We start with an illustration of the various techniques useful for frequency-stabilization of laser sources at different levels of accuracy, both from the point of
view of the frequency error sensing and of the feedback-loop implementation. Then,
we consider several methods to be used in connection with commercially available
He-Ne laser sources. We discuss in particular the Zeeman splitting and the crossed
polarization modes as frequency-sensing method, and the piezoelectric versus ther-
mal-expansion controls of the cavity length. Several commercial He-Ne tubes of dif
ferent manufacturers were tested, as regard to, e.g.:(i) modal distribution at various discharge current, (ii) preferred polarization directions, (iii) frequency
range of Zeeman splitting, (iv) frequency-pulling effects. In order to achieve sta
bilization in the range of 1 part on 10^{10}, it was not necessary to resort to iodi-
ne-cell references, and the acuracy of Ne atomic line has been found adequate.
We select a side-arm tube as the best device for thermal control of cavity length,
whose response time was satisfactorily fast when well designed PID controller and
thermal element were employed. Details of operation, including discharge currents,
thermal transient analysis and frequency compensation scheme will be discussed.

Introduction

Frequency stabilization is useful in a variety of measurement applications of
laser. The most stringent requirement is obviously that of length and frequency
metrological standard [1], where frequency stability $f/\Delta f$ and reproducibility of
the order of 10^{12} are attained. In other optoelectronic applications a medium
level stability is required, in the range $10^9 \div 10^{10}$; examples of such applications
are: (i) high-resolution atomic spectroscopy, (ii) optical-heterodyne detection,
(iii) long-distance interferometry. Hereafter we compare different approaches to
medium-level frequency-stabilization and present the results obtained in the development of a two-mode, thermal control, frequency stabilization scheme.

1-Laser sources

We have tested various samples of He-Ne laser tubes, made by different manufacturers. The following parameters have been taken into account: (i) mode pattern as a function of the discharge current, (ii) frequency-pulling effects, (iii) mode's polarization directions, (iv) effects of Zeeman splitting and associated frequency range. A summary of the main features of the considered sources is reported in Fig.1.

Laser tube		Cavity configuration	Cavity length (mm)	Output power at I_s= 5 mA (mW)
Spectra Physics 155	≠ 1	int.	270	0.4
Spectra Physics 155	≠ 2	"	270	0.5
Jodon 211	≠ 3	"	180	1.2
Quante-L.T. PF 280	≠ 4	"	260	1.4
Quante-L.T. PF 280	≠ 5	"	260	1.6

Fig.1- Main features of the considered commercial laser tubes

Some tubes (samples 1, 2 and 3) have shown Zeeman effect with both normal and longitudinal magnetic field, with differents beating frequencies near 10^5 Hz, without showing any preferred preferred polarization direction. The other two samples (4 and 5) have on the contrary shown Zeeman effect into a more restricted range of frequencies, in conjunction with two strongly preferred mode's polarization directions. A comparison of frequency intervals of Zeeman beating frequency is made in Fig.2.

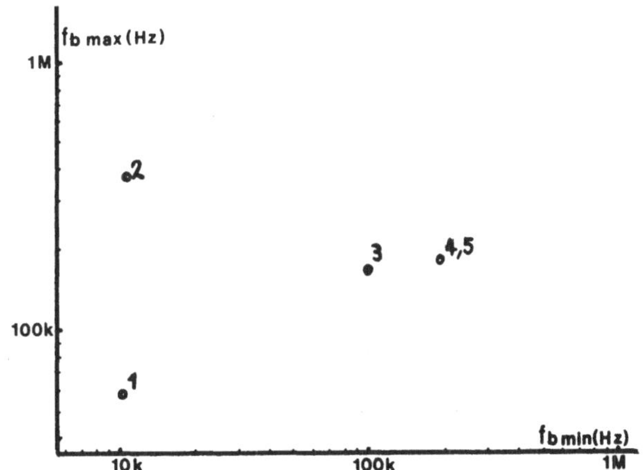

Fig.2- Zeeman beating frequency range of the different samples

When operated at the nominal current ($\sim$5mA) all the tubes have shown a three-mode laser oscillation, while at an operating current ranging from 2.5 to 3.5 mA they were well into the two-mode regime, as shown in Fig.3.

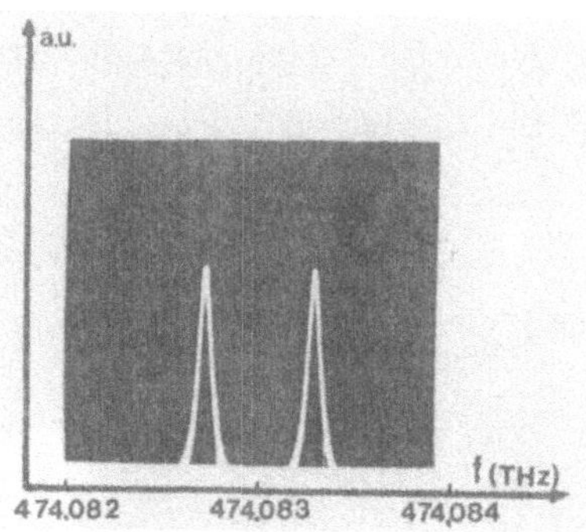

Fig.3- Typical mode structure in the current range 2.5÷3.5mA

During warm-up in the two-mode region the beating frequency between the two orthogonally polarized oscillating modes was monitored by means of a pin photodiode and well defined frequency-pulling effects were observed (see Fig.4).

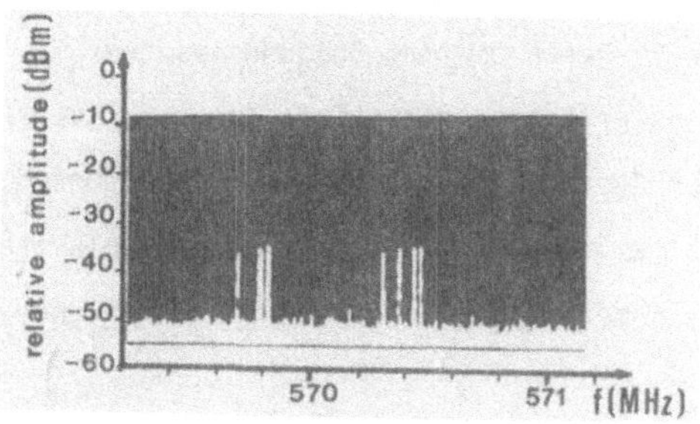

Fig.4- Frequency-pulling effects: this exposure, taken over 10 s, shows the variation of the beating frequency between two adjacent longitudinal modes during warm-up of the sample no.5

Mode structure beyond 2.5mA was not investigated, because of the extremely high values required for the ballast resistor, and consequently for the high-voltage power supply.

2-Stabilization configurations

We discuss here only those approaches capable to yield a frequency stability in the range $10^9 \div 10^{10}$, i.e. active stabilization of optical frequency [11]. About passive stabilization we only mention that it allows to reach a frequency stability of 10^8 at best [3]. Many techniques can be used both for frequency-error signal pick-up and for cavity length actuation. Among the frequency-dependent effects useful to pick-up an error signal we can quote, beyond the well-known absorbing-cell technique: (i) the dip at the top of the gain curve (Lamb dip)[2], (ii) the frequency splitting of modes due to e.m. field induced birefringency (Zeeman and Stark effects), (iii) the orthogonality of polarization of adjacent longitudinal modes (gain-hole effect). Among the techniques usable to actuate the cavity length variation necessary to control the oscillation frequency we can consider: (i) the piezoceramic actuator and (ii) dissipation of thermal power.

2.1-The Zeeman-stabilized laser

A way to obtain a frequency-dependent signal is to take advantage of the magnetic quantum number m_1 degenerancy suppression due to an applied magnetic field B. This is the well-known Zeeman effect, leading to a gain curve peaks separation, for orthogonal polarizations [4], $\Delta f = eB/(4\pi m)$. The modifications suffered by the gain curve as a consequence of the presence of an external magnetic field are shown in Fig.5.

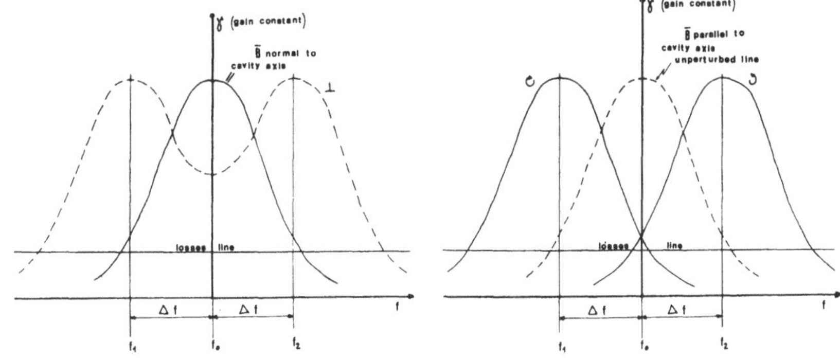

Fig.5- Modification of the gain curve for orthogonal polarizations when is applied: a) a magnetic field normal to the cavity axis and b) an axial magnetic field

The Zeeman-induced birefringency causes different amount of frequency pulling to the two orthogonally polarized oscillating modes[5],and the resulting beating frequency can be used as the error signal in the feedback loop. A possible configuration for a Zeeman-stabilized laser is depicted in Fig.6, where is used a thermal actuation of the cavity length variation.

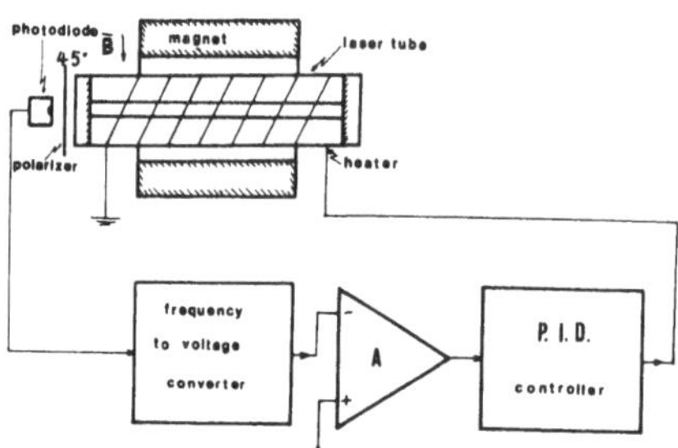

Fig.6- A possible implementation of a Zeeman stabilized laser

This scheme gives stabilities of the order of 10^9 [6,9,10] , and with some modifications it is theorically possible to achieve frequency stability as high as 10^{13}, the limiting factor being a 10^5 stability required for the applied magnetic field [8].

2.2-The two-mode stabilized laser

In laser sources with a restricted range of Zeeman beating frequency and/or with strongly preferred polarization directions, such as our samples no.4 and 5, we can use the hole-burning effect. This has as a consequence the orthogonality of polarization direction of the adjacent longitudinal modes, so that they can be separately detected, allowing their amplitude difference to be used as a frequency-sensitive signal. We describe now the realization of a stabilized He-Ne laser employing the above mentioned frequency-dependent effect, and the proposed scheme can be seen in Fig.7 and in Fig.8.

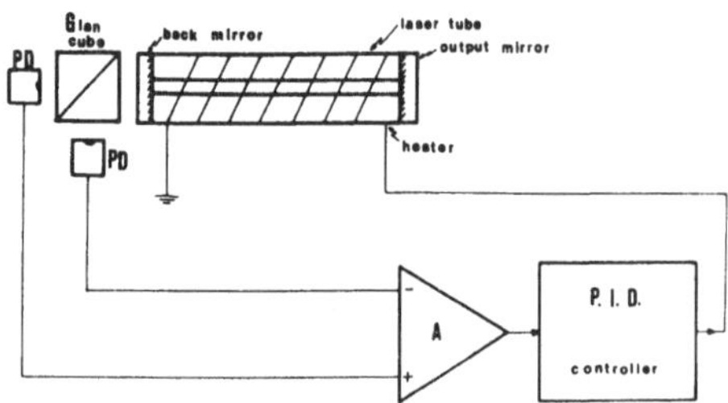

Fig.7- Configuration of the two-mode stabilized laser

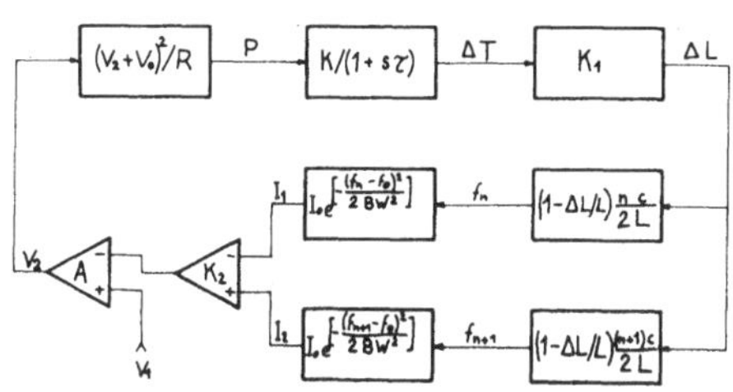

Fig.8- Block-scheme of the system reported in Fig.7, showing the various transfer characteristics involved

First, we illustrate the thermal time constant characteristics and the discharge current requirements to obtain laser action well into the two-mode region of operation. The thermal time constant has been measured to be $\tau=480$ s (sample no. 5). The frequency-dependent signal is directly supplied by the difference of the intensities of the two modes, through the gain line-shape that is assumed to be invariant, being the discharge current electronically stabilized. The error signal is then fed, through a PID filter and a power booster, to an electrical heater dissipating a power of the order of that of the gas discharge itself. The heating element has been realized with a resistive wire wounded along the capillary tube. This winding introduce another thermal time constant $\tau_2=870$ ms. The PID controller parameters has been choosen to optimize medium-term stability. The optical frequency stability at short-medium term has been measured to be 5×10^9, with an integration time of 1 s.

3-Conclusions

From a comparison between the two considered techniques of frequency stabilization usable in conjunction with commercially available He-Ne low-power laser tubes, we observe that the two approaches are almost equivalent from the point of view of the frequency stability obtainable in practice: $f/\Delta f$ in the range of 10^9.

So, the choice between them has to be based on different considerations: i.e. the availability of a reasonable wide range of Zeeman beating frequency and a cavity exhibiting a perfect axial symmetry are facts that make such a tube a candidate for Zeeman-effect based stabilization. On the other hand, a restricted range of the Zeeman beating frequency, the difficulty to be operated in the single-mode regime and/or the existence of preferred directions of polarization in the optical cavity, lead to the conclusion that it is preferred a frequency stabilization by means of of the two-mode amplitude difference technique.

When both stabilizing schemes seems to be applicable, considerations such as the permitted amount of optical feedback have to be taken into account. The maximum optical feedback that can be sent back to the source is 10^{-9} for the two-mode configuration and 10^{-7} for the transverse Zeeman scheme, as reported in Ref.7.

References

[1] F. Bertinetto, B. I. Rebaglia, M. Liverani, S. Gualini, ALTA FREQUENZA, Vol. XLIV, n.10, Ottobre 1975, pp. 279E-283E, 569-573.

[2] W. E. Lamb, Phys. Rev. 134, p. 1429 (1974).

[3] H. Ogasawara, J. Nishimura, Appl. Opt., Vol. 21, n. 7, pp. 1156-1157 (1982).

[4] L. D. Landau, E. M. Lifšic, Mec. Quant. Teoria non Rel. - Boringhieri, Torino 1969

[5] S. Donati, V. Speziali et al.,RI 78/2 ,Ist. di Elettronica,Pavia, Feb. 1978

[6] N. Umeda, M. Tsukiji,H. Takasaki,Appl. Opt., Vol. 19, n. 3 pp. 442-450.

[7] N. Brown, Appl. Opt., Vol. 20, n. 21, pp. 3711-3714.

[8] K. Nemeš, M. Valič, Proc. 5th Int. Cong. Laser 81, Springer-Verlag 1982

[9] T. Baer, F. V. Kowalski,J. L. Hall, Appl. Opt., Vol. 19, n.18, pp.3173-3177.

[10] H. Ogasawara, J. Nishimura, Appl. Opt., Vol. 22, n. 5, pp. 655-657.

[11] M. Sargent III, M. O. Scully in: F. T. Arecchi,E. O. Schulz-Dubois ed. by, Laser Handbook Vol. 1, North Holland Publishing Co., Amsterdam 1972.

Raman Conversion of Dye Lasers:
A Coherent High Power Light Source in the VUV and th IR

P. Lokai, K. Hohla, D. Basting

Lambda Physik Gmbh, Goettingen, W.Germany

Pulsed dye lasers have shown a tremendous increase in peak power and in average power since the availability of excimer laser pumps. In addition the development of new and efficient dyes has broadened the spectral region of dye lasers: 0.3-0.9 μm can be reached with peak power in excess of 4 MW and linewidth is reduced to 0.005 cm^{-1}.

For years not only spectroscopists demand for an extension of this tuning range into the VUV and into the IR. Various nonlinear schemes have been tested, mostly on a laboratory scale. Due to the high peak power of dye lasers the Raman conversion process is an alternative for both wavelength regimes /1-4/. In comparison to other methods, stimulated Raman scattering (SRS) in gases is of extreme experimental simplicity, the nonlinear material (gas) has high damage threshold and very good optical quality without absorption at the generated wavelength. H_2 gas is widely used as a Raman medium because it exhibits a large Raman shift (4155 cm^{-1}) and narrow transition linewidth ($\approx$775 MHz). Different techniques have been applied to achieve high conversion efficiency in the SRS process: the use of resonators to provide feedback of the Stokes radiation /6/, multiple pass configuration /5/ and the use of hollow dielectric waveguides to guide the pump and Stokes readiation /4/.

We have investigated the conversion efficiency of a dye laser to the UV as well to the infrared spectral range. For both ranges we compared two devices: a) the normal standard Raman cell and b) a capillary waveguide structure (cf. Fig. 1). The input dye laser was an excimer pumped Lambda Physik FL 2002 E dye laser with output energy of approximately 40 mJ in a 15 nsec pulse. The linewidth of the pump laser was 0.04 cm^{-1}.

For the capillary experiment the dye laser was focussed into the high precision bore quarz tube. Coupling losses were minimized by achieving an approximate mode match between the near Gaussian spatial mode of the dye laser and the low loss EH_{11} mode of the waveguide. The fundamental line and the various Stokes orders were separated by a CaF_2 prism or germanium filter.

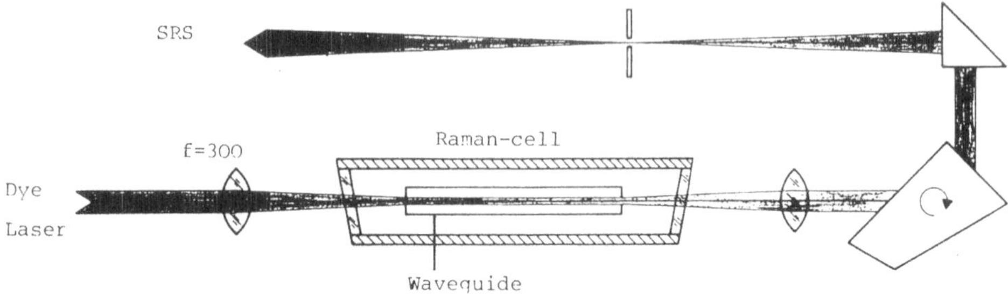

Fig. 1 Schematic view of a waveguide-type Raman cell

For the UV spectral range we got the well known result that the SRS
is a powerful shifting method for the wavelength range below 220 nm.
And even in the range from 220 to 260 nm it compares not to bad with
the normal doubling scheme using KPB.

Comparing the two Raman shifting devices it turned out that the
conventional device is superior to the waveguide. This is mostly due
to enhanced losses in the tube, because for th UV the capillary will
not work as a waveguide. Quite different is the situation in IR
spectral range. We therefore adressed our work mostly to the extension of the Raman scheme to the IR spectral range.

To the first Stokes order we measured a conversion efficiency of 17 %
which is similar to that obtained using the conventional Raman cell
(cf. Fig. 2). Of special interest are the efficiencies of the second
and third Stokes components. The second Stokes order could reach an
efficiency of more than 10 % and the third Stokes component more than
2 % (compared to the dye laser). Using efficient dyes like Rhodamine
6G, Sulforhodamine 101, DCM, Rhodamin 700 and Pyridine 2 the wavelength range from 575 to 750 nm is covered with energies of more than
40 mJ/pulse. Thus using the first three Stokes components the spectral tunability in the IR has been extended up to 9 μm with erngies
of $>$ 0.3 mJ per pulse.

Up to now the longest wavelength we observed is 9.5 μm, but this is
only due to the transmission limit of the optical material in use.

The dependence of the Ramanshifted output energy on the input dye
laser is shown in Fig. 3. One observes clearly a threshold behaviour
of the S3 line at an input energy of 15 mJ. To maximize the output
one has to adjust the pressure as indicated by Fig. 4. Optimum output
is obtained at 18 bar for S3, while S2 reaches this maximum already
at 8 bars.

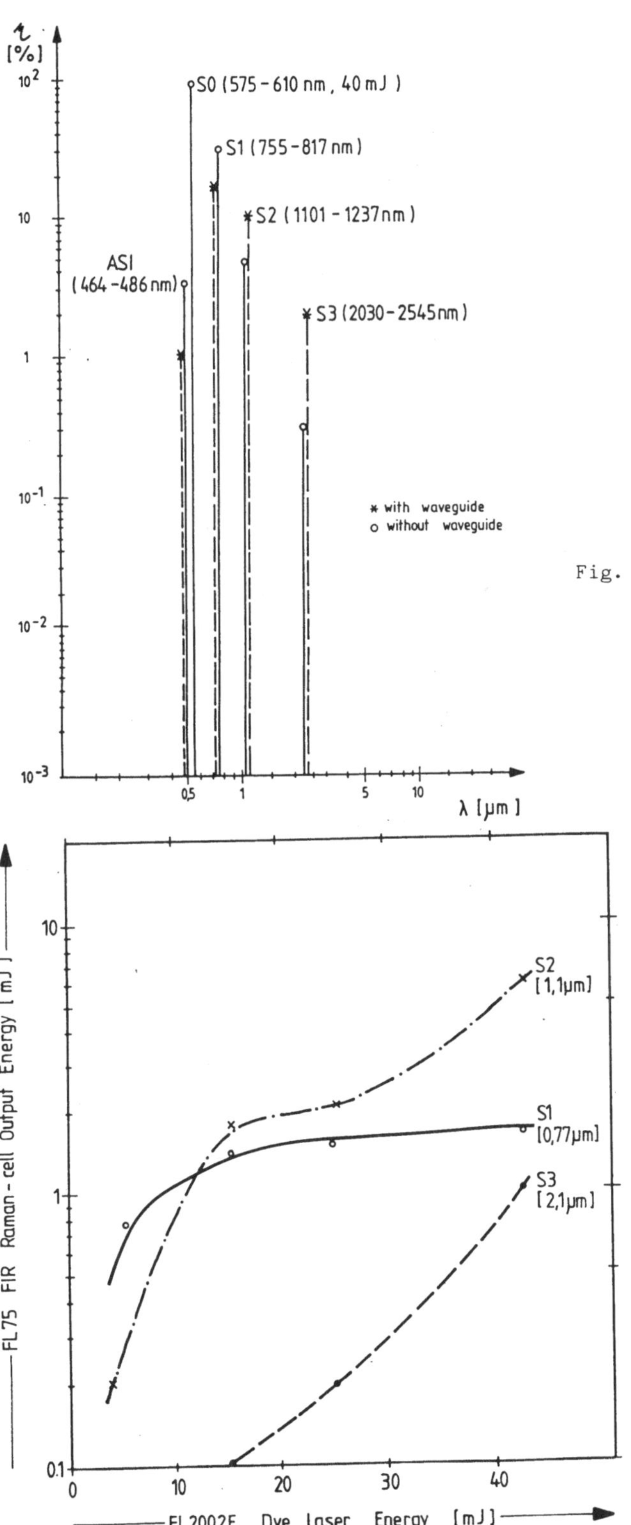

Fig. 2. Comparison of the SRS efficiency in H_2 in conventional Raman cell and with capillary waveguide design

Fig. 3. Energy for the Stokes S1, S2, S3 as function of pump energy at H_2 pressure 15 bars

The linewidth of the generated IR radiation is an important parameter that determines the usefulness of the SRS method. If the interaction of the various waves (S0, S1, S2) is completely parametric, the spectral linewidth of the scattered radiation should be equal to that of the pump source. The bandwidth of the various higher order Stokes- and Antistokes lines were investigated by starting from UV (IR) dyes and monitoring this bandwidth by FP-etalons. It turned out that up to 0.04 cm^{-1} the bandwidth of the dye laser is transformed as well to the higher order Stokes lines as to the higher order Antistokes lines.

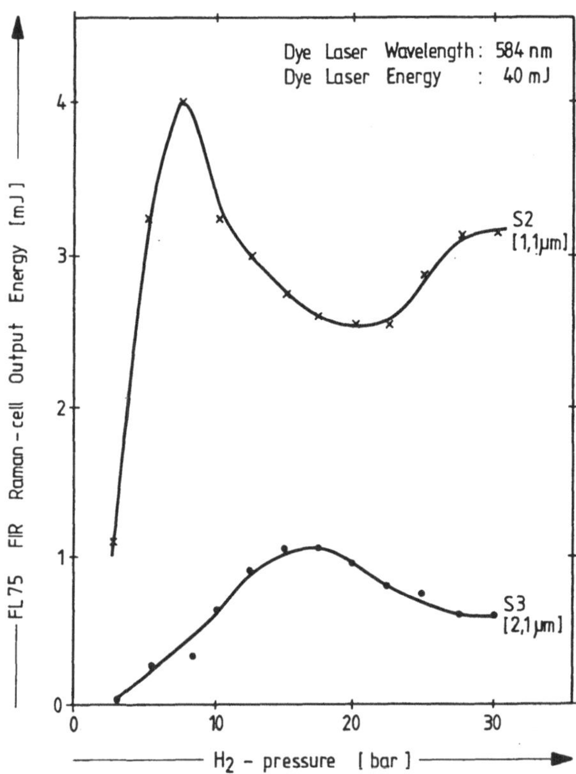

Fig. 4. Energy for the Stokes S1, S2, S3 as function of pump energy at H$_2$ 15 bars

In summary the Raman conversion of dye laser radiation has proved to be a nearly ideal high power tunable light source in the VUV and with IR (cf. Fig. 5) with excellent spectral properties (0.04 cm^{-1})

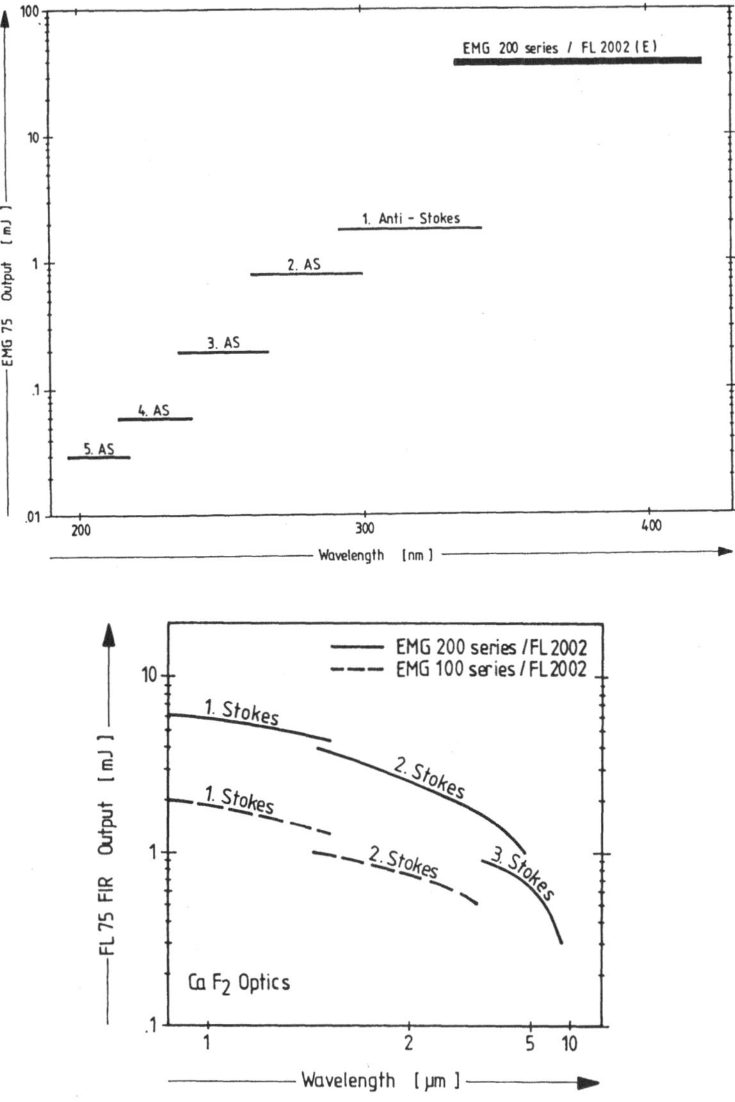

Fig. 5 a,b,Extending of the tunability of dye lasers in the UV and IR

References

/1/ V. Wilke, W. Schmidt, Appl. Phys. 18, 177 (1979).

/2/ H. Schomburg, H.F. Döbele, B. Rückle, Appl. Phys. 30, 131, (1983).

/3/ D.J. Brink, D. Proch, D. Basting, K. Hohla, P. Lokai, Laser and
    Opto-elektronik 3, 41, (1982).

/4/ W. Hartig, W. Schmidt, Appl. Phys. 18, 235 (1979).

/5/ A.J. Berry, D.C. Hanna, Opt. Comm. 45, 211 (1978).

/7/ W.R. Trutna, R.L. Byer, Appl. Opt 19, 301 (1980).

Q-switching and Cavity Dumping of CO$_2$ Waveguide Lasers

B.J. Stocker
Philips Research Laboratories,
Redhill, Surrey RH1 5HA, UK.

Introduction

For certain optical radar applications a pulsed source of infra-red radiation at 10.6 μm is required, at repetition rates greater than can be obtained from a TEA laser. A Q-switched CW carbon dioxide laser offers such a source, particularly if the compact waveguide type can be used. There are three main methods of Q-switching the laser, using either a rotating mirror, an electro-optic crystal, or a saturable absorber. However, there is relatively little information on the application of these methods to waveguide CO$_2$ lasers.

The lasers, which were constructed in our laboratories, have an alumina waveguide of up to 2 mm internal diameter, 2 mm wall thickness, cooled by flowing water in a jacket also constructed of alumina. A longitudinal d.c. discharge of about 2 mA is maintained along the waveguide in a gas mixture of He, CO$_2$ and N$_2$, with in certain cases the addition of Xe, at a total pressure of about 100 torr.

Active Q-switching

Initial experiments used a laser tube with KCl Brewster angle windows and external concave mirrors. Such a laser, capable of giving 3 W CW., was Q-switched by rotating one mirror at up to 30,000 rpm (500 Hz) giving pulses of 330 W peak power 100 ns pulse width.

For most applications faster repetition rates are required and electro-optic Q-switching is preferable[1]. To overcome the losses due to the electro-optic crystal a laser with greater gain length is required and those used have two or three discharge sections, each of 100 mm active length. Figure 1 shows one with three sections.

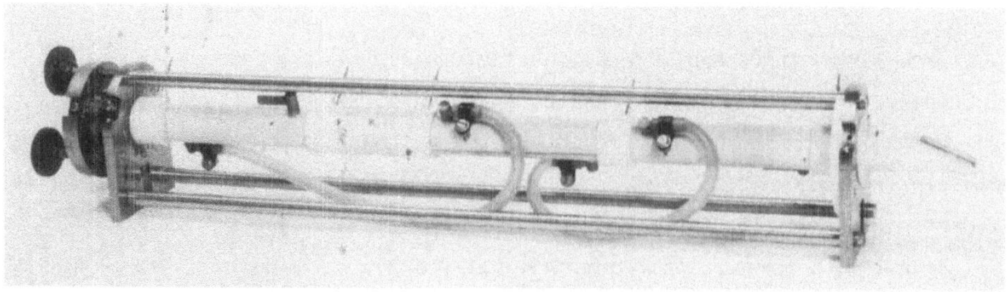

Fig. 1 Three-section laser tube

At one end of the laser there is an internal plane mirror, mounted on a bellows, with adjusting screws. The other end has a ZnSe Brewster angle window. The laser is operated from three power supplies, those for the left and centre sections being of negative polarity and the right positive. Figure 2 shows the laser tube mounted on a rail with mounts for an external, partially transmitting, concave mirror and for a water-cooled CdTe electro-optic crystal. A CdS quarter-wave plate is mounted between the laser window and the CdTe crystal.

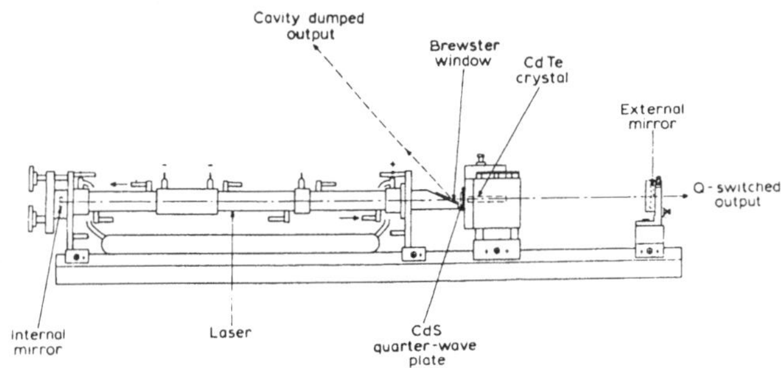

Fig. 2 Q-switched laser assembly

With no voltage applied to the electro-optic crystal the quarter-wave plate prevents lasing, because the polarized radiation from the laser window returns via the plate to the window in the orthogonal plane of polarization to be partially reflected out of the cavity. On applying a voltage pulse, the electro-optic crystal cancels the effect of the quarter-wave plate and the photon density in the laser cavity builds up to give the Q-switched pulse.

However the CdS plate is only quarter-wave at 10.6 μm wavelength and it is necessary to prevent lasing on the $00°1{\rightarrow}02°0$ transition between 9.2 and 9.6 μm. Other workers have used a grating but we chose to coat the external mirror to give 50% transmittance at 9.3 μm but 10% at 10.6 μm where Q-switching was to occur. This successfully prevents CW lasing in competition with the Q-switched operation at 10.6 μm.

The electro-optic crystal is driven from a Velonex generator giving pulses of up to 2.7 kV into a 200 ohm load across the crystal.

Using a flowing gas mixture of He, CO_2 and N_2 the laser has given Q-switched pulses of 660 W peak power, 110 ns pulse width. The corresponding CW power with the electro-optic crystal and quarter-wave plate removed was 10W.

If the voltage pulse is terminated at the peak of the Q-switched pulse a cavity dumped pulsed is obtained, by reflection off the Brewster angle window. To make this pulse of greater amplitude the external mirror was replaced by one of 1% transmittance at 10.6 μm but 40% near 9.3 μm. Figure 3 shows typical Q-switched and cavity dumped pulses observed on photon drag detectors. Cavity dumped pulses of 3.2 kW peak power 12 ns pulse width have been obtained using flowing gas. A

sealed-off gas mixture of 100 torr of 16 He, 4 CO_2, 2 N_2, 1 Xe gave 2.2 kW peak power.
These pulses were unchanged for repetition frequencies up to 30 kHz.

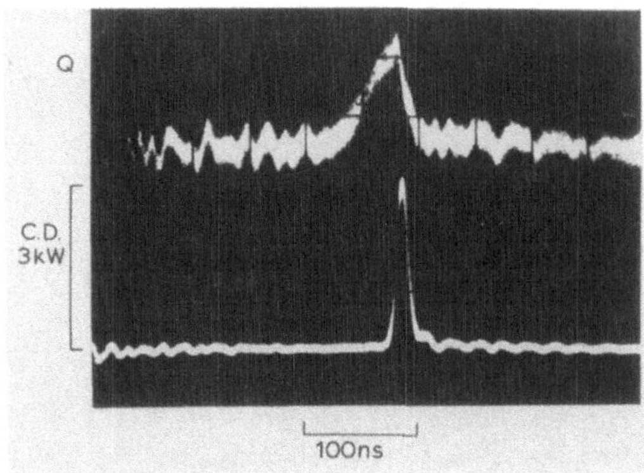

Fig. 3 Q-switched (Q) and cavity dumped (CD) pulses

Passive Q-switching

As indicated previously an alternative method of Q-switching is by means
of a saturable absorber, thus obviating the need for power dissipation in the supply
for the electro-optic crystal and parallel load. Cavity dumped operation is not
possible, however. The usual gas used for switching at 10.6 µm is SF_6[2]. To
obviate the need for a prism or grating in the cavity the example of Krupke[3] was
followed of mixing the SF_6 with a greater pressure of C_2F_3Cl. This absorbs over the
range 9.0 to 9.7 µm, preventing lasing on the $00°1 \rightarrow 02°0$ transition (also prevented by
the coating on the external mirror) and also more weakly from 9.7 to 10.4 µm, covering
the R branch of the $00°1 \rightarrow 10°0$ transition. Unfortunately, when introduced to a small
cell, with KCl Brewster angle windows, in the cavity (in place of the electro-optic
crystal) CW lasing was observed at about 10.3 µm, with no evidence for Q-switching.

The external mirror was therefore replaced by a grating and positive lens,
to maintain the laser on the P20 line at 10.6 µm. To observe the Q-switched pulse
the internal mirror was replaced by one of 10% transmittance. With pure SF_6 in the
cell Q-switched pulses were obtained, but the SF_6 pressure and grating angle were
extremely critical. Figure 4 shows typical pulses obtained, of about 300W peak
power, $5\frac{1}{2}$ kHz repetition frequency, but the pulses showed appreciable jitter. These
operational problems are believed to be related to the rapid variation in the SF_6
absorption coefficient as the laser is tuned across the pressure-broadened line width

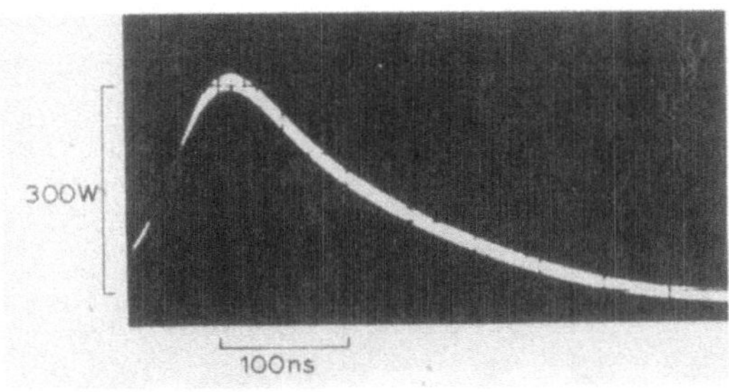

Fig. 4 Passive Q-switched pulse

Conclusion

Q-switching of a compact CO_2 waveguide laser can provide a source of high energy pulses at repetition frequencies in the kilohertz range. Unfortunately the techniques of Q-switching by a rotating mirror or a saturable absorber are limited and the best results are obtained using a CdTe electro-optic crystal, at the expense of additional energy losses.

In the final arrangement an additional mirror is included to reflect the cavity dumped output parallel to the Q-switched output. Also a gas reservoir is attached to the laser tube and contains $Mg(ClO_4)_2$ dessicant, to control the water vapour pressure and allow a reasonable life on one gas filling.

Acknowledgement

The work on active Q-switching was carried out under contracts from the Procurement Executive, Ministry of Defence, sponsored by DCVD.

References

(1) WAYNE, R.J, BUCZEK, C.J. and CHENAUSKY, P.P.
     IEEE J.Quantum Electron. QE-13, 23D (1977)

(2) WOOD, O.R. and SCHWARZ, S.E.
     Appl. Phys. Letters 11, 88 (1967)

(3) KRUPKE, W.F.
     Appl. Phys. Letters 14, 221 (1969)

EQ-switched and Cavity Dumped Tunabel CO$_2$ Laser and the Generation of Very Short Optically Pumped Far Infrared Laser Pulses

H.J.A. Bluyssen and A.F. van Etteger

Research Institute for Materials, University of Nijmegen

Toernooiveld, 6525 ED Nijmegen, The Netherlands

INTRODUCTION

Very short far infrared (FIR) radiation pulses are useful for purposes of time resolved and saturation spectroscopy in solid-state and molecular physics. In a previous paper[1] we have shown how very short FIR radiation pulses can be generated by optically pumping of molecular FIR laser with pulses from an electrically pulsed Q-switched (EQ-switched) CO$_2$ laser. Due to limitations caused by the rotating mirror which was used as the Q-switching element, pulsewidth, peakpower and repetition rate could not be varied independently.

In this paper we describe the operation and characteristics of a tunable CO$_2$ laser with an electrically pulsed discharge which can be either Q-switched or cavity dumped by an intracavity CdTe electrooptic modulator and its use for the generation of very short FIR laser pulses in CH$_3$F, CH$_3$OH and HCOOH.

In the EQ-switched mode the width of both the pumppulse and the FIR laser pulse can be varied by means of the length of the voltage pulse on the CdTe crystal.

EQ-SWITCHED AND CAVITY DUMPED CO$_2$ LASER

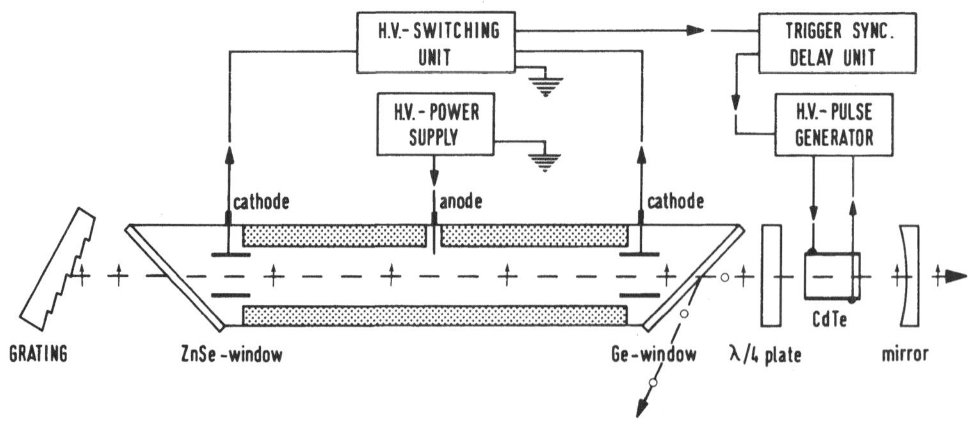

Fig. 1. Schematic diagram of EQ-switched and cavity dumped CO$_2$ laser.

A schematic diagram of the EQ-switched/cavity dumped tunable CO_2 laser including the complete electronics is shown in Fig. 1. The longitudinal discharge current of the low pressure ($\sim$ 20 Torr) plasma tube is pulsed giving rise to current densities in the plasma tube of up to 200 mA in pulses of 100 μs and up to a repitition rate of 1 kHz instead of the 20 mA in a cw discharge. In addition to a $\lambda/4$ plate the electrooptic modulator crystal is placed inside the laser cavity near the spherical mirror and is driven by a high voltage pulsegenerator delivering 4.5 kV pulses of variable length (50 ns- 1 μs), a risetime and decaytime of $\sim$ 40 ns and a repitition rate of up to 1 MHz. The high voltage pulsegenerator is synchronized with an adjustable delay to the current generator. If the voltage over the CdTe crystal is absent the linearly polarized radiation is coupled out via the Brewster angle window. If the voltage is present a radiation field can be build up in the cavity. In the EQ-switched mode the spherical mirror has a transmission of about 20% and radiation is coupled out during the phase the radiation field is building up. In the cavity dumped mode the partially transmitting spherical mirror is replaced by a totally reflecting one. The transmission is coupled out via the 76% reflecting intracavity Ge Brewster angle window during the decaytime of the voltage pulse which determines the width of the laser pulse.

Due to the high current density CO_2 laser pulses of a few tens of nanoseconds wide can be generated up to peak powers of 10 kW for the strongest of about 50 different observed lines. The shortest pulses with the highest peak power were obtained in the cavity dumped mode. Results are given in Fig. 2 which shows the peak power of 17 different CO_2 laser lines of the 9P-branch; the other three branches 9R, 10P and 10R show similar results. The peak power was determined from the average output power, the repetition rate and the pulsewidth. The pulseshape was observed by means of a Molectron type P3-01 pyroelectric detector with a risetime of 5 ns. A typical result is shown in Fig. 3. Values from 60 ns up to 100 ns have been measured for the pulse width of the laser lines of Fig. 2. It should be noted that in the cavity dumped mode the pulse width is mainly determined by the decaytime of the voltage pulse over the electrooptic modulator crystal and therefore cannot be varied. On the other hand in the EQ-switched mode the pulsewidth is determined by the gain of the medium and the losses of the cavity. Moreover the tail of the pulse and therefore its width can be affected by the length of the voltage pulse over the CdTe crystal. This is shown in Fig. 4 where the pulse width of the 9P20 CO_2 laser line is given for various lengths of the voltage pulse.

Finally it should be noted that due to the maximum duty cycle of about 10% of the current generator and the current pulse duration of about 100 μs the repetition rate of the system is limited to about 1 kHz.

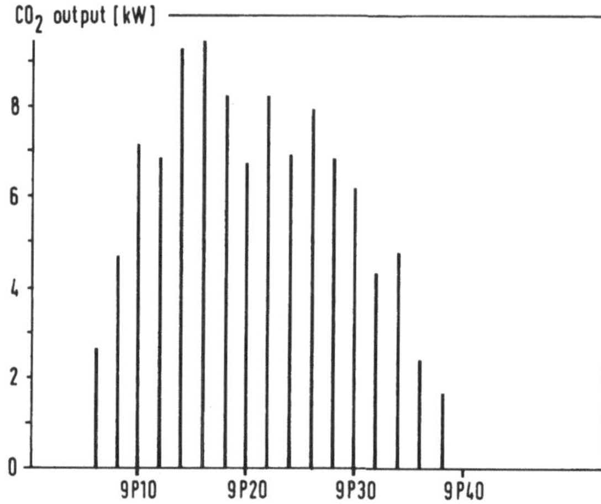

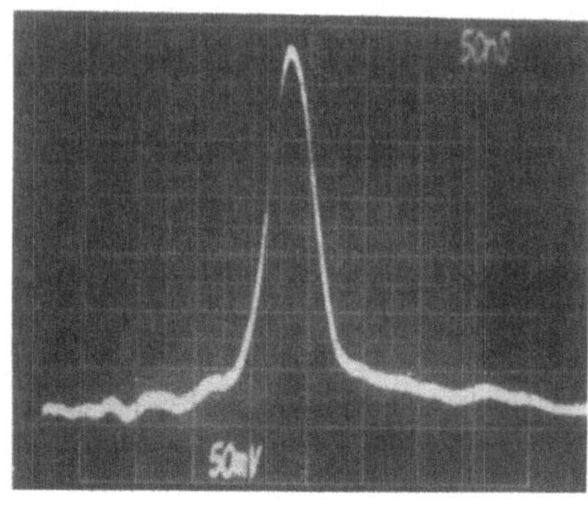

Fig. 2. Output power of CO_2 laser lines of the 9P branch in the cavity dumped mode

Fig. 3. CO_2 laser pulse shape in the cavity dumped mode

GENERATION OF VERY SHORT FAR INFRARED LASER PULSES

The EQ-switched and cavity dumped CO_2 laser pulses have been used to generate very short far infrared (FIR) radiation pulses by optically pumping of molecular gases such as CH_3F, CH_3OH and HCOOH.

In general the CO_2 laser radiation excites a vibrational transition in the molecule giving rise to a population inversion of two rotational levels like in the symmetric top molecule CH_3F, which was used by Chang and Bridges[2] to show for the first time operation of a far infrared optically pumped molecular laser. Since then hundreds of laser lines in tens of different molecules have been observed in the wavelength region from 10 µm up to 2000 µm.

A very important and interesting molecule for the generation of FIR radiation by optically pumping is CH_3OH and its deuterated components such as CH_3OD, CH_2DOH, CD_3OD etc. Due to the internal rotation these molecules have a much more complex vibrational rotation spectrum than CH_3F and therefore the number of available laser lines is much larger. Untill this moment a number of 330 different lines have been observed in this molecule which were for the greater part optically pumped by a cw CO_2 laser. For our 1 m long FIR waveguide laser both quartz and metallic waveguides were used with diameters of 13 and 25 mm. The pumppulses are focussed by an f = 50 cm ZnSe lens onto the 1 mm coupling hole of the input mirror. The FIR pulses transmitted through the 2 mm coupling hole of the output mirror were detected with a liquid He cooled n-GaAs detector of which the response time is limited by the lifetime of the electrons in the conduction band. The detector signals are fed to a Tektronix 7912 AD program-

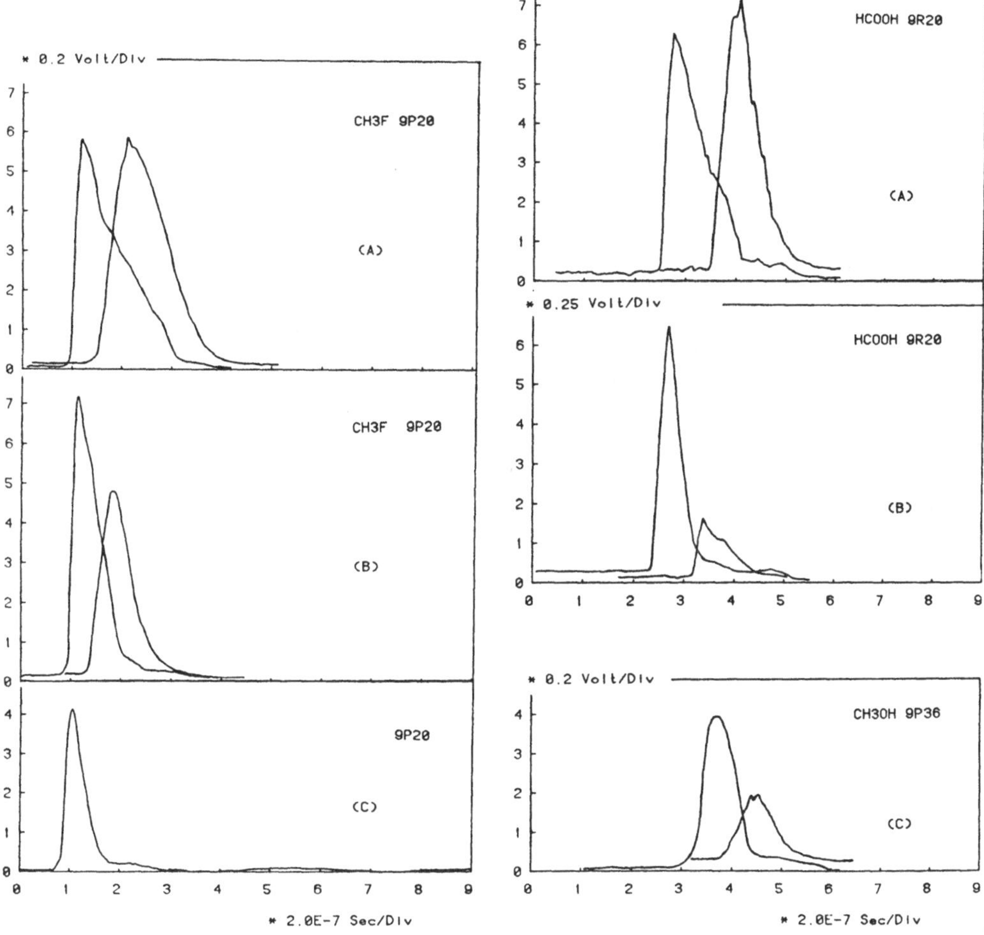

Fig. 4. FIR laser pulses in CH_3F at
λ=496 µm and 9P20 CO_2 laser
pumppulse. Width of the pulses
is varied by the voltage pulse
on the CdTe crystal

Fig. 5. FIR laser pulses in HCOOH
at λ=426 µm and in CH_3OH
at λ=118.8 µm

mable digitizer and analyzed by a Tektronix 4052 Desktop Graphic Computer and Display.

Fig. 4 shows typical results if FIR radiation pulses at a wavelength of λ = 496 µm generated in the molecular gas CH_3F at a pressure of about 350 mTorr in addition to the 9P20 (CO_2 laser line) pumppulse. The width of the pumppulse was varied from 200 ns (A) to 120 ns (B) and 80 ns (C) by decreasing the length of the voltage pulse over the CdTe crystal. For the characteristic times of the CH_3F laser pulses i.e. risetime t_r, decaytime t_d and pulse builduptime t_b, which is the time between the start of the pumppulse and the onset of the FIR pulse, we find for the case A: t_b = 80 ns, t_r = 70 ns and t_d = 240 ns, for case B: t_b = 80 ns, t_r = 70 ns and

44

t_d = 140 ns. For the last case (C) no FIR radiation pulse was observed which implies that the width of the pumppulse must be greater than the pulse builduptime t_b.

Both t_b and t_r depend on the gain of the FIR laser medium $g(\nu)$ = $B.h\nu$. ΔN_{FIR} where B is the transition rate, $h\nu$ is the FIR quantum of energy and ΔN_{FIR} is the population inversion density which is build up during the time t_b. ΔN_{FIR} depends on the intensity and width of the pumppulse, the pumptransition rate and the pressure of the molecular gas p. In a previous paper[1] it was shown that for p > 300 mTorr the gain and both t_b and t_r are independent of the pressure. So only by increasing the intensity of the pumppulse one can decrease t_b and t_r as will be shown in Fig. 5. The decaytime t_d depends on the molecular rotational relaxation time $(\pi\Delta\nu)^{-1}$, where $\Delta\nu$ is the homogeneous linewidth of the transition, which is approximately 40 MHz at 1 Torr for the polar laser molecules that have been studied to date[3]. For p = .35 Torr, $(\pi\Delta\nu)^{-1} \approx$ 30 ns, while the decaytime for case B of Fig. 4 is much longer. This may be due to the lifetime of the electrons in the conduction band of the GaAs detector. Fig. 5 shows results of FIR pulses from HCOOH at λ = 426 μm at a presure of about 1 Torr in addition to the 9R20 pumppulse. Due to the higher peak power of the pumppulse in case B, the pulse builduptime of the FIR pulse is shorter than for case A as was already mentioned above. On the other hand the shorter pumppulse of B generates a weaker FIR pulse than in case A. After the onset of the rapidly increasing FIR pulse, ΔN_{FIR} is quickly depleted and the gain is decreased with respect to the initial small signal gain. If the pumppulse is short, pumping stops before the maximum of the FIR pulse can be reached.

Finally Fig. 5C shows a FIR pulse at λ = 118.8 μm generated in CH_3OH and its 9P36 pumppulse as an example of a large number of possible FIR pulses in CH_3OH and its deuterated components.

In conclusion we have described the generation of very short FIR radiation pulses by optical pumping using an EQ-switched tunable CO_2 laser as the pumpsource. By variation of the length of the voltagepulse on the intracavity electrooptic modulator, the width of the pumppulse and the FIR pulse can be varied. For generation of FIR pulses the width of the pumppulse must be at least equal to the pulsebuilduptime of the FIR pulse. Therefore the cavity dumped pulses which are less than 100 ns wide could not be used to generate FIR laser pulses.

Acknowledgements: It is a pleasure to thank R. van Dongen, F. Schut and J. Gerritsen for their technical assistance.

REFERENCES

1. H.J.A. Bluyssen, A.F. van Etteger, J.C. Maan and P. Wyder, IEEE J. Quantum Electron., vol. QE-16, 1347 (1980.
2. T.Y. Chang, T.J. Bridges and E.G. Burkhardt, App.Phys.Lett. 17, 249 (1970).
3. D.T. Hodges, Infrared Phys., vol. 18, 375 (1978).

Dual-Beam Large Aperture CO_2 Amplifier

J.Badziak, M.Borzęcki, A.Chojnacka, Z.Dźwigalski, A.Kalbarczyk,
Z.Kurzyński, L.Perliński, J.Teter

S.Kaliski Instytute of Plasma Physics and Laser Microfusion
Warsaw Poland

I. Introduction

The construction of high-power laser systems require to use
final amplifiers of large aperture. For example it is necessary
10 + 20 cm active aperture to obtain energy of 100 J in nanosecond
pulse from CO_2 laser amplifier. Such devices have been built in
several scientific laboratories /eg. /1 + 5// . One of the possible
solution is to construct double-sided CO_2 electron-beam controlled
amplifiers. This construction ensures, among the others, to obtain
large volumes of an active medium at relatively low overal dimensions
of the device. The first solution of this type was developed at the
Los Alamos Scientific Laboratory /1/. In this paper we present construction and initial results of testing of the electron-beam controlled CO_2 amplifier with a double active volume, built at the IPPLM.

II. Construction of the amplifier

The amplifier consists of two basic parts: two-sided electron
gun and two amplifier chambers. The view of the amplifier is shown
in Fig. 1. The electron gun provided with a two-sided cold cathode
is supplied from two parellelly operating five-stage Marx generators
/5 x 50 kV/. Amplifier chambers containing an active medium are made
of glass fibre reinforced polyester resin. Thanks to thier good
electric insulation properties they are characterized by relatively
small dimensions and weight as compared to commonly used metal chambers. Both active media have the same geometry of maximum dimensions
20 x 20 x 120 cm^3. Spacing of main electrodes is variable from 10 to
20 cm. Each active medium is pumped from a two-stage Marx generator.
A specially developed trigger pulse cable generator ensures a proper
synchronization of the electron gun and systems supplying active
media.

III. Results

It is known that the energy deposited in amplifier active medium with the non-self-sustained discharge is decided first of all by parameters of the electron gun i.e. by useful current density of the beam /transmitted current/ and its time duration. Fig. 2 presents transmitted e-beam current I_{bt} and its duration /at half-amplitude/ $\tau_{0,5}$ versus spacing d of electron gun electrodes for a copper pointed cathode. There are two reasons of decrease $I_{bt} = f/d/$: decrease of electric field intensity between gun electrodes with the increase

Fig. 1. View of the dual-beam CO_2 laser amplifier.

of thier spacing and a considerable large beam divergence. On the other hand, reasons why $\tau_{0,5}$ increases versus d are not completly clear. The useful current decays when instability, created in the expanding cathode plasma, develops /total current canalizes/. Probably, the development of the above instability depends on the density of the current taken from the cathode. This current, for a metal pointed cathode, grows versus time and the cricital value, necessary for instability developement, is reached quicker in case of smaller electrode spacing. Better results /I_{bt} almost 3 times greater and $\tau_{0,5}$ longer by 25% / are obtained for a titanium pointed cathode. It seems to be caused by the fact that practically only top lands operate in case of the titanium points while in case of the copper points there operate side surfaces as well.

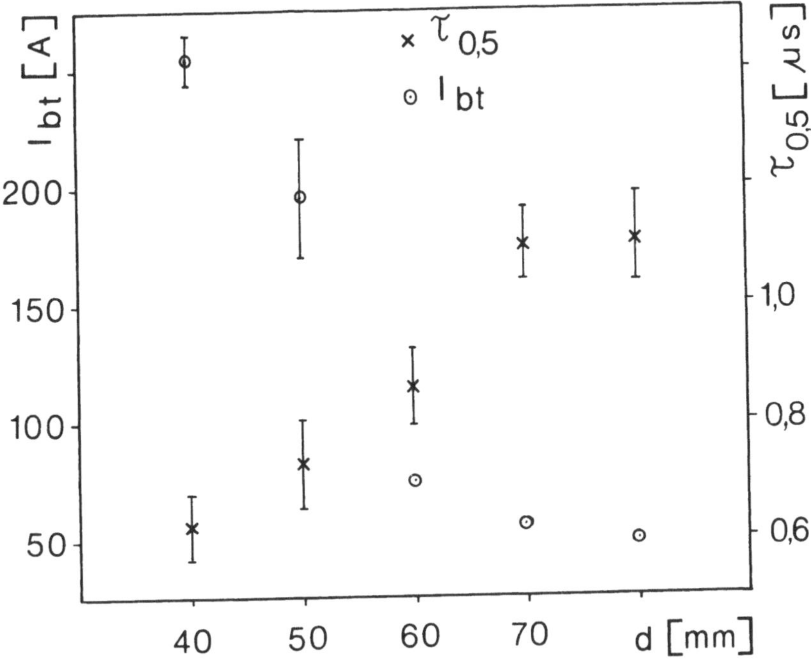

Fig. 2. Transmitted e-beam current and its pulse duration as a
function of the gun electrodes distance for copper edges
of the cathode.

Preliminary measurements of the small signal gain coefficient
g_o were carried out for the spacing between the main electrodes equal
to D = 10 cm and D = 15 cm. Charging voltage of the two-stage Marx
generator was varied from 30 to 60 kV. Results presented below were
obtained at capacity of one stage equal to 3,2 uF.
The value of g_o was measured by probing the medium with cw CO_2
laser radiation /6/. Furthermore, voltage variation with time on
medium electrodes and the current flowing throught the medium were
recorded.
Characteristics taken for given parameters of the amplifier
are presented in Figs. 3 and 4. Fig. 3 shows a dependence of g_o upon
the charging voltage of Marx generator supplying the active medium
for different pressures of CO_2: N_2: He mixture. Fig. 4 presents the
dependence g_o = f/E/p/ at pre-determined electron gun parameters and
different pressures of CO_2: N_2: He = 1:1:3 mixture. One can see that

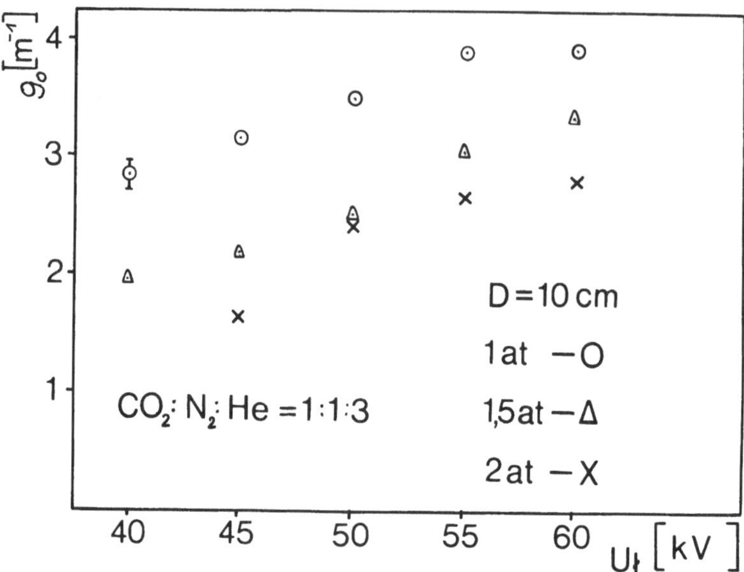

Fig. 3. Small signal gain coefficient versus supply voltage of pumping generator.

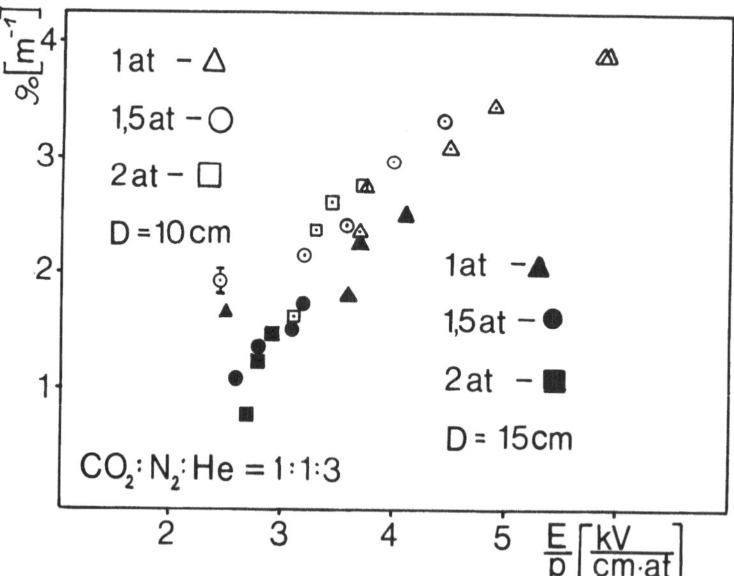

Fig. 4. Small signal gain coefficient versus electric field.

g_o grows with the increase of E/p. At applied E/p values we still do not enter the range of saturation of the dependence $g_o = f/E/p/$. This permits to increase g_o value still in an easy way by application of higher E/p. Moreover, one can see from Figures, that the discharge passes from the self-sustained regime to the non-self-sustained one /low variations of E/p at high p/ with the increase of pressure.

Testing of the amplifier is being continued.

References

1. LASL Progress Report LA-6982-PR, 1977.
2. Annual Progress Report on Laser Fusion Program, ILE Osaka, 1979.
3. V.A.Adamovich, V.N.Ansimov, E.A.Afonin, V.Yu.Baranov, at.al., Applied Optics, vol.19, 918, 1980.
4. YU.Y.Avilov, V.Yu.Baranov, V.A.Burcev, at.al., Voprosy atomnoi nauki i tehniki - Termojadernyi sintez, 2/10/, 3, 1982.
5. U.Bizzarri, G.Messima, L.Picardi, A.Vigmati, Optics and Laser Technology, 210, August 1981.
6. A.Kalbarczyk, Z.Kurzyński, M.Loth, IPPLM Report, No 23/80/ 43.

A Study of Gas-Throat Mixing Gasdynamic Laser Driven by Combustion of C_6H_6 and N_2O

K.Kitagawa
First Research Center, Japan Defense Agency
Nakameguro 2-2-1, Meguro, Tokyo, 153, Japan

Conventional gasdynamic lasers (GDLs) driven by combustion of gas fuels and oxidizers (G-F/O) have abilities to produce high energy of more than several tens of kilowatt. The lasing performance of the GDLs will be improved considerably by changing the conventional type to the mixing type GDLs (MGDLs) and the G-F/O to liquid fuels and oxidizers (L-F/O) because of the high optical gain of MGDLs and the small size of L-F/O.

Various mixing schemes have been reported for MGDLs such as side wall injection /1/, screen mixing /2,3/, throat mixing /4,5/ and gas-throat mixing /6/. For these mixing techniques, the throat mixing GDL can generate the highest gain /5/ and the gas-throat mixing GDL is expected to provide the high optical gain due to its rapid and homogeneous mixing capabilities /6/.

Since there exist no fuels and oxidizers which produce only hot N_2 gas by chemical reaction, donor gases in MGDLs driven by combustion must contain contaminants such as O_2, CO, CO_2 and H_2O. In these contaminants oxygen has a rather long relaxation time compared to the other gases /7/, and when C_6H_6 and N_2O are burned at a low equivalent ratio oxygen becomes a dominant contaminant in the combustion gases and hence, we have selected C_6H_6 and N_2O for our system.

Fig. 1 shows the experimental apparatus. The fuels and oxidizers in the run tanks are pressurized by 210 kg/cm^2-N_2 gas and supplied to the combustor through a needle and solenoid valve (see Fig. 1). The benzene and nitrous oxide are fired by a torch which used $CO+H_2$ and N_2O and ignited by an electric spark plug. The total mass flow rate of F/O and the combustion pressure are designed to be 500 g/sec and 100 kg/cm^2, respectively. Liquid carbon dioxide (L-CO_2) pressurized at 120 kg/cm^2 is injected to the nozzle and mixed with the combustion gases at the throat. The maximum time of combustion, which is limited by the amount of available L-F/O, is designed to be 5 sec.

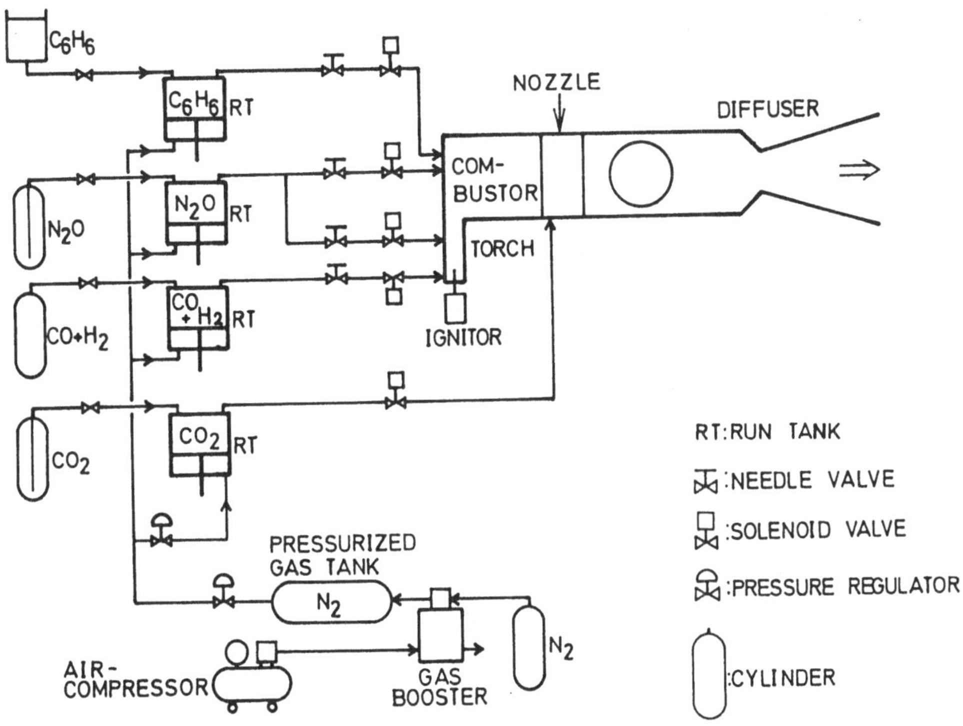

Fig.1. Experimental appatus of MGDL driven by combustion of C_6H_6 and N_2O. The pressurized gas tank contains 420 kg/cm^2-N_2 gas which is compressed from commercially available N_2 cylinders by a gas booster.

The gas-throat mixing nozzle is consisted of 73-conical nozzle units which are placed as the screen nozzle. When the donor gas goes through the nozzle throat, L-CO_2 is injected vertically to the primary flow. The cross section of the nozzle units is shown in Fig. 2. Two different area ratio of the conical nozzle are provided for our experiments, one of the area ratio is 44 and the other is 78.

At the throat of the nozzle, combustion gases are choked by the gas-throat formed by the injected laser gases. Since the gas-throat has very rapid mixing capability, primary and secondary flows will be fully mixed instantaneously at the throat section /6/. Then the mixture of donor and laser gases are expanded and accelerated to the supersonic flow in the divergent section of the nozzle.

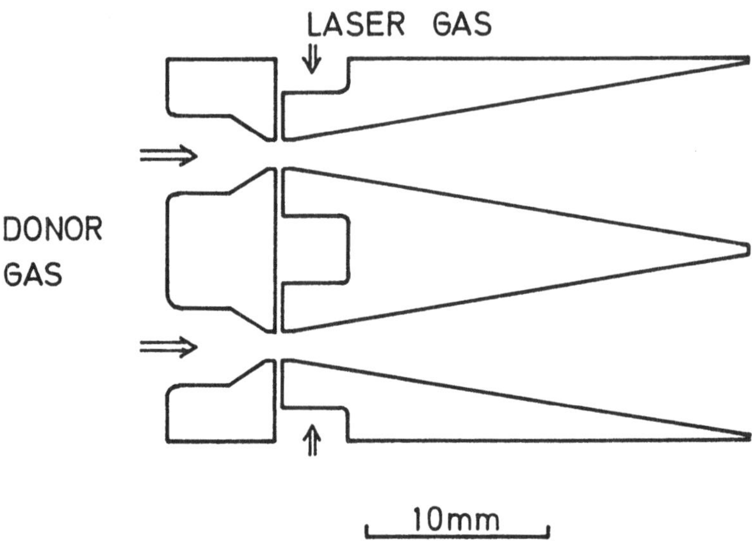

Fig.2. Cross section of gas-throat mixing conical nozzle units. These conical nozzle are placed as the screen nozzle.

In our preliminary experiments, stable and soot-free combustion have been obtained up to a pressure of 120 kg/cm^2. The maximum gain obtaine was 4.9 %/cm which is the highest value yet reported in the literature Subsequent experiments for laser output measurements are now proceedin and the detailed results will be presented.

References.
/1/ B.R.Bronfin, L.R.Boedeker, and J.P.Cheyer: Appl. Phys. Lett. 16 (1970) 214
/2/ P.Cassady, J.Newton, and P.Rose: AIAA Journal. 16 (1978) 305
/3/ W.Schall, P.Hoffman, and H.Hügel: Appl. Phys. Lett. 48 (1977) 688
/4/ V.N.Croshko, N.A.Fomin, and R.I.Soloukhin: Acta Astronautica. 2 (1975) 929
/5/ R.Bailly, M.Péalat, and J.P.E.Taran: Rev. Phys. Appl. 12 (1977) 17
/6/ K.Kitagawa: Appl. Phys. Lett. 39 (1981) 527
/7/ R.L.Taylor, and S.Bitterman: Rev. Mod. Phys. 41 (1969) 26

Optoelektronische Komponenten
und Sensoren

Optoelectronic Components
and Sensors

Optoelektronische Komponenten und Sensoren

K. Berchtold und J. Angerstein
TELEFUNKEN electronic, D-7100 Heilbronn, Deutschland

Der Artikel beschreibt die Trends der Entwicklung
moderner optoelektronischer Bauelemente; auf spezielle Systemlösungen wird eingegangen, um dort
die Schlüsselfunktion beschriebener Bauelemente
aufzuzeigen.

1   Einleitung

Die Bauelemente der Optoelektronik stellen einen technisch und
wirtschaftlich interessanten Markt dar, der jährliche Wachstumsraten
von weit über 15 % aufweist; die Methoden der Optik und die immer
weiter getriebene Perfektion der Bauelemente an der Schnittstelle
Optik/Elektronik bieten damit ein so hohes Innovationspotential dar,
daß in nahezu alle Bereiche moderner Technik neue Lösungen auf der
Basis optoelektronischer Komponenten eindringen. Hier werden Verfahren und besondere Systemlösungen beschrieben, die sich die heute
fortgeschrittene Perfektion der Material- und Bauelementetechnologie
zunutze machen. Dabei finden auch Teilgebiete der Sensortechnik
Erwähnung, wo heute optoelektronische Komponenten - als direkte
Fühler oder mittelbar unter Verwendung von Fasern - durchaus kostengünstige Lösungen aufzeigen.

2   Wichtige Ausgangsmaterialien

Ähnlich wie in /1/ sind in Bild 1 wichtige Materialien eingeordnet,
wobei

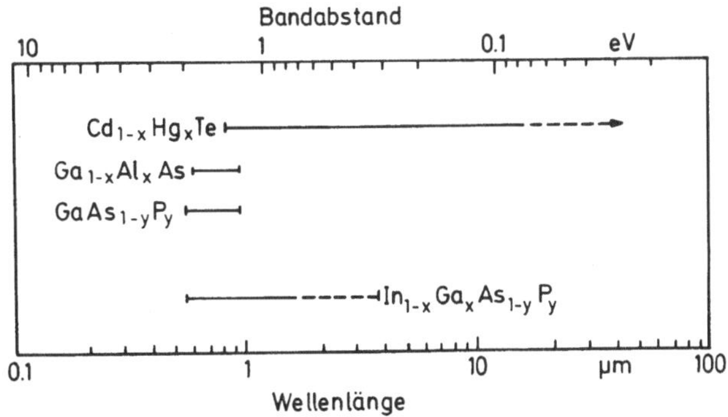

Bild 1.   Bandabstand und Wellenlängenbereich wichtiger optoelektronischer Materialien

als Ordnungsparameter Bandlücke und Wellenlänge dienen. Im folgenden werden zunächst Materialien näher betrachtet, die heute und wohl auch in Zukunft, sei es als Sender oder Empfänger, in den verschiedenen Wellenlängenbereichen der angewandten Optoelektronik verwendet werden (s. Bild 1). Es kann festgestellt werden, daß eine Auswahl bestimmter Materialien im Laufe der Entwicklung bereits stattgefunden hat; so stellt CdHgTe als Empfänger im infraroten Bereich das dominierende Material dar, während die Bleiverbindungen (PbSnTe, PbSnSe) für Empfänger nicht im selben Maße weiterentwickelt wurden (s. 3.3). Auch wird in Zukunft die quaternäre III-V-Verbindung InGaAsP das vorwiegend in Gebrauch befindliche Material der Nachrichtentechnik bei 1.3 µm bis 1.6 µm sein, so daß sich mit dem noch vor einigen Jahren auch aussichtsreichen GaAlAsSb nur noch wenige Entwicklungsarbeiten beschäftigen.

Für die vielfältigen Anwendungen in der Optoelektronik hat sich - über das Silizium hinaus, das als Standarddetektormaterial verwendet wird, hier aber nicht weiter betrachtet werden soll - eine Reihe verschiedener Materialien als geeignet für die entsprechenden Anwendungen herausgestellt. Diese Materialien werden aber in nahezu allen Fällen nicht als homogene Kristalle bearbeitet, sondern es werden epitaktische Schichten mit modifizierten Eigenschaften auf das einkristalline Substrat aufgebracht. Die Vorteile der Anwendung von Epitaxieschichten bei Sendern und Empfängern im Vergleich zur Benutzung von homogenem Ausgangsmaterial sind höherer Freiheitsgrad bei der Festlegung von Zusammensetzung und Dotierung, höhere Reinheit und bessere Kristallqualität, so daß daraus im allgemeinen bessere Bauelementeeigenschaften (Quantenausbeuten bei Sendern bzw. Empfängern) resultieren. Es bietet sich daher an, die unterschiedlichen Bauelementetechnologien nach Material und nach Art des Epitaxieverfahrens zu klassifizieren. Bei einer für die Funktion des Bauelementes (Sender, Empfänger) vorgegebenen Wellenlänge wird man danach fragen, welche Substrate zur Verfügung stehen und ob eine Anpassung der Gitter von Substrat und Epitaxieschicht bei Wahl der Bandlücke (Wellenlänge) möglich ist. Die folgende Übersichtstabelle gibt die derzeit in der Technik am häufigsten verwendeten Herstellverfahren für III-V-Bauelemente wieder.

Substrat	InP	GaAs			GaP
Epitaxie	LPE (MO)	LPE (MO) (MBE)	VPE		LPE
Material	InGaAs	GaAlAs GaAs:Si	GaAsP	GaAsP :N	GaP:N GaP:Zn:O
Bauelement	Sender Empfänger	Rote bis Infrarote LED's und Laser	Rote LED	grün gelb orange	Grüne LED Rote LED's
Wellenlänge nm	1300 ... 1700	660 800 900	660	565 590 630	565 690

Tabelle 1. Substrate und Herstellverfahren wichtiger optoelektronischer Bauelemente

Die Tabelle zeigt, daß es auch heute eine Reihe von konkurrierenden Verfahren gibt, um bestimmte Bauelementeigenschaften zu erzielen. So sind die Verfahren für rote LED's stark unterschiedlich (näheres s. 2.1). Es können hierzu beim Vergleich von Flüssigphasen- und Gasphasenepitaxie folgende Feststellungen getroffen werden:

- Die Flüssigphasenepitaxie führt bei LED's zu höheren Quantenausbeuten (s. 2.1) infolge besserer Materialgüte; grüne LED's werden heute vorwiegend so hergestellt.

- Die Gasphasenepitaxie erlaubt niedrige Herstellkosten (höherer Scheibendurchsatz, bessere Steuerbarkeit). Auch können nachfolgende Prozesse wie Planartechnik (z. B. für monolithische Anordnungen) mit höherer Ausbeute durchgeführt werden (s. 3.1 Monolithische LED-Zeilen; man wird daher Gasphasenverfahren (VPE- bzw. MO-CVD-Epitaxie) anstreben, wenn es die Anforderungen an die Bauelementeeigenschaften zulassen (dünne Schichten für Hetero-Strukturen, ausreichende Helligkeit bei Rotdioden).

- Die Gasphasenepitaxie erlaubt ein höheres Maß an Flexibilität (Zusammensetzung, Dotierung). Einen weiteren Schritt in dieser Richtung stellt die MBE-Methode dar, womit spezielle Strukturen gezielt hergestellt werden können.

2.1 GaAsP - GaAlAs-System

Rote LED's

Technologien für die Herstellung roter LED's sind in Bild 2 dargestellt.

Technologien für rote LED's

Bezeichnung	Material	Substrat	η (%)	GF(mlm/A)	λ(nm)
Standard	$GaAs_{0.6}P_{0.4}$	GaAs	0,4	350	660
High - efficiency	$GaAs_{0.4}P_{0.6}$:N	GaP	0.4	1500	630
Zn – O	GaP: Zn, O	GaP	6	700	690
Intensivrot	$Ga_{0.65}Al_{0.35}As$	GaAs	3	4000	650

Bild 2.   Technologien für rote LED's

Rot emittierende Lumineszenzdioden auf der Basis von GaAsP auf GaAs-Substrat werden mit Hilfe der Gasphasenepitaxie weltweit in sehr großen Mengen produziert. Die entsprechenden Fertigungsmethoden sind daher stark ausgereift; dies wiederum bedingt eine hohe Materialperfektion und Homogenität, wodurch beispielsweise Reihenanordnungen (s. 3.1) mit kleinen Toleranzen der elektro-optischen Parameter realisiert werden können.

Rote Emission ist auch im System GaAlAs möglich und erlaubt hier gerade eine besonders effektive, strahlende Rekombination. Diese LED's werden heute so hergestellt, daß auf einem p-leitenden GaAs-Substrat zwei GaAlAs-Schichten flüssig-epitaktisch abgeschieden werden. Den Schichtaufbau zeigt Bild 3.

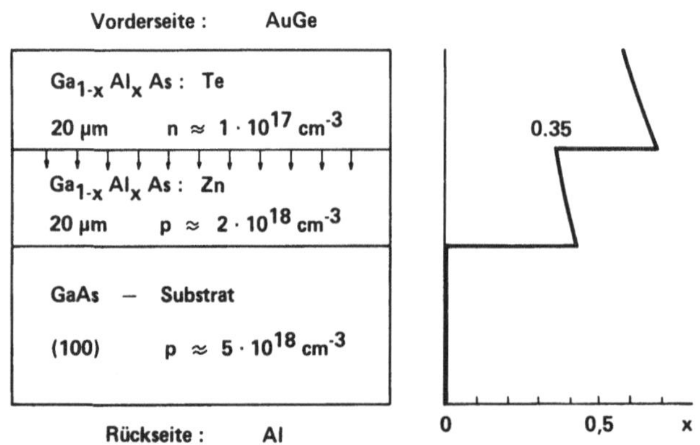

Bild 3.   Schichtaufbau für intensiv-rote LED's

In letzter Zeit hat nun die GaAlAs-Technologie entscheidende Fortschritte gemacht / 2, 3 /. GaAlAs-Dioden sind etwa 10 mal heller als Standard-Rotdioden (660 nm) und 3 mal heller als "high-efficiency-red" ("orange" auf GaP-Substrat).

GaAlAs-Laser

Derzeit werden GaAlAs-Laser mit Hilfe der Flüssigphasenepitaxie hergestellt. Um den Durchbruch der Lasertechnologie im kostengünstigen Einsatz bei hohen Stückzahlen zu sichern, sind auf der Seite der Chipherstellung neue Verfahren langfristig erforderlich, die eine rationellere Herstellung ermöglichen. Es ist zu erwarten, daß die Gasphasenepitaxie aus metallorganischen Verbindungen (MOCVD) sich gegenüber der Flüssigphasenepitaxie bei der Laserherstellung durchsetzen wird. Bei MO-CVD-Epitaxie handelt es sich um einen pyrolithischen Abscheidevorgang, der für das GaAlAs-System bei (im Vergleich zur herkömmlichen Gasphasenepitaxie) niedrigen Prozeßtemperaturen (700 °C) abläuft und somit dünne Schichten und scharfe Übergänge bei hoher Gleichmäßigkeit über die Scheibe herzustellen gestattet. Es werden erhebliche Anstrengungen unternommen, die MOCVD-Epitaxie auch beim InGaAsP-System zu verwenden / 5 /.
Wenn auch bekanntlich die Entwicklung des Halbleiterlasers und speziell die des GaAlAs-Lasers ursprünglich sehr stark vorangetrieben wurde, weil er für den Einsatz der Glasfaserübertragung unentbehrlich schien, so hat heute auf ganz anderen Gebieten (optische Speicher, zu denen auch die "Compact-Disc" gehört) die Markteinführung des GaAlAs-Lasers in sehr großen Stückzahlen bereits begonnen. Schon auf der Laser 79 wurde über eine Vielzahl unterschiedlicher Strukturen berichtet und Aufbau und Eigenschaften beschrieben; daher sollen im folgenden zusammenfassend die unterschiedlichen Merkmale von "gain"-geführten und "index"-geführten Lasern im Hinblick auf die Anwendung in der Speichertechnik behandelt werden / 4 /.
In Bild 4 sind die beiden Strukturen schematisch dargestellt; den Vergleich der beiden Strukturen gibt Tabelle 2 wieder, wobei die für die Speichertechnik maßgebenden Parameter angegeben sind.

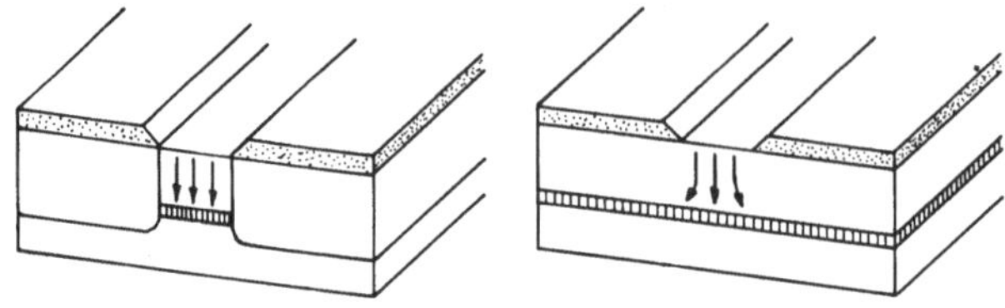

Bild 4. Strukturen eines "index"-geführten (a) und eines "gain"-
geführten (b) Lasers in schematischer Darstellung

Parameter	index	gain
Ausgangs-leistung	5 mW	> 20 mW
Schwellstrom	20 mA	80 mA
Astigmatismus	< 5 μm	< 20 μm
Wellenfront	sphärisch	nichtsphärisch
Spektrum (Moden)		
longitud.	1	vielmodig
trans.	1	1
Modulationsband-breite	1 GHz	350 MHz
Signal-Rausch-verhältnis	(-)	(+)

Tabelle 2.   Vergleich unterschiedlicher Laserstrukturen

Es kann aus heutiger Sicht festgestellt werden, daß für Laser hoher
Ausgangsleistung (> 20 mW) z. B. als Schreiblaser, "gain"-geführte
Strukturen wie der V-Nut-Laser, vorteilhaft eingesetzt werden kön-
nen, zumal sich der Astigmatismus durch Zylinderlinsen ohne großen
Kostenaufwand beseitigen läßt.

2.2 InGaAsP-System

Heute steht ohne Zweifel fest, daß aufgrund der geringen Dämpfung und der geringen Materialdispersion der Faser im Wellenlängenbereich zwischen 1 µm und 1.6 µm sich die Übertragung in diesem Bereich durchsetzen wird. Als Ausgangsmaterial sowohl für Sende- wie auch für Empfangselemente steht der quaternäre Mischkristall InGaAsP zur Verfügung, wobei InP als Substrat dient 6/.
Die Technologie des InGaAsP-Systems entspricht im wesentlichen Punkten der Flüssigphasentechnologie des GaAlAs-Systems. Wie erwähnt, wird auch hier langfristig das MO-CVD-Verfahren angestrebt. Im Vergleich zum GaAlAs-Laser zeigt der InGaAsP-Laser eine wesentlich größere Temperaturabhängigkeit des Schwellstroms.
Dieses sogenannte T_o-Problem beim InGaAsP-Laser wird dazu führen, daß Laser vorwiegend im Fernnetz eingesetzt werden, wo aufwendigere Temperaturstabilisierungen mit Peltierkühler wirtschaftlich tragbar sind. Der erwähnte Aufwand ist möglicherweise dann zu umgehen, wenn kostengünstige, index-geführte Laser mit Schwellstromdichten um 10 mA zur Verfügung stehen. Die Chancen für LED's, als Planar- oder Kantenemitter im Teilnehmerbereich eingesetzt zu werden, sind groß, wenn auch die geringeren Bandbreiten von LED's im Vergleich zu Lasern deren Einsatzbereich begrenzen.
Bei den Empfängern werden sich wohl Pin-Dioden generell gegenüber Lawinendioden im Teilnehmerbereich durchsetzen, zumal mit Hilfe von GaAs-MESFET's rauscharme und schnelle Pin-FET-Kombinationen zur Verfügung stehen.
Den Aufbau einer typischen Pin-Empfangsdiode zeigt Bild 5.

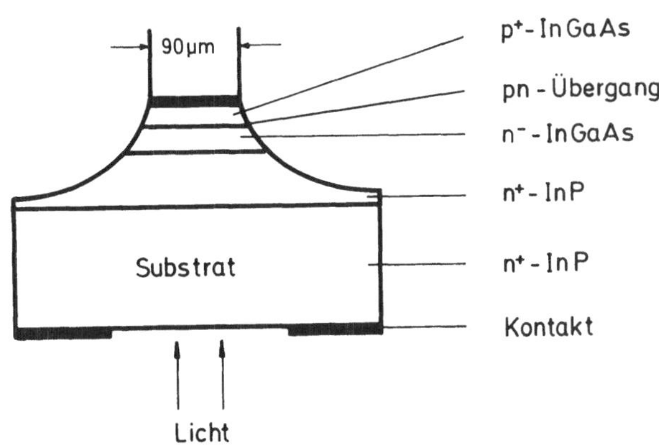

Bild 5.   Aufbau einer InGaAs/InP Pin-Diode 6/.

Die Absorption findet in der InGaAs-Epitaxieschicht statt, so daß die Dioden bis etwa 1.7 µm empfindlich sind. Die Dunkelströme liegen typisch unter 10 nA. Bei 1.3 µm beträgt der Quantenwirkungsgrad 70 %. Auch angesichts der Forderung nach einer Mehrfachausnutzung der Faser mit unterschiedlichen optischen Trägerwellen ist heute die Frage noch nicht entschieden, ob die Übertragung bei 1.3 µm (auf grund der geringen Materialdispersion in der Faser) oder bei 1.5 µm (aufgrund der geringen Verluste) stattfindet, oder ob nicht letzten Endes der gesamte Bereich im Wellenlängenmultiplex genutzt wird.

3   Spezielle Systemlösungen auf der Basis optoelektronischer
     Komponenten

Im folgenden werden spezielle Systemlösungen beschrieben, die auf der Verfügbarkeit optoelektronischer Techniken beruhen.

3.1 LED-Zeilen für die Elektrophotographie

Es konnte gezeigt werden, daß die Belichtung mit monolithischen LED-Bildpunktzeilen in der Elektrophotographie bei Photoleitertrommeln für genügend schnelle Entladung sorgt (< 2.5 µW/mA bei 660 nm). Die räumliche Auflösung reicht bei ≤ 100 µm Raster auch für hochwertige Drucker und Kopiergeräte aus. Als parallel arbeitende Anordnung ohne mechanischen Scanner mit der Möglichkeit der direkten Helligkeitssteuerung über den Flußstrom stellt die LED-Zeile ein nahezu ideales Belichtungsverfahren dar. Die Anordnung bei der Elektrophotographie zeigt Bild 6. Bild 7 deutet die Möglichkeiten konkurrierender Verfahren bei unterschiedlichen Photoleitern an.

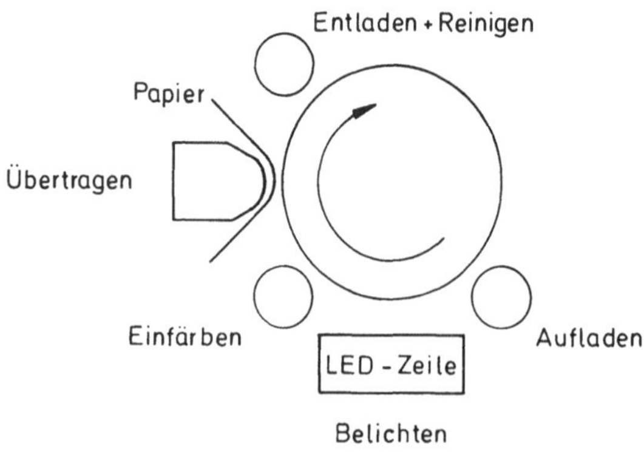

Bild 6.   Prinzipskizze des Verfahrens der Elektrophotographie

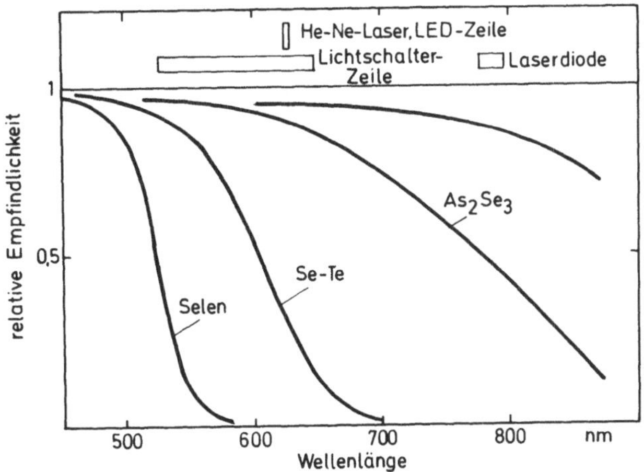

Bild 7.   Empfindlichkeitsbereiche unterschiedlicher Photoleiter

Es werden monolithische Module mit etwa 200 Bildpunkten (ca. 50 x 50 μm^2 aktive Fläche) zu längeren Zeilen zusammengesetzt. Die weitgehend zur Perfektion getriebenen GaAsP- auf GaAs-Substrat-Technologie (Gasphasenepitaxie, Planartechnik) ermöglicht die erforderliche hohe Gleichmäßigkeit ($\pm$ 50 %) der Belichtungsintensität über die Zeilenlänge.

3.2 Sensoren auf der Basis optoelektronischer Komponenten

Jede Detektordiode oder auch Solarzelle stellt einen Sensor, nämlich für elektromagnetische Strahlung, dar. Auf der Basis dieser direkten Messung kann eine Vielfalt von Meßwertaufnehmern gebaut werden, die alle auf dem Lichtschrankenprinzip aufbauen; so gelangen die unterschiedlichsten optischen Sensoren für die Messung einer Position zur Anwendung. Derartige Sensoren können durchaus auch kostengünstige Massenanwendungen z. B. in Zündmodulen für Automobile, erschließen; auch stellt beispielsweise die Laserabtasteinheit in optischen Speichern einen Sensor im erweiterten Sinne dar. Hier sollen jedoch im folgenden nur Sensoren auf der Basis optischer Fasern beispielhaft betrachtet werden. Ihre Wirkung beruht auf dem Nachweis geringer Phasenunterschiede (Interferometer) oder Amplitudenunterschiede, wobei die Faser selbst sowohl zur Lichtleitung wie auch zur Umsetzung der direkten Sensorgrößen (Druck, Temperatur etc.) dient /7, 8/ . Faseroptische Sensoren werden häufig verwendet, wo ihre Unempfindlichkeit gegenüber hohen Temperaturen vorteilhaft ist. So lassen sich z. B. in Glasfasern Dotierstoffe wie seltene Erden einbauen, die die Absorption in der Faser bei bestimmten Wellenlängen (z. B. 850 nm) beeinflussen. Die Temperaturabhängigkeit dieser Absorption (im Bereich von 30 °C bis 900 °C) kann als Sensorprinzip fungieren, wodurch sich robuste Temperatursensoren aufbauen lassen. Als Beispiel eines optischen Sensors als kostengünstige Lösung in der Automobilelektronik soll ein Füllstandssensor dienen. Das Prinzip ist in Bild 8 dargestellt. Die Realisierung erfolgt mit Hilfe herkömmlicher Infrarotsender (GaAs) und -empfänger (Si).

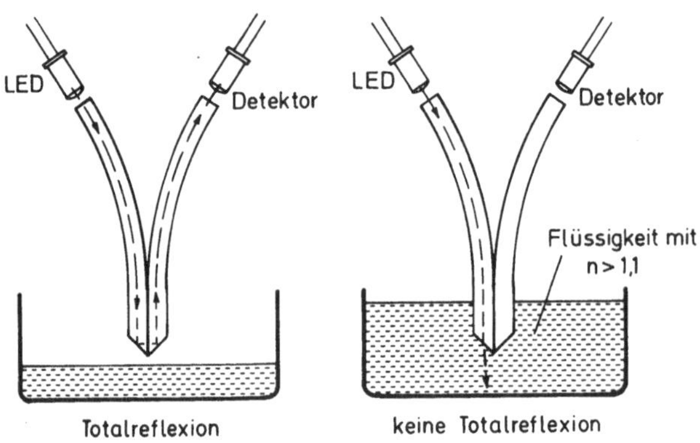

Bild 8.   Prinzipskizze eines Füllstandsensors

3.3 Trends in der Infrarottechnik (CdHgTe)

Die Geräte der Wärmebildtechnik im atmosphärisch transparenten Bereich von 3 μm bis 5 μm und 8 μm bis 14 μm sind heute noch durch den Einsatz von monolithischen Detektorzeilen (CdHgTe als Photoleiter mit parallelen Kanälen, ca. 200 Bildpunkte) geprägt 9 . In Bild 9 sind die infrage kommenden Detektormaterialien für die genannten Wellenlängenbereiche zusammengestellt. Im unteren Teil ist die Transparenz der Atmosphäre aufgetragen. Die schraffierten Bereiche deuten die sogenannten optischen Fenster an.

Schon seit langem war der Weg in Richtung der Verwendung von Detektormatrizen höherer Bildpunktanzahl mit direkt angeschlossener Signalverarbeitung ("focal plane array") vorgezeichnet. Infrage kamen auch Si-Infrarot-CCD's (Dotierungen von Si mit Ga oder In); heute muß festgestellt werden, daß neben technologischen Schwierigkeiten der Mangel an kostengünstigen Kühlern (20 K bis 30 K) für den

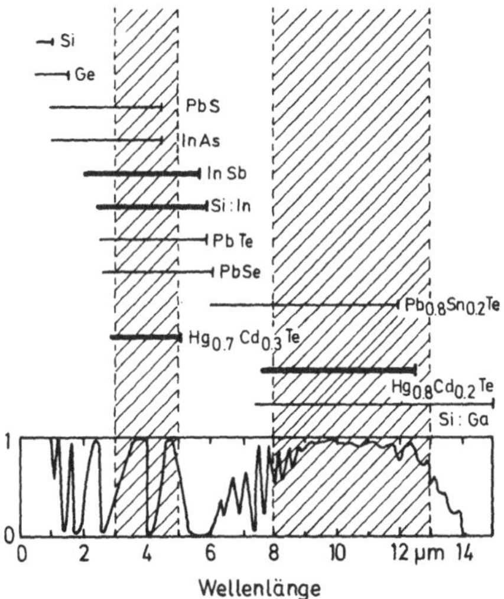

Bild 9.   Detektormaterialien für die Wärmebildtechnik

Nichteinsatz dieser Entwicklungen verantwortlich ist. Es ist heute
abzusehen, daß sich CdHgTe-Detektormatrizen in photovoltaischer Form
bei "focal-plane-arrays" durchsetzen werden, wobei die Bildverarbeitung in Si-CCD-Arrays erfolgt, die in hybrider Form an die Detektoren innerhalb derselben Montageebene angebracht werden.
Die folgenden beiden Bilder (Bild 10 und Bild 11) zeigen die Methoden auf, wie die Schnittstelle Detektorarray-Signalverarbeitung
gestaltet werden kann.

Der hohe Aufwand, der für derartige Entwicklungen erforderlich ist,
läßt eine Konzentration auf ein aussichtsreiches Material als sinnvoll erscheinen (CdHgTe), insbesondere als sich hierbei ein gewisser
Vorsprung in der Materialbeherrschung sich herauskristallisiert hat.

IR-CCD (schematisch)

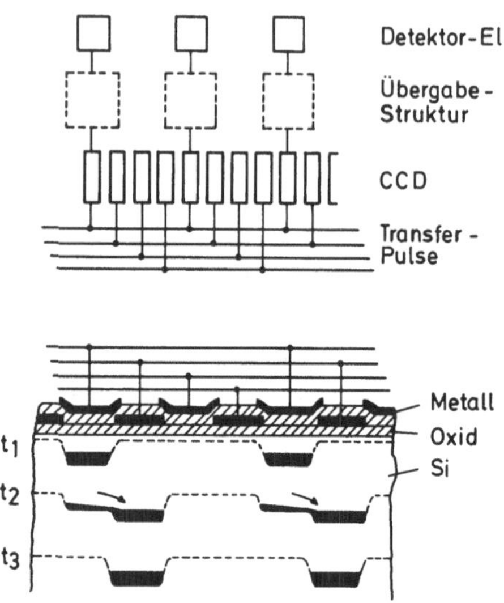

Bild 10.   Anordnung von Detektorarray und Signalverarbeitung

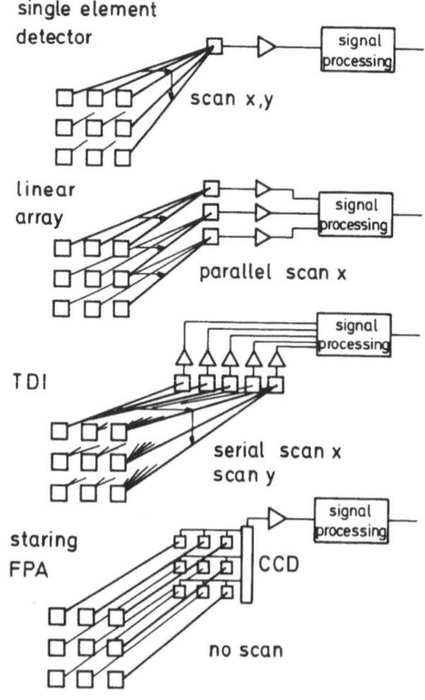

Bild 11.   Unterschiedliche Scan-Mechanismen in der Wärmebildtechnik

4 Ausblick

Die perfekte Beherrschung der Materialtechnologie stellt einen
Schlüssel für neuartige Systemlösungen in der Informationstechnik
dar. So konnten wir verfolgen, wie verfügbare Entwicklungen (z. B.
GaAlAs-Laser) nicht erwartete Einsatzbereiche in Massenanwendungen
(z. B. optische Speicher) finden und dort kostengünstige Lösungen
ermöglichen. Es ist zu erwarten, daß auch in Zukunft ähnliche Ten-
denzen, z. B. beim InGaAs-System auftreten, so daß der schnelle
Fortschritt auf dem Gebiet optoelektronischer Komponenten entschei-
denden Einfluß auf die zukünftige Situation der Mikroelektronik
ausüben wird.

5 Literatur

/1/ J. Hesse, K. Berchtold, J. Angerstein
    Conf. Proc. Laser 79, Munich
    "Trends in Optoelectronic Components"

/2/ C. Cohen
    Electronics, Nov. 20 (1980), 74

/3/ J. Varon et al.
    IEEE, ED-28 (1981), 416

/4/ E.E. Wagner, J. Angerstein, K. Petermann
    Proc. SPIE; Techn. Conf. Europe; Geneva, Apr. 18 - 22; (1983)

/5/ K.W. Benz, H. Haspeklo, R. Bosch
    Journal De Physique
    Colloque C5, Supplément au no 12, Dez. 1982 p C5 - 393

/6/ O. Krumpholz, H.P. Vollmer
    NTZ, Nov. 82, 35, 692

/7/ G.D. Pitt
    Elektr. Nachrichtenwesen 57, 1982, 102

/8/ E. Snitzer, J.R. Dunphy
    G. Meltz
    Int. Conf. on Fiberoptic Rotation Sensors
    Nov. 1981, Cambridge MA

/9/ H. Maier, J. Hesse
    in "Crystals, Growth, Properties and Applications" Vol. 4;
    Springer Berlin-Heidelberg 1980

Recent Advances in High-Speed Photon Detectors

BRANKO LESKOVAR
Lawrence Berkeley Laboratory
University of California
Berkeley, California 94720, USA

Recent progress of some fast high-gain photon detectors using photoemission and secondary emission processes is reviewed and summarized. Specifically, performance characteristics are presented, of the new Amperex XP 2020, RCA 8854, and Hamamatsu
R 647-01 conventionally design photomultipliers. Also, characteristics are presented
of the ITT F 4129 and Hamamatsu R 1564U extended lifetime microchannel plate photomu-
tipliers as well as certain special made photomultipliers intended for application in
positron emission tomography, high energy physics and plasma diagnostic experimental
systems. Finally, microchannel plates as photon detectors for ultra violet and x-ray
wavelengths are discussed.

Introduction

Fast high-gain standard and microchannel plate photomultipliers, are among the fast-
est and most sensitive devices for detecting the incidence of photons on a target surface. These devices have gained wide acceptance in research instrumentation, particularly in radiation detection (1), atomic and molecular subnanosecond flourescence decay studies, (2), optical ranging experiments (3), optical communication systems (4)
and plasma diagnostics (5). The detection of signals in practical systems in these
areas, requires photon detectors with high quantum efficiency, high gain, fast time
response and high data-rate capabilities. They should also have good output pulse-
height and time resolution. In many cases large sensitive areas are required. Also,
position-sensitive detection or imaging of incident radiation patterns is sometimes
necessary. In practically all these applications a minimum amount of noise or spurious signal should be present in the detector output.

For more than fifty years research applications have used the phenomenon of photoemission to convert absorbed incident radiation into an electron stream which is then
amplified by a secondary emission system. The detection process begins with a cathode
from which the incident radiation exites the emission of photoelectrons. The emitted
electrons are directed to a surface which has been treated to have high secondary electron emission. The secondary electrons produced by this first dynode are then directed
to another secondary emitter. The process is repeated for as many times as are required to amplify the initial electron stream by the desired amount, after which it
is collected by an anode. Finally, the output current from the electron multiplier
feeds external circuitry to provide the output signal. Statistical variations inherent in the excitation of photoelectrons by the incident photons and the statistical
nature of the secondary emission process cause the output signal to vary from one
pulse to the next, even with a constant number of incident photons. The resulting
distribution in output pulse-height limits both the pulse-height and time resolution
of the detector.

Previous studies have shown that photomultipliers with dynodes having cesium-activat
gallium phosphide secondary emitting surfaces exhibit better pulse-height and time
resolution capabilities than photomultipliers employing conventionally activated dynodes (6-7). Also in previous papers it has been shown that microchannel plate high-
gain photomultipliers exhibit significantly better time resolution than conventional
electrostatically focused photomultipliers (8-13).

Furthermore, it has recently been shown that new high-gain photon detectors employir
microchannel plates in cascade for electron multiplication will, under optimized operating conditions, exhibit the highest pulse-height resolution ever obtained (14).

Based on the above mentioned work, further effort has been made to investigate and review the time and pulse-height resolution of some new generation commercially available photomultipliers.

The measurements of the characteristics of these photomultipliers were made with a measuring system which has previously been described in Reference 16. The system has a time resolution of approximately 25 ps, FWHM.

Specifically, performance characteristics have been studied of the new Amperex XP 2020, RCA 8854 and Hamamatsu R 647-01 photomultipliers which use conventional multiplier structures. Furthermore, the characteristics have been investigated of a new generation of ITT F 4129 and Hamamatsu R 1564U extended life microchannel plate photomultipliers. Finally, characteristics have been reviewed of special made photomultipliers to be used in positron emission tomography, high energy physics and plasma diagnostics experimental systems.

The electrostatically focused Amperex XP 2020 photomultiplier uses a 12 stage discrete dynode structure with a semi-transparent bialkali (S24) photocathode having a useful diameter of 45 mm. Its peak spectral response is at 400 nm with a quantum efficiency of 26%. The photomultiplier has copper berillium dynodes instead of $AgMgOC_S$ dynodes which were used previously in a similar device. The tube design is optimized to have a small single electron time spread and for high repetition rates in counting operations.

The RCA 8854 photomultiplier is a variant of the RCA 4522. It has a high gain GaP (C_S) first dynode followed by thirteen BeO dynodes. The new RCA designation for the photocathode is 35 ET (formerly 118), which has a peak response at 400 nm and a quantum efficiency of 27%. It's spectral response extends from 200 nm to 600 nm. The maximum useful photocathode diameter is 114 mm. This photomultiplier is designed for experimental research instrumentation where good pulse-height resolution and large photocathode areas are important.

The Hamamatsu R 647-01 is a 13 mm-diameter, 10 stage, head on, flat face plate type photomultiplier with a bialkali photocathode having an S-11 response. The photocathode has its peak response at 420 nm and a useful diameter of 9 mm. Because of its small size, the photomultiplier is particularly suitable for high spatial resolution positron emission tomography, and also for high energy physics and nuclear chemistry experimental systems. The electron multiplier utilizes a box type structure and grid dynodes.

The new ITT F 4129 photomultiplier has an S-20 photocathode with a maximum usable diameter of 18 mm and three microchannel plates in cascade for the electron multiplication. The plates are in a Z-configuration to reduce the positive ion feedback. The three plates are identical, having 12 µm diameter channels with length to diameter ratios of 40. Proximity focusing is used for the input and collector stages. In ITT F 4129f device a protective film is provided between the photocathode and the microchannel plate which leads to a significant improvement in quantum efficiency, stability and life expectency as well as in the total elimination of the afterpulses.

The Hamamatsu R 1564U photomultiplier has a bialkali photocathode with a usable diameter of 18 mm, and two microchannel plates in cascade for electron multiplication. The anode is matched to a 50 Ohm connector. The entrance part of the first microchannel plate is covered with a thin aluminum film to prevent the bombardment of the photocathode by positive ions. As in the ITT photomultiplier this results in a significant increase in photocathode life, and stability of quantum efficiency.

The life of a microchannel plate photomultiplier without protective film is determined by a decrease in the photocathode quantum efficiency because of photocathode positive ion bombardment. The second determining factor is change in the channel wall secondary emission coefficient due to electron scrubbing, especially in high gain region of a channel. This is not effected by the protective film. The life of a device is defined as a total charge density accumulated at the anode at which device losses 50% of its overall responsitivity.

Results and Discussions

The results of the measurements of characteristics of the Amperex XP 2020, RCA 8854, Hamamatsu R 647-01, R 1564U and ITT F 4129 photomultipliers are summarized in Table 1. Also, some of the results are presented in Figs. 1-4.

Table 1. Summary of Characteristics Measurements of Some New Generation Conventionally Designed and Microchannel Plate Photomultipliers. Full Photocathode Illumination.

	Amperex XP 2020	RCA 8854	Hamamatsu R 647–01	ITT F 4129	Hamamatsu R 1564U
DC Gain	$>3 \times 10^7$	3.5×10^8	$>10^6$	1.6×10^6	5×10^5
Supply Voltage Between Anode and Cathode (V)	2200	2500	1000		3400
Microchannel Plate Voltage (V)				2500	
Rise Time (ns)	1.5	3.2	2	0.35[a]	0.27
Electron Transit Time (ns)	28	70	31.5	2.5[a]	0.58
Impulse Response, FWHM, (ns)	2.4	4.0	3.5	0.52[a]	
Single Photoelectron Time Spread, FWHM, (ns)	0.51	1.55	1.2	<0.20[a]	0.09
Multiphotoelectron Time Spread, FWHM, (ns)	0.12[b]		0.40[c]	0.10[d]	
Peak-to-Valley Ratio of Pulse-Height Spectrum with Optimized Operating Conditions		1.9:1		2.47:1[e]	
Dark Pulse Count[f] (cps)	450	155	54	1800	
Quantum Efficiency[g] %	26	27	28	20	15
Photocathode Diameter (mm)	44	114	9	18	18

[a] These characteristics were measured for prototype packaged photomultipliers.

[b] Measured using 2500 photoelectrons per pulse.

[c] Measured using 100 photoelectrons per pulse.

[d] Measured using 800 photoelectrons per pulse.

[e] Optimized operating condition for F 4129 was $V_M = 2710V$. In this case the photomultiplier gain was 6.6×10^7.

[f] Dark pulse summation is defined by: $\displaystyle\sum_{1/8\ \text{photoelectron}}^{16\ \text{photoelectrons}}$ = counts per second.

[g] Quantum efficiency values for proximity focused microchannel plate photomultiplier decrease significantly during operating time for devices without protective ion barrier film between the photocathode and the input of the microchannel plate. Table shows quantum efficiency initial values.

Witn full photocathode illumination, and with a light pulse produced by a 200 ps electrical pulse, the rise time, impulse response (FWHM), and single photoelectron time spread (FWHM), were 1.5 ns, 2.4 ns, and 0.51 ns, respectively for the XP 2020, (16). Figure 1 shows two single photoelectron time spectra spaced 4 ns apart. Measurement of the dark pulse spectrum showed that the photomultiuplier pulse-height resolution is not good enough to show the one, two and three photoelectron peaks. This is typical for all conventionally designed photomultipliers using the first dynode with low gain. The high frequency counting measurements show that the XP 2020 can be operated at a higher pulse repetition rate than 55 MHz (the voltage divider must be capable of providing the current needed for its operation). Figure 2 shows, the anode output pulses at repetition rate of 55 MHz. The upper trace is the anode output pulse, the lower trace is the driving electrical pulse. The light source was a light emitting diode, type XP 21, driven by pulses having a width of 3 ns and approximately 10 V amplitude. The diode was operated in an avalanche mode.

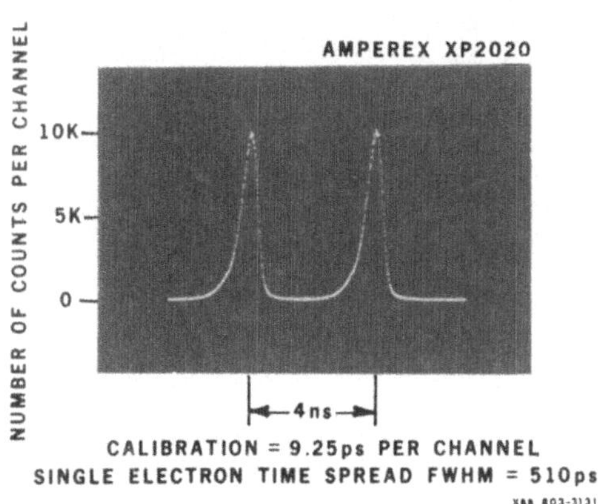

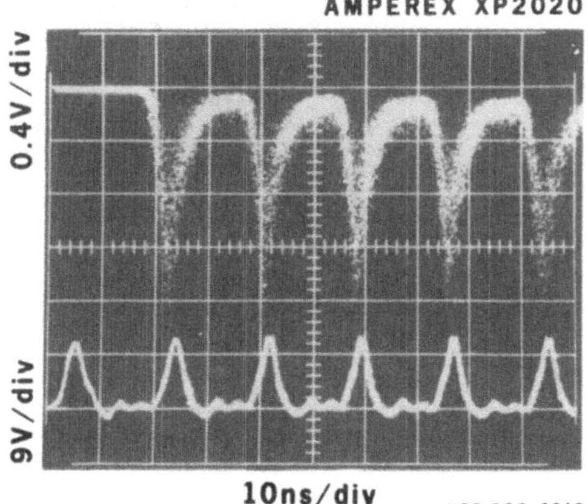

Fig. 1 Single photoelectron time spread of the XP 2020 photomultiplier with full photocathode illumination.

Fig. 2 Anode output pulses (upper trace) of XP 2020 photomultiplier using input pulse repetition frequency of 55 MHz with 2% duty cycle.

The rise time, impulse response, and single photoelectron time spread, for the large photocathode area 8854 photomultiplier were 3.2 ns, 4 ns and 1.55 ns, respectively. This device utilizes a negative-electron-affinity $GaP(C_s)$ secondary emission surface on the first dynode of the electron multiplier resulting in a high pulse-height resolution (17). Figure 3 shows the photomultiplier pulse-height spectrum indicating the peak to valley ratio of 1.9:1. With this high resolution of the single and multi-photoelectron peaks an effective separation of spurious single photoelectron peaks (thermionic electrons and others) and the desired multiphotoelectron peaks can be achieved. The 8854 has shown resolution of three distinct photoelectron peaks.

The rise time, impulse response, and single photoelectron time spread, for the very small photocathode area R 647-01 photomultiplier were 2 ns, 3,5 ns, and 1.2 ns respectively, (18). The device is particularly suitable for application in positron emission tomography systems where high spatial resolution is required. Also, the device is well suited for some of the more recently developed systems where time-of-flight information of the positron annihilation γ-rays are used in combination with conventional projection data to improve the signal-to-noise ratio for image reconstruction.

The F 4129 microchannel plate photomultiplier has a rise time, impulse response and single photoelectron time spread of 0.35 ns, 0.52 ns, and 0.2 ns, respectively. The device exhibits excellent timing capabilities and is very much less sensitive to ambient magnetic fields than the best conventionally designed photomultipliers, (7,20). This is mostly due to the small thickness of microchannel plates (approximately 2 mm), very strong applied electric field (5-10 kV/cm) and proximity focusing used between the photocathode and microchannel plate. Measurements have also shown that the photomultiplier operating characteristics can be optimized to yield a peak-to-valley ratio of 2.47:1. Figure 4 shows a typical single photoelectron pulse shape from the device.

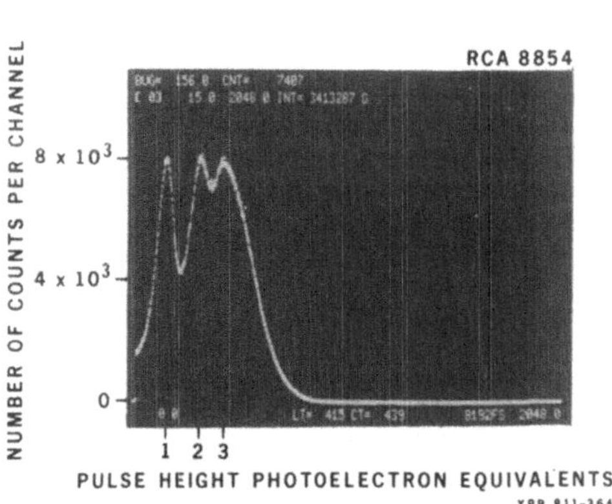

Fig. 3 Typical pulse-height resolution of the 8854 photomultiplier.

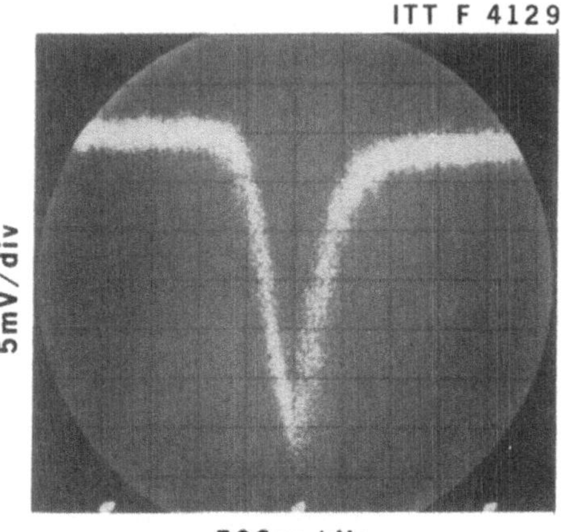

Fig. 4 Typical single photoelectron pulses from the F 4129 operated at V_M = 2400 V, using 200 ps light impulse excitation.

The life of the new generation F 4159f device with a 7 nm thick ion barrier film is significantly increased when compared with the life of the F 4159, (19).

The rise time, impulse response, and single photoelectron time spread for R 1564U microchannel plate photomultiplier were approximately 0.27 ns, 0.58 ns, and 0.09 ns, respectively, (21).

Because of a thin Al film (thickness of approximately 13 nm) which covers the front surface of the first microchannel plate, the device has exhibited a significant increase in life and stability of quantum efficiency of the photocathode as compared to devices without the Al film. Measurements showed that devices without the Al film exhibit a half gain degradation after a total output charge of 10^{-3} C/cm², (22). The photomultiplier with Al film showed a constant gain up to an accumulated output charge of 10^{-2} C/cm² which results from the gain degradation of the plate itself.

In addition to the above mentioned commercially available photomultipliers, there are a number of special made devices developed for particular applications. The Hamamatsu R 1548 is a dual rectangular photomultiplier with a bialkali photocathode having its maximum response at 420 nm, (23). Maximum useful area of the photocathode is (10 mm x 20 mm) for each of the two channels. The anode pulse rise time and single photoelectron time spread are approximately 1.8 ns and 1.0 ns, respectively. The device has a gain of 2×10^6. It was developed for positron emission tomography. The Hamamatsu R 1449 has 508 mm diameter quasi-hemispherical glass window with bialkali photocathode having its peak response at 420 nm and has a gain of 10^7 at 2000 V applied voltage. The anode pulse rise time, impulse response, and single photoelectron time spread are 18 ns, 30 ns and 7 ns, respectively, (22). The photomultiplier was developed for high energy physics experiments such as proton decay studies.

Several developments in high-speed photon detectors have been made for plasma diagnostic, (5), where excellent impulse response properties and high linearity of output current are important. One of the latest of these is the ITT MCP-1 microchannel plate photomultiplier, (24). The device consists of a semitransparent multialkali photocathode, microchannel plate and tapered coaxial 50 Ohm anode. The anode is mated to a special vacuum feedthrough made by EG&G. The device has an accelerating screen between the microchannel plate and anode to improve the time response by isolating the plate from the anode. If the screen is held at ground potential, the anode signal does not start to rise until the electron burst passes the screen. The gain of this multiplier is approximately 10^4. Its impulse response measured with 810 nm laser pulser is shown in Fig. 5. Output pulse rise time and impulse response were 273 ps and 248 ps, respectively.

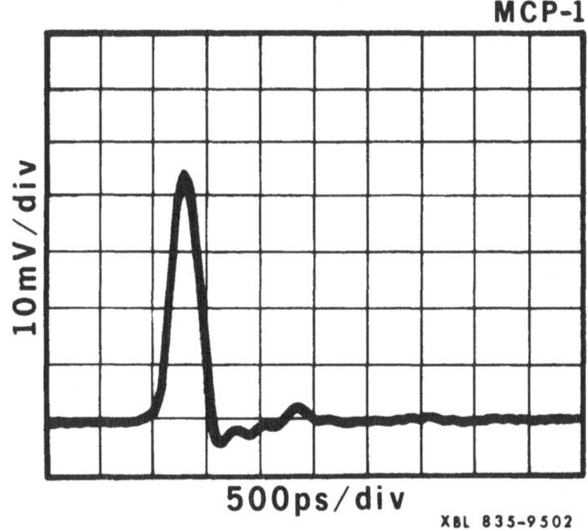

Fig. 5. Impulse response of MCP-1 microchannel plate photomultiplier operated with accelerating voltage of 1 kV.

Microchannel plates can operate efficiently as photon detectors in the extreme ultra violet and x-ray wavelengths in windowless configuration. The quantum efficiency of the plate itself, in which the lead glass and electrode surfaces prepared by the manufacturer act as photocathode, lies in a 1-10% range for most x-ray wavelengths as shown in Fig. 6. The standard means of enhancing these values is to deposit a material of relatively high photoelectric yield on the plate surface and channel walls. Lithium fluoride, magnesium fluoride and cesium iodite have been used for this purpose, (25). An increase of 65% in the quantum efficiency for 1.98 keV photons can be obtained using a Mg F_2 coated plate.

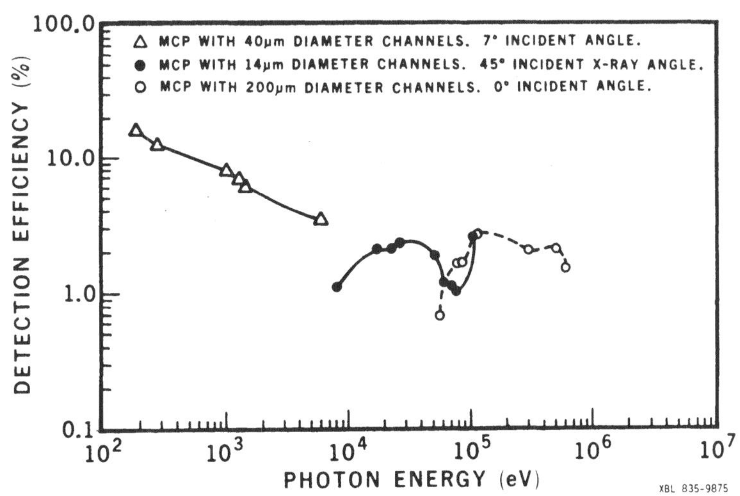

Fig. 6 X-Ray detection efficiencies.

74

Acknowledgments
This work was performed as part of the program of the Electronics Research and Development Group of the Lawrence Berkeley Laboratory, University of California, Berkeley and was partially supported by the Director's Office of Energy Research, Office of Health and Environmental Research, U.S. Department of Energy under Contract DE-AC03-76SF00098.

References
1. G.F. Knoll, Radiation Detection and Measurement, John Wiley & Sons, New York (1979).
2. B. Leskovar, C.C. Lo, P.R. Hartig, K.H. Sauer, Rev. of Sci. Instr. 47, No. 9, 1113-1121 (1976).
3. W.J. Carrion, Proc. of the Society of Photo-Optical Instrumentation Engineers, 134, 151-154, March 1978, Fort Walton Beach, Florida.
4. R.M. Gagliardi, S. Karp, Optical Communications, John Wiley & Sons, New York (1976).
5. P.R. Lyons, L.D. Looney, J. Ogle, R.D. Simmons, R. Selk, B. Hopkins, L. Hocker, M. Nelson, P. Zagarino, Proceedings of the Los Alamos Conference in Optics, Santa Fe, New Mexico, April 7-10, 1981, SPIE, 288, 404-411 (1981).
6. B. Leskovar and C.C. Lo, IEEE Trans. Nucl. Sci., NS-19, No. 3, 50-62 (1972).
7. C.C. Lo and B. Leskovar, IEEE Trans. Nucl. Sci., NS-28, No. 1, 698-704 (1981).
8. B. Leskovar and C.C. Lo, Nucl. Instr. and Methods, 123, No. 1, 145-160 (1975).
9. B. Leskovar, Nucl. Instr. and Methods, 128, 115-119 (1975).
10. B. Leskovar, Physics Today, 30, (11), 42-49 (1977).
11. B. Leskovar and C.C. Lo, IEEE Trans. Nucl. Sci. NS-25, No. 1, 582-590 (1978).
12. B. Leskovar, and C.C. Lo, IEEE Trans. Nucl. Sci., NS-26, No. 1, 388-394 (1979).
13. B. Leskovar, Proc., of the 4th Intl. Congress - Laser 79 Opto-Electronics, Munich, West Germany, 581-586. Published by IPC Science and Technology Press Ltd., England (1979).
14. B. Leskovar, Proc. of the 5th Intl. Congress, Laser 81 - Optoelectronics, Munich West Germany, June 1981. Published by Springer-Verlag, Vol.: Optoelectronics in Engineering, 381-386 (1982).
15. C.C. Lo, P. Lecomte and B. Leskovar, IEEE Trans. Nucl. Sci., NS-24, No. 1, 302-311 (1977).
16. C.C. Lo and B. Leskovar, IEEE Trans. Nucl. Sci., NS-28, No. 1, 659-665 (1981).
17. C.C. Lo and B. Leskovar, IEEE Trans. Nucl. Sci., NS-29, 1, 184-190 (1982).
18. C.C. Lo and B. Leskovar, Performance Studies of Hamamatsu R 647-01 Photomultiplier, Lawrence Berkeley Laboratory Report, LBL-15494, December 15, (1982).
19. The F 4126, F 4128, F 4129 Series of Photomultipliers Incorporating Microchannel Plate, Electro-Optical Product Division of ITT Corp. Fort Wayne, Indiana (1980).
20. D.H. Ceckowski, E. Eberhard, and E. Carney, IEEE Trans Nucl. Sci., NS-28, 677-682 (1981).
21. The Microchannel Plate Photomultipliers Tubes (MCP-PMTs), Hamamatsu TV Co., Ltd., Hamamatsu, Japan (1982).
22. T. Hayashi, Nucl. Instr. and Methods, 196, No. 1, 181-186 (1982).
23. T. Yamashita, M. Ito and T. Hayashi, Proceedings of the International Workshop on Physics and Engineering in Medical Imaging, 209-211 (1982).
24. D.F. Simmons, B.A. Smith, and P.B. Lyons, Development of High-Speed Microchannel Photomultiplier, Los Alamos National Laboratory Report. Presented at the LANL Conference on Optics, April 11-15, 1983.
25. K.W. Dolan, J. Chang, Proc. SPIE. 106, 178 (1977).

Ge-Photoelemente für radiometrische Messungen im nahen Infrarot

K.D. Stock und K. Möstl
Physikalisch Technische Bundesanstalt
Postfach 33 45, 3300 Braunschweig

1. Summary

Ge-photoelements can expand the wavelength range for radiometric measurements covered by silicon detectors to about 1500 nm without severe loss of accuracy /1/. This work deals with measurements of the responsivity of large area Ge-photoelements (Judson J16, 10 mm ∅, USA) as function of current and voltage e.g. linearity measurements in the case of short circuit and open loop operation. Surprisingly the dynamic application range limited by the beginning nonlinearity on the upper side and the noise on the lower side is greater for open loop than for short circuit operation. The nonlinearity caused by the I/U-characteristics of the p-n-junction can be distinguished from other instrinsic loss mechanisms. The temperature dependence of the responsivity also shows one part caused by the p-n-characteristic and another one which can be explained by the temperature dependence of the absorption coefficient.

2. Ersatzschaltbild und Innenwiderstand

Nach Abb. 1 teilt sich ein im Photoelement durch Strahlung generierter Strom I_{PH} in einen außen meßbaren Strom I (durch den Serienwiderstand R_S und den Lastwiderstand R_L) und einen Dioden-Verlust-Strom I_D (durch den nichtohmschen Diodenwiderstand R_D). Im Gegensatz zu Si-Photoelementen ist das Verhältnis I_D/I_{PH} bei den untersuchten großflächigen Ge-Photoelementen nicht mehr vernachlässigbar klein.

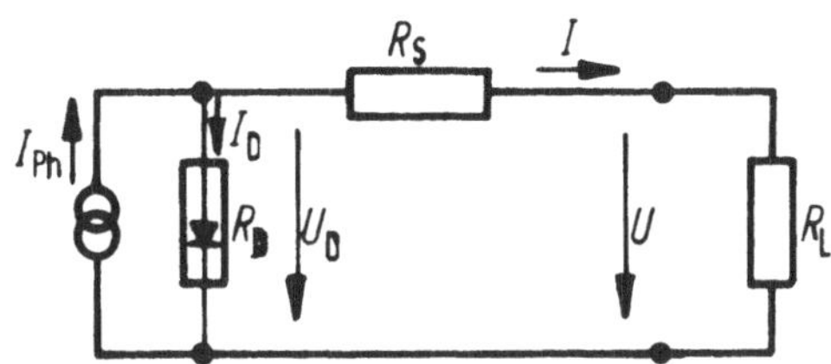

Abb. 1. Ersatzschaltbild eines Photoelementes mit externem Lastwiderstand bei konstanter Bestrahlungsstärke

Das Stromteilungsverhältnis I_D/I ist bestrahlungsstärkeabhängig und führt zu einer
Veränderung der Empfindlichkeit s (s nach DIN 5031). Die Vergrößerung von R_L verstärkt
diesen Effekt. Daher wird bei der Absolutbestimmung radiometrischer Größen i.a. der
ideale Kurzschlußbetrieb (R_L =0) angestrebt. Abb. 2a zeigt Ergebnisse von Kennlinien-
messungen im Durchlaßbereich des Photoelementes. Die ungewöhnliche Darstellung der

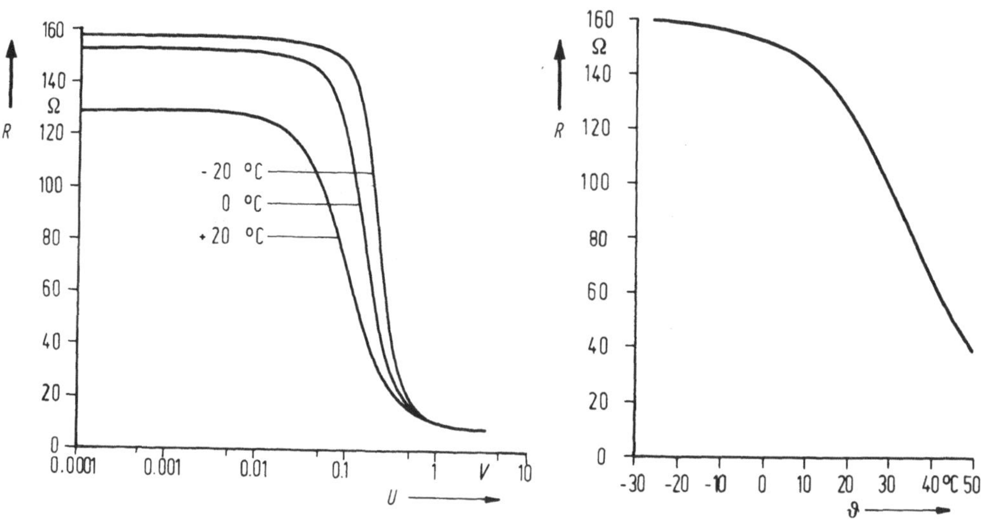

Abb. 2. Innenwiderstand $R = R_D + R_S$ eines Photoelementes
a. als Funktion der Spannung U für verschiedene Temperaturen
b. als Funktion der Temperatur im "ohmschen" Bereich, d.h. - 0,2 mV $\leq U \leq$ + 0,2 mV

I/U Kennlinien in der Form R(U) =U/I(U) ist für die Diskussion des in Abschnitt 3
erwähnten Kennlinienanteils bei der Nichtlinearität und der Temperaturabhängigkeit
von s ausgesprochen hilfreich. Z.B. läßt sich erkennen, daß sich R mit wachsendem
U einem konstanten Anteil, dem Serienwiderstand R_S (etwa 6,5 Ohm),nähert.

3. Linearität der Empfindlichkeit im Kurzschluß
Unter Linearität eines Strahlungsempfängers versteht man die Proportionalität
zwischen Bestrahlungsstärke und Ausgangssignal. Die Abweichung von dieser Proportionalität wird als Nichtlinearität bezeichnet. In der PTB wird die Nichtlinearität
eines optischen Strahlungsempfängers in einem automatisierten Verfahren gemessen,
das dem von Bischoff /2/ benutzten Additionsverfahren entspricht. Hieraus läßt sich
die für die Praxis wichtige Abhängigkeit der Empfindlichkeit vom Photostrom berechnen. Die durchgezogene Linie in Abb. 3 zeigt s für ein Ge-Photoelement. Dabei ist s
auf die Empfindlichkeit s_0 normiert, die bei einem Photostrom bei I=1 μA ermittelt
wurde. Selbst im idealen Kurzschluß (R_L =0) liefert das bestrahlungsstärkeabhängige
Stromverteilungsverhältnis I_D/I einen Beitrag zur Nichtlinearität des Photoelementes

(Kennlinienanteil, siehe Abschnitt 2). Dieser Anteil ist in Abb. 3 als gestrichelte Linie eingetragen. Im Kurzschlußbetrieb überwiegen aber die mit I anwachsenden Verlustmechanismen (z.B. Rekombinationen) den Kennlinienanteil. Sie begrenzen den

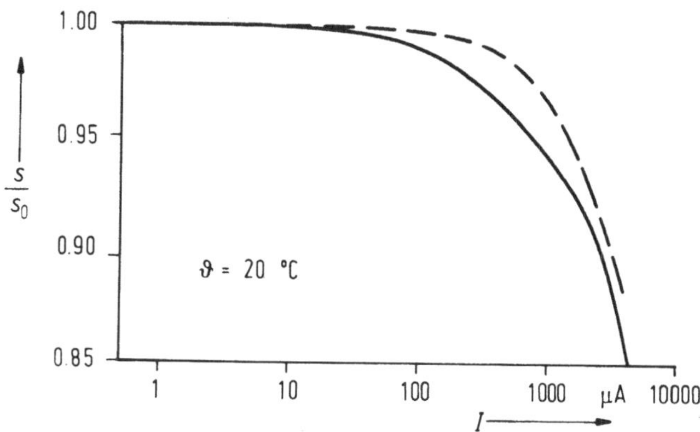

Abb. 3. Relative Empfindlichkeit eines Ge-Photoelementes als Funktion des außen meßbaren Stromes I (Kurzschluß). Durchgezogene Kurve: Messung. Gestrichelte Linie: berechnet aus der Kennlinie (Kennlinienanteil)

Einsatzbereich des Photoelementes nach oben hin. Nach unten hin wird der dynamische Einsatzbereich durch das Rauschen des Elementes begrenzt. Setzt man eine 1-prozentige Schranke für die Abweichung (1- s/so) im oberen Bereich und für die relative Standardabweichung im unteren Bereich, so erhält man einen dynamischen Einsatzbereich von etwa 2,3 Dekaden.

4. Linearität im Leerlauf

Die untere Grenze des dynamischen Einsatzbereiches kann wesentlich gesenkt werden, wenn das Element im Leerlauf betrieben wird ($R_L = \infty$, Spannungsmessung). Zwar setzt die Nichtlinearität hier bei kleineren Bestrahlungsstärken ein (Abb. 4), jedoch ist bei dem beschriebenen Ge-Photoelement der dynamische Einsatzbereich größer als im Kurzschlußbetrieb. s_0 ist hier die Empfindlichkeit bei U=1 μV. Die in Abb. 4 errechnete Skala für den Strom, den das Photoelement unter sonst gleichen Bedingungen im Kurzschluß liefern würde, ermöglicht den Vergleich mit Abb. 3. Im Gegensatz zum Gebrauch von Si-Photoelementen kann also für radiometrische Messungen mit Ge-Photoelementen der Leerlaufbetrieb unter den in Abschnitt 6 genannten Einschränkungen seine Berechtigung haben. Benutzt man die Definition des dynamischen Einsatzbereiches aus Abschnitt 3, so erhält man den im Leerlauf vergleichsweise großen Bereich von 3,4 Dekaden.

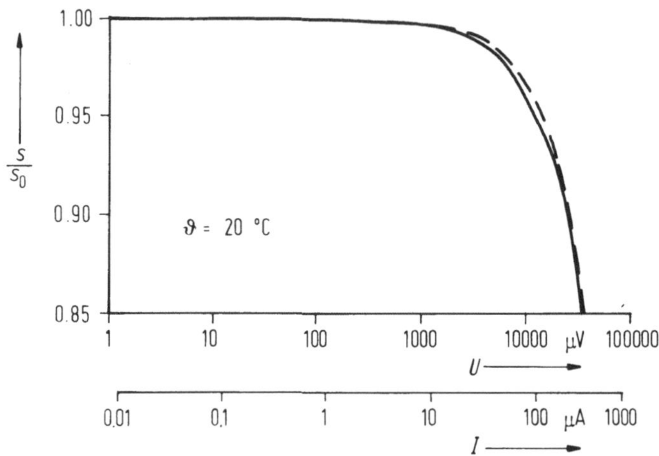

Abb. 4. Relative Empfindlichkeit eines Ge-Photoelements als Funktion der außen
meßbaren Spannung U (Leerlauf). Gestrichelte Kurve: Kennlinienanteil

5. Temperaturverhalten der Empfindlichkeit im Kurzschluß

Die Abb. 5 zeigt das Absinken der Empfindlichkeit s mit ansteigender Temperatur ϑ
im Kurzschlußbetrieb. Bei Zimmertemperatur beträgt der Temperaturkoeffizient etwa

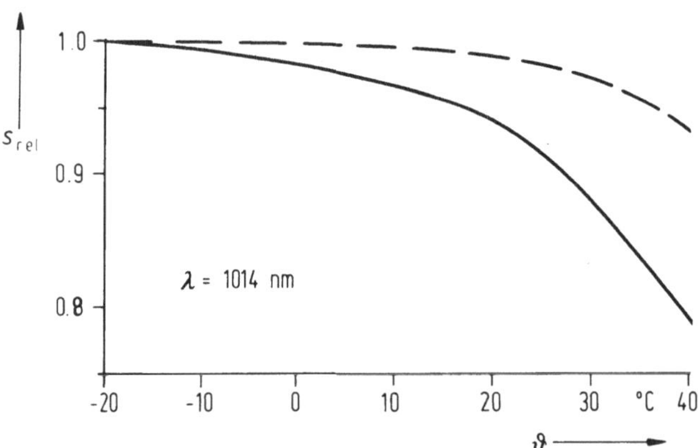

Abb. 5. Relative Empfindlichkeit eines Ge-Photoelementes als Funktion der
Temperatur ϑ (Kurzschluß); durchgezogene Kurve: Messung. Gestrichelte Kurve:
Berechnet aus der Temperaturabhängigkeit des Innenwiderstandes

0,4 %/°C. S_{rel} in Abb. 5 ist auf die Empfindlichkeit bei der Temperatur ϑ = -20°C
normiert. Die gestrichelte Linie in Abb. 5 zeigt den aus der Temperaturabhängigkeit

der Kennlinien berechneten Empfindlichkeitsabfall. Er ist im Vergleich zum Leerlauf (vgl. Abschnitt 6) relativ gering. Wie bei Si-Photoelementen /3/ ist die Temperaturabhängigkeit der Empfindlichkeit vor allem dann zu beachten, wenn die verwendete Strahlung im wesentlichen bereits vor der Verarmungszone absorbiert wird. Die entstandenen Ladungträger müssen erst über Diffusion in die Verarmungszone gelangen, um zum Meßsignal beizutragen. Der mit der Temperatur wachsende Absorptionskoeffizient /4/ sorgt über eine Verlängerung dieses Diffusionsweges für eine erhöhte Rekombination und damit für eine Verringerung des Meßsignals, bzw. der Empfindlichkeit.

6. Temperaturverhalten im Leerlauf

Im Gegensatz zum Kurzschlußbetrieb (Abschnitt 5) ist im Leerlauf die Temperaturabhängigkeit des Innenwiderstandes R die Hauptursache der Änderung der Empfindlichkeit s. Hier ist die gemessene Spannung gleich dem Spannungsabfall, den der generierte Strom I_{Ph} an R_D (Abb. 1) erzeugt. Der Verlauf von $R_D(\vartheta) = R(\vartheta)-R_S$ als Funktion der Temperatur ϑ ist aus Abb. 2b zu entnehmen. Bei Zimmertemperatur beträgt die Temperaturabhängigkeit der Empfindlichkeit etwa 2,3 %/$^{\circ}$C. Diese starke Temperaturabhängigkeit stellt hohe Anforderungen an die Temperaturkonstanz bei Messungen im Leerlauf.

7. Schlußbemerkung

Die Abweichungen von der Homogenität der Empfindlichkeit ist innerhalb einer zentralen Empfängerkreisfläche von 7 mm $\emptyset$ kleiner als $\pm$ 3 % /1/. Bei Absolutmessungen mit diesen niederohmigen Ge-Photoelementen ist zu beachten, daß ein maximaler Lastwiderstand R_L von nur 0,13 Ohm zulässig ist, wenn eine maximale Abweichung der absoluten Empfindlichkeit von z.B. 0,1 % vom idealen Kurzschlußbetrieb gewünscht wird /5/. Diese für Absolutmessungen gültige Forderung ist allein eine Folge des Anspruchs an ein möglichst kleines Stromverteilungsverhältnis I_D/I, d.h. des Ersatzschaltbildes (Abschnitt 2). Im Gegensatz dazu gelten für Relativmessungen die Einsatzbeschränkungen aufgrund der Diskussion der Nichtlinearitäten (Abschnitt 4 und 5). Gewährleistet man eine den Ergebnissen von Abschnitten 5 und 6 entsprechende Temperaturkonstanz, so weisen die untersuchten Ge-Photoelemente für radiometrische Relativmessungen in der Kombination von Kurzschluß- und Leerlaufbetrieb einen dynamischen Einsatzbereich von mehr als 4 Dekaden auf.

Literatur
/1/ STOCK,K.D. und MÖSTL, K.: Proc. 10th Int. Techn. Comm. Photon Detectors Berlin (1982) 40
/2/ BISCHOFF, K.: Z. Instrumentenkunde 69 (1961) 143
/3/ MÖSTL,K. und STOCK,K.D.: PTB Mitteilungen 92 (1982) 11
/4/ DASH,W.C. und NEWMAN,R.: Physical Review 99 (1955) 1151
/5/ STOCK,K.D. und MÖSTL,K.: PTB Jahresbericht (1982) 161

Si: In Infrared Detectors at Low Temperatures: a New Mode of Operation

H. HARJUNMAA
University of Helsinki, Department of Physics
Siltavuorenpenger 20 D, SF-00170 Helsinki 17
Finland

INTRODUCTION

In this paper, new experiments on the photochronic detection method /1,2/ are reporte
The method is hereby briefly summarized: A lightly doped homogeneous single crystal
silicon sample, fitted with two ohmic contacts, is refrigerated to a temperature wher
essentially all charge carriers are bound (for instance, with phosphorus doping, belc
25 K). The crystal is exposed to the radiation to be measured. After a suitable exposure time has elapsed, a voltage is applied between the contacts, and, after a short
delay, a breakdown occurs in the crystal. If a series resistor is used, the breakdown
is non-destructive. The voltage is cut off and the cycle is repeated over and over
again. The breakdown delay serves as a measure of the intensity of the infrared radiation falling on the crystal.

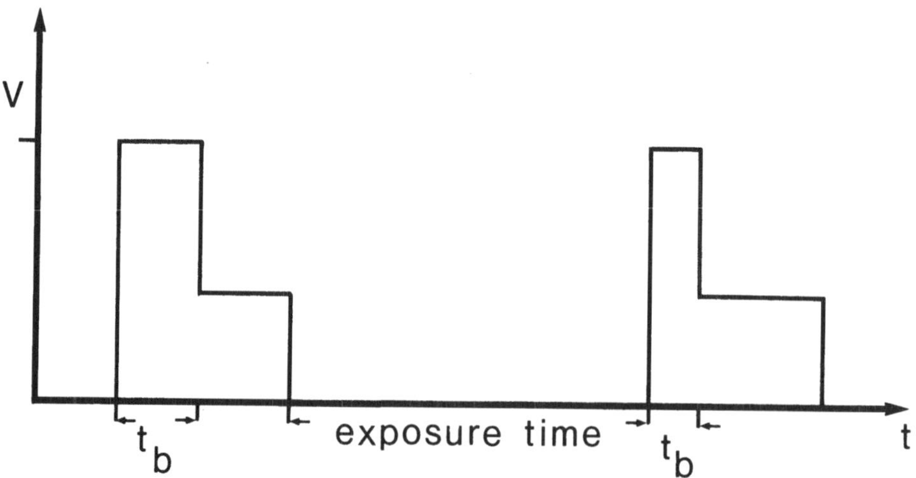

Fig. 1. Typical operation sequence of the Si:In photochronic detector. Exposure time
τ = 10 ms, breakdown delay in the dark $t_{b,dark}$ = 100 μs, illuminated $t_{b,illum}$ = 60 μ:
temperature = 10 K, voltage = 30 V

EXPERIMENTS

In this work, the spectral response (i.e. the wavelength dependence of the radiation-induced change in the breakdown delay) of the Si:In photochronic detector was measured. The measuring system comprised a globar radiation source, a grating monochromator and an optical switchbox, the purpose of which was to enable switching of the monochromatic radiation between the sample and a pyroelectric reference detector. The sample was cooled using a closed-cycle helium refrigerator; the rest of the apparatus was at room temperature. A radiation chopper was used with the reference detector, but not with the Si:In detector.

A typical operation sequence is shown in fig. 1. In the case illustrated, radiation shortens the breakdown delay. With short enough exposure times, the opposite effect is observed. The signal of the photochronic detector is defined as

$$S = \frac{t_{b,illum} - t_{b,dark}}{t_{b,dark}} \qquad (1)$$

Signals in the range $S = -0.99...+100$ have been observed. To preserve linearity, one usually works in the range $S = -0.5...+1$. The breakdown delay varies slightly from breakdown to breakdown with an amount Δt_b. Accordingly, the noise of the photochronic detector is defined as

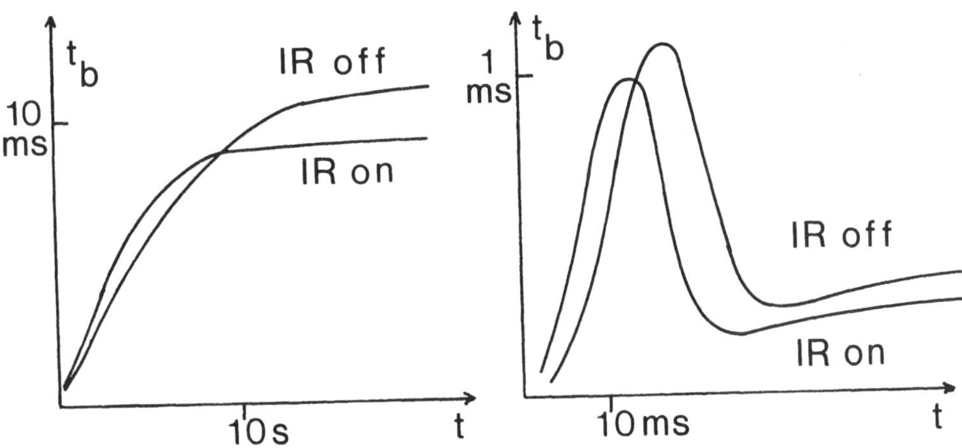

Fig. 2. Breakdown delay as a function of exposure time. a) Si:P, b) Si:In

82

$$N = \frac{\Delta t_b}{t_b}$$

(2)

Typically, N is of the order of 1%.

The breakdown delay depends on the exposure time in the manner depicted in fig. 2.
The measured spectral response of the Si:In photochronic detector, normalized with
respect to the reference detector signal, is presented in fig. 3.

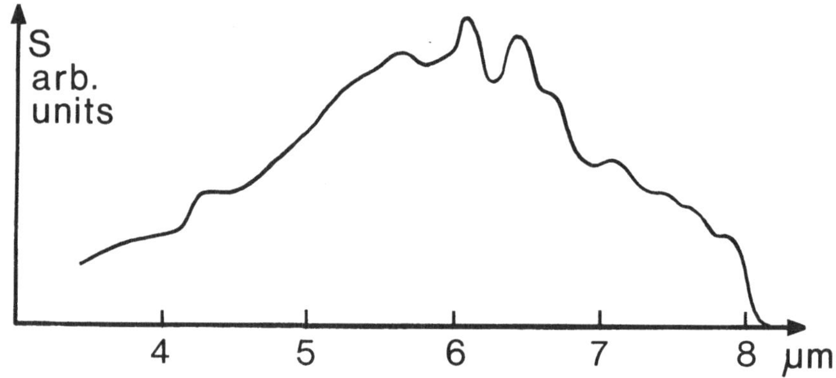

Fig.3. Spectral response of Si:In photochronic infrared detector. Temperature = 10 K,
sample thickness = 0.35 mm, lateral illumination. Room temperature resistivity of the
sample = 110 Ω cm

ANALYSIS

The spectral response has a cut-off wavelength of 8.0 μm. Clearly, then, the effect
of photons is to cause photoionization of the non-excited impurity atoms. In order
to explain the phenomenon of delayed breakdown and its sensitivity to photons, a simplified energy-level scheme with two long-lived excited states of the impurity atoms
is invoked /3/. Phosphorus-doped silicon was chosen as the object of theoretical efforts, because its (τ, t_b)-curve is simpler, and because there is no reason to sus-
pect a fundamental dissimilarity between it and Si:In, as their behavior is otherwise
essentially similar. (The spectral response of the Si:P photochronic detector has not
been measured, as it lies in an experimentally more difficult wavelength range). The
two long-lived excited states in question are the split-off upper 1S states of the
phosphorus atom (see fig. 4).

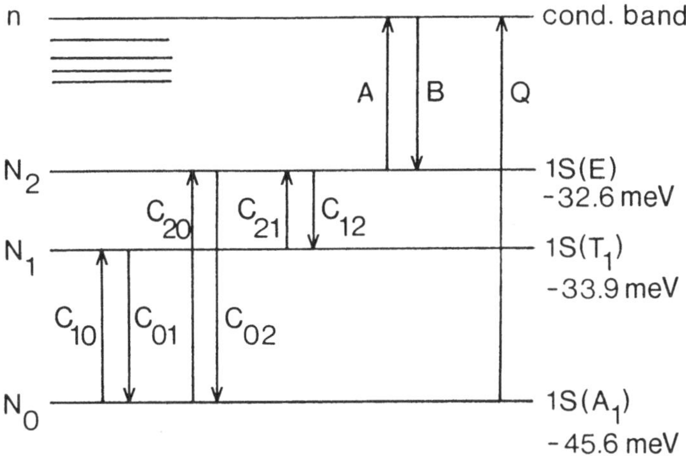

Fig. 4. Energy levels of phosphorus in silicon

The following response mechanism is envisaged: During exposure, the electric field is off. An incoming photon ionizes a phosphorus atom, and, after a photoconductive life-time τ_e, the electron recombines, via phonon cascade, into the upper one (1S(E)) of the long-lived excited states. While in the conduction band, the electron may cause impact de-excitation of one or several excited neutral phosphorus atoms. After the electric field is switched on in order to effect a breakdown, the impact-caused transitions will take the electrons predominantly upwards in the energy level system. Impact ionization will occur, and the ensuing positive feedback causes an avalanche breakdown, the sooner, the larger is the population of the excited states.

The fairly long delay and the sharp breakdown are properties, verified numerically, of the system of rate equations written according to fig. 4:

$$\frac{dn}{dt} = AN_2 - Bn(n + N_A) + QN_0$$

$$\frac{dN_2}{dt} = Bn(n + N_A) + C_{20}N_0 - C_{02}N_2 + C_{21}N_1 - C_{12}N_2 - AN_2$$

$$\frac{dN_1}{dt} = C_{10}N_0 - C_{01}N_1 + C_{12}N_2 - C_{21}N_1$$

$$N_0 = N_D - N_A - n - N_1 - N_2$$

(3)

where N_D = number density of donors

N_A = number density of acceptors

A, B, C_{ij} = transition coefficients

Q = photoionization rate.

The photoconductive lifetime is short (10^{-7} s) and, consequently, irradiation tends to increase the population of the 1S(E) state and shorten the delay. When there is an appreciable population on the $1S(T_1)$ state, however, impact de-excitation may cause a lengthening of the delay.

In experiments with Si:P /3/, the lifetime of the 1S(E) state was found to be of the order of minutes at 18 K, the lifetime of the $1S(T_1)$ state being about one third that of the 1S(E) state.

The dependence of the breakdown delay on the populations of the excited states is non linear. To alleviate this nonlinearity problem, a variation of the photochronic mode of operation is proposed. In this variation, the exposure time is controlled so a to keep the breakdown delay constant. As long as the exposure time is longer than the photoconductive lifetime and shorter than either of the excited state lifetimes, the detector actually counts up to a fixed number of incoming photons, and the frequency of breakdowns effected by the control circuit is proportional to the radiation intensity.

References
(1) HARJUNMAA, H, LEHTO, A, Appl. Opt. 18 (1979) 959
(2) LEHTO, A, HARJUNMAA, H, US Pat. 4 228 354
(3) LAWAETZ, P, PROCTOR, W.G, to be published in J. Phys. C.

Digitale Restlichtkamera im Tieftemperaturenbetrieb

B. AULBACH, W. PFEIFFER
Technische Hochschule
Schloßgraben 1, D 6100 Darmstadt

Einführung

Zur optischen Untersuchung des Entladungsaufbaus in komprimierten Gasen werden hochempfindliche Kamerasysteme eingesetzt /1/. An die Ultrakurzzeitkamera werden Anforderungen wie die Einzelphotonenempfindlich-
keit und Belichtungszeiten im Nanosekundenbereich gestellt. Ein Konzept
mit tastbarem Doppelkanalplattenbildverstärker und nachgeschalteter
Restlichtkamera erfüllt diese Forderungen (Fig. 1). Die Entwicklung eines speziell für den Einzelbildbetrieb konzipierten Kamerasystem ermög-
die für Meßzwecke notwendige Verbesserung der Übertragungseigenschaften.

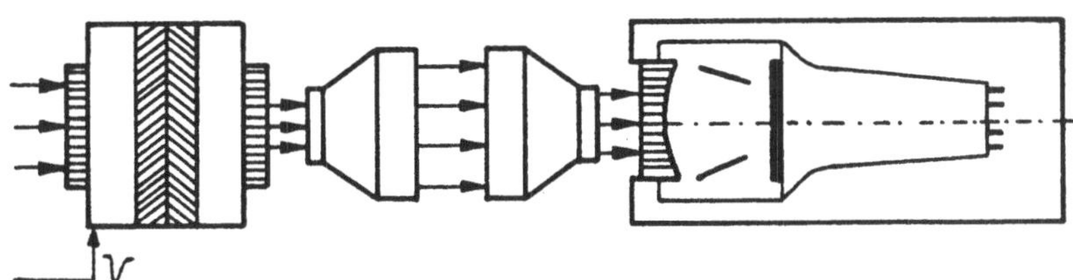

Fig. 1. Schematischer Aufbau der Ultrakurzzeitkamera

Funktionsprinzip der digital gesteuerten Restlichtkamera

Verschiedene Anforderungen und Randbedingungen, die von den durchgeführten Experimenten einerseits und dem Funktionsprinzip eines Multi-
diodenvidikons /2/ andererseits vorgegeben sind, haben zu folgendem
Konzept für den Aufbau einer speziellen Videokamera geführt:
- Das Target des SIT-Vidikons wird zweidimensional mit einer Bildpunkt-
  matrix von 256 x 256 Bildpunkten abgetastet.
- Zur Reduktion des Dunkelstroms wird das Target der Kameraröhre auf
  bis zu $-50\ ^{\circ}$C gekühlt. Die Kühlung ermöglicht durch die ausgedehnte
  Speicherzeit die Bildintegration auf dem Target und ein verlangsam-
  tes Auslesen des Bildinhalts (SLOW-SCAN-Abtastung).

- Die durch die kapazitive Trägheit des Targets bedingte Nichtlinearität wird durch verschiedene Targetspannungen während der Lösch- bzw.
  Lesezyklen behoben.

Das 256 x 256 Bildpunktformat bedeutet im vorliegenden Fall keine Einschränkung, da das Nyquistkriterium aufgrund der eingeschränkten Auflösung der Aufnahmeeinrichtung (ca. 10 lp/mm im dynamischen Betrieb)
nicht verletzt wird.

Die SLOW-SCAN-Abtastung zweidimensionaler Bildvorlagen erfordert die
Tastung des Elektronenstrahls an jeder Stelle des gelesenen Bildfen-
sters. Die Tastzeit wird in den Bereich der Bildpunktverweilzeit einer
Echtzeitkamera (100... 200 ns) gelegt. Wird diese erheblich überschrit-
ten, treten bei Strahlstromstärken, die eine gute Fokussierung bzw.
Strahlannahme sicherstellen, Bloomingeffekte auf . Blooming bedeutet
hier, daß innerhalb weniger Fernsehzeilen die gesamte Bildinformation
ausgelesen wird. Das Videosignal wird mit geringer Bandbreite verarbeitet. Trotz geringem schaltungstechnischem Aufwand wird so ein S/R-Ver-
hältnis von über 1:100 erreicht.

In Fig. 2 ist das Blockschaltbild der Digitalkamera skizziert:
Die Kamera ist über eine optoelektronische Datenübertragungsstrecke
mit einem Bildspeichersystem verbunden. Die zentrale Ablaufsteuerung
erzeugt die X-/Y-Adressen für die Ablenkspulen und den Strahltastim-
puls am Gitter G_1 für das Auslesen der Bildpunkte. Das Videosignal
wird schmalbandig verstärkt und analog/digital gewandelt. Der Bildpunkt wird unter einer von der Steuerung vorgegebenen Adresse in den
Frame-Buffer geschrieben. In Abhängigkeit der verschiedenen Betriebsarten Löschen, Integrieren bzw. Lesen schaltet die Steuerung die Tar-
get-, Kathoden- und Hochspannung der gekühlten SIT-Röhre. Nach Beendigung der Löschzyklen wird eine automatische Offsetkorrektur durchgeführt, um den Nullpunkt des A/D-Wandlers auf den Hintergrundwert des
Bildes abzugleichen.

<u>Eigenschaften und Anwendungsbeispiele</u>
Das Dunkelstromverhalten des verwendeten SIT-Vidikons bei Kühlung zeigt
Fig. 3. Der dem Dunkelstrom proportionale Ladungsverlust wird nach einer Speicherzeit T_S gemessen. Der Maximalwert des Dunkelstroms ist auf
10^3 Skalenteile (Vollaussteuerung des A/D-Wandlers) normiert. Die minimale Zykluszeit der im SLOW-SCAN-Kamera liegt bei 3 s. Die Kurven zeigen eine Reduktion des Dunkelstroms (T_S = const.) um etwa den Faktor 2
pro 10 $^{\circ}$C Temperaturerniedrigung. Bei einer gegebenen Temperatur wird
die Speicherzeit unter Berücksichtigung des maximal zulässigen Hinter-

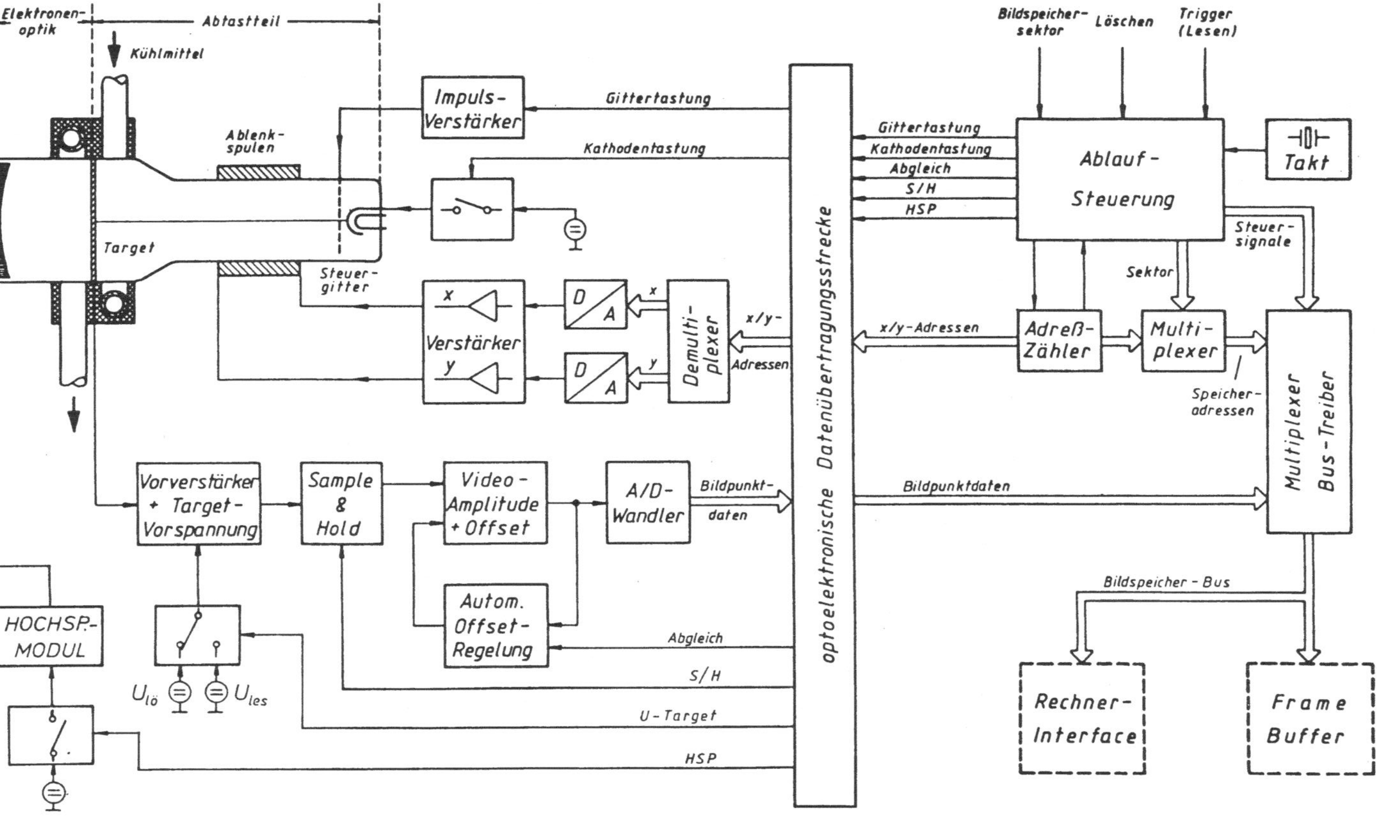

Fig. 2. Blockschaltbild der digital gesteuerten Restlichtkamera

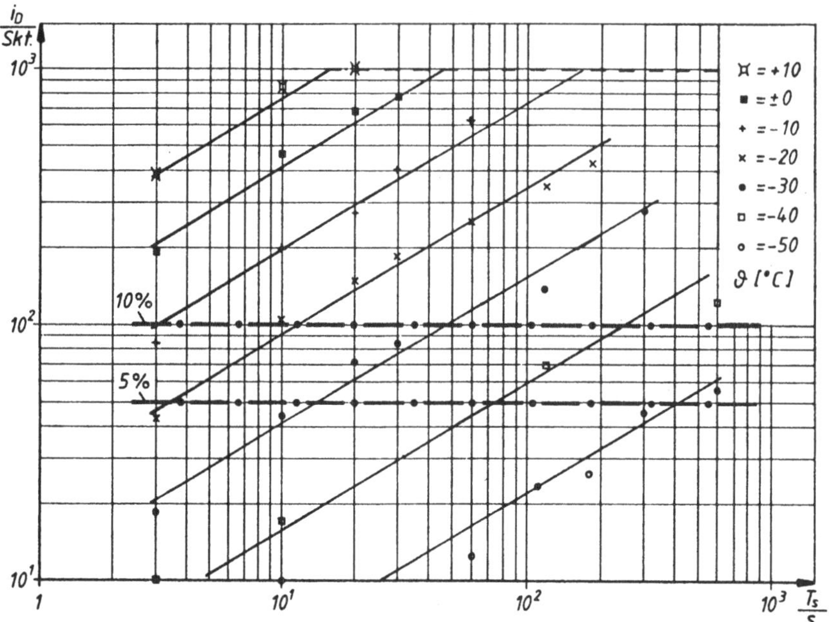

Fig. 3. Eigenschaften der SIT-Röhre TH 9659 bei Kühlung ($i_D = f(\vartheta, T_s)$) mit Grenzkurven für 5 % und 10 % Hintergrundrauschen

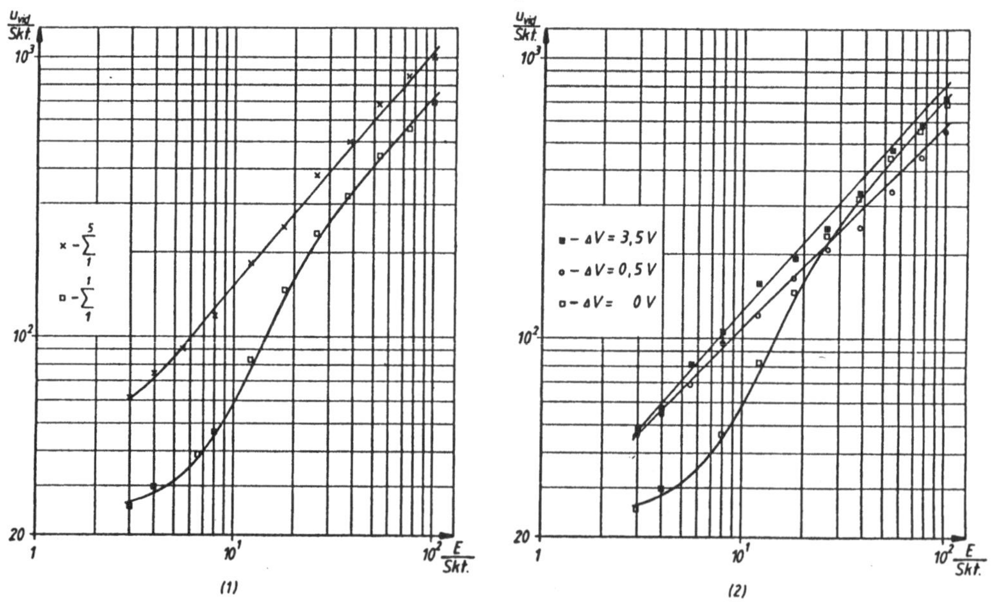

Fig. 4. Linearisierung des Videosignals durch
(1) Addition des mehrfach gelesenen Bildinhalts
(2) Erhöhen der Targetspannung während des Lesezyklus

grundrauschens ermittelt, das während der Integrationszeit durch den Dunkelstrom aufgebaut worden ist. Eine hochauflösende Messung bei Integrationszeiten über 10 s ist nur möglich, wenn die Kameraröhre auf mindestens -30 $^{\circ}$C abgekühlt wird.

Die Nichtlinearität kann durch Bildpunktaddition des mehrfach ausgelesenen Targets bzw. durch die Wahl verschiedener Target-Kathodenspannungen bei den Lösch- bzw. Lesezyklen eliminiert werden /3/. Durch die Kühlung braucht bei keinem der genanntem Korrekturverfahren das Ansteigen des Hintergrundrauschens berücksichtigt werden. Die Bildaddition in Fig. 4.1 zeigt bereits eine weitgehende Linearisierung des Videosignals nach 5 Schritten. Eine vereinfachte Versuchsdurchführung erlaubt die Erhöhung der Targetspannung während der Lesezyklen. Die Meßergebnisse zu Fig. 4.2 zeigen, daß eine brauchbare Linearisierung bereits dann auftrit, wenn die Lesespannung um den Wert des durch Hochenergieelektronen überhöht aufgeladenen Targets vergrößert wird. Eine weitere Steigerung des Nutzsignals wird zwar beobachtet, wenn die Spannungsdifferenz weiter erhöht wird, aufgrund der vergrößerten Targetkapazität und der schlechteren Strahlstromannahme müssen dann allerdings bis zu 100 Löschzyklen durchgeführt werden.

Fig. 5 zeigt zwei Anwendungsbeispiele. Die obere Abbildung zeigt ein Vorentladungsbild in Stickstoff bei maximaler Verstärkung der Ultrakurzzeitkamera. Bei reduzierter Verstärkung und Mehrfachbelichtung des Targets (hier n = 20) wird durch die aufintegrierten Photoelektronen bereits die Kontour der Leuchtfigur sichtbar.

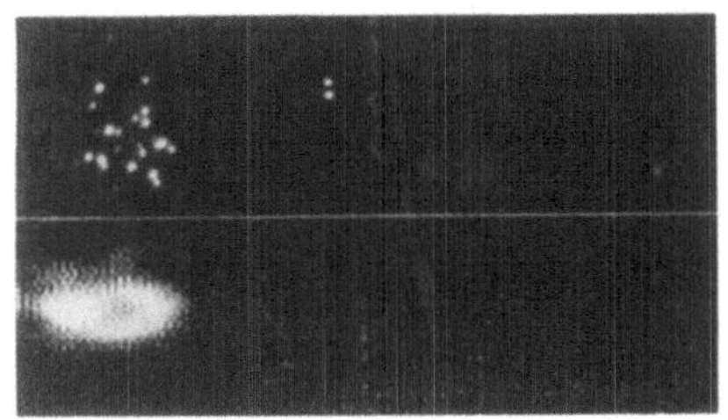

Fig.5. Anwendungsbeispiele:
(1) Einzelphotonenaufnahme (bei max. Verstärkung der Anlage)
(2) Bildintegration (bei reduzierter Verstärkung der Anlage)

(1) PFEIFFER, W., AULBACH, B., LIEBEROTH-LEDEN, B., and WITTMER, D.: 4th IEEE Pulsed Power Conf. Albuquerque (1983) invited paper
(2) CROWELL, M.H., LABUDA, E.F.: BSTJ 48 (1969) 1481
(3) HONEYCUTT, R.K., BURKHEAD, M.S.: Astron. Obs. with Television-Type Sensors, Symp. at Univ. British Columbia (1973) 229

Pyroelektrische Polymerfolien als Sensoren zur Leistungsmessung von Impulsgaslasern

H. Gündel, W. Bohmeyer, W. Kabel und H. Volkmann
ZIE der AdW, Berlin, DDR

R. Danz
IPOC "Erich Correns" der AdW, Teltow, DDR

1. Einleitung

Pyroelektrische Detektoren haben in den letzten Jahren auf unterschiedlichen Gebieten Anwendung gefunden. Sie gehören zu den thermischen Detektoren, und ihr Anwendungsgebiet reicht vom Fernen Infrarot (FIR) bis
zum Vakuumultraviolett (VUV) /1,2/.
Im Vergleich zu herkömmlichen Detektoren (Golay-Zellen, Thermoelementen
Detektoren nach dem Wirkprinzip des inneren Fotoeffekts u. a.) haben si‹
eine gute Nachweisempfindlichkeit und kurze Ansprechzeit, sind diesen
darüberhinaus aber in der Einfachheit der Konstruktion sowie der billigen Herstellbarkeit überlegen. Im Gegensatz zu Fotoleitungsdetektoren,
die bei Temperaturen des flüssigen Stickstoffs oder Heliums betrieben
werden, arbeiten sie bei Zimmertemperatur.
Als Werkstoff für pyroelektrische Wandlerelemente wurden bisher hauptsächlich ferroelektrische Materialien eingesetzt, wobei sowohl einkristalline Materialien (z. B. Triglycinsulfat (TGS)) als auch ferroelektr:
sche Keramiken (z. B. Bleizirkonate (PZ)) Verwendung fanden.
Daneben wurden auch Polymermaterialien bekannt, die aufgrund ihrer pola-
ren Primärstruktur Polarisationseffekte aufweisen. Unter den Polymeren
hat Polyvinylidenfluorid (PVDF) den größten pyroelektrischen Koeffizienten.
Im folgenden wird über den Aufbau, die spezifischen Eigenschaften und
Besonderheiten solcher Detektoren für die Energie- bzw. Leistungsmessung an verschiedenen Impuls-Lasertypen berichtet.

2. Funktionsweise pyroelektrischer Detektoren

In pyroelektrischen Materialien führt eine Temperaturänderung dT zu eine
Veränderung der makroskopischen Polarisation von $dP = \lambda\, dT$. Die Größe λ
heißt pyroelektrischer Koeffizient. Dadurch entsteht eine Oberflächenladung von

$$dQ = A\, dP = A \cdot \lambda\, dT \qquad (1)$$

A bezeichnet die Fläche des Wandlerelements.

Eine solche Temperaturänderung kann z. B. durch Exposition mit dem Licht
eines Lasers bewirkt werden. Der Wärmeübergang vom Material des Sensors
auf die Umgebung erfolgt mit einer thermischen Zeitkonstante τ_{th}.
Für den Fall hinreichend kurzer Strahlungsimpulse

$$t_L << \tau_{th} \qquad (t_L - \text{Halbwertsbreite des Laserimpulses})$$

gilt dann

$$dT = B \cdot P_L \, dt \qquad (2)$$

P_L ist die Laserleistung.
Der Proportionalitätsfaktor B hängt vom Absorptionsvermögen und der Wär-
mekapazität des Sensormaterials ab.
Damit läßt sich die Änderung der Oberflächenladung direkt mit der einge-
strahlten Laserleistung in Verbindung bringen /3,4/.
Es wird die influenzierte Ladungsmenge oder deren zeitliche Änderung
registriert.
Deshalb sind zwei Betriebsarten zu unterscheiden:
Differenzierender Betrieb RC $<<$ t_L
Die influenzierte Ladung fließt in diesem Fall momentan über den Last-
widerstand R ab und erzeugt einen Spannungsabfall von

$$U_R = R \, I_p = R \, \frac{dQ}{dt} = A \cdot B \cdot \lambda \cdot R \cdot P_L \qquad (3)$$

Das Meßsignal U_R ist damit der Laserleistung proportional (vergl. Fig. 1)

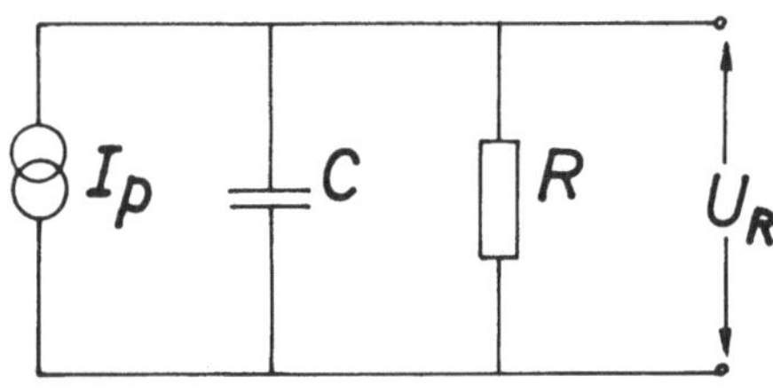

Fig. 1 Ersatzschaltbild des pyroelektrischen Leistungsmessers

Integrierender Betrieb RC $>>$ t_L
Die Spannung am Kondensator ergibt sich zu

$$U_C = \frac{1}{C} \int dQ = \frac{A \cdot B \cdot \lambda}{C} \int P_L \, dt = \frac{A \cdot B \cdot \lambda}{C} \cdot E_L \qquad (4)$$

Die Meßspannung U_C ist damit der Energie des eingestrahlten Laserimpulses proportional.

3. Besonderheiten von Detektoren auf PVDF-Basis

Die Auswahl des geeigneten Materials für einen pyroelektrischen Detektor ergibt sich aus den spezifischen Einsatzbedingungen, wie der geforderten Detektorfläche, der Arbeitsfrequenz, der Detektivität, nicht zuletzt aber auch aus wirtschaftlichen Gesichtspunkten.
Der pyroelektrische Koeffizient von PVDF liegt um etwa eine Größenordnung unter dem des als Vergleich heranzuziehenden TGS. Das hat entsprechend eine ebenfalls um eine Größenordnung geringere spezifische Detektivität zur Folge. Für die Messung der Leistung von Impulslasern spielt dieser Gesichtspunkt jedoch fast immer eine untergeordnete Rolle /5/.
Ein hoher spezifischer Widerstand und eine kleine Dielektrizitätskonstante erleichtern die Realisierung der geforderten RC-Konstante und damit den meßtechnischen Nachweis der influenzierten Ladung erheblich.
PVDF wird als dünne, mechanisch feste Folie angeboten. Damit können im Prinzip Detektoren beliebig großer Fläche ohne die bei Einkristallen oder Keramiken notwendigen aufwendigen Bearbeitungsschritte realisiert werden. Handelsübliche Folien liegen in Dicken von 5 ... 30 μm vor.
PVDF ist chemisch inert, insbesondere auch völlig unempfindlich gegen Wasserdampf. Daher entfallen aufwendige Kapselungen bzw. der Einsatz von Fenstern.

4. Aufbau der Detektoren

PVDF ist ein fluoriertes Polyäthylen. Die Pyroelektrizität steht mit der Existenz polarer CF_2-Gruppen im Polymer in Verbindung, die zunächst ungeordnet sind. Daher ist eine Formierung der Folien bei erhöhter Temperatur und Feldstärken bis zu 300 MV/m erforderlich. Der pyroelektrische Koeffizient ist bis zu Temperaturen um 100°C konstant. Bei Dauertests über zwei Jahre wurden keine Alterungserscheinungen beobachtet. Zum Abgriff des Meßsignals ist eine Metallisierung der Oberfläche notwendig. Im Hochvakuum aufgedampftes Aluminium ergibt gut haftende, wischfeste Schichten.
Im Betrieb des Detektors als Energiemesser ist eine wellenlängenunabhängige und über die Empfängerfläche homogene Absorption erwünscht.
Dies wird durch eine geeignete Lackschicht erreicht, die eine konstante Absorption bei Wellenlängen zwischen 0,2 und 25 μm sichert. Die maximale Belastbarkeit dieser Schichten liegt bei 20 MW/cm^2. Wegen der langen Wärmeübergangszeiten muß die Absorptionsschicht im Leistungsmesser-Regime weggelassen werden. In diesem Fall muß eine Volumenabsorption in der Polymerschicht erreicht werden.
In Fig. 2 sind drei Varianten der Flächenaufspannung gezeigt, die sich durch die realisierbaren thermischen Zeitkonstanten unterscheiden.

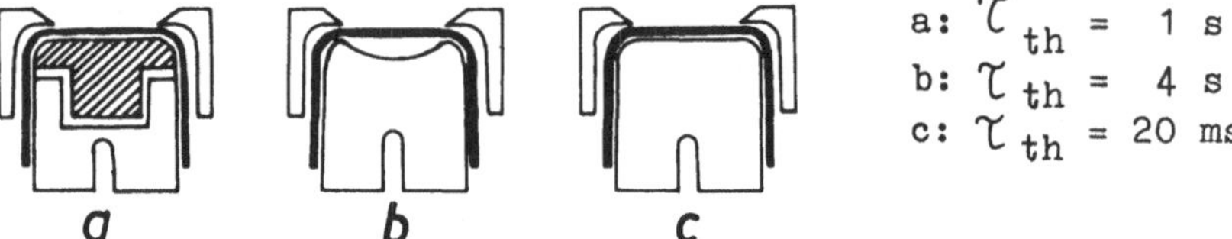

Fig. 2 Aufbau der Folienhalter

Die Auswahl der geeigneten Variante ergibt sich aus den Einsatzbedingungen. Die Registrierung des Meßsignals im Leistungsmesser-Regime erfolgt mit Hilfe eines Oszillografen. Für den Einsatz des Energiemessers wurde ein Spitzenwertmeßgerät entwickelt, das die direkte Anzeige der Laserimpulsenergie gestattet.

5. Anwendungsbeispiele

In den nachfolgenden Figuren werden die Möglichkeiten der vorgestellten pyroelektrischen Detektoren bei Energie- und Leistungsmessungen an sehr unterschiedlichen Lasertypen demonstriert. Dabei wird eine Wellenlängenbereich zwischen 337 nm (N_2-Laser) und 28 μm (H_2O-Laser) überstrichen.

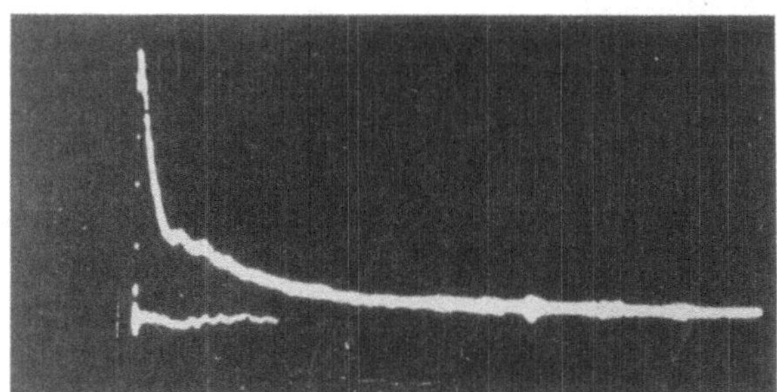

Fig. 3 Zeitlicher Verlauf eines CO_2-Laser-Impulses
Zeitablenkung 500 ns/Einheit, RC = 30 ns

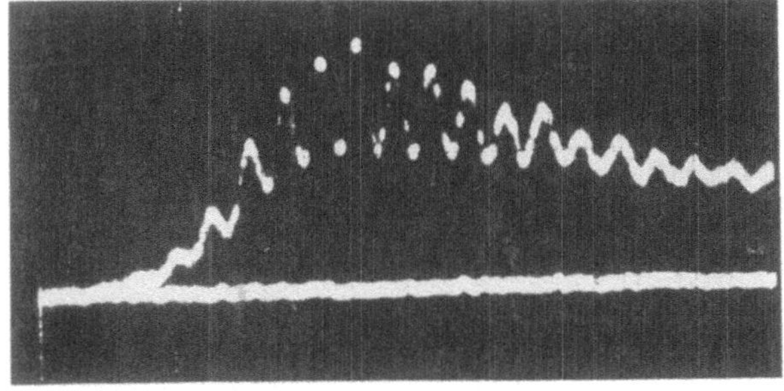

Fig. 4 Wie Fig. 3; Zeitablenkung 20 ns/Einheit, RC = 5,1 ns
(Die hohe Zeitauflösung gestattet den Nachweis des mode-locking.)

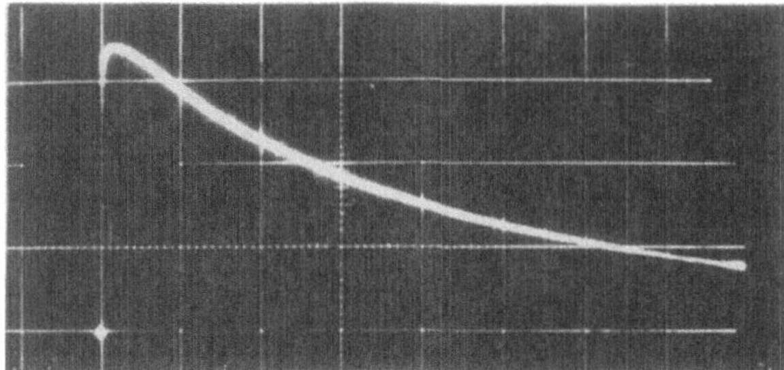

Fig. 5 Zeitlicher Verlauf des Meßsignals im Energiemeßregime ohne
Lackschicht (CO_2-Laser) Zeitablenkung 100 μs/Einheit; RC = 2,5 ms
Der Spitzenwert ist der Laserimpulsenergie proportional, der Abfall
erfolgt mit der thermischen Zeitkonstante

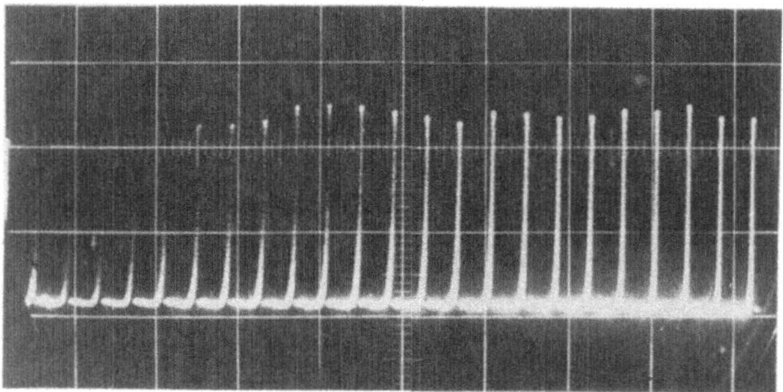

Fig. 6 Impulsenergie eines N_2-Lasers (337 nm) bei einer Folgefrequen
von 20 Hz. Die Impulsdauer eines Einzelimpulses beträgt 10 ns.
Zeitablenkung 0,1 s/Einheit, Vertikale Ablenkung 0,5 mJ/Einheit

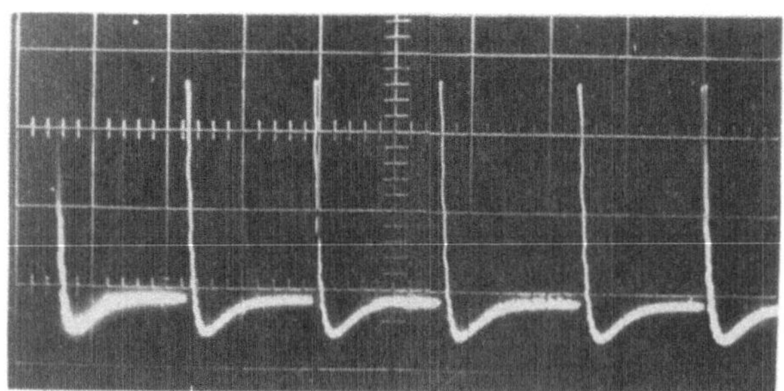

Fig. 7 Strahlungsmessung am H_2O-Laser (28 μm)
Zeitablenkung 1 s/Einheit; Vertikale Ablenkung 1mJ/Einheit

Literatur
(1) HADNI,A.: J.Phys.E.Sci.Instrum. 14 (1981) S. 1233-1240
(2) STOTLAR,S.C. u.a.: SPIE 190 LASL Optics Conference (1979) S.347-3
(3) KULESHOW,V.P., MALYUTA,D.D.:Prib. i. Techn. Eksp. 24 (1981) 205
(4) BOHMEYER,W. u.a.: Exp. Technik d. Physik 30 (1982) S. 65-74
(5) PORTER,S.G., Ferroelectrics 33 (1981) S. 193 - 206

Optronisches und lasertechnisches Messen und Prüfen

Optronisches Messen und Prüfen

Lasertechnisches Messen und Prüfen

Holographische Interferometrie

Lasers and Optoelectronics in Measuring and Testing

Optronic Measuring and Testing

Lasers in Measuring and Testing

Holographic Interferometry

Optische Prüfverfahren auf dem Vormarsch

H. ROTTENKOLBER
Rottenkolber Holo-System GmbH
Erhardstrasse 12, D-8000 München 5

1. Allgemeines

Die ersten Worte des Genesis umreißen eindrucksvoll die Bedeutung des
Lichtes für unsere Welt: "Da sprach Gott: Es werde Licht. Und es ward
Licht. Und Gott sah, daß es gut war. Aber Licht wäre nichts, gäbe es
nicht Augen, es zu sehen. Das vom Auge aufgenommene Lichtsignal wäre
nichts ohne Weiterverarbeitung der optischen Signale, zum Beispiel im
menschlichen Gehirn. Damit aber sind die <u>Grundelemente optischer Prüf-</u>
verfahren auch schon umrissen:

- Lichtquellen und Optik als "Sender"
- Optik und photosensitives Material als "Empfänger"
- Weiterverarbeitung der Lichtsignale

Weshalb sind optische Prüfverfahren so einzigartig ? Das liegt an den
besonderen <u>Eigenschaften der Grundelemente.</u> Sie arbeiten weitgehend

- trägerheitslos
- berührungsfrei
- mehrdimensional

Durch Entwicklungen der letzten zwanzig Jahre wurde das Instrumentarium der optischen Prüftechnik wesentlich erweitert:

- 1960 "erfindet" Theodore Maiman ein "neues" Licht, den Laser.
- die Opto-Elektronik und der Computer erfahren durch die Halbleiter-
  Technologie einen gewaltigen Aufschwung.

Immer sind es die faszinierenden <u>Eigenschaften des Lichtes,</u> welche den
Einsatz der optischen Prüfmethoden bestimmen:

- Art der Lichtausbreitung
- die Lichtgeschwindigkeit
- die Lichtwellenlänge oder Farbe
- die Interferenz

Die Verbesserung der abbildenden Eigenschaften von Linsen und Objektiven macht die Erforschung neuer Gebiete möglich. Nach dem Blick in den Makrokosmos mit Hilfe von Fernrohren drang der Mensch in den Mikrokosmos mit Hilfe von Mikroskopen vor. Das <u>Mikroskop</u> ist heute ein unentbehrliches Prüfgerät in der <u>Medizin</u> wie in der <u>Werkstoff-</u>prüfung. Die Aufbereitung von mikroskopischen Bildern mit Hilfe des heutigen Entwicklungsstandes der Elektronik war die erste Anwendung der <u>digitalen Bildverarbeitung.</u> Sie erlaubt das rasche Erkennen und Auszählen von Teilchen bestimmter Größe und Kontur. Während die Lichtmikroskopseite weitgehend ausentwickelt sein dürfte, sind auf der Seite der Bildverarbeitung noch erhebliche Entwicklungsreserven, die parallel zur Entwicklung im Computerbereich ablaufen werden.

Mit der Erfindung lichtempfindlicher Schichten durch Daguerre in der Mitte des vorigen Jahrhunderts wird die Entwicklung der <u>Photographie</u> und der <u>Kinematographie</u> in Gang gesetzt. Eine völlig neue Dimension, die Zeit, wurde erforschbar. Neben dem Zeitraffer ist vor allem die <u>Hochgeschwindigkeitsphotographie</u> eine wertvolle optische Prüfmethode. Zunächst vor allem im <u>militärischen Bereich</u> zur Erfassung von rasch ablaufenden Vorgängen eingesetzt, gewinnt die Nutzung im zivilen Bereich immer mehr an Bedeutung. Der <u>Sportler</u> gewinnt aus der zeitlichen Auflösung ebenso neue Erkenntnisse wie die <u>Unfallforschung</u> aus dem Studium der Bewegungen eines Dummy bei einer Aufprallstudie. Die Möglichkeiten der Änderungen des Maßstabes Zeit sind meiner Meinung nach aber noch längst nicht ausgeschöpft. Luft- und Satelitenaufnahmen, langfristige Wetterprognosen und Umweltdiagnostik, in Verbindung mit anderen analytischen Methoden sind nur einige der Perspektiven.

Die Bedeutung der <u>stereoskopischen Photographie</u> bei der Landvermessung wurde schon früh erkannt. Die Auswertung stereoskopischer Bilder mit Hilfe der digitalen Bildanalyse wird teilweise durchgeführt. Über die Landvermessung hinaus wird der Wunsch nach <u>optischer Konturvermessung</u> immer stärker, so zum Beispiel im Turbinenbau, in der Gießereitechnik oder bei Preßformen. Neben die Stereographie tritt die sogenannte Moiré-Technik. Hier wird ein Linienmuster auf den Prüfkörper projiziert und durch das Muster beobachtet. Man erhält Höhenschichtlinien, deren Abstand von der Geometrie der Prüfapparatur abhängt. Obwohl die Moiré-Technik keine neueren Entwicklungen voraussetzt, ist sie erst in den letzten Jahren in den Vordergrund getreten. Die automatische Bildanalyse fehlt in diesem Zusammenhang noch vollständig.

Bis ins vorige Jahrhundert waren die natürlichen Lichtquellen, wie
Sonne und Sterne, das menschliche Auge und das menschliche Gehirn
das einzige Instrumentarium optischer Prüfmethoden. Dabei wurde die
geradlinige Ausbreitung der Lichtstrahlen und die Fähigkeiten des
Auges, Hell-Dunkel-Signale unterscheiden zu können, ausgenutzt. Die
Farbempfindlichkeit wurde nicht genutzt. Die Lichtgeschwindigkeit,
die Phasenlage des Lichtes und die Polarisation können vom menschlichen Auge nicht erfaßt werden.

2. Art der Lichtausbreitung

Die <u>Sonnenuhr</u> ist eines der ältesten optischen Prüfgeräte der Mensch-
heitsgeschichte. Die Seefahrer aller Zeiten benutzten den <u>Sextanten</u>
zur Orientierung an den Fixsternen. Der <u>Theodolit</u> dient zur Vermessung
auf dem Lande. Moderne Vermessungsgeräte sind mit einem He-Ne-Laser
als aktive Lichtquelle ausgestattet. Die digitale Darstellung der
Vermessungsergebnisse ist die vorläufige letzte Stufe dieser Anwendung.

Licht ist manipulierbar. Man kann die Richtung von Lichtstrahlen durch
Spiegel oder beim Übertritt in ein anderes Medium, zum Beispiel von
Luft in Glas, ändern. So entstehen optische Komponenten wie Spiegel,
Linsen und Prismen, die zu Abbildungssystemen wie Fernrohren oder
Mikroskopen zusammengefügt werden können. Die Qualität der Abbildung
hängt entscheidend von der <u>Qualität der optischen Komponenten</u> ab.
Diese kann nur durch geeignete Prüfmethode erfaßt werden. Nach der
von <u>Foucault</u> entwickelten Methode werden noch heute Spiegel geprüft.
Dabei werden Abweichungen von einer Sollkontur des Spiegels in
Helligkeitswerte umgesetzt. Dasselbe Prüfprinzip erweiterte Töpler
bereits 1846 zu dem nach ihm benannten Schlierenverfahren. Lichtstrahlen, welche ein inhomogenes Brechzahlfeld durchlaufen, erfahren
eine Änderung ihrer ursprünglichen Richtung. Diese Richtungsänderung
wird in proportionale Helligkeitswerte umgesetzt. Ein inhomogenes
Brechzahlfeld kann bei <u>Wärme- und Stoffaustauschvorgängen</u> und in
<u>schlierigem Glas</u> ebenso gegeben sein, wie bei Druckänderungen. In
der <u>Strömungsforschung/Gasdynamik</u> liegt dann die Hauptanwendung des
Schlierenverfahrens. Auf demselben Prinzip, jedoch weniger aussage-
fähig, beruhr das Schattenverfahren von Dvorak. Durch die Entwicklung
des Lasers werden die vorgenannten Verfahren immer mehr zugunsten
von interferometrischen Verfahren verdrängt.

Jahrelang wurden optische Sensoren nur als Lichtschranken eingesetzt.
Sie waren einfache Ja-Nein-Informanten. Dann lernten sie zählen. Mit
der Aneinanderreihung zu Dioden-Zeilen und unter Verwendung geeigneter Lichtquellen können sie heute Wegänderungen messen. Füllstands-
messungen in Behältern und Dickenmessung an Blechen sind nur erste
Applikationen. Das Ergebnis steht digital zur Verfügung und führt bei
entsprechender Schwellwertvorgabe zur automatischen Überwachung.
Optische Sensoren sind die Wegbegleiter der auf dem Vormarsch befindlichen Robotergeneration. Die Roboter lernen sehen.

Die Videotechnik tritt immer mehr an die Stelle des menschlichen Auges.
Bilder können gespeichert werden und stehen für einen Soll-Ist-Ver-
gleich als digitale Werte bereit. Bei einfachen Überwachungsaufgaben
ist das Prinzip ebenso einsetzbar wie in Verbindung mit anderen optischen Geräten. Die Prüfung von Schaltkreisen unter dem Mikroskop
läßt sich damit weitgehend automatisieren. Die Entwicklung der Videotechnik wird in den nächsten Jahren in Richtung höherer Bildauflösung
gehen.

3. Die Lichtgeschwindigkeit

Bereits vor 300 Jahren hat Olaf Römer die Lichtgeschwindigkeit gemessen. Sie ist so groß, daß man optische Prüfmethoden als nahezu
trägheitslos bezeichnen kann. Aber auch die Ausbreitungsgeschwindig-
keit des Lichtes selbst kann ein Prüfkriterium sein. Sowohl die Erfindung des Lasers als auch die Entwicklung "schneller Elektronik"
machten dies möglich.
Mit einem Laser können Blitze von wenigen Nanosekunden Dauer und
ausreichender Energie abgeschossen werden. Nach dem Radarprinzip
wird die Laufzeit des Blitzes ermittelt und die Entfernung errechnet.
Moderne Panzer sind mit solchen Entfernungsmeßgeräten ausgerüstet. Im
Rahmen des Apollo-Programms konnte mit bis dahin unvorstellbarer Genauigkeit die Entfernung Erde-Mond gemessen werden.

Aus der Akustik ist der sogenannte Doppler-Effekt bekannt. Danach
ändert sich die Frequenz, das ist die Tonhöhe, je nachdem, ob sich die
Schallquelle auf den Empfänger zu oder sich von ihm wegbewegt. Mit
Hilfe moderner Elektronik läßt sich der Doppler-Effekt auch bei Licht
ausnutzen. In einer Flüssigkeits- oder Gasströmung mitgeführte Partikel werden mit Laserlicht, meist He-Ne-Laser, beleuchtet. Das Laserlicht wird an den bewegten Partikeln rückgestreut und von einem orts-

festen Empfänger registriert. Die geringe Frequenzverschiebung des an
den bewegten Teilchen rückgestreuten Lichtes gegenüber dem Beleuch-
tungslicht wird ermittelt und in die Geschwindigkeit des bewegten
Teilchens umgerechnet. Die als Doppler-Anemometrie bezeichnete
Methode ist heute ein unverzichtbares Instrument der Strömungsforschung

Aus dem zeitlichen Verlauf der Geschwindigkeit in eine Zeitintervall
kann der in diesem Zeitintervall zurückgelegte Weg berechnet werden.
Nach diesem Prinzip arbeitet der Laser-Abstandsmesser, der zur Fein-
steuerung an Werkzeugmaschinen eingesetzt wird.

4. Die Lichtwellenlänge

In der zweiten Hälfte des siebzehnten Jahrhunderts entdeckte Newton
die Zerlegbarkeit des weißen Sonnenlichtes in die Spektralfarben. Der
Regenbogen wurde erklärbar. Es dauerte aber noch zwei Jahrhunderte,
ehe man das Licht als elektromagnetische Welle begreifen lernte; das
sichtbare Sprektrum ist nur ein kleiner Ausschnitt. Röntgen entdeckte
die nach ihm benannten Strahlen, welche tief im kurzwelligen Bereich
über das UV-Licht hinaus einzuordnen sind. Im langwelligen Bereich
schließt sich Infrarot und fernes Infrarot an das sichtbare Spektrum
an. Optik und optische Prüfmethoden reichen heute von Röntgenstrahlen
bis ins ferne Infrarot.

Die Spektroskopie ist die älteste optische Prüfmethode, welche auf
der Lichtwellenlänge beruht. Es war ein Münchner namens Fraunhofer,
der bereits 1814 die nach ihm benannten Linien im Sonnelicht entdeckte.
Die chemischen Elemente, und, später entdeckt, die Bausteine des Atoms,
haben typische Spektrallinien, durch die sie sich identifizieren
lassen. Generationen von Physikern waren und sind mit der Erforschung
dieses Phänomens befaßt. Mit der Erfindung des Lasers hat die Spektroskopie entscheidende neue Impulse bekommen. Die Spekralanalyse,
heute bereits weitgehend automatisiert, ist aus keinem Werkstofflabor
wegzudenken. Sie wird als Lidarsystem in der Umwelttechnik, bei der
Satellitenbeobachtung der Erdoberfläche ebenso eingesetzt wie in der
Erforschung ferner Galaxien.

Röntgenstrahlen sind aus der medizinischen Diagnostik nicht mehr wegzudenken. In der Anfangsphase vorzugsweise zur Beurteilung der Knochen-
struktur eingesetzt, fanden die kurzwelligen Strahlen sehr rasche An-

wendung auch bei der Untersuchung der Weichteile. Die Mediziner
lernten auch geringe Kontrastmodulationen lesen. Gehirntumore wurden
ebenso erkennbar wie TBC-Schatten in der Lunge. Heute werden ver-
suchsweise Techniken der _Bildanalyse_ eingesetzt, um die Interpretation zu verbessern. Eine über die Bildverbesserung hinausgehende
Automatisierung im Medizinbereich ist in nächster Zukunft kaum zu
erwarten. Dem Arzt wird die individuelle Beurteilung und Entscheidung der Symtome vorbehalten bleiben müssen.

In der _Verbrechensbekämpfung_ wird die Röntgenüberwachung z.B. in
Flughäfen eingesetzt. Hier, wie im Medizinbereich, ist die individuelle Beurteilung des Röntgenbildes wohl nicht zu ersetzen.

Weniger spektakulär ist der Einsatz von Röntgenstrahlen im _Technik-
bereich_. Hier werden als Ergänzung zur Ultraschalltechnik _Risse und
Lunker_ sichtbar gemacht. _Stahlcord-Fahrzeugreifen_, wegen ihrer
höheren Lebensdauer geschätzt, werden mit Röntgengeräten überwacht.
Prüfkriterium ist das gleichmäßige Muster des Stahlcords. Eine
automatische Erkennung von Musterabweichungen ist sicher in naher
Zukunft zu erwarten. Schließlich kann man _Eigenspannungen_ mit Hilfe
von Röntgenstrahlen erkennen.

Im Gegensatz zur Röntgendiagnostik steht die _UV-Diagnostik_ etwas
stiefmütterlich zur Seite. Bestimmte Stoffe reflektieren vorzugsweise UV-Licht. Diese Stoffe dienen zum "Anfärben" z.B. von Geld-
scheinen oder sie setzen sich in Haarrissen von technischen Oberflächen an. Mit UV-Licht können einfache _Fälschungen_ oder _Haarrisse_
erkannt werden. Darüber hinausgehende Anwendungen sind mir nicht
bekannt.

Im Gegensatz dazu hat die _Infrarot-Diagnostik_ ein breites Anwendungs-
feld. Einfache Lichtschranken übernehmen, unbehindert von sichtbarem
Licht, _Zählfunktionen._ Infrarot-_Temperaturmeßgeräte_ mit Digitalanzeige
werden sowohl im Walzwerk als auch im Triebwerksbau benutzt.

Die Entwicklung der Halbleiter-Technologie des letzten Jahrzehnts
hat entscheidende Impulse gegeben. Im militärischen Bereich werden
Infrarot-Detektoren zur Kontrolle der _Flugbahnen von Raketen_ und zur
Überwachung von Vorgängen am Boden und in der Luft verwendet.

Die Umsetzung von _Wärmebildern_ in Bilder des sichtbaren Lichtes wurde in den letzten Jahren erheblich verbessert. An <u>Bauwerken</u> konnten Schwachstellen mit hohen <u>Wärmeverlusten</u> sichtbar gemacht werden. Im Bereich des <u>Maschinenbaus</u> kommt es häufig durch Reibung oder chemische Reaktionen zu unerwünschten Überhitzungen, welche mit Infrarotgeräten sichtbar gemacht werden. Diese Geräte sind bereits soweit entwickelt, daß bestimmten Temperaturen Grauwerte oder Farben zugeordnet werden können. Abhängig von Einsatzgebieten lassen sich ohne weiteres Signale beim Überschreiten von Schwellwerten angeben.

Komplementär zur Röntgendiagnostik wird Infrarotdiagnostik in der <u>Medizin</u>, vor allem zur Erkennung von <u>Brustkrebs</u>, eingesetzt. Die Grundidee beruht auf den durch höheren Stoffwechsel in Krebszellen bedingten höheren Körpertemperaturen.

5. Polarisation von Lichtwellen

Störende Spiegelungen an Glasscheiben werden vom Photographen mit einem sogenannten Polarisationsfilter unterdrückt. Dies ist möglich, weil das an der Glasscheibe reflektierte Licht zum Teil polarisiert wird, d.h. das reflektierte Licht schwingt in einer Ebene. Das Polarisationsfilter hat die Eigenschaft, nur in einer Ebene schwingendes Licht durchzulassen. Wird die Durchlaßebene des Polarisationsfilters senkrecht zur Schwingungsebene des reflektierten Lichtes eingestellt, wird dieses nicht durchgelassen - die Spiegelung verschwindet. Bestimmte Harze ändern unter mechanischer Zug- oder Druckeinwirkung ihre Polarisationseigenschaften (Doppelbrechung, Durchlaßebene). Es besteht ein mathematischer Zusammenhang zwischen der Änderung der Polarisationseigenschaften und dem mechanischen Spannungszustand, wobei die Stoffeigenschaften des Harzes und die Durchstrahlungsdicke zu berücksichtigen sind. In der Spannungsoptik wird ein Modell des Prüfkörpers aus Harz in ein Lichtbündel aus polarisiertem Licht gestellt und mit einem Polarisationsfilter beobachtet. Wird auf das Modell eine mechanische Spannung aufgebracht, so kann über das Polarisationsfilter die Änderung der Polarisationseigenschaften beobachtet werden.
Spannungsoptische Bilder sind Interferenzmuster. Die Liniendichte gibt Aufschluß über den Spannungszustand.
Die Interpretation des spannungsoptischen Bildes muß noch vom Spezialisten vorgenommen werden. Die Bildverarbeitung wird zu einer automatischen Aussage und damit zu einer breiteren Anwendung führen.

6. Lichtinterferenz

Der Nachweis der Wellennatur des Lichtes geht wesentlich auf Thomas
Young zurück. Er wies durch seinen Doppelspalt-Versuch nach, daß
unter bestimmten Voraussetzungen "Licht plus Licht gleich Dunkel"
möglich war. Werden zwei oder mehr Lichtwellen überlagert, kommt
es zur ganzen oder teilweisen Auslöschung oder Verstärkung dieser
Lichtwellen. Diese Erscheinung bezeichnet man als Interferenz.

Lichtinterferenz ist die Grundlage zahlreicher optischer Prüfverfahren, welche sich durch hohe Anzeige-Empfindlichkeit auszeichnen.
In der Zeit vor Erfindung des Lasers standen Lichtquellen zur Verfügung, deren Licht mehr oder weniger darbig war. Ferner waren definierte Phasenbeziehungen innerhalb der von solchen Lichtquellen ausge-
sandten Lichtwellen nicht gegeben. Eindeutige Interferenz zweier oder
mehrerer Lichtwellen ist aber nur möglich, wenn Licht einer einzigen
Wellenlänge, d.h. einer Farbe, gegeben ist (zeitliche Kohärenz).
Durch engbandige Filter kann dieser Bedingung mehr oder weniger
genügt werden. Eindeutige Interferenz ist aber auch nur möglich,
wenn die zur Interferenz gebrachten Lichtwellen in einer eindeutigen
Phasenbeziehung zueinander stehen (räumliche Kohärenz). Ehe das Laserlicht zur Verfügung stand, mußte diese letztgenannte Bedingung durch
hochpräzise optische Anordnungen erfüllt werden. Laserlicht befreit
weitgehend von diesen Zwängen, da es der Natur seiner Erzeugung nach
diese Bedingungen selbst erfüllt. Der Laser ist für interferometrische Anordnungen die geeignete Lichtquelle schlechthin.

Das interferometrische Meßprinzip beruht darauf, daß die Phase des
Lichtes in einer Teilwelle moduliert wird und durch Überlagerung mit
einer zweiten Teilwelle in Form von Interferenz sichtbar wird. Die
Modulation der Phase geschieht durch Modulation des sogenannten
optischen Weges, der bekanntlich das Produkt aus geometrischem Weg
und er Brechzahl ist. Daraus ergibt sich die Nutzungspalette.

Ende des letzten Jahrhunderts wurde das Interferometer von Mach und
Zehnder bekannt, das auf der Konzeption von Jamin beruht. Das Gerät
von Mach/Zehnder wird bis heute eingesetzt in der Erforschung von
Vorgängen des Wärme- und Stoffaustausches und in der Strömungsforschung. Es hat etwa dasselbe Einsatzgebiet wie das Töplersche
Schlierenverfahren, dem es aber wegen seiner quantitativen Aussage-
fähigkeit weit überlegen ist. Erst durch den Einsatz der sogenannten

holographischen Interferometrie in den letzten Jahren wurde das aufwendige Mach-Zehnder-Interferometer allmählich abgelöst.

Bei dem <u>Michelson-Interferometer</u> werden die beiden Teilstrahlen in sich selbst zurückgeführt. Es wurde vorzugsweise zur <u>Überprüfung von ebenen Oberflächen</u> eingesetzt. Eine Modifikation ist das Interferometer nach <u>Twyman-Green</u>, welches die Untersuchung auch von <u>Linsen, Objektiven oder gekrümmten Oberflächen</u> zuläßt. Die Analogie zum Foucault-Test liegt auf der Hand, wiederum aber als Vorteil der Interferenzverfahren die Möglichkeit der quantitativen Aussage. Etwa denselben Einsatzbereich haben <u>Vielstrahl-Interferometer</u> nach dem Prinzip von <u>Fizeau</u>, welche vor allem als <u>technische Mikroskope</u> Bedeutung haben.

Die hier erwähnte klassische Interferometrie wurde mit einem Schlag revolutioniert: Laserlicht ist in bisher nicht bekanntem Maße interferenzfähig.

Die körnige Struktur einer von Laserlicht beleuchteten Fläche ist ein Indiz für dessen Interferenzfähigkeit. Diese als "Speckle" bezeichnete Erscheinung wurde Anfang der siebziger Jahre als Prüfmethode konzipiert und seitdem weiter entwickelt. Vom optischen Aufwand und den Umgebungsbedingungen her weniger anspruchsvoll, hat die <u>Speckle-Methode</u> allerdings einen gravierenden Nachteil: die Grobkörnigkeit und damit schlechte Auflösung der Speckle-Bilder, die bis heute auch durch elektronische Nachverarbeitung nicht entscheidend verbessert werden konnte.

Fünf Jahre nach Erfindung des Lasers ging ein Zauberwort durch die Welt:Holographie.Zunächst bestaunt als dreidimensionale Photographie, als Laserphotographie. Man kann durch das "Hologrammfenster" sozusagen um die Ecke sehen. Im Hologramm sind, im Gegensatz zu klassischen Photographie, die Phasenlagen der von einem Objekt ausgehenden Wellenfronten "eingefroren". Von einer Arbeit von Denis Gabor aus dem Jahre 1948 ausgehend, hatten Leith und Upatnieks von der Michigan University in Ann Arbor die Sensation möglich gemacht. Wenig später hielt die H olographie Einzug in die Interferenz-Meßtechnik als <u>holographische Interferometrie.</u> Dabei werden zwei Wellenfronten von dem gleichen Objekt überlagert, wobei mindestens eine Wellenfront im Hologramm gespeichert ist. Da die Speicherung zeitlich versetzt erfolgt, werden Formänderungen an dem Prüfobjekt oder Phasenänderungen in transparenten

Objekten oder Medien als Interferenzfeld sichtbar. Die Beschaffenheit
der Oberfläche des Prüfobjektes oder transperenter Objekte spielt dabei keine Rolle: Das heißt, die Tür zur breiten technischen Anwendung
von hochpräzisen Interferenzmethoden war aufgetan.

Die holographische Interferometrie dringt in die klassischen Anwendungen vor: <u>Wärme- u. Stoffaustausch, Strömungsforschung, Qualitätskontrolle optischer Komponenten.</u> Die neuen Anwendungsbereiche: <u>Zer-
störungsfreie Werkstoffprüfung von Kunststoffen</u> und Qualitätskontrol-
len von Reifen, statische <u>Verformungsermittlung</u> als Konstruktionsun-
terstützung neben der klassischen Spannungsoptik, der Dehnmaßstreifen
und der Modalanalyse. <u>Schwingungsanalyse</u> als hervorragende Methode zur
Sichtbarmachung von Schwingungsbildern. Es war ein langer Marsch vom
Labor zur praktischen Anwendung. Dies wird deutlich am Beispiel holographischer Reifenprüfung: Es werden Formänderungen kleiner als ein
Mikrometer an so unoptischem Material wie schwarzem Gummi in einer so
brutalen Umgebung wie einer Reifenproduktionshalle innerhalb von
3 Sekunden angezeigt.

Die Interpretation der Interferenzbilder wird derzeit noch von
Spezialisten vorgenommen. Bildverarbeitung und Aufbereitung im
Computer sind die notwendigen nächsten Schritte auf dem Weg zur
breiten Anwendung dieser Technologie.

7. Zusammenfassung

Durch die Entwicklung der letzten beiden Jahrzehnte haben optische
Prüfmethoden einen gewaltigen Aufschwung erfahren.Zwei Ereignisse
waren hierfür entscheidend:

- die Erfindung des Lasers

- die Entwicklung der optischen Sensor- und Computertechnik.

Da die letztgenannte Entwicklung noch in vollem Gange ist, kann der
<u>heutige Stand der Technik</u> bestenfalls als eine <u>Momentaufnahme</u> ver-
standen werden. Diese Entwicklung führteund führt zu automatisier-
baren Prüfabläufen und befreit von Fehlinterpretationen. Dadurch
werden Regelkreise möglich, deren erste, noch recht einfache Aus-
wirkung, der sehende Roboter ist.

Bereits die Anwendung des derzeitigen Standes der Technik im Bereich der optischen Sensoren und der Bildverarbeitung für die verschiedensten optischen Prüfmethoden wird zu einer weiteren gewaltigen Expansion optischer Prüfverfahren führen. Die Erschließung immer neuer Einsatzgebiete ist nur noch eine Frage der Zeit. Der Vormarsch optischer Prüfverfahren geht unaufhaltsam voran.

Echtzeit-Bildverarbeitung im industriellen Einsatz

M. PRAMMER
Institut AEE, Technische Universität Wien
Gußhausstraße 27 - 29, A-1o4o Wien

R. HORAK
Fa. Optik-Elektronik GesmbH
Eumigstraße 2, A-2351 Wr. Neudorf

Zusammenfassung

Die Automatisierung der visuellen Qualitätskontrolle erfordert
Bildverarbeitungssysteme, die im Takt der Fertigung, also in Echt-
zeit, arbeiten können. Der Industrieeinsatz stellt hohe Anforderungen
an Meßgenauigkeit, Datendurchsatz, Fehlerquote und Verfügbarkeit.
Das vorgestellte System wurde zur Prüfung von Stückgut der Halb-
fertigteil- und Finalproduktion entwickelt. Als Aufnehmer werden
Digitalkameras eingesetzt, bestehend aus CCD-Zeilensensor, Analog/
Digital-Umsetzer sowie Steuer- und Kalibrierlogik. Der angeschlossene
Rechner basiert auf einer Multi-Prozessor-Architektur und kann
2 MByte/s pro Kamera auswerten. Die Prüfung basiert auf einem Muster-
vergleich mit einer im Teach-in-Modus gewonnenen Referenz. Die Haupt-
schwierigkeiten liegen in der Wahl signifikanter Erkennungsmerkmale
und der Festlegung einer Toleranzbreite.

Einleitung

Der zunehmende Einsatz bildverarbeitender Systeme für die Produktions-
sichtkontrolle wird durch folgende Faktoren bestimmt:
- Die Kosten für visuell/manuelle Inspektion von Stückgut erreichen
  10 % der gesamten Bearbeitungskosten (/1/). Optische Prüfplätze
  erfordern volle Konzentration auf eine repititive, ermüdende Tätig-
  keit.
- Die Assemblierung durch Handhabungsautomaten bedingt hundert-
  prozentige Kontrolle aller Einzelteile. Eine bei manueller Montage
  implizit ausgeschiedene defekte Schraube wird von einem Roboter
  eingedreht und beschädigt das komplette Produkt (/2/). Der Prüf-
  takt muß also dem Fertigungstakt entsprechen (typ. 3 Stück/s),
  womit menschliche Prüfer überfordert sind.

Entsprechend dem Bedarf der Industrie werden beträchtliche Forschungs-
anstrengungen auf diesem Gebiet unternommen (/3/ - /5/). Abb. 1 zeigt
die prinzipielle Integration des visuellen Prüfsystems in eine
Fertigungsstraße. Das Fließband transportiert das Stückgut in die

Prüfstation. Es wird geeignet beleuchtet und von Kameras abgetastet.
Die digitalisierten Kamerasignale werden von einem Prozeßrechner
ausgewertet. Der Rechner steuert einen nachfolgenden Sortierer, der
im einfachsten Fall defekte Teile ausscheidet. Weiters kann der
Computer Rückmeldungen an die Fertigung liefern und Produktions-
statistiken erstellen.

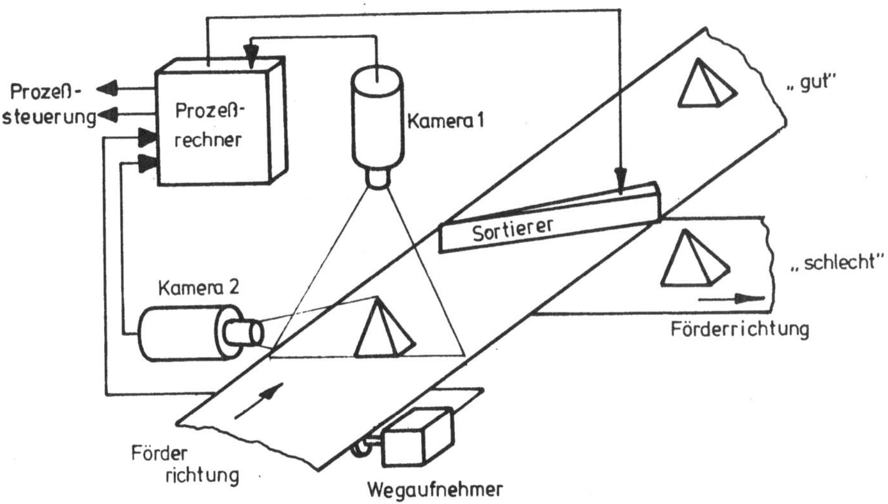

Abb. 1. Konzept eines visuellen Prüfsystems

Der kontinuierliche Vorschub legt eine zeilenweise Abtastung quer
zum Band nahe. Voraussetzung ist die rotationsrichtige Einbringung
des Prüfgutes. Ein Meßfühler ermittelt die Transportgeschwindigkeit,
um einen eindeutigen Bildzusammenhang herzustellen.

Das vorgestellte System mußte folgenden Forderungen genügen:
- Meßgenauigkeit: Das Verhältnis von Sichtfeld zu Bildpunkt soll
    besser als 1000 : 1 sein. Der Abbildungsmaßstab muß eindeutig und
    driftfrei sein, da jede Neukalibrierung eine teure Fertigungs-
    unterbrechung bedeutet.
- Kontur- und Oberflächenkontrolle: Es sollen verschiedene Längen-
    und Breitenmaße bestimmt und die Güte der Oberfläche beurteilt
    werden. Die feinstrukturierte Oberfläche erfordert eine Grauskala
    von mindestens 50 Stufen pro Bildpunkt.
- Durchsatzrate: Bei einer Taktrate von 3 Stück/s verbleiben bei
    zwei Kameraeingängen 0,15 s Bearbeitungszeit pro Bild.

<u>Sensoren</u>

Die erforderliche Auflösung und Stabilität wird nur von CCD-Zeilen-
sensoren erreicht. Abb. 2 zeigt den Aufbau einer Digitalzeilenkamera.
Das Sensorsignal wird bereits in der Kamera mit 6 - 8 bit und 2 MHz
Abtastrate digitalisiert und Byte-parallel auf RS-422 Leitungen ge-
sendet. Zur Stabilisierung des Dunkelstromes wird der CCD-Sensor
über ein Peltier-Element gekühlt. Eine Kalibrierungslogik führt bei
jedem Zeilendurchlauf einen Abgleich mittels Abtastung einer Referenz-
fläche durch und ermöglicht Vergleichsmessungen bei unterschiedlichen
Lichtbedingungen.

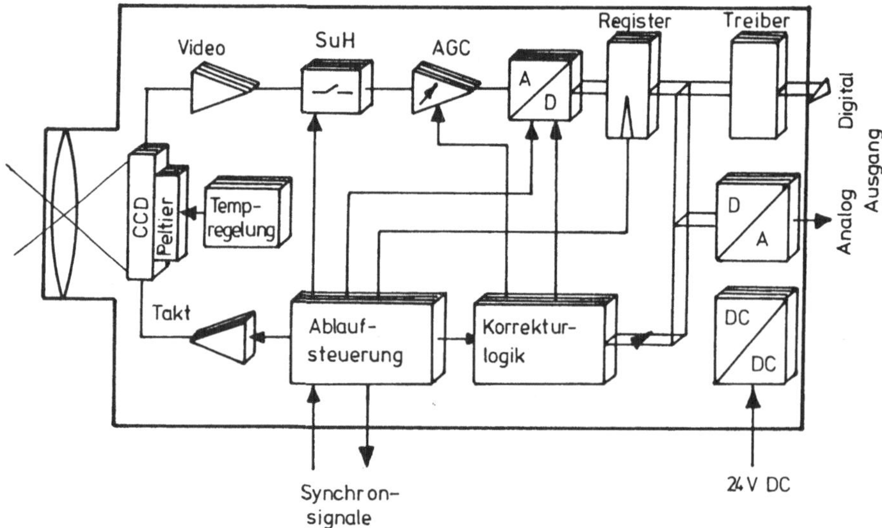

Abb. 2. Blockschaltbild einer Digitalzeilenkamera

Die Kameras erfüllen folgende Sensorfunktionen:
- Konturkontrolle: Das leere Bildfeld erscheint im Gegenlicht weiß,
  das Meßobjekt dunkel. Breitenabmessungen ergeben sich durch Aus-
  zählen der Bildpunkte zwischen Helligkeitssprüngen. Längenmaße
  werden in Verbindung mit dem Transportwegsensor ermittelt.
- Lichtschranke: Das Auftreten dunkler Bildpunkte innerhalb einer
  Zeile zeigt den Eintritt des Meßobjektes in die Station an.
- Oberflächenkontrolle: Auf die Oberfläche des Prüflings wird
  Weißlicht mit 40° Einfallswinkel gerichtet. Diese Schrägbeleuchtung
  macht durch Schattenwurf Kratzer, Risse und Erhebungen sichtbar.

Prozessor

Die hohe Prozeßgeschwindigkeit wird durch überlappende Verarbeitung ("Pipelining") auf einem Multi-Prozessorsystem erreicht (Abb. 3).

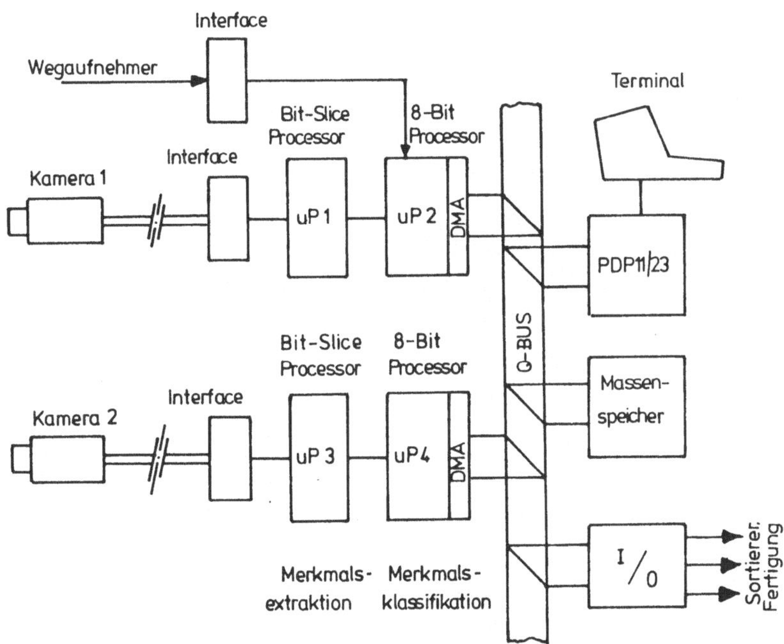

Abb. 3. Multi-Prozessorsystem

Für jede Kamera ist ein Verarbeitungskanal vorhanden. In der ersten Stufe führt ein bipolarer Bit-Slice-Prozessor die Merkmalsextraktion durch. Benachbarte Bildpunkte werden zu Feldern von 4 x 4 und 16 x 16 Elementen zusammengefaßt, in denen die Helligkeitsverteilungsstatistik errechnet wird. Ein Merkmalsvektor von ca. 4 KByte wird an einen 8-bit-MOS-Prozessor weitergereicht, der ihn mit einer Referenzmatrix korreliert und daraus einen Fehlervektor errechnet. Die Fehlervektoren aller Kanäle werden zur PDP 11/23 - CPU übertragen, die die Entscheidung über Akzeptanz oder Zurückweisung fällt und entsprechend die Ausgänge aktiviert.

Algorithmus

Das Dimensionsproblem ist eindeutig definierbar. Das Bild wird durch ein Medianfilter mit 8 Pixel Schlepplänge aufbereitet und mittels Schwellwertvergleich binärisiert. Die gefundenen schwarz/weiß-Sprünge

bilden die Objektkontur. Es werden folgende Meßwerte bestimmt:
minimale, mittlere und maximale Breite, Position und Verdrehung,
mittlere Länge sowie Schärfe der Konturschnitte. Für jedes Maß sind
Grenzwerte festgelegt, deren Über- und Unterschreiten zur Rück-
weisung führen. Aufeinanderfolgende gleichartige Fehler werden zur
Fertigung rückgemeldet.
Die Prüfung der Oberflächentextur erfordert wesentlich mehr Rechen-
und Bedienungsaufwand. Die Meßfläche wird in 1000 Felder gerastert.
Die Komponenten des erwähnten Fehlervektors sind Erwartungswert,
mittlere Schwarz- und mittlere Weißvarianz in jedem dieser Felder.
In einem vorbereitenden Lerndurchgang ("Teach in") werden ausgewählte
Gut- und Schlechtmuster geprüft, wobei dem System vom Bediener die
Art des Fehlers mitgeteilt wird. Die erhaltenen Merkmalsvektoren
werden gesammelt und ihre Abstandsmaße zueinander bestimmt. Der
Aufbau der Referenzmatrix ergibt sich aus der Forderung, die Abstandsmaße der Gutmuster zueinander minimal und zu den Schlechtmustern
maximal werden zu lassen.
Daneben besteht die Möglichkeit, während des Betriebes einzugreifen.
Wird eine offensichtliche Fehlentscheidung des Systems vom Operator
erkannt, kann er vom Terminal aus das Programm anhalten und das
letzte Bild im internen Speicher einfrieren lassen. Der Rechner versucht in einem iterativen Suchprozeß seine Sollparameter so zu verändern, daß die korrekte Entscheidung zustande kommt.

Diese Entwicklung ist ein Gemeinschaftsprojekt der Technischen
Universität Wien, der Firma Optik-Elektronik GesmbH und dem Forschungs-
förderungsfonds für die gewerbliche Wirtschaft Österreichs
(Proj. Nr. 3/3617-I/P).

Literatur

/1/ NEVINS, J.L. et al: Productivity, Technology and Product System
   Productivity Research, Report R-928, Vol.2, Draper Laboratory 1976.

/2/ GEISSELMANN, H.: Optische Sensorsysteme schließen Automatisierungs-
   lücken. ELEKTRONIK 13/82, S.134 - 141.

/3/ CHIN, R.T., HARLOW, C.A.: Automated Visual Inspection: A Survey.
   IEEE Transactions on Pattern Analysis and Machine Intelligence,
   Vol. PAMI-4, 6, Nov. 1982.

/4/ PORTER, G.B., MUNDY, J.L.: Visual Inspection System Design.
   Computer, Mai 1980, S. 40 - 80.

/5/ KAZMIERCZAK, H. (Hrsg.): Erfassung und maschinelle Verarbeitung
   von Bilddaten. Springer Verlag 1980.

Erkennen metallischer Gefügestrukturen durch Korrelation mit artifiziellen Mustern an einer lichtoptischen Diffraktionsanlage[+]

W. EIKHORST[++] und S. BOSECK
Universität Bremen, Fachbereich Physik/Elektrotechnik-Kybernetik
Postfach 33 04 40, 2800 Bremen 33

Die Gütebestimmung von Metallproben wird vom metallographisch vorgebildeten Fachmann durch den Vergleich mit Richtreihen vorgenommen (1a,b). Diese Methode ist relativ langwierig und zudem mit einem subjektiven Fehler behaftet. Deshalb versuchen wir eine automatische Analyse von Metallschliffen vorzubereiten (2). Diese beruht auf dem Vergleich artifizieller Muster mit dem Fourier-Leistungsspektrum des Objekts, hier dargestellt durch das Schliff-Ätz-Bild des metallischen Gefüges.

Zur Methode: Da unsere Untersuchungen nicht am Originalbild durchgeführt werden, erstellen wir zunächst mit Hilfe einer lichtoptischen Diffraktionsanlage (Abb. 1) auf der Basis der Fraunhoferschen Beugung ein zweidimensionales Fourier-Leistungsspektrum. Zu diesem Zweck wird ein Negativ des Metallschliffbildes (in unserem Fall von Perlit; ca. 200-

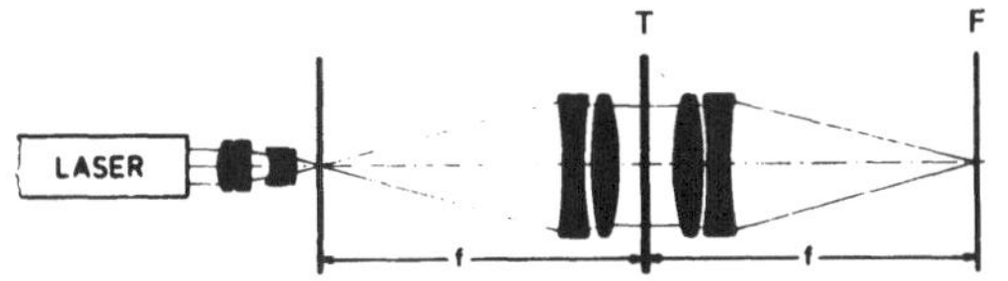

Abb. 1. Schematischer Aufbau der lichtoptischen Diffraktionsanlage

fach vergrößert) an der Stelle T (Abb. 1) in die Diffraktionsanlage (2) eingebracht und das in F erzeugte Leistungsspektrum über eine (nicht dargestellte) Zwischenoptik und einen Optischen-Spektrumanalysator (OSA)[+++] ausgelesen. Eine weitere Verarbeitung, sowie Speicherung finden in einem Apple II Plus-Rechner statt.

Zur vergleichenden Analyse des Fourier-Leistungsspektrums eines Objekts mit dem einer verwandten Bildvorlage ist es sinnvoll, eine Datenreduktion durchzuführen, indem man sukzessiv immer nur eine Diagonale des Beugungsbildes untersucht. Dies wird durch Ausblenden eines Spaltes in der Fourierebene erreicht. Das gesamte zweidimensionale Fourier-Leistungs-

---

+   gefördert mit Mitteln der Stiftung Volkswagenwerk
++   Teil einer Dissertation
+++ Hersteller b & m spectronic

spektrum wird dann durch schrittweises Drehen des Objektes erschlossen.

Durch diese methodische Einschränkung ergibt sich eine starke Verminderung der Datenmenge, wenn man ein Verfahren findet, die interessanten Richtungen im Beugungsbild aufzusuchen. Dies ist sowohl durch eine richtungsspezifische Intensitätsmessung (3) als auch durch die hier durchgeführte Korrelation des Beugungsspektrums der gesuchten Richtung mit den Beugungsspektren des schrittweise gedrehten Objektes möglich.

$$KF = \sum_{N=1}^{400} \frac{X_{(N)}}{\sqrt{\sum_{M=1}^{400} X_{(M)}^2}} \times \frac{Y_{(N)}}{\sqrt{\sum_{M=1}^{400} Y_{(M)}^2}}$$

KF = Korrelationsfaktor; $X_{(N)}$ = Spektrum 1; $Y_{(N)}$ = Spektrum 2

<u>Zum Experiment:</u> Dieser Vorgang sei anhand des Beugungsbildes eines Strichgitters mit 4 LP/mm (Abb. 2) veranschaulicht. Das Strichgitter wird in Winkelschritten von insgesamt 20° gedreht und die zugehörigen Beugungsspektren werden mit dem Spektrum in Abb. 2 korreliert.
Bei $\varphi = -1^O$ findet optimale Übereinstimmung statt (Abb.3).Die Abweichun< von der Ideallage 0° ist durch eine ungenaue Winkeleinstellung und durc: die Breite des Spaltes zu erklären.

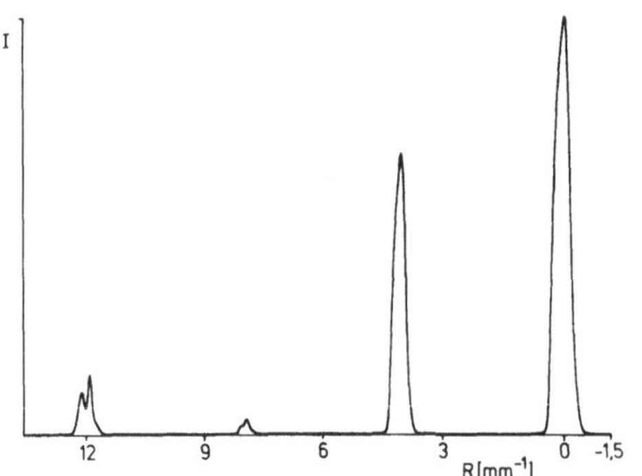
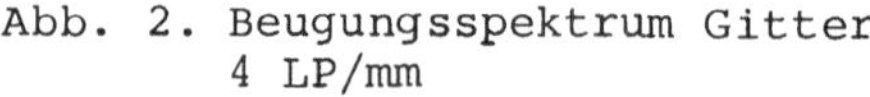
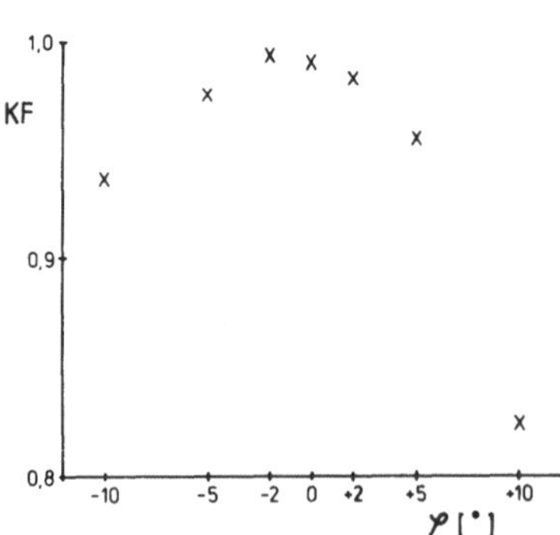

Abb. 2. Beugungsspektrum Gitter
4 LP/mm

Abb. 3. Korrelation in Abhängigkeit vom Neigungswinkel

Hat man eine Verzugsrichtung aufgefunden, so kann man versuchen, Periodizitäten zu identifizieren.
Hierzu werden die in den Abb. 4 und 8 eingetragenen Bereiche dieser Perlit-Schliffbilder gebeugt (Abb. 5 u. 9). Die Leistungsspektren werden entlang der gekennzeichneten Richtungen durch den OSA aufgezeich-

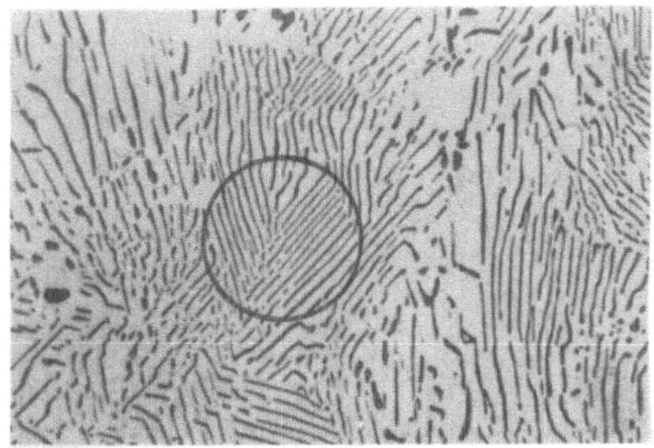

Abb. 4. Perlit 1 (Schliff-Ätz-Bild)

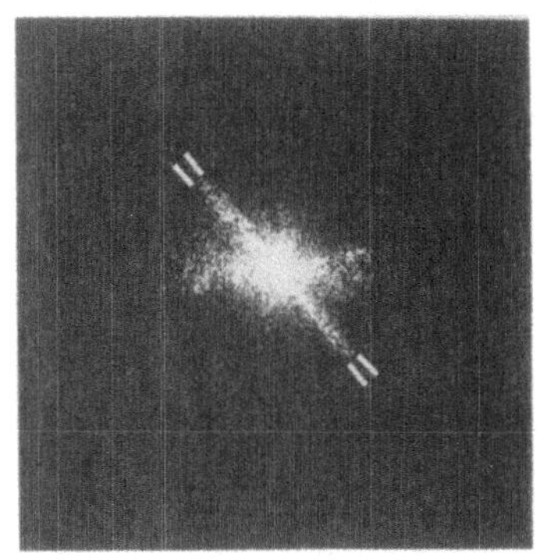

Abb. 5. Perlit 1 (Beugungsbild)

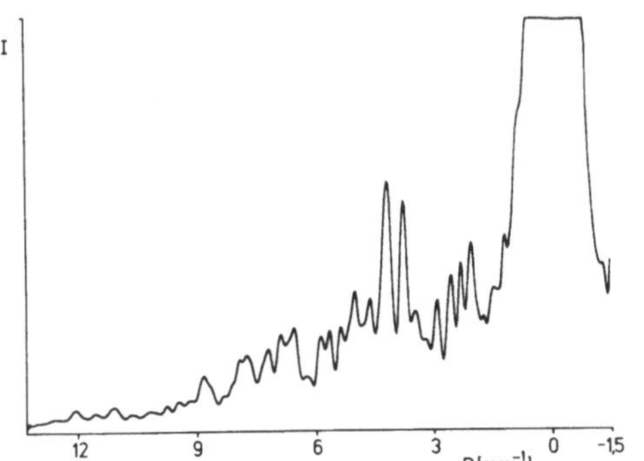

Abb. 6. Beugungsspektrum Perlit 1/1
(s. Abb. 5)

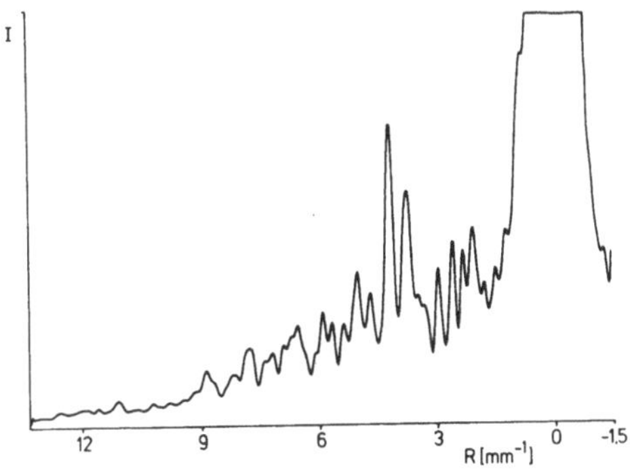

Abb. 7. Beugungsspektrum Perlit 1/2 (s. Abb. 5)

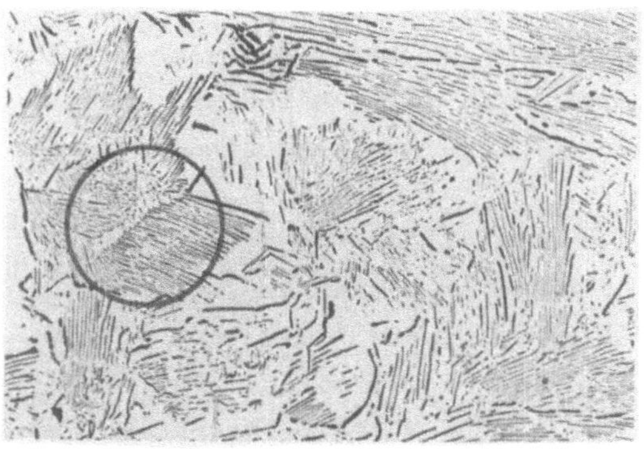

Abb. 8. Perlit 2 (Schliff-Ätz-Bild)

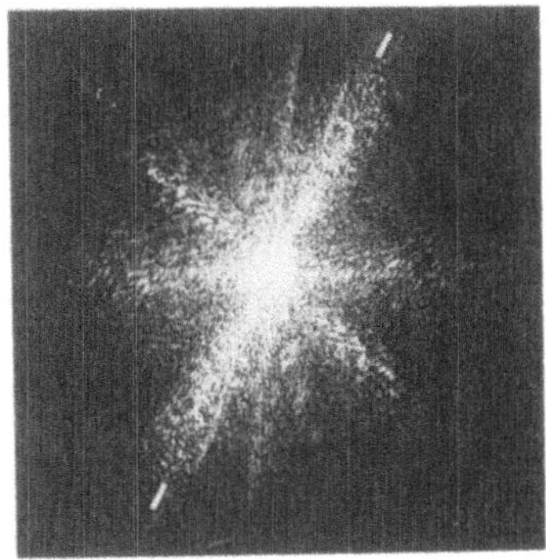

Abb. 9. Perlit 2 (Beugungsbild)

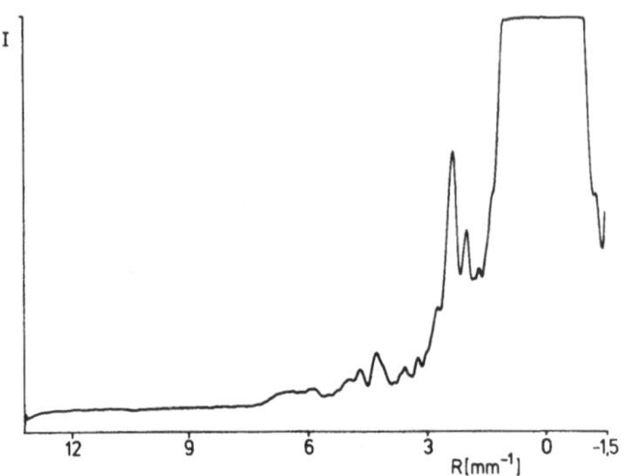

Abb. 10. Beugungsspektrum Perlit 2
(s. Abb. 9)

net (Abb. 6, 7 u. 10). Die Abb. 6 und 7 sind dabei beides Beugungsspek-
tren des Perlit 1 (Abb. 4), die geringfügig voneinander differierende
Richtungen des Beugungsbildes (Abb. 5) repräsentieren. Hinzu kommen die
Leistungsspektren zweier Strichgitter mit 4 bzw. 1,6 LP/mm.
Im weiteren wird das Spektrum Perlit 1/1 (Abb. 6) als Ausgangsspektrum
angesehen und eine Korrelation mit den übrigen Spektren durchgeführt
(Ergebnisse Tab. 1).

Tabelle 1. Korrelation von Perlit mit Perlit bzw. realen Strichgittern

Perlit 1/1 x Perlit 1/2	0,993	Perlit 1/1 x 4 LP/mm	0,427
x Perlit 2	0,835	x 1,6 LP/mm	0,462

Zufriedenstellend sind die hohe Korrelation zur selben, geringfügig ver
drehten Perlitstruktur und die erheblich geringere Korrelation zur frem
den Perlitstruktur.
Unbefriedigend ist die Korrelation mit den Strichgittern, da der Perlit
1 bei einem mittleren Lamellenabstand von ca. 4 LP/mm mit dem Strichgitter von 1,6 LP/mm das höhere Ergebnis liefert.
Dies ist offensichtlich auf die Schmalbandigkeit der Peaks (Abb. 2) in
den Beugungsbildern der Strichgitter zurückzuführen. Die Peaks in den
Perlitbeugungsbildern dagegen umfassen ein breites Frequenzband, da der
Abstand der Lamellen schwankt.
Zur Vermeidung dieses Fehlers wurde die Korrelation mit drei artifiziellen Gitter-Beugungsspektren durchgeführt. Diese "Filter" haben an den
Stellen der üblichen Beugungspeaks Balken mit einer Breite von ± 10 %
der Gitter-Ortsfrequenzen. Die Höhe der Peaks nimmt mit dem Faktor 2/
Harmonische ab. Das "Filter" in Abb. 11 ist auf den mittleren Lamellenabstand des Perlits 1 (4 LP/mm), das "Filter" in Abb. 12 auf den mittleren Lamellenabstand des Perlits 2 (2,1 LP/mm) ausgerichtet. Das dritte "Filter" mit 1,6 LP/mm ist nicht graphisch dargestellt.

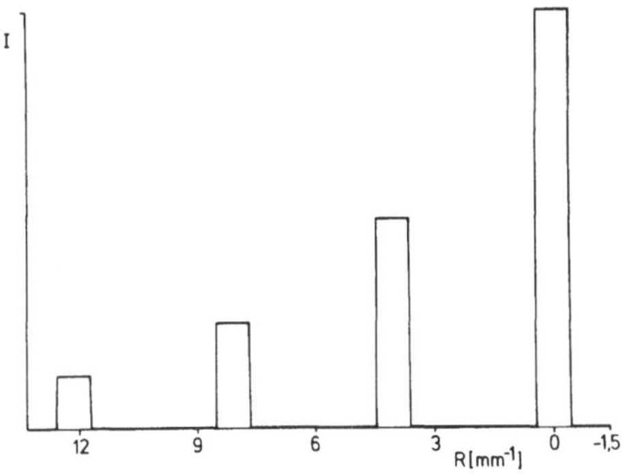

Abb. 11. Artifizielles Spektrum
        4 LP/mm

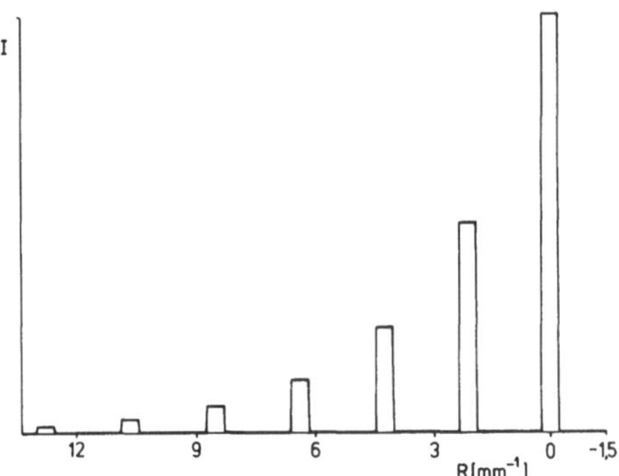

Abb. 12. Artifizielles Spektrum
        2,1 LP/mm

Folgende Tab. 2 zeigt die Korrelationsergebnisse mit diesen artifiziellen Beugungsspektren.

Tabelle 2. Korrelation von Perlit 1/1, Perlit 2 und dem 4 LP/mm Strich-
           gitter mit artifiziellen Filterfunktionen

Perlit 1/1	x Filter 4	0,547	Perlit 2	x Filter 1,6	0,413
	x Filter 2,1	0,498	4 LP/mm	x Filter 4	0,595
	x Filter 1,6	0,335		x Filter 2,1	0,376
Perlit 2	x Filter 4	0,258		x Filter 1,6	0,004
	x Filter 2,1	0,508			

Es zeigt sich, daß diese einfachen Filterfunktionen bereits zu verwertbaren Korrelationsergebnissen führen.
Eine Verfeinerung der "Filter", unter Berücksichtigung der Streulicht-
funktion, dürfte stärkere Differenzierungen erlauben. Am Ende sollte es
möglich sein, für jede zu analysierende Metallstruktur einen artifiziellen Filtersatz zu finden, der schrittweise die bisher benutzten Richt-
reihenelemente ablöst.

Literatur
(1a) KULMBURG, A.: Microchim. Act. Suppl. 5 (1974) 181-206
(1b) STAHL-EISEN-PRÜFBLATT 1570-71, 2. Ausg. (1971)
(2) EIKHORST, W., S. BOSECK und H. VETTERS: Forschungsbericht des Se-
     nators für Wissenschaft und Kunst der Freien Hansestadt Bremen
     (1980)
(3) EIKHORST, W., S. BOSECK und H. VETTERS: BEDO 15 (1982) 209-216

Sensorsystem auf der Basis eines Laserscanners zur Kantenfindung in der Materialbearbeitung

TH. KREIS, H. KREITLOW, W.JÜPTNER und G. SEPOLD

Bremer Institut für angewandte Strahltechnik, BIAS
Ermlandstr. 59, D 2820 Bremen 71

Einleitung

In der Materialbearbeitung mit Hochleistungslasern, z.B. beim Härten, Umschmelzen oder Entgraten wird eine automatisierte Erkennung der Werkstückform, besonders der Werkstückkante, benötigt, um daraus die Steuerdaten für die Laserstrahlführung zur Erzielung gleichmäßiger und fehlerfreier Bearbeitungsergebnisse zu gewinnen. Die Bewegungen des Bearbeitungskopfes, der die Optik für die Strahlformung des Materialbearbeitungslasers enthält, kann durch einen in mehreren Koordinaten rechnergesteuerten Tisch, auf dem das Werkstück befestigt ist, durch einen in mehreren Bewegungsfreiheitsgraden schwenk- und verschiebbaren Halter für den Bearbeitungskopf, oder durch einen Industrieroboter realisiert werden.

Um nahezu beliebig im Raum verlaufende Kanten in allen drei Raumrichtungen zu erkennen, wurde ein Sensorsystem entwickelt, welches mit Hilfe des Lichtschnittverfahrens arbeitet. Dabei wird ein Muster erzeugt, das mit Hilfe der digitalen Bildverarbeitung ausgewertet wird, um die Lage der Kante bezüglich eines sensorfesten Koordinatensystems zu bestimmen. Aus diesen Daten werden im On-Line-Betrieb die Steuersignale für den Vorschub erzeugt, nach dem Vorschub wird von der dann erreichten Position aus der nächste Kantenpunkt bestimmt. Dieses Vorgehen erfordert schnelle Algorithmen, ein solcher wird im folgenden vorgestellt /1,2/.

Formbestimmung mit dem Lichtschnittverfahren

Beim Lichtschnittverfahren wird ein Lichtfächer unter schiefem Winkel zur optischen Achse einer TV-Kamera auf das zu vermessende Objekt projiziert /3,4/. Für sämtliche beleuchteten Oberflächenpunkte des Werkstücks ist dann einerseits bekannt, auf welchem Strahl durch den Kamerafokus sie liegen, andererseits, daß sie in der durch den Lichtfächer definierten Fläche liegen. Aus diesen Informationen lassen sich eindeutig alle drei Koordinaten der beleuchteten Oberflächenpunkte, insbesondere der gesuchten Kantenpunkte, bezüglich eines kamerafesten Koordinatensystems bestimmen, Bild 1.

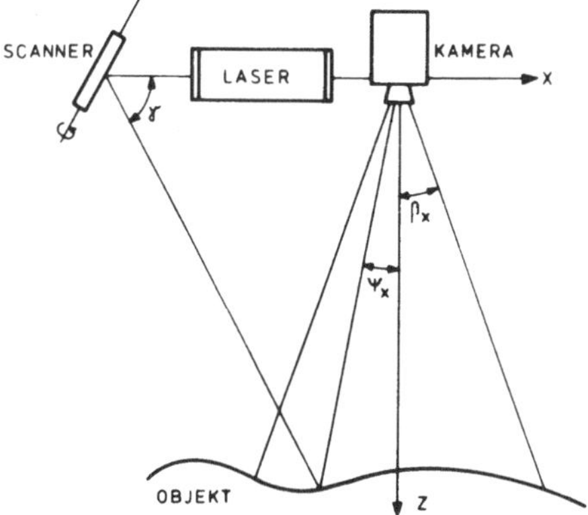

Bild 1. Prinzip des Lichtschnitt-
sensors

Zur Erzeugung des Lichtfächers stehen verschiedene Lichtquellen in Verbindung mit Schlitzblenden, Zylinderlinsen sowie mechanischen und akustooptischen Ablenkern zur Verfügung. Im vorliegenden Fall wurde ein Laserstrahl mit mechanischem Scanner verwendet. Zur Aufnahme des Lichtschnittmusters ist ein zweidimensionaler Detektor notwendig, für den gefordert wird, daß er kein Nachziehen bei bewegten Bildern, sowie nur geringe geometrische Verzerrungen im Bildfeld zeigt. Derartige Detektoren stehen als Plumbicon- oder CCD-Halbleiter-Kamera zur Verfügung. Bei der Digitalisierung des Bildes kommt man mit einer Quantisierung in nur zwei Graustufen aus, da durch die Verwendung eines Laserstrahls hoher Kontrast im Bildfeld vorliegt, was auch im Hinblick auf eine hohe Verarbeitungsgeschwindigkeit von Vorteil ist. Weiter werden dadurch Probleme mit der Tiefenschärfe bei dreidimensionalen Objekten vermieden. Insbesondere bei Verwendung eines Linienfilters vor der Kamera wird ein hoher Störabstand im Detektorsignal aufgrund der hohen Intensität und der Monochromasie erreicht.

Kantenbestimmung

In dem von der Kamera aufgenommenen digitalisierten Bild werden in einem Vorverarbeitungsschritt mit Hilfe von Methoden der digitalen Signalverarbeitung Störungen und Rauschen unterdrückt, Lücken im Lichtschnittverlauf aufgefüllt und das Muster auf eine Breite von nur einem Bildpunkt verdünnt. Aus den einzelnen Bildpunkten können jetzt die dreidimensionalen Koordinaten der dazugehörigen Objektpunkte bestimmt werden, was allerdings einen für die On-Line-Steuerung der Bearbeitung zu hohen Rechenaufwand erfordert. Daher wurde ein Algorithmus entwikkelt, der den Kantenpunkt direkt im digitalisierten Bild bestimmt. Dieser Algorithmus benutzt nur die Integer-Arithmetik, nur eine Multiplikation pro Kantenpunkt und keine Division oder trigonometrische Funktion. Dadurch wird eine hohe Auswertegeschwindigkeit erreicht, und es werden Kanten mit in nahezu beliebigem Winkel aneinanderstoßenden Flächen sowie beliebiger Verrundung erkannt.

Das aus K Punkten bestehende, vorverarbeitete Lichtschnittmuster wird nach folgendem Algorithmus ausgewertet:

1. Schritt: Als Startpunkt wird der Punkt (i,j) der Lichtschnittspur mit größtem i gewählt. Der Laufindex m und die Größen T_m und S_m werden zu
$$m = 0, \quad T_0 = 0, \quad S_0 = 0$$
gesetzt.

2. Schritt: Für m = 1,...,K wird der jeweils nächste Punkt der Lichtschnittspur gesucht und
$$S_m = S_{m-1} + |i_m - i_{m-1}|$$
$$T_m = T_{m-1} + |j_m - j_{m-1}|$$
gesetzt.

3. Schritt: Der Index n>0 wird gesucht, für den erstmals
$$T_K S_n (S_K - S_n) \leqslant S_K T_n (T_K - T_n)$$
erreicht wird.

Dann sind (i_n, j_n) die Koordinaten des Kantenpunktes im digitalisierten Bild für diesen Lichtschnitt. Nur für diesen Punkt werden seine kartesischen Koordinaten im sensorfesten Koordinatensystem berechnet und diese Daten bei einem Teach-In-Vorgehen im Steuerrechner gespeichert oder im Falle des On-Line-Vorgehens direkt in die Steuerdaten für den Laserbearbeitungskopf weiterverarbeitet /5/. Der Algorithmus führt nicht zum Ziel bei einer Verzweigung der Kante oder bei abgeschatteter Kante. Ein an einer Kante erzeugtes und vorverarbeitetes Lichtschnittmuster zeigt Bild 2, die Berechnungsgrößen des Algorithmus sind in Bild 3 aufgetragen.

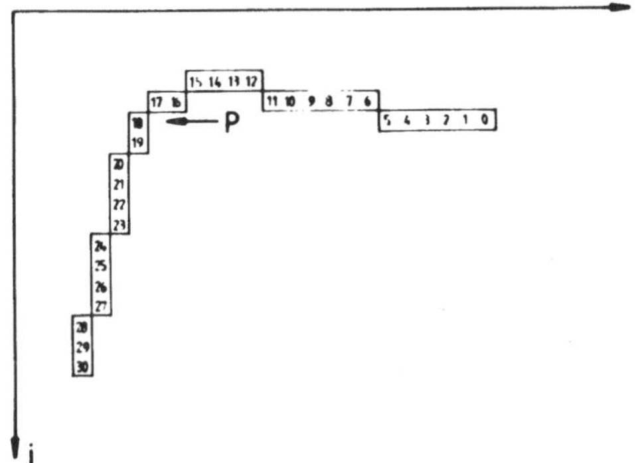

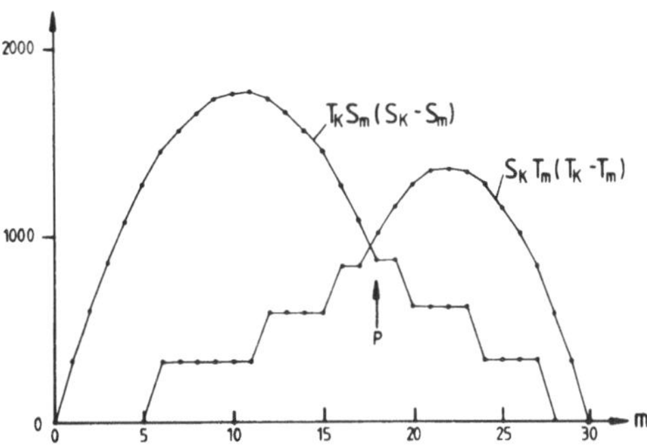

Bild 2. Digitalisiertes Muster
an einer Kante

Bild 3. Berechnungsgrößen für
das Muster aus Bild 2.

Anbringung des Sensors am Laserbearbeitungskopf

Für eine Kantenfindung und Bearbeitung des Werkstücks im On-Line-Betrieb kann der Sensor am Bearbeitungskopf befestigt werden. Dieser
kann sich im allgemeinen in drei Translations- und drei Rotationsfrei-
heitsgraden bewegen, Bild 4. Der Sensor dagegen muß die Translationen
und die Rotation R_z mitmachen, darf jedoch nicht die Rotationen um die
x- und die y-Achse mit ausführen, da andernfalls die zu erkennende
Kante aus dem Bildfeld des Sensors verschwinden würde. Dies erfordert
eine Anordnung der Gelenke des Bearbeitungskopfes, die einen Schwenk
um R_x oder R_y ohne eine dazu notwendige Rotation um R_z ermöglicht.

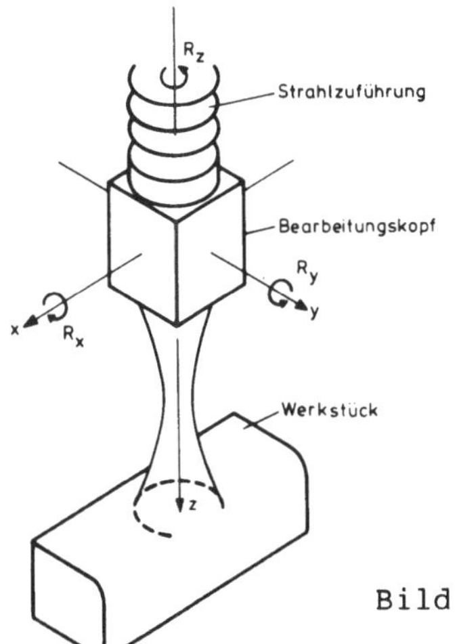

Bild 4. Bewegungsfreiheitsgrade
des Bearbeitungskopfes

Maximale Krümmung der Kantenspur

Um den Strahl des Bearbeitungslasers und den Bearbeitungskopf nicht ins Bildfeld zu bekommen und um beim On-Line-Prozeß Zeit zur Aufbereitung der Steuerdaten zu gewinnen, muß der Sensor einen Vorhalt vor dem Bearbeitungskopf haben. Bei der oben vorgeschlagenen Anbringung ist gewährleistet, daß das auszuwertende Bildfeld immer in Vorschubrichtung vor dem Bearbeitungskopf liegt. Bei zu kleinen Krümmungsradien der Kantenspur im Raum kann jedoch die Kante seitlich aus dem Bildfeld verschwinden. Um dies zu vermeiden, darf entweder der Abstand zwischen Sensor und Kopf einen von der maximalen Krümmung bestimmten Wert nicht überschreiten oder es können nur Werkstücke mit einem durch den Vorhalt festgelegten minimalen Krümmungsradius bearbeitet werden. Der Zusammenhang zwischen dem minimalen Krümmungsradius R_{min}, dem Vorhalt d und der Breite B des quadratischen Auswertefeldes ist, Bild 5:

$$(d + B/2)^2 = B*R_{min} - B^2/4$$

Der minimale Krümmungsradius hängt in erster Linie von der Breite des Auswertefeldes ab, die über die Höhe des Sensors über dem Bearbeitungskopf variiert werden kann.

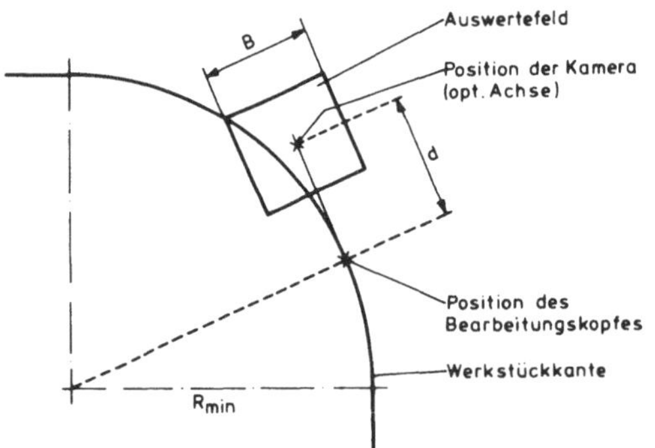

Bild 5. Abhängigkeit des Krümmungsradius vom Auswertefeld

Zusammenfassung

Ein Sensorsystem auf der Basis eines Laserscanners wurde entwickelt, welches es ermöglicht, bei nahezu beliebig geformten Objekten die zu bearbeitenden Kanten zu erkennen und zu verfolgen. Es zeichnet sich dadurch aus, daß die Erkennung der Kante und die Bearbeitung im On-Line-Verfahren durchgeführt werden kann. Dies wurde dadurch erreicht, daß schnelle und flexible Algorithmen implementiert wurden. Das System ist in der Lage, eine Kante in beliebiger Lage im Raum zu verfolgen, solange sichergestellt ist, daß eine Kante vorhanden ist. Die Winkel und Abrundungen der Kante können nahezu beliebig sein. Weiter wurden die Befestigung des Sensors am Bearbeitungskopf sowie Grenzen der Auswertung behandelt.

<u>Literatur</u>

/1/ KREIS,TH., H.KREITLOW und W.JÜPTNER: Acta Imeko (1982) 229

/2/ KREIS,TH., H.KREITLOW und W.JÜPTNER: Proc. 2nd Int. Conf. on Robot Vision and Sensory Controls, IFS Publ. Ltd. (1982) 9

/3/ VANDERBRUG,G.J., J.S.ALBUS und E.BARKMEYER: Proc. of 9th ISIR (1979) 213

/4/ FOITH,J.P.: IITB-Mitteilg. (1977) 32

/5/ SPUR,G., H.RITTINGHAUSEN und H.SINNING: Zeitschr. f. wirtsch. Fert. 75 (1980) 293

Synthetic Horizontal Interlace, a Technique for Doubling the Horizontal Resolution of Imaging Automatic Trackers

Stanley P. Buchanan

Martin Marietta Aerospace,Orlando, FL   32855/USA

Introduction

Television format imaging devices are a major subsystem of numerous military fire control systems that function as optical trackers. Beyond the sensor, fundamental components consist of optics, stabilization platform, and video processing electronics optimized to respond to the electromagnetic radiation produced by the target and its background. This paper presents a conceptually simple hardware/software scheme for improving the system horizontal resolution by a simple modification of that portion of the video processing electronics that functions as the imaging automatic tracker electronics. The baseline system described here was tested in the live TOW firings at the U.S. Army's MICOM Test Range in 1981.

TV-TOW System Description

A TOW (Tube Launched, Optically Tracked, Wire Guided) anti-tank missile is effective at ranges greater than 3km. It is equipped with a Xenon beacon emitting in the near infrared ($\sim.8\mu$m). The beacon and intended target are tracked by an operator from a control station and deviations from the desired trajectory are sensed to produce steering commands which are transmitted up the wire to the missile. This manually based design was replaced with an automatic tracking and guidance system for demonstration of the following requirements:

1. The missile beacon must be acquired very rapidly (less than 200 msec from its entrance to the FOV) or the missile will abort.

2. Tracking must continue through a degrading signal to noise ratio as the TOW travels downrange.

3. After initial acquisition, the tracking subsystem must be able to reacquire the beacon if a breaklock occurs. This requires the tracker to automatically change its internal tracker parameters.

The system configuration consisted of 12° field of view missile capture, TV camera, boresighted together with 2.5° FOV target tracker camera. These cameras input scene video to the two, independent point trackers, designated the Capture and Narrow FOV trackers, respectively. A TV area correlator (TVAC) provided scene stabilization on the selected target. After missile launch, the point trackers develop error signals proportional to the displacement of the TOW beacon image from the center of the TV FOV. These error signals are used to steer the TOW missile into the center of the FOV.

See Figure 1 for the system block diagram. The Capture Tracker is responsible for initially acquiring the beacon, while the Narrow FOV tracker guides the TOW missile to target impact after transfer. All tracking is automatic after the target is intially acquired.

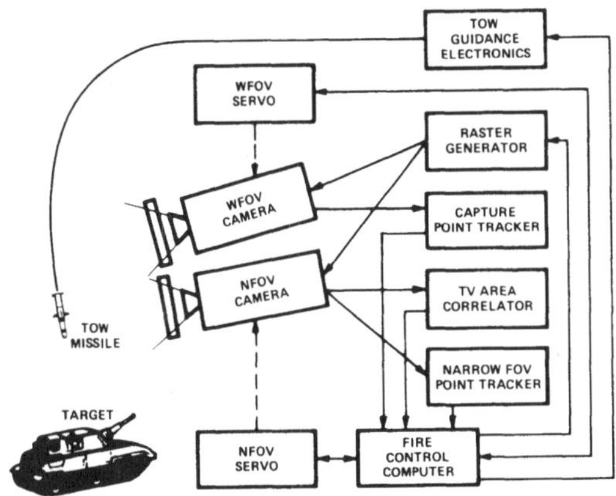

Figure 1.

Functional Block Diagram of Overall System

Upon missile launch, the capture tracker goes into a lobing search mode which covers the entire wide FOV except for small stability boundaries at the edges. When the tracker detects the TOW beacon video, the gated portion of the video is significantly decreased to enhance discrimination of the beacon from obscurations and background clutter within the FOV.

After a short interval, in which it is expected the TOW has been commanded into the center of the FOV, the narrow FOV tracker is activated and assumes responsibility for acquiring, tracking and guiding the TOW. With its reduced FOV, the beacon signal to noise ratio is increased to make possible tracking the dimming beacon until target impact.

<u>Contrast Tracker/Baseline Description</u>

The contrast trackers are slight modifications to previously described image automatic trackers /1/ which operate on the beacon "modulation" less background. A Z80 microprocessor within the tracker provides the flexibility necessary for controlling the tracker modes as well as processing the sensor video. This flexibility was the key to enabling the tracker to do double duty as both an acquisition sub-system and a tracking sub-system. Previous TOW operational modes required a human operator to manually move the camera and acquire the TOW before automatic tracking via a non-imaging scanned array was possible.

It is believed that the successful demonstration of the system described here is the first time a TOW has been automatically tracked and guided using a TV sensor.

Resolution Improvements with Horizontal Interlace

The standards for 525 line commercial television (EIA RS 170, etc.) were utilized in this system for availability and convenience. Military systems continue to use such standards but many are now converting to 875 line format for increased tracking accuracy. The vertical interlace problems of these formats are easily overcome by averaging error data over a frame (2 fields @ 60 Hz = 1 frame at 30 Hz) to achieve 1 TV Line (.12% in 875 line systems) of accuracy. Accuracy in the horizontal direction at these higher line rates has been limited by the drift of analog techniques used or by digital counter processing (pixel clock) rates. The utilization of horizontal interlace (alternating the phase of high speed pixel count clock, see Figure 2), while maintaining an offset in the microprocessor for the "interlaced horizontal field" allows a doubling of horizontal resolution. This now makes the tracker error uncertainty identical in both coordinate axes. This is achieved without going to higher speed clocks which would require a change in logic from the currently low power schottky type devices to the higher power consumptive families such as ECL.

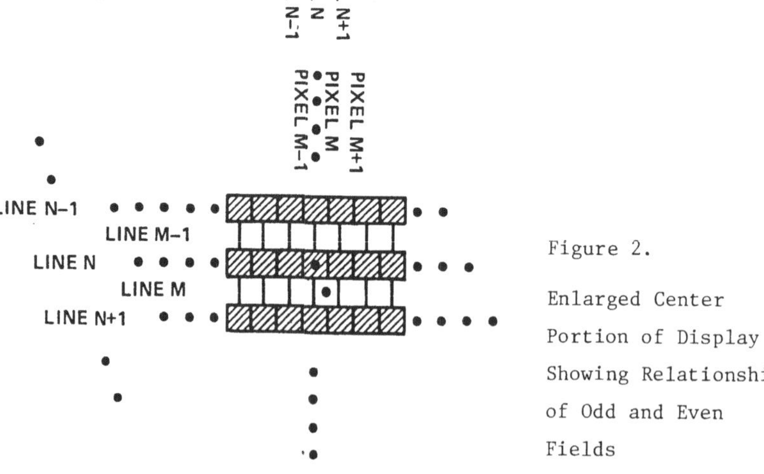

Figure 2.

Enlarged Center Portion of Display Showing Relationship of Odd and Even Fields

1. Buchanan, S. P., "Automatic Tracking, Improved Performance for Electro-Optical Imaging and Target Acquisition Systems", Optics and Laser Technology (February 1980), IPC Business Press.

Ink Layer Thickness Monitor (I.T.M.O.)

K.J. SCHELL and J. TIJHOF
Joh. Enschedé en Zonen
P.O. Box 114, 2036 LS Haarlem, Holland

1. <u>Introduction and summary</u>

In the production of securities, such as banknotes, cheques, bonds, etc., intaglio printing plays an important role.

Intaglio printing starts with an engraving. This engraving is filled with a high viscous ink. The excess amount of ink is wiped off with the aid of an in liquid immersed wiping roller and the image is then transfered under high pressure on paper. This transfer process is called intaglio printing.

The excess amount of ink wiped off by the wiping roller is important for the economy of intaglio printing. To reduce this amount and meanwhile keeping the printing quality, we developed a monitor. The monitor is based on the absorbency of intaglio ink in the near infra red region. Disturbing reflections are surpressed by using polarized light and Brewster angle. The relation between the intensities - of the light reflected by the surface and of the light passing through the ink layer - is a measure for the ink thickness. The paper presented will start with a short survey of intaglio printing. Then the relations between the intensities and the ink layer thickness will be shown. Use is made of the Fresnel laws of reflection and the Lambert relation for absorption.

The design and the development of the ink layer monitor finally led to a working model and tests were made on a production press. The results of these tests lead to the final conclusion:
- Variations strongly affecting the printing quality
  are shown by the monitor but variations lightly
  affecting the printing quality are not shown.

The human eye has again proved its performance in quality control.

2. Intaglio Printing

Fig. 1 shows the cylinder on which the intaglio plate is clamped, fig. 2 shows inking train and chablone rollers. A thick layer of ink is chabloned through the inking train in three colours on the intaglio plate. The thick layer is wiped in the engravings of the intaglio plate through an, in liquid immersed, wiping roller. The ink efficiency is only a few % and therefore carefull monitoring of the applied ink layer before wiping can reduce the ink consumption!

wiping   plate
cylinder   cylinder

Fig. 1

chablone   inking train
rollers

Fig. 2

3. Method of thickness measuring

Fig. 3 shows the method for measuring the thickness before wiping.

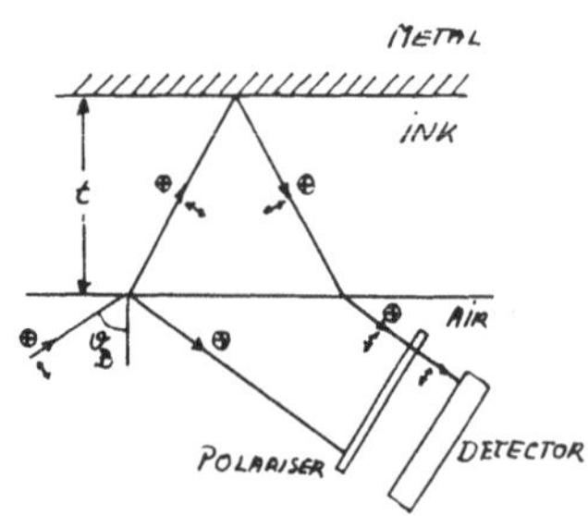

Fig. 3

⊕ Light polarised with electric vector perpendicular to plane of incidence.

↕ Light polarised with electric vector parallel to plane of incidence.

A beam of unpolarised light strikes the air/ink interface at an angle of incidence θ_B (Fig. 3); part of the light is reflected. The remaining light passes through the ink film, is reflected at the ink/metal interface, passes through the ink film again and finally emerges parallel to the light reflected from the air/ink interface. The intensity of this "emergent light" is modified by (a) absorption in the ink (b) absorption on reflection at the ink/metal interface and (c) loss by reflection at the ink/air interface. The emergent light contains with $2t/\cos.\theta$ information about the thickness of the ink film.

However, a photodetector positioned to receive the emergent light would also, in general, receive the light reflected at the air/ink interface, and since the intensity of this "reflected light" can be expected to greatly exceed that of the emergent light, the output of the photo-detector would show little sensitivity to ink film thickness. It is necessary, therefore, to remove the effect of the reflected light and at a certain angle of incidence known as the Brewster Angle (θ_B), the reflectance is zero for light polarised with the electric vector parallel to the plane of incidence. Since unpolarised light may be considered to be composed of equal quantities of light polarised in two perpendicular planes, and since the ink can be expected to behave as a dielectric, the reflected light shown in Fig. 3 does not contain light polarised with the electric vector parallel to the plane of incidence at θ_B. If the polariser is then oriented so that it transmits only light polarised with the electric vector parallel to the plane of incidence, the detector does not respond to the reflected light. In the ideal case then (shown in Fig. 3) only light which passes through the ink affects the detector, whose output is then a function of the ink thickness.

4. <u>Calculation of the sensitivity</u>

The sensitivity is defined as $\dfrac{I_{det} - I_c}{I_{det}}$.

I_{det} stands for the light intensity falling directly on the detector.
I_c stands for the intensity of the light on the detector after refraction at and absorption through the ink layer

130

The sensitivity depends on:

4.1. Wavelength distribution of the light source. A halogen-tungsten
     lamp is used and the distribution is following Planck's law with
     T = 3000 °K.

4.2. The transmission characteristic of the ink. Fig. 4 shows the trans-
     mission measured for ink nr. 6235 at thickness of 0.0015" and 0.008"
     There is a window between 1 and 2.8 μ. The exponential Lambert-Beer
     absorption law is assumed to be valid.

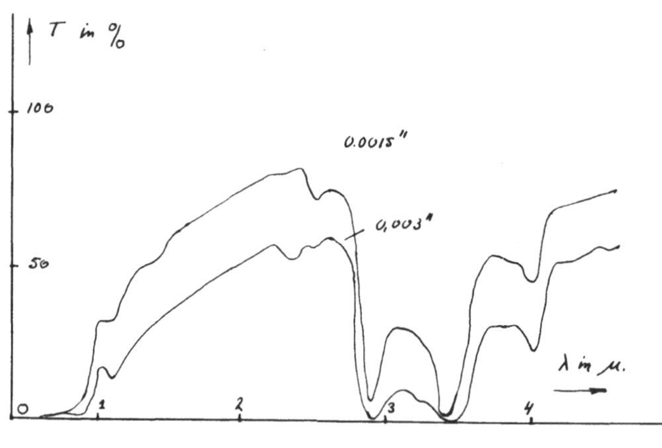

Fig. 4

4.3. Reflection laws of Fresnel for the air/ink and ink/metal interface.
4.4. The spectral sensitivity in the near infra red region of the photo-
     detector.

A numerical calculation done with the aid of a computer ultimately
shows in fig. 5 the relation between sensitivity and thickness for
ink nr. 6235. Curve 1 is calculated with an intensity after reflection of 15% at the Brewster Angle and a transmission of 100%
through the polariser (worst-case approach). Curve 2 accounts for
the spectral sensitivity of the detector and nr. 3 is calculated
with a 7% reflection at the air/ink interface.

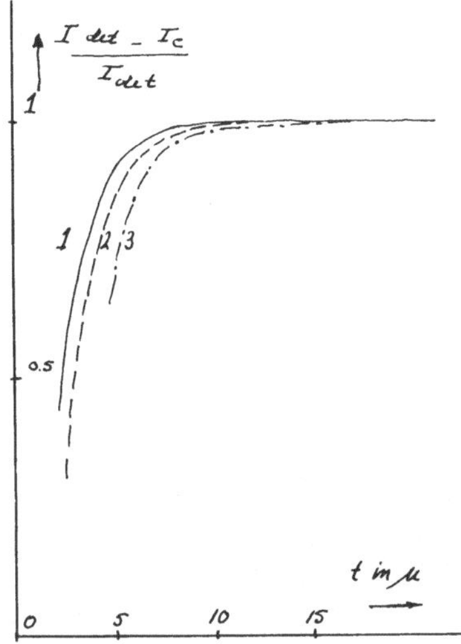

Fig. 5

The measuring range lies between 5 and 12,5 micron.

5.   <u>Results and interpretation</u>

Fig. 6 shows a real time signal extracted from the monitor.

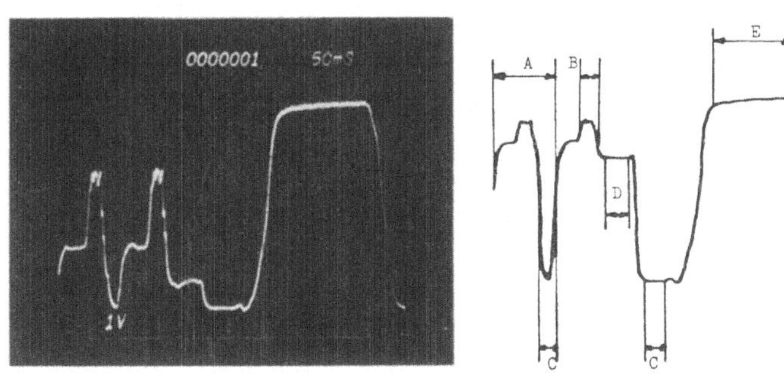

Fig. 6

A = heigth of the banknote
B = intaglio gravure
C = intaglio plate without ink
D = intaglio plate with ink
E = gap

The upper part of Fig. 7 shows two samples of a banknote.

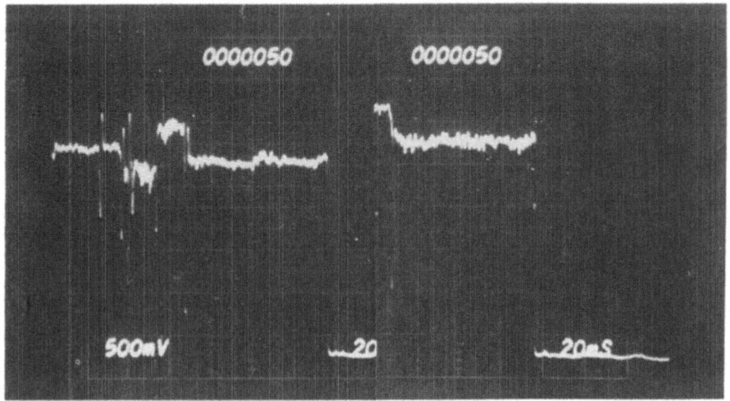

Fig. 7

The difference in optical density is 0,18 (approx. 20% of the max.
density). The corresponding difference in ITMO density is approx.
0,3 V. But noise and spikes, due to the multicolour chablone inking
are prohibiting a more accurate measurement.

ITMO was intended as a tool for the printer so that early information
was available in order to make the necessary corrections to avoid
small changes (approx. 0,04) in density. Fig. 7 clearly shows that
the method is not sensitive enough; variations strongly affecting
the printing quality are shown by the monitor, but variations lightly
affecting the printing quality are not shown.

Berührungslose und zerstörungsfreie Messung der thermischen Abstrahlung/Absorption und der Rauheit der technischen Oberflächen mittels Lichtleiter

Dr. Ing. Charley Buxbaum, 5400 Baden, Schweiz

Die vorgeschlagene Methode für die Untersuchung der Beschaffenheit von technischen Oberflächen ermöglicht nicht nur die Rauheit im breitesten Bereich zu quantifizieren, sondern auch die Eigenschaften der Mikrostruktur der Oberfläche im IR-Strahlungsbereich zu messen. Unter dem Begriff "Mikrostruktur" ist hier eine Struktur der Oberfläche zu verstehen, welche bis zu der feinsten Porosität einer hochglänzenden Metalloberfläche geht.

Die Funktionseignung der leistungsmässig hochgezüchteten Werkstücke ist in grossem Mass von der Oberflächenbeschaffenheit abhängig. An den Oberflächen nämlich spielt sich die Interaktion der Bauelemente mit der Umwelt ab: *Gleiteigenschaft, Kontamination, chemische Resistenz, Hochspannungsfestigkeit, Elektronenemission, thermischer Haushalt usw.* werden meistens durch die letzte Fertigungsoperation der Bestandteile bestimmt. Zwischen dieser Operation und der eigentlichen Anwendung des Bauelementes muss die Qualität der Oberfläche überprüft werden. In der modernen Fertigung ist der Erfolg dieser Qualitätskontrolle unter anderem auf die Objektivität, Reproduzierbarkeit, Geschwindigkeit und Störungslosigkeit des Fertigungsablaufes angewiesen. Je nach Automatisierungsgrad der Fertigung wird die Qualitätskontrolle vermehrt auf Vergleiche mit Musterstücken beschränkt. Auf Messungen mit Auswertung der Daten in absoluten Zahlen wird vermehrt verzichtet.

Im Bild 1 sind zwei völlig unterschiedliche Oberflächen ersichtlich. Laut Perthometer besitzen sie aber fast denselben Rauheitswert. Der Qualitätsunterschied, welchen wir zwar mittels Mikroskop beobachten können, wurde erst mittels Ausmessen des thermischen Abstrahlungsvermögens ε_T quantifiziert: Das ε_T der platinierten Probe ist halb so gross als dasjenige der anderen Probe. Im Bild 2 ist ein umgekehrtes Beispiel dargestellt. Die Oberflächen, im Vakuum aufgeheizt und pyrometrisch gemessen, besitzen praktisch dieselben ε_T-Werte, die perthometrisch gemessene Rauheit variiert aber um Faktoren.

Die gleichzeitige Bestimmung der Rauheit und der thermischen Strahlungseigenschaften (Reflexion, Absorption, Abstrahlung) liefert somit wertvolle Information über das makroskopische und mikroskopische Bild der Oberflächen in welchen auch die Korngrösse, Porosität, Oxydation, Hygroskopie, Erosion, Verstaubung, Farbänderung usw. inbegriffen sind.

Die Werte der thermischen Abstrahlung ε_T einer Oberfläche sind mittels Reflexionsmessungen R_λ der IR-Strahlung erfassbar: $\varepsilon_T = 1-R_\lambda$. Die R_λ hängt sowohl von den optischen Eigenschaften der Oberfläche (Farbe) als auch von dem Wellenspektrum der Strahlungsquelle ab. Die wahren Temperaturen der beiden Strahler : T_1-Temperatur der untersuchten Oberfläche und T_2-Temperatur der primären Strahlungsquelle relativieren die scheinbare Temperatur T der gemessenen Oberfläche nach der Formel: $T = \sqrt{T_1 T_2}$. Mit der Einstellung der T_2 sind deshalb auch bei sonst kalten Proben Oberflächentemperaturen von über T = 1000 K simulierbar.

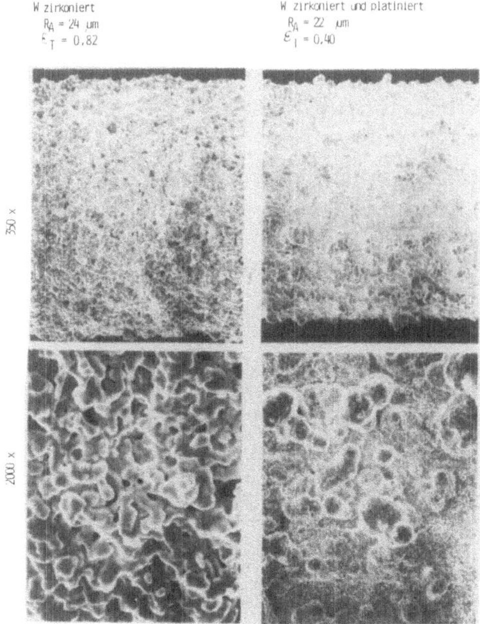

Bild 1 - Drahtoberfläche mit variablem
therm.Abstrahlungsvermögen und
konst. Rauheit

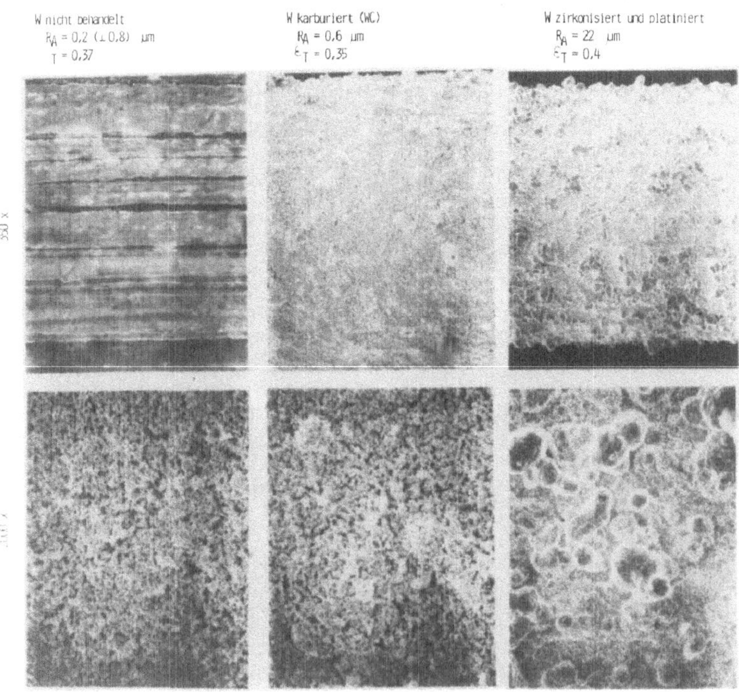

Bild 2 - Drahtoberfläche mit variabler Rauheit und konst. thermischen Abstrahlungsvermögen

Thermische Abstrahlung ε_T (Absorption A_T)

Die geläufige Methode für die ε_T-Bestimmung basiert auf den Heizleistungs- und Temperaturmessungen der im Vakuum aufgeheizten Proben. Die Proben müssen zuerst die geeignete Form besitzen (z.B. aufheizbarer Draht). Das Messprozedere erstreckt sich entsprechend über Stunden.

Die Reflexionsmessung dagegen ist zerstörungsfrei und für "on line"-Kontrolle geeignet. Einmal geeicht nach bekannten Proben liefert sie zudem sowohl lokale als auch integrierte ε_T-Werte:

$$\varepsilon_T = 1 - R_\lambda = A_T$$

Im Bild 3 ist die Reflexion als Funktion der im Vakuum ermittelten ε_T-Werte in relativen Einheiten dargestellt. Die Charakteristik wurde durch Messungen auf Oberflächen aus W, Zr und deren Karbiden gewonnen und ist für Metalle mit vergleichbarem Grauheitsgrad gültig. Oberflächen anderer Qualität - wie Farbaufstriche, organische Stoffe, Pulverschichten, Farbmetalle u.d.ä.- liefern Reflexionskurven mit ähnlichem Durchlauf, quantitativ aber verschoben. Für die Bestimmung der wahren ε_T- und A_T-Werte eines unbekannten Materials ist wiederum eine Eichung notwendig.

Rauheit R_R

Die Reflexionsintensität einer senkrecht bestrahlten Oberfläche ist winkel-abhängig und ist durch eine Glockenkurve darstellbar. Die Steilheit der Schenkel der Glockenkurve hängt stark von der Oberflächenstruktur ab. Während die Intensität des Reflexionsmaximums $R_{\lambda M}$ mit zunehmender Rauheit sinkt, nimmt die Intensität des Streulichtes zu. Wird dabei das $R_{\lambda M}$ nicht berücksichtigt und nur die *Schenkelsteilheit* R_S = f (Winkel) verschiedener Glockenkurven als Funktion der Oberflächenbeschaffenheit betrachtet, kann ein *Stoff- und farben unabhängiges Mass für die Oberflächenbeschaffenheit* gewonnen werden. Sollten sich verschiedene Oberflächen mit derselben Rauheitsart nur durch den Grad der Rauheit unterscheiden, stellt auch die zu jeder der Oberfläche gehörende R_S die quantitative Aenderung der Rauheit R_R dar.

Im Bild 4 ist ein Beispiel dargestellt, in welchem R_R von unterschiedlich behandelten Oberflächen auf Cu-Draht ($\emptyset$ = 0,4 mm) mit den perthometrisch gemessenen Rauheitswerten R_A verglichen wurden. Als Bezugspunkt 1 ist die *Rauheit* R_A = 0,5 μm zu verstehen. Die breite Streuung der R_R-Werte stellt die Unhomogenität der Rauheit der Proben dar. Für Vergleichsmessungen in der Fertigungslinie (on line) bringt die optische Oberflächenbetrachtungs-methode neben den anfangs aufgezählten Vorteilen zusätzlich noch die Möglichkeit mit , sowohl lokale als auch integrierte Rauheit des Werkstückes direkt zu messen.

Nach demselben Prinzip wie Rauheit ist z.B. auch die *Korngrösse* von losen Stoffen wie *Staub, Pulver, Granulate usw.* bequem messbar. Im Bild 4 ist auch die Korrelation zwischen der Korngrösse in μm der Farbpulver (FeO_2-Pigment) und den R_R-Werten dargestellt. Die R_R-Werte wurden diesmal mit dem VOLPI-Finishtest[*] gewonnen. Als Bezugspunkt 1 wurde die *Korngrösse* des braunen Farbpigmentes von 120 μm gewählt.

---

[*] Gerätebezeichnung der Fa. VOLPI AG., Urdorf/ZH, Schweiz

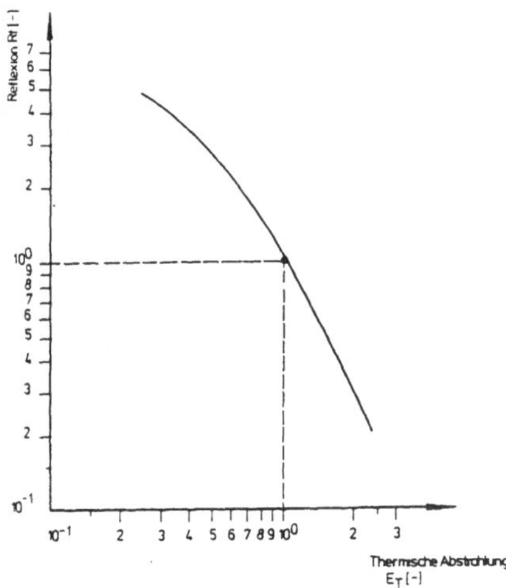

Bild 3 - Aenderung des Reflexionssignals als Funktion
der thermischen Abstrahlung E_T in relativen
Werten. Proben: Oberflächen unterschiedlich
behandelten Mo- u. W-Drähte mit einheitlichem
Durchmessen 0,61 mm. Bezugspunkt: $E_T = 0,4$

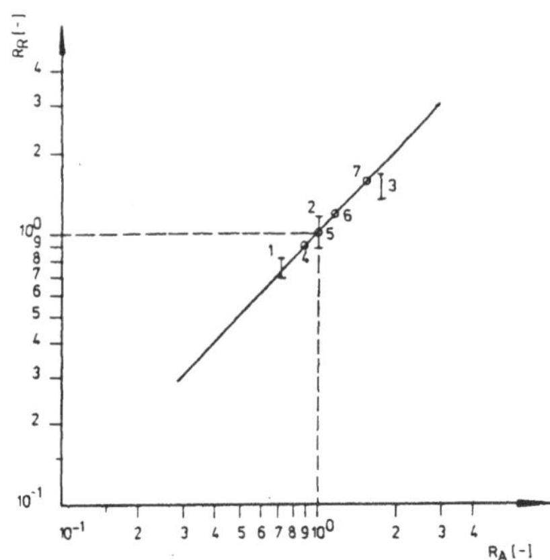

Bild 4 - Vergleich in relativen Einheiten der mittels Reflexionsmessungen (R_R) und Perthometer (R_A) gewonnenen Rauheitswerte von unterschiedlich behandelten Kupferdrähten ($\emptyset$=0,4mm): 1-blankes Cu,
2-chemisch aufgerauhtes Cu, 3-sandgestrahltes Cu.
Bezugspunkt: R_A=0,5µm, R_R=0,8V. Messpunkte 4-7
beziehen sich auf die Bestimmung der R_R von Pul-
ver-Farbpigment mit Korngrössen: 4-105µm,5-120µm,
6-140µm, 7-130µm.

Das Gerät für die kombinierte Messung von ε_T und R_R

Die auf dem Reflexionsprinzip beruhende Bestimmung von ε_T und R_R wurde mit
vier optisch getrennten Lichtwellenleitern (LL) und geeigneter Optik gelöst.
Die Einordnung der LL in dem Messkopf der Fa.VOLPI ist im Bild 5 schematisch dargestellt. Der die IR-Strahlung sendende LL_1 ist sowohl für die ε_T-
als auch R_R-Messart bestimmt. Er ist senkrecht zur Probe gerichtet. Parallel
mit LL_1 ist in derselben optischen Achse ein zweiter Lichtwellenleiter -
LL_2 - plaziert, welcher die senkrechte Reflexion der Probenoberfläche für
die Bestimmung der ε_T empfängt. Zwei weitere Lichtwellenleiter - LL_3 und
LL_4 -, welche zusammen mit LL_1 für die Bestimmung der Rauheit R_R dienen,
sind um kleine Winkel von der optischen Achse des LL_1 abgeneigt.

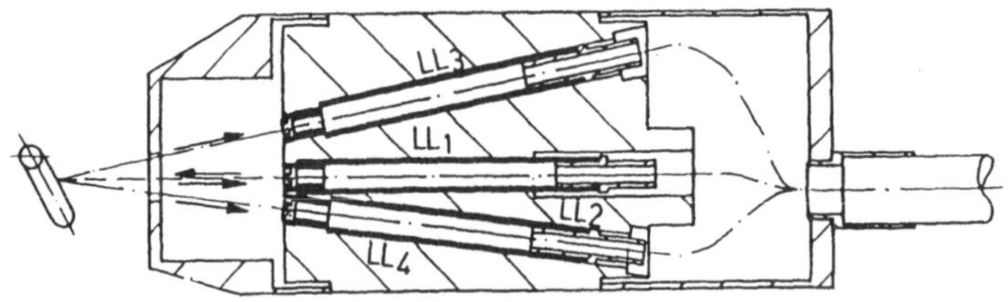

Bild 5 - Messkopf des VOLPI-Finishtest-Gerätes mit vier optisch
getrennten Lichtleitern LL_1 - LL_4, welche auf eine Bolzen-
probe gerichtet sind.

Ich danke der Fa. VOLPI herzlich für die zur Verfügung gestellten Unterlagen.

Örtlich und zeitlich aufgelöste Aufnahme und Auswertung schwacher transienter Leuchterscheinungen

D. WITTMER, W. PFEIFFER
Technische Hochschule Darmstadt
Schloßgraben 1, D 6100 Darmstadt

Einleitung

Mit Hilfe getasteter Bildverstärker ist es möglich, Leuchterscheinungen örtlich zweidimensional und zeitlich aufgelöst zu untersuchen. Durch Verwendung geeigneter Bildverstärker sowie einer nachgeschalteten Restlichtkamera wird hierbei eine hohe Lichtverstärkung erreicht. Durch Speicherung des Videobildes auf Videoband und anschließende Digitalisierung mit verminderter Abtastrate ist eine rechnergestützte Auswertung möglich (BILD 1).

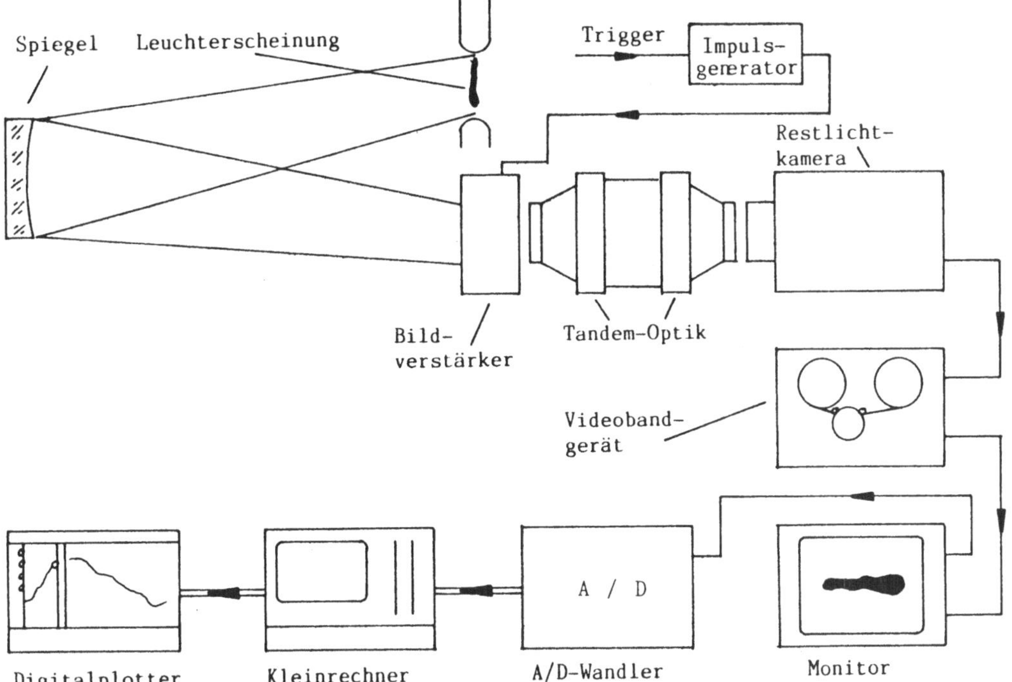

Bild 1. Versuchsaufbau zur örtlich und zeitlich aufgelösten Aufnahme und rechnergestützten Auswertung von Leuchterscheinungen

Bildaufnahmeeinheit

Zentraler Bestandteil der Bildaufnahmeeinheit ist der tastbare Bildverstärker, der einerseits die Funktion des Verschlusses und andererseits einen Teil der Bildverstärkung übernimmt /1/. In der hier beschriebenen Versuchseinrichtung kommen zwei verschiedene Bildverstärker-Typen zum Einsatz. Die Bildverstärkerkaskade (Bild 2) besteht aus drei hintereinandergeschalteten nähefokussierten Bildverstärkerdioden hoher Betriebsfeldstärke (PROXIFIER), wobei die erste getastet wird. Mit einem entsprechenden Spannungsimpuls (Bild 3) sind Verschlußzeiten bis hinab zu 5 ns möglich. Der Bildverstärker mit Mikrokanalplatte (Bild 4) arbeitet ebenfals nach dem Nahfocus-Prinzip, besitzt jedoch intern eine Mikrokanalplatte zur Sekundärelektronen-Vervielfachung. Bei Tastung der Photokathode mit einem negativen Spannungsimpuls (Bild 5) sind Verschlußzeiten bis <2 ns möglich. Bild 6 zeigt vergleichend die spektrale Eingangsempfindlichkeit der beiden Bildverstärker bei gleicher S 20 - Photokathode. Bedingt durch die feldunterstützte Elektronen-Emission bei hohen Betriebsfeldstärken zeigt die Bildverstärkerkaskade ein deutlich besseres Verhalten im langwelligen Bereich als der Bildverstärker mit Mikrokanalplatte. Grundsätzlich wird die Bildverstärkerkaskade eingesetzt, wenn eine hohe örtliche Auflösung im dynamischen Betrieb (>20 Lp/mm) oder eine hohe Empfindlichkeit im langwelligen Bereich erforderlich ist. Sind hingegen kürzeste Verschlußzeiten notwendig, findet der Bildverstärker mit Mikrokanalplatte Verwendung. Die Verstärkung ist bei beiden Typen vergleichbar. Ebenso haben beide Bildverstärker die Eigenschaft, daß sie, auch im dynamischen Betrieb, sehr verzeichnungsarm sind.

Das auf dem Leuchtschirm des verwendeten getasteten Bildverstärkers kurzzeitig erscheinende Bild wird mit Hilfe einer speziellen Restlicht-Kamera aufgenommen. Diese Kamera enthält statt der sonst üblichen SIT-Röhre eine PROXICON-Röhre, die aus einem NEWVICON mit vorgeschaltetem Bildverstärker (PROXIFIER) besteht (Bild 7). Diese Röhre zeichnet sich durch sehr geringe Verzeichnung aus, so daß eine nachfolgende rechnergesteuerte Entzerrung völlig entfallen kann. Die Kopplung zwischen dem Ausgang des getasteten Bildverstärkers und dem Eingang der Kamera geschieht über eine lichtstarke Tandem-Optik.

Da die Information des getasteten Bildes in nur einem einzigen Video-Halbbild gespeichert ist, ist zu deren Auswertung ein Speicher, entweder in Form eines Bildspeichers oder in Form eines Videobandgerätes mit Standbildbetrieb, notwendig. Derzeit wird ein Videobandgerät benutzt, welches auf einen Triggerimpuls hin ein einzelnes Halbbild aufzunehmen vermag.

Auswertung

Zur rechnergestützten Auswertung ist eine vorherige Digitalisierung nötig. Dies wird durch ein Digitalisierungsverfahren mit verminderter Abtastrate erreicht, wobei das zu digitalisierende Video-Standbild spaltenweise abgetastet wird. Nach diesem Vorgang kann die digitale Bildinformation durch entsprechende, auf den jeweiligen Anwendungsfall zugeschnittenen Programme verarbeitet werden /2/. Hierzu steht ein Kleinrechner zur Verfügung, auf dem Programme zur statistischen Verarbeitung, zur Glättung oder zur Linearisierung des Bildinhalts implementiert sind. Bild 8a zeigt als Beispiel einer quasi-dreidimensionalen Ausgabe über Digitalplotter das Ausgangsbild eines getasteten, jedoch für die Tastung ungeeigneten Bildverstärkers bei gleichmäßiger Eingangsbeleuchtung. Deutlich sichtbar ist hier der Verstärkungsabfall zur Mitte hin, der durch eine zu geringe Kathodenleitfähigkeit entsteht. Bild 8b zeigt unter sonst gleichen Bedingungen das Ausgangsbild eines gut tastbaren Bildverstärkers, der Verstärkungsabfall zur Mitte hin ist völlig verschwunden. Bild 8c zeigt das gleiche Ausgangsbild, das jedoch zusätzlich mit Hilfe eines digitalen Filters geglättet wurde.

Zur endgültigen Speicherung der interessierenden Videobilder können grundsätzlich die Diskettenlaufwerke des Rechners eingesetzt werden. Da eine Diskette jedoch nur zwei Bilder aufnehmen kann, ist dieses Verfahren allerdings sehr aufwendig. Es empfiehlt sich daher auch zur Archivierung die analoge Speicherung auf Videoband.

Zusammenfassung

Zur örtlich und zeitlich aufgelösten Aufnahme von schwachen Leuchterscheinungen eignen sich tastbare Bildverstärker nach dem Nahfocus-Prinzip, und zwar Bildverstärker mit Mikrokanalplatte ebenso wie Bildverstärkerkaskaden. Zur Aufzeichnung auf Videoband findet eine Restlichtkamera mit sehr geringer Verzeichnung Verwendung. Die gesamte Anlage ermöglicht Belichtungszeiten von wenigen Nanosekunden bzw. Auflösungen von mehr als 20 Lp/mm. Die Verarbeitung der sehr großen Datenmenge erfolgt durch einen Rechner und der Darstellung auf einem Digitalplotter.

140

Literatur
(1) PFEIFFER,W., AULBACH,B., LIEBEROTH-LEDEN,B., WITTMER,D.: 4th IEEE
    Pulsed Power Conference, Albuquerque (1983) to be published
(2) PFEIFFER,W. und AULBACH,B.: 7th Int. Conf. on Gas Discharges and
    their Appl., London (1982) 499

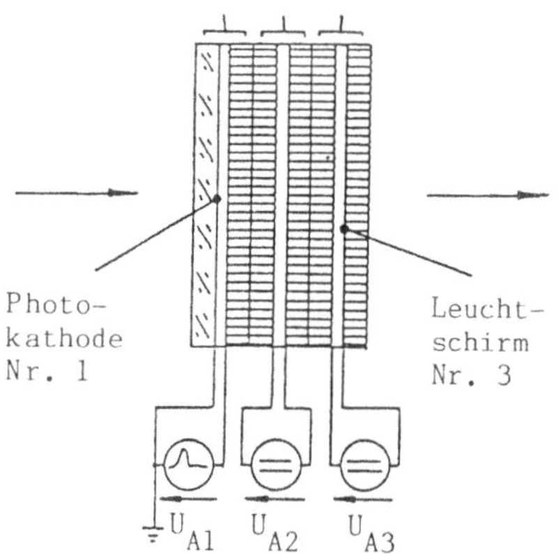

Bild 2. Bildverstärkerkaskade,
bestehend aus drei Dioden

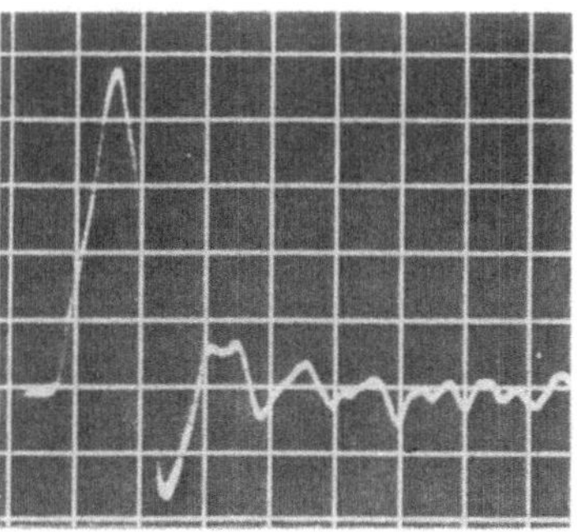

Y: 1.7 kV/div; t: 10 ns/div

Bild 3. Spannungsimpuls zum
Helltasten der Kaskade

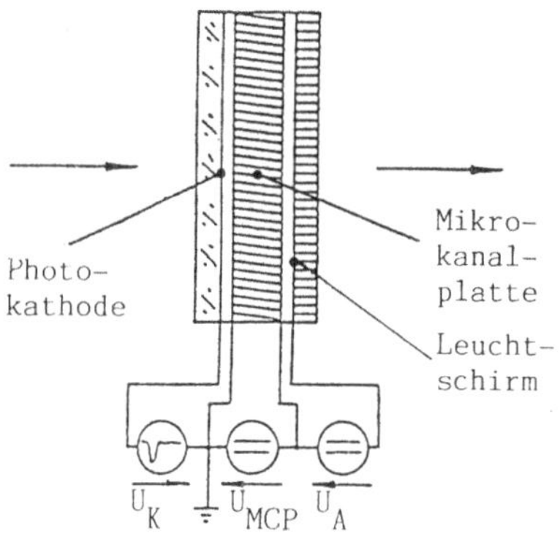

Bild 4. Bildverstärker mit
Mikrokanalplatte (MCP)

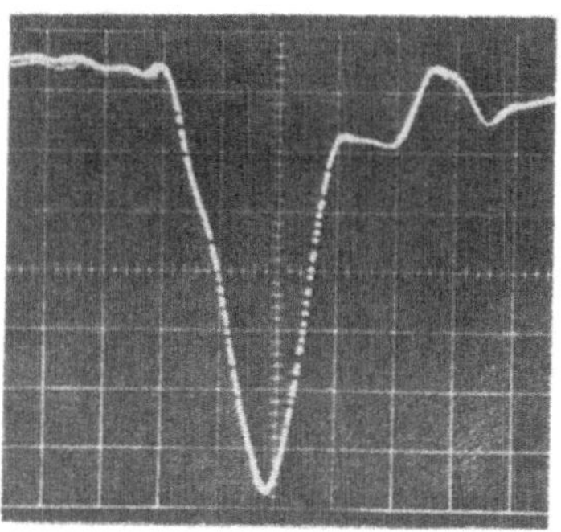

Y: 50 V/div; t: 1 ns/div

Bild 5. Spannungsimpuls zum
Helltasten des Bildverstärkers
mit Mikrokanalplatte

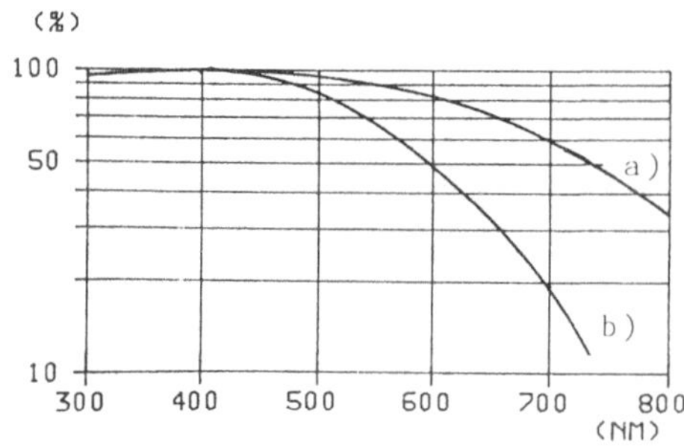

Bild 6. Spektrale Eingangs-
charakteristik

a) Bildverstärkerkaskade
b) Bildverstärker mit MCP

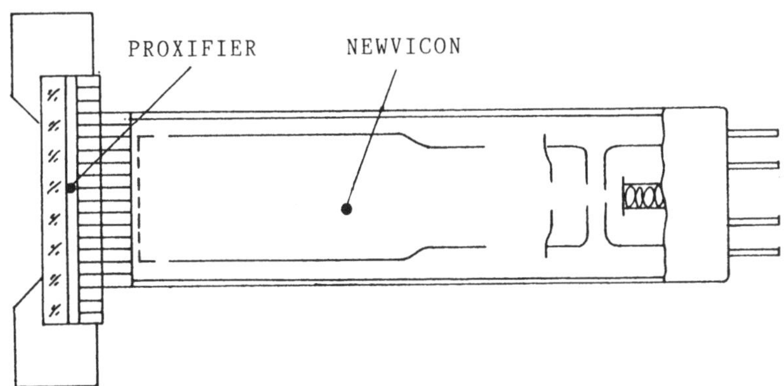

Bild 7. Aufbau einer PROXICON-Bildaufnahmeröhre

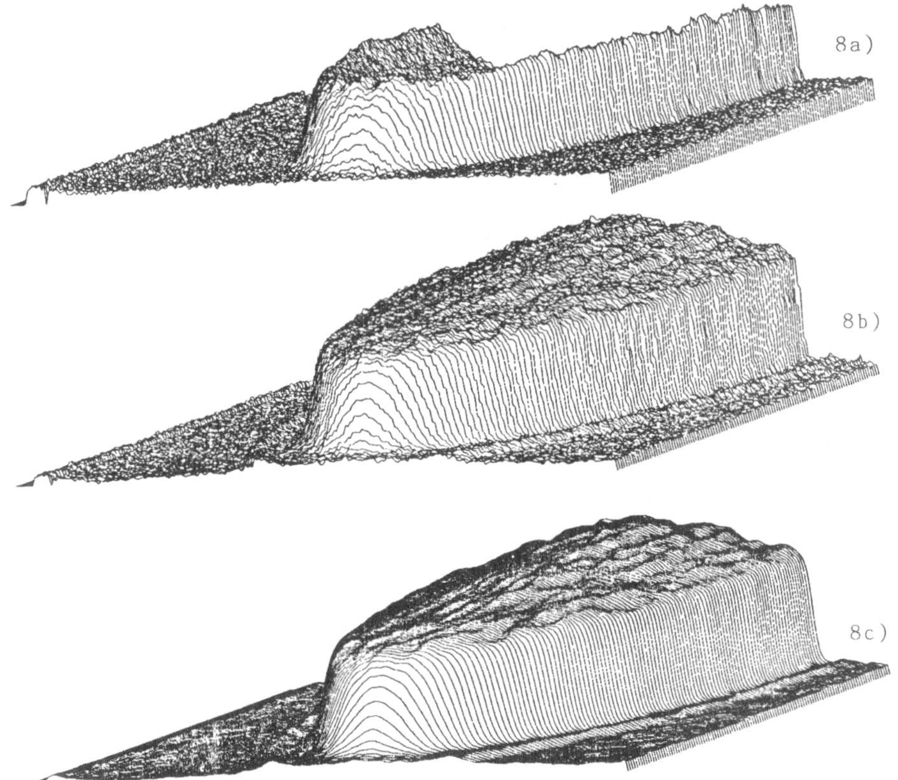

Bild 8. Räumlich zweidimensionale Intensitätsverteilung getasteter
Bildverstärker bei gleichmäßiger Eingangsleuchtdichte

    a) Bildverstärker mit geringer Kathodenleitfähigkeit
    b) Bildverstärker mit hoher Kathodenleitfähigkeit
    c) Zusätzliche digitale Filterung des Bildes in b)

Two-Wavelength Heterodyne Interferometry

A. F. Fercher and H. Z. Hu
University of Essen, Physics Dept.
P.O. Box 1o3 764, D-43oo Essen 1

Two-wavelength interferometry has been described in the literature as a means to
obtain reduced sensitivity interferometry. So far only two-wavelength techniques
which use the Moiré effect /1/ and holographic techniques /2, 3/ have been described. In this paper we show that the two-wavelength technique extends the applicability of the heterodyne interferometer and furthermore that it allows to
measure the contour of optically rough surfaces by interferometry.

In optical testing by interferometry the wave transmitted and/or reflected by an
optical element is compared with a reference wave of well-defined form. This is
done by suitably interfering the two waves to obtain an interferogram. The optical
path difference (OPD) between object wave and reference wave is then derived from
the interferogram which displays the phase differences between the two waves. In
the heterodyne interferometer this is performed by frequency-shifting the test beam
and/or the object beam in order to produce a difference frequency ω_H. In the interferogram the fringes then move in a direction perpendicular to themselves. The OPD
is determined with the help of two detectors: A fixed reference detector at a
reference point R and a scanning detector at measurement points M, N, ... These
detectors measure the time of arrival of the fringes and the phase difference between M and R is equal to the ratio of the time of arrival of the fringes at these
points divided by the period $2\pi/\omega_H$. The result is unique if the phase difference
between neighbouring measurement points M and N is smaller than 2π, i.e. if the
distance $\overline{MN}$ is smaller than the fringe spacing. In all other cases the phase
difference would only be determined modulo 2π. Hence we can only determine the
phase difference relative to a point sufficiently near to the measurement point. To
measure the phase differences in the interferogram uniquely the separation of the
sampling points must not exceed the fringe spacing. This is a rather limiting
condition at least if the surface of the test object is complex, as is the case,
for example, with silicon wafers, diamond turned optical components, aspheric sur-

faces and in general with all optical surfaces having steeply sloped zones with a short lateral width. The sampling distance might have to be very small in some areas but we do not know in advance where. Therefore an interferometric technique which measures arbitrary phase differences absolutely at any point M is desirable. The two-wavelength technique meets this requirement. The phase difference can be measured in an absolute way independently of its value at neighbouring points.

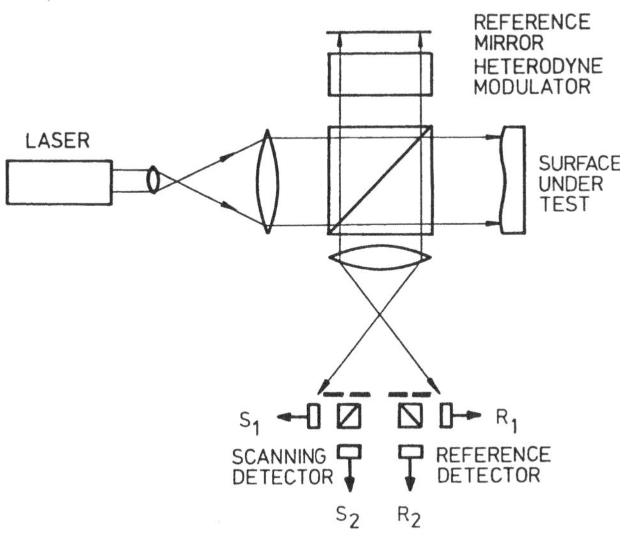

Fig.1. Two-wavelength heterodyne interferometer.

In Fig. 1 the two-wavelength technique is illustrated for the case of a Michelson interferometer. However, the two-wavelength technique is applicable to any interferometer. The interferometer is illuminated with light containing two wavelengths λ_1 and λ_2. Hence two interferograms appear simultaneously. At the reference point R we have the detector signals:

$$R_1(t) = I_1 \left[1 + \cos(\alpha_1 + \omega_H t) \right]$$

$$R_2(t) = I_2 \left[1 + \cos(\alpha_2 + \omega_H t) \right]$$

(1)

α_1, α_2 being the phase differences between object waves and reference waves of λ_1 and λ_2 respectively at the reference point R and ω_H being the heterodyne frequency.

At the measurement point M we have the waves

$$S_1(t) = I_1\left[1 + \cos(\alpha_1 + \mu_2 + \omega_H t)\right]$$

$$S_2(t) = I_2\left[1 + \cos(\alpha_2 + \mu_2 + \omega_H t)\right] \tag{2}$$

μ_1 and μ_2 being the additional optical phase differences between object and reference waves of λ_1 and λ_2 respectively at the measurement point M relative to the reference point R. Let $d(M)$ be the additional OPD at the measurement point M relative to the OPD at the reference point. Then we have

$$\mu_1 = \frac{2\pi}{\lambda_1}\, d(M) - 2n_1\pi$$

$$\mu_2 = \frac{2\pi}{\lambda_2}\, d(M) - 2n_2\pi \tag{3}$$

and n_1, n_2 and d are three unknowns. In addition we now measure the phase difference Δ of the signals $S_1(t)$ and $S_2(t)$

$$\Delta(M) = \alpha_1 - \alpha_2 + \frac{2\pi d(M)}{\Lambda} - 2\pi(n_1 - n_2) \tag{4}$$

Provided that the additional OPD $d(M)$ in the interferogram nowhere exceeds $\Lambda = \dfrac{\lambda_1\lambda_2}{|\lambda_1\lambda_2|}$ we have $n_1 = n_2$ and hence there are three equations to determine the three unknowns. Thus the OPD may be determined at an arbitrary point M in the interferogram without any reference to a neighbouring point with the precision of conventional interferometry, provided $|d(M) - d(R)| < \Lambda$. If on the other hand the OPD in neighbouring points M, N, ... is taken into account, this condition is reduced to $|d(M) - d(N)| < \Lambda$ and thus very rough surfaces can be measured by this interferometric technique.

Heterodyne interferometry in general allows to determine OPD with high precision, i.e. with a standard deviation of λ/N and say $N = 100$. If this high precision is not needed, then it is possible to abandon the reference detectors and only use the scanning detectors. The precision is then Λ/N. In this case we have the straightforward application of the heterodyne principle to low-sensitivity two-wavelength interferometry. The technique presented here does neither make use of the Moiré-effect nor does it need a hologram. Hence we have a technique which makes possible direct interferometric measurements on rough surfaces. The only condition that must be fulfilled by the surface is that the OPD between neighbouring measurement points must not exceed Λ. Fig. 2 presents first results of measuring the surface profile of an optically rough object with the two-wavelength heterodyne interferometer.

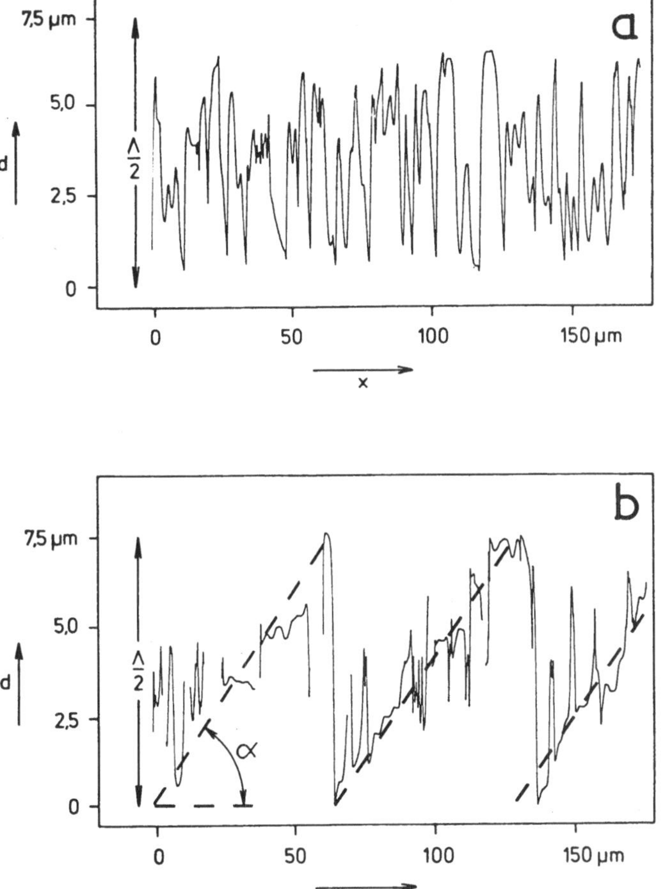

Fig. 2. Voltmeter output displays the surface profile d(x) of an optically rough but macroscopic plane surface (a). In b the surface has a tilt angle α.

Only the scanning detectors had been used. The signals $S_1(t)$ and $S_2(t)$ were fed into a lock-in-amplifier operating in the vector voltmeter mode. According to Equ. (4) the output of the voltmeter is proportional to d(M). The object under test was a macroscopically plane grinded piece of glass. In these measurements the scanning detectors remained in a fixed position and the object was moved. The 647 nm and 676 nm lines of a Kr-laser were used with an effective wavelength of $\Lambda = 15$ μm. Fig. 2a shows the voltmeter signal obtained, when the normal of the object surface was perpendicularly to the interferometer axis and the object was moved in x-direction parallel to the interferometer axis. In Fig. 2b the object was first tilted to obtain an angle α between the surface normal and the interferometer axis. Then the object was moved again in x-direction. Besides the microstructure the voltmeter output shows the mean increase of d in accordance with the tilt α.

We have shown that a direct interferometric measurement of the surface profile of rough surfaces is posible by two-wavelength heterodyne interferometry. An important benefit of this technique is that focusing is not needed. The heterodyne technique has the merit of being very fast. Consequently we expect that the technique described above finds applications as non-contacting profilometer in mechanical engineering and also in the optical production technology.

References

/1/   C.POLHEMUS: Appl. Opt. 12 (1973) 2o71
/2/   J.C. WYANT: Appl. Opt. 1o (1971) 2113
/3/   F.M. KÜCHEL and H.J. TIZIANI: Opt. Commun. 38 (1981) 17

Laser Densitometry of High Optical Densities

H. Zorc

"Rudjer Bošković" Institute

Zagreb, Yugoslavia

Introduction

The application of lasers has recently become widespread in the field of science, technology, industry and even in household. Numerous experimental laser methods made
considerable progress, while at the same time entirely new fields have been discovered,
e.g. holography.

This work deals with application of lasers in densitometry, i.e. in measuring of optical
densities which formerly could not be obtained by conventional methods. More precisely,
it deals with monochromatic laser densitometry, since monochromatic lasers were
used (Nd-YAG and ruby lasers).

One of numerous applications of lasers is also in densitometry of absorption glasses
and anti-laser safety filters. Calorimetric method is commonly used for determination
of optical density of absorption glasses, while optical density of interference filters
can only be measured by using optical methods. It is also the only applicable method
for measuring of combined absorption/interference filter density.

Measuring method

Basically, the whole thing is very simple – laser light falls onto a specimen, passes
through it and falls onto a detector, where transmitted energy is being registered.
Since the range between the incident and transmitted energy is rather large (up to 12
orders of magnitude), it is clear that only the energy which passes through the specimen
should be measured, not any other. Schematic diagram is shown in Fig. 1.

All other forms of energy, like mechanical shock, electric noise, parasitic light caused
by multiple reflections, flash light, etc., represent disturbances that can considerably
influence sensitivity of the method. Special attention should be paid to multiple reflection
light, flash light and possible induced emission of light in the specimen itself.

Multiple reflections are commonly described as "light crawling", since the light seems

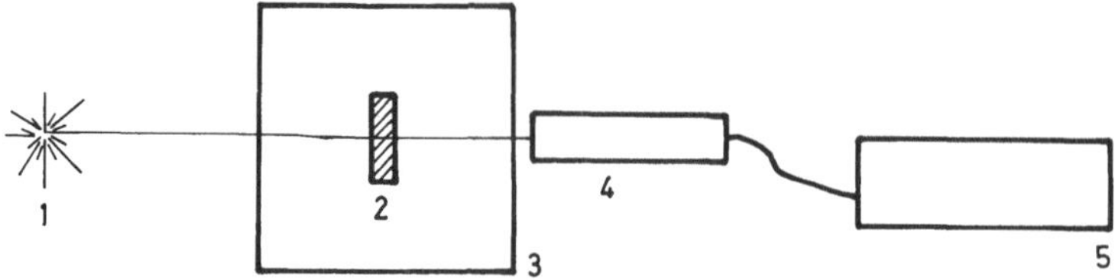

1. Laser beam
2. Specimen
3. Test chamber
4. Detector
5. Console (energy meter)

Fig. 1. Schematic diagram of the laser densitometry used in the paper.

to get around obstacles and thread among them. This problem can be solved by disabling the detector to receive any kind of light, except the one that passes <u>through</u> the specimen. The problem is simply solved if there is normal light incidence, but if incident angle is continuously varied, things become more complicated.

Elimination of flash light is far more serious problem, because it is necessary to measure only the density at laser wavelength. The use of coloured glass is often insufficient especially if their optical densities are lower than those of the specimen. The 0.5 mm thick Si turned out to be a good filter for flash light elimination in the use of Nd-YAG laser, and RF 645 glass with selective beam splitter in the use of ruby laser.

The induced radiation emission in the specimen can also be considerable, relating to the transmitted energy. This problem is solved by putting a band-pass filter between the specimen and the detector.

Except for the parasitic light, which can greatly influence the measurement and even make it completely impossible, a very important factor is also type of the used laser. When a Q-switch laser is used the transmitted energy becomes function of the incident energy. Consequently, instead of density measurement we have also nonlinear effects measured. Our experiences have shown that Q-switch lasers are not advisable for such measurements.

Application

The above described measuring method – laser densitometry – was applied to optical density measurements of anti-laser safety filters. The Nd-YAG and ruby lasers with 1 J output energy (duration of a pulse was 1 ms) were used. Laser Precision Corp.

RJ 7100 instrument with corresponding detectors (with sensitivity up to 10^{-12} J) was used. Maximal optical density obtained by this method on absorption/interference filters was 12, relating to the applied instrumentation. This method proved to be very accurate and reliable on great number of the tested filters manufactured by LAIR.

Summary

The method of monochromatic laser densitometry and problems connected with its application have been described. Solutions to the problems have been given and the use of laser energy up to 1 J, with 0.1 – 1 ms pulse duration, has been suggested. This is very important because of nonlinear effects. Attention should be paid to energy density that falls onto a specimen, depending on its damage threshold.

Reference

For further information please contact H. Zorc, "Rudjer Bošković" Institute, LAIR, P.O.Box 1016, 41001 Zagreb, Yugoslavia.

Self-calibrating Automatic Laser Focus Control System

R.E. KUNZ

GRETAG Limited
Althardstrasse 70, CH-8105 Regensdorf, Switzerland

Introduction

Present laser-based techniques often require precisely controlled focused laser
beams. Methods and arrangements for focusing laser light on various objects usually
are tailored to very specific situations, imposing severe constraints on the properties of the objects or on the focused laser beam itself.

In this paper we are proposing a novel laser focus control system combining two complementary methods to achieve accuracy, speed, and ruggedness without sacrificing
versatility. After listing some major system requirements, we will describe the
basic principles of each method. From this, their distinct advantages as well as the
benefits from their combination should become apparent.

Design considerations and operating principles

The design of our system was determined to a large extent by the **system requirements**
outlined below:

1. Performance:
   a) sufficient accuracy for adjustments in the range from $<1\,\mu$m to >1 mm,
      adaptable to the actual application;
   b) fast focus detection for real-time applications;
   c) aim is to set (minimize) spot diameter, not just to position the object.

2. Reliability:
   a) immunity against external disturbances such as mechanical, thermal, and
      electrical influences;
   b) not affected by changing polarization of light;
   c) independent of angular distribution of light;
   d) cope with coherent optical noise (speckling).

3. Versatility:
   a) for planar and non-planar surfaces;
   b) for a wide class of surface properties;
   c) neither special "marks" nor any previous processing of surface required;
   d) for cw and pulsed radiation;
   e) allow for changes of laser beam parameters.

4. Compactness:
   a) no auxiliary light sources;
   b) no additional elements in front of the focusing lens;
   c) simple, rigid arrangement, few parts, easily assembled and aligned.

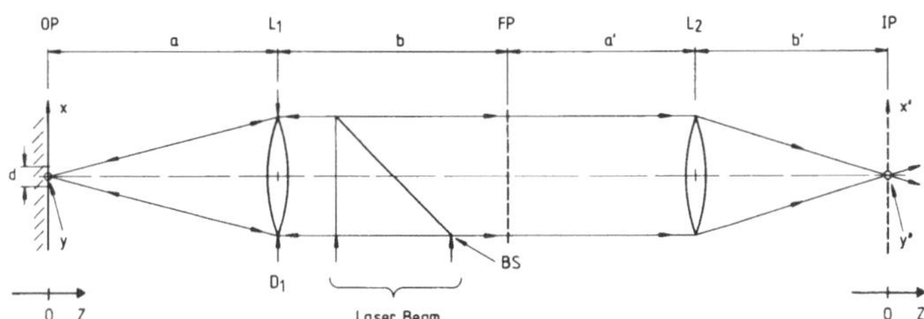

Fig. 1.
Basic elements
and definitions.

A focus detection system based on analyzing the radiation reflected by the illuminated surface presents an appropriate way to fulfill these requirements. For clarity
and brevity, we refer the discussion to the arrangement shown in Fig. 1. In this
example, a laser beam is focused on the object via mirror (beam splitter) BS and
lens L_1 (focal length f_1). The light reflected from the illuminated spot
(diameter d) in the object plane OP is analyzed by two different methods. The first
one uses an optical system L_2 to form an image of the illuminated spot and probes
the light distribution near its image plane IP. The second one probes the light
distribution in the Fraunhofer region of the scattering area, as e.g. in the
Fraunhofer plane FP of L_1 ($a \cong f_1 \cong b$ in Fig. 1). We will denote these methods by
ISP (**image space probing**) and FRP (**Fraunhofer region probing**), respectively.

First, we will discuss the realization of a powerful ISP system using spatially filtered multiple imaging of the illuminated spot. An example of such a **multiple image
filtering** (MIF) arrangement is shown in Fig. 2a. Grating G in front of cylindrical
lens L_2 allows to split the reflected wave u_r into secondary waves $u_{r\,1,\,2,\,3}$
forming multiple images at IP. Spatial filters (e.g. V-shaped masks $M_{1,\,2,\,3}$ shown
in Fig. 2b) allow to probe the image forming waves simultaneously at different
locations z' (cf. Fig. 1). Variation of the focusing conditions (e.g. of object position z) influences the relative powers in the filtered images, i.e. photocurrents
$I_{1,\,2,\,3}$ of photodetectors Det. 1, 2, 3, respectively (cf. Fig. 2c). Using differential amplification, signed focus error signals as e.g. (I_2-I_1) or $(I_2-I_1)/I_3$ may be
readily obtained and used for fast focus control.

For achieving high sensitivity and stable operation of this "optical comparator", it
is crucial to maximize **common mode rejection** (CMR), i.e. to balance the different
optical channels with respect to external changes. A major aspect is the symmetry of
the arrangement. We have been able to conceive a highly symmetrical and well
balanced system by splitting the primary wave u_r as homogeneously as possible, i.e.
independent of location and polarization at the beam splitting element. This is
accomplished by using diffractive optical elements such as gratings or **holographic
optical elements** (HOEs) instead of ordinary beam splitters or, even worse, of using
different parts of the scattered wave as secondary waves. Sustaining the high degree
of symmetry through the whole system, including photodetectors and electronics,

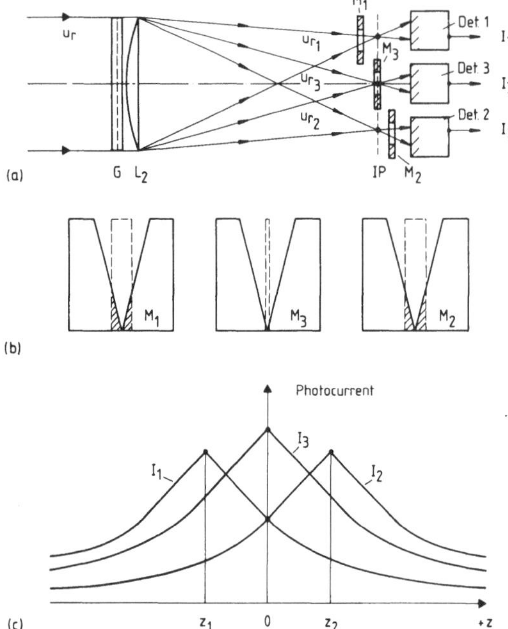

Fig. 2.
Principle of MIF method:

(a) Sketch of MIF arrangement;

(b) Masks M_1, M_2, M_3 (solid lines) and correspondent light distributions (broken lines). Hatched portions indicate blocked light;

(c) Idealized variation of detector output currents I_1, I_2, I_3 versus position z of object (cf. Fig. 1).

results in a very high CMR of external disturbances such as changes in polarization and angular distribution of the scattered light as well as mechanical, thermal, and electrical influences.

Another important feature is the flexibility to adapt the system's characteristics to the actual requirements of a given application. Introducing anamorphic optical elements and matching spatial filters increases versatility and ease of alignment, whereby CMR may be enhanced even further. We would like to point out that the spatial filtering may be effectuated in many different ways, e.g. using filters with complicated transmission functions instead of simple masks, or not requiring separate filters at all. If a HOE is used to perform the essential operations on the scattered light, extremely compact systems may be realized with a minimum number of components, featuring maximum flexibility and reliability at the same time.

The MIF system described above is able to provide fast control of focus position for cw light sources as well as for pulsed radiation, in contrast to other methods. Furthermore, it may also be utilized for automatically focusing incoherent light with only slight modifications. Besides of representing a compact focus detection system for focusing laser light, this type of arrangement lends itself also to applications e.g. in non-contact profilometry and sensor technology.

After having finished the basic design of this MIF system, we came across a publication /1/ and a patent /2/, both describing optical non-contact profilometers using

MIF techniques. However, the symmetry and the additional degrees of freedom of our system represent substantial advantages for practical applications.

As is typical for accurate imaging in real space, precise alignment and long-term stability of both optical arrangement and properties of the light source are vital requirements. Therefore, to assure high precision over long time periods, and to compensate or allow for light source parameter changes, calibration of the imaging setup by a method not suffering from the same limitations is of utmost importance.

A second method working in Fourier space, i.e. an FRP method as defined before, is appropriate for accomplishing this task since it is able to effectively measure the diameter of the illuminated spot without being affected by its actual position. In fact, observing the speckles in the backscattered radiation offers the possibility to realize a system which is not critical with regard to alignment.

As is well-known (cf. Ref. /3/), coherent light scattered from "optically rough surfaces" shows a statistical intensity modulation, the average width of the features (speckles) in the Fraunhofer region being inversely proportional to the width of the illuminated spot. Figure 3 (photographs taken from a TV monitor) shows the decrease in speckle size near the Fraunhofer plane FP as a function of defocusing (cf. Fig. 1; $a \cong f_1 = 30$ mm, $1/e^2$-diameter of HeNe laser beam $\cong 8$ mm; $z = 0$ (a), $50\,\mu$m (b), $100\,\mu$m (c), $200\,\mu$m (d), $400\,\mu$m (e), $800\,\mu$m (f)). This illustrates the principle of characterizing the state of focusing by **speckle pattern evaluation** (SPE), realizable e.g. by video signal processing. The most significant characteristic of this method is that it is based on width sensing rather than on position sensing.

Typical applications of such a system include the calibration of focusing units with critical alignment, and the monitoring of proper focusing conditions in various laser-based systems. Moreover, it can be used to determine the variation of the illuminating beam's width as a function of defocusing, thus providing information on beam parameters as e.g. depth of focus, and on eventual beam aberrations.

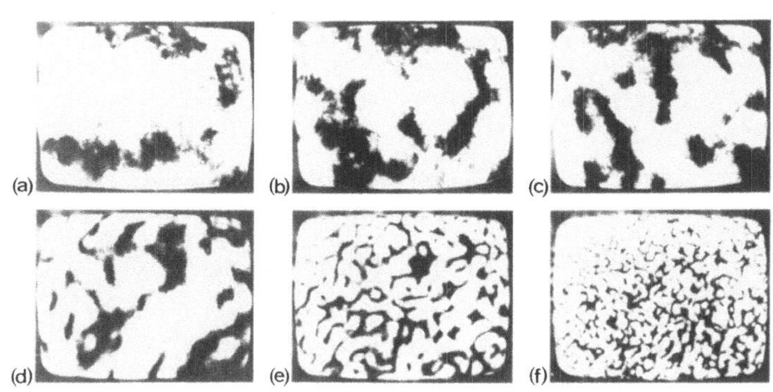

Fig. 3.
Illustration of speckle size variation with object position z (cf. Fig. 1).

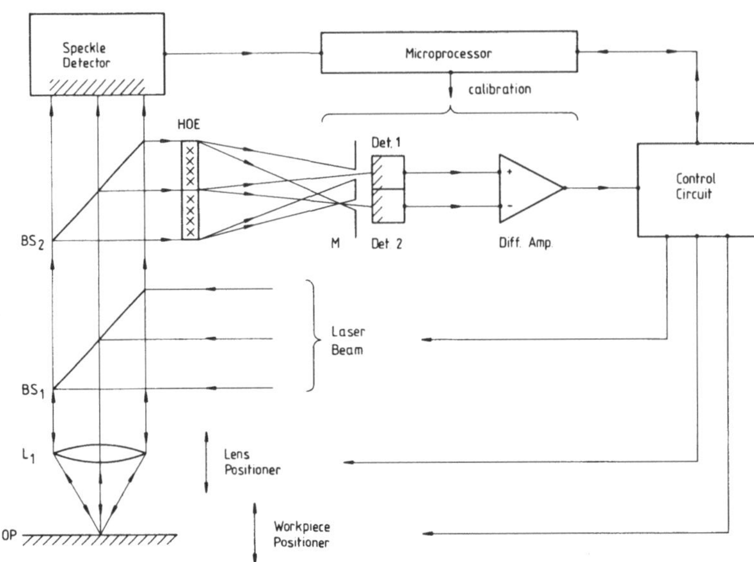

Fig. 4.
Self-calibrating
automatic laser
focus control
system.

Although focus detection may be performed based on either ISP or FRP methods exclusively, a particularly important application is the realization of a self-calibrating system as shown in Fig. 4. The MIF arrangement, shown here with a single HOE performing all essential optical transformations, consists of merely two parts, the HOE and the masked detector pair. It provides fast focus control via a differential amplifier and a control circuit. Calibration is done by the SPE subsystem using a microprocessor. A remarkable feature of this system is its ability to control not only the workpiece position, but also the lens position as well as the laser beam parameters.

In conclusion we would like to point out that the two complementary methods, combined in an automatic, self-calibrating focus control system, allow to master wanted and unwanted changes of geometrical or optical parameters, maintaining accuracy, speed, and versatility in critical, laser-based equipment and processes.

It is a pleasure for me to acknowledge J. Knus, U. Murbach, and M. Tuor for their assistance in setting up the necessary laboratory experiments.

References

/1/ Y. FAINMAN, E. LENZ, and J. SHAMIR, "Optical profilometer: a new method for high sensitivity and wide dynamic range", Appl. Opt. 21, 3200-3208 (1982).

/2/ F.-M. TEISSIER, "Procédé et dispositif de palpage optique", French Patent No. 94.871 (8 Dec. 1969).

/3/ R.K. ERF, Ed., "Speckle Metrology" (Academic Press, New York, 1978), and References cited therein.

Farbige Weißlichtholographie

R. Leßing
Spindler u. Hoyer GmbH u. Co.
Königsallee 23
3400 Göttingen

Führt man bei der Herstellung eines Hologramms den Referenzstrahl, der
den kohärenten Untergrund bildet, so, daß er dem Objektstrahl entgegenläuft, bilden sich in den Interferenzmaxima stehende Wellen mit
einem Abstand von $\lambda/2$. Es entsteht also nicht nur ein für die Holografie typisches Interferenzmuster, vielmehr ist dieses auch noch "geschichtet moduliert". Dieser Effekt kann bei hinreichend großem Auflösungsvermögen des Aufnahmematerials gespeichert werden, wobei natürlich der Brechungsindex n des Materials eine Rolle spielt: Die
Schichtungsmaxima haben jetzt einen Abstand Δ von $\lambda/2n$. Mit λ
(z.B. aus einem He-Ne-Laser) = $0.633\,\mu m$ und n = 1.5 kommt man zu
$\Delta = 0.21\,\mu m$ entsprechend einem Mindestauflösungsvermögen des Materials
von 4740 L/mm. Silberhalogenidfilme (z.B. 8E75 HD der Firma Agfa,
Leverkusen, können dieses Auflösungsvermögen gerade erreichen. Bei
Verwendung kürzerer Wellenlängen (Ar^+ Laser $0.5145\,\mu m$, $0.488\,\mu m$) ist
das Filmauflösungsvermögen zu gering, so daß die Speicherung von
stehenden Wellen dieser Wellenlänge in Silberhalogenidfilmen nicht
sehr effizient sein kann.

Erfolgversprechend scheint die Verwendung eines Farbstofflasers zu
sein, der (z.B. mit DCM betrieben) eine Wellenlänge von $0,66\,\mu m$
emittiert, was zu einem Modulationsabstand von $0.22\,\mu m$, entsprechend
einer Ortsfrequenz von 4545 L/mm führt.

Die Rekonstruktion des holografisch gespeicherten Objektes kann jetzt,
da es moduliert aufgenommen wurde, mit weißem Licht erfolgen.
Figur 1 zeigt den Rekonstruktionsmechanismus.

Wird der belichtete Silberhalogenidfilm entwickelt, so entstehen an
den Stellen der Schichtumgsmaxima teildurchlässige Silberschichten,
(Lippmann-Schicht). An den Stellen der Minima bleibt das Silberhalogenid
erhalten. Beim Bleichen des Hologramms wird allein das Silber entfernt,

156

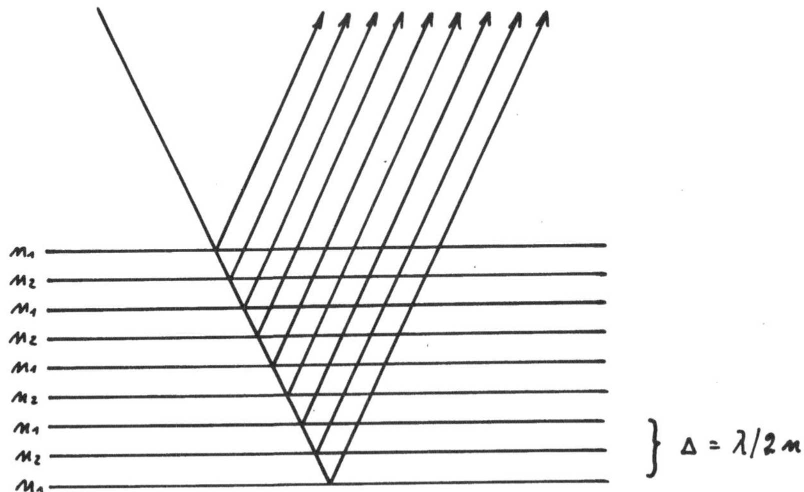

Figur 1   Rekonstruktion einer Wellenlänge aus einem λ/4 Schichtsystem mit Hilfe des Bragg-Effektes.
Die Rekonstruktionswellenlänge ist die gleiche wie bei
der Aufnahme der Schicht.

so daß eine λ/4 Schichtung unterschiedlicher Brechungsindices übrig-
bleibt. Einmal nämlich der Emulsionsträger (Gelatine) allein, zum
anderen angereichert mit verbleibendem Silbersalz.
Messungen des Phasenhubs von auf 8E75 HD hergestellten holografischen
Gittern (1) ergeben einen je nach phototechnischer Behandlung der
Platte unterschiedlichen Wert bis zu λ/2. Der Brechungsindexunter-
schied kann jetzt ermittelt werden. Setzt man n_1 (Gelatine) mit 1.5 an,
so ist n_2 (Gelatine + Silbersalz) hiernach = 1.55. Es würde sich bei
einer Gesamtschichtdicke des Films von 7.5 µm eine maximale Rekonstruktionseffizienz von 42% ergeben. Tatsächlich liegen die Verhältnisse bei der Lippmann-Holografie etwas anders:
Bedingt durch die Nähe der Auflösungsgrenze des Materials kommt man
nur auf Beugungseffizienzen von ca. 30%. Dies entspricht zwei Brech-
werten von 1.5 bzw. 1.535. Die Halbwertsbreite kann hieraus rechne-
risch ermittelt werden, sie beträgt 35 nm, also etwa die gleiche,
wie ein metallisches Interferenzfilter. Dies würde zur einfarbigen
lichtstarken Rekonstruktion von Hologrammen mit weißem Licht ausreichen. Bedenkt man jedoch, daß die Rekonstruktionswellenlänge
nahezu die gleiche ist, die bei der Aufnahme verwendet wurde, (es
mußte ja wegen des Auflösungsvermögens des Films dunkelrotes Laserlicht sein), so erscheint der rekonstruierte Gegenstand ebenfalls im

tiefroten Wellenlängenbereich, wo das menschliche Auge recht unempfindlich ist, und somit der subjektive Bildeindruck unbefriedigend
wird. Will man nicht zu anderem holografischen Aufnahmematerial übergehen (z.B. Dichromaten), gibt es eine Möglichkeit, die Sichtbarkeit
von Lippmann-Bragg-Hologrammen zu steigern. Die gemessene Effizienz
der Rekonstruktion ist mit 30% groß genug, um den Hologrammgegenstand
hell erscheinen zu lassen. Es braucht nur die Rekonstruktionsvorlage
verkleinert zu werden. Kurzzeitig gelingt dies durch Aufheizen und
damit Dehydrieren der Filmemulsion des fertigen Hologramms. Die Gelatine schrumpft etwas, der Schichtungsabstand und damit die Rekonstruktionswellenlänge werden kleiner. Allerdings nimmt nach dem Abkühlen
die Gelatine durch die Wasseraufnahme aus der Luft allmählich wieder
ihre alte Schichtdicke an, und die Verhältnisse sind wie vorher.

Abhilfe schafft hier eine Expansion der Emulsionsschichtdicke vor
der Belichtung des Hologramms. Setzt man die unbelichtete Platte einer
10%igen Glycerinlösung aus (Dauer ca. 1 Min.) so wird die Emulsion
nach dem Trocknen um ca. 17% vergrößert. Diese vorbehandelte Platte
unterliegt jetzt dem schon bekannten Belichtungsprozess und dient zur
Speicherung der Lippmannschichtung mit $\Delta_o = \lambda_o / 2n$ ($\lambda_o = 0.66\,\mu m$).

Da die geometrische Schichtdicke größer ist als vorher, werden nun
auch mehr Lippmann-Lagen gespeichert (40 gegen vorher 34). Der anschließende photochemische Prozeß (Entwickeln, Wässern, Bleichen,
Wässern) läßt das Glycerin wieder aus der Emulsion, wodurch sie nach
dem Trocknen wieder ihre alte Dicke von 7,5 μm einnimmt. Selbstver-
ständlich schrumpft dabei auch der Abstand Δ_o der Lippmann-Lagen um
dem vorher eingestellten Expansionswert von 17%, so daß jetzt
$\Delta_1 = \lambda_1 / 2n$ wird, mit $\Delta_1 = \Delta_o (1-0.17)$ und also $\lambda_1 = \lambda_o (1-0.17)$. Die Rekonstruktion des Hologrammobjekts erfolgt im grünen Wellenlängenbe-
reich. Bedingt durch die nach wie vor 40 gespeicherten Lippmann-
Lagenpaare erhält man eine Rekonstruktionseffizienz von 34% bei einer
Halbwertsbreite von 30 nm. Neben der "besseren" Lage des Rekonstruk-
tionsmaximums ist die Effizienz höher und die Halbwertsbreite kleiner,
so daß eine brillante Darstellung des Objekts erwartet werden kann.
Fig. 2 zeigt eine mögliche Art, ein solches Hologramm herzustellen.

Das vom Laser kommende Licht wird mit einem Mikroskopobjektiv aufgeweitet und mit einem Raumfilter (20 .. 30 μm Ø) von unerwünschten
Strukturen gereinigt. Die etwas schräggestellte Hologrammplatte wird

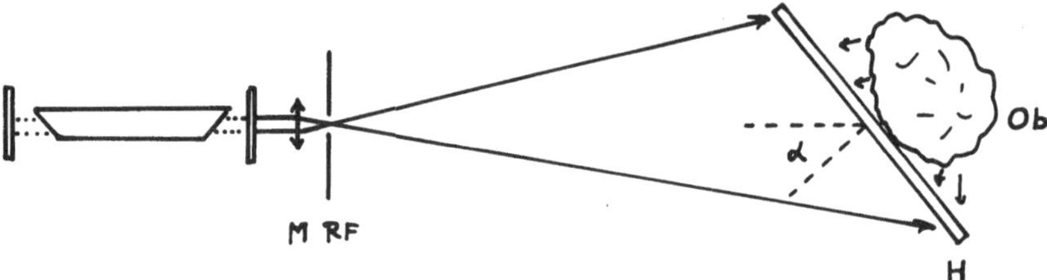

Fig. 2   Herstellung eines "Einstrahl-Weißlichthologramms". M: Mikro-
skopobjektiv 40:1, RF: Raumfilter, H: Hologrammplatte,
Ob: Objekt. Der Abstand zwischen RF und M ist ca. 1m.
Der Winkel α ist vorzugsweise der Brewsterwinkel. (E-Vektor ∥

von diesem Licht direkt bestrahlt (Referenzstrahl). Das durch die
transparente Platte durchgelassene Licht wird vom Objekt, welches
möglichst hell sein sollte,* auf die Platte zurückreflektiert, wobei
nun die Lippmann-Schicht entsteht. Objekt und Platte können, ja sollten sich berühren, um einerseits gegenseitige Bewegung auszuschließen,
andererseits größtmögliche Bildschärfe zu erzielen. Nach einigen Sekunden Belichtungszeit (bei 10 mW Laserleistung, Strahl Ø am Hologramm
ca. 20 cm) kann entwickelt werden: (A) 10g Pyrogallol + 20g Na_2SO_2 in
1 l Wasser, (B) 60g Na_2CO_2 in 1 l Wasser. (A) und (B) sind zu gleichen Teilen kurz vor dem Entwickeln zu mischen. Nach dem Wässern wird
gebleicht: 5g $K_2Cr_2O_7$ + 5 cm^3 H_2SO_4 in 1 l Wasser, und danach die
Platte getrocknet (keinen Spiritus zur Schnelltrocknung verwenden).
Die Rekonstruktion des Objekts erfolgt mit einer Richtleuchte in
Reflexion.

Belichtet man unterschiedliche Teile des Geschehens (z.B. Objekt und
Hintergrund) nacheinander auf verschiedene unterschiedlich expandier-
te Platten, so ist eine kontrasterhöhende mehrfarbige Rekonstruktion
möglich. Dies gelingt auch auf einer Platte, wenn man zwischen den
Belichtungen die oben geschilderte Filmschichtexpansion durchführt. -

* Gut geeignet sind blanke Münzen.

Für anregende Diskussionen danke ich Herrn Professor H.J. Gutjahr,
Fachhochschule Köln, Abtlg. Kohärente Optik.

Literatur: 1) D. MALLWITZ, R. LESSING:
"Entwicklung eines linearen Meßsystems"
Forschungsbericht T81-031 BMFT (1981)

Optimierung der automatisierten Auswertung von Specklegrammen beim Einsatz eines schnellen Fouriertransformators

H. KREITLOW, TH. KREIS und W. JÜPTNER

Bremer Institut für angewandte Strahltechnik, BIAS
Ermlandstr. 59, D 2820 Bremen 71

Einleitung

Die Specklefotografie ist eine kohärentoptische Methode zum berührungslosen Bestimmen von Verformungen im Mikrometerbereich an rauhen Oberflächen und wird bevorzugt zur zerstörungsfreien Dehnungs- und Spannungsanalyse eingesetzt. Die von einem aufgeweiteten Laserstrahl beleuchtete rauhe Oberfläche des Objekts wird in zwei Belichtungen bei verschiedenen Belastungszuständen über eine abbildende Optik in einer hochauflösenden Fotoplatte aufgenommen. Zur Auswertung wird die entwickelte und als Specklegramm bezeichnete Fotoplatte punktweise mit einem konvergenten Laserstrahl beleuchtet. Dadurch erscheint auf einer hinter dem Specklegramm stehenden Mattscheibe ein Beugungshalo, der mit Young'schen Streifen moduliert ist, deren Abstand und Winkellage mit der Verformung des dem durchstrahlten Fleck korrespondierenden Objektpunktes unmittelbar in Zusammenhang stehen: Der Betrag des Veränderungsvektors ist umgekehrt proportional zum Streifenabstand, die Richtung dieses Vektors ist orthogonal zu den Streifen. Zur Ermittlung des Dehnungs- und Spannungsfeldes muß eine hohe Anzahl dicht liegender Meßpunkte ausgewertet werden, was nur durch ein automatisiertes Auswerteverfahren wirtschaftlich möglich ist.

Halbautomatische Auswertesysteme, die auf Videotechnik basieren und vom Benutzer die Positionierung einer Auswertelinie orthogonal zu den Young'schen Streifen erfordern /1/, oder die Drehung des gesamten Musters, bis die Streifen parallel zu den Fernsehzeilen liegen /2/, sind vorgeschlagen und erprobt worden. Auch wenn die Bestimmung der Streifenabstände und Berechnung der Verschiebungen bei diesen Systemen automatisch geschieht, ist durch die notwendige interaktive Voreinstellung der zeitliche Aufwand für die vollständige Auswertung eines Specklegramms noch erheblich und die Genauigkeit auf einige Prozent bzw. Winkelgrade beschränkt. Neben dem Umstand, daß nicht die gesamte im Streifenmuster enthaltene Informationsfülle ausgenutzt wird, können bei diesen Systemen durch Linsenfehler /3/ oder durch die Verschiebung der Streifenmaxima aufgrund der variierenden Intensitätsverteilung im Beugungshalo /4/ fehlerhafte Ergebnisse erzeugt werden.

Zur Vermeidung dieser Nachteile und zur vollautomatisierten Auswertung von Specklegrammen wurde die digitale zweidimensionale Fouriertransformation des Young'schen Streifenmusters und die Berechnung des Verformungsvektors aus dem Amplitudenspektrum vorgeschlagen /1,5/.

Automatisierte Specklegrammauswertung mit Fouriertransformatoren

Das auf der Mattscheibe erzeugte Muster wird auf die Empfängerfläche einer Video- oder Matrixkamera abgebildet, über einen Analog-Digital-Wandler quantisiert und digitalisiert und die digitalen Werte im Random-Access-Memory des angeschlossenen Bildverarbeitungssystems gespeichert, Bild 1. Über festverdrahtete Hardware oder geeignete Software wird die zweidimensionale Fouriertransformation und daraus das Amplitudenspektrum erzeugt, Bild 2. Aus der Lage des ersten Nebenmaximums außerhalb des Gleichanteils werden Betrag und Winkel der Komponenten des Verschiebungsvektors parallel zur Specklegrammebene bei der Aufnahme berechnet. Diese Daten werden für sämtliche Punkte eines für die Dehnungsmessung geeigneten Auswerterasters bestimmt und zwischengespeichert. Durch numerische Differentiation nach den Ortskoordinaten läßt sich hieraus das Dehnungsfeld und daraus das Spannungsfeld ermitteln.

Auflösung im Specklegramm und im Streifenmuster

Für eine Spannungsanalyse wird ein dichtes Auswerteraster benötigt, da die Genauigkeit der berechneten Spannungen vom Abstand der zur numerischen Differentiation herangezogenen Auswertepunkte abhängt. Die räumliche Auflösung bei der Auswertung des Specklegramms ist durch den Durchmesser des Auswertelaserstrahls beschränkt. Je kleiner der Durchmesser des Strahls, um so besser ist zwar die Auflösung, um so größer sind jedoch die Speckles im Beugungsmuster. Da aber die Speckles als breitbandiges stochastisches Rauschen auftreten, führen sie im Spektrum nicht zu lokalen Maxima, die die Bestimmung der von den Young'schen Streifen erzeugten Nebenmaxima erschweren, Bild 3.

Da die Verschiebungsvektoren in allen Richtungen gleich empfindlich gemessen werden soll, empfiehlt sich eine Digitalisierung in ein quadratisches Bildpunktraster. Die Anzahl der Bildpunkte entlang einer Zeile im Bildfeld ist entsprechend dem Abtasttheorem /6/ nach unten begrenzt durch die doppelte Anzahl der erwarteten Young'schen Streifen innerhalb des Bildfeldes. Eine Erhöhung der Bildpunktanzahl vermindert den Einfluß des Specklerauschens, führt jedoch nicht zu einer genaueren Bestimmung des Maximums, da sich durch eine Erhöhung der Abtastrate zwar die noch erkennbaren Ortsfrequenzen erhöhen, die kleinen Ortsfrequenzen aber nicht genauer bestimmen lassen.

Einfluß des Beugungshalo auf das Spektrum

Durch die radial abfallende Intensität des Beugungshalo werden die Maxima der Young'schen Streifen verschoben, so daß Fehler bei der Bestimmung der Abstände der Streifenmaxima auftreten. Diese Verschiebung vergrößert sich mit abnehmender Sichtbarkeit der Streifen /4/. Dagegen werden die für die Auswertung benötigten Maxima im Amplitudenspektrum nicht verschoben: Die $\cos^2$-förmigen Young'schen Streifen erzeugen im Spektrum scharfe Impulsfunktionen, während der Halo einen verbreiterten Gleichanteil erzeugt. Der Multiplikation von Halo und Streifenmuster im Ortsraum entspricht eine Faltung der jeweiligen Transformierten im Ortsfrequenzraum. Diese Faltung der Transformierten von Halo und Streifenmuster führt zu einer Verbreiterung der Impulsfunktionen, nicht jedoch zur Verschiebung von deren Maxima, Bild 4. Die Beugungshalo hat sogar den positiven Effekt, daß sie bei der Fouriertransformation wie eine Fensterfunktion wirkt und Frequenzanteile unterdrückt, die durch ein Auswertefeld entstehen, das nicht an die Perioden der Young'schen Streifen angepaßt ist, (aliasing).

Redundanz der zweidimensionalen Fouriertransformation

Bei der zweidimensionalen Fouriertransformation werden zuerst alle Zeilen des Bildes transformiert, sodann dieses Ergebnis entlang aller Spalten transformiert. In der Software für die Transformation kann neben der Verwendung des FFT-Algorithmus, der die Anzahl der notwendigen komplexen Multiplikationen drastisch reduziert /6/, die Anzahl der notwendigen Rechenschritte durch Ausnutzung der vorhandenen Redundanz weiter verringert werden: Da das Bild als reelle Matrix vorliegt, lassen sich jeweils zwei Zeilen in einem Schritt transformieren, eine als Real-, die andere als Imaginärteil. Das Ergebnis wird in den hermiteschen und antihermiteschen Teil aufgespalten. Da das zweidimensionale Amplitudenspektrum punktsymmetrisch zum Ursprung ist, braucht nur die rechte oder linke Hälfte der spaltenweisen Transformationen durchgeführt werden und das Nebenmaximum braucht nur in dieser Hälfte des Spektrums gesucht werden.

Bestimmung des Ortes des Nebenmaximums

Zur Bestimmung der Koordinaten des Nebenmaximums können einfache Such-
algorithmen die Ortsfrequenz $(u,v) \neq (0,0)$, für die die Amplitude $S(u,v)$
maximal wird, ermitteln. Dieses Vorgehen liefert nur eine geringe Auflösung, die für eine Ermittlung genauer Dehnungswerte durch numerische
Differentiation im allgemeinen nicht ausreicht. Eine genauere Bestimmung erhält man, indem der Schwerpunkt einer Umgebung um das gefundene
Maximum gesucht wird, d.h. man bildet die gewichtete Summe aller Ortsfrequenzen, wobei die Gewichte die Intensitäten des Spektrums sind.
In vielen Fällen ist auch diese Genauigkeit noch nicht ausreichend.
Eine Verbesserung bietet die Erweiterung des in /4/ für eine Dimension
vorgeschlagenen Verfahrens auf zwei Dimensionen. Hierbei wird wieder
das Maximum im Spektrum gesucht, sodann in einer Umgebung des Maximums
im kontinuierlichen Spektrum des diskreten Bildes mit feinerer Unter-
teilung eine genauere Bestimmung der Lage des Maximums vorgenommen.

Berechnung des Verschiebungsvektors

Die Auswertung eines Punktes eines einzigen Specklegramms erlaubt die
Bestimmung der Komponenten des Verschiebungsvektors parallel zum
Specklegramm bei der Aufnahme. Wird in dieser Ebene ein x-y-Koordina-
tensystem definiert, so ist der Verschiebungsvektor

$$(d_x, d_y) = \frac{L \lambda}{M} (u,v)$$

Hierbei sind (u,v) die aus dem Spektrum ermittelte Ortsfrequenz des zu
den Young'schen Streifen gehörigen Nebenmaximums, L der Abstand des
Specklegramms von der Mattscheibe, λ die Wellenlänge des Auswertela-
sers und M der zusammengesetzte Abbildungsmaßstab bei Aufnahme des
Specklegramms sowie bei der Aufnahme des Streifenmusters.

Berechnung des 3D-Verformungsfeldes einer Oberfläche

Sollen nicht nur die Bewegungen innerhalb einer Ebene der Objektoberfläche bestimmt werden, so müssen mindestens zwei Specklegramme aus
unterschiedlichen Richtungen gleichzeitig aufgenommen werden. Da aus
jedem Specklegramm zwei Komponenten einer Projektion des Verschiebungsvektors gewonnen werden, erhält man die drei Komponenten des gesuchten Verschiebungsvektors aus einem überbestimmten Gleichungssystem
mit Hilfe der Methode der kleinsten Fehlerquadrate /7/:

Der zu bestimmende Veränderungsvektor im Objektpunkt P sei $\vec{d}(P) =$
(d_x, d_y, d_z). Jede Beobachtungsrichtung sei durch den Einheitsvektor

$\vec{b}_i = (b_{xi}, b_{yi}, b_{zi})$ charakterisiert. Aus dem aus der m-ten Beobachtungsrichtung aufgenommenen Specklegramm erhält man die m-te Projektion $\vec{d}_m$ des Veränderungsvektors, die sich schreiben läßt als

$$\vec{d}_m = P_m \, \vec{d}$$

mit

$$P_m = - \underset{\sim}{b}_m \, \underset{\sim}{b}_m$$

und

$$\underset{\sim}{b}_m = \begin{pmatrix} 0 & -b_{zm} & b_{ym} \\ b_{zm} & 0 & b_{xm} \\ -b_{ym} & b_{xm} & 0 \end{pmatrix}$$

Mit den $r \geqslant 2$ Projektionen $\vec{d}_m$ läßt sich dann $\vec{d}$ berechnen nach

$$\vec{d} = \left(\sum_{m=1}^{r} P_m \right)^{-1} \left(\sum_{m-1}^{r} \vec{d}_m \right)$$

Aus dem Veränderungsvektorfeld $\vec{d}(P)$ für die Punkte P der Objektoberfläche lassen sich die Bewegungen des starren Körpers, die Translationen und Rotationen, oder aus den partiellen Ableitungen in den verschiedenen Raumrichtungen die Dehnungen und Rotationen bestimmen. Dabei muß berücksichtigt werden, daß keine Ableitungen in Normalenrichtung zur Oberfläche möglich sind. Dies ist keine grundsätzliche Einschränkung, da es an freien Oberflächen keine Kraftkomponenten in Normalenrichtung gibt/8/. Die Matrix der partiellen Ableitungen läßt sich dann in ihren symmetrischen Teil, die Dehnungskomponenten, und ihren schiefsymmetrischen Teil, die Rotationskomponenten, zerlegen /7/.

Zusammenfassung

Für Systeme zur vollautomatisierten Auswertung von Specklegrammen mit Hilfe der zweidimensionalen Fouriertransformation wurde der Einfluß verschiedener Parameter wie Wahl der Auflösung im Specklegramm und im Young'schen Streifenmuster, Specklerauschen, sowie Beugungshalo untersucht. Weiter wurden Möglichkeiten zur Erhöhung der Auswertegeschwindigkeit durch schnelle Rechenverfahren aufgezeigt. Darüber hinaus wurden Verfahren zur Ermittlung dreidimensionaler Verformungsfelder sowie von Dehnungen und Spannungen hergeleitet.

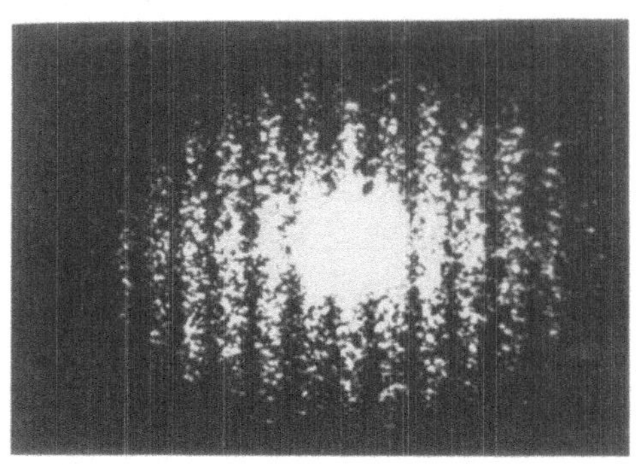

Bild 1. Young'sche Streifen

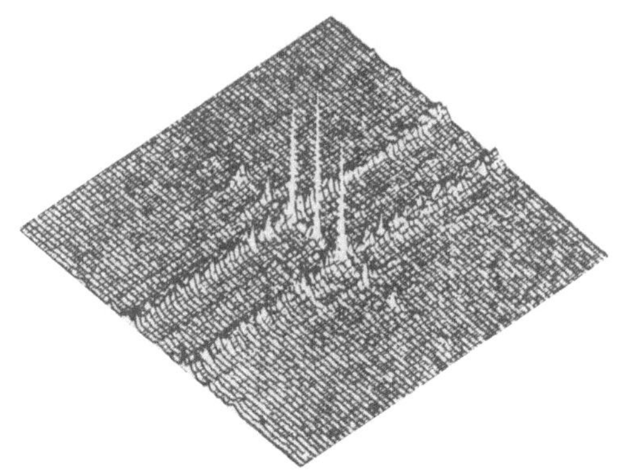

Bild 2. Amplitudenspektrum

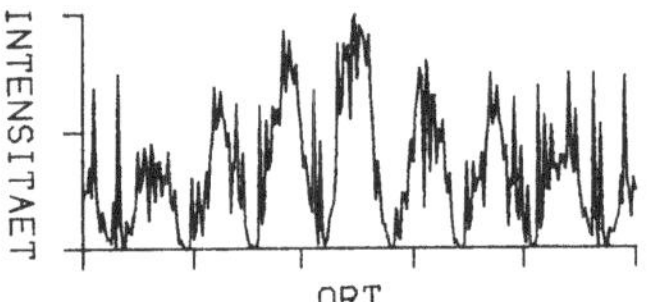

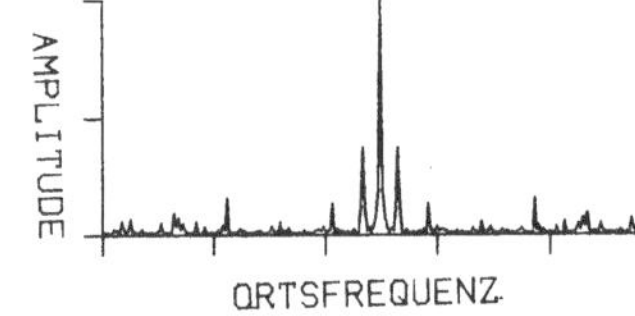

Bild 3. Einfluß des Specklerauschens auf das Spektrum

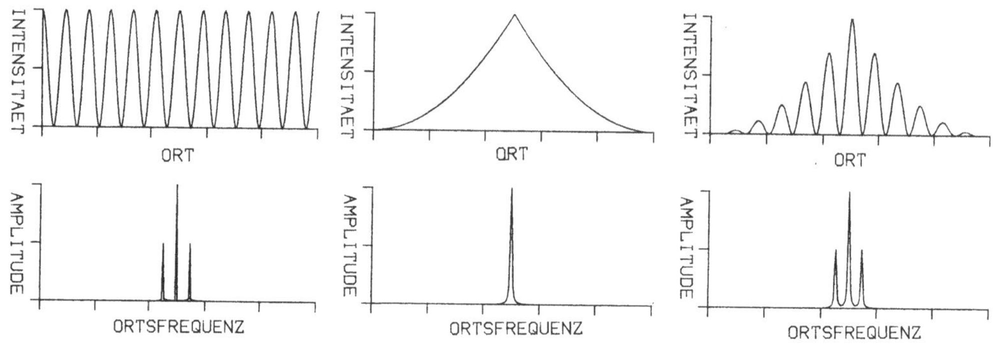

Bild 4. Einfluß des Beugungshalo auf das Spektrum

Literatur

/1/ KREITLOW,H. und TH.KREIS: SPIE-Proc. 210 (1979) 18

/2/ INEICHEN,B., P.EGLIN und R.DÄNDLIKER: Appl. Optics 19 (1980) 2191

/3/ ARCHBOLD,E., A.E.ENNOS und M.S.VIRDEE: SPIE-Proc. 136 (1977) 258

/4/ KAUFMANN, G.H.: Appl. Optics 20 (1981) 4277

/5/ BRUHN,H. und A.FELSKE: VDI-Ber. 399 (1981) 13

/6/ STEARNS,S.: Digital Signal Analysis, Hayden Book Co. (1975)

/7/ PRYPUTNIEWICZ,R.: Appl. Optics 17 (1978) 3613

/8/ DÄNDLIKER,R., B.INEICHEN und F.M.MOTTIER: in Eng. Uses of Coherent
    Optics, Cambridge Univ. Press (1976) 99

Ein Speckle-Verfahren zur Schwingungsanalyse

B. RUTH, D. HAINA und W. WAIDELICH
Gesellschaft für Strahlen- und Umweltforschung mbH
Abteilung für Angewandte Optik
Ingolstädter Landstr. 1, D-8042 Neuherberg

Einleitung

Die bisherigen optischen Methoden zur berührungslosen Schwingungsanalyse, wie das Holographie-Verfahren oder die Heterodyn-Technik, eignen sich nur für kleine Amplituden /1,2/. Messungen mit der Speckle-
Methode dagegen erfassen Amplituden bis zu 0.2 mm /3,4/. Ein Nachteil
der bekannten Verfahren ist, daß Amplitude und Frequenz gleichzeitig
nur mithilfe einer Referenz bestimmt werden können.

Methode

Das von einer Oberfläche gestreute Laserlicht bildet eine granuläre
Struktur, die Speckles. Eine Objektbewegung führt bei geeigneter optischer Anordnung zu einer Bewegung des Speckle-Musters. Bei dem Aufbau in Fig. 1 bildet die Linse den Strahlfleck des Lasers mit der
Vergrößerung M in die Ebene des Pinholes S ab. In diesem Fall bestimmen Objektabstand und Blendendurchmesser die mittlere Speckle-Größe.

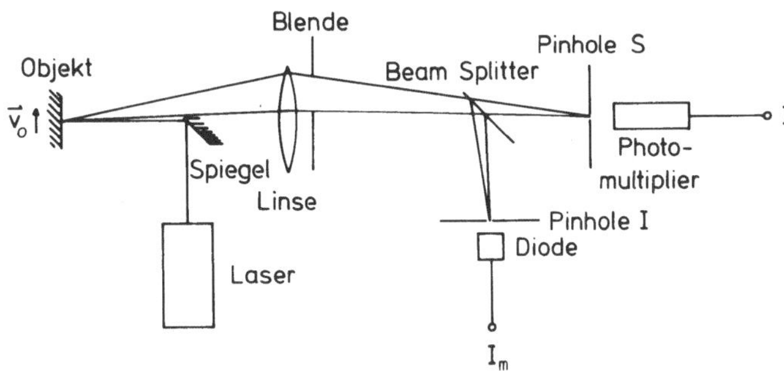

Fig. 1. Optischer Aufbau zum Messen der Speckle-Bewegung mit einer
Abbildung.

Mit der Voraussetzung, daß die optische Achse des Abbildungssystems
senkrecht auf der Objektoberfläche steht und das Objekt nur eine

transversale Geschwindigkeit hat, gilt für die Speckle- und Objektgeschwindigkeit $\vec{v}_s$ bzw. $\vec{v}_0$

$$\vec{v}_s = - M \vec{v}_0 \tag{1}$$

unabhängig von der Richtung von $\vec{v}_0$ /5,6/. Der Laserstrahl wird in
die optische Achse eingespiegelt, damit der Strahlfleck zusammen mit
seinem Bild auch bei einer geringfügigen Variation der Objektebene
auf der optischen Achse bleibt. Der Vorteil dieser Anordnung besteht
also darin, daß die einfache Beziehung (1) unabhängig von der Richtung der Objekt-Geschwindigkeit gilt. Nachteilig dagegen ist die Notwendigkeit, das Objekt und alle Komponenten zu justieren.

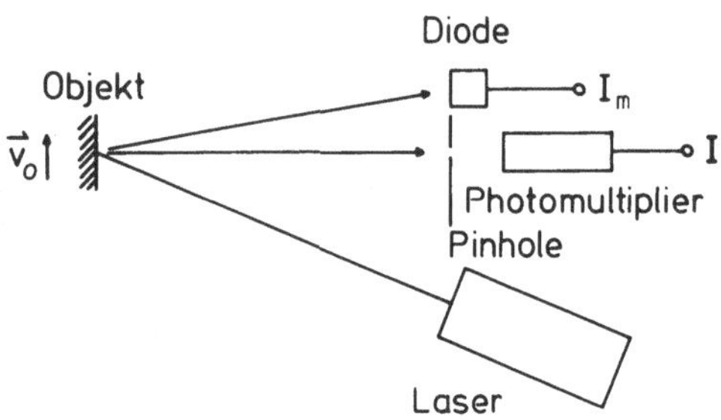

Fig. 2. Optischer Aufbau zum direkten Messen der Speckle-Bewegung.

Bei dem einfacheren Aufbau in Fig. 2 läßt sich der Objektabstand in
einem größeren Bereich variieren, ohne die Speckle-Bewegung wesent-
lich zu beeinflussen. Denn die Speckle-Größe ist in einem gewissen
Bereich unabhängig vom Objektabstand, da sie vom Verhältnis zwischen
Objektabstand und Strahlfleckgröße bestimmt ist und ein zunehmender
Abstand durch einen gleichzeitig vergrößerten Strahlfleckdurchmesser
kompensiert wird. Daher ist bei diesem Verfahren eine zusätzliche
Geschwindigkeitskomponente in der Objekt-Normalen nicht störend.
Falls die Beobachtung senkrecht zu der Objektoberfläche erfolgt, gilt

$$\vec{v}_s = C \; \vec{v}_0 \tag{2},$$

wobei C ein Tensor ist und im wesentlichen von dem Verhältnis der
Abstände zwischen Laser und Objekt bzw. Detektor und Objekt abhängt.
Die Richtung der tangentialen Objektgeschwindigkeit bezüglich der
Laser-Einstrahlung geht ebenfalls in C ein, eine Vernachlässigung
dieser Abhängigkeit führt aber nur zu einem Fehler von maximal 1 %,
wenn der Winkel zwischen Einstrahlrichtung und Beobachtungsrichtung
kleiner als 20° ist /7/.

In beiden Verfahren führt also die Objekt-Bewegung zu einer Bewegung des gesamten Speckle-Musters. Zur Bestimmung der Speckle-Geschwindigkeit innerhalb kleiner Zeitintervalle wird nun die Speckle-Intensität I mit einem Photomultiplier hinter einem Pinhole registriert. Eine elektronische Schaltung, die in /7/ beschrieben ist, zählt die Schnittpunkte von I mit einer Schwelle, die proportional zur gleichzeitig gemessenen mittleren Intensität ist. Dazu dienen die Dioden in Fig. 1 und 2. Die Zählrate ist proportional zum Mittelwert des Geschwindigkeitsbetrages und liegt während des folgenden Zeitintervalls in digitaler und analoger Form vor. Eine zeitabhängige Geschwindigkeit führt daher zu einer Treppenfunktion.

Das Signal der Meßanordnung in Fig. 3 kann nun mit einem Oszillographen beobachtet oder mit einem Transientenrecorder gespeichert werden.Mit einer entsprechenden Triggerung ist es damit auch möglich, nicht-periodische oder auch einmalige Vorgänge zu registrieren. Für Untersuchungen von periodischen Bewegungen eignet sich der angeschlossene Frequenzanalysator. Dabei ist die Frequenz trotz der Treppenfunktion direkt feststellbar und die Amplitude wird entsprechend der Meßintervall-Länge und der betrachteten Frequenz modifiziert /7/ Bei kleinen Amplituden ist ein Mittelungsprozess durch die Frequenzanalyse notwendig, da sonst die Amplitude aufgrund des statistischen Charakters der Speckles einem zu großen Fehler unterliegt.

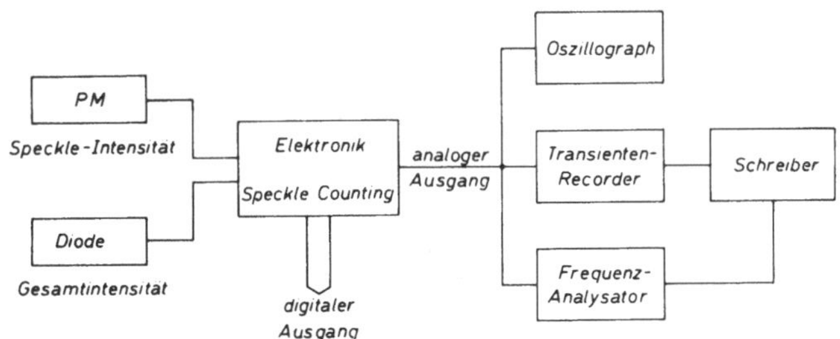

Fig. 3. Meßanordnung zur Schwingungsanalyse.

Ergebnisse

Für Beispielsmessungen wird die optische Anordnung in Fig. 2 benutzt. Die Leistung des HeNe-Laser beträgt 9 mW und der Strahl, der im Winkel von 5° zur Beobachtungsrichtung auf das Objekt fällt, hat einen Durchmesser von 1 mm. Die mittlere Speckle-Größe ist 100 µm, so daß der Kontrast bei einer Blende vom Durchmesser 50 µm gut erhalten

bleibt. Zur Untersuchung von niedrigen Frequenzen wird ein Zeitintervall von 10 ms gewählt. In Fig. 4 ist das Spektrum des analogen Aus-
gangssignals bei einer eindimensionalen Schwingung eines Testobjek-
tes dargestellt.Da die Anlage den Betrag der Geschwindigkeit mißt,
erscheint der Hauptbeitrag im Spektrum bei der doppelten Frequenz.
Die Peaks bei Vielfachen der einfachen und doppelten Frequenz zeigen, daß es sich um keine reine Sinusschwingung handelt. Die ermittelte Frequenz stimmt genau mit der des Testobjekts überein. Eine
phasengerechte Addition aller Amplitudenbeiträge führt bis auf einige Prozent zu der Schwingungs-Amplitude des Test-Objekts.

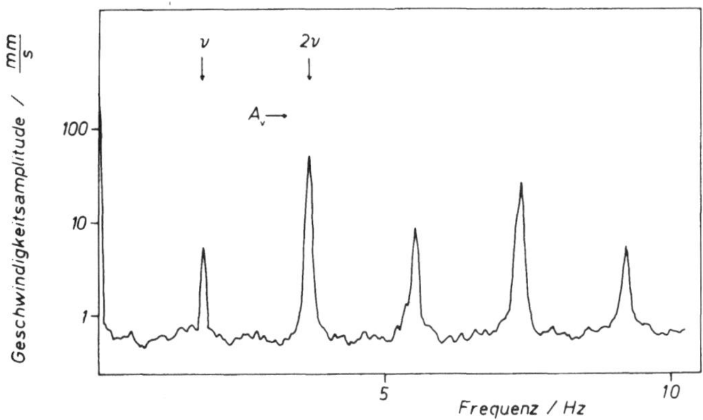

Fig. 4. Spektrum einer Schwingung. ν = 1.84 Hz; A = 12.5 mm
A_r: Errechnete Amplitude bei einer reinen Sinusschwingung.

Vibrationen von Objekten sind im allgemeinen keine eindimensionale
Bewegungen. In einem weiteren Beispiel führt daher das Test-Objekt
eine überlagerte Schwingung in x- und y-Richtung mit beliebigen Frequenzen aus. Das Spektrum ist in Fig. 5 dargestellt.

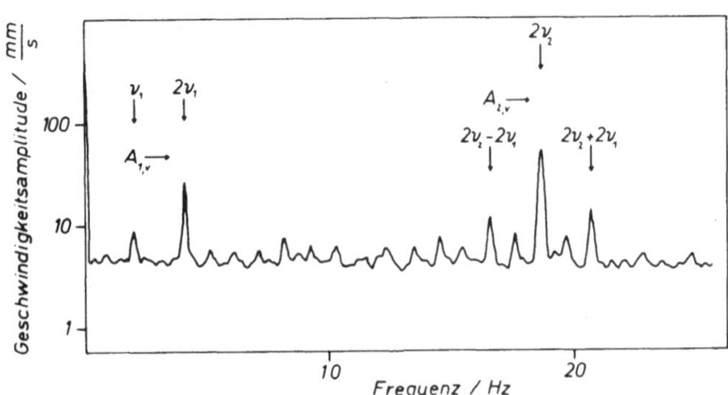

Fig. 5. Spektrum von zwei überlagerten Schwingungen.
ν_1 = 2.0 Hz; A_1 = 12.5 mm; ν_2 = 9.4 Hz; A_2 = 5.0 mm

Beide Frequenzen lassen sich eindeutig und in Übereinstimmung mit
den vorgegebenen Objektfrequenzen anhand der größten Peaks ermitteln. Daneben treten auch kleinere Peaks bei Differenz-Frequenzen
auf, die auf Schwebungen zurückzuführen sind.

Diskussion

Das Verfahren ermöglicht die Analyse von beliebigen Objekt-Schwin-
gungen mit Amplituden im mm- und cm-Bereich. Aufgrund der optischen
Anordnung führt die transversale Objektbewegung zu einer Speckle-Bewegung, deren Geschwindigkeitsbetrag durch die Detektoren und die
Auswerte-Elektronik ermittelt wird. Überlagerte Schwingungen lassen
sich anhand einer Frequenzanalyse idendifizieren, falls die beteiligten Frequenzen nicht zufällig Vielfache voneinander sind. Dann treten Spezialfälle auf, wie z.B. bei der kreisförmigen Schwingung, die
auf diese Weise nicht analysiert werden kann.
Die möglichen Amplituden sind nach unten auf etwa 100 μm begrenzt,
da die Speckle-Größe aufgrund der verfügbaren Blenden nicht unter
etwa 100 μm sinken darf. Bei höheren Frequenzen steigt das Schrot-
Rauschen des Photomultipliers, das sich aber wieder durch den Einsatz
stärkerer Laser reduzieren läßt. Für große Amplituden und kleine Frequenzen gibt es keine Einschränkungen.
Mögliche Anwendungen liegen bei Berücksichtigung des erfassbaren Am-
plituden- und Frequenz-Bereichs bei der Schwingungsanalyse von Mo-
toren.

Literatur
/1/ Powell, R.L., and Stetson, K.A., 1965, J.opt.Soc.Am., 55, 1593.
/2/ Ueha, S., Shiota, K., Okada, T., Tsujiuchi, J., Sakamoto, Y.,
    and Mori, E., 1975, Jap. J. appl. Phys., 14, Suppl. 14-1, 335.
/3/ Tiziani, H.J., 1971, Optica Acta, 18, 891.
/4/ Tiziani, H.J., and Klenk, J., 1981, Appl. Opt., 20, 1467.
/5/ Yamaguchi, I., Komatsu, S., and Saito, H., 1975, Jap. J. appl.
    Phys. 14, Suppl. 14-1, 301.
/6/ Yamaguchi, I., and Komatsu, S., 1977, Optica Acta, 24, 705.
/7/ Ruth, B., Haina, D., and Waidelich, W., Optica Acta (im Druck).

Erfassung von Deformationen an einem untertägigen Salzpfeiler mit Hilfe kohärentoptischer Meßverfahren

M.-A. Beeck, B. Garbe und W. Geier', U. Hunsche und I.Plischke"
'Institut für Meßtechnik im Maschinenbau, Universität Hannover,
 Nienburger Str. 17
"Bundesanstalt für Geowissenschaften und Rohstoffe, Stilleweg 2,
 D-3000 Hannover

1. Einleitung

Die Kenntnis des Kriechverhaltens von Salzgestein ist für eine Stand-
sicherheitsanalyse untertägiger Hohlräume notwendig. In den letzten
Jahren wurde, gestützt auf Laborversuche, ein Gesetz /1/ formuliert,
daß dieses Verhalten beschreibt. Ziel der derzeitigen Untersuchungen
ist die Überprüfung, ob dieses Kriechgesetz auf große Gesteinsvolumen
übertragbar ist. Für vertretbare Versuchszeiten (Tage bis Monate)
müssen aufgrund der kleinen Kriechrate (µm/Tag) hochauflösende Meßtechniken eingesetzt werden. Dazu benutzt man bisher mechanische Meßuhren, induktive Wegaufnehmer und Extensometer. Das vom Kriechvorgang
bewirkte dreidimensionale Verschiebungsvektorfeld der Oberflächenver-
formung kann so nur in einer Komponente integral zwischen zwei ausge-
zeichneten Punkten des Versuchsobjektes erfaßt werden.

In dieser Arbeit wird die Einsatzmöglichkeit optischer Meßmethoden,
die flächenhaft und zeitgleich arbeiten, zur Verschiebungsmessung
untersucht. Bei den Meßobjekten handelt es sich einerseits um eine
Salzwürfelprobe, deren zeitliches Verformungsverhalten im Labor in
einem Druckversuch untersucht wurde, andererseits um einen Versuch,
der an einem Salzpfeiler unter Tage durchgeführt wurde. Das dreidimensionale Verschiebungsfeld setzt sich aus einer in-plane-Verschiebung in der x-y-Ebene, zu messen durch Specklefotografie /2/, und einer
out-of-plane-Verschiebung in z-Richtung, zu messen durch holografische
Interferometrie /3/, zusammen (siehe Bild 1).

Die Ergebnisse der optischen und der herkömmlichen Meßmethoden
werden verglichen und die Vorteile der optischen Methoden beschrieben.

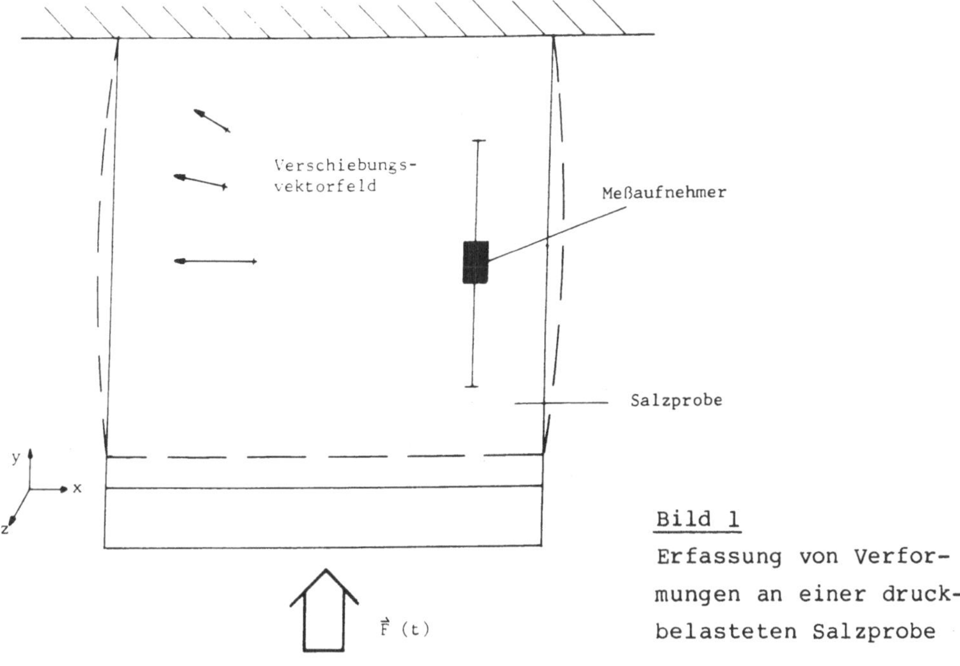

Bild 1
Erfassung von Verfor-
mungen an einer druck-
belasteten Salzprobe

2. Specklefotografie an druckbelasteten Salzwürfelproben

Ein Probenwürfel aus Salzgestein mit 75 mm Kantenlänge wurde mit einer
definierten einaxialen Druckspannung belastet und die dadurch bewirkte
zeitliche Verformung mittels Specklefotografie zeitseriell gemessen
(Bild 2).

Bild 2
Versuchsaufbau für
die Druckbelastung
der Salzprobe

172

Die Belastungseinrichtung bestand aus drei Edelstahlsäulen von 2200 mm
Länge und 50 mm Durchmesser. Sie bildeten ein gleichseitiges Dreieck,
und eine massive Stahlplatte am oberen Ende wirkte als Widerlager. Im
Schwerpunkt des Dreiecks befand sich ein hydraulischer Stempel, der
von unten die definierte Druckkraft erzeugte. Eine zweite Stahlplatte
(Grundplatte) zwischen Stempel und Probe wurde an den Säulen geführt.
Die Wegdifferenz zwischen Grundplatte und Widerlager wurde durch
Meßuhren mit 10 µm Auflösung registriert.

Der optische Aufbau war auf einer verwindungssteifen Platte montiert,
die an zwei Säulen der benutzten und einer Säule einer benachbarten
Belastungseinrichtung befestigt war. Die optische Zielachse der Meß-
kamera lag in Richtung des mittigen Normalenvektors der zu vermessen-
den Würfelfläche. Als Lichtquelle wurde ein 15mW He-Ne-Laser benutzt.
Die numerische Apertur des Kameraobjektivs und der Abbildungsmaßstab
bestimmen die minimal meßbare Verschiebung zu 7 µm, die Auflösung ist
nach Überschreiten dieses Wertes besser.

Da die große Eindringtiefe des Lichtes in das Salzgestein die Kohärenz
des Lichtes und damit den Specklekontrast verminderte, wurde die Probe
mit weißer Farbe präpariert. Es wurde die Verformung in den ersten 24
Stunden nach einer Spannungserhöhung von 14 auf 18 MPa zeitseriell
registriert. Der zeitliche Ablauf und die Zeitdifferenzen zwischen den
Doppelbelichtungen sind in Bild 3 grafisch dargestellt.

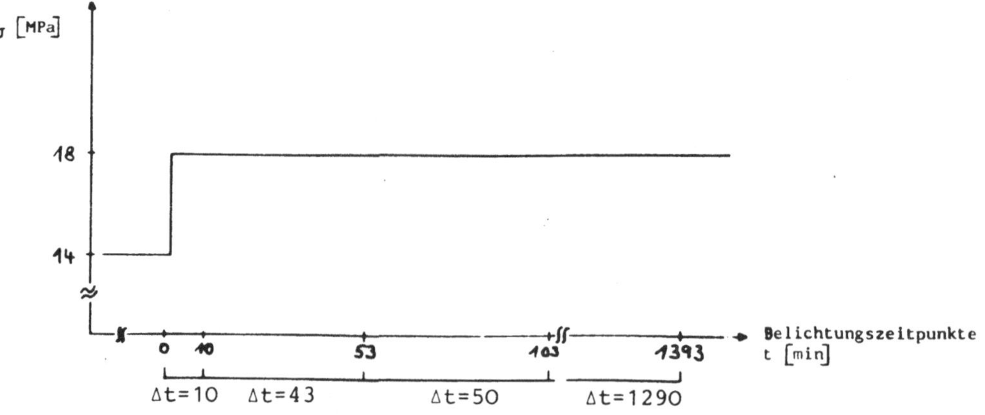

Bild 3   Δt /min/ : Zeit zwischen den Doppelbelichtungen

Zeitlicher Ablauf der Verschiebungsmessungen am Kriechprüfstand

Beim Auslesen der Specklegramme ergab jeder Punkt der Oberfläche ein
gut moduliertes Young'sches Streifensystem, so daß eine flächenhafte
Auswertung vorgenommen werden konnte. Zur Überprüfung der Korrelation
der herkömmlichen mit der optischen Meßmethode wurden korrespondierende
Meßwerte verglichen. Dabei waren die mit den Meßuhren aufgenommenen
Werte mit den Verschiebungsdifferenzen zwischen Grundplatte und Widerlager zu vergleichen. Diese Ergebnisse sind in Tabelle 1 gegenübergestellt.

	Meßuhren	Speckle-fotografie	Verformungsgeschw. aus Meßuhren
Δt = 10 min	126 µm	79 µm	12,6 µm/min
Δt = 43 min	38 µm	7 µm	0,9 µm/min
Δt = 50 min	26 µm	24 µm	0,5 µm/min
Δt = 1290 min	66 µm	64 µm	0,05 µm/min

Tabelle 1
Vergleich der mit zwei Verfahren gemessenen vertikalen Verformungen
eines druckbelasteten Salzwürfels

Zusätzlich wurden die mittleren Verformungsgeschwindigkeiten für jede
Messung berechnet. Die offensichtlich starken Abweichungen der ersten
beiden Messungen lassen sich aus der Versuchsdurchführung erklären. Vor
und nach dem Ablesen der Meßuhren mußte das Filmmaterial gewechselt und
belichtet werden. Dadurch wurde das Zeitintervall, über das tatsächlich
gemessen wurde, bei den Specklemessungen verkürzt. Die großen Verformungsgeschwindigkeiten bei den ersten beiden Messungen verstärkten den
Effekt. Bei der vierten Messung über 21,5 Stunden waren diese beschriebenen Zeitdifferenzen vernachlässigbar. Beide Meßverfahren liefern hier
übereinstimmende Meßwerte.
In Bild 4 ist die aus den Specklegrammen bestimmte in-plane-Verschiebung der Würfelproben flächenhaft dargestellt. Die in den Auswertpunkten
angreifenden Vektorpfeile geben in Betrag und Richtung die gemessenen
Oberflächenverschiebungen wieder und können zur Bestimmung der Deformationen herangezogen werden. Im zweiten Zeitbereich (Δt = 43 min) ist
eine besonders starke Verformung in x-Richtung festzustellen, die durch
die herkömmlichen Meßverfahren nicht erfaßt werden konnte.

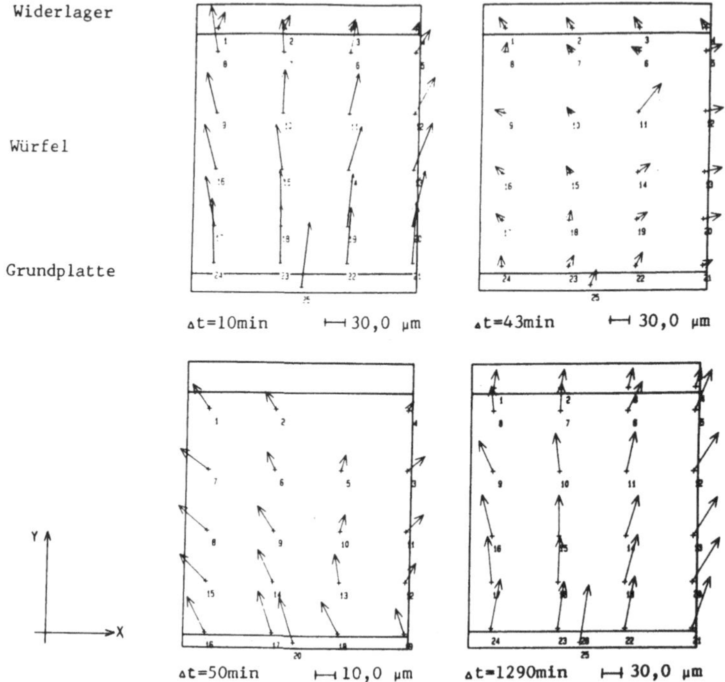

Bild 4

Zeitserielle Darstellung der in-plane-Verschiebung einer
druckbelasteten Salzprobe

Der aufgrund der flächenhaften Meßwerterfassung bei zumindest gleich-
bleibender Meßgenauigkeit höhere Informationsgehalt dieser Messungen
ermöglicht eine weitgehende Überprüfbarkeit von Rechenmodellen zur
Kriechverformung durch das Experiment.

3. Kohärent-optische Verformungsmessungen an einem untertägigen Salzpfeiler

Der Versuchsort befindet sich auf der 490 m-Sohle des Salzbergwerks
ASSE II. Der Versuchspfeiler ist an Firste und Sohle mit dem Gebirge
verbunden und hat eine quadratische Grundfläche von 1,5 m Kantenlänge
und eine Höhe von 3 m (Bild 5).

Bild 5
Salzpfeiler mit Meßfeld

Um definierte Spannungen zu erzeugen, wurde der Pfeiler in der Mitte
waagerecht geschlitzt und ein Stahlkissen eingebracht. Einzelheiten
des Versuchsaufbaus sind bei /4/ dargestellt. Mit dieser Versuchsanordnung wurden Kriech- und Lastwechselversuche durchgeführt.

Ziel dieser Untersuchungen soll es sein, festzustellen, inwieweit das
aus Laborversuchen experimentell ermitelte Kriechgesetz auf große
Gesteinsbereich übertragbar ist, oder ob ein Maßstabseffekt existiert,
wie er aus Festigkeitsuntersuchungen an manchen Gesteinen bekannt ist.
Die Lastwechselversuche dienten darüber hinaus der Bestimmung der
elastischen Eigenschaften des Salzpfeilers.

Um die zur konventionellen Messung von Verformungen erforderlichen
induktiven Wegaufnehmer und Meßuhren an den beiden Blockhälften be-
festigen zu können, wurden auf jeder der vier Blockseiten 18 Meßbolzen
verankert. Ferner wurden auf allen vier Seiten Extensometer bis zu
einer Teufe von 9 m in das Gebirge hinein installiert (s. Abb. 9).

Als Vorbereitung für die optischen Messungen wurde der Pfeiler mit
retroreflektierender Farbe gestrichen, um die Belichtungszeit zu
verkürzen. Als Lichtquelle wurde ein Argon-Ionen-Laser mit 1 Watt Ausgangsleistung benutzt, so daß eine Fläche von 1,5 m² erfaßt werden
konnte. Zur Kühlung des Lasers diente ein geschlossener Wasserkreis-

lauf mit einem Ausgleichsbehälter von 800 l, so daß ein kontinuierlicher Betrieb über 12 Stunden möglich war. Der optische Aufbau befand sich auf einem verwindungssteifen Dreibein (s. Bild 6), das sich in 4 m Abstand vom Pfeiler befand. Aufgrund des stark verkleinerten Abbildungsmaßstabs verringerte sich die Meßempfindlichkeit gegenüber den Laborversuchen auf 40 µm.

In der Pfeilermitte wurden zusätzlich zwei Dehnungsmeßstreifen horizontal und vertikal geklebt, um eine dritte Meßmethode für den Vergleich der Meßwerte zu erhalten.

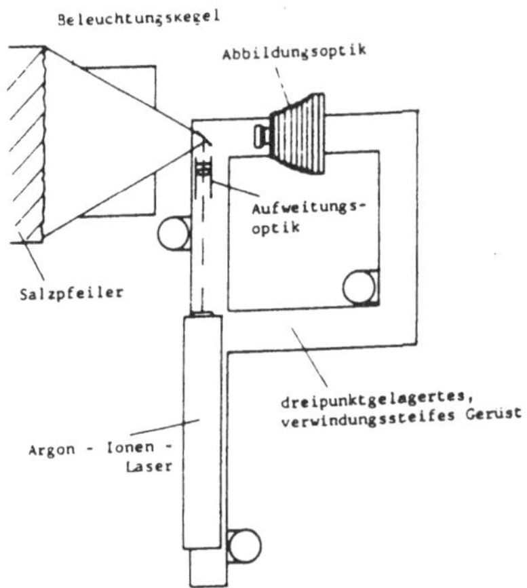

Bild 6
Prinzipielle Versuchsanordnung für die untertägige Specklefotografie

Die optischen Verformungsmessungen wurden während der Lastwechselversuche durchgeführt, bei denen die vertikale Spannung am Versuchspfeiler mit Hilfe des Druckkissens stufenweise zwischen 9,5 und 3 MPa variiert wurde. Während die Specklefotografie wie im Labor gute Ergebnisse lieferte, konnte die out-of-plane Verschiebung mittels holografischer Interferometrie nicht gemessen werden. Durchgeführte Frequenzanalysen von Körper- und Luftschallmessungen am Pfeiler zeigten, daß Schwingungsamplituden in der Größenordnung von λ auftraten und somit die Aufzeichnung von Hologrammen verhinderten. Analysen des Luft- und Körperschalls auf der 800 m-Sohle ergaben wesentlich geringere Schwingungsamplituden, so daß dort ein erfolgreicher Einsatz der holografischen Interferometrie zu erwarten ist.

Die Specklegramme ließen sich wieder flächenhaft auswerten. Phasenschwankungen in der oben angegebenen Größenordnung bewirken nur geringe

Schwingungen des Speckleschwerpunktes und beeinflussen nicht die
Gesamtmessungen. Bild 7 zeigt in vektorieller Darstellung die
Verschiebungen in Abhängigkeit von den Laststufen. An der Messung
Δσ = − 4 MPa wird nun die Aussagekraft der Methode demonstriert.

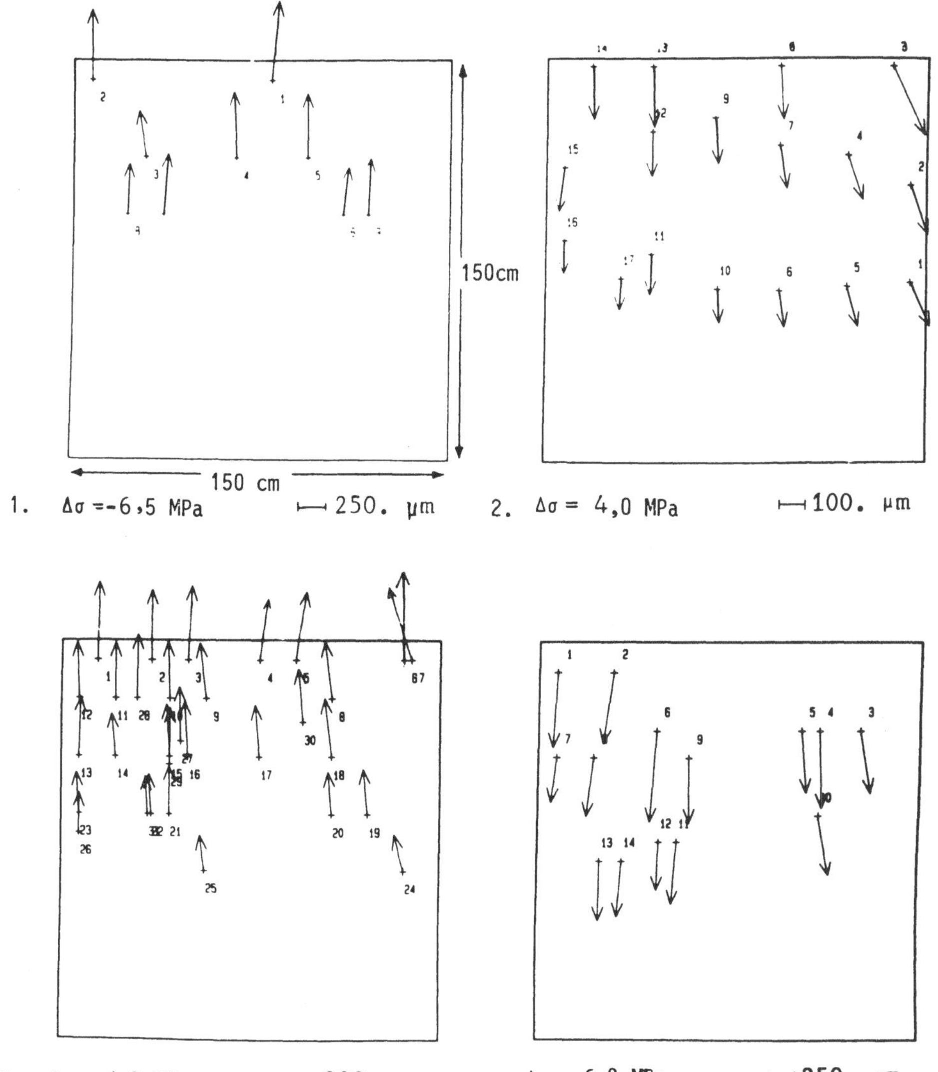

Bild 7
Zeitserielle Darstellung der in-plane-Verschiebung des Meßfeldes auf
den untertägigen Salzpfeiler bei Lastwechselversuchen. Die Belastung
wurde in der folgenden Reihenfolge variiert:

9,5 − 3,0 − (5,0) − 3,0 − 7,0 − 3,0 − 9,0 MPa

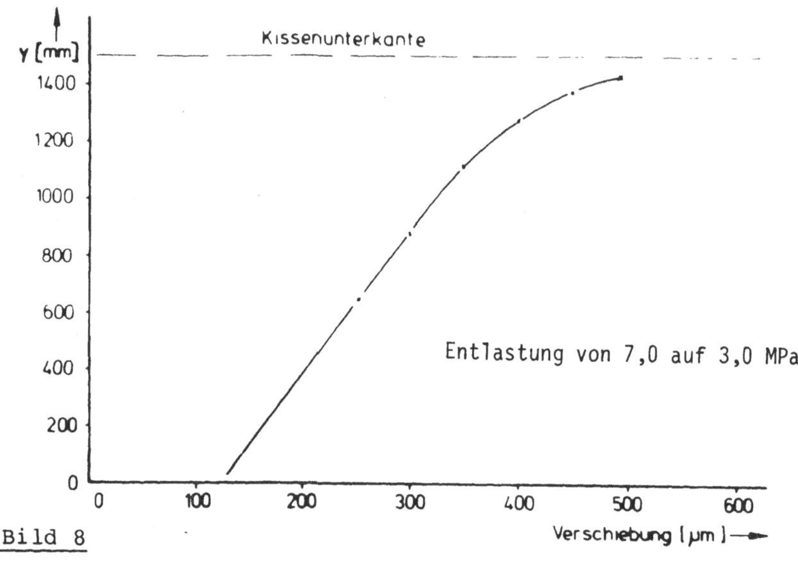

__Bild 8__

Die über die horizontale Koordinate gemittelten vertikalen Verschiebungen des Pfeilers bei einer Lasterniedrigung, bestimmt aus dem Specklegramm

In Bild 8 ist die aus der Messung erhaltene Verschiebung als Funktion
der Höhenkoordinate y des Pfeilers dargestellt. Die Werte aus dieser
Kurve wurden in Tabelle 2 mit denen der in Bild 9 gekennzeichneten
Meßstrecken verglichen.

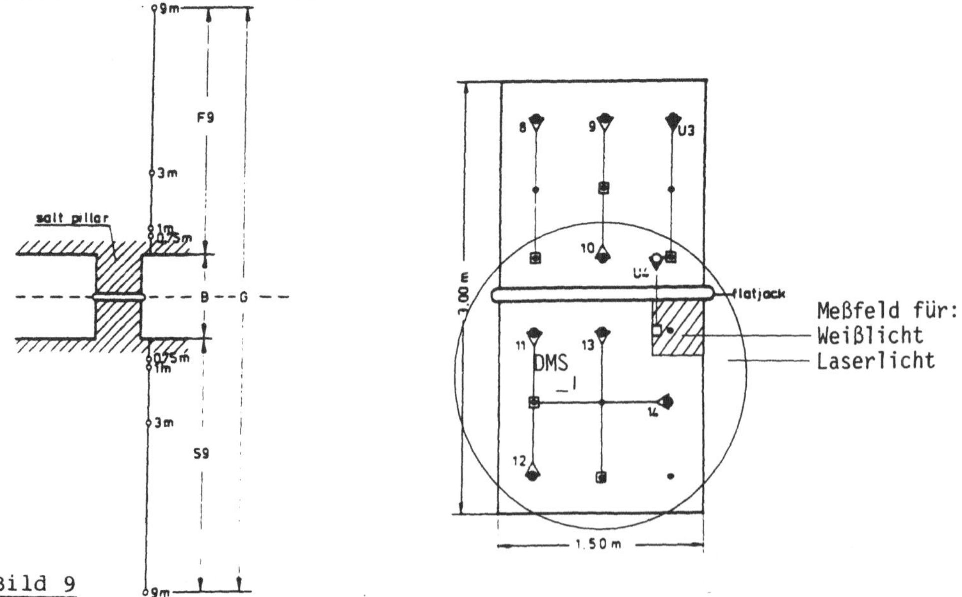

__Bild 9__

Anordnung der Extensometerfestpunkte und der Meßstrecken auf der optisch
untersuchten Pfeilerseite. Zum Vergleich wurden folgende Wegaufnehmer-
bzw. Extensometermeßstrecken herangezogen: 11, 13 bzw. B.

Meßstrecken (y-Koordinaten am Pfeiler)	induktive Wegaufnehmer (WA) (μm)	DMS (μm)	Speckle-fotografie (μm)
200 - 1200 (WA13)	212	181	219
700 - 1200 (WA11)	116	91	120
Extensometermeß-strecke 0,5 · B	93	–	110

Tabelle 2

Vergleich der mit verschiedenen Verfahren gewonnenen vertikalen Verformungen am Salzpfeiler bei Lasterniedrigung (7,0 - 3,0 MPa)

Aufgrund der Nichtlinearität der Verformung über die Höhe ist es wichtig, gleiche y-Koordinatendifferenzen zu benutzen. Diese Bedingungen
können die Dehnungsmeßstreifen (DMS) nicht erfüllen, daher ist die
Übereinstimmung nicht so gut. Die Übereinstimmung zwischen Meßuhren
und Specklemessung ist dagegen sehr gut. Zusätzlich liefert eine Extrapolation der Verschiebungskurve auf den Wert für y = 0 in guter Übereinstimmung den Wert der Extensometermessung. Das heißt, daß durch die
Specklefotografie die Ergebnisse der herkömmlichen Meßmethoden vollständig bestimmt werden können. Sie liefert jedoch Zusatzinformationen
über die vertikale und horizontale Verformung über die gesamte Fläche.

4. Vergleich von Weißlicht- und Laserspeckle-Fotografie

Parallel zu den oben beschriebenen Messungen mit einem Laser wurde ein
Versuch mit einem Elektronenblitz als Weißlichtquelle durchgeführt. Es
wurde ein Fläche von ca. 350 x 350 mm² unterhalb des Druckkissens am
Pfeiler ausgeleuchtet, die mit einer Plattenkamera fotografiert wurde.
Die in der retroreflektierenden Farbe enthaltenen Glaskörper von ca. 30 μm
Durchmesser wirken bei Beleuchtung als Punktlichtquellen. Kann die ver-
wendete Optik und der Film diese Struktur auflösen, so ergibt die Ab-
bildung dieser Lichtpunkte eine speckleähnliche Struktur im Negativ.
Werden Doppelbelichtungen durchgeführt, so erhält man analog zur kohä-
renten Abbildung Young'sche Interferenzstreifen /5/.

Die Durchführung des Versuchs ließ nur eine prinzipielle Bewertung der
Ergebnisse zu. In Tabelle 3 sind die mittels Laserspeckle- und Weiß-
lichtspeckle-Fotografie zeitgleich gewonnenen Meßwerte dargestellt. Die

Ergebnisse stimmen einigermaßen überein und lassen hoffen, daß die Weißlicht-Speckletechnik ebenfalls gute Ergebnisse liefern kann. Erst durch weitere Meßreihen können Aussagen zur Meßunsicherheit getroffen werden.

(MPa)	- 6,5	2,0	- 2,0
Laserspeckle	657	157	228
Weißlichtspeckle	715	202	232

Tabelle 3
Vergleich der vertikalen Verformungen aus Laserspecklefotografie und Weißlichttechnik

5. Zusammenfassung

Ausgehend von Verformungsmessungen an Salzwürfelproben in einem Kriech-prüfstand wurde die Specklefotografie erstmals im Bergwerk unter den dort herrschenden Umgebungsbedingungen eingesetzt. An einem Salzpfeiler, der an Firste und Sohle mit dem Gebirge verbunden war, wurden in einem Lastwechselversuch Verschiebungsmessungen durchgeführt.

Die über induktive Wegaufnehmer integral längs einer vorgegebenen Meß-strecke erfaßten Verschiebungen wurden durch die aus den Specklegrammen ermittelten Werte innerhalb der Meßgenauigkeit bestätigt. Mit der Specklefotografie kann aufgrund der flächenhaften Meßwerterfassung zu-sätzlich eine quantitative Darstellung des in-plane-Verformungsverhal-tens einer gesamten Seitenfläche des Pfeilers gewonnen werden.

Darüber hinaus wurde für die untertägigen Messungen die Speckletechnik mit Weißlicht als Beleuchtung erprobt. Die Ergebnisse lassen erwarten, daß bei nahezu gleicher Meßempfindlichkeit die teuren bzw. versorgungs-technisch aufwendigen Laser damit ersetzt werden können.

Dank an das Land Niedersachsen, das dieses Vorhaben mit Forschungs-mitteln förderte, und an die Gesellschaft für Strahlen- und Umwelt-forschung für die Hilfe bei der Durchführung der Messungen unter Tage.

Literatur

/1/ ALBRECHT, H., U. HUNSCHE (1980): Gebirgsmechanische Aspekte bei
    der Endlagerung radioaktiver Abfälle in Salzdiapiren unter
    besonderer Berücksichtigung des Fließverhaltens von Steinsalz.-
    Fortschr. Miner. 58 (2), p. 212 - 247.

/2/ ENNOS, A. E. (1975): Speckle Interferometry.- In: Laser Speckle
    and Related Phenomena
    Editor: Daipty, J.C.,
    Topics of Applied Physics, 9, p. 203 - 253.
    Springer-Verl., Berlin, New York, Heidelberg.

/3/ BURGHARDT, COLLIER, LIN (1971): Optical Holografy, Academie
    Press, New York, San Francisco, London.

/4/ HUNSCHE, U., I. PLISCHKE, H.-K. NIPP, H. ALBRECHT (1983):
    An in-situ creep experiment using a large rock-salt pillar.-
    Proc. Sixth Int. Symp. on Salt, May 24 - 28, 1983, Toronto,
    Canada, in press.

/5/ FORNO, C. (1975): White-light speckle photography for measu-
    ring deformation, strain, and shape.- Optics and Laser
    Technology, p. 217 - 221.

Vergleichende Holographie

B. Breuckmann, M. Schiller, W. Thieme
M.A.N. Neue Technologie
Dachauer Straße 667, 8000 München 50

Problemstellung

In dem sich ständig verschärfenden Wettbewerb werden zunehmend höhere
Anforderungen an die Produkte hinsichtlich Qualität und Lebensdauer
gestellt. Einhergehend damit steigen auch die Anforderungen an die
Meßtechniken, die zur Qualitätsbeurteilung und zur zerstörungsfreien
Werkstückprüfung herangezogen werden. In den Fällen, wo Form oder
Verformungsverhalten ständig mit großer Genauigkeit vermessen oder
kontrolliert werden müssen, bietet sich als hochwertige Meßtechnik die
holografische Interferometrie an. Ihre Vorteile liegen dabei in der
hohen Empfindlichkeit und der flächenhaften Meßwerterfassung. Nachteilig ist, daß die benötigte Information häufig nicht direkt aus dem
Hologramm abgelesen werden kann, sondern verdeckt vorliegt. So werden
z. B. Fehler, die bei einer Belastungsprüfung nur zu geringen Formänderungen führen, im Streifenmuster der Gesamtverformung untergehen.
Solche unauffälligen, nicht unbedingt harmlosen Fehler sind nur durch
eine aufwendige Punkt-für-Punkt-Auswertung des Interferogramms zu ent-
decken. Mit Hilfe der vergleichenden Holografie jedoch kann dieses
Problem umgangen werden.

Vergleichende Holografie

Das Verfahren der vergleichenden Holografie wurde 1980 von D.B.Neumann
/1/ vorgeschlagen. Dieses Verfahren gestattet es, uninteressante Grundmuster wie z. B. das der Gesamtverformung bereits während der Aufnahme
auf optischem Weg zu eliminieren und zwar dadurch, daß die Eigenschaften (Form und Verformungsverhalten) des Testbauteils direkt mit denen
eines fehlerfreien Masterbauteils verglichen werden. Als Meßergebnis
erhält man ein Interferogramm, in dem die Gleichanteile unterdrückt
sind, nur dort, wo Abweichungen vorliegen, taucht ein dafür charakteristisches Streifenmuster auf.

Im M.A.N. Holografielabor wurde das vorgeschlagene Verfahren auf seine praktische Durchführbarkeit untersucht und modifiziert; das Vergleichs- interferogramm wird jetzt auf folgende Weise erhalten: Zunächst werden die beiden Hologramme von Anfangs- und Endzustand eines fehlerfreien Masterbauteils auf einer Platte mittels zweier Referenzstrahlen auf- gezeichnet (Abb. 1a). Für die eigentliche Vergleichsaufnahme wird diese Platte mit den konjugierten Referenzstrahlen beleuchtet. Die abgebeugten Objektwellenfelder dienen zur Beleuchtung des Testbauteils, das sich jetzt an der Stelle des Masters befindet (Abb. 1 b). Von An- fangs- und Endzustand wird bei entsprechend geschalteten Referenzstrah- len ein Doppelbelichtungshologramm angefertigt. Stimmen Master- und Testbauteil überein, so sind die Objektwellen nach der Reflexion am Testbauteil wieder in Phase, das Doppelbelichtungshologramm weist kein Streifenmuster auf. Abweichende Eigenschaften der Bauteile dagegen werden durch ein charakteristisches Streifenmuster angezeigt.

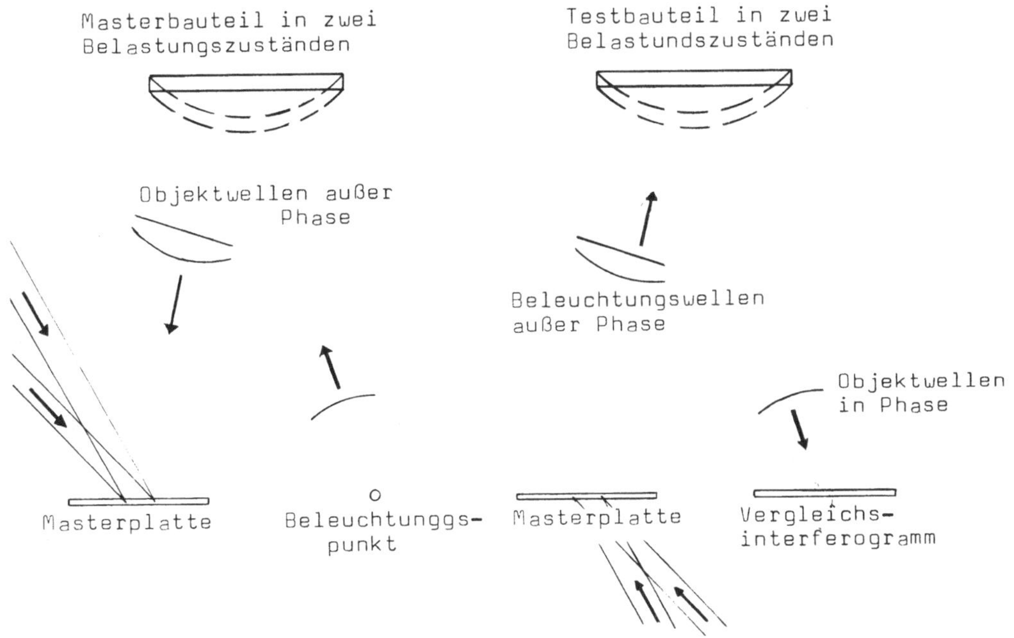

Abb. 1 a: Herstellung der Master- platte. Die Hologramme von An- fangs- und Endzustand werden mit zwei verschiedenen Referenzstrah- len aufgenommen.

Abb. 1 b: Herstellung des Ver- gleichsinterferogramms. Nach Reflexion an einem fehlerfreien Testbauteil sind die Beleuch- tungswellen wieder in Phase.

Im Gegensatz zu Neumann, der die Verwendung von zwei Masterplatten
vorschlägt, macht die Zwei-Referenzstrahlmethode das Verfahren praxis
freundlich:
- Der optische Aufbau bleibt fest ohne zwischenzeitliche Reposentio
  nierung von Hologrammplatten
- Die geringe Empfindlichkeit gegenüber optischen Inhomogenitäten
  erlaubt die Verwendung handelsüblicher Platten
- Die abgebeugten Objektwellen erzeugen auf dem Testbauteil das
  Grundmuster, es verschwindet im Fall einer Dejustierung und gestattet somit eine permanente Verfahrenskontrolle.

Mit diesem modifizierten Verfahren wurde die Durchführbarkeit der
optischen Mustereliminierung demonstriert. Die Abb. 2 a, b zeigen
zwei Interferogramme eines Balkens unter Biegebelastung. Das erste
ist ein übliches Doppelbelichtungshologramm mit einem dichten Streifenmuster. Das zweite ist ein Vergleichshologramm, hier ist das
Grundmuster bis auf wenige Streifen eliminiert. Die nicht ganz vollständige Eliminierung ist vermutlich auf Spiel in der Biegevorrich-
tung zurückzuführen und ist behebbar.

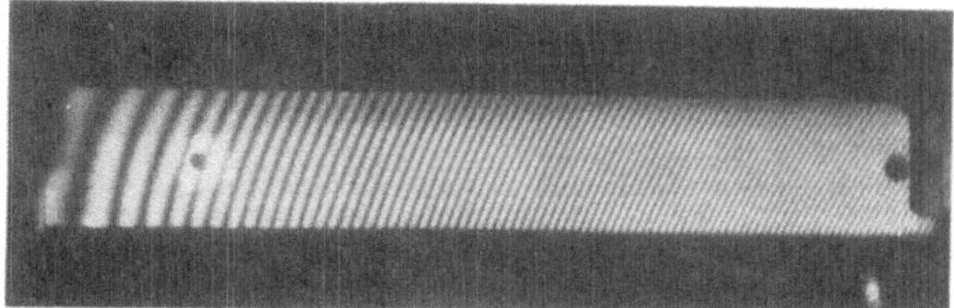

Abb. 2a: Doppelbelichtungshologramm der Biegeverformung

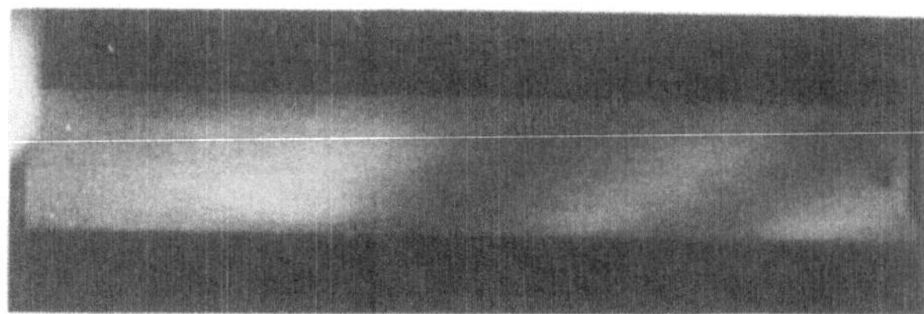

Abb. 2b: Vergleichshologramm, die Biegeverformung des Testbauteils
         ist weitgehend eliminiert

<u>Vorteile</u>

Die Unterdrückung der Gleichanteile führt zu einer deutlichen
Hervorhebung des fehleranzeigenden Streifenanteils. Art, Lage und
Größe werden auf einen Blick erkennbar. Dieses extrem starke Hervor-

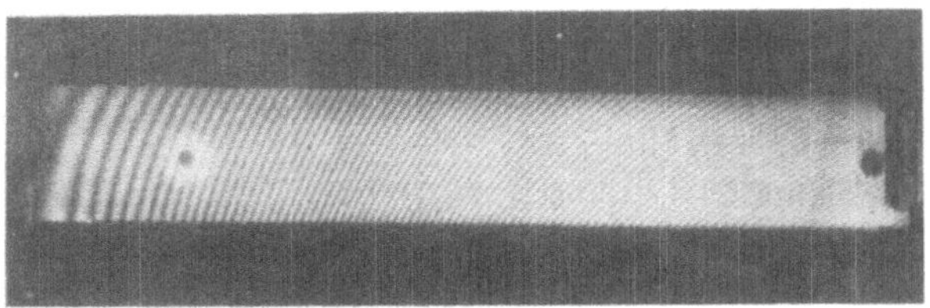

Abb. 3a: Doppelbelichtungshologramm der Biegung plus einer zusätz-
         lichen Belastung

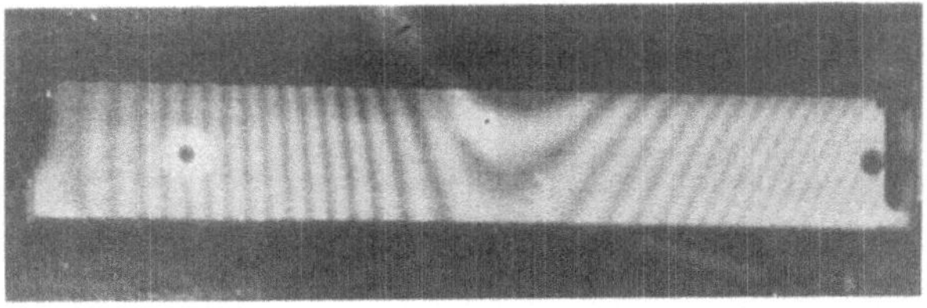

Abb. 3b: Zugehöriges Vergleichshologramm, Größe und Angriffspunkt
         der zusätzlichen Belastung sind auf einen Blick erkennbar

treten von Abweichungen vom gewöhnlichen Verformungsbild wurde
experimentell simuliert, indem auf den Biegebalken eine zusätzliche
Belastung aufgebracht wurde. Abb. 3 a zeigt ein Doppelbelichtungshologramm der Gesamtverformung. In einem solchen Fall ist ohne eine
genaue Auswertung eine Aussage über die zusätzliche Belastung unmöglich. Abb. 3 b zeigt das Vergleichshologramm, Größe und Angriffspunkt der zusätzlichen Belastung sind klar zu erkennen.

Im Gegensatz zur üblichen Holografie bleiben bei der Vergleichsholo-
grafie auch hohe absolute Verformungen unsichtbar. Das erlaubt es,
im Fall der zerstörungsfreien Werkstückprüfung die Belastung ohne
Rücksicht auf die Größe der Verformung in den Bereich zu legen, in
dem Fehler am signifikantesten hervortreten.

Die freie Wahl der Belastungsparameter hebt die Empfindlichkeit
der holografischen Prüftechnik um sicherlich mehr als eine Größenordnung an. Zusammen mit der leichten Auswertbarkeit ist die Prüftechnik soweit verfeinert worden, daß man auf ein breites Einsatz-
spektrum in Qualitätssicherung und Werkstückprüfung hoffen darf.

/1/ D. B. Neumann, Comparative Holography, Tec. Digest, Topical
    Meeting on Hologram Interferometry and Speckle Metrology,
    Opt. Soc. Am., 1980, PP MB 2-1.

Abbildung von akustischen Platten- und Volumenwellen mit Hilfe der holographischen Interferometrie zur verbesserten Fehlerbeschreibung bei der Ultraschallprüfung

A. Krüger, H.-A. Crostack
Universität Dortmund / Fachgebiet Qualitätskontrolle
D 4600 Dortmund

1. Einleitung

In der zerstörungsfreien Werkstoffprüfung mit Ultraschall bestehen
Schwierigkeiten sowohl hinsichtlich der genauen Beschreibung von Fehlstellen nach Form und Größe als auch hinsichtlich der Trennung von
Fehler- und Störanzeigen. In besonderem Maß gilt dies für die Erfassung oberflächennaher Fehler und bei der Prüfung von dünnwandigen Bauteilen und Verbundwerkstoffen. Für die hierzu erforderliche umfassen-
dere Aufnahme und Analyse des durch den Fehler beeinflußten Ultra-
schallfeldes bietet sich eine großflächige optische Abbildung mit
Hilfe der holographischen Interferometrie an /1,2/. Dadurch werden die
Vorteile beider Techniken miteinander verknüpft. Da die Amplituden der
Ultraschallschwingungen im Nanometerbereich, d.h. bei einigen tausend-
steln der Lichtwellenlänge liegen, ist eine deutliche Steigerung der
Empfindlichkeit der holographischen Interferometrie erforderlich.
Sofern dies erreicht wird, läßt sich aus den erhaltenen Bildern, die
das Schallfeld mit hoher Ortsauflösung wiedergeben, aufgrund des weit-
aus umfangreicheren Informationsinhaltes eine wesentlich genauere Fehlerbeschreibung sowie eine bessere Trennung zwischen unterschiedlichen
Anzeigen vornehmen. Diese Technik erbrachte bereits wesentliche Verbesserungen bei der zerstörungsfreien Prüfung von Oberflächenbeschich-
tungen auf Bindefehler, Risse und Poren /3,4/. Ziel der vorliegenden
Untersuchungen ist es, die Eignung des Verfahrens zur Prüfung dünnwandiger Bauteile zu ermitteln.

2. Versuchsaufbau

Die zur Abbildung hochfrequenter Ultraschallschwingungen erforderliche
Empfindlichkeitssteigerung der optischen Holographie läßt sich durch
stroboskopische Beleuchtung der Objektoberfläche im Takt der Ultra-
schallschwingung und einen Sprung in der relativen Phase zwischen Ob-
jekt- und Referenzstrahl während der Aufnahme erreichen. Praktisch

wird die stroboskopische Beleuchtung durch einen akustooptischen Auskoppler im Laserresonator und der Phasensprung durch einen elektrooptischen Kristall im Referenzstrahl realisiert, Bild 1.

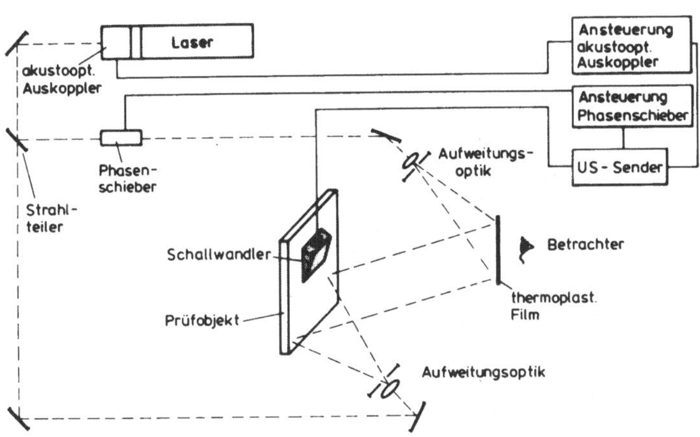

Bild 1. Versuchsaufbau für die holographische Ultraschallabbildung

Beide werden synchron zum Ultraschallsignal gesteuert /1,2/. Der übrige Aufbau ist konventionell. Mit dieser Aufnahmetechnik können Schwingungsamplituden bis herab zu ca. 2 Å abgebildet werden. Die Auswertung kann dabei sowohl von einem Betrachter als auch einem Rechner vorgenommen werden. Zur Anregung der Ultraschallsignale werden CS-Impulse /5/ von 20 - 100μs Dauer bei Wiederholfrequenzen von 0,4 kHz eingesetzt. Die benutzten Schallfrequenzen liegen bei 0,3 bis 3 MHz. Nach einer geeigneten Leistungsverstärkung werden diese Signale einem resonanten PZT-Wandler zugeführt, der über einen Plexiglaskeil unter einem vorwählbaren Winkel in den Probekörper einschallt.

3. Abbildung von geführten Ultraschallwellen

Zur Prüfung dünnwandiger Bauteile, wie z.B. Bleche und Rohre, werden in der US-Technik üblicherweise geführte Wellen - Platten-, Stab- oder Rohrwellen - verwandt. Bei entsprechender Anregung unter Einsatz gesteuerter Signale (CS-Technik) und bei Einschallung unter definiertem Winkel ist die Erzeugung bestimmter Wellenmoden möglich, die zu einem empfindlichen Fehlernachweis führen. Da die Wellenart außerdem stark von der Geometrie des Körpers abhängt, führen nicht nur Fehlstellen

188

sondern auch Formänderungen zu Reflektoranzeigen. Eine Unterscheidung
zwischen diesen unterschiedlichen Anzeigen ist bei der reinen Ultraschallprüfung jedoch nicht möglich, so daß z.B. die Prüfung von
Schweißnähten mit belassener Decklage und Wurzel auf Schwierigkeiten
stößt.

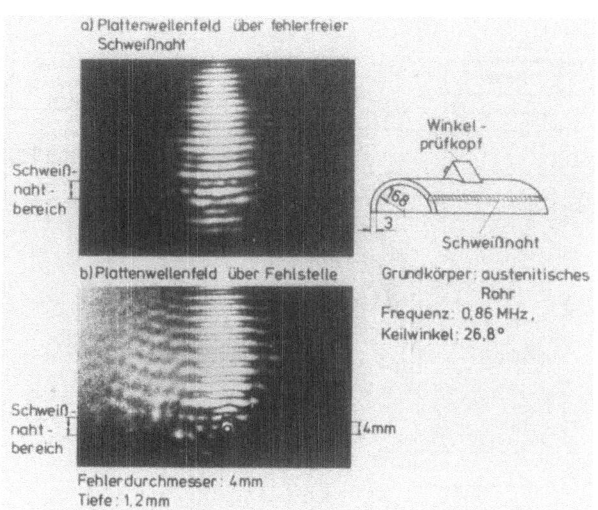

Bild 2. Holographische Abbildung von Plattenwellenfeldern

Hier erbringt die holographische Schallfeldabbildung deutliche Vorteile, Bild 2. Die in Umfangsrichtung über ein austenitisches Rohr
(Durchmesser: 168 mm, Wanddicke: 3 mm) laufende Welle wird durch die
in Längsrichtung verlaufende Schweißnaht nur wenig verzerrt und gedämpft, Bild 2a. Die Anregung der Plattenwelle erfolgt mit einer Frequenz von 0,86 MHz unter einem Einstellwinkel von 27° (Wellenmode a_1).
Die hellen Streifen stellen die Maxima der Auslenkung dar, während die
dazwischenliegenden dunklen Bereiche Minima wiedergeben. Im Gegensatz
zum fehlerfreien Fall wird das Schallfeld im Bereich eines künstlich
in die Schweißnaht eingebrachten 4 mm großen Fehlers (Flachbodenbohrung, 1,2 mm unterhalb der Oberfläche endend) deutlich gestört. Bild
2b. Der Fehler äußert sich dabei durch Ausbildung eines eigenständigen
Schwingungsmusters. Lage und Größe des Fehlers werden unmittelbar
angezeigt und die Trennung zwischen Wellenfrontstörungen durch die
Schweißnaht und durch den Fehler ist infolge der hohen axialen und la-
teralen Auflösung ebenfalls möglich. Dabei kommen selbst Fehler, die
kleiner sind als die Schallwellenlänge, noch deutlich zur Anzeige, wie
dies für eine 2 mm große Flachbodenbohrung (Tiefe: 1,2 mm) in Bild 3

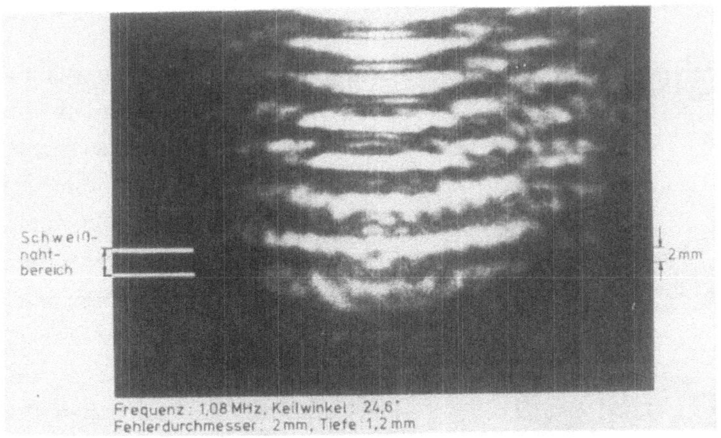

Bild 3. Nachweis eines Fehlers in einer Schweißnaht

wiedergegeben ist. Der Fehler ruft hier eine ringförmige Verzerrung
der Wellenfront hervor, die sich in weiten Bereichen äußert, obgleich
sein Durchmesser nur etwa ein Drittel der Wellenlänge beträgt. In Abhängigkeit von der Tiefenlagen konnten bisher Fehler bis herab zu ca.
10% der Schallwellenlänge gefunden werden. Neben der Auffindung der
Fehler ist besonders interessant, daß die abgespaltenen Wellenarten
großflächig abgebildet werden. Hiermit eröffnet sich die Möglichkeit
zu grundsätzlichen Untersuchungen über das Reflexions-, Beugungs- und
Streuverhalten von Ultraschallwellen an Fehlstellen. Mit ähnlichem
Erfolg lassen sich auch dünnwandige gewalzte Bleche auf Lunker und
Doppelungen (flächige durch den Walzprozeß bedingte Materialtrennungen)
prüfen. Bild 4 gibt das Ergebnis der Prüfung eines 2,5 mm dicken Ble-
ches mit natürlichen Fehlstellen wieder. Während oben in Bild 4 zum
Vergleich das Wellenfeld der 1 MHz-Plattenwellen über einem fehlerfreien Bereich zu erkennen ist, sind unten in Bild 4 und in Bild 5
die Störungen sowohl über der Lunkerzeile wie über der Dopplung aus-
zumachen. An die von links heranlaufende ungestörte Schallwelle, die
in Bild 5 nur kurz ist, da hier der Schallsender dicht ans Fehlerge-
biet gesetzt ist, schließt ein Gebiet mit starken Amplitudenüberhöhun-
gen (erhöhten Helligkeiten) als Anzeige für eine im Blech befindliche
Lunkerzeile an. In dem folgenden Gebiet ändert sich das Schallfeld mit
zunehmendem Abstand wieder, wobei der "Schallschatten" der Fehlstelle
seinen Einfluß behält. Typisch für die beiden sich daran anschließenden Doppelungen, Bild 5, ist ein stark strukturiertes Schallfeld mit
eigenständigem Schwingungsmuster, deren Beginn und Ende jedoch gut ab-

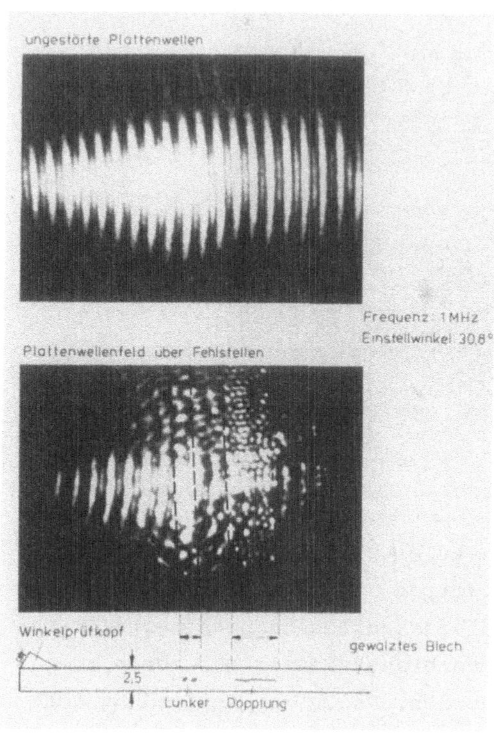

Bild 4. Abbildung von Plattenwellen zur Detektion von Lunkerzeilen und Dopplungen

schätzbar ist.

Neben der Prüfung dünnwandiger Komponenten mit Plattenwellen eignet sich diese Technik, wie erste Untersuchungen zeigen, auch zur Abbildung von Volumenwellen, die allerdings in ihrer Amplitude schwächer sind. In Bild 6 ist das an der Oberfläche erscheinende Echo einer 10 mm breiten Nut in einem Stahlblock bei Anschallung mit einer Transversalwelle unter 42° abgebildet. Bei so tiefen Fehlern ist natürlich nicht unmittelbar eine Aussage über Fehlerform und -größe möglich, da das Schallfeld entsprechend aufgeweitet ist. Jedoch erlaubt diese hochauflösende flächige Erfassung der

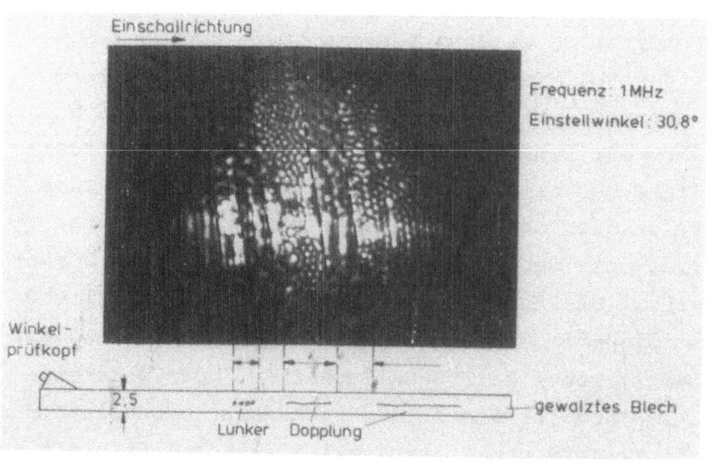

Bild 5. Flächenhafte Darstellung von Dopplungen

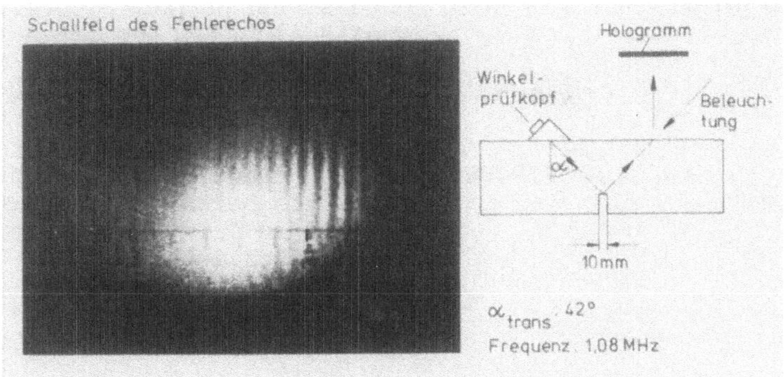

Bild 6. Holographische Abbildung von Volumenwellen

Schallfeldverteilung an der Oberfläche die Rekonstruktion des am Fehler reflektierten Schallfeldes. Bei Kenntnis der Fehlertiefe läßt sich damit ähnlich wie bei der akustischen Holographie eine Rückrechnung von der Oberfläche auf Größe und Form des Reflektors in der Tiefe durchführen. Dies ist Gegenstand laufender Arbeiten.

4. Zusammenfassung

Durch eine spezielle holographische Aufnahmetechnik wird die Abbildung von Platten- und Volumenwellen bei hohen Ultraschallfrequenzen möglich. Durch die hierbei gewonnenen flächigen Schallfeldbilder ergeben sich gegenüber der normalen Ultraschallprüfung eine Reihe von Vorteilen. So werden beim Einsatz von Plattenwellen, Lage, Form und Größe von Fehlstellen in Form eines Bildes angezeigt, wobei sich Fehleranzeigen gut von Formanzeigen z.B. aus Schweißnahtüberhöhungen trennen lassen. Hierzu trägt neben der flächigen Darstellung das hohe axiale und laterale Auflösungsvermögen, das weit unterhalb der Schallwellenlänge liegt, bei. Dies erlaubt auch den Einsatz relativ niedriger Frequenzen, die den Ultraschall weniger stark streuen; ein besonderer Vorteil bei grobkörnigen Werkstoffen. Darüberhinaus belegen die Untersuchungen, daß auch die Abbildung von tiefer liegenden Fehlstellen durch Volumenwellen möglich ist, so daß aus der holographisch gemessenen Schallfeldverteilung an der Oberfläche eine Rückrechnung auf die Fehlergeometrie ermöglicht wird.

192

/1/ CROSTACK, H.-A., W.R.FISCHER;   Untersuchungen zur Steigerung der
                                     Empfindlichkeit der optischen Holo-
                                     graphie zur Abbildung von Ultra-
                                     schallfeldern; Materialprüf. 23
                                     (1981) 11, S. 384/387

/2/ CROSTACK, H.-A., W.R. FISCHER,   Ein holographisches Verfahren zur
    A. KRÜGER   optischen Detektion von Ultraschall-
                                     feldern in der zerstörungsfreien
                                     Werkstoffprüfung
                                     Optoelektr. in der Technik
                                     Vortragsband Int. Kongress LASER'81
                                     (1982), Springer-Verlag, S. 73/78

/3/ CROSTACK, H.-A., W.R. FISCHER   Einsatz von Ultraschall in Verbin-
                                     dung mit holographischer Interfero-
                                     metrie zur zerstörungsfreien Prü-
                                     fung thermisch gespritzter Schichten
                                     Materialprüfung 24 (1982), 2, 49-54

/4/ CROSTACK, H.-A., A. KRÜGER,   Non-destructive testing of thermally
    W.R. FISCHER, H.-D. STEFFENS   sprayed coatings by using optical
                                     holography to receive ultrasonic
                                     waves
                                     10. Int. Conf. "Therm. Spritzen"
                                     2.-6.5.83 in Essen, DVS-Berichts-
                                     band 80, 1983, S.28-30

/5/ CROSTACK, H.-A.   Beitrag zur Verbesserung der Ultra-
                                     schallprüfung beim Ermitteln von
                                     Fehlern in schwer prüfbaren Werk-
                                     stücken
                                     Habilitationsschrift, Universität
                                     Dortmund, 1978

Kompensation von Bildmodifikationen und dreidimensionale Deformationsberechnung aus Doppelpulshologrammen

E. MÜLLER und V. HRDLICZKA
Institut für Werkzeugmaschinenbau
ETH Zürich, CH-8092 Zürich

Einleitung

Basierend auf den theoretischen Grundlagen der Holographie und der holographischen Interferometrie wurde eine Auswertungsmethode für holographische Interferogramme entwickelt, die eine quantitative Bestimmung des Verschiebungsfeldes für die folgenden drei Fälle erlaubt:

- ein einzelnes holographisches Interferogramm,
- gleichzeitig mit verschiedenen Sichtpunkten aufgenommene holographische Interferogramme zur Berechnung dreidimensionaler Verschiebungsfelder
- ein mit anderem Laserlicht rekonstruiertes holographisches Interferogramm (Pulslaser)

Die Verwendung von Pulslasern mit kurzer Hologrammbelichtungszeit reduziert die Anforderungen an die Stabilität des holographischen Aufbaues drastisch. Die mit einem Rubin-Pulslaser belichteten Hologramme werden in der Regel nicht mit einem teuren Rubin-Dauerstrich- oder einem Dye-Laser, sondern mit einem preisgünstigeren He-Ne-Laser, rekonstruiert. Dadurch entstehen wegen der geänderten Lichtwellenlänge bei der Hologrammrekonstruktion Bildmodifikationen, die eine exakte Auswertung der Pulshologramme verhindern.

Im ersten Teil des Artikels werden die durch Rekonstruktion mit anderer Lichtwellenlänge verursachten Bildmodifikationen behandelt, und im zweiten Teil folgt die Bestimmung eines Verschiebungsfeldes durch die Auswertung dreier gleichzeitig aufgenommener Interferogramme.

Bildmodifikationen

Änderungen der Einfallsrichtung des Rekonstruktionstrahles auf das Hologramm oder die Verwendung einer anderen Lichtwellenlänge beeinflussen den Beugungswinkel bei der Erzeugung des Objektstrahls und bewirken dadurch Bildmodifikationen (1,3). Das rekonstruierte Bild erscheint vom wirklichen Objektstandort verschoben, verzerrt und in der Grösse verändert. Zudem sind die rekonstruierten Objektpunkte astigmatisch.

Verschiebung, Verzerrung und Grössenänderung des rekonstruierten Objektbildes können durch rechnerische Kompensation der Sichtwinkeländerung korrigiert werden. Die Berechnung des astigmatischen Abbildungsbereiches erlaubt eine Abschätzung der erreichbaren Bildqualität sowie eine Berechnung des ursprünglichen Sichtpunktes.

Die Sichtwinkeländerung bei der Hologrammrekonstruktion ist in der Regel bereits bei kleinen Wellenlängenänderungen gut sichtbar und kann durch den Vergleich einer Punktlichtwelle mit der effektiv erzeugten Welle berechnet werden (1).

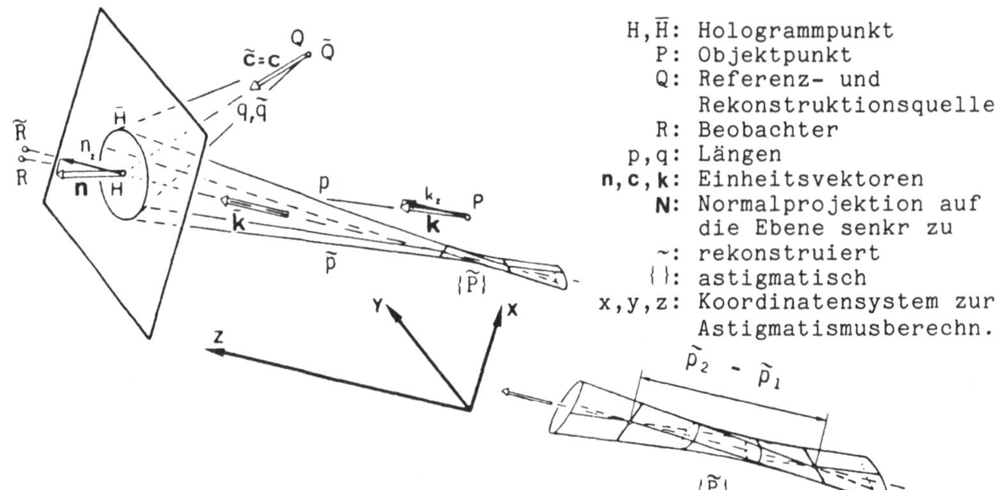

Fig. 1. Modifikation der Sichtrichtung und astigmatische Abbildung, erzeugt durch Änderung des holographischen Aufbaues bei der Rekonstruktion.

Die resultierende Phasendifferenz im Hologrammpunkt H muss für benachbarte Hologrammpunkte $\bar{H}$ stationär sein. Diese Bedingung liefert die Grundgleichung der modifizierten Sichtrichtung und die Gleichung für eine reine Laserwellenlängenänderung ($c=\tilde{c}$).

$$N[\frac{1}{\tilde{\lambda}}(\tilde{k} - \tilde{c}) - \frac{1}{\lambda}(k - c)] = 0, \quad |\tilde{k}| = |\tilde{k}| = 1, \qquad N\,k = N[(1 - \frac{\lambda}{\tilde{\lambda}})\,c + \frac{\lambda}{\tilde{\lambda}}\tilde{k}].$$

Die Sichtwinkeländerung erzeugt eine Verzerrung sowie eine Grössenänderung des rekonstruierten und modifizierten Bildes.

Durch Nullsetzen der zweiten Ableitung der Phasendifferenz erhält man eine Gleichung, die den Astigmatismus des rekonstruierten Objektpunktes bestimmt. Für eine reine Änderung der Laserwellenlänge bei der Rekonstruktion kann die Gleichung in Komponenten eines speziellen Koordinatensystems umgeschrieben werden. Die zwei Eigenwerte dieser Gleichung ($\tilde{p}_1$ und $\tilde{p}_2$ in Fig. 1) begrenzen den astigmatischen Abbildungsbereich des rekonstruierten Punktes {P̃} und müssen für jeden Objektpunkt einzeln berechnet werden (2).

Fig. 1 zeigt zusätzlich, dass der Sichtpunkt des rekonstruierten Hologrammbildes R nicht mit dem entsprechenden Sichtpunkt des ursprünglichen Objektes übereinstimmt. Analog zur Umrechnung der rekonstruierten und modifizierten Bildpunkte in ursprüngliche Objektpunkte kann aus der Lage des Objektsichtpunktes der entsprechende astigmatische Sichtpunkt {R} des Objekts bestimmt werden (4,5).

Durch eine Wellenlängenänderung werden nicht nur die Objektpunkte, sondern auch die Interferenzstreifen modifiziert. Die Berechnung der optischen Weglängenänderung D im Lokalisationspunkt der Interferenzstreifen führt auf die Gleichung $D = \mathbf{u} \cdot \mathbf{g} = \mathbf{u} \cdot (\mathbf{k} - \mathbf{h})$ mit $\mathbf{u}$ = Verschiebungsvektor und $\mathbf{g}$ = Empfindlichkeitsvektor. Dieses Resultat zeigt, dass durch die Rekonstruktion mit anderer Lichtwellenlänge die Interferenzstreifen analog zu den Objektpunkten modifiziert werden, und dass für die Deformationsberechnungen die unmodifizierten Beleuchtungs- $\mathbf{h}$ und Beobachtungsrichtungen $\mathbf{k}$ bekannt sein müssen (2).

In Fig. 2 sind zwei Beispiele holographischer Aufbauten und die jeweils
zu erwartenden Bildmodifikationen dargestellt.

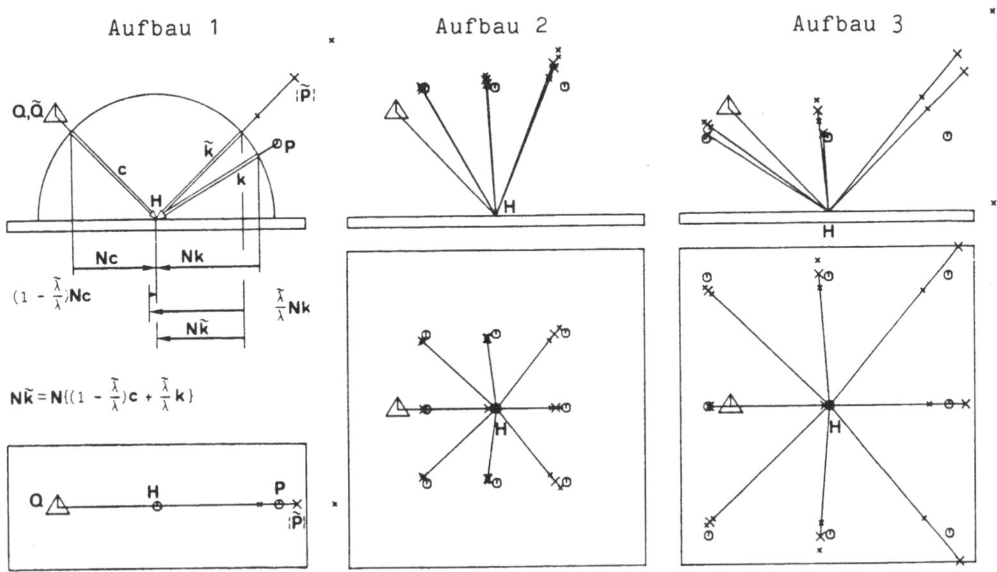

Fig. 2. Modifikation der mit einem Rubinlaser ($\lambda = .6943\,\mu$m) aufgenommenen
und mit einem He-Ne-Laser ($\tilde{\lambda} = .6328\,\mu$m) rekonstruierten Objektpunkte P. Dargestellt sind je zwei Normalprojektionen eines holographischen Aufbaues. Links ist die Berechnung der modifizierten Sichtrichtung $\tilde{k}$ für einen einzelnen Objektpunkt, in der
Mitte und rechts für mehrere Objektpunkte dargestellt. Gut
sichtbar sind Bildverzerrungen und Bildverschiebungen wie auch
die astigmatischen Abbildungsbereiche (× ×—×).

Bestimmung eines räumlichen Verschiebungsfeldes

Aufgenommen werden gleichzeitig drei holographische Interferogramme mit
einem Rubin-Doppelpulslaser beim Schnittaustritt eines Stahlhalters.
Jedes Interferogramm wird einzeln rekonstruiert und photographiert
(Fig. 3). Der holographische Aufbau und die Sichtpunkte der photogra-
phischen Aufnahmen werden ausgemessen. Aus den Sichtpunkten der photo-
graphischen Aufnahme werden die entsprechenden Sichtpunkte (Mittelpunk-
te der astigmatischen Bereiche) des Objektes bestimmt.

Eingetragen ins Photo sind
die Ordnungen der Interferenzstreifen, die Fixpunkte
und der mittlere Objektpunkt
P. Acc1 und Acc2 bezeichnen
die Beschleunigungsaufneh-
mer, die am Spannelement befestigt sind.

Fig. 3. Photo des rekonstruierten Hologrammes H2.

196

Die räumliche Definition der Objektgeometrie mit einem Rechner-Programm
erlaubt es nun, eine Ansicht des Objektes mit Hilfe des Plotters zu erstellen, die dem von der Modifikation befreiten Objektbild entspricht.

Die Ordnungen der Interferenzstreifen werden mit Hilfe der beiden in
Fig. 3 sichtbaren Federelemente bestimmt. Diese sind auf einem starren
Halter befestigt und stützen sich auf das Aufspannelement ab. Die Ver-
wendung von grossen Hologrammplatten (4x5 in.) erlaubt verschiedene
Sichtrichtungen und somit eine Überprüfung der in der Photo eingetrage-
nen Ordnungszahlen der Interferenzstreifen.

Die Photos der Hologrammbilder werden nun in bestimmten Punkten (Fixpunkte und Bildpunkte) digitalisiert und die Koordinaten zusammen mit
den Ordnungen der Interferenzstreifen gespeichert.

Da die Lagen der Fixpunkte sowohl auf dem Plotterbild als auch auf dem
Photo bekannt sind, können die digitalisierten Punkte mit Hilfe der
Fixpunkte in die entsprechenden Plotterbilder eingezeichnet werden.

Das Einzeichnen der digitalisierten Punkte entspricht einer Globalkor-
rektur der Bildmodifikationen. Unterschiede zwischen der Objektkontur
und den digitalisierten und eingezeichneten Punkten sind ein Mass für
die Bildverzerrung. Fig. 4 zeigt zwei dem Photo in Fig. 3 entsprechende
Plots, links mit eingezeichneten digitalisierten Bildpunkten, rechts
mit von der Modifikation befreiten Bildpunkten.

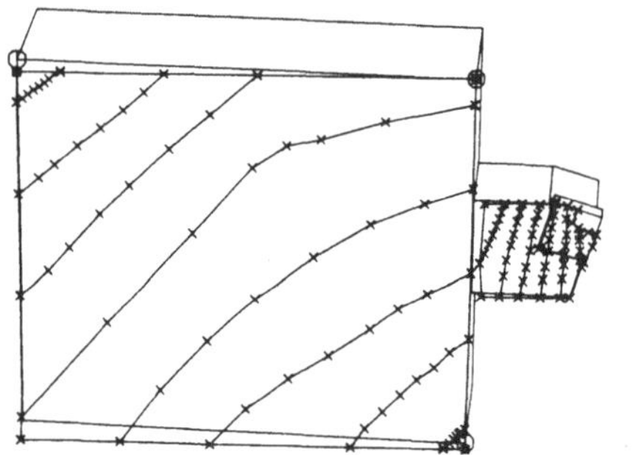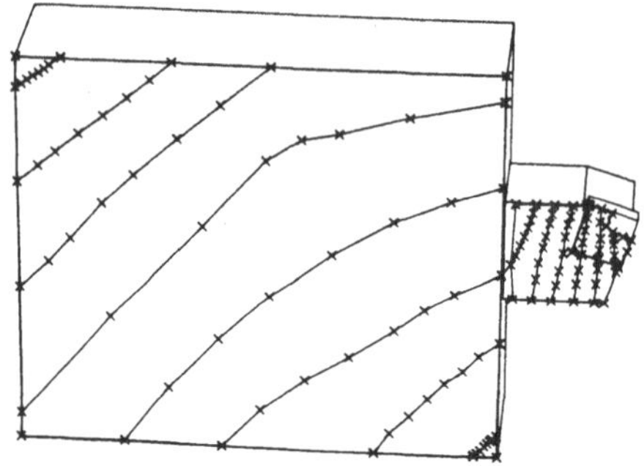

Fig. 4. Plotterbild des Zerspanungswerkzeuges mit Spannelement aus der
        Ansicht des Hologrammes H2. Eingezeichnet sind links die digi-
        talisierten Bildpunkte und rechts die von der Modifikation be-
        freiten Bildpunkte. Ein Vergleich zeigt, dass ohne Modifika-
        tionsberechnung, trotz Einzeichnen mit Hilfe von Fixpunkten,
        ein grosser Digitalisierungsfehler vorgetäuscht wird.

Aus dem Bild rechts in Fig. 4 können nun die den Bildpunkten entsprechenden Objektpunkte durch Projektion des Bildes auf das Objekt bestimmt werden. Die Deformationsberechnung liefert anschliessend aus jedem Messpunkt eine Projektion des Verschiebungsvektors auf die jeweilige Empfindlichkeitsrichtung des holographischen Aufbaues.

Um einen dreidimensionalen Verschiebungsvektor zu bestimmen, müssen die
drei Projektionen des Verschiebungsvektors im gleichen Objektpunkt bekannt sein. Diese Forderung verlangt einen Übergang von Bildpunkten auf
Gitterpunkte. Die Interferenzstreifenordnungen der umliegenden Bildpunkte bestimmen in Funktion des Bildpunktabstandes vom Gitterpunkt,
den Wert der Interferenzstreifenordnungen des Gitterpunktes.

Die Darstellung der dreidimensionalen Verschiebungsvektoren in Fig. 5
zeigt deutlich die Starrkörperbewegung und die Deformation des Objektes
zwischen den beiden Belichtungen bei der Hologrammaufnahme.

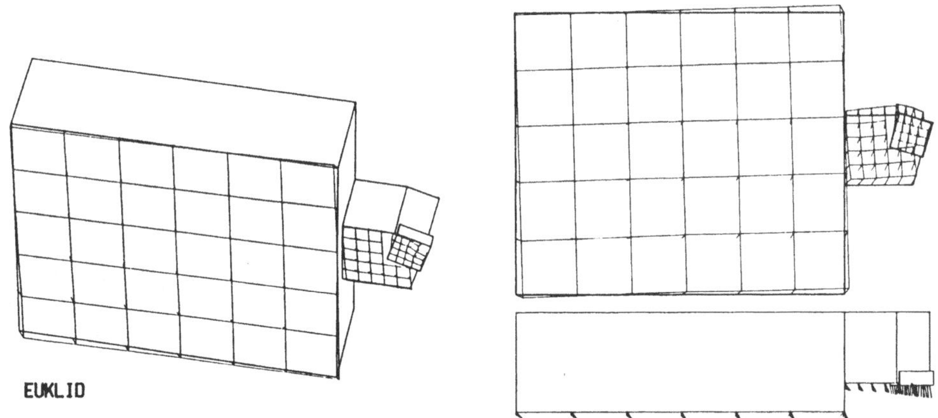

Fig. 5. Ansicht des Objektes mit Gitterpunkten und eingezeichnetem
        dreidimensionalem Verschiebungsfeld (2000fach überhöht). Links
        eine Zentralprojektion und rechts zwei Normalprojektionen.

Eine Überprüfung der holographisch interferometrischen Messung wurde
mit Hilfe der beiden am Aufspannelement befestigten Dreiachs-Beschleu-
nigungsaufnehmer (Fig. 3) durchgeführt (2).

Schlussfolgerungen

Die vorgestellte Auswertungsmethode für holographische Interferogramme
erlaubt eine Kompensation der bei der Rekonstruktion erzeugten Bildmodifikationen und eine genaue Auswertung holographischer Interferogramme. Das Verfahren lässt sich gut automatisieren und liefert anschauli-
che Darstellungen der errechneten Verschiebungsvektoren.

Literatur
(1) Schumann W., Dubas M.: Holographic Interferometry. Springer Verlag
    Berlin (1979).
(2) Müller E.: Auswertung holographischer Interferogramme. (Evaluation
    of Holographic Interferograms). Diss. ETH 7246 (1983).
(3) Dändliker R., Hess K., Siedler Th.: Astigmatic Pencil of Rays Re-
    constructed from Holograms. Israel Jour. of Tech. 18,240-246
    (1980).
(4) Schumann W.: Duality Property in Holographic Imaging. J.Opt.Soc.Am.
    71, 525-528 (1981).
(5) Müller E., Hrdliczka V., Cuche D.: Computer-based Evaluation of Ho-
    lographic Interferograms. SPIE 398-07 (1983).

Die Holographie, ein Hilfsmittel zur genaueren Bestimmung von Verformungen und Dehnungen an Bauteilen im Vergleich zu Berechnungsmethoden der Festigkeits- und Schwingungslehre

G. SCHÖNEBECK
Kraftwerk-Union
Wiesenstraße 35, D 4330 Mülheim/Ruhr

Einleitung

Die moderne Festigkeitslehre hat leistungsfähige Berechnungsverfahren zur Dimensionierung von Bauteilen entwickelt. Trotzdem muß besonders bei nicht einfachen Bauteilformen die Haltbarkeit durch Versuche kontrolliert werden. Das gilt um so mehr, wenn die Bauteile lebenswichtige Funktionen für die Maschine oder das System übernehmen müssen. Trotz dieser Vorsorge kommen immer wieder Schadensfälle durch zu Bruch gehende Bauteile vor. Hier hat die holografische Interferometrie in Verbindung mit dazu speziell entwickelten holografischen Methoden wirksam helfen können. Weiterhin konnte mit der holografischen Interferometrie und den nachfolgend skizzierten Verfahren festgestellt werden, daß bisherige Verfahren der Festigkeitslehre, auch die sogenannten exakten Lösungen, das wirkliche Verformungsfeld nicht immer vollständig wiedergeben. Die holografische Interferometrie mit den entsprechenden Verfahren erlaubt, die wirklichen, vollständigen Verformungen exakt zu messen. Das ist die Voraussetzung für eine einwandfreie Spannungsermittlung. Man kann so Vergleiche ziehen und die Fehlergrößen der Berechnungsverfahren abschätzen. Der nächste Schritt ist dann das Verbessern der Berechnungsverfahren mit erneuter holografischer Kontrolle usw.

Überblick

Nachfolgend werden Verfahren zusammenfassend skizziert, die aus der Praxis der holografischen Interferometrie im Laufe einiger Jahre entstanden sind. Mit diesem Überblick soll dem Anwender geholfen werden, die für seine Zwecke richtigen Verfahren auffinden und auswählen zu können.

1. Eine allgemeine Methode zur Bestimmung räumlicher Verschiebungsfelder [1, 2, 3]

Mit dieser Methode werden mit nur einer Hologrammplatte über Spiegel verschiedene Ansichten des Bauteiles holografisch aufgenommen. Bild 1 zeigt das Prinzip. Damit ist es auch möglich, Dehnungen von Verschiebungen zu unterscheiden, ohne einen festen Bezugspunkt haben zu müssen. Nach der Gleichung

$$\overline{n}_\eta^{\,o} \cdot \overline{D} = \left| \overline{n}_\eta \right| = \frac{\lambda}{2} \frac{1}{\cos \frac{d\eta}{2}} \cdot N_\eta \quad \text{mit } \eta = 1;\ 2;\ 3$$

erfolgt die Auswertung aus einem Hologramm (Bild 2) mit verschiedenen Ansichten.

Bild 3 zeigt das Ergebnis. Diese exakte Methode mit Spiegeln ist Voraussetzung auch für manche der folgenden Verfahren.

2. Modellwerkstoff Silicongummi [4, 5]

Um Dehnungsspitzen im oberflächennahen Bereich z.B. bei Kerbproblemen auffinden zu können, wurde Silicongummi als Modellwerkstoff verwendet. Weder mit Dehnungsmeßstreifen, spannungsoptischen Methoden, noch mit Speckle-Methoden sind diese sehr eng begrenzten Spitzen aufzufinden, weil sie wegintegriert werden. Sie waren deshalb bisher auch völlig unbekannt. Bild 4 zeigt einen Bereich eines Turbinenschaufelfußes mit Anriß. Das Interferenzbild wurde ausgewertet und in Bild 5 dargestellt. Der kleinste Krümmungsradius der gemessenen Querkontraktion und damit auch der Bereich der größten Dehnung längs der Randlinie liegt dort, wo auch der Schwingungsriß aufgetreten ist. Er liegt außerhalb des nach der Kerbtheorie als gefährdet angesehenen Kerbbereiches.

3. Modellwerkstoff Gips, Porcelin [4,5]

Als Hilfsmittel bei Entwicklungsaufgaben für Berechner und Konstrukteure wurde Gips und ein gipsähnlicher Kunststoff, nämlich Porcelin eingeführt. Diese Stoffe verhalten sich so elastisch wie Stahl (siehe Punkt 5). Das Bild 6 gibt einen Überblick über eine Eigenschwingungsform einer Schaufel aus Stahl, Gips und Porcelin. Die Eigenschwingungsfrequenzen stehen zueinander in einem festen Verhältnis entsprechend dem Verhältnis ihrer Massen und Elastizitätsmodule.

4. Anwendung auf Torsionsprobleme

4.1 Elastische Achse [6]

Bei Torsionsproblemen ist es oft erwünscht, die elastische Achse eines Körpers zu kennen. Die Berechnung ist aber nur in wenigen Sonderfällen möglich. Unter Verwendung von Verfahren 1 wird das Momentanzentrum der Torsionsbewegung gemessen. Die Verbindung dieser Punkte ist die elastische Achse der Körpers. Bild 7.

4.2 Prandtl'sches Seifenhautgleichnis [7]

Dieses Gleichnis besagt, daß das Volumen unter einer bei Überdruck ausgelenkten Membran (wie bei einer modernen Traglufthalle) ein Maß für die Drillsteifigkeit der überspannten Grundfläche (Profil) ist. Der Tangentenanstieg an die Membran ist direkt ein Maß für die Größe der Schubspannung. Mittels der Holografie werden die Höhenschichtlinien der durch Δp ausgelenkten Membran gemessen. Damit sind die sonst nicht berechenbaren unbekannten Potentialfunktionen holografisch ermittelt. Bild 8 zeigt eine Aufnahme mit dem Referenzkreis. Die selbstgegossene Membran hat eine Dicke von weniger als 1 μm. Weitere Torsionsprobleme werden in [8] beschrieben.

5. Das Stufenhologramm [9, 10]

Bei genaueren Festigkeitsberechnungen muß man auch mit genaueren Werten des Elastizitätsmoduls E rechnen. Außerdem braucht man diesen E Modul auch für die Modellwerkstoffe (Gips, Porcelin). Mit dem neuen Verfahren des Stufenhologrammes kann man auch bei kleinster Belastung den E-Modul um den Nullbereich präzis messen. Das Prinzip besteht darin, daß bis zu ca. 20 überlappende Doppelbelichtungshologramme auf einer Hologramm platte nacheinander aufgenommen werden. D.h. der holografische Meßbereich für Verformungen wird bis um das ca. 20-fache erweitert. Bild 9 zeigt das Prinzip, Bild 10 die Bestimmung des E-Moduls von Gips bei verschiedenen Belastungen. Man sieht, daß Gips bis zum Bruch einen konstanten E-Modul besitzt ($E = 5140 \pm 20$ N/mm^2). D.h. Gips ist vollelastisch bis zum Sprödbruch.

6. Querverformung von Biegestäben [11]

Wird z.B. ein einseitig eingespannter prismatischer Stab durchgebogen oder zum Schwi gen gebracht, so erleidet er auch eine Querverformung senkrecht zur Stabachse. Bei de schwingenden Stab mit rechteckigem Querschnitt nach Bild 11 ist die Querverformung an den gekrümmten, d.h. nicht parallelen Interferenzstreifen zu erkennen. Die weißen Zonen sind die Knotenbereiche. Die exakten Berechnungsverfahren kennen diese Art von Deformation nicht. In Bild 12 ist die Auswertung dargestellt. Erst wenn diese Deformationen mit in die Rechnungen einbezogen werden können, kann man auch die Eigenfrequenzen von Stäben treffsicherer berechnen, als es bisher möglich war. Die bis jetzt fehlenden Informationen über diese Sekundärdeformationen sind auch ein Grund für die mangelnde Genauigkeit von Schwingungsrechnungen, besonders für höhere Eigenschwingungsformen.

Abschlußbemerkung

Die Punkte 1 bis 6 zeigen als Beispiele eine Auswahl der Möglichkeiten, die die moderne holografische Interferometrie der konstruktiven Technik bieten könnte, wenn Holograf und Berechner bzw. Konstrukteur besser zusammenarbeiten würden. So aber bleiben viele Möglichkeiten ungenutzt, weil einerseits die Holografen oft nicht die Probleme der Technik genau genug kennen und andererseits Berechner und Konstrukteure mit den holografischen Interferogrammen nichts anzufangen wissen. Das dürfte auch ein wesentlicher Grund dafür sein, weshalb holografische Techniken noch nicht so Eingang in die Praxis gefunden haben, wie sie es eigentlich verdient hätten. Zumal holografische Techniken besonders bei Entwicklungsaufgaben schneller, sicherer und billiger sind al z.B. Finite-Elemente-Methoden.

Literatur

[1] SCHÖNEBECK, G.: VDI-Berichte 313 (1978) 155
[2] SCHÖNEBECK, G.: Diss. München (1979)
[3] SCHÖNEBECK, G.: Laser 79 (1979), Proceedings 576
[4] SCHÖNEBECK, G.: VDI-Berichte 366 (1980) 79
[5] SCHÖNEBECK, G.: 11. National Conference, B.H.E.L. Hyderabad (1982) FV 30
[6] SCHÖNEBECK, G.: VDI-Berichte 399 (1981) 39
[7] SCHÖNEBECK, G.: Laser 81 (1981)
[8] JUCKENACK, D.: Konstruktionstechnik 5, TU-Berlin (1983)
[9] SCHÖNEBECK, G.: SPIE-Conference, Genf (1983) Vortrag 398-17
[10] SCHÖNEBECK, G.: VDI-Berichte 480 (1983) 73
[11] SCHÖNEBECK, G.: Noch nicht veröffentlicht

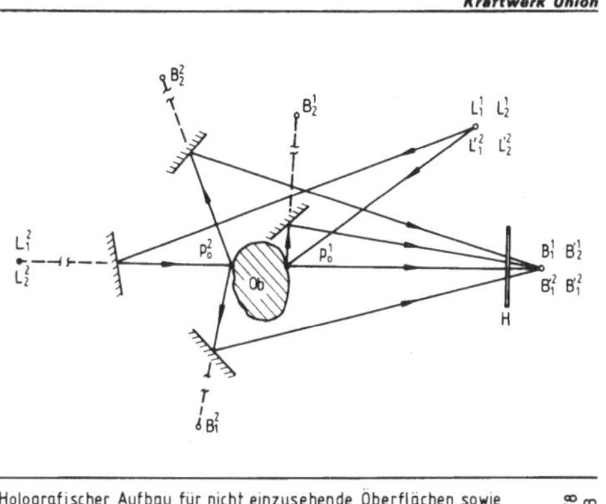

Bild 1, Prinzip der "Allgemeine Methode"

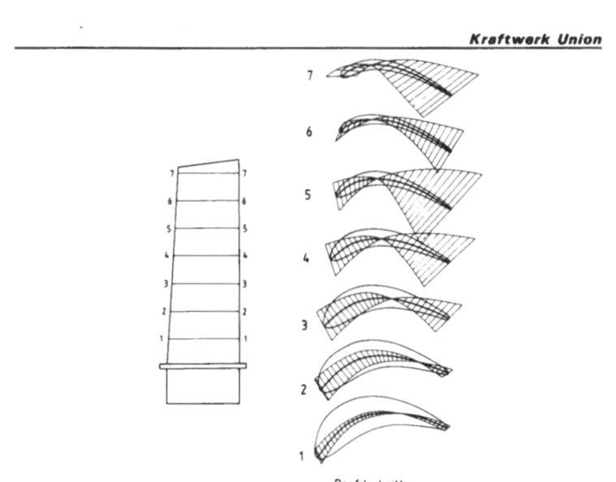

Bild 2, 1 Hologramm, 3 Ansichten

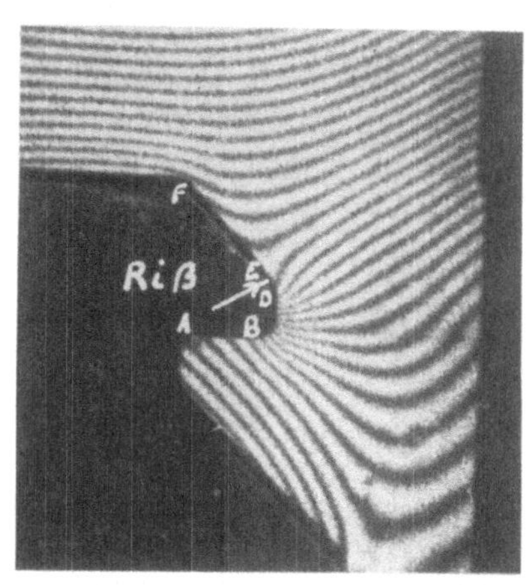

Bild 3, Auswert.der "Allgem.Methode"

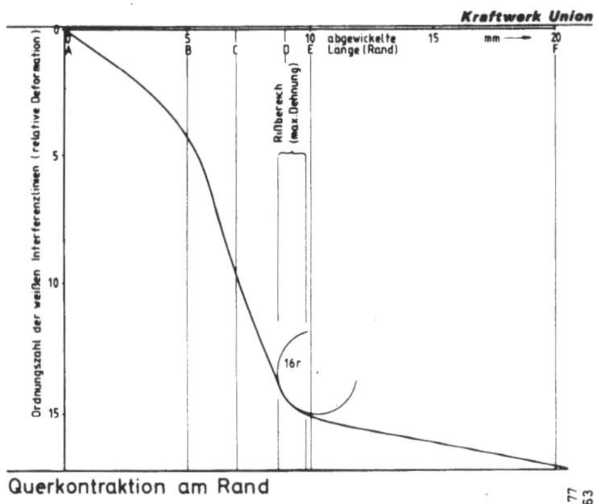

Bild 4, Höhenschichtlinien d.Querkontraktion

Bild 5, Ausw.der Querkontraktion

Bild 5, Modellmaterialien

202

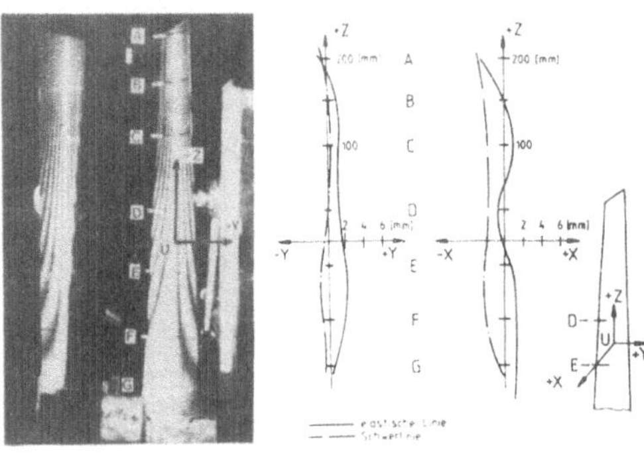

Interferenzbild   Diagramm
Bild 7, elastische Linie

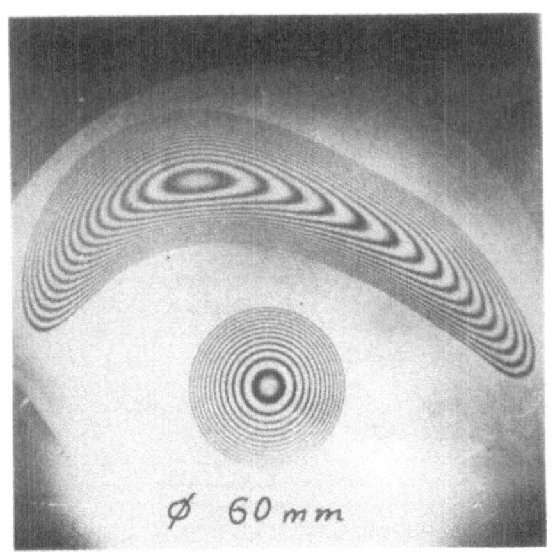

Bild 8, Seifenhautgleichnis (Hologramm)

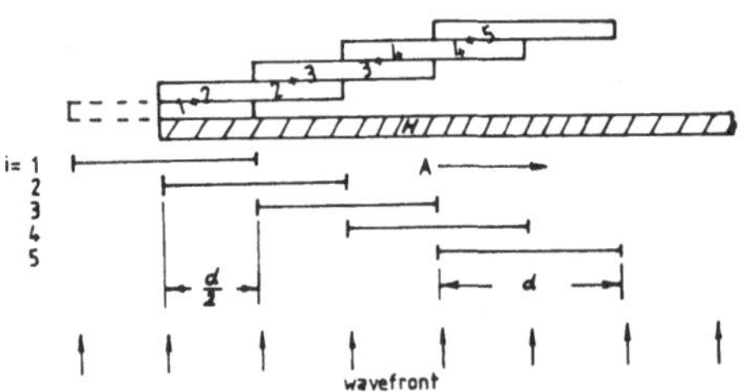

Überlappende Doppelbelichtungshologramme (1+2), (2+3), (3+4) usw.

d = stufenweise verschiebbarer Spalt
Bild 9, Stufenhologramm (Prinzip)

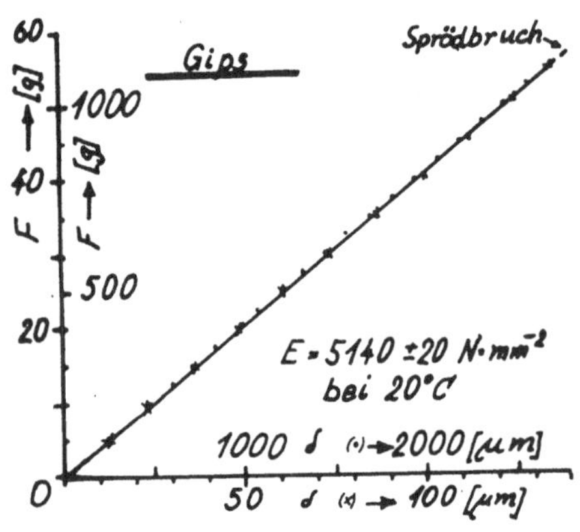

Bild 10, E-Modulbestimmung (Gips)

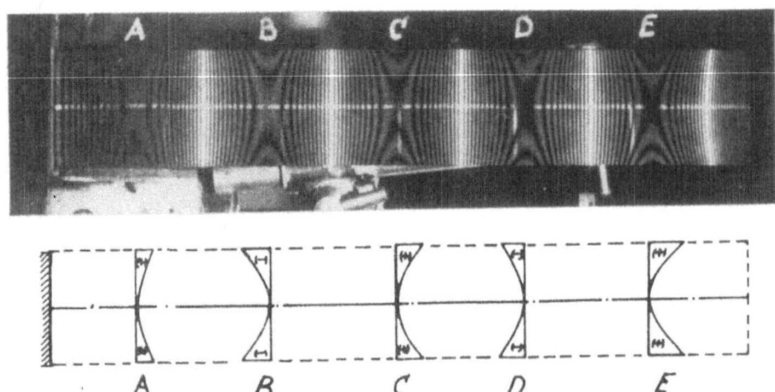

Bild 11, Querverformung   oben: Holografisches Interferenzlinienbild
Bild 12, beim Biegestab   unten: Auswertung

Holographische Schwingungsanalyse an rotierenden Bauteilen mit Hilfe eines objektlagen- und schwingungsphasenbezogenen Lasertriggerverfahrens

J.GELDMACHER, H.KREITLOW, E.VOGT* und B.DIRR*
Bremer Institut für angewandte Strahltechnik, BIAS
Ermlandstraße 59, D-2820 Bremen 71
*Institut für Mechanik, Universität Hannover,
Appelstraße, D-3000 Hannover 1

1.Einleitung

Die Entwicklung neuartiger holografischer Verfahren /1,2/ in den letzten Jahren ermöglicht die Untersuchung des dynamischen Verhaltens rotierender technischer Bauteile durch die Analyse ihrer Schwingungsformen. Hierzu ist es notwendig, den Einfluß der Drehbewegung auf die Interferenzmusterbildung während der Hologrammplattenbelichtung zu eliminieren. Dies gelingt durch Anwendung der vier folgenden Methoden:

1. der holografische Meßaufbau wird optimiert so, daß er bezüglich der Messung von In-Plane-Drehbewegungen des Bauteils möglichst wenig empfindlich ist,
2. das drehende Bauteil wird durch ein mit halber Objektdrehzahl rotierendes Prisma eines optischen Bild-Derotators stationär abgebildet,
3. der aus Hologrammplatte und Referenzwelle bestehende Teil des Interferometers führt eine Drehbewegung mit genau der gleichen Drehzahl und Drehrichtung wie auch das rotierende Bauteil aus,
4. die Belichtung der Hologrammplatte wird durch gezieltes Ansteuern des Lasers durchgeführt.

Über die Entwicklung und den Einsatz des Verfahrens unter 4. wird berichtet.

2.Objektlagen- und schwingungsphasenbezogenes Lasertriggerverfahren

Bei diesem Verfahren gelingt die holografische Speicherung von Schwingungsinterferogrammen rotierender Objekte, wenn die Doppelbelichtung der Hologrammplatte in der gleichen ausgewählten räumlichen Objektlage durchgeführt wird und die zu vergleichenden Schwingungsphasen die holografische Meßempfindlichkeit nicht überschreiten. Der zeitliche Abstand der Belichtungsimpulse beträgt dabei mindestens eine volle Umlaufzeit des Objektes.

Die wesentlichen Voraussetzungen für den erfolgreichen Einsatz dieses Meßverfahrens sind:

- Zur Triggerung der Laserimpulse muß die jeweils ausgewählte Objektlage auf Bruchteile eines Mikrometers erfaßt und das daraus abgeleitete Steuersignal dem Laser zugeführt werden.
- Die Verzögerungszeit zwischen Lasertriggerung und Hologrammplattenbelichtung darf sich in Abhängigkeit von der Drehzahl nur soweit ändern (Mikro- bzw. Nanosekunden), daß die dadurch holografisch gemessenen Objektdrehungen noch keine Interferenzmuster erzeugen. In Bild 1 wird diese zulässige zeitliche Änderung in Abhängigkeit von der Drehzahl für Objekte mit einem Durchmesser von 200, 400 und 800 mm bei einer Aufbaugeometrie S=B=(0,0,1000mm) nach Bild 2 dargestellt.
- Zur Berücksichtigung der holografischen Meßempfindlichkeit muß nach Auslösen des ersten Laserimpulses die jeweilige Schwingungsphase nach jeder Objektumdrehung mit der Phase zum Zeitpunkt des 1.Laserimpulses verglichen werden. Der zweite Laserimpuls wird dann ausgelöst, wenn bei gleicher Raumlage des rotierenden Objektes die Differenz der Schwingungsphasen innerhalb der holografischen Meßempfindlichkeit liegt.

Die Vorteile des Verfahrens gegenüber den anderen o.g. Verfahren sind:
- Der holografische Meßaufbau unterscheidet sich nicht von dem für Schwingungsanalysen an stationären Objekten. Die Justierung des Meßaufbaus ist daher einfacher.
- Bei großen Objekten sind keine Bildhologramme erforderlich.
- Es können auch feststehende Komponenten, z.B. Sichtfenster von Meßkammern, geringer optischer Qualität im Strahlengang der Objektwelle zwischen Objekt und Hologramm angeordnet werden.
- Das rotierende Objekt kann aus beliebigen Beobachtungsrichtungen holografiert werden.

3.Durchführung und Ergebnisse

Das Verfahren wurde zur Schwingungsanalyse an einem auf einer Schwungscheibe rotierenden Schaufelmodell (einseitig eingespannter Biegebalken mit b:h:l=28mm:3mm:225mm) in einem Prüfstand unter Betriebsbedingungen erfolgreich eingesetzt. In Bild 3 ist schematisch der verwendete holografische Meßaufbau dargestellt. Als Lichtquelle diente ein Rubin-Riesenimpulslaser mit temperaturstabilisiertem Etalon und Resonanzreflektor sowie einer Ausgangsleistung von etwa 20 mJoule. Das über einen Piezokristall angeregte Objekt konnte in diesem Prüfstand bei Drehzahlen bis 5500 U/min und einem Meßkammerunterdruck von wenigen Torr rotieren, wobei die Beleuchtung und Beobachtung des Objekts durch ein im Meßkammerdeckel eingefügtes Plexiglasfenster geringer optischer Qualität (Ebenheit und Parallelität) mit den Abmessungen $100 \times 400 mm^2$ erfolgte.

Die für die Steuerung des Lasers notwendigen Signale wurden über einen am Objekt befestigten DMS (Schwingungsphasenkontrolle) mit nachgeschalteter Telemetrieübertragung sowie über eine auf der Objektachse angebrachte Codiereinheit (Drehlagenerfassung) gewonnen, siehe Bild 4.

Bei den holografischen Messungen wurde das Schaufelmodell in den zuvor berechneten und im Campbell-Diagramm, Bild 5, als Funktion der Drehzahl dargestellten unteren Eigenfrequenzen angeregt. In den Bildern 6 bis 10 sind einige der bei unterschiedlichen Drehzahlen gemessenen Schwingungsformen als Interferogramme mit den zugehörigen quantitativen Auswertungen zu sehen.

Zur Absicherung eines ausgewählten Rechenmodells /3/ für die Berechnung von Eigenformen wurden die damit ermittelten Schwingungsformen mit den holografisch gemessenen verglichen. Die Berechnung der Eigenformen erfolgte über Energieausdrücke nach dem Verfahren von Rayleigh-Ritz: Mit einem allgemeinen Ritzansatz wird ein Funktional der potentiellen Energie aufgestellt, welches minimiert wird. Dies führt auf ein allgemeines Eigenwertproblem, dessen Lösung die Konstanten im Ritzansatz bestimmen.

Eine Analyse der Rechenergebnisse und der holografischen Interferogramme zeigte Abweichungen der berechneten von den gemessenen Schwingungsformen, die durch eine Überlagerung der erzwungenen Eigenformen und einer zusätzlichen Grundform mit doppelter Drehfrequenz -angeregt durch das Erdschwerefeld- erklärt werden können.

4. Zusammenfassung

Es wurde gezeigt, daß das holografische objektlagen- und schwingungsphasenbezogene Lasertriggerverfahren zur Messung von Schwingungsformen an rotierenden Bauteilen auch unter Betriebsbedingungen geeignet ist. Dabei konnte anders als bei anderen holografischen Verfahren ein Sichtfenster geringer optischer Qualität bzgl. Ebenheit und Parallelität der Fensterflächen zur Beobachtung des Objekts in einer evakuierten Meßkammer verwendet werden. Ein Vergleich der holografisch gemessenen Schwingungsformen mit den berechneten Eigenformen ergab Hinweise für die Verbesserung des ausgewählten Rechenmodells.

Literatur

/1/ Beeck,M.-A.u.H.Kreitlow: Proc.Laser 79,IPC (1979),408-419
/2/ Geldmacher,J.,H.Kreitlow,P.Steinlein u.G.Sepold:SPIE-Proc.398(1983)
/3/ Vogt,E.,J.Geldmacher,B.Dirr u.H.Kreitlow:erscheint demnächst in
      Experimental Mechanics

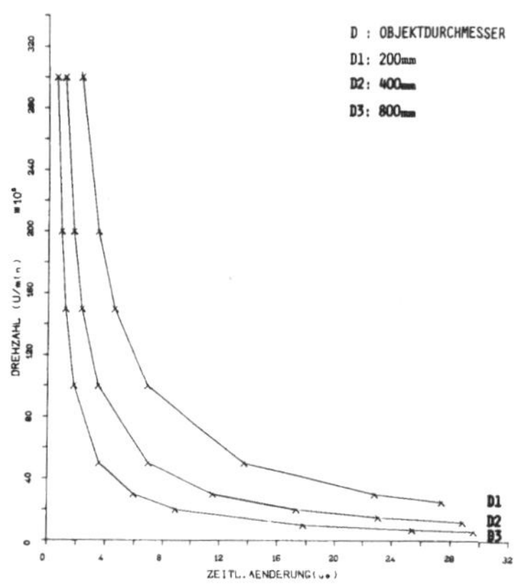

Bild 1. Zul. zeitl. Abweichung der Hologrammbelichtung

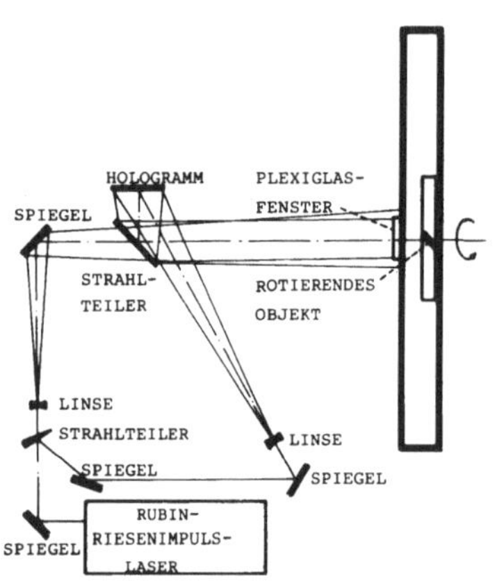

Bild 2. Aufbaugeometrie des eingetzten holografischen Meßaufbaus

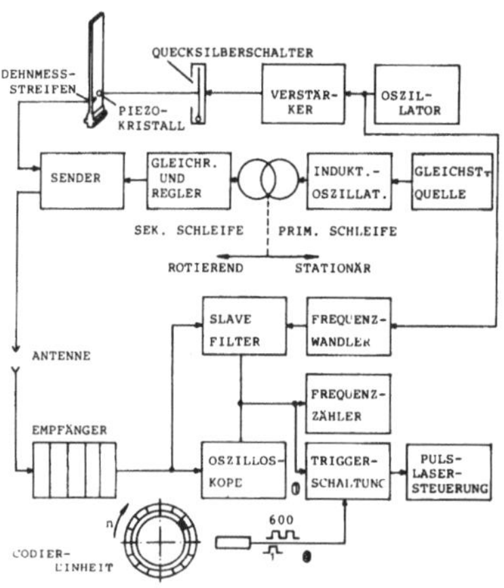

Bild 3. Schematische Darstellung d. holografischen Meßaufbaus

Bild 3. Erfassung der Signale zur Lasersteuerung

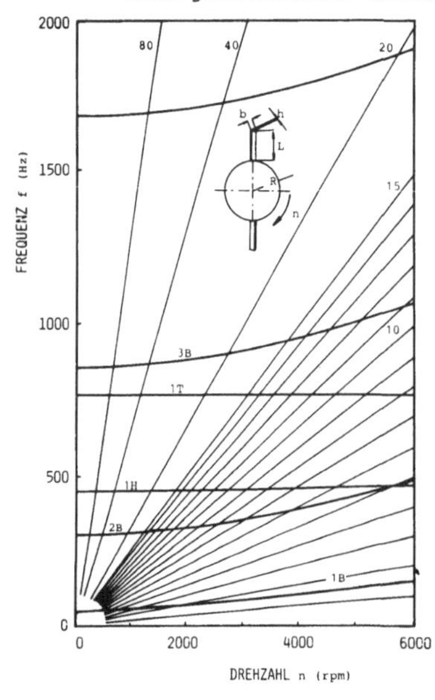

Bild 5. (links) Campbell-Diagramm des untersuchten Schaufelmodells

Bild 6. (rechts) Vergleich von berechneten mit holografisch gemessenen Schwingungsformen

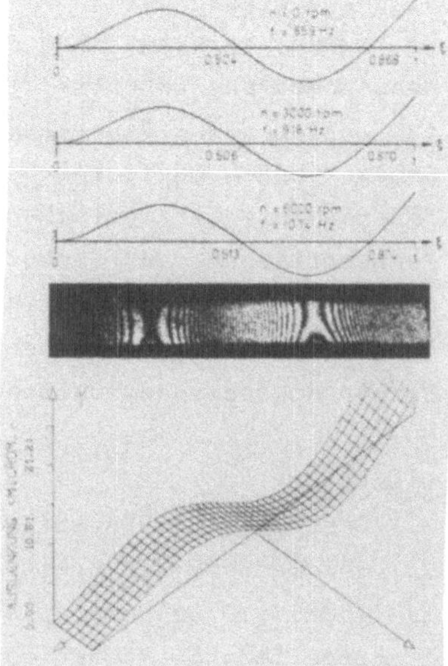

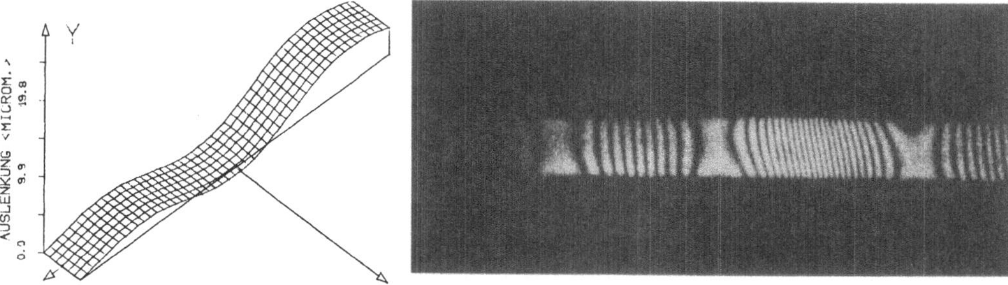

Bild 7. Schwingungsinterferogramm eines rotierenden Schaufelmodells bei
n=4000 U/min, f=1850 Hz sowie dessen quantitative Auswertung

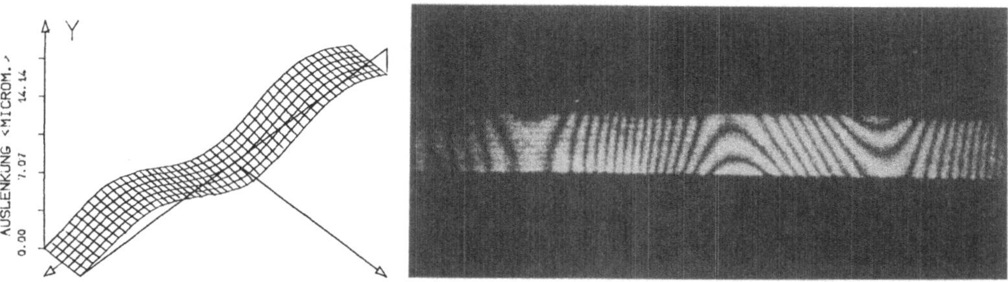

Bild 8. Schwingungsinterferogramm eines rotierenden Schaufelmodells bei
n=5000 U/min, f=1903,1 Hz sowie dessen quantitative Auswertung

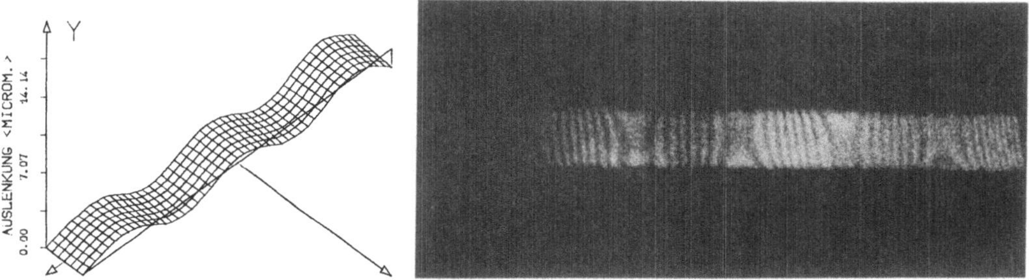

Bild 9. Schwingungsinterferogramm eines rotierenden Schaufelmodells bei
n=3000 U/min, f=4356 Hz sowie dessen quantitative Auswertung

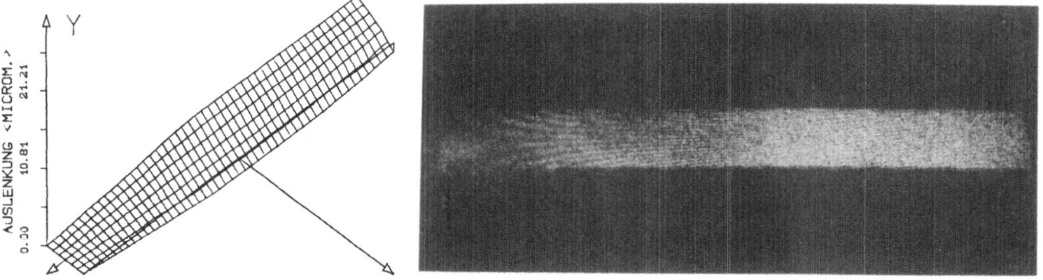

Bild 10.Schwingungsinterferogramm eines rotierenden Schaufelmodells bei
n=3000 U/min, f=777,0 Hz sowie dessen quantitative Auswertung

Angewandte Holographie bei den Zweiphasenströmungen der chemischen Verfahrenstechnik

J. J. TIMKÓ
Forschungsinstitut für Technische Chemie, UAK
Budapest/Ungarn

Die Zweiphasenströmungen der chemischen Verfahrenstechnik sind dynamische, stochastische Prozesse, die mit viel Parametern beeinflusst werden.

Mit holographischer Interferometrie können Strömungen, Wärme- und Stoffaustausch abgebildet werden, aber von den unwiederholbaren Eigenschaften eines dispersen Systems – z.B. der Bildung von Tropfen oder Blasen – geben Interferogramme fast keine Information. Deshalb wurde eine impulsholographische Methode entwickelt / 1, 2, 3 /, um das strömende disperse System näher kennen zu lernen. Mit diesem "in-situ" anwendbaren Verfahren können Grösse, Grössenverteilung, Häufigkeit und räumliche Lage von Partikeln oder Tropfen auch bestimmt werden.

Die Impulsholographie hat den grossen, unersetzlichen Vorteil, die äusserst rasch ablaufende Vorgänge trägheitsfrei und ohne störende Eingriffe zu erfassen. Die Auswertung ist in beliebiger Zeit später möglich und gibt räumliche Informationen von dem ganzen Prozess.

Das – in dem Holographischen Laboratorium des Forschungsinstitutes für Technische Chemie der Ungarischen Akademie der Wissenschaften entwickeltes – Messverfahren hat eine patentierte Anordnung / 1, 2, 3/ bei den Aufnahmen, wo gleichzeitig zwei Hologramme entstehen können.

Als Beleuchtungsquelle dient ein Impulsrubinlaser, mit 20 ns Blitzdauer. Die Hologramme werden auf Agfa-Gevaert "Holotest" Filme oder Platten aufgenommen. Dem abbildenden Prozess gemäss werden die Aufnahmen entweder mit einer Streulinse aufgeweitetem Laserstrahl, oder mit parallelem Laserbündel durchgeführt. Dabei kann die Geradeaus-Methode oder der optische Aufbau mit getrenntem Referenzbündel verwendet werden.

Die Wiedergabe erfolgt durch einem kontinuierlichen HeNe-Gaslaser und geschlossenem TV-System. Der optische Aufbau gibt einen gewissen Spielraum auch bei der Abbildung. Das räumliche Bild erscheint als eine Reihe von

nacheinanderfolgenden, scharfgestellten Ebenen. So können überlagerte Tropfen
oder Blasen getrennt werden. Von dem Monitor wird die obengenannte Bildreihe
photographiert und ausgewertet. Die Auswertung kann auch automatisch, elektro-
optisch vorgenommen werden.

Dazu dient ein Analysengerät, mit einer Rechenmaschine gekoppelt.

Das Messverfahren wurde in dem Laboratorium und auch bei Versuchen in
Industriegrössen mehrfach verwendet.

Die Methode war bisher bei Zerstäubung /mechanische- und pneumatische Düsen,
Zentrifugalscheiben/; Sedimentation /mit und ohne Flokkuliermittel/; Wirbel-
schicht; Kristallbildung- und auflösung; Luftfiltrieren und Blasenbildung
erfolgreich verwendet.

Eine interessante Zweiphasenströmung wurde in einem Apparat neulich untersucht, wo Blasenbildung dazu diente, die Transportvorgänge zu beschleunigen.

Als Versuchsapparat diente eine Kolonne in halbindustrieller Grösse; Ø 450 mm.
An die obere Wasserfläche wurde senkrech ein Wasserstrahl mit einer Geschwindigkeit von 22 m/sec geführt. Der Aufprall des Wasserstrahles auf der Oberfläche
hat die Wirkung, das die Luft wie ein Sack miteingezogen wird. Dieser "Luftsack"
bricht in kleine Blasen zusammen. Die Blasen werden in der Richtung der Achse
immer grösser und endlich schäumt das ganze Wasser, erfüllt von Blasen. Die
Blasen steigen dann am Wandrand empor an die Oberfläche. Die erzeugten Blasen
haben ein Durchmesser von 10– 100 μm. Der Stoffaustausch, laut Messungen der
Technischen Universität in Budapest – wo der Apparat auch konstruiert wurde –
übertrifft die in der Fachliteratur erwähnten Werte. /4/

Die Aufgabe war – nach der Möglichkeit – festzustellen, wie die Oberfläche
des Wasserspiegels sich beim Aufprall des Wasserstrahles benimmt, wie die
Luftblasen entstehen, ob und wie die Blasen grösser werden. Das alles wird
in einem Schaum holographiert, der sich mit Schnellfilmen nicht durchblicken
liess. Dazu wurde Durchlicht- und Auflichtholographie verwendet.

Durchlichtholographie hat sich bewährt, wo der in die ruhende Flüssigkeit
eindringende und zerfallende Flüssigkeitsstrahl noch weniger turbulent war,
als in dem Schaumzustand. Die Hologramme zeigen sogar die wellenförmige
Oberfläche des Strahles, wo die Rände abreissen.

210

Die gleichzeitig aufgenommene Schnellfilm-aufnahmen zeigen die ganze Erscheinung
sehr deutlich, aber sie geben von den sich bildenden Blasen keine räumliche
Information. Von den Hologrammen können aber die einzelnen Blasen auch ausgewertet werden, ihre räumliche Lage ihre Grösse und Grössenverteilung. Diese
Daten bestimmen das Verfahren eindeutig.

Der turbulente Schar von Tropfen und Blasen wurde mit allen erwähnten Versuchs-
anordnungen abgebildet. Dabei könnte festgestellt werden, dass die Durchlicht-
hologramme klarere Konture von den Tropfen und den Blasen zeigen als die Auf-
lichthologramme, aber mit Auflichthologrammen lässt sich ein grösseres Volumen
durchblicken. Die Koalescens der einzelnen Blasen und ihr Aufstieg neben der
Wand des Apparates wurde von den Hologrammen nicht nur im ganzen Querschnitt
sichtbar; sondern die erwähnten Blasenparametern könnten teilweise auch
quantitativ ausgewertet werden.

Diese Experimente haben die Anwendungsreihe der Methode vollständig gemacht,
da hier Gasdispersion in Flüssigkeit abgebildet wurde.

Mechanische Düsen dienten schon früher als Versuchsobjekte. In dem Gebiet der
Energiewirtschaft hat die holographische Methode dazu geholfen von Brennstoff-
düsen erzeugte Flüssig-Gas Dispersionen im kalten Zustand und brennend näher
kennen zu lernen /5/. In diesem Falle war die flüssige Phase das dispergierte
Mittel.

Wenn die Ergebnisse beider Experimente verglichen werden, kann auch festgestellt werden, dass die Grenzflächen der disperser Phase die optimale
Anordnung bei Aufnahme und Wiedergabe stark beeinflussen.

Das "in-situ" verwendbare holographische Messverfahren hat bisher in allen
Vervendungsgebieten /6/ wo in dem chemischen Prozess Zweiphasenströmungen eine
Rolle spielten neue Information über das disperse System gegeben und Theorie
und Praxis einander näher gebracht.

Literatur

/1/ TIMKÓ et al. Hung. Patent 175.498.
                "Eljárás és berendezés diszperz rendszerek fizikai jellem-
                zőinek holográfiás meghatározására." 1981. nov. 4.

/2/ U.S.A. Patent 4.278.319.
            "Process and apparatus for the determination of the physical
            characteristics of dispersed systems by holography."
            Jul. 14. 1981.

/3/ U.K. Patent. GB 2.042.754B.
            "Process and apparatus for the determination of the physical
            characteristics of dispersed systems by holography."
            Oct. 27. 1982.

/4/ KENYERES, Techn. Univ. Budapest, Inst. f. Landw. Chem. Techn.
            "HTPS-System" 1982.

/5/ TIMKÓ, J.J.   "The Investigation of Transport Phenomena by Applied
            Holography."
            FLOW VISUALIZATION II. /Editor: W. MERZKIRCH/
            Hemispere Publishing Corporation, Washington New York
            London, 1982. /pp. 535-541./

/6/ TIMKÓ, J.J.   "Angewandte Holographie in der chemischen Verfahrenstechnik
            - als vielversprechende Messmethode."
            OPTOELEKTRONIK IN DER TECHNIK /Editor: W. WAIDELICH/
            Springer-Verlag, Berlin Heidelberg New York,1982./pp.63-68./

Weitere Information:

    Holographisches Labor des Forschungsinstitut für Techn. Chemie der UAK
    H-1502 Budapest, Pf. 98.

Echtzeitholographie mit dem Dauerstrichlaser bei Schwingungsuntersuchungen

R. Feiertag und M. Trundt
Elektromechanische Konstruktionen, Universität -
GHS Wuppertal, Fuhlrottstr. 10, 5600 Wuppertal 1

Die Holographie als berührungsloses optisches Meßverfahren
zur rückwirkungsfreien Messung kleinster Verformungen zeigt
ihre besonderen Vorteile bei Schwingungsmessungen, da man
dabei die gesamte Objektoberfläche mit einer einzigen
Messung erfassen kann. Es ist dadurch einfach möglich, die
Maxima und Minima des Schwingungsbildes zu lokalisieren und
die Amplituden quantitativ anzugeben. Durch Frequenz- und
Amplitudenmodulation des Schwingerregers kann sehr einfach
und rasch eine Frequenzanalyse des Untersuchungsobjektes
durchgeführt werden. /1/

Obwohl die Vorteile dieses Verfahrens auf der Hand liegen,
ist es erstaunlich, daß dieses Verfahren in der Praxis noch
relativ wenig genutzt wird. Gerade in der Feinwerk- und
Gerätetechnik, wo oft störende mechanische Schwingungen
kleiner Amplituden auftreten, ist ein rückwirkungsfreies
Schwingungsmeßverfahren sehr wichtig.

Bisher beschränkt sich die Anwendung der Holographie zur
Schwingungsanalyse mit Impulsen im wesentlichen auf Objekte
größerer Abmessungen, wie Autoreifen, Turbinenschaufeln,
Tragflügel oder Getriebegehäuse. /2, 3/ Dies ist darauf zurückzuführen, daß die praktischen Einsatzmöglichkeiten der Echtzeitholographie mit Dauerstrichlaser zur Messung mechanischer
Schwingungen bis heute noch nicht genügend erprobt sind, obwohl gegenüber der Doppelpulsholographie der gerätetechnische
Aufwand für Holographie mit dem Dauerstrichlaser wesentlich
geringer ist.

Mit einem Versuchsaufbau aus einem 1W-Dauerstrichlaser, der
an einer schwingungsisolierten T-Nutengußplatte über Luft-
polster gedämpft aufgehängt ist, wird die Schwingungsiso-
lierung gegenüber der Umwelt durch ein pneumatisches Regel-
system erreicht. Die Hologramme werden mittels einer Holo-
sofortbildkamera auf thermoplastischem Film gespeichert.
Die Dokumentation der Versuchsergebnisse erfolgt mit Video-
recorder oder Fotokamera.

Im wesentlichen wurden zwei Hologrammtechniken erprobt.
Die eine Methode besteht darin, ein Hologramm des ruhenden
Objektes anzufertigen und auf thermoplastischem Film zu
speichern. Dieses Hologramm wird dann mit dem schwingenden
Objekt überlagert rekonstruiert. Durch Überlagerung der
beiden Objektzustände in den Umkehrlagen der Schwingung entsteht ein Interferenzmuster.

Bild 1 zeigt das Hologramm einer Lautsprechermembran bei
einer Anregungsfequenz von 800 Hz. Es sind deutlich zwei
zentrumsymmetrische Schwingungsbereiche mit je 8 Schwingungs-
maxima zu erkennen. Das symmetrische Interferenzmuster entsteht durch eine gleichmäßige Auslenkung der Membran.

Bild 1:
Echtzeithologramm einer
Lautsprechermembran
bei 800 Hz.

Die andere Methode ist die Speicherung mehrerer Schwingungs-
perioden auf einem Hologramm. Bild 2 zeigt das gleiche Objekt
wie Bild 1 auch bei einer Anregungsfrequenz von 800 Hz. Das
Schwingungsmuster ist das gleiche wie bei Bild 1. Der Unterschied liegt zum einen in dem besseren Kontrast des Hologramms
und zum anderen darin, daß doppelt so viele Interferenzlinien
erscheinen.

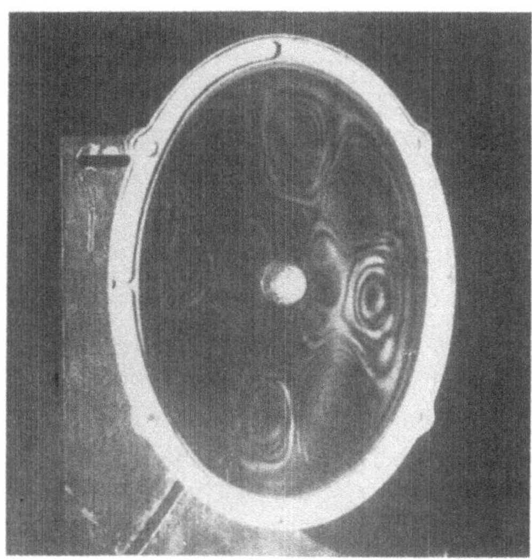

Bild 2:
Lautsprechermembran bei
800 Hz.

Der Vorteil der ersten Methode liegt darin, daß man das Bezugshologramm für mehrere Messungen bei verschiedensten Frequenzen
heranziehen kann. Dabei ist allerdings zu bedenken, daß sich
die Hologramme durch den Einfluß umweltbedingter Störfaktoren
permanent verändern. Je länger man das Bezugshologramm stehen
läßt, desto schlechter wird die Qualität und desto mehr
Interferenzlinien erscheinen infolge des ständig vorhandenen
Störpegels. Es ist also stets abzuwägen, wie lange man ein
Bezugshologramm verwenden kann. Im Gegensatz hierzu liefern
Nullhologramme schwingender Objekte zwar stets kontrastreiche
Interferenzbilder, wobei der Einfluß äußerer Störgrößen infolge
der Überlagerung beider Umkehrlagen der Objektschwingung aus-
geschaltet wird.

Von Nachteil ist allerdings, daß für jede Messung ein neues
Hologramm angefertigt werden muß und die doppelte Linienzahl
auftritt, wodurch schneller die Auflösungsgrenze erreicht
wird. Diese Methode ist deshalb immer dann vorzuziehen, wenn
man genaue Schwingungsbilder bei einer Frequenz darstellen will.

Es hat sich gezeigt, daß besonders dünnwandige Versuchsobjekte sehr gut für holographische Messungen geeignet sind.
Versuchsreihen von Lautsprechermembranen ergaben aussagekräftige Ergebnisse. Es ist bemerkenswert, daß es kein alternatives Meßverfahren gibt, mit dem Schwingungen weicher Membranen optisch dargestellt werden können. Die Existenz solcher Schwingungsmuster wurde etwa bereits um 1880 von
Friedrich Chladni mit Hilfe von Bärlappsamen gezeigt, den
exakten Nachweis konnte aber erst die Holographie erbringen.
Holographisch ist es möglich, auch fehlerhafte oder zu weiche
Einspannungen der Membran zu erkennen. Materialfehler machen
sich durch Unregelmäßigkeiten des Interferenzmusters bemerkbar. Auch das Übertragungsverhalten, Eigenfrequenzen, obere
und untere Grenzfrequenz von Lautsprechermembranen können
auf diese Weise sichtbar gemacht und bestimmt werden.
Interessant ist z. B. auch, daß bei den untersuchten Laut-
sprechern keine obere Grenzfrequenz gefunden werden konnte.
Selbst im Ultraschallbereich von 30 kHz zeigte die Membran
deutlich ein sichbares Interferenzmuster. Solche Ergebnisse
konnten bisher mit herkömmlichen Meßverfahren nur mit großem
Aufwand gewonnen werden.

Neben Lautsprechermembranen kann auch das Schwingungsver-
halten von Antrieben, Lüftermotoren, Pumpen, Transformatoren
und anderen elektromechanischen Geräten untersucht werden.
Bei diesen Objekten kann man aussagekräftige Ergebnisse erhalten, denn das Interferenzmuster ermöglicht es, jede Verformung der Oberfläche quantitativ zu bestimmen.
Besonders vorteilhaft ist die Holographie bei der Untersuchung von Gehäusen. Da es sich bei Gerätegehäusen meist
um dünnwandige, zu Schwingungen neigende Blech- und Kunst-
stoffkonstruktionen handelt, läßt sich das Verfahren zur

Optimierung der Gehäuseform heranziehen, da mit Hilfe des
Interferogrammes genau angegeben werden kann, welche Bereiche
zu Schwingungen angeregt werden und wo die stärksten Bean-
spruchungen auftreten.

Die Untersuchungsergebnisse sind besonders bei geringen
Amplituden nicht immer eindeutig. Daher ist es wichtig, den
Einfluß der umweltbedingten Störfaktoren zu erkennen und
diese möglichst gut zu beseitigen. Bei dem hochempfindlichen
Meßverfahren der holographischen Interferometrie machen sich
Temperatur- und Luftfeuchtigkeitsschwankungen, Luftbewe-
gungen, Raum- und Körperschall als Störungen des Interferenzmusters bemerkbar. Bei der Auswertung ist es häufig schwierig,
diese überlagerten Störgrößen zu erkennen und die entsprechenden Interferenzlinien zu eliminieren. Deshalb sollte das
Versuchslabor möglichst klimatisiert und schallgedämmt sein.

Besonders vorteilhaft an der Echtzeitholographie ist, daß die
Erfolge von Dämpfungsmaßnahmen sofort nachprüfbar sind. Damit
können Optimierungen von Blech und Kunststoffgehäusen durch
Einschweißen von Rippen, durch Aufkleben von Dämmatten oder
durch Ausschäumen in kurzer Zeit durchgeführt und der Erfolg
nachgewiesen werden.

Literatur

(1) Wernicke, G.; Osten, W.: Holografische Interferometrie
    Physik-Verlag, 1982 Weinheim

(2) Schönebeck, G: Eine allgemeine holografische
    Methode zur Bestimmung räumlicher Verschiebungen
    Dissertation, München, 1979

(3) Steinbichler, H: Beitrag zur quantitativen
    Auswertung von holografischen Interferogrammen,
    Dissertation, München 1973

Messungen der Verformungen durch Schwerkrafteinfluß an Großbauteilen mit einem speziellen holographischen Interferometer

W.JÜPTNER, J.GELDMACHER, G.SEPOLD und D.PLATHNER*
Bremer Institut für angewandte Strahltechnik, BIAS
Ermlandstraße 59, D-2820 Bremen 71
*Institut für Radioastronomie im Millimeterbereich, IRAM
Voie 10, Domaine Universitaire de Grenoble, France

1.Einleitung

Die Konstruktion des aus hyperbolisch geformten Reflektoren zusammengesetzten 30-Meter-Radioteleskopspiegels für den Submillimeterbereich zur Beobachtung bis zu Entfernungen von 10 Milliarden Lichtjahren erfordert eine bisher auf diesem Gebiet nicht gekannte Formbeständigkeit. Die einwandfreie Funktion dieser astronomischen Meßgeräte ist nur gewährleistet, wenn die Formabweichung der Reflektoroberfläche (größer $1m^2$) unter Betriebslast weniger als 100 Mikrometer beträgt. Die Verformung, d.h. die Abweichung von der Fertigungskontur der Reflektoroberflächen, während des astronomischen Einsatzes kann durch Windbelastung, durch thermische Belastung aufgrund von Temperaturschwankungen sowie durch Schwerkraftbelastung als Folge unterschiedlicher Reflektorstellungen bezüglich des Erdschwerefeldes hervorgerufen werden.
Zur Einhaltung der hohen Anforderungen an die Formbeständigkeit wurden neue Reflektorstrukturen entwickelt, die auf Verformung bei den genannten Belastungsarten geprüft wurden. Die untersuchten Reflektoren mit einer Fläche von $1000x1250mm^2$ bestanden aus einer Basisstruktur, über der im Abstand von etwa 50mm vierzehn Mini-Reflektoren befestigt waren, die die Reflektoroberfläche bildeten, Bild 1. Während die Verformungen bei Temperatur- und Windbelastung (Luftdruckänderung) mit herkömmlichen holografischen Interferometern und Techniken gemessen werden können, ist für die Verformungsmessung bei Schwerkraftbelastung ein spezielles drehbares holografisches Interferometer erforderlich. Über diese neuartige Verformungsmeßtechnik wird berichtet.

2.Holografischer Versuchsaufbau

Der zu entwickelnde holografische Meßaufbau muß hinsichtlich einer konstanten Aufbaugeometrie die folgenden Anforderungen erfüllen:

- Durch Drehung des holografischen Meßaufbaus um 90° kann die Reflektorstruktur in eine senkrechte Position (keine Schwerkraftbelastung in Oberflächennormalenrichtung) und eine waagerechte Position (maximale Schwerkraftbelastung in Oberflächennormalenrichtung) gebracht werden. Durch den interferometrischen Vergleich beider Objektzustände, z.B. mit dem holografischen Doppelbelichtungsverfahren, soll der Einfluß der Schwerkraft auf die Reflektorform gemessen werden.

- Die Torsionsteifigkeit des Meßaufbaus muß so groß sein, daß er sich bei der notwendigen Drehung um 90° zwischen den beiden Objektbelichtungen nicht verwindet, andernfalls würde dies ebenfalls zur Simulation einer Objektbewegung führen.

- Die Biegesteifigkeit des Meßaufbaus muß so groß sein, daß seine Verformung durch die Schwerkraft sehr klein ist gegen die zu messende Schwerkraftverformung des Objekts.

- Aufgrund der notwendigen Größe des Meßaufbaus ist eine Isolierung gegenüber Umgebungseinflüssen -Luftschall,Bodenerschütterungen- in vertretbarem Maße nicht möglich. Darüberhinaus können wegen der langen Zeiten zwischen den beiden holografischen Belichtungen vor und nach der Drehung des Meßaufbaus Relativbewegungen der optischen Komponenten auftreten, die eine Ganzkörperbewegung des Objekts im Meßergebnis simulieren. Dieser Meßfehler ist durch optische Maßnahmen zu vermeiden.

Der nach diesen Forderungen entwickelte und eingesetzte Meßaufbau zur holografisch interferometrischen Messung der Schwerkraftverformung, Bild 2, kann in drei Funktionseinheiten aufgeteilt werden: Grundrahmen, Laser und holografisches Interferometer.

<u>Grundrahmen</u>: Dieser Teil des Meßaufbaus wurde aus Aluminiumhohlprofilen X-95 erstellt. Er diente zur Aufnahme und zur gegenseitigen Positionierung von Reflektor, Interferometer und Laser sowie zur Drehung dieser Komponenten ohne Änderung der gegenseitigen Lage. Die Größe des Rahmens wurde so gewählt, daß eine vollständige Ausleuchtung und Beobachtung des Objekts gewährleistet war. Aufgrund der günstigen Materialeigenschaften der verwendeten Aluminiumhohlprofile konnte das Eigengewicht bzw. die Eigenverformung unter Beibehaltung einer hohen Biege- und Torsionssteifigkeit klein gehalten werden.

<u>Laser</u>: Als Lichtquelle wurde ein Rubin-Riesenimpulslaser mit temperaturstabilisierten Etalons zur Modenselektion im Einzelpulsbetrieb mit einer Ausgangsleistung bis zu 20 mJ bei einer Pulslänge von etwa 40 Nanosekunden eingesetzt. Mit dieser Lichtleistung können Objektflächen von über 2 m^2 Größe vollständig ausgeleuchtet werden, wie Vorversuche zeigten.

<u>Interferometer</u>: Die vom Impulslaser ausgekoppelte Lichtwelle wurde aufgeweitet und über einen Spiegel zur Objektbeleuchtung umgelenkt, s.Bild 3. Aus diesem Objektwellenfeld wurde über einen Referenzspiegel, der starr mit dem Reflektor verbunden war, die Referenzwelle ausgekoppelt, in Richtung der Hologrammplatte umgelenkt und dort mit der Objektwelle überlagert.

Mit Hilfe dieses Interferometeraufbaus nach dem Referenzspiegelverfahren gelingt die Eliminierung von Ganzkörperstörbewegungen (gleiche Bewegung aller Objektoberflächenpunkte) aus dem Meßergebnis /1/. Die durch die Störbewegung verursachte Lichtwegänderung der Objektwelle wird durch den gekoppelten Referenzspiegel in gleicher Größe und Richtung auf die Referenzwelle übertragen. Führen jedoch Oberflächenpunkte gegenüber dem Referenzspiegel unterschiedliche Bewegungen durch, werden diese durch das Verfahren nicht kompensiert. Bei solchen Verformungen handelt es sich i.a. um Verformungen des Objektes, die wie bei den vorliegenden Messungen die zu messenden Größen sind. Das Referenzspiegelverfahren erlaubt somit trotz Unterdrückung von Störbewegungen die Messung von Verformungen.

3. Versuchsdurchführung und Ergebnisse

Die Messungen wurden sowohl bei senkrechter als auch waagerechter Ausgangslage des Reflektors durchgeführt, wobei die Drehung des Meßaufbaus aufgrund der einfachen Handhabung manuell durchgeführt werden konnte. Zur Kontrolle der Eigenverformung des Meßaufbaus wurde neben dem Objekt auch ein Teil des Grundrahmens holografisch belichtet, so daß mögliche Biege- oder Torsionsbewegungen im Rahmen holografiert werden konnten. Aufgrund dieser Information konnte in Vorversuchen der Grundrahmen durch zusätzliche Maßnahmen wie Diagonalverstrebungen soweit verbessert werden, daß die Forderungen hinsichtlich Biege- und Torsionssteifigkeit erfüllt wurden. Dies wird durch die sehr gute Reproduzierbarkeit der holografischen Meßergebnisse bestätigt.

Störungen aus der Umgebung aufgrund fehlender Schwingunsisolierung konnten durch den Einsatz des holografischen Referenzspiegelverfahrens aus dem Meßergebnis eliminiert werden.

Schon die qualitative Betrachtung der erzielten holografischen Interferogramme weist auf unterschiedliches Verformungs- und Bewegungsverhalten der einzelnen Mini-Reflektoren hin. Noch deutlicher macht dies eine quantitative Auswertung, die in jeweils in 9x9 Rasterpunkten der Mini-Reflektoroberflächen durchgeführt wurde, Bild 4. Es zeigt sich, daß sich die Formabweichungen aus zwei Komponenten zusammensetzen:

1. aus einer Kippbewegung der Mini-Reflektoren, hervorgerufen durch eine Durchbiegung der Basisstruktur infolge der Schwerkraft. Die Größe und Richtung dieser Bewegung ist abhängig von der Durchbiegung am Befestigungsort des jeweiligen Mini-Reflektors,
2. aus der Verformung der Mini-Reflektoroberfläche infolge der Schwerkraft.

Der wesentlich größere Anteil an der Formabweichung wird durch die Kippbewegung hervorgerufen. Dies zeigt beispielhaft das Interferogramm und dessen Auswertung, Bild 5. Eine reine Kippbewegung des Mini-Reflektors würde parallele und gradlinige Interferenzstreifen im Interferogramm erzeugen. Die zusätzliche Veformung bewirkt die leichte Krümmung im Streifenverlauf.

4. Zusammenfassung

Die durchgeführten holografisch interferometrischen Messungen hatten zum Ziel, Verformungen von großflächigen, hyperbolisch geformten Teleskopreflektoren infolge einer Schwerkraftbelastung zu bestimmen. Dazu wurde für dieses Meßproblem ein drehbarer nichtschwingungsisolierter Meßaufbau für große Objekte entwickelt und erfolgreich eingesetzt. Neben der Bestimmung der Reflektoroberflächenverformung zeigten die Meßergebnisse auf, daß Verbesserungen der Formbeständigkeit vor allem durch eine stabilere Basisstruktur zu erreichen sind.

Durch die Messungen konnte darüberhinaus nachgewiesen werden, daß die holografische Interferometrie als Meßverfahren für Schwerkraftverformungen an Bauteilen größer 1 m^2 erfolgreich einsetzbar ist.

Literatur

/ 1/ Jüptner,W.,   Untersuchungen zur Fehlerbewertung an
      Kreitlow,H.,   Reaktorbauteilen mit Hilfe der holographischen Interferometrie,
      Fischer,B. und   fischen Interferometrie,
      Geldmacher,J.   Abschlußbericht zum BMFT-Forschungsvorhaben RS 327, Mai 1980

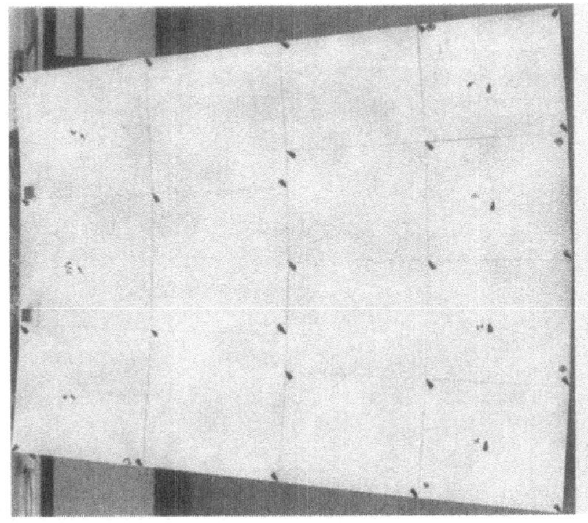

Bild 1. Reflektor eines 30-Meter-
Radioteleskopspiegels

Bild 2. Holografischer Aufbau zur
Messung der Schwerkraft-
verformung

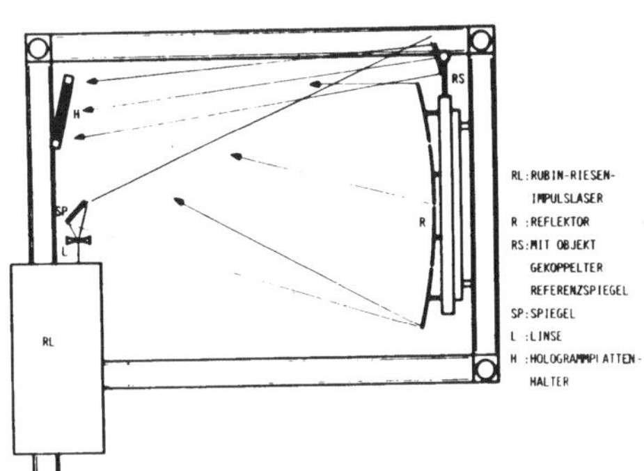

Bild 3. Holografisches Interferometer
nach dem Referenzspiegel-
verfahren

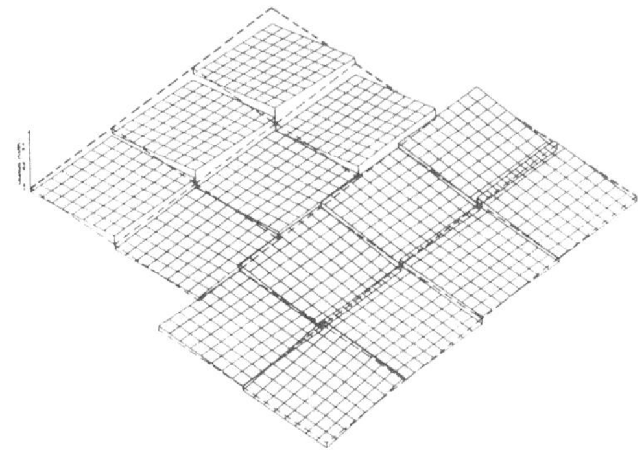

Bild 4. Verformung des Reflektors
bei Schwerkraftbelastung

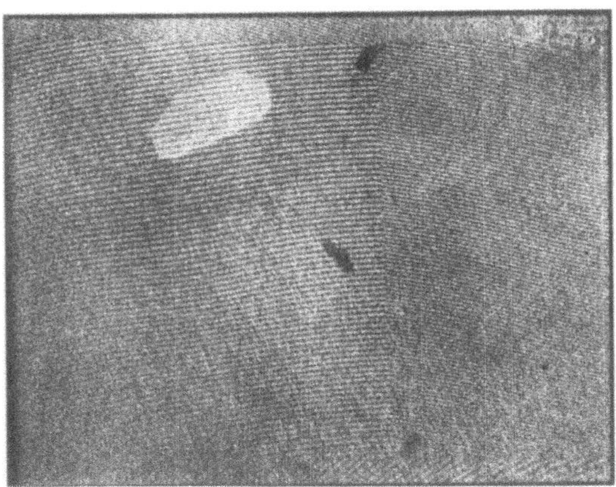

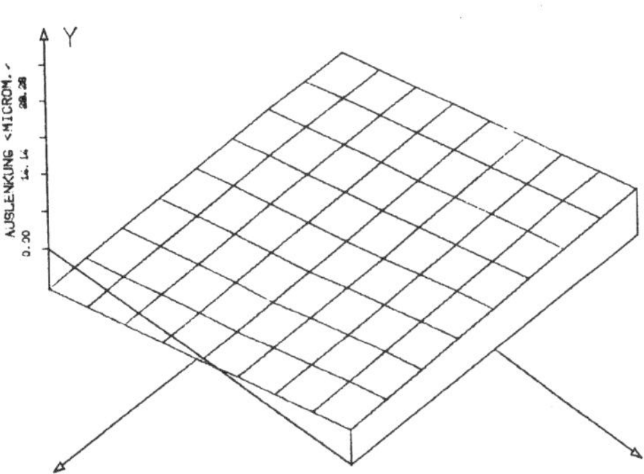

Bild 5. Holografisches Interferogramm und quantitative Auswertung eines
Mini-Reflektors bei Schwerkraftverformung

Laser in der Materialbearbeitung

CO_2-Laser: Systeme und Diagnostik

CO_2-Laser: Anwendung

Absorption, Gasdurchbruch, Schockwellen

Nd-YAG-Laser: Systeme und Anwendung

Laser in Material

CO_2-Lasers: Systems and Diagnostics

CO_2-Lasers: Application

Absorption, Gas Breakdown, Shock Waves

Nd-YAG-Lasers: Systems and Application

The Future of High-Power Laser Welding

Jeffrey P. Carstens
United Technologies Research Center

INTRODUCTION

Laser welding has been widely adopted in a number of industries for production joining tasks. The majority of applications to date are for spot- or seam-welding of thin materials, typically less than 5 mm, employing pulsed or cw solid-state lasers with up to 1 kW of average power, or CO_2 lasers with average power up to 5 kW. Applications involving higher-power lasers (>5 kW) have been fewer in number, and are just now beginning to move into production use.

The purpose of this report is to consider several aspects of the potential use of high-power CO_2 lasers for industrial welding applications. As an initial perspective, the advantages of laser welding are discussed and quantified based on a generalized model of laser welding performance. Consideration is then given to the current and expected near-future technical capabilities of laser welding equipment, both with respect to lasers and the beam transfer systems necessary to direct the beam on the workpiece. Finally, the prospect of high-power CO_2 laser welding systems for heavy industrial welding tasks is illustrated by some recent welding system design concepts generated at United Technologies Research Center.

LASER WELDING ADVANTAGES

Welding Performance

To allow a first-order quantitative assessment of laser welding advantages, an empirical model has recently been developed at UTRC to correlate high-power laser welding performance. A generalized estimate of laser welding performance based on this correlation is shown in Fig. 1. Data supporting this correlation come from various UTRC lasers operating up to 4 kW (Ref. 1), 10 kW (Ref. 2), 15 kW (Ref. 3) and 18 kW (Ref. 4). Data on welding up to 90 kW (Ref. 5) were also considered. Material thickness reported in Refs. 1-5 ranged from 2 mm to 38 mm; welding speed varied from 10 to 250 mm/sec. The data used were restricted to include results only from low-alloy steels. This restriction gave a better correlation, and deals with the material of greatest interest for the majority of industrial welding applications. Except for the 90 kW data, all beams were from unstable resonators with magnification of 2.0. Also, in all cases, laser power is that power entering the (two-mirror) focus head. Factors causing scatter in the correlation include different plasma removal techniques, different focusing optics, and different amount of weld underbead established.

The welding performance shown in Fig. 1 can be extended to heavier-section material without increasing laser power by the use of multipass welding techniques. Typical configurations for one-, two-, three- or four-pass welds from both sides are shown in Fig. 2. For welds involving more than two passes, a narrow v-groove and filler wire will be required for the outer passes. Single-pass filler welds of depths up to 1.5 cm have been made at UTRC with filler wire, without seriously degrading the performance indicated on Fig. 1. Therefore, in projecting the performance data in Fig. 1 to a multipass weld, it is assumed that each pass of laser welding requires the same energy input per unit length as a butt weld of the same thickness as the depth of the pass.

Weld Speed

Using this model, a quantitative estimate of welding time (beam or arc time) can be made for laser welding and compared with conventional techniques. The results of this exercise are shown in Fig. 3. Arc times for conventional welding techniques assume flat butt welds with prepared grooves, and are taken from Ref. 6 for submerged arc welding (SAW) and from Ref. 7 for narrow-gap gas metal arc welding (GMAW). These welding techniques were chosen for comparison because they have the highest welding speed of conventional techniques.

The laser welding speed advantage for single-pass welding is seen to be a strong function of laser power. For thicknesses appropriate to single-pass laser welding, the 5 kW laser is about 50% faster than conventional welding, while the 15 kW laser is three to five times faster and the 30 kW laser allows welding speeds five to ten times faster than SAW.

For thicker-section welding, the time required by the GMAW technique rises substantially, due apparently to the restrictions on the electrode size by the narrow gap. In the range of 3 to 6 cm of weld thickness, the four-pass laser weld with 15 kW of laser power enjoys a speed advantage of 25 to 35 times the conventional method. There may be few welding environments where this full speed advantage can be utilized. However, significant increases in productivity are clearly available.

Heat Input to Weld

A second major advantage to laser welding that may be analyzed using laser welding performance data shown in Fig. 1 is the total amount of heat input to the weld. To calculate this for laser welding, knowing the laser welding power and speed, all that is required is the absorption efficiency in the laser welding process. Previous studies at UTRC (e.g., Ref. 8) have shown that the absorption efficiency of laser keyhole welding of low-alloy steels is typically about 90%. Since the power referred to in the performance analysis is the power going into the focus head, a small energy loss is allowed for the two focusing mirrors, and 85% of the laser power is assumed to be absorbed by the weld metal. For conventional welding techniques, approximately 70% of the power from the welding head will be absorbed in the molten pool. Based on these assumptions, a direct comparison can be made of the heat input to the weld.

This comparison is shown for single-pass laser welds and four-pass laser welds in Fig. 4. The reduction in heat input to the weld with laser welding is very significant at all weld thicknesses. This reduction in thermal energy input to the weld has a first-order impact on many applications where postweld processing required by distortion (e.g., straightening or grinding) can be significant in terms of processing time and cost.

Weld Configuration

A third distinct advantage of laser welding is the narrow width of the laser weld nugget itself. This narrow width, and the rapid freezing rates implied by the narrow width, allows the joining of two butting surfaces with little regard for the weld configuration (e.g., T or butt weld). For example, as shown in Fig. 5, the energy requirements for making a two-pass tee-weld with filler wire are essentially the same as those required for an autogenous two-pass butt weld with the same total area to be joined.

The very thin weld, which is characteristic of laser welds made at weld speeds higher than 20-25 mm/sec, also allows laser welding to be used in configurations where normal welding techniques would be unacceptable due to the size of the weld nugget relative to the initial cross section of the pieces being welded. This concept is illustrated by the cross section on the right of Fig. 5. Here, a fully-penetrating weld was made utilizing the entire depth of the thick element to the right, without penetrating or violating the outer surface of the sheet metal on the left-hand side of the weld. This advantage can be particularly important to designers when considering welding as an alternate approach to the fabrication of small parts.

Weld Quality

A fourth basic advantage to laser welding is the achievement of levels of weld quality not attainable with other welding techniques.

The best example of unique laser weld quality is the demonstration of fusion zone purification with laser welding (Ref. 9). It has been demonstrated in a variety of high-strength, low-alloy steels that purification of the metal can take place in the welding process, leaving fewer nonmetallic impurities in the weld metal than exist in the base metal. This phenomenon has led to enhanced toughness of the weld nugget (as demonstrated, for example, by increased Charpy impact strength) relative to the base metal. This attribute can be of great importance for adopting laser welding in areas where weld toughness is a prime consideration.

LASER SYSTEM CAPABILITIES

Industrial Laser Development

As noted above, the utility of multikilowatt lasers for welding applications is affected by the available laser output power level. Current developments in industrial laser systems at United Technologies Research Center indicate that power levels for industrial applications will not be a restricting factor in the immediate future. These developments will be briefly reviewed.

A modular laser system has been developed at United Technologies Research Center (UTRC) (Ref. 10). The concept, which is illustrated in Fig. 6, employs a basic laser module, capable of 3 kW of continuous output, which can be assembled in multiple units to provide higher output levels. Each module has its own gas flow loop, heat exchanger and discharge region. A separate set of discharge electrodes is provided for each module. The cavity optics are arranged on optical platforms at either end of the modular laser; the platforms are connected by an optical truss member which extends the length of the laser underneath the discharge regions.

UTRC has manufactured a considerable number of these modular laser systems, including some for production use involving continuous operation on a 16- to 20-hr-per-day schedule.

During 1982, a four-module system, shown in Fig. 7, was fabricated for use in the UTRC laser test laboratory in East Hartford. The output end of the laser and the aerodynamic output window, or aerowindow, are clearly visible. As discussed in Ref. 10, the aerowindow allows the laser beam to exit from the low-pressure laser cavity into atmospheric pressure through a tailored gas jet which isolates the cavity environment from the atmosphere while eliminating beam distortion and reliability problems incurred with solid windows. In operation, a transfer duct is placed over the end of the aerodynamic window to duct the beam into a beam transfer cabinet. At this cabinet, the beam power and beam profile are monitored, and the beam is then switched to one of two workstations.

This laser has been operated in its initial tests as an unstable resonator with a beam magnification of 2.0. With this optical configuration, the laser has been operated for brief periods of time at power levels as high as 20 kW, and has been used continuously for welding tests at 15 kW. These numbers substantially exceed the nominal 12 kW power rating which is applicable to continuous use in production welding.

Specific factors limiting the power output of this type of laser have not yet been firmly established. However, it appears that powers in excess of 20 kW can be obtained by adding additional modules to the system.

Beam Transfer System Development

As laser welding is applied to larger workpieces, the desirability of moving the beam about a stationary workpiece will increase. The potential beam misalignment going into the focus head is a function of the number of moving mirrors in the train and their average pointing accuracy, as indicated in Fig. 8. If the beam going into the focus mirror is misaligned by an amount, $\emptyset$, this results in a wander of the focal point relative to the location of the weld, indicated as ϵ in Fig. 8. Since laser welds are typically very narrow, with a minimum dimension frequently of the order of 1.2 mm or less, it is imperative that the focused beam pointing direction lateral to the weld path stay constant, typically to within less than ±0.5 mm. The angular error, introduced into the beam by a moving mirror whose angular alignment is off by an amount α_m, is equal to $2\alpha_m$. Since these errors are additive as the beam moves through the system, the total allowable error is given by the sum of twice the average moving mirror angular error times the number of moving mirrors in the system, as noted in Fig. 8.

As an example, assuming an allowable aiming error of ±0.5 mm and a 50 cm focal length allows an angular error $\emptyset$ of 1 milliradian. The misalignment error allowed with one moving mirror is therefore 0.5 mrad, which is easily achievable. For three moving mirror axes, the misalignment error on each axis must be less than 0.17 mrad, which exceeds the accuracy normally attainable on large-gantry systems. Accordingly, alternate means are desired to reduce the accuracy requirements of multiple moving mirror systems.

One such system, based on work done at UTRC, involves detecting the beam misalignment and eliminating it with the final mirror in the train just prior to the focusing mirror. A schematic diagram of this system is shown in Fig. 9. The turning mirror, which will be adjusted to eliminate angular variations in the incoming beam, has a focusing sampling grating applied to its front surface. This grating (discussed in Ref. 11) samples an extremely small percentage (typically 0.01%) of the incoming high-power beam and focuses it on a quadrant detector located at a selected angle to the input beam (a 45° angle is shown in Fig. 9). The system characteristics are such that, if the turning mirror is controlled to keep the sampled beam focused on the center of the quadrant detector (which is fixed relative to the downstream focusing mirror), the reflected beam will maintain parallelism with its initial alignment direction. With the beam corrector system in place, the mirror motion accuracy requirements are greatly decreased, eliminating the extremely high precision of mirror motion otherwise required.

FUTURE LASER WELDING SYSTEMS

The preceding developments will be combined in various ways in the future to give new capabilities to high-power laser welding systems. As the necessary laser and beam control subsystems are developed, welding applications can be expected to proliferate into more difficult environments. This growth is described conceptually by looking at two potential future laser welding systems.

A flexible, multiaxis welding system, for use in factories in which many different types of welds are required, is shown in Fig. 10. This system is a general-purpose machine for a large welding shop where welding is done on a diverse number of workpieces. The focus head supporting gantry runs on accurate guideways from one end of the enclosure to another, and workpieces up to 3 m x 3 m x 3 m in size are brought into the enclosure from either end on a separate set of rails. Thus, while a weld is being made in the enclosure, the next workpiece is being fixtured or tack-welded on a rail car outside the enclosure at the opposite end. As soon as welding is finished on the first workpiece, it is shuttled out of the enclosure and the new workpiece is shuttled in the other end. Four degrees of welding motion freedom are available: the three orthogonal axes shown in Fig. 10 and the rotational axis in the focus head shown in Fig. 11.

The focus head carries a number of subsystems, as shown in Fig. 11, in addition to the previously-discussed beam-correcting system. A seam-tracker is required to locate and accurately position the focus head. Such seam-trackers are standard items now. In the future, however, their sensory capability will be increased to measure both joint gap and mismatch, to allow control of the required rate of filler addition.

An even more advanced laser welding system is shown schematically in Fig. 12. Here the laser generator itself is mounted on a truck to allow movement from one work site to another. This may involve moving the laser several hundred meters or several hundred kilometers. Again, a beam-correcting station at the focus head would allow the use of relatively inaccurate mirror actuation schemes, as shown by the cantilever gantry in the sketch. Such a system would have a great deal of flexibility in terms of moving to different portions of a large job. An example of this is the welding of copper-nickel sheet onto the hull of a large ship as part of a retrofit operation to reduce or eliminate fouling of the hull. This application has been studied in detail by UTRC (Ref. 12). It should be noted that present industrial lasers, of the type made at UTRC, are adequately rugged in their basic design to withstand the dynamic loadings involved in being transported on a truck from one weld site to another.

As can be seen from these examples, present-day developments in lasers and beam transfer systems show that there will be little restriction to applications of laser welding in terms of workpiece size or location as long as the weld seam itself is accessible to the extent necessary to make the weld.

CONCLUDING REMARKS

The benefits of high-power laser welding are seen to be significant in terms of speed, heat input to the weld, weld configuration flexibility, and weld quality. Current developments in laser and beam control technology indicate the basic feasibility of achieving these benefits with flexible, efficient laser welding systems.

REFERENCES

1. Duhamel, R. F.: Single-Module Laser Welding Performance. UTRC Report No. R81-112134-21, December 1, 1981.

2. UTRC Brochure: Industrial Applications of High-Power Lasers. June 1981.

3. Metzbower, E. A. and D. W. Moon: Laser Beam Welding of ASTM A-36 Steel. To be published in Lasers in Metallurgy. K. Mukherjee and J. Mazumder, Eds. TMS-AIME, Chicago, Illinois, February 1981.

4. Duhamel, R. F. and C. M. Banas: Laser Welding of Steels and Nickel-Base Alloys. Paper presented at Second International Conference on Applications of Lasers in Materials Processing, Los Angeles, California, January 24-26, 1983.

5. Banas, C. M.: Laser Welding to 100 kW. UTRC Report R76-912260-2. Work performed under Navy Contract N00173-76-M-0107. February 1977.

6. Modern Welding Technology by Howard B. Cary. Prentice Hall, Inc., 1979.

7. Hunt, J. F.: Welding Steam Headers by Narrow-Gap GMAW. Welding Design and Fabrication, December 1982.

8. Banas, C. M.: CO_2 Laser Materials Processing. From Physical Processes in Laser-Materials Interactions. M. Bertolotti, Ed. Plenum Publishing Corp., 1983.

9. Breinan, E. M. and C. M. Banas: Fusion Zone Purification During Welding with High-Power CO_2 Lasers. UTRC Report R111087-2, April 1975.

10. Carstens, J. P.: High-Power CO_2 Laser Materials Processing Technology at United Technologies Research Center. Paper given at VDI Technologie Zentrum on "Materials Processing with CO_2 High-Power Lasers," Stuttgart, April 1-2, 1982.

11. Mottier, F. M.: Diffraction-Limited Focusing Grating for High Energy Laser Diagnostics. Paper No. 365-30, presented at 26th Annual International Symposium of the SPIE, August 23-27, 1982.

12. Banas, C. M.: Laser Weld Attachment of Cu-Ni Alloy to Ship Steel. Final Report to International Copper Research Association on Project 241, March 1981.

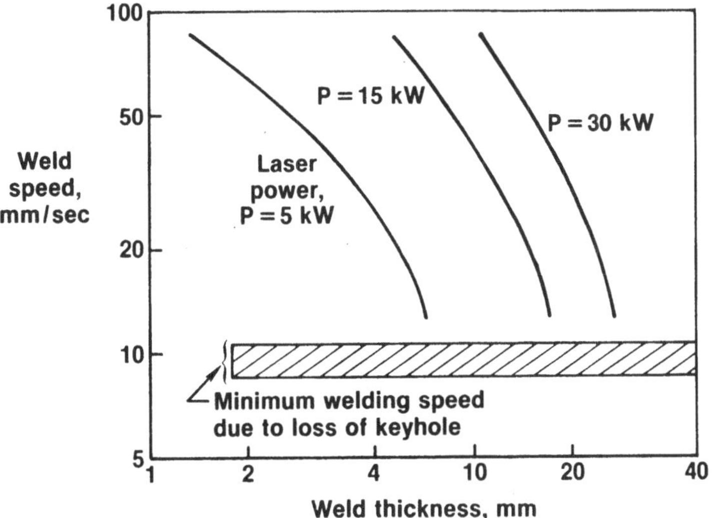

Fig. 1 Weld Speed vs Power

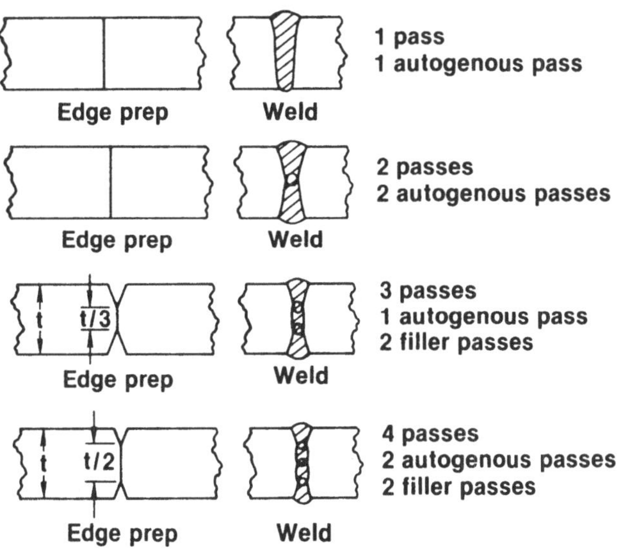

Fig. 2 Multi-Pass Weld Configurations

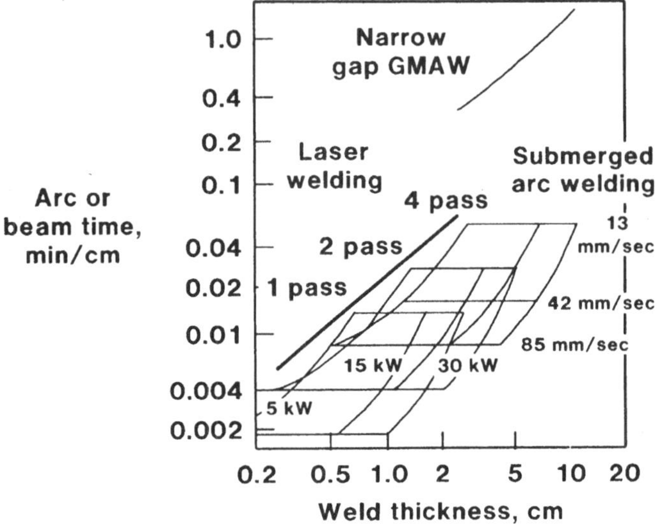

Fig. 3 Welding Time Comparison

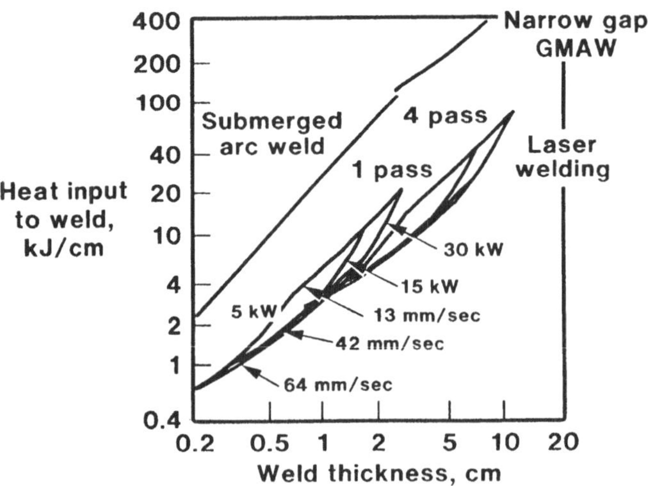

Fig. 4 Heat Input to Weld

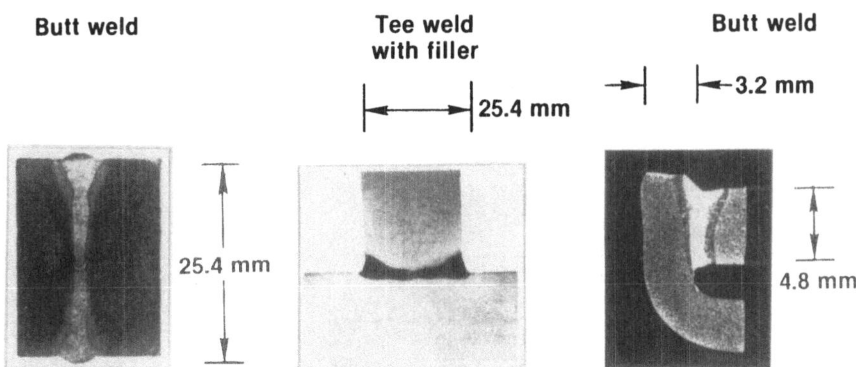

Fig. 5 Laser Weld Configurations

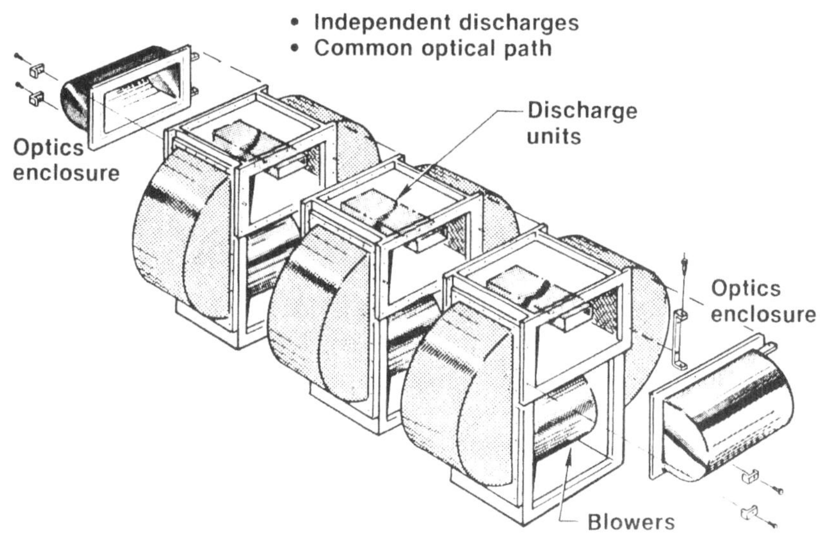

Fig. 6 Modular Laser Concept

Fig. 7 4 Module 12 kW Laser

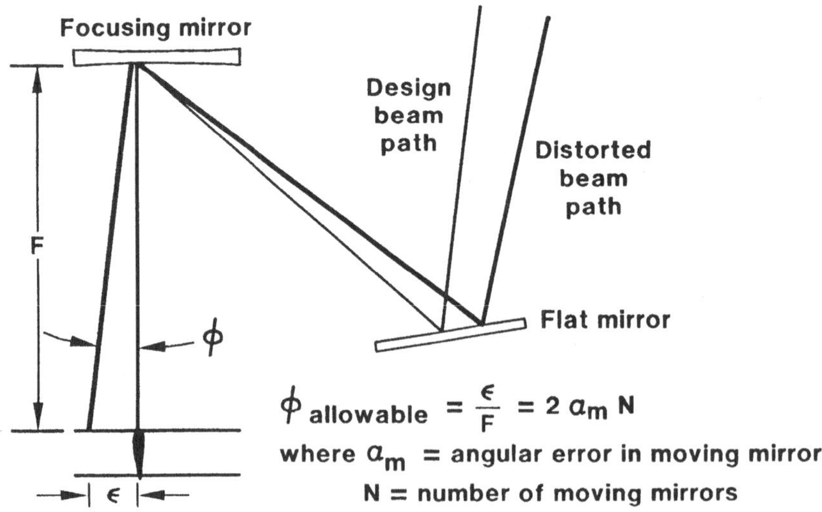

$$\phi_{allowable} = \frac{\epsilon}{F} = 2\,\alpha_m\,N$$

where α_m = angular error in moving mirror

N = number of moving mirrors

Fig. 8 Beam Pointing Error Geometry

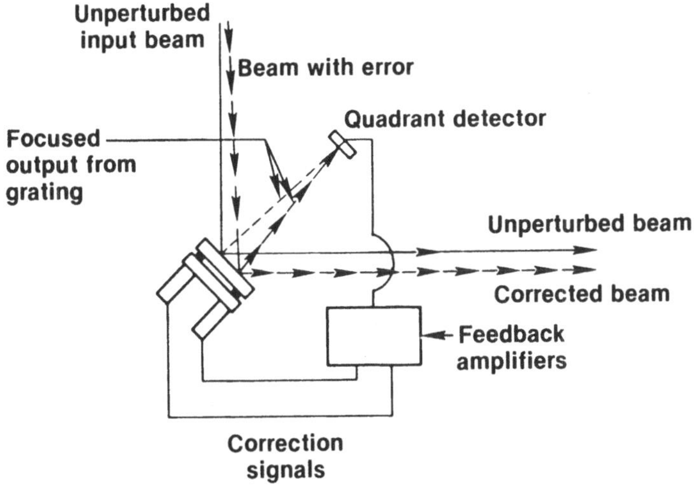

Fig. 9 Beam Correcting Mirror Using Focusing Grating on Mirror Surface

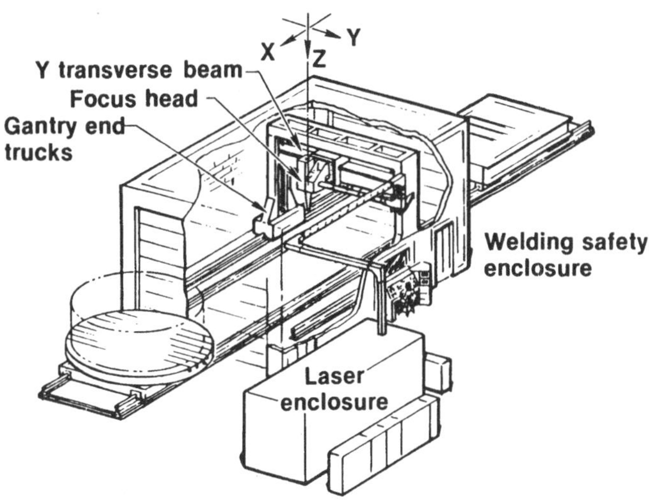

Fig. 10 Overall View of Large Laser Welding System with Moving Gantry

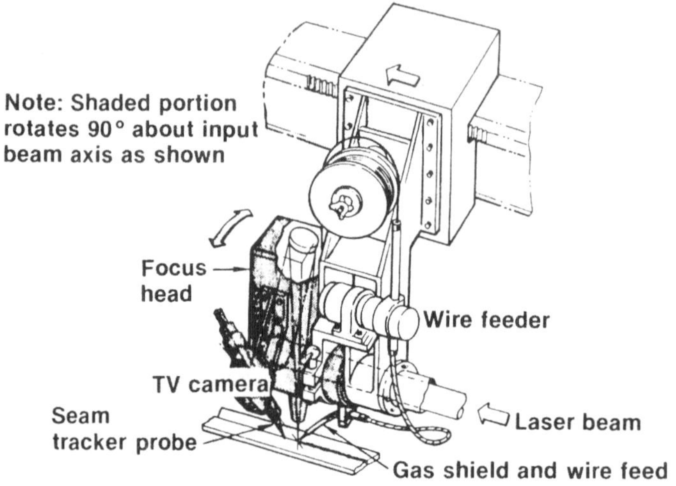

Fig. 11 Focus Head for Laser Welding Gantry

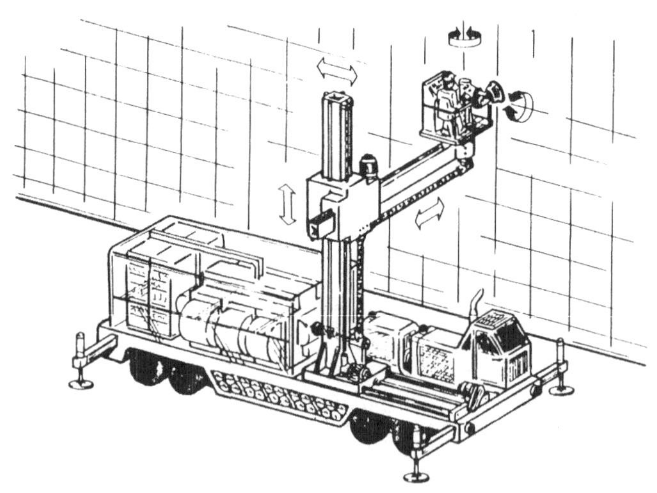

Fig. 12 Mobile Laser Welder

Criteria to Match Future CO$_2$-Laser Systems to Material Processing

L.BAKOWSKY, P.LOOSEN, E.BEYER, G.HERZIGER
Institut für Angewandte Physik, Technische Hochschule Darmstadt
Schloßgartenstr. 7, D-6100 Darmstadt

Introduction

Applications of high power CO2 lasers to industrial material processing often show a discrepancy between processing results and the limits of processing quality and velocity predicted by theoretic calculations. Quality and velocity of laser processing are limited, because the laser parameters are not yet adapted to the characteristic phenomena of the interaction of high power laser radiation with matter. In particular two points turn out to be of interest
- the influence of laser induced plasma on quality and efficiency of the machining
- the coupling between workpiece and laser system during machining due to optical feedback.

Higher processing velocity and quality will be reached, if the parameters of a new generation of high power CO2-laser systems are matched to the special requirements of laser hardening, welding, cutting, etc. For a survey of these requirements it is useful to distinguish between two intensity regimes: Low-intensity processing is preferrably the domain of laser hardening, high-intensity laser radiation is preferrably used for material removal processes, such as drilling , cutting, etc.

1. Laser Hardening

An illustration of laser hardening is shown in the first picture. The surface is heated up by the absorbed laser power and cools down by heat conduction into the bulk material. Compared to conventional heat treatment, laser processing shows order of magnitudes higher cooling and heating rates. Analytic treatment is performed by three dimensional heat equation with processing parameters depending on temperature and materials. Today the majority of unsolved problems is concerning the workpiece. For examples the stability and efficiency of absorbing coatings and the transformation processes taking place at the very high heating and cooling rates are to be mentioned. Concerning the laser systems the main problem is to adjust the local and temporal intensity distribution to the

238

transformation process. The aim is to heat up the surface rapidly and to keep the
temperature constant during the transformation time. This is achieved by
selecting the mode configuration, the use of beam integrators and special
intensity modulation in time.
Therefore future systems for applications in the low-intensity region have to be
mode-controlable in space and time.

2. Drilling and Cutting

For drilling and cutting it is necessary to melt, vaporize and remove the
material. To succeed the losses by heat conduction, much higher laser intensities
are required than in the transformation hardening case. Characteristic phenomena
in this intensity range are the appearance of a laser induced plasma and an
intensity-dependent absorption. Figure 2 shows the process schematically.
Focussing a laserbeam of power P and divergence θ by an optical system yield an
intensity on the surface

$$I = P/\pi \cdot r_F^2 \quad ; \quad r_F = \theta \cdot f \qquad (1)$$

In case of a Gaussian intensity distribution the surface temperature may be
calculated from the general heat equation /1/

$$T(0,0,0,t) = A \cdot \frac{P}{r_F} \cdot \frac{1}{\pi^{3/2} \sqrt{2} \kappa \cdot \rho \cdot c} \ \text{arc} \ \tan \sqrt{\frac{8 \kappa t}{r_F^2}} \qquad (2)$$

ρ :density
c:specific heat
κ :thermal conductivity
A:absorptivity

For $t > 10^{-4}$ s the arctan in equation 2 may be equated to $\pi/2$. This leads to a
time independent minimum laser power to melt or vaporize the material

$$P_{V/M} = \frac{\sqrt{2\pi} \cdot K}{A} \cdot T_{V/M} \cdot r_F \qquad (3)$$

Figure 3 shows $P_M (r_F)$ (lower curve) and $P_V (r_F)$ (upper curve) for copper,
aluminum, and steel. Following these curves it seems to be impossible to cut or
melt copper with todays commercial $CO2$-laser systems.
There is a significant point however, which has not been taken into acccount for
the calculation of the curves. In the high-intensity range the absorption is no

longer constant but becomes an intensity dependent parameter. Figure 4 shows measurements A(I) for steel irradiated with a Nd-YAG laser with pulse durations t_p =7μs and t_p =50 ns. As the laser intensity on the target's surface approaches a critical value I the absorption instantaneously rises from A=0.4 to A $>$ 0.9.

In further experiments the pulse duration has been varied over a wide range. Comparing the meassured threshold intensities with equation 2 yield that the rise of the absorption is connected with the vaporization limit and the appearance of a laser induced plasma in front of the surface /2/. Transmitting the Nd-YAG experimental results to material processing with CO2-lasers a reduction of the minimum laser powers in figure 3 up to a factor 100 is expected (figure 5).

If the intensity is chosen greater than the threshold intensity for a laser induced gas breakdown I_B (figure 5), the plasma starts to absorp the laser power. With further increasing intensity the plasma layer takes off and shields the target completely (figure 6).The plasma layer expands towards the laser at multiple sound velocity /5,6/. Due to the associated pressure liquid target material is pressed out of the processing zone and the processing result is determined by the plasma expansion. However in a small intensity intervall the laser induced plasma acts as an absorption increasing element. In this intensity intervall material removal can be performed with maximum efficiency and accuracy. As a consequence of the previous mentioned phenomena continous laser systems are not suited for processing in the high intensity range. Therefore the use of pulse modulated systems is advisable. Pulse amplitude and pulse duration are to be adapted to the intensity dependent absorption and the plasma influence. Such an adapted pulse-code is shown in figure 7 /7/. The intensity of the first pulse is greater than the critical value for abnormal absorption. The pulse duration of the following pulse train is long enough to vaporize the material, the repetition rate is adjusted to the lifetime of the plasma layer, so that the disturbance becomes minimum. The pulse amplitudes are matched to the material removing process in order to reach high efficiency.

3. State of the Art of High Power CO2-Lasers

Pulse modulation and stable intensity profile are achieved with slow flow axial laser systems. Because of the heating of the laser gas by electrical input power specific laser power is limited to P/L=50 W/m. The total maximum laser power is limited to P $\simeq$ 1000 W by mechanic stability reasons. The systems usually emit in the fundamental mode (TEM00), minimum focus radii are $r_F \simeq$ 50μm. Corresponding maximum intensities on targets are I $\simeq 10^2$ W/cm^2.

In the case of fast flow axial laser systems cooling of the laser gas is increased by two orders of magnitude. The maximum specific laser power is P/L= 1000 W/m.

Technical problems concerning gas supply and circulating pump limit the maximum power to $P \simeq 5$ kW. The lasers emit in a stable local intensity distribution (low order mode), but turbulences of the gas flow and the high gain of the laser medium cause amplitude fluctuations up to 100%.

There are less difficulties combined with gas supply and circulation ,if electrical discharge, gas flow, and laser radiation are transverse to each other. With such transverse flow systems the highest laser powers are reached today ($P \leq 20$ kW). The maximum intensity reached on the workpiece surface is yet smaller than in the case of slow flow axial systems, because of local and temporal intensity fluctuations due to system instabilities.

Figure 8 shows figure 3 with the typical operation fields of the three systems. The additional drawn curve (-.-) corresponds to an intensity $I=10^{7}$ W/cm^{2}.

A modulation of the laser radiation as shown in figure 7 is not yet possible with todays high power systems.

Conclusion

The physical background of laser processing in the low intensity range, e.g. heat treatment and hardening, is fairly well understood today. To optimize the laser systems to such applications, the intensity profil is to be adapted to the special geometry in order to reach an optimum temperature distribution for the transformation process. In the high intensity range ($I > 10^{6}$ W/cm^{2}) three phenomena influence laser machining mainly

-intensity dependent absorption

-laser induced plasma in front of the workpiece

-optical feedback between workpiece and laser system

Abnormal absorption of highly reflecting materials without disturbance by the expanding plasma layer is only achieved, if the laser intensity is modulated. Pulse amplitude, pulse duration, and repetition rate are to be adapted to plasma and processing dynamics.

Literature
/1/ J.F.Ready, Industrial Applications of Lasers , Academic Press (1978)
/2/ H.G.Treusch,K.Wissenbach, DPG 47. Physikertagung, Regensburg 1983, K60
/3/ D.C.Smith, J.Appl.Phys. 48, 6 (1977)
/4/ E.Beyer,L.Bakowsky,R.Poprawe,G.Herziger at these proceedings
/5/ Y.P.Raizer, Sov.Phys.JETP 21, 1009 (1965)
/6/ R.Poprawe, E.Beyer, L.Bakowsky, G.Brumme, G.Herziger at these proceedings
/7/ G.Herziger, 3. Internationale Tagung Laser und ihre Anwendungen Dresden 1977

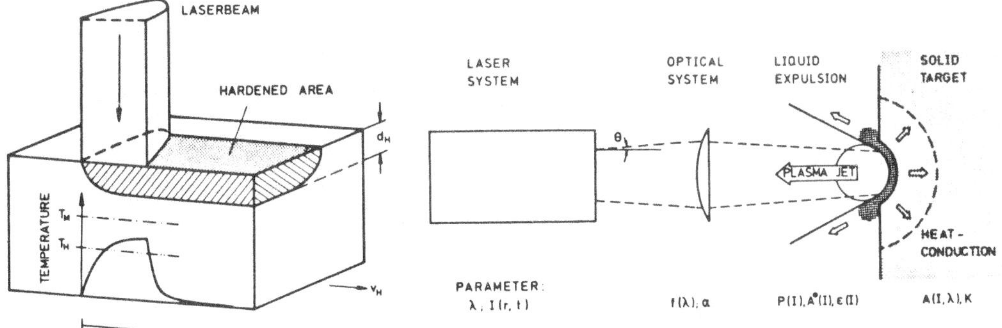

Fig.1: Scheme of laser hardening

Fig.2: Scheme   of   laser   material
processing in the high intensity range

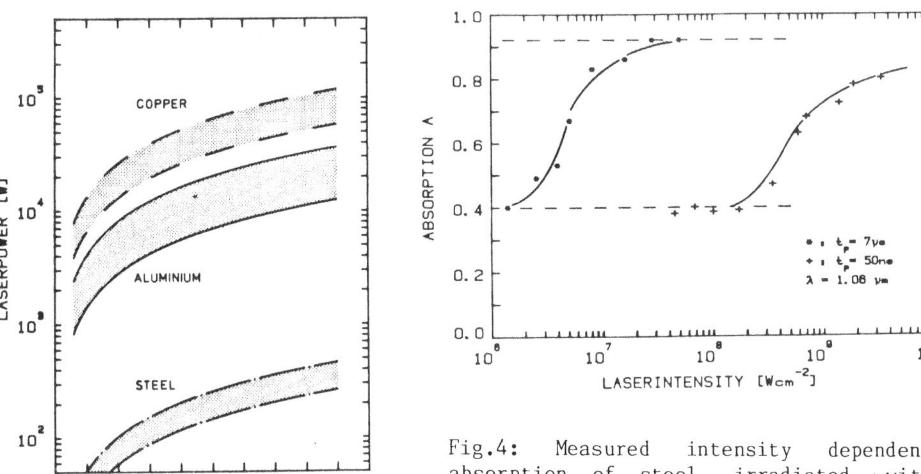

Fig.4:   Measured   intensity   dependent
absorption of steel   irradiated with
Nd-YAG laser pulses (tp = 7 μs and tp =
50 ns)

Fig.3: Calculated minimum laser power to
melt (lower curves) and vaporize (upper
curves) copper (- -),aluminium (—),and
steel(-·-) as a function of focus radius

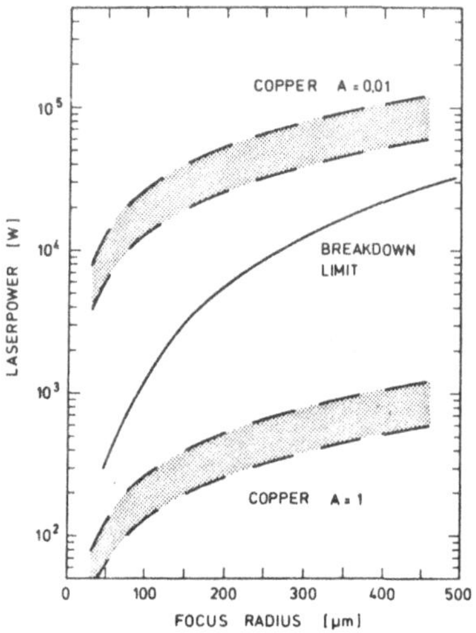

Fig.5: Calculated minimum laser power to melt and vaporize copper as a function of focus radius at A=0.01 and A=1

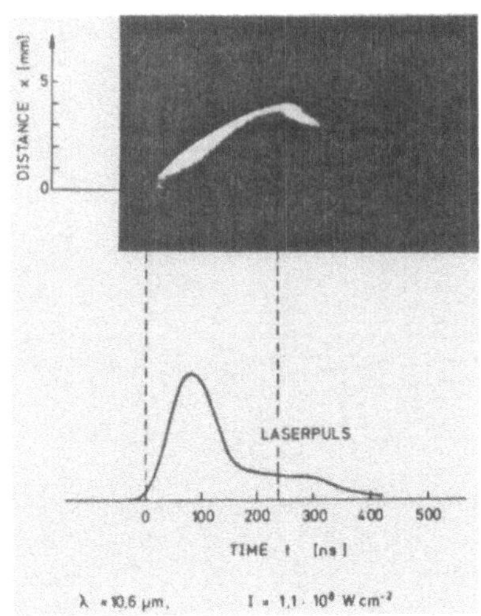

Fig.6: Streak picture of the expanding plasma layer in front of the target

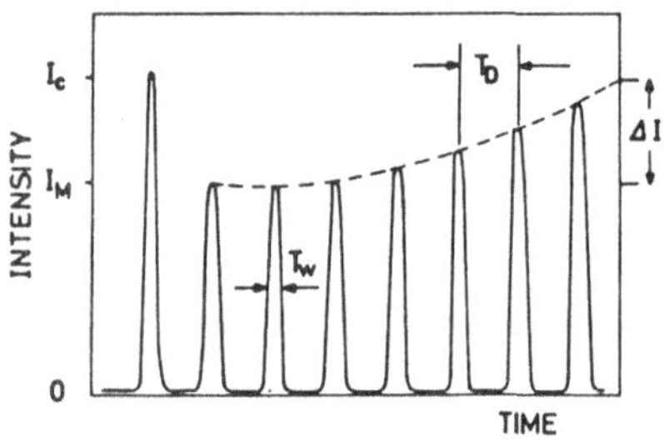
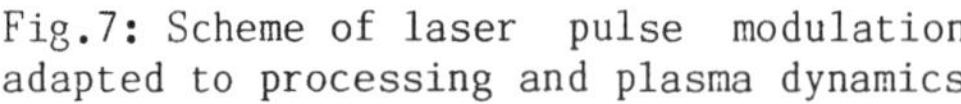

Fig.7: Scheme of laser pulse modulation adapted to processing and plasma dynamics

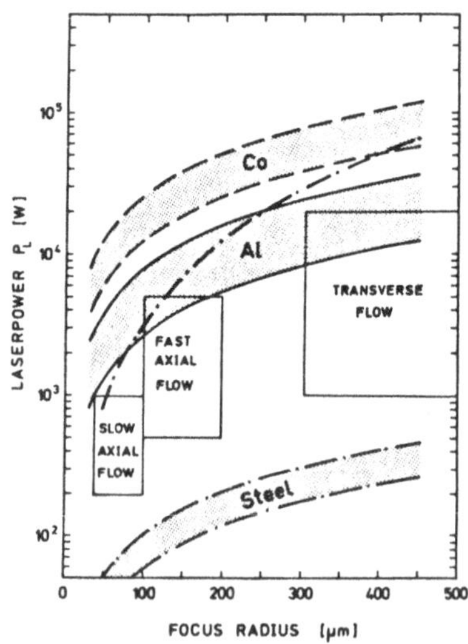

Fig.8: Operation fields of the three high power CO2-laser systems compared with the calculated curves of figure 3 (- · - :I=10^7 W/cm^2)

An Analysis of the Effect of Mode Structure on Laser Material Processing

M. Sharp, P. Henry, W.M. Steen, G.C. Lim.
Imperial College, London, England

INTRODUCTION

High power laser welding is now an accepted material processing technique. In gaining this acceptance a large amount of research has been done on the effect of various parameters e.g. traverse speed, laser power, etc /1,2/. One area which has received little attention is the effect of the mode structure of the laser beam on the weld profile.

There are two main reasons why such an investigation is important. First, many of the larger power machines do not produce a Gaussian, and so the significance of a non-Gaussian beam is important in assessing the use of these machines. Secondly, since the mode structure of the output from a machine may vary over a period of time, it is necessary to have a knowledge of the effects of the mode structure, so that the reproducibility of the weld can be guaranteed.

As a starting point of this investigation, the profile of a focussed beam operating in a TEMmn mode is shown to have the same profile as the unfocussed beam when aperture effects are neglected and perfect optics are assumed. From this it can be deduced that the focussed beam has the same intensity profile as the raw beam for any intensity profile.

A new definition of spot size is introduced which allows a consistent comparison of the spot sizes of the different modes. Using a finite element numerical model of the laser welding processing an investigation is made into whether the effect of the mode on the weld profile is due solely to the varying effective spot sizes of the modes, or whether the actual mode structure itself has an effect. A theoretical investigation is made into the effect of a finite lens aperture on the spot size, and it is also shown how lens aberration can also be accounted for. Finally an experiment is described to validate the theoretical results outlined in this paper.

Focussing of Laser Beams

The TEMmn modes for a laser with cylindrical symmetry are given in (3) as

$$A_{mn}(\tfrac{r}{w}, \theta) = (\sqrt{2}\, r/w)^n L_m^n(2r^2/w^2) \exp(-r^2/w^2)\, e^{in\theta} \qquad (1)$$

where (r, θ) are polar coordinates, W is the characteristic radius of the beam, Lm is the Laguerre polynomial of order n and degree m, and Amn is the amplitude of the TEM mn mode. The intensity is given by:

$$I(r, \theta) = |A_{mn}|^2 \qquad (2)$$

Standard texts on optics (4) derive the following equation for the focussing of a monochromatic beam of light:

$$Q_{mn}(p, \phi) = \int_0^a \int_0^{2\pi} A_{mn}(\tfrac{r}{w}, \theta)\, e^{\frac{i\pi kr^2}{2}\left(\frac{1}{s} - \frac{1}{f}\right)}\, e^{ikrp\cos(\theta - \phi)}\, r\, dr\, d\theta \qquad (3)$$

Where the geometry involved is illustrated in fig1. f is the focal length of the lens, and a is the aperture radius. Performing the integration over θ yields

$$Q_{mn}(p, \phi) = e^{in\phi} \int_0^a A_{mn}(\tfrac{r}{w})\, e^{\frac{i\pi kr^2}{2}\left(\frac{1}{s} - \frac{1}{f}\right)} J_n(zr)\, r\, dr \qquad (4)$$

Where $Z = kp/s$. At focus, $s = f$ and the complex exponential term disappears. Aperture effects can be neglected by taking a as infinity. In this case it can be shown that:

244

$$\mathcal{Q}_{mn}(p,\phi) = A_{mn}\left(\tfrac{p}{w_f},\phi\right) \tag{5}$$

where

$$w_f = \frac{\lambda f}{\pi w} \tag{6}$$

is the characteristic radius of the focussed beam. Since the Amn form an orthogonal set of functions and (3) defines a linear integral operator on the function Amn, then it can be shown that

$$\mathcal{Q}(p,\phi) = A\left(\tfrac{p}{w_f},\phi\right) \tag{6}$$

for all beams with unfocussed amplitude A (r,θ)

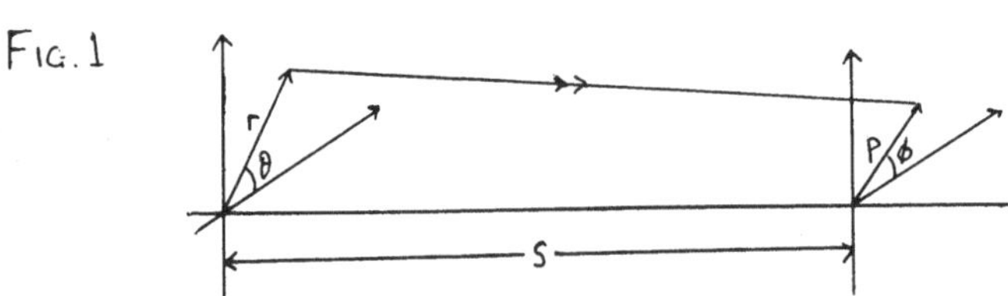

Fig. 1

<u>Spot Size</u> There have been many definitions introduced to describe the spot size of the laser beam. The characteristic radius, w, is not a good measure of the spot size since the effective spot size of the higher order modes is much greater than the characteristic radius. The important point about the characteristic radius is that it is a function of the geometry of laser cavity alone (3) and is therefore independent of the mode structure. The one draw back with the characteristic radius and other definitions is that the power contained within the defined radius is dependant on the mode itself. To rectify this the following definition of spot size is introduced. A laser beam of total power P_t and intensity distribution (r,θ) has radius, R, by

$$\int_0^R \int_0^{2\pi} I(r,\theta)\, r\, dr\, d\theta = (1 - e^{-2})\, P_t \tag{7}$$

The factor $(1-e^{-2})$ is chosen so that the beam radius R is equal to the characteristic radius, w, for the Gaussian mode. In order to describe the relative sizes of the different modes, a mode factor, K, is defined by

$$R = Kw \tag{8}$$

Mode factors for the lower order modes are given in table 1.

Mode factors of TEMmm modes.

	− N −		
	0	1	2
0	1.00	1.32	1.56
1	1.65	1.88	2.08
2	2.12	2.31	2.48

<u>Aperture Effects and Aberrations</u> It has been shown that the focussed beam profile is that of the raw beam if aperture effects are neglected, and so the spot size will be given by the characteristic radius multiplied by the mode factor. The integral in (4) was calculated numerically to show the effect of a finite aperture on the spot size for low order modes. The results are shown in Table 2.

Increase in spot size due to aperture.

a/w	m=0,n=0	m=0,n=1	m=1,n=0	m=1,n=1
0.5	2.44	2.87	1.64	2.04
1.0	1.37	1.61	1.55	1.28
2.0	1.00	1.00	1.00	1.00

It can be seen that if the aperature radius is greater than twice the characteristic radius of the beam then the effect is negligible.

Aberrations can be accounted by including an extra factor

$$\exp\left(i k W(r, \theta)\right) \qquad (9)$$

in (4) W is the aberration function /5/ and describes the deriation of the actual wavefront from a spherical wavefront. For expample

$$W(r, \theta) = C r^4 \qquad (10)$$

describes spherical aberration. C depends on the amount of aberration present Since the beam extends over a wider range for higher order modes, it would be expected that higher order modes would be effected more by aberrations.
The effects of mode structure on laser welding.Having detailed the relationship between the raw beam and the focussed beam, it is now possible to investigate the effects of the mode structure using a three dimensional finite element model/g/ adapted to accept any axisymmetric mode.

It is well known that the weld width increases and penetration decreases as the spot size of the laser beam increases. It is therefore to be expected that the weld width would be increased and the penetration decreased by using a higher order mode, since the effective spot size would be greater.
In order to see if the acutal mode structure itself has an effect the model wasrun for a TEMol* mode and for a Gaussian mode of the same effective radius, as well as a Gaussian beam of the same characteristic radius. (Table3).

Effects of mode structure on weld profile.
(all lengths in mm)

		P = 1600 W W = 0.25		P = 2000W W = 0.5	
		W	D	W	D
TEM$_{01}^{*}$	W	0.8	1.7	2.0	1.4
TEM$_{00}$	W	--	--	1.8	2.0
TEM$_{00}$ 1.32	W	0.8	2.2	2.3	2.0

It can be seen that the TEMo1* beam has less penetrating than a Gaussian of the same effective radius. This suggests that the mode structure does have an effect apart from the increase in size, but it is felt that further investigation is necessary.

Experiment

Experiments are in progress at Imperial College to validate the ideas expressed in this paper. The experiments simply consists of placing ALL Laser beam analyser in the path of a beam form a control laser 2kw CO_2 laser, so that the made structure of the raw beam can be monitored while welding is carried out. The ALL beam analyser can also be used to measure the focussed beam profile, and this will be sued to check the relationship between the raw and focussed beams.

<u>References</u>

1. Duley W.W. CO_2 Lasers effects and Applications Academic Press 1976

2. Ready J.F. Effects of High Power Laser Radiation Academic Press 1971

3. Kogelniic H. & LI,T Proc IEEE <u>54</u> 1312 1966

4. Born M. and Wolf E. Principles of Optics Pergamon

5. Longhurst R.S. Geometrical and Physical Optics Longmans 1967

6. Henry P et al Proc. ICALEO 1982 LIA.

Diagnostic of High Power CO₂-Lasers

P.LOOSEN, L.BAKOWSKY, G.HERZIGER, F.RÜHL
Institut für Angewandte Physik, Technische Hochschule Darmstadt
Schloßgartenstraße 7, D 6100 Darmstadt

1.Introduction

The result of material processing by CO2-lasers is mainly determined by the quality of the laser radiation.   The essential parameters determining the quality of the CO2-laserradiation are power, modestrukture and their time dependence.   Processes resulting in alterations of power and modestructure are shown in fig.1.

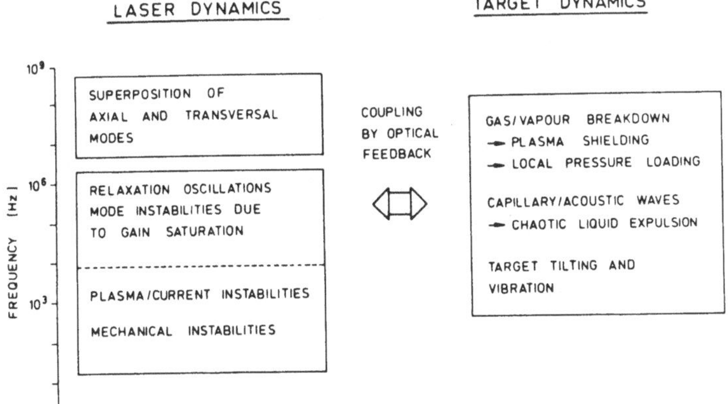

Fig.1 Feedback induced fluctuations of laser power and mode structure and its corresponding phenomena on the target.

The strong coupling of laser and workpiece during processing is an essential feature which distinguishes the laser from customary tools.   Laser and workpiece are representing a system of coupled resonators with strong interactions between laser- and targetdynamics (fig.1). The target controls the laserradiation by capillary or acoustic waves in the processing area, stimulating the laser to relaxation oscillations, mode instabilities/1/ or chaotic temporal fluctuations.   The laser changes over from continuous operation to uncontrolled pulse operation.   At higher intensities the processing result is influenced by a laser induced plasma at the target suface with optical (plasmashielding/8/, lenseffect) and mechanical effects (chaotic pressure loading of the surface/8/). The laser induced plasma is caused firstly by the mentioned pulse superelevation of the relaxation oscillations and secondly by mode instabilities. In addition CO2-lasers exhibit power superelevations by mode-lock phenomena due to the strong non-linearity of the laser medium.

These considerations may indicate, that beam diagnostic has to be performed in the processing arrangement, analysing of the free running laser beam is not sufficient for determining the quality of a laser for material processing. Time and space resolution of the diagnostic system has to be adapted to the dynamical characteristics of the processing. In the following the time constants of the mentioned power fluctuations in CO2-lasers are estimated to determine the requirements on time and space resolution of the diagnostic system.

2. Time constants of the power fluctuations in CO2-lasers

1. The highest fluctuation frequencies in the region of 2-200 MHz are due to superposition of axial and transversal modes. Fig.2 shows the frequencies of some axial and transversal modes for a long-radius resonator, as it is often used for high power lasers. Typical frequency spacing between adjacent axial modes are about 10-200 MHz, between adjacent transversal modes about 2-20 MHz. In the case of

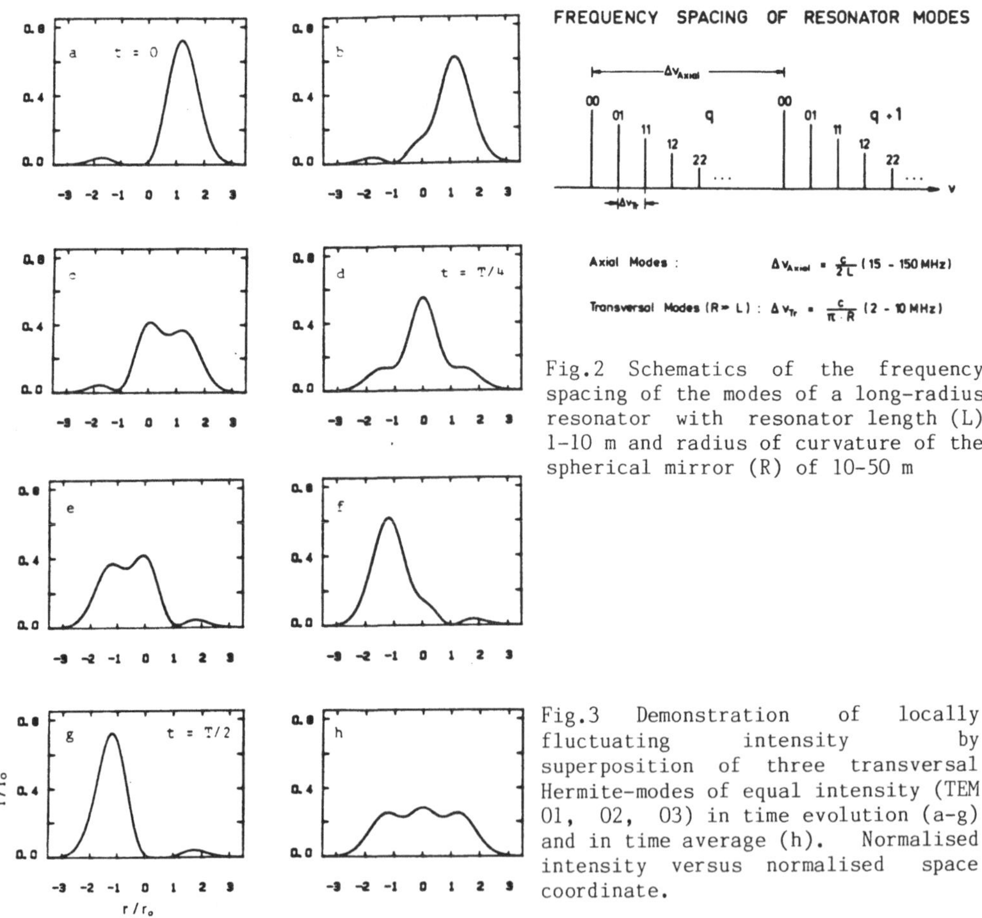

Fig.2 Schematics of the frequency spacing of the modes of a long-radius resonator with resonator length (L) 1-10 m and radius of curvature of the spherical mirror (R) of 10-50 m

Fig.3 Demonstration of locally fluctuating intensity by superposition of three transversal Hermite-modes of equal intensity (TEM 01, 02, 03) in time evolution (a-g) and in time average (h). Normalised intensity versus normalised space coordinate.

mode-locking of these modes there results a temporal fluctuation of the total power (superposition of axial modes) or a spatial fluctuation of the center of weight of the intensity distribution (superposition of transversal modes) with the indicated frequencies. Fig.3 gives a simple example of the superposition of transversal modes in the time evaluation and in the time average.

2.   Power fluctuations in the region of 50 KHz to 5MHz are caused by two phenomena. Firstly relaxation oscillations appear, well known from solid-state lasers, secondly high power laser systems show mode instabilities due to non-linear gain saturation/1/.   The frequencies of relaxation oscillations depend on the normalised pumping power P_N, the lifetime of the upper or lower laser level t_L, and the resonator lifetime t_R /2,9/.

$$f = \left(\frac{P_N - 1}{t_L * t_R} \right)^{1/2}$$

In the case of laser excitation with a suitable step function the laser power relaxes with the indicated frequency to the stationary value. In the case of continuous excitation, variation of the pumping power of a few percent can cause a total modulation of the laser power/9/. Mode instabilities are induced by nonlinear saturation of the gain/1/. At certain values of the laserparameters pumping power, resonatorgeometry and resonator losses the field distributions become unstable and, stimulated by diminishing disturbances, the laser jumps irregulary between the possible field distributions.

3.Power fluctuations in the range between Hertz and Kilohertz occur due to variations of pumping power and resonator losses. Variations of the pumping power are mainly caused by plasma instabilities and current ripples. Variations of resonator losses are attributed to mechanical and thermal instabilities of the laser structure, leading to mirror misalignment, resonator length alterations etc..

<u>3. Detectors for CO2-laserradiation</u>

Detectors for CO2-laserradiation can be classified into quantum detectors (Ge-Cu, Hg-Cd-Te detectors) and thermal detectors (bolometer, pyroelectric detectors).
The highest time-resolution down to the ns-region can be reached with quantum detectors. Risetime of Hg-Cd-Te detectors is 20-50 ns at a power detection threshold of about 0.01 mW, risetime of Ge-Cu detectors is about 1 ns at a power detection threshold of about 0.1 mW. Due to the little band gap the detectors exhibit a strong noise at room temperature. They have to be cooled with liquid nitrogen (77K, Hg-Cd-Te detectors) or with liquid helium (4K, Ge-Cu detectors). Bolometer detectors consist of a thin temperature dependent resistor layer. The

resistance of this layer is proportional to the irradiance. The risetime is limited
to some ms due to the thermal time constants. Bolometer detectors are operated at
room temperature. Pyroelectric detectors, also operated at room temperature,
generate a current proportional to the temporal temperature change of the
pyroelectric material. At frequencies above the thermal cut-off frequency (typicaly
10 Hz) this current is proportional to the laser irradiance, therefore continuous
radiation has to be chopped with frequencies above the thermal frequency. The upper
cut-off frequency is determined by the external electric circuit to typically 0.1-1
MHz at a power detection threshold of about 1 mW.

Depending on the ratio of detector area to beam area, time resolved measurements of
the total power or of the power at a point of the beam can be carried out. In order
to obtain a space resolved measurement beam scanning or deflection systems have to
be used.

4. Methods for space resolved measurement of CO2-laserradiation

The different methods proposed for space resolved measurements /3/ can be reduced to
two essential principles. Either laserradiation is deflected across a stationary
detector (mechanical/6/ or acoustooptical/5/) or the laserradiation is projected on
a detectorline or a detectorarray. Combinations of these principles are also
possible. In the following two simple but efficient deflecting methods are
discussed out of the various proposals: the rotating mirror and the rotating wire.

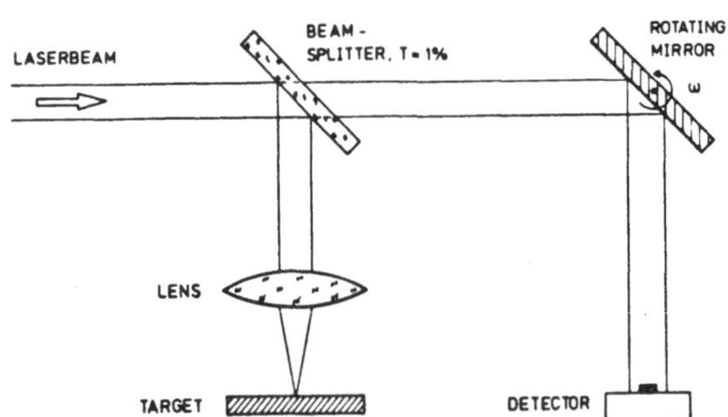

Fig.4 Schematic set-up for beam profile measurement with rotating mirror

Fig 4. shows a method of beam deflection by rotating mirror. A fraction of the
laser beam (typically some percent) is coupled out with a beam splitter and
deflected with a rotating mirror across a detector. By displaying the detector
signal with an appropriate time base on an oscilloscope the space and time
dependence of the intensity can be examined. The method can be used during
processing, it delivers straight-lined cuts through the intensity profile, there is
no distortion at the projection upon the detector. The power handling capacity of

the beam-splitter limits this method to maximal intensities of about 500 W/cm2. Special attention has to be paid to the polarisation dependence of the beam splitter transmission in the case of statistically polarised lasers.

The rotating wire method is shown schematically in fig.5 . A wire rotating through the cross section of the beam expands and deflects the beam at his highly reflecting surface. The power P_D (fig.5) is absorbed by a detector positioned at the distance L and the angle alpha to the axis. By appropriate choice of the parameter distance L, angle alpha, wire radius R_W the power P_D can be determined in a way making further beamsplitting or attenuation unnecessary. By using several detectors positioned symmetrically to the beam axis, it is possible to obtain simultaneously intensity profiles along several perpendicular oriented lines. The surface of the rotating wire must consist of a highly reflecting material such as silver or copper. Using materials like iron e.g. leads to a distortion of the measurement due to the angle and polarisation dependent reflectance at reflecting angles greater than about 60 degrees. The quantitative interpretation of the measurements has to consider the distortions occuring in deleterious detector positions and the shape of the cut lines through the beam diameter which are cirle segments.

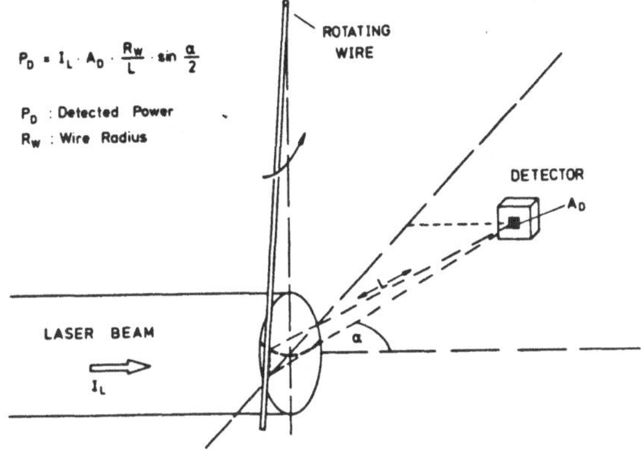

Fig.5 Schematic set-up for beam profile measurement with rotating wire

5. Results of measurements at different CO2-lasers

In the following some applications of the discussed time and space resolved diagnostic technics are presented. Results of measurements at some CO2-lasers are shown as far as they are representative of the different laser systems.

Time resolved measurements at a fast axial flow 1KW-laser are summarised in fig.6. The power is detected at a point of the beam diameter. Fig.6a shows the signal of a Ge-Cu detector with a time resolution in the ns-region. The laser power is modulated with the frequency of the transversal mode distance (20 MHz). With a time resolution in the ms-region (fig. 6b) power fluctuations are observed which are mainly induced by instabilities of the discharge plasma. These measurements have been taken with a pyroelectric detector, the laser radiation has been chopped with a frequency of about 1 KHz. Space resolved measurements have been taken at a 500 W slow flow TEM00-mode laser with the rotating mirror and a pyroelectric detector . Fig.7a shows the Gaussian intensity distribution of the free running beam. During processing a considerable alteration of the intensity profile takes place (fig.7b). The target in the lens focus induces feedback modulation of the laser power with a

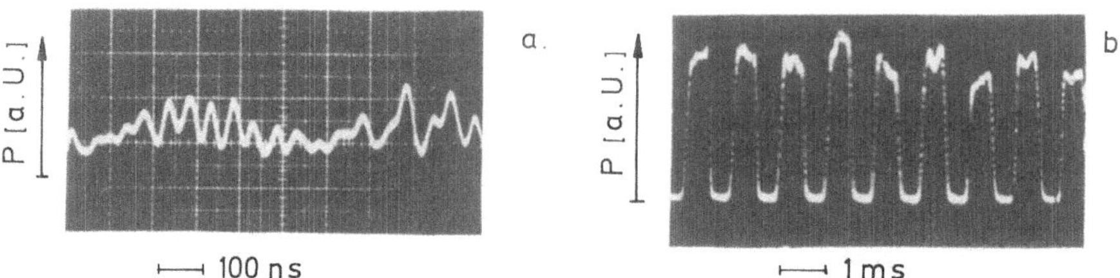

Fig.6 Time resolved measurements of the power of a fast axial flow
1 KW CO2-laser
a. with high time resolution (Ge-Cu detector), fluctuations
in the MHz-region due to superposition of transversal modes
b. with low time resolution (pyroelectric detector), fluctuations in the KHz region due to discharge instabilities

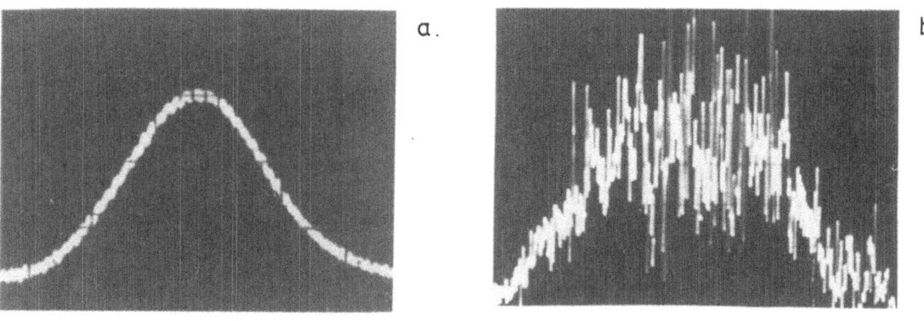

Fig.7 Space resolved measurements of the intensity distribution of
a slow flow TEM 00 CO2-laser
a. free running beam
b. with feedback modulation caused by the target in the lens
focus

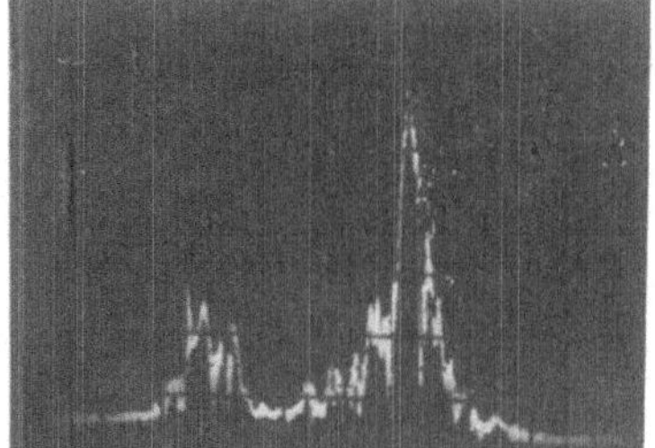

Fig.8 Space resolved measurement of the
intensity distribution of a transverse
flow 12 KW CO2-laser with an unstable
resonator

frequency of about 150 KHz. A result of the discussed rotating wire diagnostic technic is shown in fig.8. The measurement has been taken at a transverse-flow 12 KW CO_2-laser with an unstable resonator. The ring-like intensity distribution exhibits a slight asymmetry.

6. Summary

Diagnostic of the free running laser beam is not sufficient to estimate the beam quality of a CO_2 laser for material processing. During processing complex interactions between laser and target occur leading to considerable alterations of the beam characteristics. Therefore investigations of the beam quality have to be performed during processing, the diagnostic requirements are defined by the special dynamics in the system formed by the laser and the target. Space and time resolved detection of the laser beam intensity distribution is possible with the discussed technics. The technics can be diversified with respect to different regards. The scan velocity can be increased to reduce the error originating from the coupling of space and time resolution. The total beam cross section can be scanned by modification of the technics for example with X-Y vibrating mirrors or Nipkov disks/4/. However the date rates of these extended technics require higher efforts of storage and computation electronics, whereas the discussed techniques give an easy survey about the quality of the laser radiation.

Literature
/1/ J.Dembowski et al.
    Resonators for High Power Lasers
    Fourth International Symposium on Gas Flow and Chemical Lasers
    Stresa, Italy (1982)
/2/ W. English
    Dissertation
    JWG-Universtät, Frankfurt (1976)
/3/ F. Rühl
    Realisierungskonzepte für On-Line Strahldiagnostiksysteme für CO2-Hoch-
    leistungslaser
    Bundesministerium für Forschung und Technologie, F+E-Vorhaben 13 N 52031
    Darmstadt, 1983
/4/ R.Rothe, W.Jüptner, G.Sepold
    Measuring Methods to Control a High-Power CO2 Laser Beam for Material Processing
    Imeco '82, Berlin (1982)
/5/ G.Nickel et al.
    IEEE - Journal of Quantum Electronics, QE-8, 1. (1972)
/6/ G.C.Lim, W.M. Steen
    Optics and Laser Technology, June 1982
/7/ E. Beyer, L.Bakowsky, G.Herziger, A. Donges, P.Loosen
    These Proceedings
/8/ R.Poprawe, L.Bakowsky, E.Beyer, G.Brumme, G.Herziger
    These Proceedings
/9/ G. Herziger et al.
    IEEE - Journal of Quantum Electronics, QE-10, 2 (1974)

Meßgerät zur On-Line-Kontrolle von Laserstrahlen bei der Werkstoffbearbeitung

R. ROTHE, P. STEINLEIN, G. SEPOLD
Bremer Institut für angewandte Strahltechnik, BIAS, 2820 Bremen 71

1. Einleitung

Hochleistungslaser werden zunehmend zum Schweißen, Schneiden, Härten und für andere Werkstoff-Bearbeitungsfälle eingesetzt. Besonders bei Lasern mit Leistungen oberhalb 1,5 kW besteht ein technisches Interesse, die Leistungsdichteverteilung über den Querschnitt des Laserstrahls während der Bearbeitung messen und beobachten zu können.

Hiermit wird dem Operateur eine einfache Justierhilfe gegeben; und im Sinne der Qualitätssicherung besteht die Möglichkeit, Zusammenhänge zwischen sich ändernder Leistungsdichteverteilung und Bearbeitungsergebnis herzustellen. Für diesen Zweck wurde nach einer neuartigen Technik ein Gerät entwickelt, das im folgenden vorgestellt wird

2. Aufbau und Funktion des Meßgerätes

2.1 Detektion

Der wesentliche Teil des Gerätes ist ein Speichenrad, dessen Speichen um 45° zur Achse geneigt und axial versetzt sind, Bild 1. Die Speichen bestehen aus Flachstäben, deren zur Achse zeigenden Flächen verspiegelt sind. Dreht sich das Speichenrad um die Achse, so beschreibt jede Speiche einen Kegelmantel. Durch einen solchen Kegelmantel werden parallel zur Drehachse einfallende Laserstrahlen auf die Drehachse gespiegelt. Dort befindet sich ein Detektor mit kleiner Apertur. Dieser empfängt Laserlicht aus einem eindeutig bestimmten Abstand von der Achse, der sich aus der Anordnung ergibt. Mit der Bewegung einer Speiche durch den Laserstrahl hindurch wird so eine Zeile des Laserstrahls vom Detektor erfaßt und in ein elektrisches Signal transformiert. Das Speichenrad ist so aufgebaut, daß sich jeweils höchstens eine Speiche im Strahlbereich befindet.

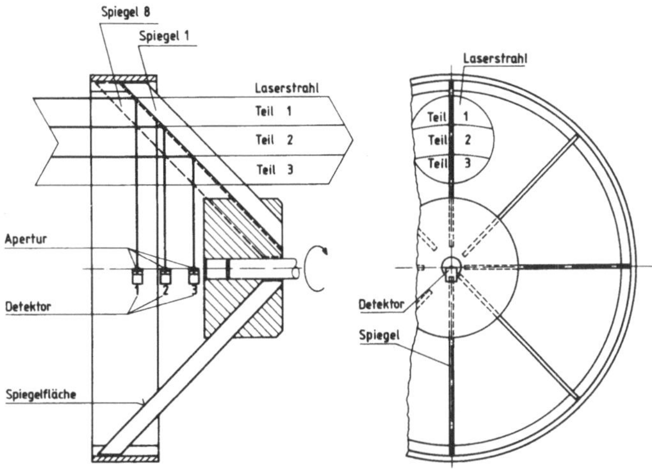

Bild 1. Meßgerät zur On-Line-Kontrolle von Laserstrahlen, Prinzipdarstellung

Durch die axial versetzte Anordnung der Speichen wird bei einer Radumdrehung eine der Speichenzahl entsprechende Zeilenzahl zeitlich nacheinander abgetastet. Die Drehfrequenz beträgt 50 Hz, so daß in einer Sekunde 50 Bilder des vollen Strahlquerschnittes aufgenommen werden. Durch Anordnung mehrerer Detektoren auf der Drehachse werden mehrere Zeilen gleichzeitig erfaßt.

2.2 Signalverarbeitung

Die Detektorsignale werden verstärkt und können beispielsweise auf einem Oszillografen dargestellt werden, Bild 2.

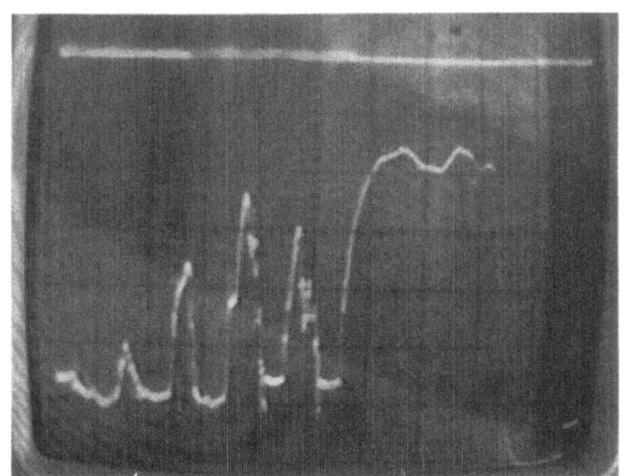

Bild 2. Oszillogramm eines Dektorsignals von sechs Speichen

Zeitablenkung:
2 ms/Skt.Zeile 1...5
0,5 ms/Skt.Zeile 6

vertikaler Maßstab:
1 V/Skt.

Laserleistung: 3 kW

Derartige Signale lassen sich z. B. zur Überwachung und Regelung des
Lasers direkt heranziehen.

Durch die gewählte Drehzahl von 50 s^{-1} bietet sich eine Weiterverarbeitung dieser Signale nach Fernsehnorm an, so daß es möglich ist, alle
bekannten Hilfsmittel einzusetzen, die eine Beurteilung der Leistungsdichteverteilung im Laserstrahl erleichtern. Diese läßt sich z. B. auf
einem Fernsehschirm in Graustufen oder Farbewerten darstellen, Bild 3.
Hier ist eine ausgeprägte Struktur eines fehljustierten Laserstrahls
gezeigt. Das Zentrum des Laserstrahls befindet sich im linken Bildteil.
Der rechte Teil ist sehr inhomogen. Die "weißen" Quadrate sind Bereiche
mit der höchsten Leistungsdichte.

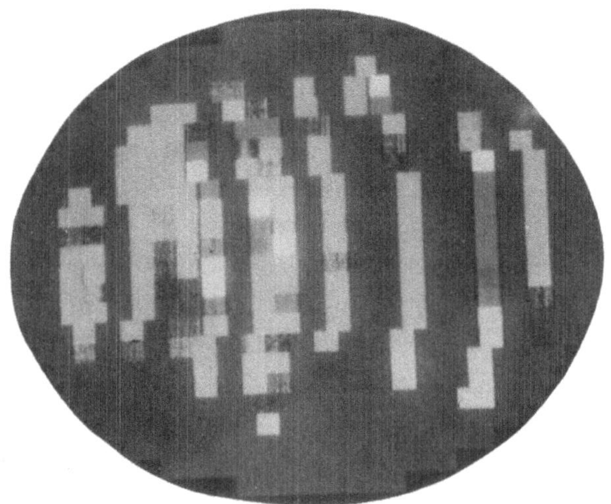

<u>Bild 3.</u> Darstellung der Leistungsdichteverteilung eines Laserstrahl-
querschnittes über "Farbwerte" (Schwarz-weiß-Kopie eines Farb-
bildes). Eine Struktur transversaler Moden ist ausgeprägt,
weiße Quadrate zeigen Leistungsspitzen an

Als weitere Darstellungsarten der Leistungsdichteverteilung wurden die
Pseudo-3D-Darstellungen erprobt und die Ausgabe als Ziffernbild auf
dem Fernsehmonitor mit zusätzlicher Einblendung der Gesamtleistung,
die durch Integration der Leistungsdichte errechnet wurde.

3. Auslegung und Eigenschaften

Ein Gerät ist zum Einsatz für 5 bis 15 kW-Laser mit maximalem Strahldurchmesser von 50 mm ausgelegt und gebaut worden, Bild 4. Es ist mit
8 Speichen und 3 Detektoren ausgerüstet. Hiermit wird der Strahlquerschnitt in 24 Zeilen unterteilt.

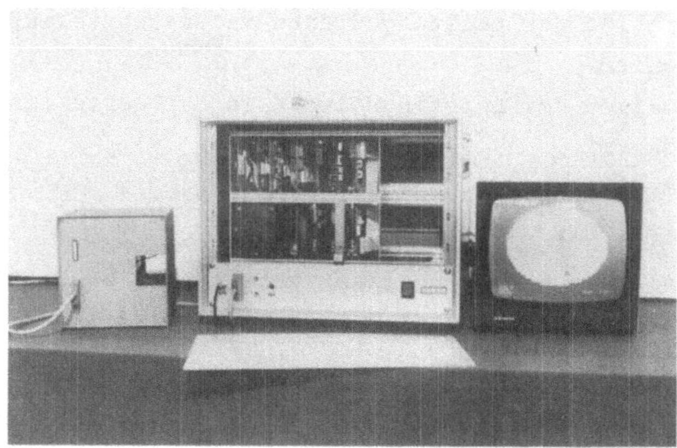

<u>Bild 4</u>. Gerät zur Messung der Leistungsdichteverteilung von CO_2-Laser-
strahlen

      Links: Meßgerät
      Mitte: Bildverarbeitungsteil, geöffnet
      Rechts: Monitor mit Querschnittsbild eines Laserstrahls
                   (Graue Bereiche sind Intensitätsmaxima)

Die Auflösung einer Zeile wird je nach Signalauswertung entweder durch
das zeitliche Auflösungsvermögen der Detektoren oder von der Speichenbreite bestimmt. Die obere Grenzfrequenz der Detektoren ist mit 100 kHz
festgelegt. Hier ist das Signal-Rauschverhältnis noch gut. Damit ist
die Auflösung einer Zeile mit 2×10^{-2} mm entsprechend 2500 Bildpunkte
gegeben. Im zweiten Auswertemodus führt die Speichenbreite 2 mm zu
24 Bildpunkten in einer Zeile, so daß hier der Strahlquerschnitt mit
$24 \times 24 = 576$ Punkten on-line aufgelöst wird.

4. Zusammenfassung

Das vorgestellte kalibrierbare Gerät zur On-Line-Messung von CO_2-Laser-
strahlen ermöglicht die Beobachtung und Auswertung der Leistungsdichteverteilung über den Querschnitt von Hochleistungslaserstrahlen. Die
Bildfrequenz beträgt 50 Hz, so daß eine Weiterbearbeitung der Signale
nach Fernsehnorm vorgenommen werden kann. Das Prinzip läßt sich auf
unterschiedliche Laserleistungsklassen und Strahldurchmesser durch Anpassen der mechanischen Größen übertragen. Ein Gerät wurde erprobt.
Hiermit war es möglich, auch kurzfristige Schwankungen der Intensitäts-

verteilung im Laserstrahl sichtbar zu machen, wobei als Darstellungsarten eingesetzt wurden:

- Oszillografenbilder der Detektorsignale
- Pseudo-3D-Darstellung der Leistungsdichteverteilung auf dem
  Oszilloskop, womit sich ihre zeitliche Änderung längs einer
  Zeile darstellen läßt
- Transformation der Leistungsdichte in Grauwerte auf einem
  Fernsehschirm
- Transformation der Leistungsdichte in Farbwerte auf einem
  Fernsehschirm
- Angabe der Leistungsdichte als Ziffernbild mit Angabe der
  Gesamtleistung.

Hiermit lassen sich Hochleistungslaser während der Werkstoffbearbeitung
steuern und regeln. Außerdem können kurzfristig Justiervorgänge
reproduzierbar optimiert werden.

Optical Feedback during Laser Material Processing

E. BEYER, A. DONGES, P. LOOSEN, G. HERZIGER
Institut für Angewandte Physik, Technische Hochschule Darmstadt
Schlossgartenstr. 7, D-6100 Darmstadt

1. Introduction

Compared to the conventional material processing the laser shows an important particu-
larity. The laser properties are changed stochastically by optical feedback. This
feedback is investigated since 1964 in several papers /1-4/. By arranging a target in
the focus area of the processing optic a part of laser radiation - dependent on reflection coefficient and target position - is reflected into the laser, being amplified
there. The processing optics, target and laser output mirror form an optical resonator, which is coupled to the laser resonator (fig. 1). The following experiments
refer to CO_2-high power laser systems. But the phenomenon of optical feedback occurs
at all types of lasers used in material processing.

2. Methods of Diagnostics

Two modes of operation have been used (fig. 1). Therefore, 1 % of the laser power was
coupled out either from the laser resonator or from the coupled resonator. A pyroelectric detector was used monitoring the signal. An experimental set-up with a chopper
shows the temporal oscillations of the laser power (fig. 2). A set-up with a rotating
mirror as well as a rotating wire shows the spatial beam distribution /5/. Temporal
and spatial oscillations of laser power superimpose to the total beam distribution.
The experimental results with different diagnostic methods, inside and outside of the
coupled resonator system (fig. 1) show reasonable agreements.

3. Experimental results

The coupling factor of the resonator system is given by the transmission of the laser
mirror. The losses of the coupled resonator depend on the reflectivity of the target
and its position in the focus area. Among other effects losses and coupling govern
the optical feedback. The laser signal appears to be undisturbed if the target is
located outside the lens focus only a few mm. Strongest laser spiking is observed if
the target is located in the focus (fig. 3). With a temporal spreading of signal
periodical intensity fluctuations with a time constant of about 45 µsec are observed

(fig. 4). Compared to the average laser intensity, the resulting intensity peaks can be strongly intensified. This temporal behaviour may be explained with the help of re laxation times of CO_2 molecules, which are shown in fig. 5. For laser power spiking the relaxation times τ_1 of the low and high laser levels are important. The pulsed disturbance of a free running laser resonator system results in two typical spiking frequencies approximated by the following equation /9,10/.

$$\omega \simeq \sqrt{\frac{P_h-1}{\kappa_1 \tau_r}} \qquad (1)$$

P_h = normalize pumprate

τ_1 = laser level relaxation time

τ_r = laser resonator relaxation time

The experimentally approximated spiking frequencies (fig. 6) are of the same order of magnitude as the calculated frequencies using equation (1).

calculation	experiment
$\tau_{S1} \simeq 6.5$ µsec	$\tau_{S1} \simeq 8.5$ µsec
$\tau_{S2} \simeq 30$ µsec	$\tau_{S2} \simeq 42$ µsec

Normally a coupled resonator system represents a damped oscillator system. Each stimu lated spiking is damped with a characteristical time constant. For continous oscilla- tions as shown in the experiments a disturbance in the resonator necessarily implies a frequency spectrum within the range of the relaxation processes. If this disturbanc is in resonance with a laser spiking frequency, a very intense peak is rising.

A disturbance of the resonator system takes always place if the target begins to melt by laser irradiation. Normally a high frequent oscillation develops in the melt by capillary waves. Thus a feedback spiking only could be observed, if the laser power is sufficient to melt the target surface. In fig. 7 laser signals of the free running beam and the intensity distribution in the coupled resonator are represented. The photon density is increasing exxentially with decreasing losses of the coupled reso- nator (fig. 7).

During welding in front of the target a plasma develops which is nearly transparent at low laser intensities. However, the target is screened by the absorbed plasma due to an overincreased intensity of a feedback spike. The exact influence of laser spiking on the laser material processing is unknown.

In fig. 8 the intensity distributions during different laser welding processes are shown. The losses of the coupled resonator are increased during the welding of

thicker targets by the development of a vapour capillary and by plasma absorption
resulting in less distinct spiking.

4. Conclusion

The optical feedback changes the characteristics of the laser oscillator by the process
on the target. Distance- and angle variation of the target, movements of the melt or
change of absorption- and refraction index of the plasma on the target surface result
in stochastic fluctuations of the laser radiation. In UV and visible range the optical
feedback can be eliminated by Faraday rotators. In IR-range there is no economically
feasible feedbacklock for high power radiation, so that for precision material processing
knowledge and control of the feedback is important.

Literature
(1) EICHLER, H., and G. HERZIGER: Z. Angew. Phys. 23 (1967) 297
(2) EICHLER, H., and W. WIESEMANN: Z. Angew. Phys. 28 (1969) 125
(3) EICHLER, H., G. HERZIGER Z. Angew. Phys. 12 (1964) 193
(4) DÄNDLIKER, R., and T. TSCHUDI: Appl.Optics 8 (1969) 1119
(5) LOOSEN, P.: these proceedings
(6) RÜHL, F.: BMFT-Report 13 N 52031, Darmstadt (1983)
(7) ENGLISCH, W.: Dissertation, JWG Universität Frankfurt (1976)

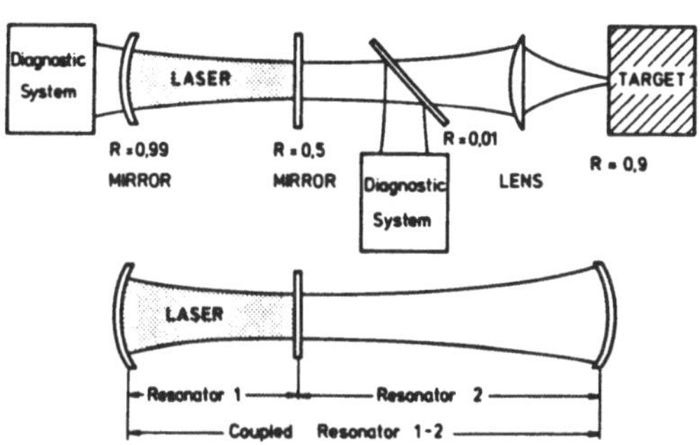

Fig. 1. Scheme of coupled resonators

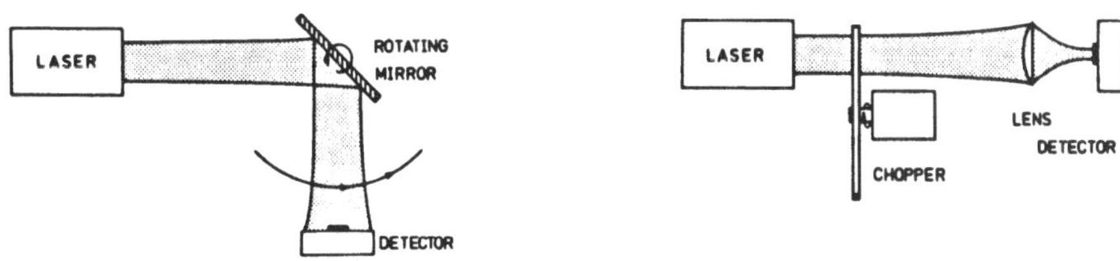

Fig. 2. Experimental set-up for temporal and spatial measurement
of laser power fluctuation

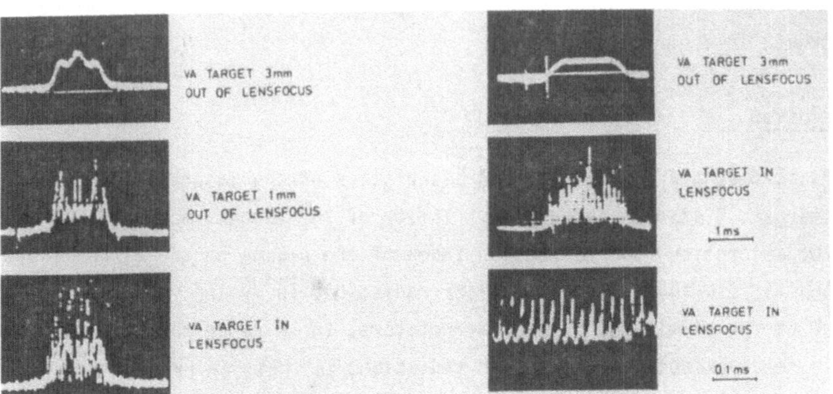

Fig. 3. Spatial beam distribution as a function of focus position

Fig. 4. Diagnostic of temporal spiking with chopper

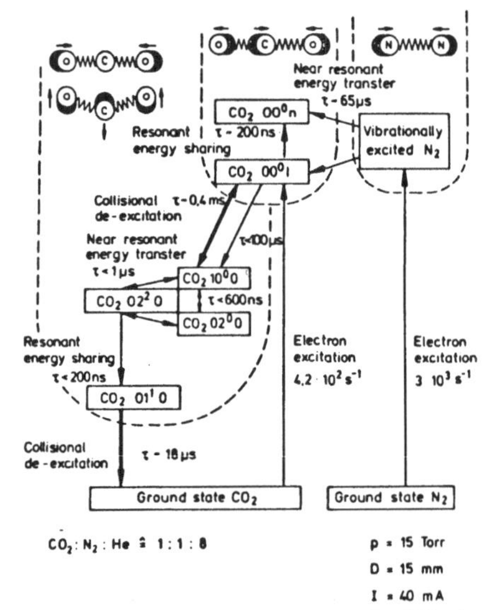

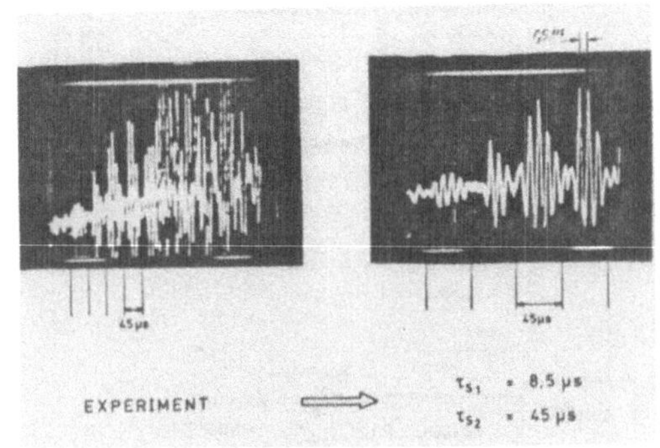

Fig. 5. Schematic diagram of the CO_2-laser mechanism showing relaxation times for a typical gas mixture

Fig. 6. Scope traces showing temporal spiking

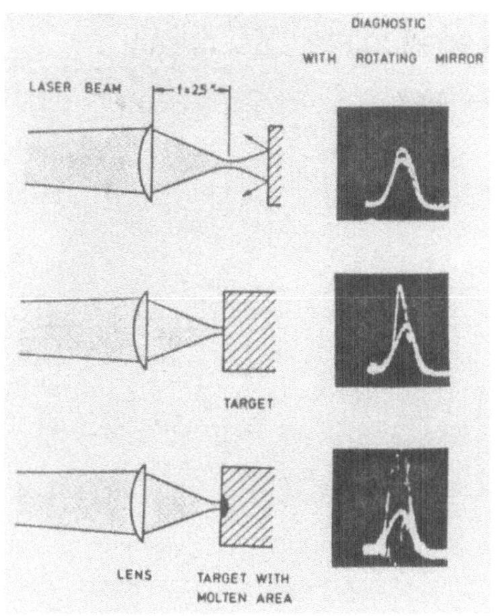

Fig. 7. Influence of target surface and position on beam distribution

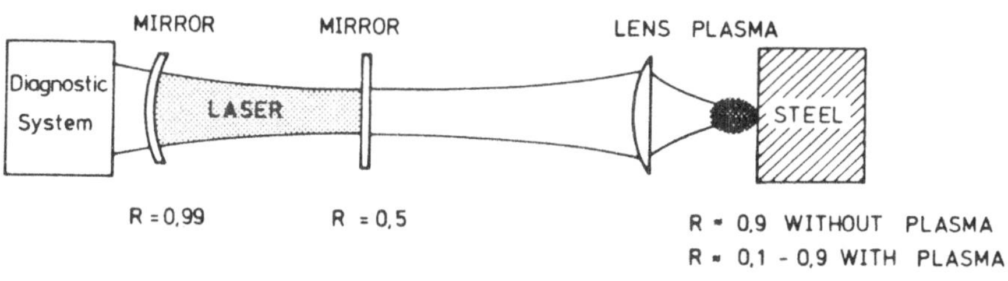

DIAGNOSTIC WITH ROTATING MIRROR

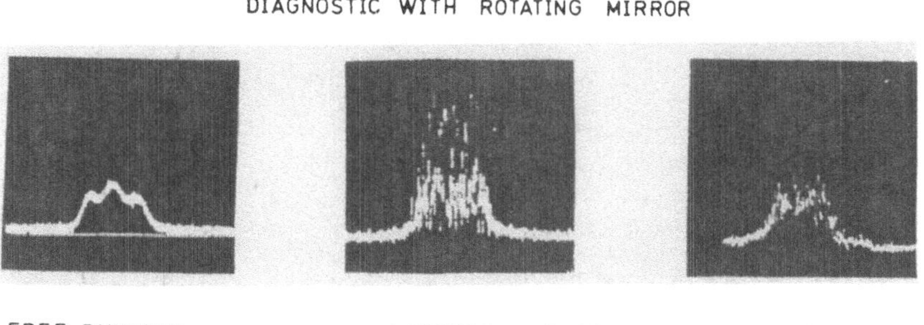

Fig. 8. Beam distribution as a function of welding process

Ein Gerät zur Analyse der Intensitätsverteilung im Laserstrahl während der Materialbearbeitung

Dr. P. Arnold

Firma ALL - Applikationslabor für Lasertechnik

Hans-Grässel-Weg 1, 8000 München 70

Herr Loosen und Mr. Sharp haben im vorhergehenden Vortrag verschiedene Detektorsysteme zur Strahlenanlyse vorgestellt, die im grundlegenden Aufbau erprobt wurden.

Basierend auf der Idee des rotierenden Drahtes - das Prinzip wurde von Dr. Steen, Imperial College, London, entwickelt - wurde ein Gerät zum Einsatz in Labor und Industrie entwickelt, das hier in den Grundzügen vorgestellt wird.

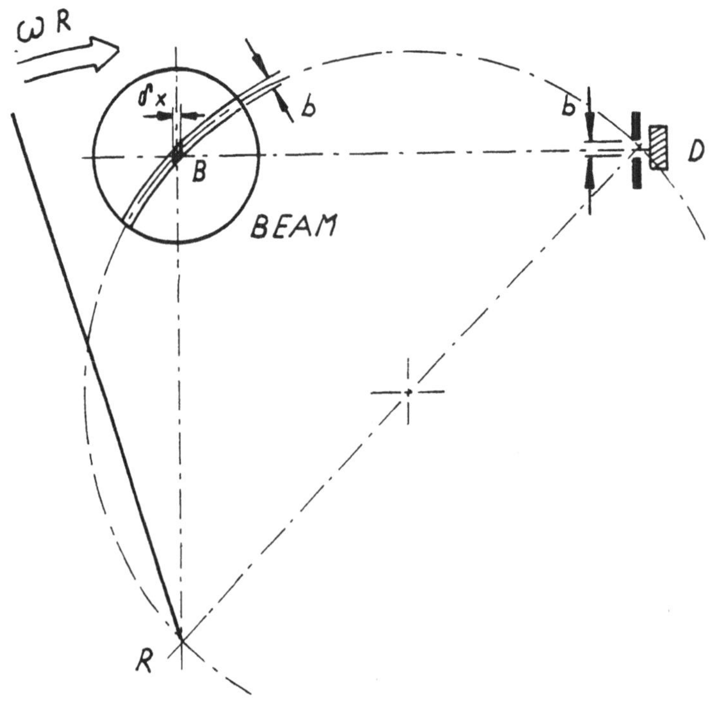

Ein dünner,hochreflektierender Draht rotiert in einer Ebene senkrecht
zum Strahl - angetrieben von einem Synchronmotor. Von dem Draht wird
während des Durchgangs durch den Strahl ein geringer Bruchteil der Intensität in einem pyroelektrischen Detektor D gestreut. Der Detektor
"sieht" die Intensitätsverteilung in dem Streifen der Breite "b". Die
Streifenbreite "b" kann mittels Blenden vor dem Detektor eingestellt
werden.
Der Detektor sieht bei Durchgang des Drahtes das momentane Flächenele-
ment dx·b. Bei üblichen Drahtdurchmessern der Größenordnung 1 mm beträgt die Breite des Flächenelements dx etwa 10µm. Aufgrund dieser guten Ortsauflösung ist das Gerät auch zur Messung im Fokuspunkt geeignet.
Der ausgeblendete Bereich des Strahls ist ein Teil des Thaleskreises
über der Basis R-D.
Unter Ausnutzung dieser Reflektionsbedingung in einer listigen Anordnung
mit zwei Detektoren, kann man bei Durchgang des Drahtes durch den Strahl
gleichzeitig eine Intensitätsverteilung in X- und Y-Richtung erhalten-
wie im folgenden Bild dargestellt.

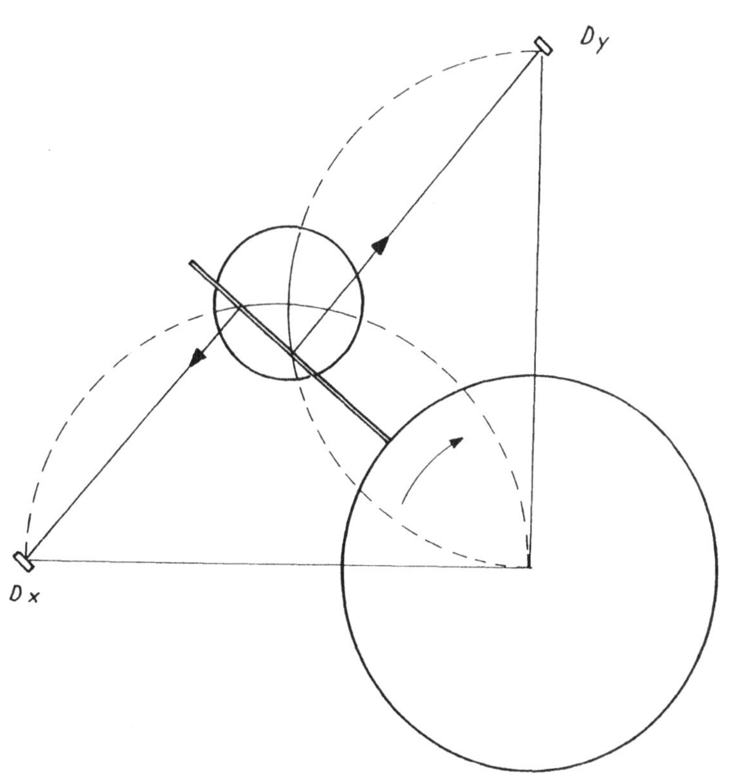

Basierend auf dieser Anordnung haben wir ein Gerät zur Strahlenanalyse
entwickelt, mit den folgenden Merkmalen:

1) klein, transportabel - geeignet für Service
2) einfache Darstellung des Signals - es genügt ein einfacher Zwei-
   strahl-Oszillograph, wie er in jedem Labor vorhanden ist.

Die Darstellung der Intensitätsverteilung erfolgt auf einem Kreisbogen.
Durch geeignete Dimensionierung wurden die Geräte so ausgelegt, daß die
Abweichung des Kreisbogens von der Sehne maximal 7% beträgt.

Die Un. linearitäten vergrößern sich mit zunehmendem Strahldurchmesser.
Aus diesem Grund werden 3 Geräte gebaut, entsprechend zunehmendem
Strahldurchmesser. Das kleinste Gerät ist geeignet für Strahldurchmesse
von 0 - 25 mm, das mittlere für 10 - 40 mm und das größte für 20 - 60mm
DieLeistung nach oben ist unbegrenzt. Die untere Leistungsdichte im
Strahl beträgt etwa 1 Watt pro mm^2.

Das Gerät enthält einen Synchronmotor, 2 pyroelektrische Detektoren
mit 2 entsprechenden Verstärkern, die in 4 Empfindlichkeitsstufen
schaltbar sind. Durch einen elektronisch verstellbaren Trigger kann
die Abbildung auf dem Oszillographenschirm immer so eingerichtet werden
daß sich das Bild in der Mitte befindet.

Der Laser-Beam-Analyser muß auf Strahlmitte eingerichtet werden. Zu
diesem Zweck ist die gesamte Anordnung von Motor, Detektor und Verstärker mittels eines Feintriebes in x- und y-Richtung verschiebbar.
Die jeweilige Position wird digital angezeigt.
Die Laser-Beam-Analyser sind sowohl für YAG- als auch für CO_2-Laser
geeignet. Bei Verwendung spezieller Fenster im Detektor kann mit einem Gerät sowohl der eine wie auch der andere Lasertyp ausgemessen
werden.

Nun zu den Anwendungsmöglichkeiten.
Das hauptsächlichste ist die Justierung des Lasers.

Auf dem nächsten Bild ist ein Beispiel einer Laserjustierung, eines
langsam geströmten CO_2-Lasers der Klasse 500 W, dargestellt.Das Gerät wird in einiger Entfernung vom Auskoppelspiegel aufgestellt.
Der Oszillograph steht am Ort der Spiegeljustage. Dies hat sich als
äußerst nützlich erwiesen, denn beim Einstllen des Spiegels kann unmittelbar beobachtet werden, wie sich die Intensitätsverteilung im

Strahl ändert. Da die Signalhöhe zur Leistung direkt proportional ist,
kann auf schnelle und einfache Weise der Laser auf höchste Leistung
bei optimalem Mode justiert werden.

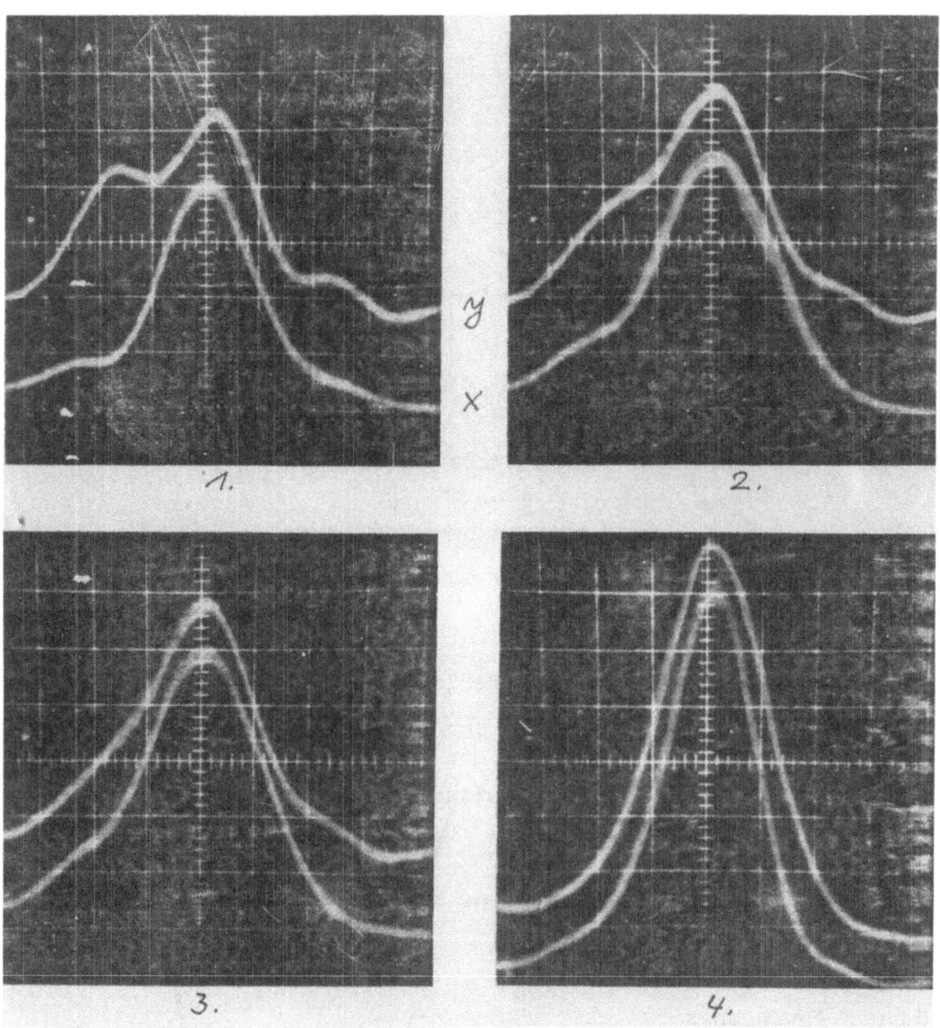

Z.B. ist zwischen der Einstellung 3 und 4 kein Unterschied auf dem
Leistungsmesser zu sehen. Eine Justierung von CO_2-Lasern allein
nach Leistung führt bestimmt nicht zum besten Ergebnis.

Der Laser-Beam-Analyser kann während der Bearbeitung im Strahl verbleiben.

Beim Schweissen zeigen sich interessante Ergebnisse. Wie Herr Loosen im vorherigen Vertrag dargestellt hat, wird durch die Rückreflexion der Laserstrahlung vom Metall der Mode entscheidend beeinflußt.

Der Laser-Beam-Analyser läßt sich noch für andere Anwendungen einsetzen, z.B. Messung im Pulsbetrieb und Messung des Fokusdurchmessers.

Im Pulsbetrieb wird der Motor ausgeschaltet und der Draht im Strahl arretiert. Ein geringer Bruchteil der Strahlung wird auf dem Detektor reflektiert. Das Detektorsignal zeigt den zeitlichen Pulsverlauf.

Bedingt durch das hochschmelzende Grundmaterial des Drahtes, seine spiegelnde Oberfläche und seine hohe Geschwindigkeit, kann mit dem Laser-Beam-Analyser der Durchmesser des Fokuspunktes gemessen werden.

Als Zusatzgerät kann ein mittlerweile sehr preiswerter Kleincomputer angeschlossen werden.
Die Signale des Laser-Beam-Analysers werden digitalisiert und gespeichert. Der Rechner kann verschiedene Operationen durchführen:

-Darstellung der aktuellen und gespeicherten Intensitätsverteilung auf dem Bildschirm.
-die Integration des Signals ergibt die Leistung
-Auswertung des Strahldurchmessers, eine Aussage über die Divergenz
-Abweichungen der Strahlmittelposition, eine Aussage über die Zeige-
 stabilität des Lasers.

Für den Einsatz des Laser-Beam-Analysers in der industriellen Produktion ergibt sich eine interessante Anwendung. Die Intensitätsverteilung des im guten Zustand justierten Lasers wird gespeichert. Das aktuelle Signal während der Bearbeitung wird ständig mit dem Soll-Signal verglichen.

Veränderungen des Lasers, die zu einer Abweichung der Intensität oder Moden-Struktur führen und damit die Qualität der Bearbeitung beeinflußen können, werden erkannt und angezeigt.

Multi-Kilowatt CO$_2$ Lasers for Material Processing[*]

K. Kakizaki, N. Nishida, and T. Takahashi

Toshiba Corp., Manufacturing Engineering Lab., Yokohama, Japan

INTRODUCTION

High-power CW CO$_2$ lasers with nominal output powers of 1,3, and 5 kW have been developed.   They are designed with an emphasis on the output beam quality. Knowledge of beam characteristics is of great importance in materials processing applications in relation to beam delivery and focusing. In this paper will be presented outline of the devices and their output power and beam characteristics.

OUTLINE OF LASERS

All the lasers described in this paper are of the transverse flow type and similar in basic designs with the laser head and discharge section as shown in Figs.1 and 2. The discharge takes place along the gas flow between individually ballasted multipin cathodes placed upstream in the flow and water-cooled tubular anodes downstream./1/

The electrode configuration in Fig.2 enables high-power uniform discharge over a large disccharge volume. Low gas pressure (35 to 45 Torr) operation assures higher gain and smaller gain saturation parameter resulting in larger mirror out-coupling and lower beam intensity on mirror/window surfaces reducing possibilities of optics damage.

Fig.1 View of 5kW laser head
with beam handling box (black,
left) mounted on platform

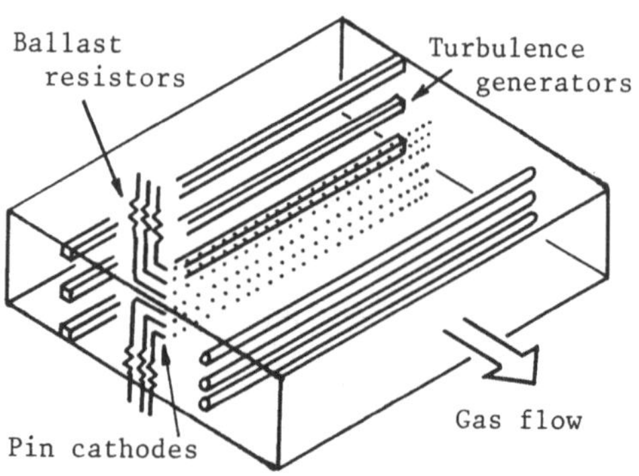

Fig.2 Schematic of discharge section

RESONATORS AND OUTPUT CHARACTERISTICS

Examples of typical resonator configurations and their beam intensity profiles for high-power CO_2 lasers are shown schematically in Fig.3. A stable resonator can sustain either TEM_{00}, $TEM_{01}*$, or highly multimode (Fig.3 (a),(b),(c) respectively) depending upon the size of a limiting aperture inside the resonator. Mode volume for TEM_{00} mode is small unless a very long (folded) resonator is employed.

The confocal unstable resonator Fig.3(d) can give a highly collimated beam out of an active medium having a large Fresnel number, and is a preferred choice in recent high-power CO_2 laser experiments.

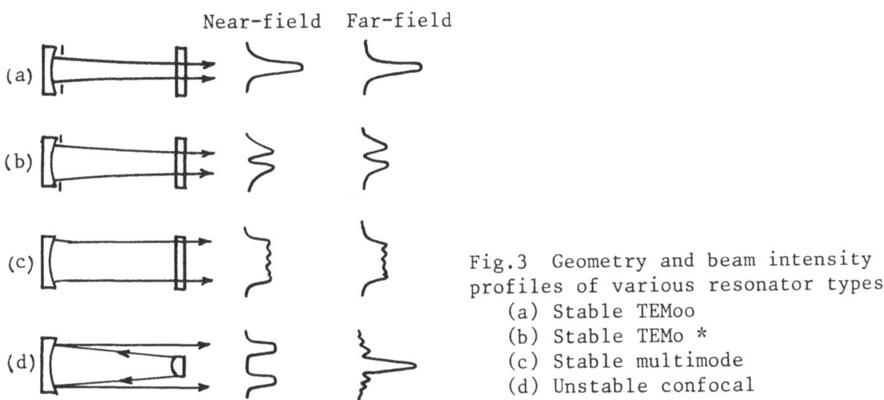

Fig.3 Geometry and beam intensity profiles of various resonator types
    (a) Stable TEMoo
    (b) Stable TEMo *
    (c) Stable multimode
    (d) Unstable confocal

Table 1 shows the resonators and output powers of the present lasers. Low order stable modes are preferred for cutting applications.

The resonators (c) and (d) can be used interchangeably with the 5kW model. The actual resonator configuration is a Z-shaped folded cavity as shown in Fig.4. It consists of a ZnSe output mirror (uncoated 83%T) and three diamond-turned copper mirrors in case of the stable resonator, and a ZnSe output window, a coupling mirror and four copper mirrors in case of the unstable resonator. Output power characteristics are shown in Fig.5. Powers over 5kW are obtained with either resonator.

TABLE 1

Model	Resonator	Output power and mode
1 kW	5-pass stable	1kW TEM_{00}, Can be pulsed
3 kW	3-pass stable	3kW Multi, 1.5kW $TEM_{01}*$
5 kW	3-pass stable	5kW Multi
"	3-pass unstable	5kW Diffraction coupled

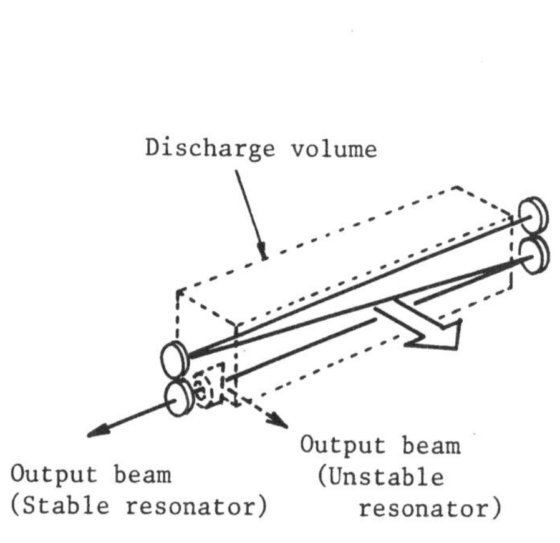

Fig.4 Resonator configuration

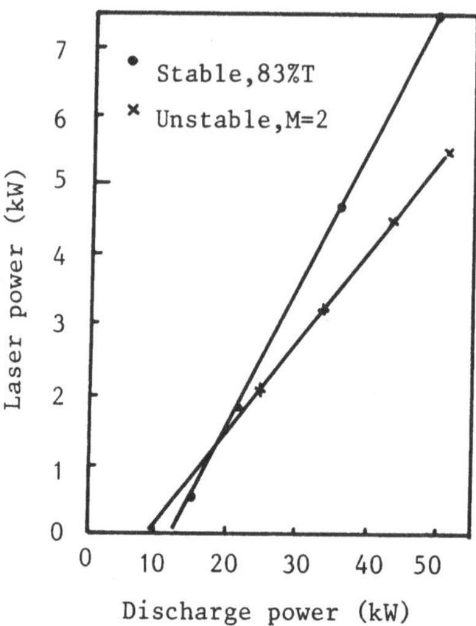

Fig.5 Output characteristics

<u>FOCUSING CHARACTERISTICS</u>

Beam intensity at the focal point may be a parameter of the greatest importance in materials processing applications. This can be calculated if near-field intensity and phase distributions are known. This is the case with the stable resonator; the field distribution is given in a Hermite-Gaussian form, and a good agreement is obtained between theoretical and experimental results. For the unstable resonator, however, there is no simple expression. A uniform-intensity, equal-phase distribution is often assumed to approximate focusing properties.

The far-field intensity distributions of the 5kW beams were determined by focal plane intensity measurements. The beam was focused with a 7-meter-focal-length ZnSe lens and ablation patterns were obtained as shown in Fig.6. The minimum spot diameter for the 5-kW stable beam was 18mm, which corresponds to the angular divergence of 2.6mrad (full angle). The diameter for the 5-kW unstable beam with the magnification M=2 was smaller, and was 5 mm at the first intensity minimum, which corresponds to the divergence of 0.7mrad.

Change of intensity profiles along the focusing path was also observed and was shown to exhibit a drastic change as suggested by Fig.3(d)(right).

Fig.7 shows the calculated angular dependence of power density I(r) and integrated fractional power L(r) for unstable resonator beams with various

Fig.6 Castings of far-field burn patterns of stable (left) and unstable(right) resonators (5kW)

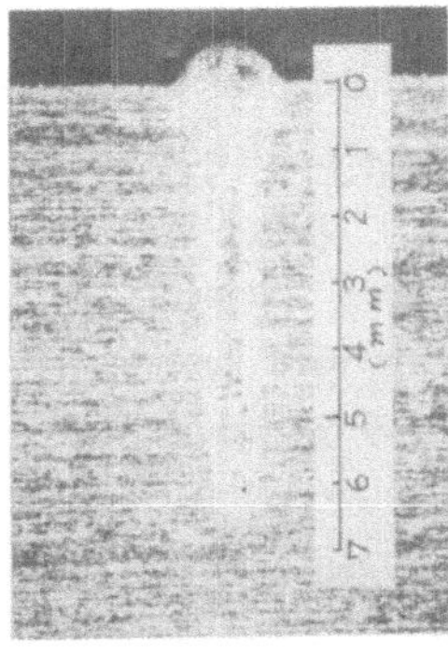

Fig.9 Cross section of bead-on-plate (5kW,10-inch FL lens, speed 2m/min)

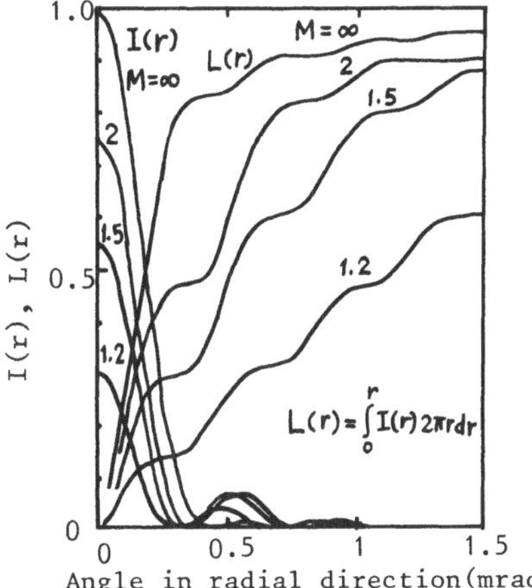

Fig.7 Intensity and integrated fractional power distribution for a stationary beam (Beam 30mm OD and total power constant)

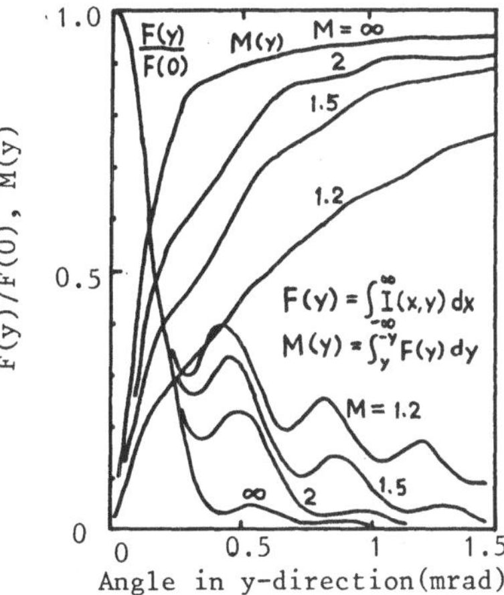

Fig.8 Intensity and integrated fractional power distribution for a moving beam (Beam 30mm OD constant)

values of M./2/ A uniform-intensity, equal-phase distribution is assumed. The calculated far-field divergence (first intensity minimum) of 0.35mrad (half angle) agrees with the experimental value mentioned previously. From this we estimate that when the 5-kW, M=2 beam is focused with a 10-inch FL lens 48% of the total power (2.4kW) goes into the central circle with 0.18mm diameter (first intensity minimum) and 34% (1.7kW) goes into the first surrounding ring with 0.4mm OD.

Fig.8 shows the power density integrated F(y) along a scanning direction(x) and the fractional integrated power M(y). F(y) corresponds to the input power into a point on a work piece when a laser beam is scanned over it. Fig.8 should be more useful than Fig.7 in estimating heat energy input in beam scanning applications, although exact processing phenomena should be analyzed by solving time-dependent heat-transfer equations.

A bead on a stainless steel plate sample is shown in Fig.9. This demonstrates a good focusability of the M=2 unstable resonator output beam.

A cutting and marking system using a 3-kW CO_2 laser with 1.5kW TEM_{01}^* mode has been installed at one of our Toshiba factories. This system consists of the laser, a 2.6m x 6.2m beam scanner, a numerical controller and a computer-aided design system connected to a host computer. Steel plates for transformer cases are marked and cut out from 2m x 6m steel plates with thickness of up to 12mm./3/

SUMMARY

A series of CO_2 lasers have been developed and their features and characteristics are described. 5-kW output beams from stable and unstable resonators are compared and shown to have a full divergence angle of 2.6mrad and 0.7mrad respectively. Far-field intensity distributions for stationary and moving beams are calculated.

REFERENDCES

/1/ A. C. Eckbreth and J. W. Davis: IEEE Jour.Quant.Electr., QE-8, 139(1972).
/2/ M. Born and E. Wolf: Principles of Optics (Pergamon, New York, 1970).
/3/ S. Fujiwara et. al.: CLEO Paper THN4 (May 19, 1983, Baltimore, MD, USA).

* This work was supported in part by the Ministry of International Trade and Industry.

Catalytic Recombination in a Multi-Kilowatt CO$_2$ Laser

A.S. KAYE
U.K.A.E.A. Culham Laboratory
Abingdon, OX14 3DB, G.B.

SUMMARY

The Culham CL10 10kW carbon dioxide laser has been fitted with a catalyst for
reversing the dissociation of CO_2. In operation at maximum power, this reduces the
dissociation from typically 35% to 15-20% and increases the output power by about
0.5kW. Variation of the catalyst throughput enables the dissociation and
recombination rates in the laser to be estimated. The dissociation rate is about 5
molecules per ion incident on the cathode; the recombination appears to occur on
sputtered cathode material and is strongly temperature dependent, resulting in a
decrease in dissociation with increasing input power.

INTRODUCTION

Gas discharge excitation of CO_2 lasers produces some dissociation of the CO_2 by
electron impact/1,2/ or collision of high energy ions in the cathode sheath. This
dissociation results in a reduction in small-signal gain and in the efficiency of
the laser /3,4/. The generation of negative ions of the dissociation products can
also adversely effect the discharge stability/1,3/.
The equilibrium dissociation is determined by a balance between various dissociation
and recombination mechanisms occurring in the laser and may only be reached after
hours of operation. Recombination may be increased by catalytic action, either on
sputtered material appearing in the laser by default/5/, or on catalysts
specifically installed for this purpose/5,6/.
In transverse flow cw lasers, the discharge current is relatively high (60amps in
CL10). As a result, processes occuring at the electrodes will play a much more
significant role than in axial flow devices dominated by the positive column.

DESCRIPTION OF EQUIPMENT

The CL10 laser has been described by Kaye et. al./7/. The laser is a transverse
flow, self sustained glow discharge device. It incorporating two of the Culham
designed CL5 5kW modules (referred to subsequently as modules A and B). The
pressure in the laser is typically 50mBar, each module has a volume of 1.7m^3 and the
gas is circulated at the rate of 4m^3 sec^{-1} per module. The nominal gas mix is

He:N$_2$:CO$_2$ in the ratio 60:25:15, with typically 1.5% of oxygen. Carbon monoxide may
alternatively be added. The gas is replenished at about 5 s.1.min^{-1}.

An additional gas circulation loop has now been installed on this laser, with a
throughput of 2.5m^{3}min^{-1}. This loop extracts gas from module B and reinjects it
into module A; the dissociation is different in the two modules with the loop in
operation. This loop incorporates a generous volume of proprietary catalyst such
that for all conditions explored, essentially total recombination of the CO$_2$
dissociation products was observed downstream of the catalyst.

The gas composition was monitored using a four channel quadrupole mass
spectrometer. The gas was sampled by continuously extracting a small volume past a
sintered disc, through which gas diffused into the mass spectrometer typically
operating at 10^{-6}mBar. The response time was typically 10secs. In practice, some
drift in sensitivity and pressure in the mass spectrometer was observed. This was
compensated by monitoring mass 28 as well as masses 44, 32 and 30. Carbon monoxide
overlaps the nitrogen peak, and this small contribution to mass 28 needed to be
taken into account in the analysis.

RESULTS WITHOUT THE CATALYST

On switching on the discharge, the dissociation increases rapidly to a maximum value
of about 50% after a few minutes. Subsequently, the dissociation slowly falls
before reaching an equilibrium value of typically 35% after 2 hours. This recovery
appears to be associated with the increasing temperature of the duct walls
downstream of the discharge. These walls are coated with sputtered cathode material

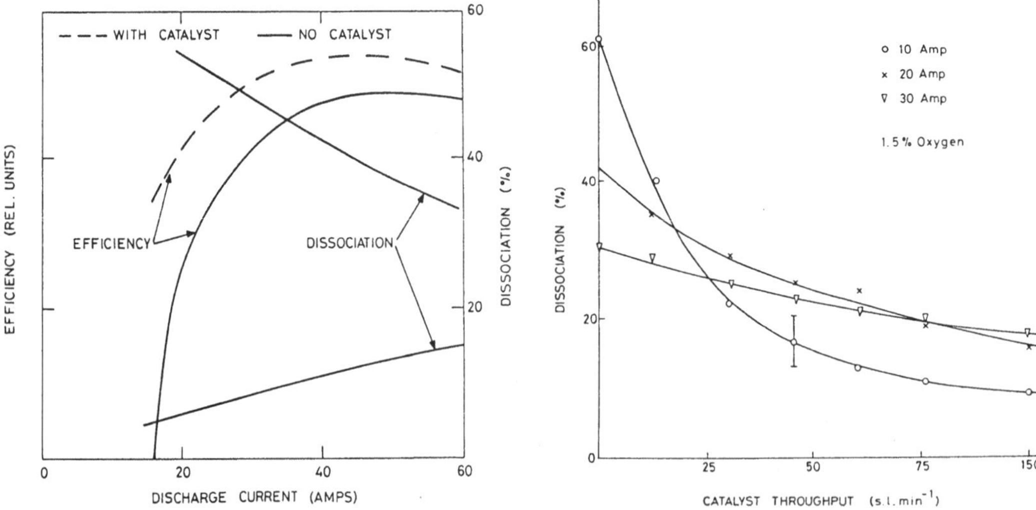

<table>
<tr><td>Fig. 1. Effect of catalyst on efficiency and dissociation.</td><td>Fig. 2. Effect of catalyst throughput on dissociation.</td></tr>
</table>

in the form of finely divided copper oxides which are well recognised catalysts at elevated temperature/8/. Increasing recombination on these surfaces reduces the dissociation. This temporal behaviour is strongly dependent on cathode material. With stainless steel or platinum, for example, the initial fall is much less severe and the subsequent recovery slight, giving an equilibrium dissociation of typically 10% after 30 minutes. This indicates that the dissociation is occurring in the vicinity of the cathode rather than the positive column.

The equilibrium dissociation and discharge efficiency vary with input power as shown in Figure (1). It is noted that at very low power, the time required to reach this equilibrium becomes increasingly long. The dissociation is observed to fall with increasing input power.

EFFECT OF CATALYST

The results here refer to module A unless otherwise stated. The equilibrium efficiency and dissociation with the catalyst in operation are also shown in Figure (1). It is observed that the efficiency is marginally increased at maximum power. Near threshold, a much more substantial increase is observed, reflecting the increase in gain, which is the main consequence of the reduction in dissociation. At high power, the dissociation is reduced by a factor 2 (1.5 in module B); at lower power, a much larger reduction is achieved, and the dissociation approaches zero. By varying the inlet pressure P to the catalyst loop, the throughput and thus the reaction rate in the loop can be varied. The dependence of dissociation, F, on loop throughout is shown in Figure (2) for a range of discharge currents, I, With increasing current, the dissociation is increasingly less affected by the catalyst.

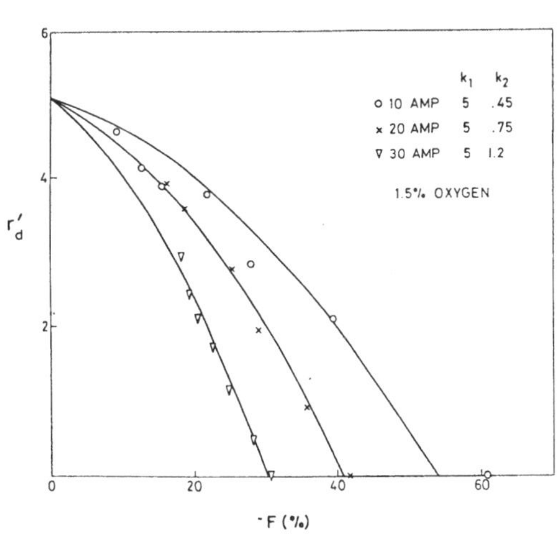

Fig. 3. Normalised nett dissociation rate in the laser.

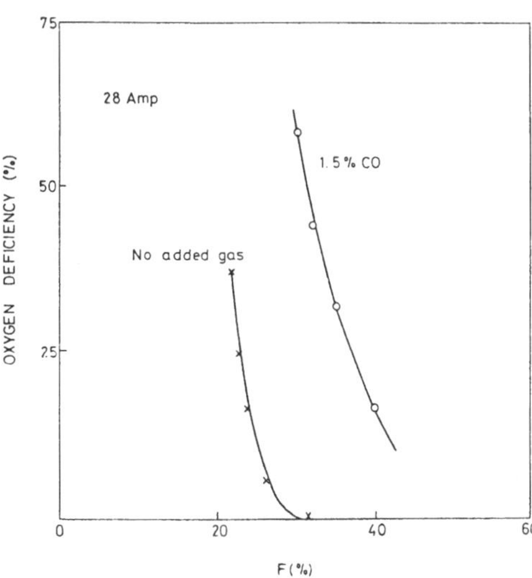

Fig. 4. Measured oxygen deficiency.

278

Now the rate of recombination in the catalyst, r_{cat}, is given by

$$r_{cat} = \frac{\nu \cdot P \cdot (F-F_0) \cdot N_A \cdot g_3}{\nu_M} \quad \text{molecules/sec} \tag{1}$$

where ν = volume throughput, F_0 = dissociation at catalyst exit (normally zero), N_A = Avogadro's No., g_3 = nominal CO_2 mole fraction and ν_M = molar volume. Define a normalised reaction rate $r' = re/I$, where I/e is approximately the rate of incidence of ions on the cathode. For equilibrium, the nett rate of dissociation in the laser, r'_d, must also be given by (1). The data of Figure (2) is replotted in the form of r'_d as a function of F in Figure (3). It is clear that r'_d falls increasingly rapidly with increasing F or I from a value at low F which is independent of current. Note that at $F = 0$, the recombination rate is zero and r'_d is the actual dissociation rate. Assume the rate of dissociation, r_{dis}, to be approximately proportional to current, and the recombination rate, r_{rec}, to scale approximately as the product of the oxygen and carbon monoxide partial pressures. Putting $F_1' = g_4/g_3$, $F_2 = g_5/g_3$ where g_4, g_5 are the nominal mole fractions CO and O_2 respectively (one of which is zero), gives

$$r'_d = r'_{dis} - r'_{rec} = k_1 - k_2 \, P^2_3 (F_1 + F)(F_2 + 0.5F) \tag{2}$$

where P_3 = nominal partial pressure of CO_2, and k_1, k_2 are rate coefficients. Curves of the form of (2) have been fitted to the experimental data of Figure (3) with the parameters shown on the figure. A good agreement with the experimental data is observed with k_1 independent of I, and k_2 increasing about linearly with I.

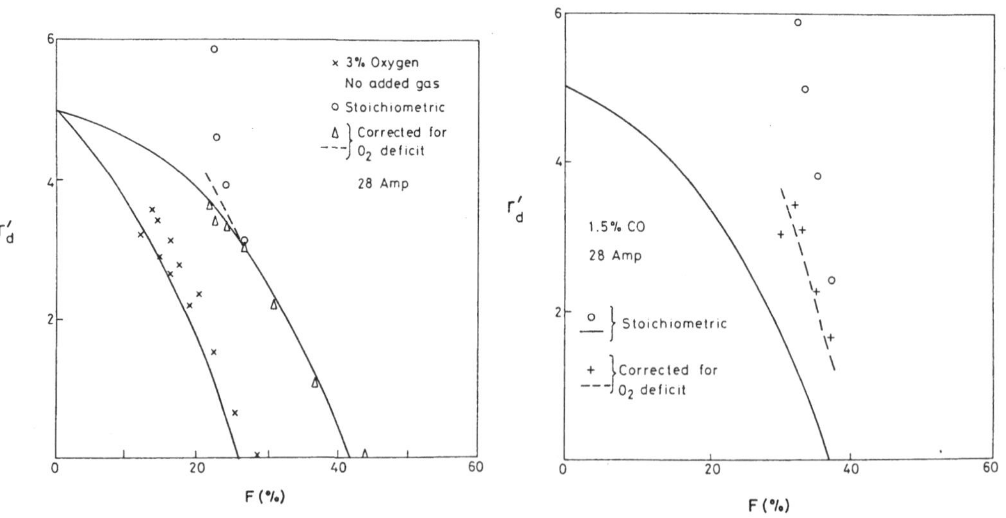

Fig. 5. Comparison of measured and predicted dissociation rates
(left) 3% oxygen & no added gas, (right) 1.5% carbon monoxide.

Using these coefficients, the nett rate of dissociation can be predicted for different gas mixtures. The results are shown in Figure (5) for 3% added oxygen, for no added gas, and for 1.5% added CO together with experimental points. Good agreement is found with the 3% O_2 data. With no added gas, the data diverges at low F; this effect is very pronounced with added CO. This results from a deficiency of oxygen, as shown in Figure (4). It is noted that this deficiency co-incides with observation of very rapid corrosion of the cathode. Using the measured oxygen partial pressure in (2) and correcting for the non-zero value of F_o in (1) brings the experimental and theoretical data into good agreement for both Figure (4) and Figure (5).

Thus a consistent fit with experimental data is obtained using a model comprising a near constant normalised dissociation rate of about 5 molecules per ion incident on the cathode, and a recombination rate scaling as the partial pressure of O_2 and CO with a rate-coefficient which increases strongly with current (about as I^2), or equivalently, with temperature downstream of the discharge. The rate of dissociation of CO_2 in the positive column can be estimated from the data of Corvin et. al./9/ to be $r'_d \sim 1.6(1-F)$. This is about a third of that required, which is consistent with a dominant role being played by the cathode - although Smith et. al./2/ have shown that the dissociation rate in typical laser mixtures can be much higher than given by the data of Corvin. The value of r'_d of about 5 could also be explained by a charge transfer cross-section of CO_2 ions of typically $10^{-15} cm^2$ at 100eV, giving several such collisions per ion across the cathode sheath. The strong current sensitivity of the recombination rate is consistent with a dominant role being played by the catalytic duct surfaces.

It is noted that the recombination in particular depends on the operating conditions over a substantial period of time and also on both the cathode material and ducting layout. The dissociation rate is also dependent on cathode material. The numerical values produced here should therefore be used with caution on other lasers.

REFERENCES

(1) WIEGAND, W.J. and NIGHAN, W.L.: Appl. Phys. Lett. 22 (1973) 583.
(2) SMITH, A.L.S. and AUSTIN, J.M.; J. Phys. D. 7, (1974) 314.
(3) BLETZINGER, P. et. al.: IEEE J. Qu. Elect. QE11 (1975) 317.
(4) ARMANDILLO, E. and KAYE, A.S.: J. Phys. D. 13 (1980) 321.
(5) ASHURLY, A.V. et. al.: Sov. J. Qu. Elect. 11 (1981) 1477.
(6) STARK, D.S. et. al.: J. Phys. E. 11 (1978) 311.
(7) KAYE, A.S. et. al.: Proc. 4th Int. Symp. Gas Flow Chem. Lasers, Stresa, 1982.
(8) KATZ, M.: Adv. Catalysis 5 (1953) 177.
(9) CORVIN, K.K. and CORRIGAN, S.J.B.: J. Chem. Phys. 50 (1969) 2570.

Untersuchung der Entladungsstabilität von 2 KW CW CO_2 Lasern mit transverser elektrischer Anregung

LI ZAIGUANG, LI SHIMIN, LIU DONGHUA, HAN NIANSENG, LI FENG,
LI JIARONG, WANG HANSHENG, CHEN ZHUHAI UND QIU JUNLIN.
Laser Institute, Huazhong University of Science and Technology,
Wuhan, China.

1. Einleitung

Die Entladungsstabilitaet von CW CO_2 Lasern ist eine wichtige Voraussetzung fuer deren Anwendung bei der Materialbearbeitung. Um eine homogene stabile Glimmentladung bei grossem Volumen zu erhalten, und die Entstehung einer Bogenentladung zu verhindern, werden die verschiedene Anregungsmethoden wie Z.B. selbstaendige Entladung oder elektronenstrahlunterstuetzte Entladung verwendet /1/.

Wir haben CW CO_2 Hochleistungslaser untersucht und aufgebaut. In diesem Lasern erfolgt die durch eine selbstaendige Entladung. Als Elektroden wurden Multinadelnkathode und Plattenanode verwendet. Mit diesem Aufbau konnte eine homogene, stabile Glimmentladung bei hohem Druck und grossem Volumen verwirklicht werden. Der Daten des CW CO_2 Lasers sind: Laserausgangsleistung 2KW, Elektrooptischer Wirkungsgrad 15%, Stabilitaet der Laserausgangsleistung $< 5\%$. Dieser Laser wird jetzt in der Lasermaterialbearbeitung eingesetzt.

2. Experimenteller Aufbau und Elektrodenanordnung

Diese Anlage besteht aus den Elektroden mit Multinadelnkathode und Plattenanode, dem Axialventilator, dem Waermetauscher mit der Wasserkuehlung, dem optischer Resonator (hochreflektierender Spiegel: Kupfer, Auskoppelspiegel: GaAs), der Vakuumsystem und Gaseinfuellsystem, der Stromquelle und der Steuerung. Das Gehaeuse ist vollschliessend aus rostfreiem Stahl.

Fig.1 zeigt der Elektrodenaufbau mit Multinadelnkathode und Plattenanode. Die Anode ist aus Kupfer. Die Multinadelnkathode ist aus Wolframdraht. Die gesamte Kathode besteht aus 224 Nadeln in vier Reihen. Die Entladungslaenge betraegt 95 cm. Um eine homogene, stabile Glimmentladung zu erhalten. ist jeder Nadel mit einem Vorwiderstand R_1 versehen. Jede Reihe ist mit einem Widerstand R_2 versehen, durch welchen Strom der Reihe eingestellt werden kann. Wenn R_2 einer dynamischer Widerstand einer Elektronenroehre ist, kann nicht nur der Strom eingestellt werden, sondern auch automatisch stabilisiert werden.

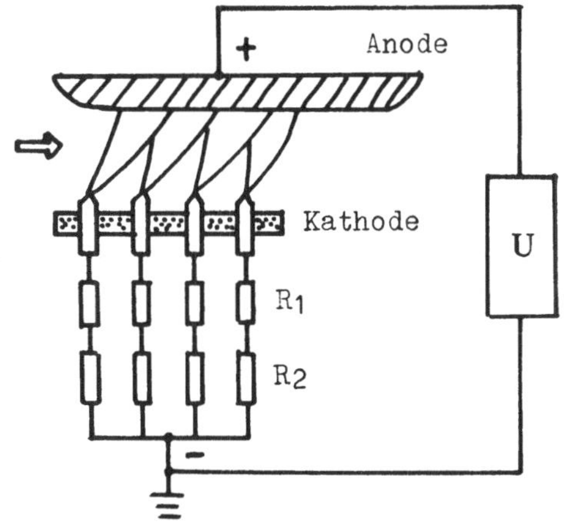

Fig.1 Schema fuer die Elektrodenkonstitution

R_1 -- begrenzender Widerstand jeder Nadel.
R_2 -- begrenzender Widerstand jeder Reihe.

3. Experimentelle Ergebnisse

Der Einfluss einiger Faktoren auf Entladungsstabilitaet wurde untersucht, Z.B. Nadelstrom, mittlere Anregungsleistungsdichte, Gasstroemungsgeschwindigkeit und Gasdissoziation.

A. Nadelstrom

In Fig.2 ist die U(I) -- Charakteristik der Entladung einer Nadelelektrode dargestellt. Unter den Entladungsbedingungen, wie sie in Fig.2 angegeben sind, haengt der Nadelstrom von der Spannung und dem

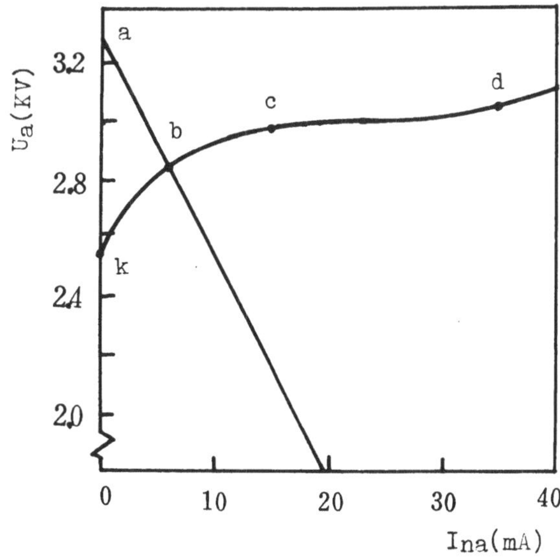

Fig.2 Die Entladungscharakteristik der Nadelelektrode.

$CO_2:N_2:He = 1:7:18$
p = 90 Torr
v = 50 msec^{-1}
s = 3.5 cm
v -- Strömungsgeschwindigkeit,
s -- Entladungsstrecke.

begrenzenden Widerstand ab. Wenn die Spannung zwischen den Elektroden
bis auf Ua = 3.3 KV zunimmt, zuendet der Entladungsraum, diese Spannung ist die Zuendspannung. Nach der Zuendung haengt der Stromwert
von dem begrenzenden Widerstand ab, und der Arbeitspunkt ist b. Wenn
der Strom auf dem Abschnitt cd der Kurve zunimmt, steigt die Spannung
langsamer an, und die Glimmentladung breitet sich auf der Nadelober-
flaeche aus. Die Stromdichte der Kathodeoberflaeche bleibt Konstant.
Oberhalb des Punktes d auf der Kurve bildet sich eine anomale Glimmentladung, die Entladung ist sehr unstabil und geht leicht in die
Bogenentladung ueber. Deshalb erfordert eine stabile Entladung einen
Nadelstrom von 15--30mA. Die Gasentladungen paralleler Nadelreihen
beeinflussen sich gegenseitig. Die Entladungscharakteristik der Nadelelektroden jeder Reihe sind ungleich. Entladung der Gasstroemungs-
richtung ist die Entladungsspannung der Nadelelektrode jeder Reihe
umso niedriger, je weiter hinten sich die Elektroden befinden. Deshalb ist die Anregungsleistungsdichte und E/N Wert fuer die Nadelelektrode jeder Reihe auch ungleich.

B. Anregungsleistungsdichte und E/N Wert

Fuer die Nadelelektroden mehrerer Reihen ist es zweckmaessig die Anregungsleistungsdichte jeder Reihe Nadelelektrode getrennt festzule-
gen, dann kann die optimale Verteilung der Leistungsdichte eingestellt werden. In Fig.3 ist die Anregungsleistungsdichte und E/N Wert
jeder Reihe dargestellt. Aus der Abbildung kann man ersehen, dass die
Anregungsleistung der 1. Reihe am kleinsten ist, hier ist es als

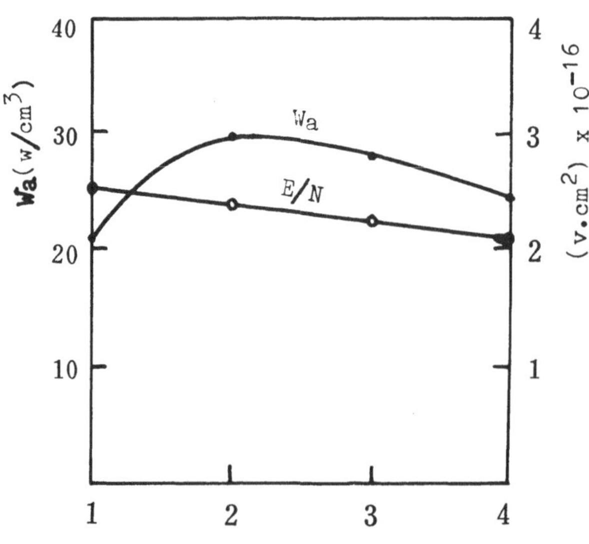

Fig.3 Verteilungsku-
rven der Anregungs-
leistungsdichte und
E/N Werte fuer die
Reihe 1 bis 4

Vorionisierung benutzt worden. Die Anregungsleistung der 2. Reihe ist am groessten, die Anregungsleistung der 3. und 4. Reihe-Nadelelektroden ist wieder niedriger. Um eine stabile Entladung zu erhalten,wurde eine mittlere Anregungsleistungsdichte 25 w/cm^3 gewaehlt.

Weil die Entladungsspannung der Nadelelektroden mit zunehmender Reihennummer abnimmt, vermindert sich E/N Wert entsprechend, wie Fig.3 gezeigt. Von der 1. Reihe bis zur 4. Reihe verringert sich der E/N Wert von 2.53×10^{-16} $V.cm^2$ bis 2.2×10^{-16} $V.cm^2$. Der mittlerer E/N Wert ist 2.34×10^{-16} $V.cm^2$. Ein hoeherer E/N Wert fuehrt zur Instabilitaet der Entladung und zur Bogenentladung.

C. Gasstroemungsgeschwindigkeit

Fig.4 zeigt die Entladungsspannung U_E und den E/N Wert unter verschiedener Gasstroemungsgeschwindigkeit. Es handelt sich wieder um die Entladung einer Reihe. Aus der Abbildung kann man ersehen, dass die Entladungsspannung mit steigender Gasstroemungsgeschwindigkeit linear zunimmt.

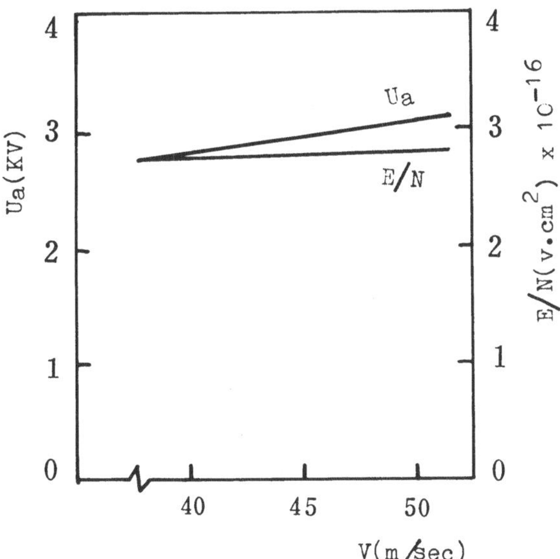

Fig.4 Entladungsspannung U_E und E/N Wert mit Gasstroemungsgeschwindigkeit.
$CO_2:N_2:He = 1:7:18$
p = 90 Torr
I_E= 1.5 A

Die Gastemperatur in dem Entladungsraum nimmt mit steigender Reihennummer zu. Entsprechend nimmt der zerlegende Material der Entladung zu. Die Entladungsstoerung der Nadelelektroden einer Reihe beeinflusst stark die Nadelelektroden der dahinter liegenden Reihe, und fuehrt zur oertlichen Bogenentladung. Deshalb sind die Anregungsleistungsdichte jeder Reihe beschraenkt. Damit die Entladungsstabilitaet erhoeht wird, muss die Gasstroemungsgeschwindigkeit zunehmen, wenn die Reihenzahl zunimmt.

284

<u>D. CO_2 - Dissoziation</u>

Im Entladungsraum werden CO_2-Molekuele durch Elektronenstoss disso-
ziert: $CO_2 + e \longrightarrow CO + O^-$

CO_2 Dissoziation fuehrt zu einer instabile Entladung. Diese Instabilitaet wird durch die negativen Ionnen hevorgerufen. Man nennt sie
Ionisierung-Instabilitaet oder Attachment-Instabilitaet /2/. Diesen
Prozess zeigt in Fig.5 Kurve a. Um die Entladung zu stabilisieren,
wurden zwei Massnahmen getroffen.

a) Zusetzen von Kohlenmonoxid (CO).

Wie die Kurve b. in Fig.4 zeigt, wenn geringe Menge CO zugesetzt werden, Z.B. $CO_2:CO:N_2:He = 1:0.4:7:19$. Es entstehen hierdurch zusaetz-
liche CO_2 Molekuele, wie die folgenden chemischen Reaktionen zeigen:

$$CO + \frac{1}{2} O_2 \longrightarrow CO_2$$
$$CO + O \longrightarrow CO_2$$

b) Stromstabilisierung durch Elektronenroehren kann die Stabilitaet weiter erhoeht, wie in Fig.5
die Kurve C zeigt.

Die Untersuchung zeigt, dass mehr
als 2KW kontinuierliche und stabile Laserleistung unter folgender
Bedingungen erreicht werden konnte: Nadelstrom 15--30 mA, mittlere Anregungsleistungsdichte 25
W/cm^3, mittlerer E/N Wert 2.35 x
10^{-16} $V.cm^2$, Gasdruck 90 Torr
($CO_2:CO:N_2:He = 1:0.4:7:19$),
Stroemungsgeschwindigkeit 50
m/sec, und Stromstabilisierung
durch Elektronenroehren.

<u>Literatur</u>

(1) E.HOAG etal: Appl. Opt. 13.
     1974.
(2) W.L.NIGHAND: "Stability of
     High-Power Molecular Laser
     Discharges" (Principles of
     Laser Plasmas. 1976).

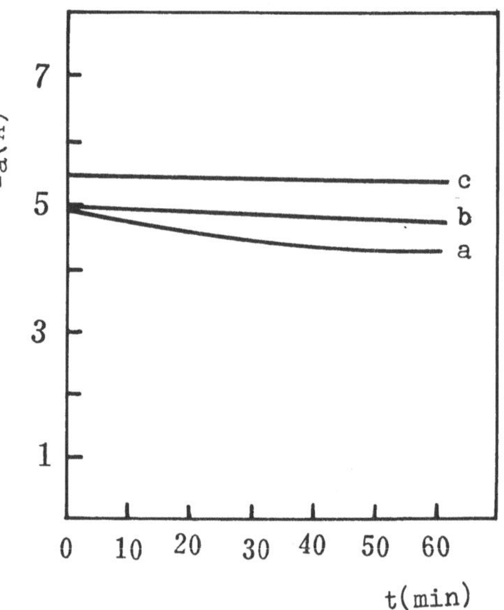

Fig.5 Die Stabilitaet der Gas-
entladung.
a. $CO_2:N_2:He = 1:7:20$
b. $CO_2:CO:N_2:He = 1:0.4:7:19$
c. $CO_2:CO:N_2:He = 1:0.4:7:19$
   bei stabilisiertem Strom

1,2 KW Gastransportlaser mit ungefaltetem Resonator

W.Abel, D.Schuöcker, B.Walter
Technische Universität Wien
Institut für Nachrichtentechnik
1040 Wien

1.Einleitung

Die wichtigste industrielle Anwendung des Hochleistungslasers stellt zur Zeit die
Materialbearbeitung, wie etwa das Laserschneiden, dar. Damit ein Laser als Werkzeug verwendet werden kann, muß er eine Reihe von Eigenschaften, wie hohe Strahlqualität, Automatisierbarkeit sowie Störungsunempfindlichkeit aufweisen. Da einer
breiten Einführung des Laserschneidens vor allem der hohe Preis der Hochleistungs-
laser entgegensteht, soll der Laser so einfach wie möglich aufgebaut sein, um
Kosten zu sparen. Die meisten am Markt erhältlichen Hochleistungslaser entsprechen
nicht allen genannten Anforderungen und wurden auch nicht immer speziell für die
Anwendung in der Materialbearbeitung konzipiert. An der Technischen Universität
Wien wurde nun im Zusammenarbeit mit der VOEST-ALPINE AG,Linz ein Hochleistungs-
Kohlendioxydlaser entwickelt, der dem erwähnten Anforderungsprofil angepaßt ist.
Dabei wurde allerdings kein grundsätzlich neues Konzept entwickelt, sondern es
wurde nur das bewährte Prinzip des Gastransportlasers im einzelnen weiterentwickelt
und optimiert.

2.Anforderungsprofil des neuen Lasers

Besonders wichtig für den Einsatz in der Fertigungstechnik ist die Qualität der
abgegebenen Strahlung. Dabei soll der transversale Strahlmodus dem Grundmodus so
nahe wie möglich kommen, um eine optimale Fokussierung des Strahls zu ermöglichen.
Ferner müssen sowohl die Leistung wie auch der Strahlmodus soweit wie möglich
zeitlich konstant gehalten werden, um eine hohe Bearbeitungsqualität zu erzielen.
Damit der Laser in Verbindung mit modernen CNC-gesteuerten Fertigungseinrichtungen
betrieben werden kann, muß er automatisierbar sein. Dabei muß das Anfahren des
Lasers sowie die Einstellung der Laserleistung durch eine CNC-Steuerung erfolgen
können. Weiters soll die Laserleistung und der Strahlmodus ständig automatisch
überwacht und bei Abweichungen von Solldaten automatisch korrigiert werden können.
Weiters muß der Laser unempfindlich gegenüber äusseren Störungen sein. So ist
ein Laser in der industriellen Fertigung sehr oft mechanischen Erschütterungen
ausgesetzt. Um dagegen unempfindlich zu sein, soll der Laser möglichst ganz aus
Metall gebaut sein, wenig bewegliche Teile enthalten und einen möglichst kurzen
Strahlengang im Resonator aufweisen. Schliesslich ist es für die Konkurrenz-
fähigkeit des Laserschneidens gegenüber anderen Schneidverfahren von größter Bedeutung, daß der Anschaffungspreis des Lasers wesentlich verringert wird, was
nur durch einen möglichst einfachen Aufbau des Lasers erreicht werden kann.

3.Konstruktive Gestaltung des neuen Lasers

Grundsätzlich wurde die bewährte Bauform eines quergeströmten Gastransportlasers
gewählt, weil damit hohe Leistungen bei kompakter und robuster Ganzmetallbauweise
erzielt werden können. Der Vakuumkessel mit Wärmeaustauscher und Gastransportein-
richtung wurde nicht selbst entwickelt, sondern von der Firma Kristalloptik-Laser-
bau zugekauft.

Das Elektrodensystem eigener Entwicklung besteht aus einer rohrförmigen wassergekühlten Katode aus Kupfer und einer großen Zahl von senkrecht zur Katode ange-
ordneten Anodenrohren aus Kupfer. Die Anzahl der Teilanoden wurde gegenüber herkömmlichen Systemen wesentlich erhöht, wobei der Abstand zwischen den einzelnen
Anoden verringert wurde. Damit wird eine gleichmässigere Anspeisung des Plasmas
mit Strom erreicht, wodurch die Stromdichte im Plasma erheblich homogener ist als
in herkömmlichen Lasern ähnlicher Bauart. Damit können im Plasma erheblich
höhere elektrische Leistungen ohne das Auftreten von Lichtbögen umgesetzt werden.
Außerdem wird infolge der kleineren Stromstärke pro Teilanode eine Wasserkühlung
der Anoden überflüssig. Um eine möglichst niedrige Betriebsspannung zu erzielen,
wird das Plasma mit einer Spannung betrieben, die nur knapp oberhalb der Brennspannung der Glimmentladung liegt und viel niedriger ist als die Zündspannung.
Die Zündung des Plasmas erfolgt dabei durch eine Hilfsentladung. Durch den geringen
Unterschied zwischen Betriebsspannung und Brennspannung der Glimmentladung können
für die Stromaufteilung relativ kleine Widerstände mit jeweils 1   verwendet
werden, die außerdem nur eine relativ geringe Belastbarkeit von 100 W aufweisen
müssen. Diese Widerstände sind auf einem wassergekühlten Kupferblock, der außerhalb des Lasers angeordnet ist, befestigt. In einer anderen Ausbauvariante wurde
der Widerstandskühler im Vakuumraum selbst angeordnet, wobei dann zur Stroman-
speisung nur zwei Vakuumdurchführungen benötigt werden. Allerdings ergaben sich
hier infolge des geringen Gasdrucks im Entladungsraum Durchschlagsprobleme, die
allerdings durch das Vergießen mit einem geeigneten Material leicht beseitigt
werden können, sodaß dann der Aufbau des Lasers infolge des Wegfallens fast aller
Vakuumdurchführungen noch weiter vereinfacht und verbilligt werden kann.
Um der Forderung nach einem möglichst einfachen, billigen und robusten Resonator
zu entsprechen, wurde auf eine Faltung verzichtet und ein hemisphärischer
Resonator gewählt, der nur aus zwei Spiegeln besteht. Der konkave Endspiegel besteht aus Silizium und weist einen Krümmungsradius von 20 Metern auf. Das Aus-
koppelfenster besteht aus Zinkselenid und weist im optimalen Fall eine Transmission
von 30% auf.Beide Spiegel haben einen Durchmesser von rund 4 cm.
Um das Justieren des Lasers automatisch mit Hilfe eines Mikroprozessors durchführen
zu können, werden alle Spiegel durch Motormikrometer verstellt. Diese sind im
Vakuumraum angeordnet, sodaß die aufwendigen mechanischen Vakuumdurchführungen
entfallen können und statt dessen nur einfache Stromdurchführungen für die

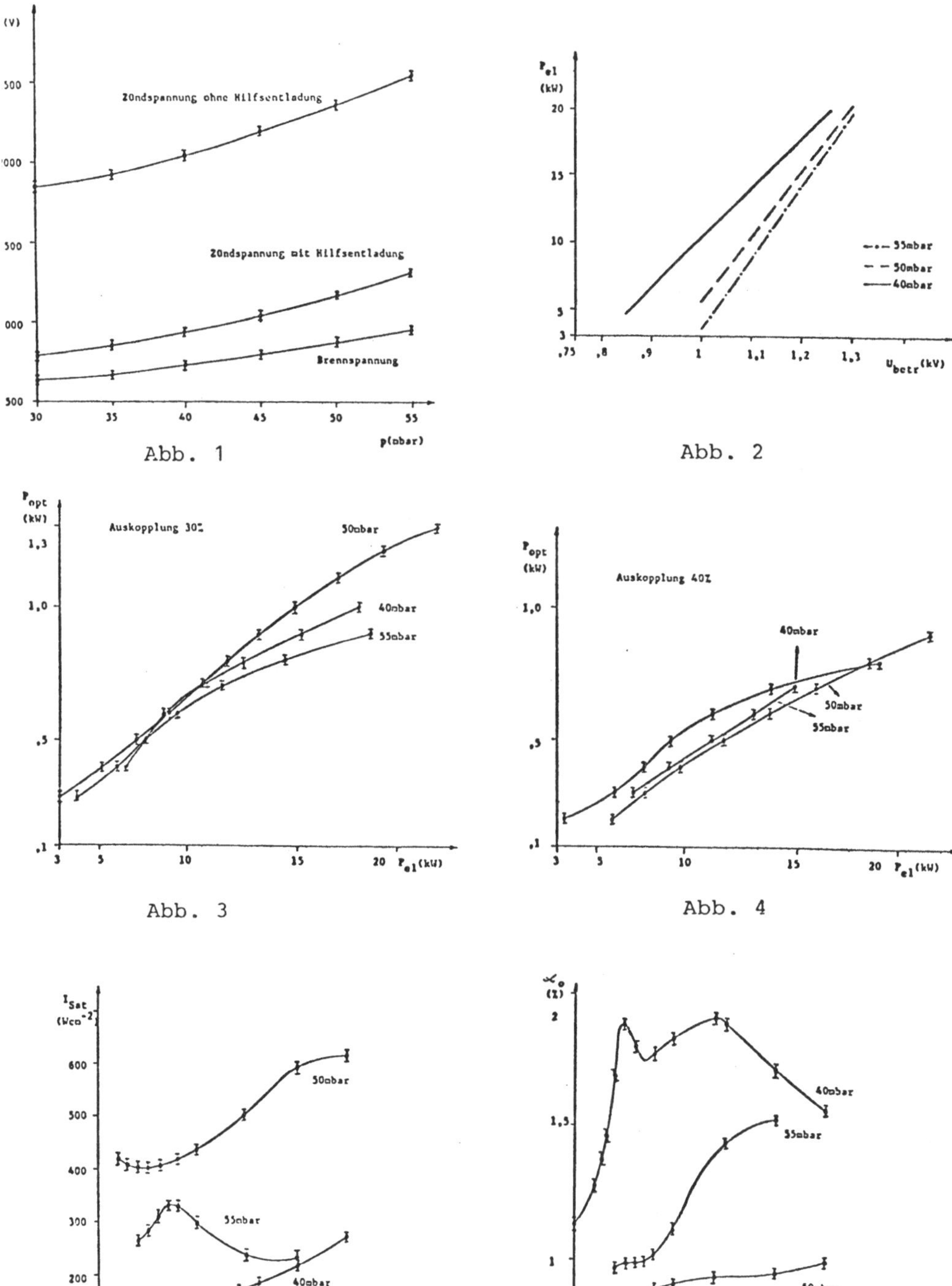

Abb. 1

Abb. 2

Abb. 3

Abb. 4

Abb. 5

Abb. 6

Anspeisung der Motormikrometer erforderlich sind. Damit wird der feinmechanische Aufwand bei diesem Laser wesentlich verringert.

4.Messungen am neuen Laser

Die Abhängigkeit der Zündspannung des Lasers vom Gasdruck ohne und mit Hilfsentladung zeigt Abb.1. In diesem Diagramm ist auch die Brennspannung der Glimmentladung enthalten. Es zeigt sich, daß durch Verwendung einer Hilfsentladung die Zündung schon bei einer Spannung erfolgen kann, die nur knapp oberhalb der Brennspannung liegt, sodaß zur Strombegrenzung nur relativ kleine Vorwiderstände erforderlich sind.

Abb.2 zeigt dann die Abhängigkeit der dem Plasma zugeführten elektrischen Leistung von der Betriebsspannung des Lasers, also der an den Vorwiderständen anliegenden Spannung, für verschiedene Werte des Gasdrucks. Die im Diagramm ersichtliche Begrenzung der elektrischen Leistung auf etwa 20 kW kommt nicht etwa durch das Auftreten von Instabilitäten, sondern durch die Leistungsfähigkeit der Stromversorgung zustande.

Die Abb.3 und 4 zeigen die vom Laser abgegebene Strahlleistung in Abhängigkeit von der dem Plasma zugeführten elektrischen Leistung und dem Gasdruck für Transmissionen des Auskoppelspiegels von 30% und 40%. Daraus kann entnommen werden, das die optimale Auskopplung bei 30% und das der optimale Gasdruck bei etwa 50 Millibar liegt.

Der Laser gibt also im optimalen Fall bei einer zugeführten elektrischen Leistung von 20 kW eine Strahlleistung von 1,2 kW ab.

Abb.5 und 6 zeigen die aus der für verschiedene Auskopplungen erzielten Strahlleistung berechneten Werte der Kleinsignalverstärkung und der Sättigungsintensität in Abhängigkeit von der zugeführten elektrischen Leistung und dem Gasdruck.

5.Schlußbemerkungen

Das auf Grund der besonderen Anforderungen der Materialbearbeitung entwickelte Konzept eines Hochleistungslasers wurde als Labormuster realisiert und zeigt einen Weg zur Entwicklung eines besonders einfachen und billigen Lasers.

Einige praktische Beispiele über Einsatzmöglichkeiten von CO_2-Hochleistungslasern beim Schweißen und Härten

A. Gukelberger

Spectra-Physics GmbH, Industrial Laser Division
Siemensstr. 20, D-6100 Darmstadt

I. Anforderungen an einen Laser im Produktionseinsatz und deren Realisierung

Nach fünfjähriger Erfahrung im Bau von transversal geströmten CO_2-Hochleistungslasern für die Materialbearbeitung im Leistungsbereich zwischen 1 und 5 kW wurde Ende 1982 der erste Vertreter einer neuen Lasergeneration der Öffentlichkeit vorgestellt.

Ziel dieser Entwicklung war es, ein besonders bedienungsfreundliches, verläßlich arbeitendes, leicht zu wartendes und wenig Stellfläche benötigendes Gerät auf den Markt zu bringen, um die Technologie des Laser-Schweißens, -Härtens und des Schneidens von Dickblech noch mehr in Produktionsbetrieben einzuführen.
Deshalb soll hier zu Beginn kurz erläutert werden, wie das oben genannte Ziel verwirklicht wurde und wodurch sich dieser neue Laser von seinem Vorgänger unter scheidet:

1. Kompakte und servicefreundliche Bauweise
Die Hochspannungsversorgung, die Zuführ- und Dosiereinrichtung für das Lasergas und der Mikroprozessor sind in einem Gehäuse unterhalb des Laserkopfes untergebracht. Dadurch wird eine sehr kompakte Bauweise erzielt, die mit einer Stellfläche von nur 2 m^2 auskommt.
Der Laserkopf selbst wird durch Ausklappen der vorderen und der hinteren Verschlußplatte für Servicearbeiten sehr leicht zugänglich.

2. Aktive Leistungsregelung

Der rückwärtige Resonatorspiegel besitzt eine Transmission von 0,2%. Dieser Lecklichtanteil wird verwendet, um die gewählte Laserleistung innerhalb einer Spanne von $\pm 2\%$ konstant zu halten. Diese Leistungskonstanz ist u.a. wichtig, um gleichbleibende und reproduzierbare Schweiß-, Schneid-, und Härteergebnisse zu erzielen.

3. Kurze Aufheizdauer

Mit bedingt durch die aktive Leistungsregelung ist der Laser bereits 2 Minuten nach dem Anschalten betriebsbereit. Dies hat eine Minimierung von Taktzeiten zur Folge und erhöht die Effizienz der Laseranlage.

4. Mikroprozessorsteuerung

Der Operator hat nur den Einschaltknopf zu drücken und die gewünschte Laserleistung einzutippen. Die Partialdrücke des Laserfüllgases sind vorprogrammiert und stellen sich automatisch richtig ein.

5. Innenliegender Resonator

Die Resonatorspiegel liegen vollständig innerhalb des Gehäuses des Laserkopfes und sind deshalb keinen mechanischen Spannungen ausgesetzt, die beim früheren Modell vom Druckunterschied zwischen Innendruck im Resonator (40 Torr) und Atmosphärendruck herrührten. Das Resonatorrohr ist aus Kohlefaser-Verbundwerkstoff aufgebaut, wobei die Laminate so gewickelt sind, daß die resultierende thermische Ausdehnung gleich Null ist. Damit entfällt die Kühlung des Resonatorrohres.

6. Keine Justierprobleme

Die Resonator und Faltspiegel sind auf wassergekühlten Invarplatten fest gegen Anschlag montiert. Lediglich der Auskoppelspiegel ist zu justieren. Um dies einfach von der Konsole aus zu ermöglichen, ist der Auskoppelspiegel in horizontaler und vertikaler Richtung motorisch verfahrbar gelagert.

7. Wesentlich reduzierte Anforderungen an das Kühlwasser

Die Kühlwassertemperatur darf den Taupunkt nicht unterschreiten und sollte $30^{o}C$ nicht überschreiten. Innerhalb dieser Spanne sind Temperaturschwankungen des Kühlwassers von $\pm 4^{o}C$ erlaubt. Besondere Anforderungen an die Beschaffenheit des Kühlwassers werden nicht gestellt. Damit wird häufig der Anschluß an einen bereits vorhandenen Kühlwasserkreislauf möglich.

II. Eigenschaften einer Laser-Schweißnaht bzw. einer Laser-Härtung und Voraussetzungen an das Bauteil

Um die Einsatzmöglichkeiten der Lasertechnologie in der Materialbearbeitung richtig einschätzen zu können, wird im Folgenden kurz auf typische Eigenschaften und Forderungen eingegangen, die an das Bauteil zu stellen sind.

a) Laser - Schweißen:

Eine Laser-Schweißnaht zeichnet sich durch ein hohes Tiefen- zu Breitenverhältnis aus. Die Wärmeeinbringung in das zu schweißende Teil ist gering, was wiederum einen äußerst geringen Wärmeverzug zur Folge hat. Der Nahtvorbereitung muß einige Aufmerksamkeit geschenkt werden, da eine Laserschweißung in der Regel ohne Zusatzwerkstoff erfolgt. Zwischenräume größer $\pm \frac{1}{10}$ mm sollten nicht auftreten. Am besten eignen sich Presspassungen.

Laser-Schweißanlagen sind relativ leicht zu automatisieren oder in einen gegebenen Produktionsablauf zu integrieren, da an Atmosphäre und nicht in einer Vakuumkammer gearbeitet werden kann.

b) Laser - Härten:

Ein Härten mit dem Laser erscheint dann sinnvoll, wenn z.B. sonst schwer zugängliche Stellen zu härten sind, wenn nur ein geringer Wärmeverzug des zu härtenden Teils zulässig ist, wenn es sich um partielles Oberflächenhärten handelt und wenn die Vorteile der Selbstabschreckung genützt werden sollen.

III. Praktische Beispiele

a) Automobilbau

1. Verschweißen von Synchronring mit Getriebegangrad.

   Bereits seit einigen Jahren werden Synchronring und Getriebegangrad bei einem europäischen Automobilhersteller produktionsmäßig in einer vollautomatischen Anlage verschweißt. Der Grund für die Fertigung von zwei separaten Teilen und der anschließenden Verschweißung liegt in der dadurch möglich werdenden kürzeren Bauweise, wodurch Platz und Gewicht eingespart werden. Die charakteristische Schweißzeit liegt bei etwa 5 Sekunden pro Teil, die Laser-Nennleistung beträgt 2500 Watt, die Schweißtiefe etwa 2 mm. Durch Konzipierung einer Doppelschweißstation wird eine fast 100%-ige und damit äußerst wirtschaftliche Auslastung des Lasers erreicht. Der besagte Automobilhersteller berichtet von einer Kosteneinsparung von 30% pro Teil.

2. Schweißen von Automatikgetriebeteilen.

   In wenigstens zehn vollautomatischen Laserschweißanlagen werden diese Teile bei Automobilherstellern in den USA, demnächst auch bei uns in Deutschland, geschweißt.

Dabei handelt es sich um Tiefziehteile, die anschließend weitergehend verzugs-frei verschweißt werden müssen. Hierzu eignen sich Laser und Elektronenstrahl, wobei in der Regel die Lasertechnologie der Vorzug gegeben wird. Dies vor allem deshalb, weil beim Laserschweißen nicht in einer Vakuumkammer ge-arbeitet werden muß.

Bei dieser Applikation liegt die Schweißtiefe bei etwa 3,5 mm, die zum Einsatz kommenden Laser haben Nennleistungen von 5000 Watt.

3. Schweißen von Schalldämpferrohren.

Bei dieser und den folgenden Applikationen handelt es sich um Beispiele, die in unserem Applikationslabor erarbeitet wurden, also noch nicht produktionsmäßig mit der Lasertechnologie gefertigt werden.

Die lasergeschweißten Teile weisen vor allem bei Al-beschichteten Rohren im Vergleich zu herkömmlichen Schweißmethoden eine höhere Festigkeit auf. Die Schweißgeschwindigkeit beträgt je nach Laserleistung und Güte der Nahtvor-bereitung zwischen 2 und 8 m/min.

4. Verschweißen von Stoßdämpfergehäuse und Federteller.

Es konnte hier gezeigt werden, daß die Schweißverbindung der geforderten Festigkeit stand hält, weit wichtiger aber ist, daß die Wärmeeinschnürung im Bereich der Schweißnaht so minimal ist, daß auf ein Nachkonen verzichtet werden kann. Die Schweißgeschwindigkeit betrug etwa 2 m/min bei einer Laserleistung von 1200 Watt.

5. Außerdem werden erfolgreiche Laser-Schweißversuche an Ventilstößelkappen, Schaltgabeln, Naben und Gelenkwellen durchgeführt.

Als genereller Trend ist festzustellen, daß aus Gründen der Kosten- und Gewichtsersparnis von Guß - auf Blechkonstruktionen übergegangen wird. Bei derartigen Konstruktionen bietet der Laser als Schweißwerkzeug häufig er-hebliche Vorteile.

6. Es wird heute bereits daran gedacht, bestimmte Karosserieteile mit dem Laser entlang einer Bahn zu verschweissen, anstatt die Teile mit dem Schweißroboter zu punkten. Die Gründe, die für das Laserschweißen sprechen, sind vor allen in der verbesserten Dichtigkeit und der erhöhten Festigkeit der Verbindung zu sehen. An den Systembauer werden bei derartigen Anlagen hohe Anforderungen gestellt.

7. Härten von Kolbenringnuten.

8. Härten von besonders beanspruchten Teilen in Zylinderlaufbuchsen.

a) Abstandshalterprofile bei Mehrscheiben - Isolierverglasungen.

Die vom Coil abgefahrenen, im Rollformer geformten und mit dem Laser verhefteten Aluminium- Abstandshalterprofile weisen eine erhöhte Formstabilität auf. Im Vergleich zum HF-Schweißen ist das Laserschweißen wirtschaftlicher. Heftgeschwindigkeiten bis zu 120 m/min bei einer Laserleistung von 1500 Watt sind möglich. Eine derartige Anlage wurde kürzlich in der Bundesrepublik in Betrieb genommen.

b) Sägenindustrie.

Bei der Produktion von Kettensägen - Schwerter werden Stellit als verschleißfester Werkstoff mit Stahlblech im Bereich des Kettenumlaufes verschweißt.
Für die Sägeblattindustrie laufen derzeit Vorversuche zum Verschweißen von Hart metallsegmenten mit dem Trägermaterial.

Die Ausführungen sollen zeigen, daß die CO_2-Hochleistungslaser im Leistungsbereich oberhalb 1 kW richtig eingesetzt heutzutage ein verläßliches und erprobtes Werkzeug vor allem für das Schweißen im begrenztem Umfang auch für das Härten darstellen und ohne Vorbehalt eingesetzt werden können.

Beitrag zum Schneiden mit CO_2-Lasern

G. SEPOLD
BIAS - Bremer Institut für angewandte Strahltechnik
Ermlandstr. 59, D-2820 Bremen 71

1. Einleitung

Wichtige Baugruppen von Laserbearbeitungssystemen sind neben Handhabungs- und Spannvorrichtungen

1. Der Laser mit Resonator und Energieversorgung, die beide in unterschiedlicher Form Einfluß auf die Strahlqualität (zeitlich und örtlich sich ändernde Leistung) nehmen,

2. die Strahlformung (Spiegel- und Halbleiteroptiken), welche die Strahlgeometrie beeinflussen und

3. der Bearbeitungskopf inklusive seiner Gasversorgung.

Nachfolgend wird hierauf eingegangen, einige Schneidergebnisse vorgestellt und auf zukünftige Entwicklungen auf dem Hochleistungs-CO_2-Laser-Gebiet hingewiesen.

2. Resonatoren und einige Schneidergebnisse

Man unterscheidet bei Hochleistungs-CO_2-Lasern zwischen längs- und quergeströmten Resonatoren, wobei die Leistung der längsgeströmten Resonatoren bis zu 5 kW beträgt, bei quergeströmten Resonatoren auch höhere Leistungen einstellbar sind. Die Frage "längs" oder "quer" ist nicht nur für die Herstellungskosten des Laserstrahlerzeugers von Bedeutung sondern auch für die Strahlqualität und -stabilität. Ein Kriterium für die Strahlqualität ist z. B. der Brennfleckdurchmesser, der wiederum von der Ausbildung der Moden abhängig ist. Je höher die Zahl der Moden,um so größer auch die Strahldurchmesser und damit um so breiter die Schweißnahtfuge und um so geringer die Schneidgeschwindigkeit, Bild 1.

Bei 10 mm Blechdicke sind vier Punkte für die erzielte Schneidgeschwindigkeit angegeben. Diese unterscheiden sich um mehr als den Faktor 1,5. Es wird vermutet, daß diese Unterschiede u. a. auf eine unterschiedliche Resonatorqualität zurückzuführen sind.

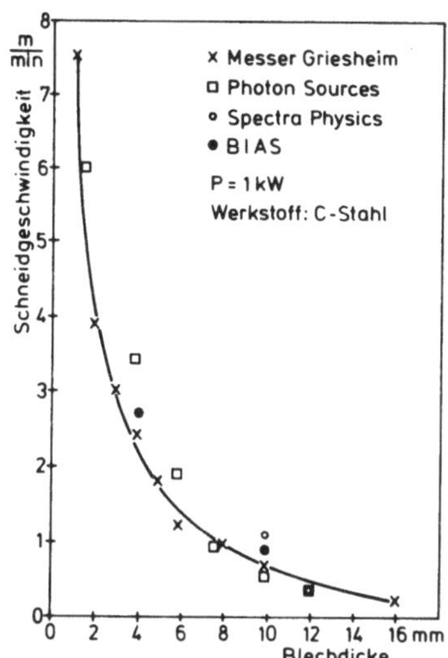

Bild 1: Abhängigkeit der Laserbrennschneidgeschwindigkeit von der
Blechdicke.
Dargestellt ist für 1 kW Schneidleistung bei gleichem Werkstoff die
durchgeschnittene ("bartfreie") Blechdicke bei variabler Schneidgeschwindigkeit.

Nachteilig bei Angaben von Firmen zur Leistungsfähigkeit ihrer Anlagen
ist häufig die Tatsache, daß über die Güte der Brennschnitte nach
DIN 2310 keine Aussagen getroffen werden, ein direkter Vergleich also
nicht möglich ist. Darüber hinaus fehlen häufig Angaben, die es ermöglichen, reproduzierbare Schnitte herzustellen. Hierzu gehören z. B.
Angaben über die Art der verwandten Düse, den Abstand der Düse zur
Blechoberfläche, den Druck am Düseneintritt, die Brennweite der verwandten Bearbeitungsoptik und ihre Ausleuchtung, um hieraus auf den
Brennfleckdurchmesser rückzuschließen.

In jedem Falle sollten jedoch eine Aufsicht auf das Schnittprofil sowie Rauheitswerte R_t angegeben werden, Bild 2.

Durch impulsförmige Betriebsweise des Laser-Netzgerätes wird die Güte
des Resonators kurzzeitig verändert, wobei hohe Spitzenimpulse gefolgt
von einem abklingenden Anteil niedrigerer Leistung entstehen. Fährt
man mit derartigen Impulsen mehrfach längs einer Linie über eine Metalloberfläche hinweg, so entsteht ähnlich wie beim Sägen ein Schnitt
relativ großer Tiefe mit sehr schmaler Wärmeeinflußzone und geringer
Rauheit, Bild 3.

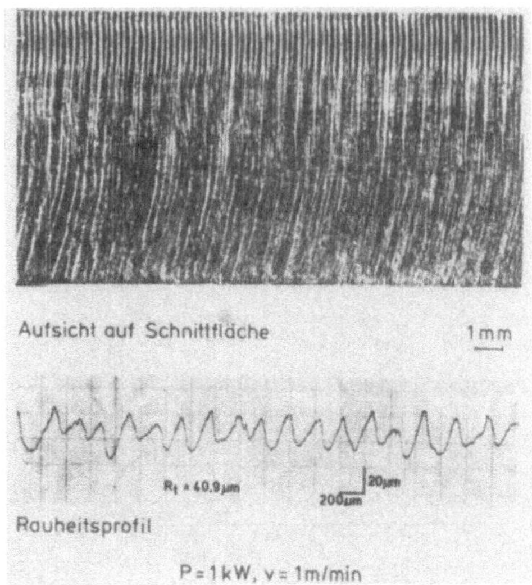

Bild 2: Laserbrennschnitt in Stahl bei Ausleuchtung des Parabol-
spiegels 40 mm; f = 150 mm; verwandte Düse: Vadura 2; Druck am
Düseneingang: 10 bar; Abstand Düse/Werkstückoberfläche 5 mm; Brenn-
fleck etwa 2 mm unter die Stahloberfläche verlegt.

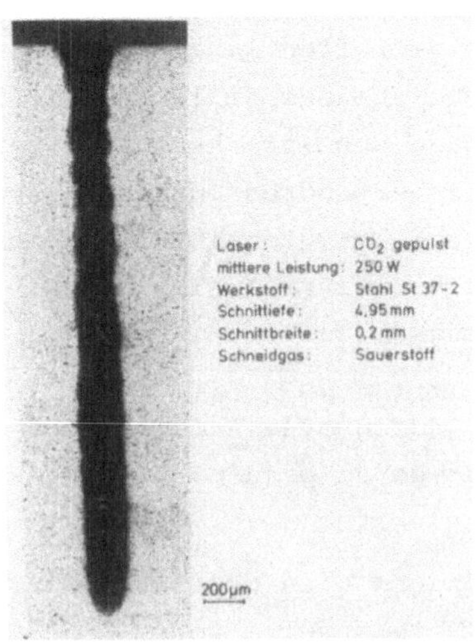

Bild 3: Schnitt mit gepulstem
Laserstrahl, 10 Durchgänge;
Frequenz: 3000 Hz; P_{max} = 1,5 kW.

Auf diese Weise lassen sich auch Werkstoffe mit hohem Reflexionsvermögen und Wärmeleitung bei relativ niedrigen mittleren Leistungen des Lasers schneiden. Weiterhin besteht die Möglichkeit, mit Hilfe dieser Technik einer Bartbildung durch Verdampfen des Metalls entgegenzutreten. Als Nachteil dieses Prozesses sind niedrige Bearbeitungsgeschwindigkeiten hervorzuheben.

3. Laserschneidsysteme

Zum Schneiden werden heute vorwiegend 500 bis 1000 W-Anlagen angeboten. Entwicklungsarbeiten weisen eine Tendenz zu höheren Leistungen auf. Polarisationsbedingte unterschiedliche Schneidergebnisse in x- und y-Richtung werden durch entsprechende Depolarisatoren korrigiert, die heute von nahezu allen Firmen angeboten werden. Deutsche Hersteller, die Laserschneidanlagen mit über 1 kW Ausgangsleistung anbieten, sind die Firmen Messer Griesheim, Rofin Sinar und Heraeus. Die zuletztgenannte Firma bietet einen quergeströmten Laser mit über 3 kW Ausgangsleistung an, Bild 4, der mit einer drehbaren Fokussieroptik (inkl. Polarisator) ausgerüstet ist.

Bild 4: Aufbau eines quergeströmten 3 kW-Lasers; Resonatorlänge kleiner 3 m; Werkfoto

Ein weiteres interessantes System wird von der Firma Spectra Physics angeboten, das über Ausgangsleistungen von über 1,2 kW verfügt, Bild 5.

Bild 5: 1,5 kW-Anlage der Firma Spectra Physics; Werkfoto

Das System ist mit einer automatischen Leistungsregelung zur Konstanthaltung der Leistung auf $\pm$ 2 % ausgerüstet. Der Resonator ist ver-
hältnismäßig klein und arbeitet nach Angaben des Herstellers im Quasi-
Mono-Mode-Betrieb. Ebenfalls neuere Modelle bieten die amerikanischen
Firmen Photon Sources und Coherent auf dem Markt an.

Aus Japan ist bekanntgeworden, daß dort Systeme mit 1,5, 3 und 5 kW
in den Markt eingeführt werden. Ein 3 kW-System wird bereits mit Erfolg in der Elektronik-Industrie zum Schneiden von über 20 m² großen,
1, 2 cm dicken austenitischen Blechen eingesetzt. Hersteller derartiger
Systeme sind die Firmen Toshiba, Mitsubishi und Hitachi.

4. Strahlführungseinheiten

Im BIAS wurden größere Bleche mit Dicken von über 30 mm laser-
brenngeschnitten. Hierzu war es erforderlich, "fliegende" Optiken zu
entwicklen mit der Schwierigkeit, auch in größeren Entfernungen (Fernfeld) ähnliche Ergebnisse wie im Nahfeld zu erzielen. Das Problem
konnte durch eine geeignete teleskopische Anordnung im Strahlengang gelöst werden. Darüber hinaus wurden Universal-Optiken entwickelt, Bild 6
womit es möglich war, höhere Schneidgeschwindigkeiten als beim Plasma-
oder Brennschneiden zu erzielen. Diese waren auch auf den Einsatz
einer neuartigen Schneiddüse zurückzuführen. Optiken und Düsen sind
Eigenentwicklungen des BIAS.

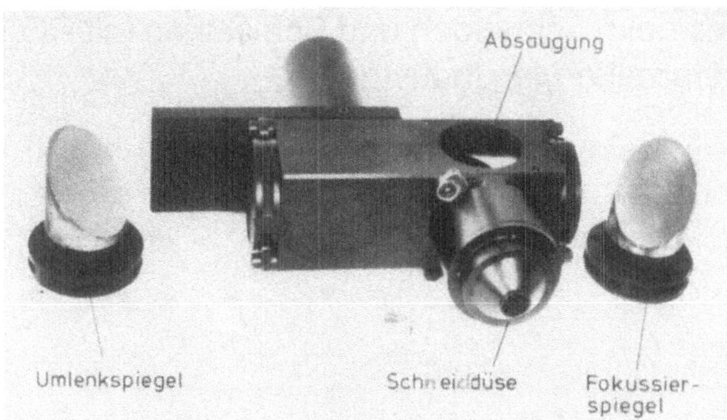

Bild 6: Bearbeitungsoptik mit leicht auswechselbarem Umlenk- und
Fokussierspiegel sowie speziellem Schneidvorsatz

5. Zusammenfassung

Im Bereich der Resonatorentwicklung hat sich der Schritt zu höheren
Leistungen von über 1 kW vollzogen. Quellen von über 5 kW müssen
zukünftig bezüglich ihrer Tauglichkeit zum Schneiden erprobt werden.
Vorhandene transmissive Optiken sollten durch Spiegeloptiken ersetzt
werden. Der Einsatz spezieller Bearbeitungsköpfe wird empfohlen.

Darüber hinaus sind problemangepaßte Handhabungssysteme für größere
Bauteile zu entwickeln ("fliegende" Optiken, CNC-gesteuerte Tische
mit Anschluß an CAD-Systeme, Roboter). Ferner sollten von Forschungsinstituten Prozeßanalysen erstellt werden, um die vorhandenen Laser-
systeme weiter zu optimieren.

Neueste Trends beim Schneiden und Schweißen mit CO_2-Lasern in der metallverarbeitenden Industrie

P. Wirth   ROFIN-SINAR Laser GmbH   Hamburg / Deutschland
I. Sarady   Technische Hochschule   Lulea   / Schweden
K. Nilsson Technische Hochschule   Lulea   / Schweden

1. Einleitung

Der Einsatz des Laserschweissens war bisher auf Anwendungen beschränkt, bei denen die Nahtgeometrie exakt und reproduzierbar festlag und daher leicht mit dem Laserstrahl zu verfolgen war. Typische Einsatzbeispiele aus der Industrie sind das Längsnahtschweissen von Rohren oder das Schweissen von Rundnähten an rotationssymmetrischen Bauteilen (Getriebefertigung). Im folgenden soll eine neue Anwendungsmöglichkeit des Laserschneidens und -schweissens vorgestellt werden, welche es erlaubt, dreidimensional geformte Blechteile über eine Stumpfnaht in nahezu beliebiger Nahtgeometrie miteinander zu verschweissen.

Die Grundidee dabei ist folgende (Bild 1):

1. Die zu verbindenden Teile werden überlappend aneinander gepresst.
2. Im Überlappbereich werden die Teile mit Hilfe des Laserstrahls besäumt und die Abfallteile entfernt.
3. Die Einzelteile werden um die Materialstärke gegeneinander verschoben und dann mit dem Laser in einer Stumpfnaht verschweisst.

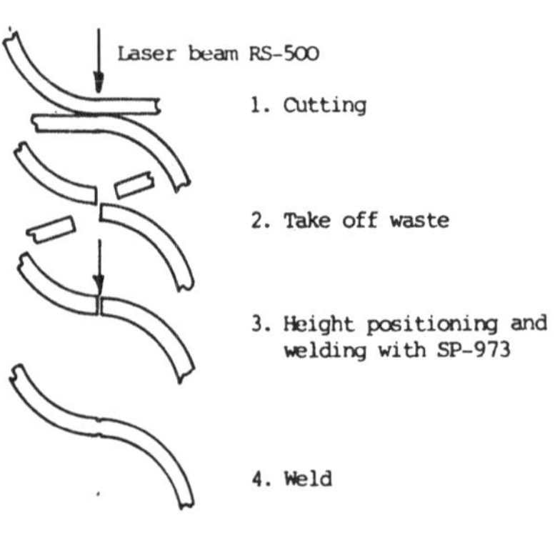

Bild 1. Prinzipskizze

Der Vorteil dieser Vorgehensweise liegt darin, daß für das Schneiden und Schweissen dieselbe Bahngeometrie und Fokussieroptik verwendet werden und dadurch die erforderliche Positioniergenauigkeit erreicht wird. Der Schnittspalt, - in der Größenordnung von o,1 mm - , welcher beim Laserschneiden entsteht, kann durch leichte Defokussierung des Schweisstrahls überbrückt werden.

Beschreibung der Anlage

Die bisherigen Versuche wurden an der Technischen Hochschule in Lulea, Schweden, am Institut für Metallbearbeitung durchgeführt.
Bild 2 zeigt den Prinzipaufbau der Anlage.

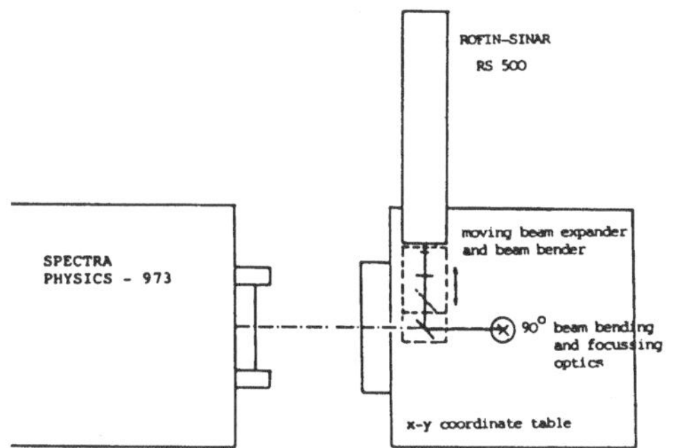

Bild 2. Anlage

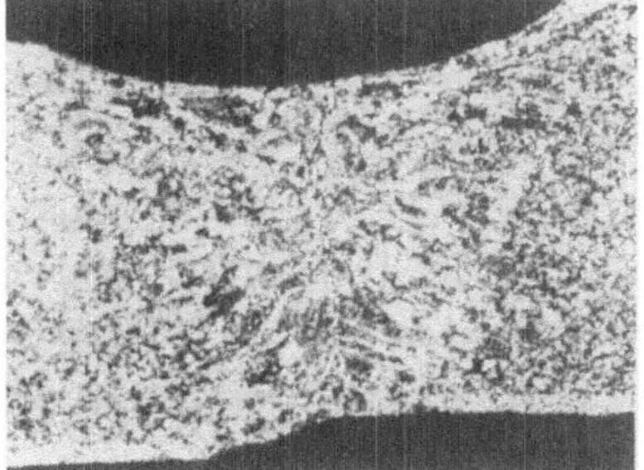

Bild 3. Laserschweissung

Zwei Laser, - ein 500 Watt CO2-Laser der Firma ROFIN-SINAR Laser GmbH und ein 2.5 kW-Laser der Firma SPECTRA-PHYSICS, werden wechselweise in die gleiche Fokussieroptik eingespiegelt. In der Fertigung soll später nur ein Laser mit 1 kW Leistung eingesetzt werden. Die Fokussieroptik zum Schneiden räumlicher Bauteile wird in Z-Richtung höhenverstellt und enthält selbst zwei rotatorische Freiheitsgrade (A, B), welche es erlauben, den Laserstrahl stets senkrecht zur Materialoberfläche zu positionieren. Unter der Optik wird das Bauteil in x,y-Richtung bewegt. Alle Achsen sind CNC-gesteuert. Zum Ausgleich von Materialtoleranzen und zur Vereinfachung der Programmierung ist die Optik mit einem kapazitivem Abtastsensor ausgerüstet, welcher für konstante Lage des Laserfokusses zur Materialoberfläche sorgt.

Tabelle 1 enthält die verwendeten Bearbeitungsparameter.

Tabelle 1. Bearbeitungsparameter

	Bearbeitungsparameter	
	Schneiden RS 500	Schweissen SP 973
Leistung	200/500 Watt	2500 Watt
Brennweite	63.5/127 mm	127 mm
Geschwindigkeit	3-5 m/min	3.5-10 m/min
Nahtbreite	o.1/o.2 mm	o.5-o.7 mm
Gas	O_2	He/N_2

Niedrig legierter Kohlenstoffstahl, Materialstärke o.8 mm
Teile wurden nicht entfettet

3. <u>Ergebnisse</u>

Die Schweissnähte wurden hinsichtlich Bruch- und Ermüdungsfestigkeit
getestet. Zum Vergleich mit den lasergeschweissten Stumpfnähten wurden lasergeschweisste und punktgeschweisste Überlappnähte herangezogen. Außerdem wurde der Einfluß der Materialbeschichtung und des
verwendeten Schutzgases untersucht, - Ergebnisse, auf die hier nicht
eingegangen werden kann.

Bild 3 zeigt eine typische Laserschweissnaht. Das Material ist ein
niedrig legierter Kohlenstoffstahl.

Versuche haben ergeben, daß hinsichtlich der Qualität der Schweiss-
naht kein Unterschied besteht, ob die Nahtkanten mit dem Laser oder
mit der Schlagschere geschnitten wurden. Auch bei größeren Spaltto-
leranzen, - ca. o.2 mm, kommt noch eine brauchbare Schweissverbin-
dung zustande.

Bild 4 zeigt den gemessenen Härteverlauf quer zur Naht.

Die Ermüdungsfestigkeit der Proben war so hoch, daß spezielle Prüflinge mit einem 8 mm-Durchmesser Loch in der Mitte gefertigt wurden.
Die Tests haben ergeben, daß die Ermüdungsrisse nicht in der Schweiss-
naht, sondern parallel dazu im Grundmaterial auftraten (siehe Bild 5).

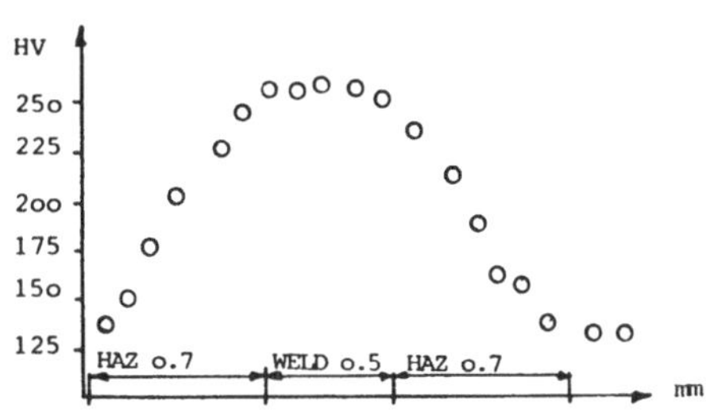

Bild 4. Härteverlauf

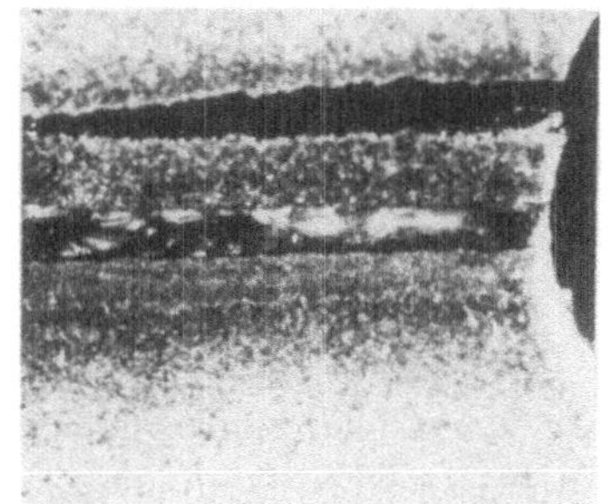

Bild 5. Ermüdungsriß neben der
Schweißnaht

In Bild 5 ist das Ergebnis der Versuche dargestellt. Es wurden
Stumpf-, Überlapp- und Doppelflansch-Nähte untersucht und dem Ergebnis von punktgeschweißten Verbindungen gegenübergestellt. Es
ergibt sich eine deutliche Überlegenheit der lasergeschweißten
Stumpfnaht gegenüber allen anderen Verbindungen.

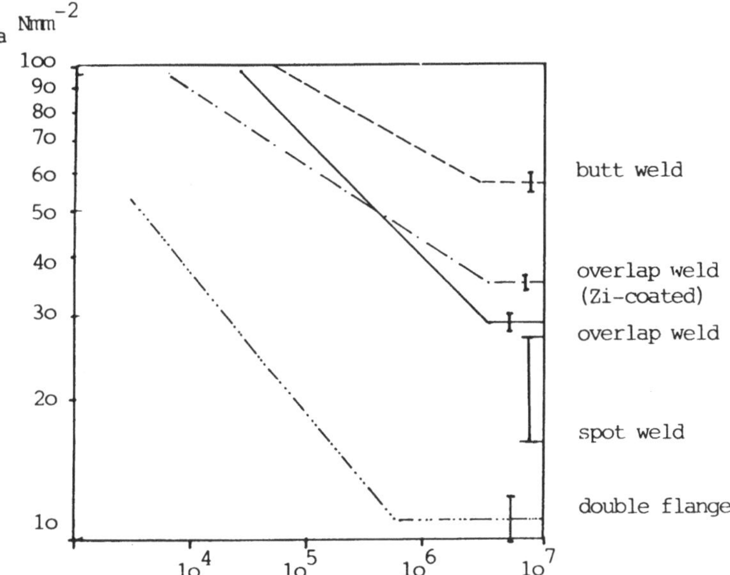

Bild 6. Ergebnis der Ermüdungstests

4. Zusammenfassung

Das dargestellte Verfahren ermöglicht es, 3-dimensional geformte
Bauteile mit hoher Oberflächenqualität (keine Überlappnaht, kein
Nahtaufwurf) miteinander zu verschweissen. Die zum Schweissen erforderliche Genauigkeit bei der Nahtführung wird dadurch erreicht,
daß das Besäumen der Teile ebenfalls mit dem Laserstrahl erfolgt.
Im Vergleich zu konventionellen Verfahren wie Punktschweissen oder
Nahtschweissen mit Lichtbogen oder Mikroplasma, ist die Wärmeeinbringung und damit der Verzug des Bauteils beim Laserschweissen wesentlich geringer.

Die lasergeschweissten Teile erfüllen alle Anforderungen hinsichtlich Biege-Wechselfestigkeit und Nahtaussehen. Damit ergeben sich
neue Möglichkeiten zur Herstellung relativ komplizierter Teilegeo-
metrien, die z.B. durch Tiefziehen in einem Arbeitsgang nicht hergestellt werden können.

Das Schneiden von metallischen und nichtmetallischen Werkstoffen mit einem 1,2 KW CW Laser

Jörg Sellner
VOEST-ALPINE AG, Linz

1)   <u>ANFORDERUNGEN AN INDUSTRIELL EINSETZBARE LASER</u>

In der Vielzahl der industriellen Anwendungen für
Laser kommen zur Zeit vorwiegend Laseraggregate mit
einer Leistung zwischen 500 - 1200 W zum Einsatz.
Die Bearbeitungsschwerpunkte liegen derzeit eindeutig
am Gebiete des Schneidens.
Nicht nur bei der Bearbeitung von Kunststoffen,
sondern auch bei der Bearbeitung von Blechen ist
derzeit noch eine Konzentration im Leistungsbereich
um 500 W zu beobachten. Der Grund hiefür liegt nicht
allein in den bis vor kurzen hohen Preisen für
KW-Geräte, sondern auch in der Tatsache, daß bei
500 W-Geräten das Verhältnis von Schnittleistung zu
Strahlenleistung noch am günstigsten war.
Seit kurzem ist hier jedoch eine rasche Entwicklung
zu KW-Geräten beobachtbar.

Für industriell einsetzbare Laser ist eine Leistungsdichte im Brennpunkt in der Größenordnung von
$1 - 5 \times 10^6$ W/cm^2 erforderlich.
Zur Erreichung dieser Leistungsdichte bedarf es nicht
nur einer geeigneten Fokussieroptik sondern auch
eines Modenbildes mit ausschließlicher Gauß-Vertei-
lung.
An die Strahlenqualität eines Lasers zum industriellen Schneiden werden somit folgende Anforderungen
gestellt:

- Grundmode

- Geringstmögliche Divergenz

- Unpolarisierter Strahl bzw. zirkular polarisierter
  Strahl

vom Laser selbst wird in Hinblick auf Schnittqualität
eine Stabilisierung der Ausgangsleistung auf
$\pm$ 1 - 2 % gefordert. Zum Einsatz des Lasers im Rahmen

einer Schneidmaschine ist es erforderlich, daß die Laserleistung kontinuierlich aus dem NC-Programm heraus ansteuerbar ist und prompt auf geänderte Sollwerte reagiert.
Vom Gesichtspunkt der Wirtschaftlichkeit her ist es erforderlich, daß der Einschalt- und Startmechanismum eines Lasers voll automatisierbar und über die CNC-Steuerung kontrollierbar bleibt.
Bei geringstem Platz-Bedarf wird von einem modernen industriell verwertbaren Laser einfache Bedienbarkeit und ein Minimum an Wartungsaufwand erwartet.

2) <u>OPTIK UND SCHNEIDDÜSE ZUM PRAKTISCHEN EINSATZ</u>

Die Strahlfokussierung von CO_2-Lasern zur Materialbearbeitung erfolgt gewöhnlich mittels plankonvexer ZnSe-Linsen und einer Brennweite von 2,5 Zoll bis 5 Zoll.

Für hohe Schneidgeschwindigkeiten in dünnen Materialien kommen – bei geeignetem Strahlenmodus – gewöhnlich 2,5 Zoll-Linsen zum Einsatz. Für Universalbearbeitung mit einem breiten Teilespektrum von metallischen und nichtmetallischen Werkstoffen empfiehlt sich – auch aufgrund größerer Werkstückdicken – der Einsatz einer 5 Zoll-Linse.
Bei kürzeren Brennweiten erweist es sich als zunehmend wichtig den Abstand zwischen Werkstückoberfläche und Schneiddüse konstant optimal zu halten. Die Erfahrung zeigt, daß zur Erreichung optimaler Schnittqualitäten und maximaler Schnittgeschwindigkeiten der Brennpunkt – bei Blechen – im oberen Drittel des Werkstückes zu liegen kommen soll.

Die theoretische Leistungsdichte bzw. Intensität im Brennpunkt beträgt bei 1,2 kW Laser 5 x 10^6 W/cm^2.

Als besonders entscheidend für die Erreichung hoher Schneidleistung und guter Schnittqualität erweist sich auch die Schneiddüse. Im Gegensatz zu Brennschneiddüsen, welche über Jahrzehnte hinweg laufend verbessert wurden und heute faktisch optimiert sind, befindet sich die Entwicklung industriell verwertbarer Laserschneiddüsen voll im Flusse.
Während beim Brennschneiden das Schneidgas den Sauerstoff als Kern umströmt, bildet beim Laserschneiden der Strahl den Kern, welcher vom Sauerstoff umströmt wird. Es ist das Ziel die Austrittsöffnung der Düse so klein wie möglich zu halten (0,2 - 0,5 mm) und die Divergenz des austretenden Gasstrahles auf ein Minimum zu begrenzen. Da zur

Erreichung einer qualitativ hochwertigen Schnittkante
der Laserstrahl unbeeinflußt durch die Düse austreten
können muß, sind Laserschneiddüsen besonders empfindlich gegen Kollisionen mit Werkstücken bzw. Werk-
stückspannsystemen. Besondere Schutzvorrichtungen
gegen derartige Kollisionen wie auch geeignete
Justagebehelfe sind erforderlich, um die Verfügbarkeit einer Schneidmaschine nicht zu verringern.

3) SCHNEIDEN VON METALLEN UND RIEFENSTRUKTUR

Beim "Laserschneiden" von Metallen ist das Auftreten
von Riefen beobachtbar.

In zahlreichen Versuchen und Proben konnte – für den
eingesetzten CO_2-Laser – ein Zusammenhang zwischen
Bahngeschwindigkeit und Riefenbreite wie folgt gefunden werden.

$$s = 0.167 \ v \qquad s = \text{Breite/Periodizität (mm)}$$
$$v = \text{Bahngeschwindigkeit (m/min)}$$

Mit zunehmender Werkstückdicke ist ein Anstieg der
Riefenhöhe selbst beobachtbar.

Da die Riefenausbildung die Rauhtiefe der Schnittkante maßgebend beeinflußt ist sie Gegenstand zahlreicher Untersuchungen verschiedenster Laserhersteller, Metallurgen und Maschinenhersteller.

Das Entstehen von Riefen und der technische Hintergrund sind noch nicht voll geklärt. Die Interpre-
tationen reichen von Wechselwirkungen zwischen dem
Schneidprozeß und der Plasmaentladung bis hin zu
Leistungsschwankungen durch Einfluß der Stromversorgungen.
Generell kann man sagen, daß das Auftreten von Riefen
eher in direktem Zusammenhang mit den eingesetzten
Geräten zu stehen als materialspezifisch zu sein
scheint.

Zur Kategorisierung von Laserschnitten empfiehlt es
sich bis auf weiteres auf die entsprechenden Richt-
linien des Brennschneidens (Ref. 10) zurückzugreifen.
Die Hauptkriterien zur qualitativen Beurteilung von
Schnitten sind die gleichmäßige Riefenfernstruktur
gemeinsam mit der Schlackenbildung an der Schnitt-
unterkante und der Wärmeeinflußzone.

Die Schnittgeschwindigkeit ist der Laserleistung direkt proportional der Blechdicke umgekehrt proportional. Diese Faustregel läßt jedoch den Einfluß des Sauerstoffes und die mit dem Sauerstoffdruck zusammenhängende Schlackenbildung außer Betracht. Die Praxis zeigt, daß bei St 37 im Dickenbereich von 2 - 6 mm eine Schlackenbildung leicht zu verhindern ist, während eine solche bei 1 mm oder 10 mm Blechdicke einer sorgfältigen Druckeinstellung bedarf. So zeigen Sauerstoffdrücke von ca. 2 - 2,5 bar bei dünnen Blechen und Sauerstoffdrücke von 0,5 - 0,8 bar bei 10 - 12 mm die besten Schnittergebnisse bei niedrig legierten Stählen.

Übersteigt das Verhältnis Schnittspaltbreite:Blechdicke den Wert 0,1, so zeigen sich Probleme in der Ausbringung des Schmelzgutes aus dem Schnittspalt. Im besonderen sind dann kugelförmige Ablagerungen (Ø 0,2 - 0,3 mm) im Schnittspalt beobachtbar. Diese behindern eine automatische Entsorgung des Schnittgutes aus der Maschine und somit eine unbemannte Fertigung. Da längere Brennweiten einen etwas breiteren Schnittspalt bewirken, empfehlen sich auch aus diesem Grunde 5 Zoll-Linsen für die flexible Fertigung bei einem breiten Blechdickenspektrum.

4)   SCHNEIDEN VON NICHTMETALLEN

Beim Schneiden von Nichtmetallen bestehen im allgemeinen folgende Bedrohungen:

a) Sichtbare Materialbeeinflussung durch Färbung

b) Freisetzung giftiger Gase

c) Schlechte "Geometrie" des Schnittes

d) Thermisch induzierte Spannungen

Besonders geeignet zum Schneiden scheinen Plexiglas und Sperrholz. Geeignet gewählte Parameter gestatten Schnittgeschwindigkeiten bis zu 10 m/min.
Dem gegenüber verhalten sich Polykarbonate gänzlich anders. Unter starker Rauchentwicklung im Schnittspalt, begleitet durch Verfärbung, gelingt nur ein schlechter Schnitt.

Auch Polyäthylen und Polypropylen lassen sich nur mit schlechtem Schnittbild und starker Rauchbildung trennen.

Dem gegenüber gelingt das Trennen von GFK-Werkstoffen ohne den Problemen a) - d). Beim Schneiden von

Tabelle 1.   Schnittgeschwindigkeiten mit 1,2 KW CO_2 Laser

Material	Dichtigkeit g/cm	$\frac{L}{kJ/cm^3}$	max. m/min	opt. 2/3 max m/min
Kiefer	0.5	0.45	11 bis 200 Watt	7
Eiche	0.85	4.6	5,2 bei 1 kW	3,5
Sperrholz	0.6	3.24	7,4 bei 1 kW	5,0
Asbest	2.5	50 – 70	0,4 bei 1 kW	0,3
Keramik	1.9 – 2.8	57 – 84	0,3 bei 1 kW	0,2
Glasverstärktes Polyester	1.45	8.9	2,7 bei 1 kW	1,8
Sperrholz	0.6	2.5	9,6 bei 1 kW	6,4
Polyäthylen PAS PE 10	0.94	15.5	1,5 bei 1 kW	1,0
Polyurethane PAS PU	1.2	1.96	12 bei 1 kW	8,0
Polyurethane Schaum	0.03	0.15	16 bei 100 Watt	11,0
Polyacetale PAS-LG	1.6	11.1	2,2 bei 1 kW	1,4
Polyamid PAS 80X	1.12	5.45	4,4 bei 1 kW	3,0
Plexiglas Lit	1.2	2.4	10 bei 1 kW	7,0
Plexiglas VA	1.2	3.8	6,3 bei 1 kW	4,2
Polycarbonate (lexan)	1.2	6.8	3,5 bei 1 kW	2,4
Hartgummi	1.4	17.7	1,4	0,9

Keramik ist besonders auf das Auftreten thermisch induzierter Spannungen zu achten, welche eine nochmalige thermische Behandlung notwendig machen könnte.

Unter Verwendung von Stickstoff konnten bei VOEST-ALPINE AG vorstehende Schnittwerte erreicht werden.

<u>REFERENZEN</u>

1. Walker R.W., CO_2-lasers: A machining success story, Photonics Spectra, Sept. 1982, p. 65 - 70.

2. Stanley L., Ream, Present and future industrial acceptance of high power CO_2-lasers, laser focus, Dec. 1982, p. 43 - 47

3. Weber, Herziger, Laser-Grundlagen und Anwendungen, Physikverlag 1972

4. Melles Griot, Optics Guide 1978, p. 9, Melles Griot Arnhem, Netherland

5. Duley W.W., p.110 - 112, Academic Press (1976), New York, San Francisco, London

6. Westerlund S., High speed cutting with curtain nozzle, colloquium on thermal cutting and flame processes, Sept. 7, 1982, Ljubljana, Yugoslavia

7. Schuöcker D., Laser cutting of bulk steel (40 mm) due to guided flow of radiation and reactive gas in the workpiece, 4th International Symposium on Gas Flow an Chemical Lasers, Stresa, Sept. 13 - 17, 1982, Italy.

8. Duley W.W., p. 151, Academic Press (1976), New York, San Francisco, London

9. Steen W.M., Lim G.C., Measurement of the temporal and spatial power distribution of a high power CO_2-laser beam, Optics and Laser Technology 82, p. 149 - 153

10. German Engineering Standard, DIN 2310

11. Hermann F.D., Schweißtechnische Praxis, Deutscher Verlag für Schweißtechnik (DVS), Düsseldorf 1979

12. Duley W.W., p. 266, Academic Press (1976), New York, San Francisco, London

13. Tikhomirov A.V., Kudrayavtsev E.P., Gas-laser cutting of materials, colloquium on thermal cutting and flame processes, Sept. 7, 1982, Ljubljana, Yugoslavia.

14.   Troitzsch, Brandverhalten von Kunststoffen, Carl Hanser Verlag, 1982, p. 27

15.   Sepold G., Teske K., Rothe R., Investigation on laser melt cutting of metals, colloquium on thermal cutting and flame processes, Sept. 7, 1982, Ljubljana, Yugoslavia.

16.   Duley W.W., p. 266, Academic Press (1976), New York, San Francisco, London

17.   La Rocca Aldo V., Der Laser als Werkzeug in der industriellen Fertigung; Spectrum der Wissenschaft, Mai 1982, p. 60 - 72.

18.   E.H. Berloffa, J. Witzmann, Laser cutting of metallic and nonmetallic materials with medium powered (1.2 kW) CW lasers, Voest-Alpine AG, Finished Products Division, Linz Austria

Laser Hardening Process Parameters

K. Wissenbach, L. Bakowsky, H.-G. Treusch, G. Herziger
Institut für Angewandte Physik, Technische Hochschule Darmstadt
Schlossgartenstr. 7, D - 6100 Darmstadt

1. Introduction

Surface heat treatment gets more and more important because of shortage of raw materials and energy. Thus, it is possible to improve the wear properties of moving parts by functional separation of base material (strength and stiffness) and surface (tribological function) in combination with simultaneous mass reduction.

The advantages of laser hardening compared to conventional hardening are (i) mostly higher hardness, (ii) less stress in the workpiece, and (iii) the production of compl hardening designs. The assumption for optimum utilization of these advantages is the quantitative knowledge and control of the physical process, which determines laser hardening. The hardness is induced by transformation of the austenitic structure to the martensitic structure. This transformation is commonly described by a time-temperature transformation diagram /1/. The diagram (fig. 1) necessarily implies cooling rates $dT/dt > 10^2$ ^{o}/s to avoid structures like bainite and perlite.

2. Basics

Fig. 2 illustrates schematically the laser hardening arrangement. The laser beam move. over the workpiece. The absorbed power density heats the surface, the necessary cooli rates for hardening are given by self-quenching in the bulk material. The hardening p cess is determined by the thermal properties of the workpiece (density ρ, specific he c_p, heat conductivity K), the absorbed power density $A \cdot I(r,t)$ and the relative veloci v between target and laser beam. Heat conduction determines the temperature distribut in the target.

With the following assumptions
- dimensions of the workpiece large compared to the dimensions of the heat source (se
  miinfinite problem)
- temperature-independent thermophysical properties
- Gaussian intensity distribution of the beam (TEM_{oo})
the temperature distribution for a workpiece moving with velocity v in x-direction is given by

$$\Gamma(x,y,z,t) = \frac{\sqrt{\rho \cdot c_p}}{4(\pi K)^{3/2}} \cdot A \cdot P_L \cdot \int_0^t \frac{1}{(t-t')^{1/2}(\rho c_p r_B^2/8K+t-t')} \cdot$$

$$\exp\left\{-\frac{z^2 \cdot \rho c_p}{4K(t-t')} - \frac{\rho \cdot c_p}{4K} \cdot \frac{(x-vt')^2+y^2}{(\rho c_p r_B^2/8K+t-t')}\right\} \cdot dt' \tag{1}$$

ρ : density A : absorption

c_p : spec. heat P_L : laser power

K : heat conductivity r_B : beam radius

The numerical computation of equation (1) in combination with the T-T-T-diagram makes it possible to calculate the hardening geometry in dependence of the process parameters.

But there is a characteristic difference between conventional- and laser hardening /3/. Fig. 3 shows a section of the Fe-C-diagram. The dot regions represent the hardening temperatures for conventional and laser hardening. In the conventional case the hardening temperatures are 30^0 - 60^0 above the GOS-line /1/. The hardening temperatures for laser hardening are 300^0 - 400^0 higher resulting in heating and cooling rates in the range of 10^3 - 10^6 ^{0}/s. Therefore, laser transformation hardening is a nonequilibrium process with nucleation rates much higher than for conventional hardening. The subsequent self-quenching results in a fine grained martensitic structure with improved hardness.

3. Experimental results

For an experimental proof of equation (1) it is necessary to control the process parameters laser power P_L intensity distribution, absorption A, beam radius r_B and hardening speed v. For practice the laser power is measured by calorimetric methods, the hardening speed is controlled by commercial translators. The measurement of absorption, beam radius and intensity distribution is not possible with conventional instruments and requires complex experimental arrangements.

Absorption A

The absorption of the workpiece is indirectly determined by measuring the incident laser power, the diffuse and the direct reflected power (fig. 4). The results for CK-85 steel of different surface treatments are shown in fig. 5. The absorption of technical surfaces varies from 10 % to 50 %. For graphite coatings the absorption increases to 88 % $\pm$ 1 %, for phosphate coatings to 77 % $\pm$ 1 %. With increasing power density the coatings are destroyed and the base materials governs the absorption.

314

Intensity distribution

The intensity distribution of the laser beam is controlled during the hardening process by the method of Steen /4,5/.

Beam radius r_B

Fig. 6 illustrates the experimental setup for the measurement of the beam radius. The spot size is determined by a power meter in combination with a copper wedge, which is moved normal to the beam. By displacing the wedge in z-direction the beam caustic is determined.

Comparison theory-experiment

Fig. 7 shows the comparison between experimental results and numerical calculations (equation (1)) for low hardened depth (laser power 200 W, beam radius 1,7 mm, hardening speed 0,5 cm/s). A cross-section of a laser hardened zone is illustrated in fig. 8 by an optical micrograph with corresponding hardness distribution. The hardened depth is about 1 mm, the maximum hardness about 830 Vickers.

The experimental and calculated hardened width are plotted in fig. 9 versus the travelling coordinate x for v = 1 cm/s. Theory and experiment show reasonable agreement. A constant hardened width is obtained after 5 mm, i.e. the heat conduction becomes steady state.

Further efforts will be done to extend the heat conduction model to finite geometries with arbitrary intensity distributions including temperature-dependence of the thermo-physical properties. Moreover approximate formulas have to be derived to limit the computational expense and to evaluate for the user rule-of-thumb formulas which reduce the number of parameter combinations. Therefore, for each application only a few experiments are necessary to determine the whole set of process parameters.

4. Summary

The presented heat conduction model allows to determine the process parameters for laser hardening for many applications. The number of parameter combinations additionally can be reduced, that laser hardening can be used in industrial production without tedious and expensive preexamination.

Literature
(1) BICKEL,E.: Die metallischen Werkstoffe des Maschinenbaues, Springer Verlag (1964)
(2) MARUO,M.,MIYAMOTO,J.,ISHIDE,T.,ARATA,Y.: Proceedings of the first international laser proceeding conference Nov 1981, Anaheim, California
(3) READY,J.F.; Industrial Applications of Lasers; Academic Press (1978)
(4) LIM,G.C.,STEEN,W.M.; Optics and Laser Technology $\underline{6}$, 149(1982)
(5) LOOSEN,P. et al.; these proceedings

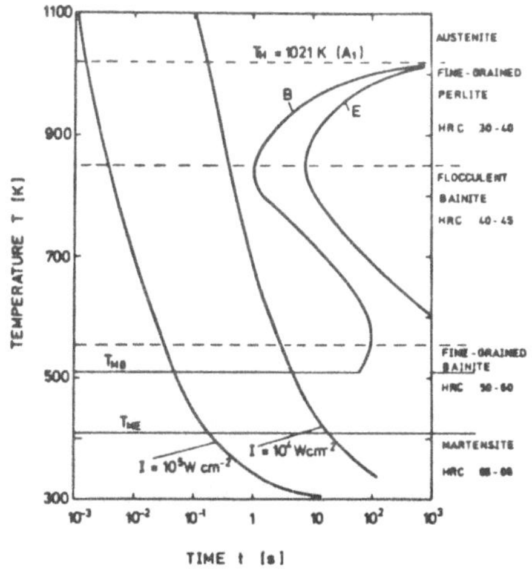

Fig. 1: TIME-TEMPATURE-TRANSFORMATION-
diagram

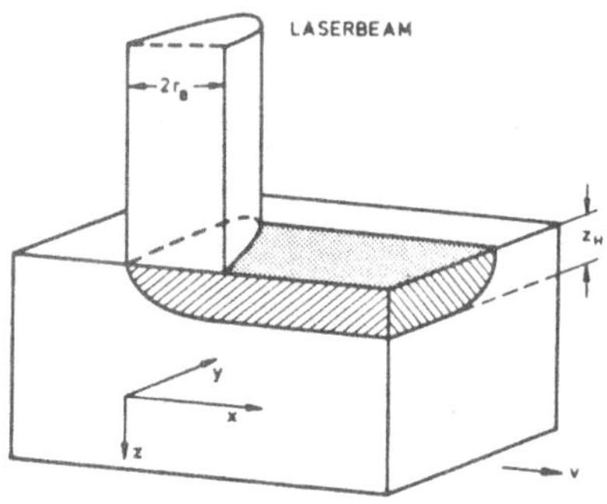

Fig. 2: Scheme of laser hardening

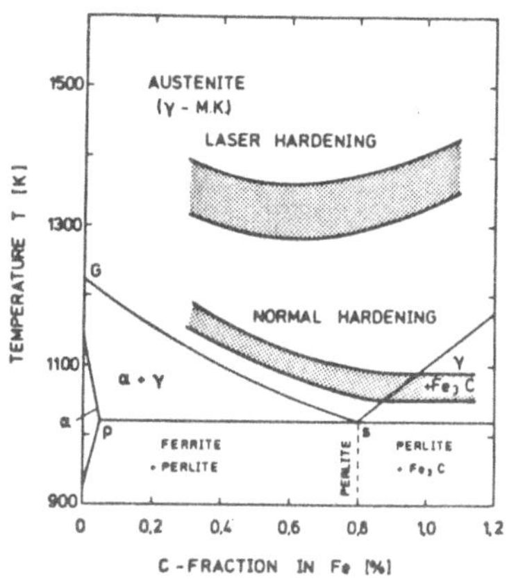

Fig. 3: Fe-Ce-diagram with hardening
temperatures for conventional-
and laser hardening

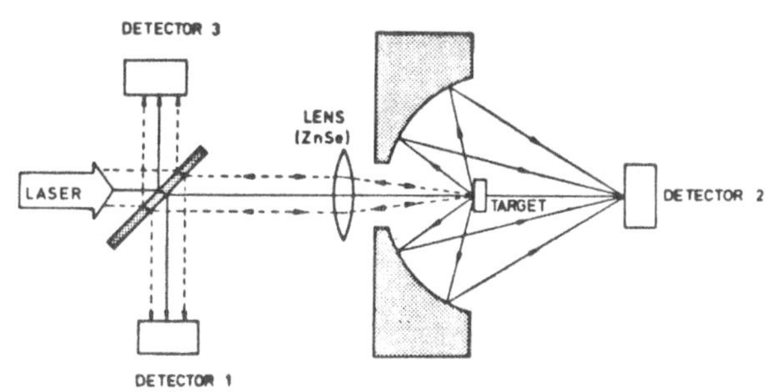

Fig. 4: Experimental setup for absorption
measurement

Fig. 5: Reflexion-coefficient for the
steel Ck-85

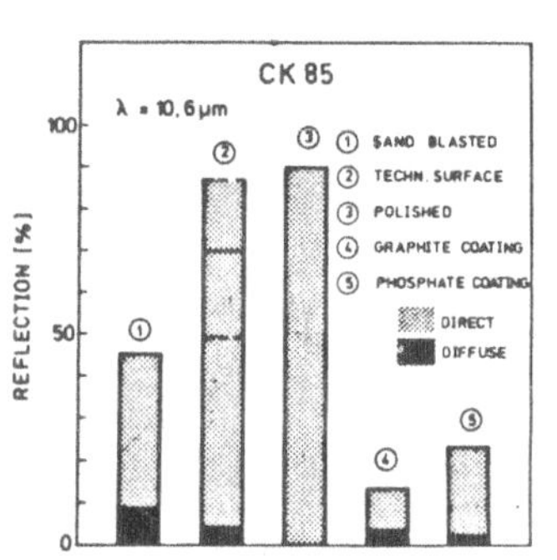

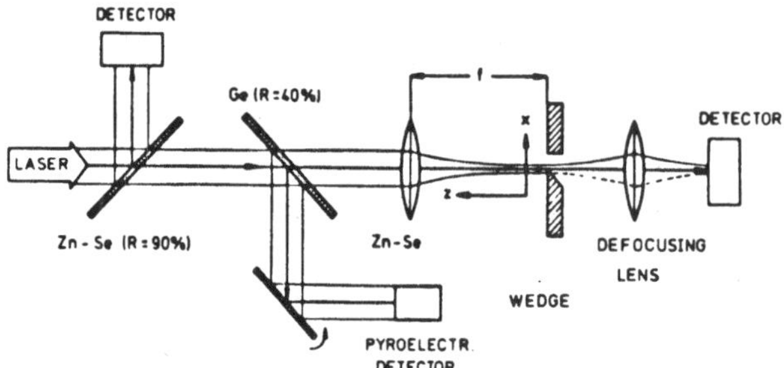

Fig. 6: Experimental setup for focus radius measurement

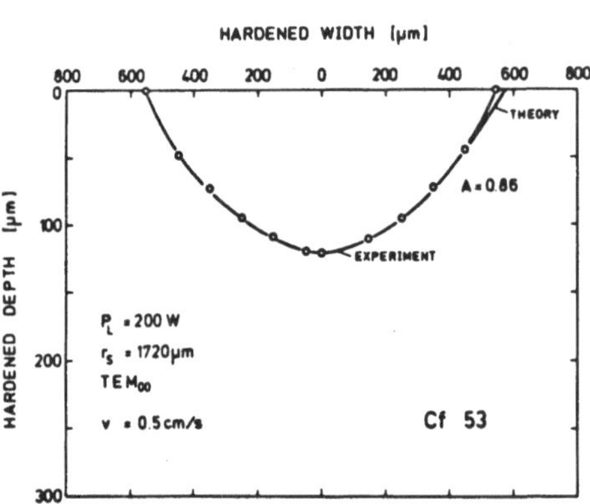

Fig. 7: Experimental and theoretical hardening geometry for the steel Cf 53

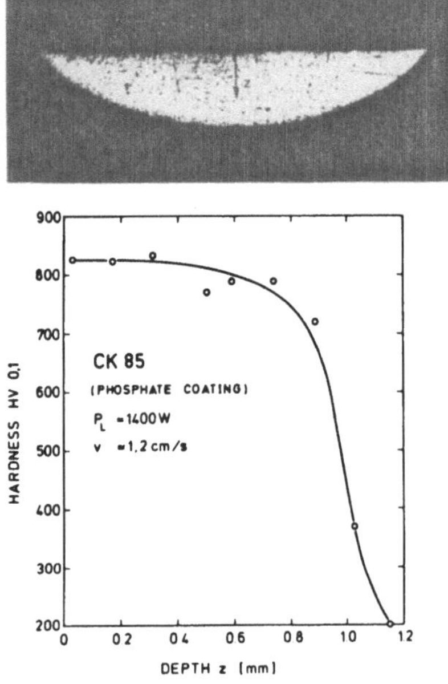

Fig. 8: Cross section of a hardened probe with corresponding hardness distribution

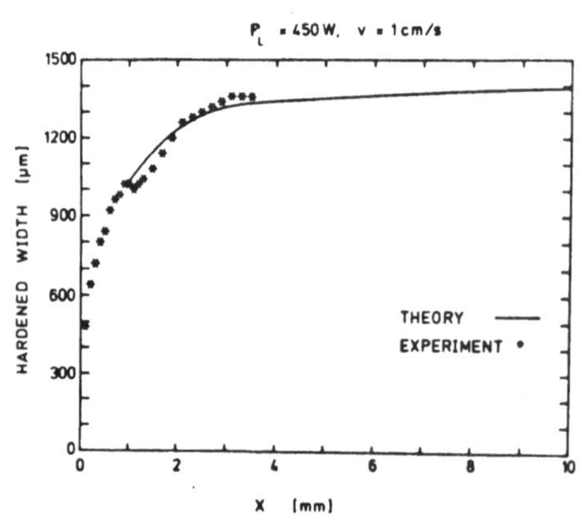

Fig. 9: Experimental and theoretical hardened width for the steel Cf 53

Aspekte des Laserstrahlhärtens

R. Becker* und G. Sepold*
R. Chatterjee-Fischer**

* Bremer Institut für angewandte Strahltechnik, BIAS, 2820 Bremen 71

** Institut für Härtereitechnik, IHT, 2820 Bremen 77

1. Einleitung

Das Härten mit dem Laserstrahl gehört zu den Kurzzeithärteverfahren,
wofür in der Regel aufgrund der hohen Energieeinbringung weniger als
eine Sekunde benötigt wird. Die Erwärmung bleibt auf eine dünne Schicht
beschränkt. Die Härtung erfolgt durch Selbstabschrecken.

Das Laserstrahlhärten wurde möglich, nachdem handelsübliche Hochleistungslaser zu erhalten waren. Heute stehen ausgereifte Anlagen mit
mehr als 10 kW Strahlleistung zur Verfügung. Das Laserhärten erfolgt
durch punktförmiges Abrastern einer Oberfläche bzw. durch Aufbringen
von Härtespuren. Das Härtungsergebnis wird beim Laserstrahlhärten im
wesentlichen durch folgende Faktoren bestimmt [1-3]:

- laserseitig: durch Leistung, Vorschubgeschwindigkeit bzw.
                              Einwirkdauer und Strahlbeschaffenheit
                              (Leistungsdichte, ihre zeitliche und räum-
                               liche Verteilung)

- werkstoffseitig: durch Werkstoffzusammensetzung, Gefügezu-
                              stand, Oberflächenbeschaffenheit

- sonstige Einflußfaktoren: Absorptionsschichten, Umgebungsmedium.

Bei den vorgestellten Untersuchungen wurde angestrebt, einige o. g.
Einflüsse auf das Härtungsergebnis zu erfassen, Zusammenhänge zu ermitteln und für die Praxis brauchbare Ergebnisse zusammenzufassen.

2. Versuchsanlage

Die Versuchsanlage bestand aus einem stationären, quergeströmten 5 kW
CO_2-Laser mit einer Härteoptik und einer Werkstück-Manipulationsan-
lage. Der Laserstrahl wurde über Spiegel in eine Härteoptik gelenkt,
die ihn auf einen rechteckförmigen Querschnitt mit konstanter Leistungsdichte transformierte, Bild 1.
Diese im BIAS entwickelte und mit Schutzgas zu betreibende Härteoptik
ermöglicht die Umformung des Strahls in Rechteckquerschnitte, so daß
die Versuche z.B. mit quadratischen Laserstrahlquerschnitten von
5 x 5 mm, 8 x 8 mm und 12 x 12 mm durchgeführt werden konnten. Der
Abstand zwischen Werkstückausgang und der Härteoptik betrug stets
3 mm. Variiert wurde die Laserleistung und die Vorschubgeschwindigkeit. Als Probenwerkstoff wurde in der Hauptsache der Stahl Ck 45
in vergütetem Zustand gewählt.

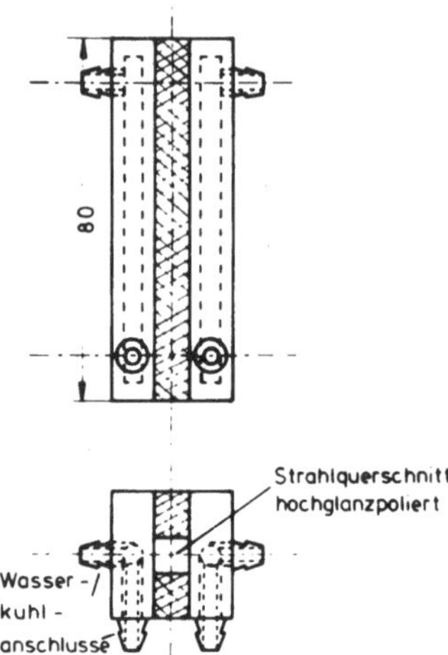

Bild 1: Schematische Darstellung des
verwendeten Strahlformers.

3. Versuchsergebnisse

3.1 Einfluß des Werkstoffoberflächenzustandes (Absorption)

Für die Energieeinkopplung in den Werkstoff ist die Oberflächenbeschaffenheit der Probe von großer Bedeutung. Es wurde eine Reihe unterschiedlich vorbehandelter Oberflächen untersucht. Das Absorptionvermögen wurde kalorimetrisch bestimmt, d.h. die Probe der Masse m
und der spezifischen Wärme c wurde mit einer Strahlleistung P bei
konstanter Leistungsdichte bestrahlt und die Temperaturerhöhung ΔT
nach der Zeit t gemessen. Zum Schutz gegen Oxidation wurde die Probenoberfläche mit Stickstoff bespült.

Oberflächenzustand	Absorptionsgrad A
geschliffen,	0,085
gefräst,	0,18
gestrahlt	0,35
Wolframpulver, aufgestreut	0,37
manganphosphatiert	0,85 .. 0,65
zinkphosphatiert	0,55
Grafitspray	0,77

Tabelle 1:

Absorptionvermögen verschiedener Stahloberflächen
(Versuchsbedingungen: Strahleinwirkzeit t = 1 s,
Leistungsdichte P = $2,5 \times 10^{4}$ W/cm^{2})

Der Absorptionsgrad wurde definiert als

$$A = \frac{m \cdot c \cdot \Delta T}{t \cdot P}$$

Die Ergebnisse der Messungen sind in Tabelle 1 zusammengestellt. Für
die durchgeführten Untersuchungen wurden Grafitsprayschichten bevorzugt eingesetzt.

3.2 Einfluß der Laserdaten auf das Härtungsergebnis

Im wesentlichen sind folgende Lasergrößen von Einfluß auf das Härtungsergebnis:

- Leistung und Leistungsdichte (-verteilung)
- Strahlgeometrie sowie
- Einwirkdauer (Vorschubgeschwindigkeit)

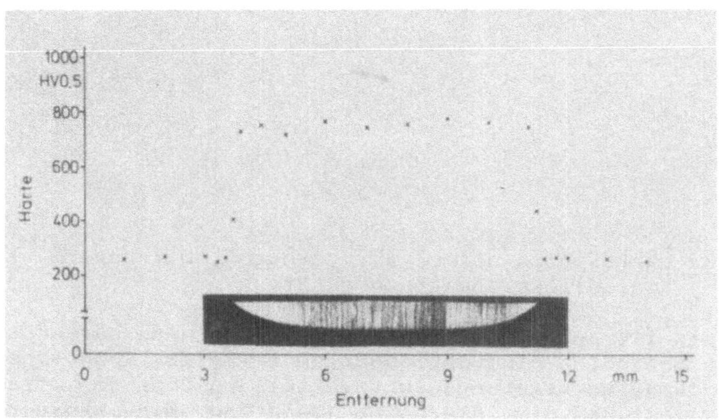

Bild 2: Gefüge und Härteverteilung (parallel zur Oberfläche)
einer Härtespur, erzeugt mit einem Strahl von
8 x 8 mm^2
P = 1420 W
v = 6 mm/s
Ck45, vergütet

Als günstigste Strahlgeometrie erweist sich ein rechteckförmiger Strahl. Bild 2 zeigt, daß mit Hilfe eines solchen Laserstrahls mit gleichmäßiger Leistungsdichteverteilung eine flache, einheitliche Härtezone erzielt werden kann. Die Härtewerte, gemessen in etwa 0,1 mm Abstand von der Oberfläche, sind über die gesamte Härtespurbreite nahezu konstant. Unter günstigen Bedingungen wurden Einhärtungstiefen bis zu 1,2 mm erreicht.

Eine gleichmäßige, sowie weitgehend rechteckige Intensitätsverteilung ist deshalb von großer Bedeutung, da örtliche Überhitzungen oder Anschmelzungen nachteilig sind und daher vermieden werden müssen. Überhitzungen oder Anschmelzungen aufgrund von ungleichförmiger Leistungsdichteverteilung führen in der Regel zu Rissen. Bild 3 gibt hierzu einige Beispiele wieder. Es enthält links eine ordnungsgemäße Härtespur mit feinnadligem Martensit. In der Mitte ist eine starke Überhitzung wiedergegeben, in deren Bereich grobnadliger Martensit und Risse vorhanden sind. Das Bild rechts weist eine Anschmelzung mit Anrissen auf.

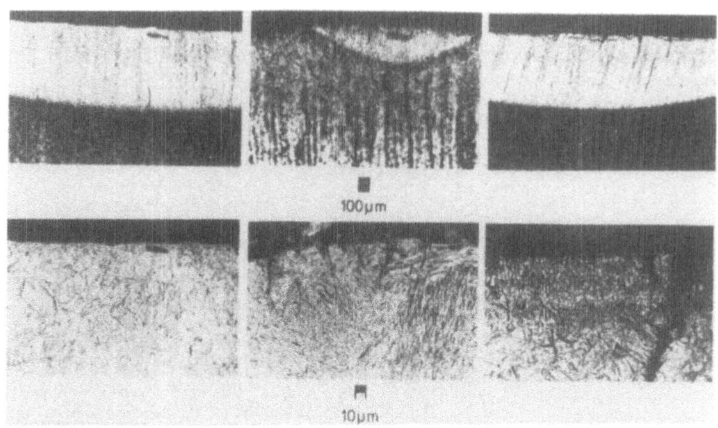

Bild 3: Härtespuren auf Ck 45, vergütet, mit unterschiedlicher Leistung gehärtet

In der Praxis ist nur die in Bild 3 links erzeugte Härtespur verwendbar. Solche Schichten sind aber dann zu erzielen, wenn eine weitgehend gleichmäßige Leistungsdichteverteilung über dem Strahlquerschnitt vorliegt und die Laserdaten (Leistung, Einwirkdauer bzw.
Vorschubgeschwindigkeit) auf den Werkstoff und die gewünschte Ein-
härtungstiefe abgestimmt sind.

Zur Abschätzung des Einflusses der Parameter-Leistungsdichte und
Einwirkdauer bzw. Vorschubgeschwindigkeit wurde eine Reihe von Experimenten durchgeführt. Ein Ergebnis ist in Bild 4 wiedergegeben.

Es wurde versucht, durch Vergleich der experimentellen Ergebnisse mit
denen physikalischer Modellrechnungen einen Zusammenhang zwischen Einhärtetiefe und Laserstrahlparameter zu ermitteln. Als grundlegendes
physikalisches Modell wurde eine Lösung der Wärmeleitungsgleichung
für einen Laserstrahl mit konstanter Intensität über dem Querschnitt
herangezogen [4]. Die Ergebnisse zeigen bisher, daß die Angabe eines
allgemein gültigen Modells nicht möglich ist. Dies dürfte auf die Abhängigkeit der thermophysikalischen Daten von der Temperatur sowie
möglicherweise auf Veränderungen des Absorptionsvermögens der Oberfläche während des Härtungsprozesses zurückzuführen sein.

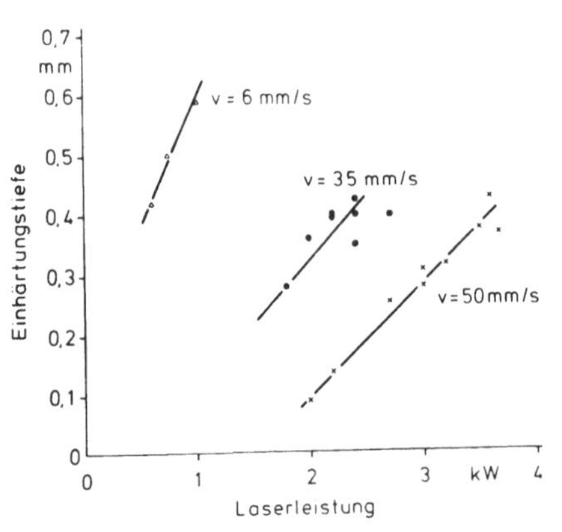

Bild 4:

Einfluß von Leistung
und Vorschubgeschwin-
digkeit auf die Ein-
härtungstiefe

CK 45 vergütet
Graphitspray
$F = 25 \ mm^2$

Für die am Werkstoff Ck 45 durchgeführten Untersuchungen wurde eine Nährungslösung berechnet. Diese Lösung hat jedoch den Nachteil, daß sie nur für den Werkstoff Ck 45 bei graphitierter Oberfläche im Leistungsbereich 1 bis 3 kW und für Strahlquerschnitte von 20 bis 150 mm^2 gilt:

$$z(P,R,t) = ((-0,032R^2 + 0,3R-1)\,\ln^2\left(\frac{P}{R^2}\right) + (0,1R^2+4))\sqrt{\ln(1+t)}$$

$$+\frac{P}{6}\cdot\ln(R-2) - 0,23R^{3/2} + 0,2P$$

z = Einhärtetiefe [mm]
P = Laser-Leistung [W]
R = äquivalenter Strahlradius = $(F/\pi)^{1/2}$ [mm]
   mit F = (quadratische) Strahlfläche
t = Einwirkzeit [s]

4. Schlußbetrachtung

Durch Laserstrahlhärten können reproduzierbare, voll-martensitische Härtespuren hergestellt werden. Voraussetzung hierfür ist eine möglichst gleichförmige Intensitätsverteilung im Laserstrahl. Als besonders vorteilhaft erweist sich eine rechteckförmige Geometrie, die es ermöglicht, breite, flache Härtungszonen in der Randschicht zu erzeugen. Hierfür wurde ein Strahlformer entwickelt und erfolgreich eingesetzt, der eine gaussförmige Leistungsdichteverteilung in eine Rechteckform mit gleichmäßiger Intensitätsverteilung transformiert.

Von besonderer Bedeutung ist auch die geeignete Auswahl von Absorptionsschichten. Von den verschiedenen untersuchten Absorptionsschichten haben sich Grafitschichten und Phosphatierungsschichten für das Härten als am günstigsten erwiesen. Dabei zeigt sich, daß sich die Absorption mit der Leistungsdichte und der Einwirkzeit des Laserstrahls ändert. Dieser Umstand ist als ein wesentlicher Faktor bei der theoretischen Vorausabschätzung eines Härteergebnisses zu berücksichtigen.

Aus einer Vielzahl von Experimenten wurde für den Werkstoff Ck 45 eine Näherungslösung für die Abhängigkeit der Einhärtetiefe von den Strahlparametern ermittelt. Es zeigt sich, daß mit den gegebenen Optiken Einhärtetiefen bis zu 1,2 mm zu erreichen waren.

5. Dank

Die Autoren danken der Arbeitsgemeinschaft Industrieller Forschungsvereinigungen (AIF) für die Unterstützung dieser Arbeit.

6. Schrifttum

1. G. STÄHLI Werkstoffe und ihre Veredlung 3 (1981) Nr. 11/12,
S. 441-446.
2. G. SEPOLD in Gas-Flow and Chemical Lasers ed. by JOHN F. WENDT, Karman Institute for Fluid Dynamies, Rhode-Saint-Genèse, Belgien
1980.
3. G. SEPOLD. R. BECKER, Schweizer Maschinenmarkt Nr. 34 (1982)
S. 72 - 75.
4. W.W. DULEY: CO$_2$-Lasers, Effects and Applications, Academy Press, London, 1976, S. 144 - 145.

Laser Surface Alloying of Ferrous Materials with Carbon

A. Walker, D.R.F. West and W.M. Steen
Department of Metallurgy and Materials Science, Imperial College of Science and
Technology, London SW7

1. INTRODUCTION

Graphite coatings have been used in laser surface melting to reduce reflectivity
and substantial amounts of carbon can be alloyed into steels from these
coatings /1,2/ eg Fig. 1. The present paper explores the potential of this
procedure for surface alloying with C.

2. EXPERIMENTAL PROCEDURE

Pure iron (C 0.016wt%) and a 1%C 1.4%Cr steel were laser treated using Control
Laser Ltd 2 kW CW CO_2 lasers typically operated at $\sim$1.7 kW. Specimens $\sim$15mm
thick and argon shielded were laser treated at various traverse speeds and beam
diameters to produce melt zones from $\sim$0.1-2.5mm in depth. DAG graphite coatings
were applied either by: (1) "painting" on the surface DAG diluted with
methanol; (2) applying a paste or slurry (in which the DAG is less diluted);
the resultant thicker coats were "protected" with adhesive tape. Successive
laser treatments (up to 12) were given, reapplying the DAG coatings between each
treatment.

Gas unsoundness and cracking were encountered in the melt zones and a procedure
with 2 lasers was used to increase the "molten" time , thus aiming to aid release of entrained gas and also to reduce the cooling rate and hence the cracking
tendency. The laser beams were operated simultaneously, one vertically and the
other at 65° to the vertical.

3. RESULTS AND DISCUSSION

3.1. Carbon alloying into S135 steel

With the aim of introducing large amounts of C, a beam of $\sim$5mm diameter was used
at 25mm s^{-1} traverse speed to produce a shallow melt zone (Figs. 2a,b) with
associated high surface/volume ratio; five successive cycles of DAG "painting"
and laser melting were applied.

A light etching region ($\sim$20 μm thick and hardness $\sim$350 HV) was present along the
interface between the melt zone and the martensitic ($\sim$900 HV) heat affected zone
(h.a.z.) (Fig. 2c); this region which consists of equiaxed grains was interpreted

as containing some martensite; it originates from solid state diffusion of
carbon leading to depression of the M_s. Dendrites of γ (containing a small
amount of martensite) extend upward from the base of the melt zone; the secondary dendrite arm spacing was ∿3 μm. The interdendritic regions contained fine
lamellar eutectic consisting of Fe_3C and retained γ (Fig. 2d). The melt zone
structure was not uniform: for example, the upper portions and the edges consisted predominantly of eutectic (Figs. 2a & b) and near the surface some primary
Fe_3C crystals were observed (Fig. 2b). The heterogeneity results from incomplete mixing of carbon in the shallow zone. Quenching to -196°C led to some
martensite formation in γ areas (Fig. 2d).

3.2. Carbon alloying into iron

Experiments with pure iron, involved conditions which produced keyholing with
associated good mixing conditions.

Figs. 3a-d show the structure produced by 12 successive DAG "painted" treatments.
The melt zone structure consists mainly of γ dendrites with some eutectic(γ +
Fe_3C). This is consistent with the hardness of ∿470 HV which increases up to ∿650
HV on cooling to -196°C due to martensite formation. The γ dendrites grew perpendicular to the melt interface. The secondary arm spacing was ∿3 μm in the
upper part of the zone (Fig. 3c). Some porosity is present, but cracking is not
a serious problem.

The use of 4 "protected" runs produced a substantially greater solution of carbon
(Figs. 4a-e). The composition was hypereutectic showing ∿40 vol% of primary Fe_3C
in a matrix of γ + Fe_3C eutectic. The hardness level is correspondingly high
(∿850 HV) and there was no significant hardness increase on cooling to -196°C,
reflecting the small proportion of γ present. Extensive porosity and cracking
occurred.

The dual beam technique (section 2) was applied to produce 4 successive melts
with "protected" graphite layers and employing 4 overlapping melt traces. The
experiment did not succeed in reducing the incidence of porosity or cracking, but
a particularly high level of carbon solution (up to ∿90 vol% primary Fe_3C) was
achieved (Fig. 5).

3.3. White iron eutectic solidification kinetics

Fine white iron structures have been previously produced by surface melting eg
/3-5/ and by solidification of powders /6/ with eutectic interlamellar spacings

(λ) of $\sim$0.5 µm /4/. Under directional solidification conditions and with $\lambda >\sim$ 2 µm, eutectic growth at a given velocity, V, occurs over a range of spacings λ_{op} with values of $\lambda_{op}\sqrt{V}$ between 11 and 29 µm$^{3/2}$s$^{-1/2}$ /7/. In laser surface melting, the average growth rate may be taken to be a substantial fraction of the traverse speed ($\sim$20 x 10^{3}µm s^{-1} in the present work); thus, taking the mean spacing $\bar{\lambda}$ as $\sim$0.7 µm, the $\bar{\lambda}\sqrt{V}$ product is significantly higher than predicted /7/.

In the present work, eutectic growth is not strictly unidirectional, but some aligned lamellae of γ and Fe$_3$C grow edgewise upwards (eg Fig. 4c). Other regions appear to show transverse sections of lamellae and may include examples of cooperative sidewise growth /8/ eg Fig. 4e. Acicular Fe$_3$C (eg Fig. 2d) can be interpreted as a divorced eutectic in very small interdendritic regions /6/.

4. CONCLUSIONS

Laser surface melting of ferrous surfaces precoated with graphite produces alloyed layers with carbon ranging up to hypereutectic levels. White iron structures can be produced containing γ + Fe$_3$C eutectics of interlamellar spacings <1 µm.

ACKNOWLEDGEMENTS are made to SERC and to Rolls Royce Barnoldswick for support and to Mr I. Hawkes and Mr M. Sharpe who developed the two-beam laser system.

REFERENCES

1. G. Christodoulou, A. Walker, W.M. Steen, and D.R.F. West, Metals Technology 1983 (in the press).
2. M. Carbucicchio, G. Meazza, G. Palombarini & G. Sambogna, J. Mat. Sci. 18, 1543 (1983).
3. T.R. Tucker, A.H. Clauer, C.T. Walters, S.L. Ream and B.P. Fairand, Lasers in Metallurgy, Eds. K. Mukerhjee and J. Mazumder, AIME, 1981, p.53.
4. I. Hawkes, W.M. Steen and D.R.F. West, Conference on Electroheat for Metals, BNCE Cambridge, 1982, paper 4.4.
5. J.H.P.C. Megaw, A.S. Bransden, T. Bell and D.N.T. Trafford, Metals Tech., 10, 2, 69 (1983).
6. L.E. Eiselstein, O.A. Ruano and O.D. Sherby, J. Mat. Sci. 18, 483 (1983).
7. H. Jones and W. Kurz, Z. fur. Metall., 72, 792 (1981).
8. M. Hillert and V.V. Subba Rao, The Solidification of Metals, Iron & Steel Institute 1968, p.204.

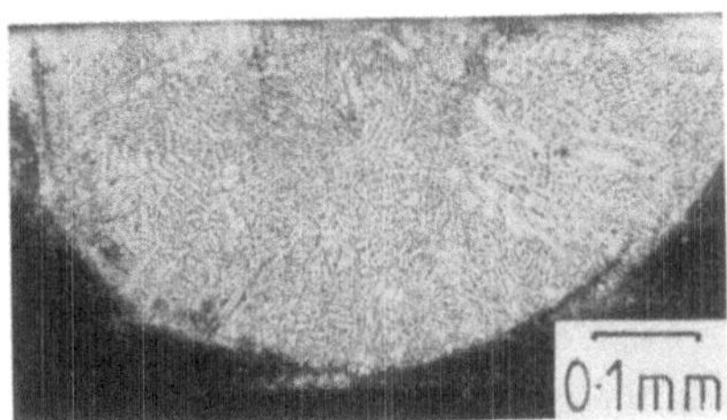

Fig. 1 Laser melted zone in DAG coated 1%C 1.4%Cr steel. Predominantly austenitic structure.

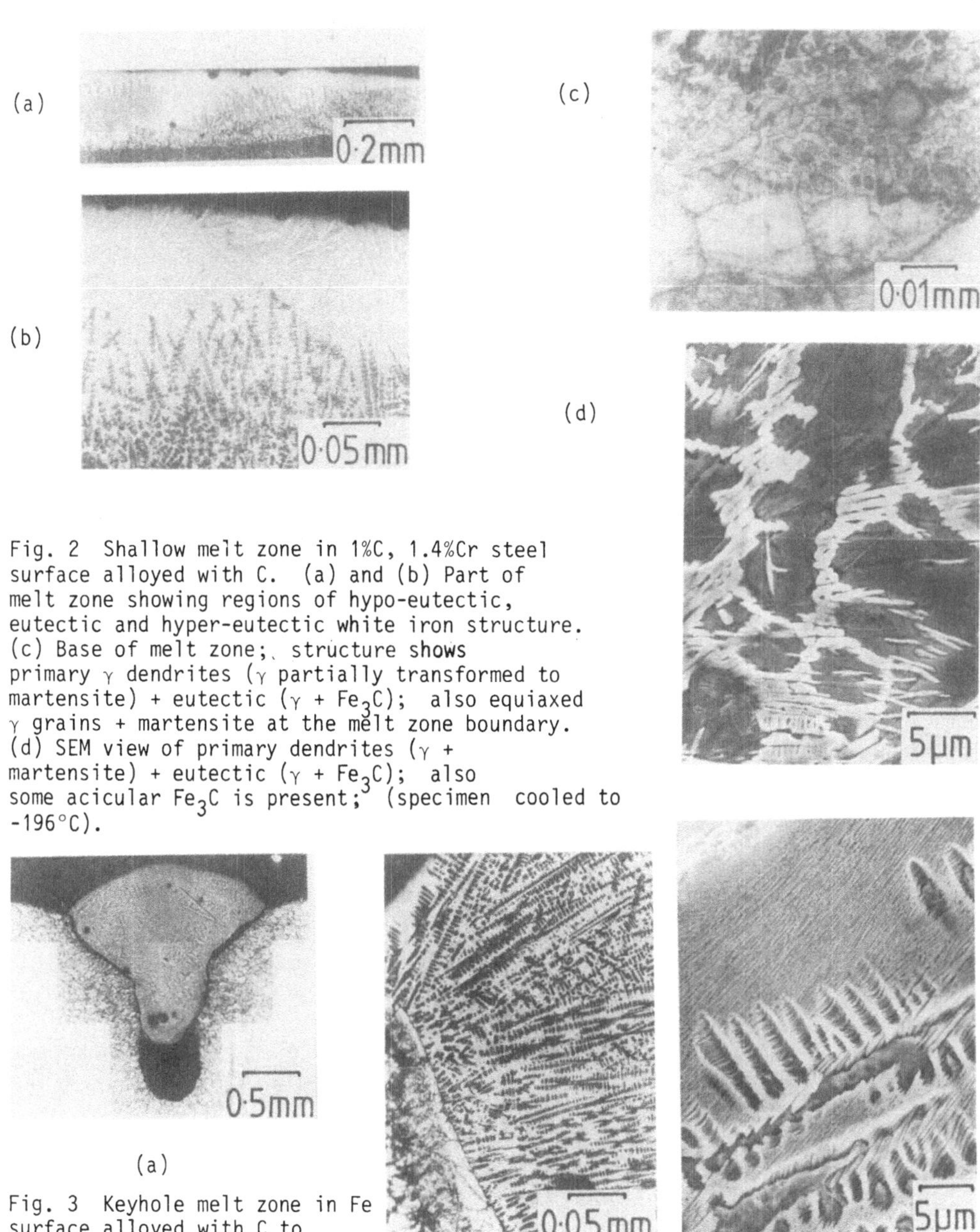

Fig. 2 Shallow melt zone in 1%C, 1.4%Cr steel surface alloyed with C. (a) and (b) Part of melt zone showing regions of hypo-eutectic, eutectic and hyper-eutectic white iron structure. (c) Base of melt zone; structure shows primary γ dendrites (γ partially transformed to martensite) + eutectic (γ + Fe_3C); also equiaxed γ grains + martensite at the melt zone boundary. (d) SEM view of primary dendrites (γ + martensite) + eutectic (γ + Fe_3C); also some acicular Fe_3C is present; (specimen cooled to -196°C).

Fig. 3 Keyhole melt zone in Fe surface alloyed with C to produce a hypo-eutectic white iron structure. (a) General view; the dark region at the base of the keyhole results from deep etching of the fine structure. (b) Part of melt zone and h.a.z. (c) SEM view of part of melt zone: primary γ dendrites (largely etched away by a deep etching technique) + eutectic (γ + Fe_3C).

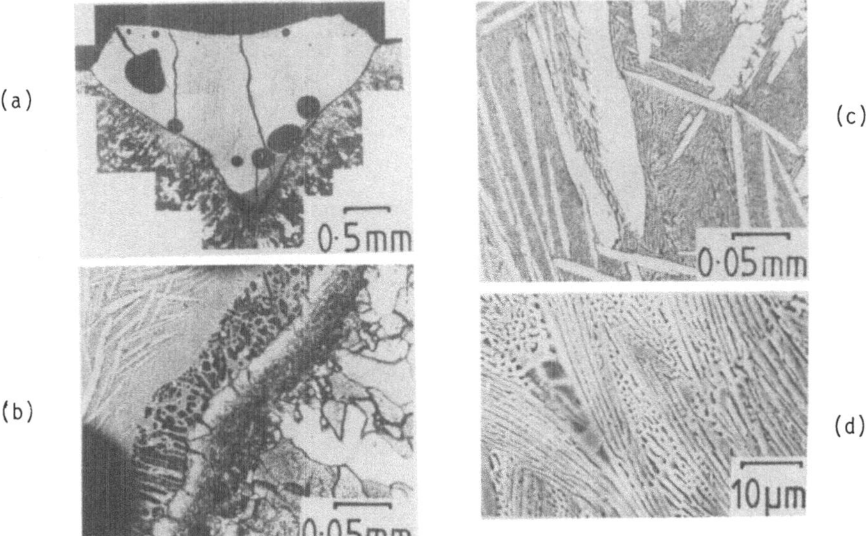

Fig. 4 Keyhole melt zone in Fe surface alloyed with C to produce a hyper-eutectic white iron structure. (a) General view, including primary Fe_3C, porosity and cracking. (b) Region spanning part of melt zone and the h.a.z.; the melt zone shows structures ranging from hypo- to hyper-eutectic; this effect is attributed to incomplete remelting of zones produced early in the series of 4 treatments. The h.a.z. shows evidence of solid state diffusion of C, leading to a γ + martensite region. (c) Primary Fe_3C + eutectic (γ + Fe_3C). (d) & (e) SEM views of the interior of blowholes (deeply etched) showing eutectic morphology. From the relative widths of the γ and Fe_3C lamellae, the volume fraction of γ in the eutectic appears to be less than that corresponding to the approximately equal proportions of the phases in eutectic of equilibrium composition; such an effect would be consistent with liquid of C content higher than the equilibrium eutectic solidifying in the "coupled zone" and hence containing more Fe_3C. However, the apparently lower volume fraction of γ may be due, at least in part, to an etching artefact since this phase is etched more deeply than the Fe_3C.

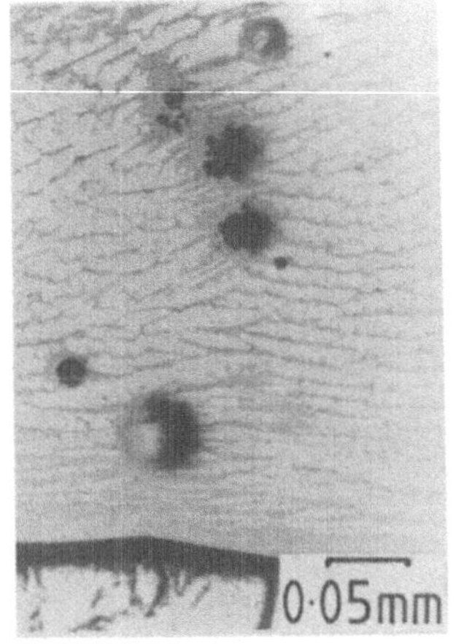

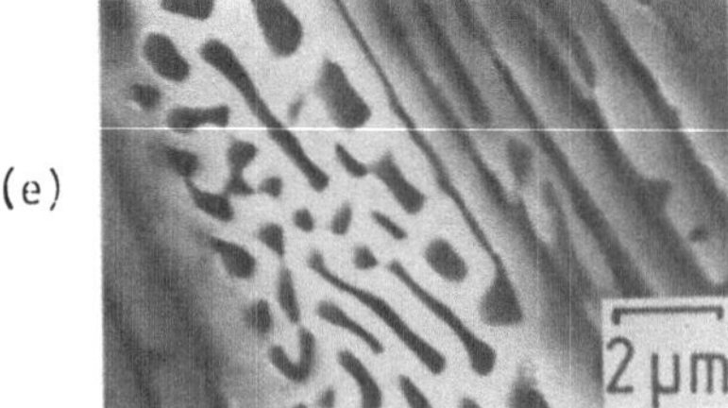

Fig. 5 Relatively shallow melt zone in Fe surface alloyed with C using a twin beam technique to produce a hyper-eutectic white iron structure. A very high volume fraction of primary Fe_3C is present.

Laser Cladding

V.M. Weerasinghe & W.M. Steen

Metallurgy Dept. Imperial College London SW7.

ABSTRACT

A process of laser cladding using argon borne powder injection is described. The
effect of the process parameters on the quality of the clad layer is discussed. The
versatility of this process is illustrated by some diverse applications. The significant process improvement obtained using a reflective shroud and shot blasted specimens is reported.

INTRODUCTION

Cladding and surface alloying are two of the many material processing applications
for which a laser has been and is being used (1 - 5). Laser cladding offers some
unique advantages:-

> (a) Controlled dilution levels
> (b) Localised heating, which reduces thermal distortion.
> (c) Controlled Shape
> (d) Good fusion bond.
> (e) Fine microstructures
> (f) Non - Contact method of application
> (g) Low residual stresses.
> (h) Minimal diffusion of the substrate elements into the clad, due to
> the short heating cycle.

The basic technique for producing uniform clad layers by laser is to overlap single
clad tracks. Intuitively this would seem to be long-winded when compared to established methods such as roll bonding, explosion cladding or powder casting, (5 - 8).
One may conclude that the laser is best suited for coating small confined areas. However, cladding rates presented here indicate that the laser may well compete favour-
ably in covering rates with conventional flat plate cladding processes.

Most metal combinations can be clad provided the clad material does not have a significantly higher melting point. Discussed here are the effects of the major process
parameters.

EXPERIMENTAL

The process used here is similar to that
currently in production by Rolls Royce
(1), namely to blow the powdered cladd
-ing material into the laser generated
meltpool as illustrated in fig.1.

Fig.1. Laser Cladding by powder injecting

Most of the work presented here is related to flat plate cladding of mild steel EN3,
with stainless steel. 316.

All specimen surfaces were degreased and shot blasted prior to cladding. Powders
with an average particle size of 60 microns were used.

The effect of the independent process variables on various clad properties is now
discussed.

DISCUSSION AND RESULTS.

(a) <u>Single track profiles</u> Three basic cross section profiles of single tracks were
observed. They are illustrated in fig2. Prifile (c) with an obtuse contact angle
is the preferred section. At higher powder feed rates and/or lower power densities
profile (a) is produced while at lower powder feed rates or higher power densities
profile (b) is produced.

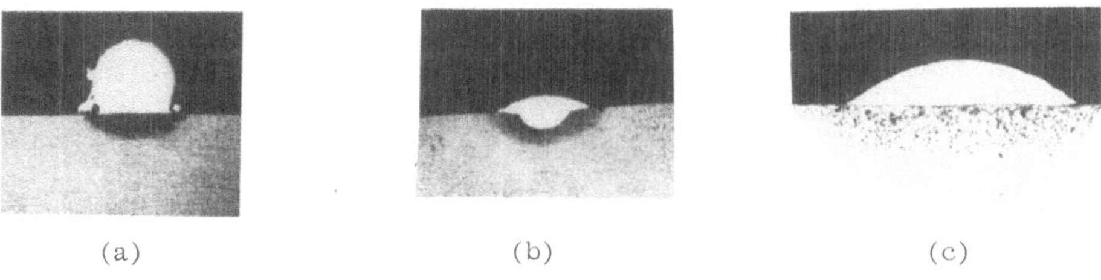

(a)   (b)   (c)

Fig.2. Three basic clad section profiles. Clad-Stainless sbustrate - Mild Steel.

(b) <u>Dilution</u> For a given power density (W/mm^2) of a given mode structure, the injected powder mass flow controls the level of dilution. This is illustrated in fig 3.
Unlike pre-placed powder cladding, dilution is independent of the cladding speed.

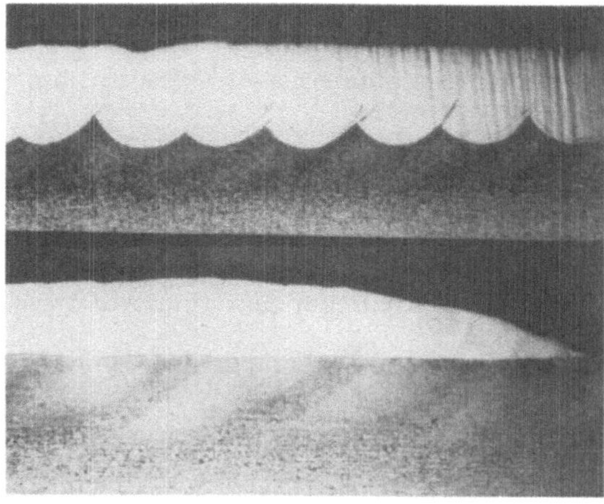

Fig. 3. Effect of powder mass flow on substrate dilution. Powder mass flow 0.09 g/s
top. 0.212 g/s bottom.

(c) <u>Porosity</u> Porosity in laser clad layers may be caused by one or a combination of
the following: cavities between two overlapped tracks (usually formed near the root
and will be referred to as inter-run porosity)solidification cavities and/or gas evo
-lution. Inter-run porosity, illustrated in fig 4 occurs when single tracks with a
section profile similar to fig.2. (a) are overlapped due to the root being shielded
from the laser.It can be eliminated totally by operating in such a way to produce section profiles as in fig 2c. It can also be eliminated by angling the powder feed
around to the line of the clad track or ultrasonically vibrating the substrate during
cladding (2) Solidification cavities occur as a porosity at or near the interface.
If the clad layer has a significantly higher melting point than the substrate the
solidification front may finish at the interface thus causing shrinkage cavities,
Fig.5.

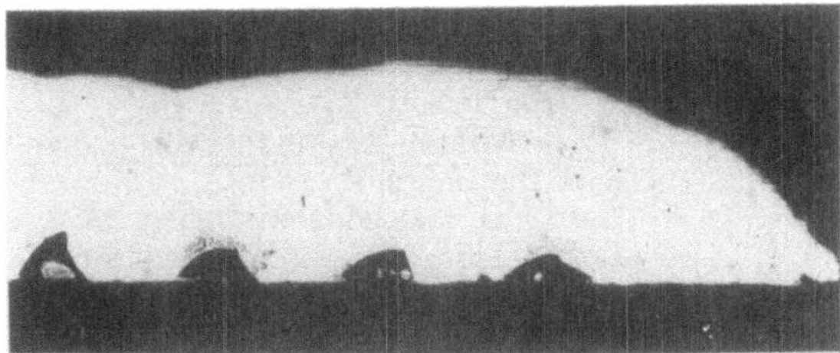

Fig. **4**. Inter-run porosity clad
– Stainless Steel 316 substrate –
Mill Steel EN3

Fig.5. Solidification void cracking
Clad – Stainless steel substrate –
Aluminium.

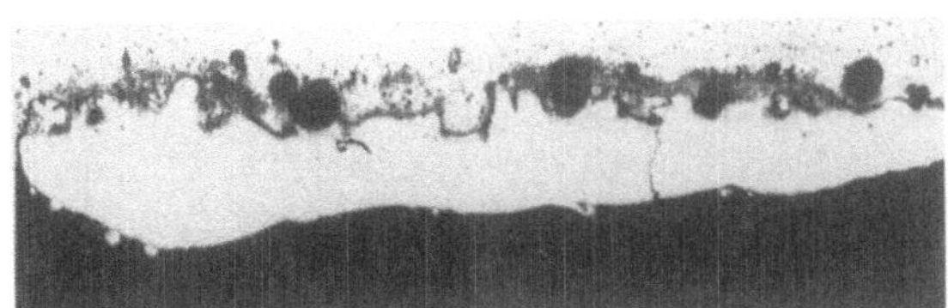

(d) <u>Cracking</u> Cracking occurs due to tensile stresses set up on solidification and also due to differential thermal expansion. It is observed in hard coatings such as iron boron (1000 VHN) but not in coatings of stainless steel (220 VHN) or stellite (580 VHN). Cracking of hard coatings such as the above stated iron **boron** has been successfully eliminated by pre-heating the substrate to 700C.

(e) <u>Surface Finish</u> The degree of overlap of adjacent tracks and the section profile of each track govern the surface finish of the clad layer. With sufficient percenta -ge overlap (usually about 50 – 60%) and with a bead profile similar to fig.2(c), a good surface finish can be obtained as illustrated in fig.6.

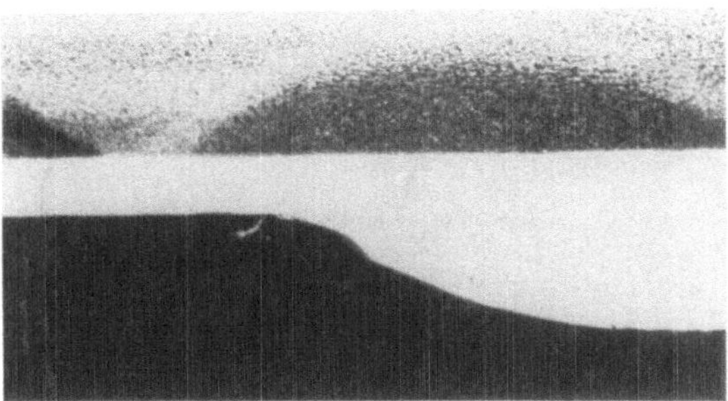

Fig. 6. Section of a stainless steel clad surface on EN3 Steel Etchant – Nital 10%
        Laser power – 1700w. with recycled reflected energy.
        Spot diameter – 5mm
        1st Coating cladding rate 18.8mm^2/s   2nd cross clad coating –
                     speed   12.5mm /s   cladding rate 10mm^2/s
                                                   speed 6.7mm /s
                                                   Powder flow – 0.2g/s

(f) <u>Cladding rates</u> From examination of bead section profiles produced at various pro- cess parameters, the following relationships are found.

$S = a - bW$	Where	S = Cladding Speed mm/s	X =	Transverse Index mm
$K = \exp(T/1.8H)$		W = Bead width mm	K =	Overlap factor (= X/W)
$SH = d$		H = Bead Height mm	C =	Area Coverage rate mm2/5
$C = KWS$		T = Clad thickness mm	a,b,d, =	Constants.

An optimum overlap factor for a particular clad thickness is found by differentiating
$C = f$ (K,T). The corresponding speed is the optimum cladding speed to produce the
particular clad thickness. Such calculations are presented in fig.7.

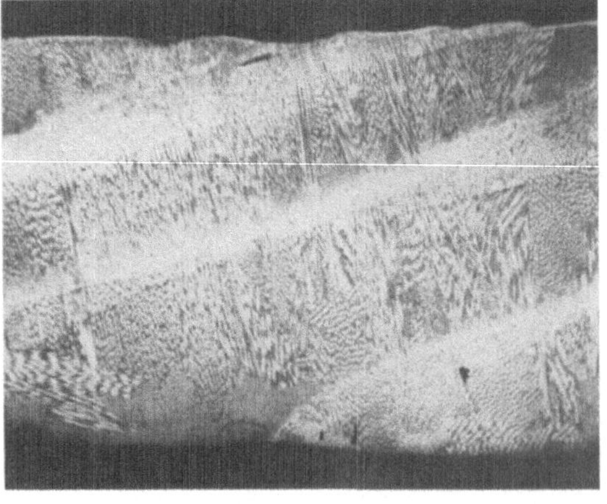

Fig. 7. Cladding rates

(g) <u>Recycling the reflected energy</u>
By using a hemispherical reflectin
device it was possible to recycle
the reflected energy and obtain a
remarkable improvement in the clad
ding rate, Fig.9. Misalligning the
axis of the device to the axis of
the laser beam allowed pre or post
heat treatment using the reflected
energy, Fig.10. By catching this
reflected energy in a calorimeter
a measure of the surface reflectiv
ity was obtained. For example a
shot blasted mild steel surface re
flected 40% of the normally incid
ent CO_2 laser radiation at 90W/mm^2
(h) <u>Process flexibility</u> – variable
composition clad layers can be mad
by this injection process. Similar-
ly clad coatings made from a mixed
powder feed are possible e.g. carb
ide particles in a stainless matri
(i) <u>Homogeneity of clad layer</u> No
macro-segregation within the clad
layer has been observed using EDAX
scans. The clad microstructure,
illustrated in fig.8, shows a fine
dendritic structure as expected
from a fast quench rate.

<u>Conclusions</u> The process has the
advantage of being a non-contact
process, causing minimal thermal
distortion, good metallurgical bonds and microstructure, and by using a reflecting device it can ach
ieve competitive cladding rates. A
further point is the great flexibility of materials which can be
deposited.

Fig. 8. Dendritic microstructure of stainless 316 clad 50X.

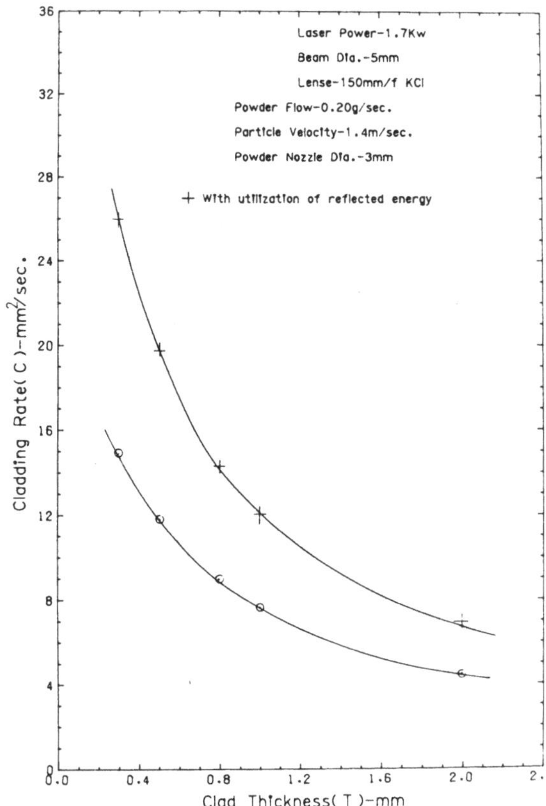

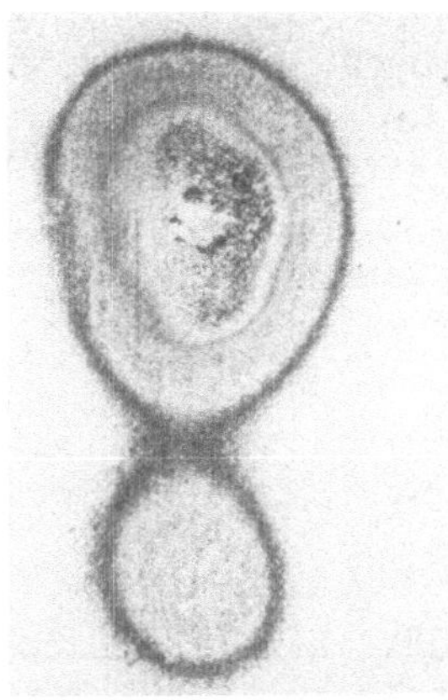

Fig.10. A beam print showing the reflected beam positioned ahead of the main beam for pre-heating.

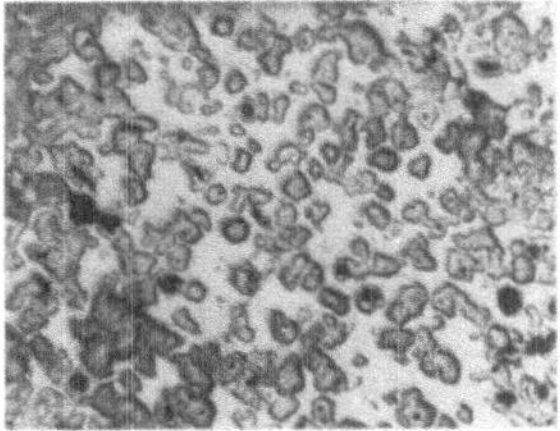

Fig.11. TiC particles in a stainless steel matrix.

Fig.9. Effect of recycling the reflected energy.

REFERNCES:

1. M.McIntyre, "Laser cladding at Rolls Royce" - 2nd Int. Conf. on Applications of Lasers in Materials processing, Jan. 1983, Los Angeles U.S.A.

2. J. Powell, W.M. Steen, "Vibro Laser Cladding" - Conf. Proc., The Metallurgical Soc of AIME Feb 1981.

3. J.D. Ayers "Particulate - TIC - Hardened Steel Surfaces by laser melt injection" Thin solid films 73, 1980, P201.

4. W.M. Steen, "Surface Coating Using a Laser" Paper 3 Conf. Advances in surface coating technology, Welding Int. London Feb. 1978.

5. B.B. Moreton. "Copper-Nickel clad steel for Marine Use" The Metallurgist, May 1981. P. 247 - 252.

6. H. Thielsch, "Stainless Clad Steels" Welding Journal, Welding research suppl. Vol. 31 March 1952. P. 142 - 150.

7. R.S. Zuchowski. E. Garrabrant. "New developments in Plasma arc Weld Surfacing" - Welding Journal (43) Jan. 1964 P. 13-20.

8. M. Riddinhough, "Hardfacing by Welding" Pub. The Louis Cassier Co. Ltd 1949.

Metallabscheidung aus Elektrolyten mit Lasern

A. K. AL-SUFI, H. J. EICHLER, I. OSTWALDT, J. SALK
Optisches Institut der TU Berlin
Straße des 17. Juni 135
1000 Berlin 12

Laser induzierte Metallabscheidung

Eine lokale Metallabscheidung aus Elektrolyten kann ohne äußere Stromversorgung durch Laserlicht bewirkt werden. Damit können z. B. Kontakte
und Leiterbahnen aus Kupfer hergestellt werden. Dazu wird ein Substrat,
auf dem bereits ein dünner Kupferfilm aufgedampft ist, in eine Kupfer-
sulfatlösung getaucht, wobei ohne Lasereinstrahlung auf das Substrat
keine Abscheidung erfolgt. Die Laserstrahlung, hier die eines Argon-
ionenlasers bewirkt eine Temperaturänderung in der Grenzschicht Metall-
Elektrolyt. Diese führt zu einer Verschiebung des Potentials Kupfer/
Kupferionen. Der bestrahlte Bereich wirkt kathodisch, die umliegende
kältere Zone anodisch. Dadurch kann eine außenstromlose Kupferabschei-
dung erfolgen. Die so erzeugten Kupferpunkte und Linien bestehen aus
einigen µm großen grobkörnigen Kristallen. Die Abscheideraten hängen
von der Wärmeleitfähigkeit des Substrats und der die Wärmeleitung be-
einflussenden Kupferfilmdicke ab. Mit Phenolharzpapier als Substrat
wurden bei punktförmigen Verkupferungen Abscheideraten bis 120 nm/s
erreicht (Bild 1). Bei linienförmiger Verkupferung von 1.5 µm Dicke
beträgt die Rastergeschwindigkeit 0.1 nm/s /1,2/.

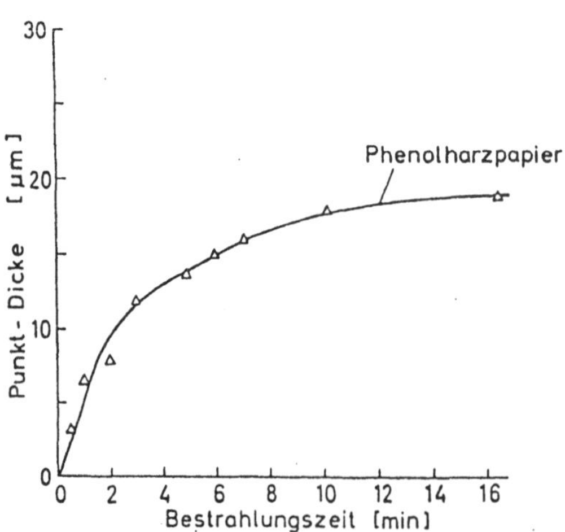

Bild 1. Abhängigkeit der ab-
geschiedenen Punktdicke einer
Kupferschicht von der Bestrahlungszeit. Argonlaserleistung
0.8 W (514,5 nm), Kupferfilmdicke 3 µm.

Eine weitere Anwendung könnte die Erzeugung von Goldkontakten auf
Kupfer sein. Punktförmige μm dicke Goldabscheidungen aus einer Goldcyanidlösung (Aurotron 439 N) benötigen allerdings sehr lange Be-
strahlungszeiten (20 min).

Laser erhöhte Metallabscheidung

Beim Abscheiden mit äußerem Strom bewirkt das Laserlicht eine wesentliche Erhöhung der Abscheiderate. Die Hauptursache für diese Erhöhung
ist eine lokale Konvektion in der Grenzschicht Metall-Elektrolyt. Sie
rührt her von der lokalen Erwärmung durch die Lasereinstrahlung und
zerstört die strombegrenzende Schicht zwischen Metall und Elektrolyt.
Dadurch tritt eine Stromerhöhung bei gleicher äußerer Spannung auf.

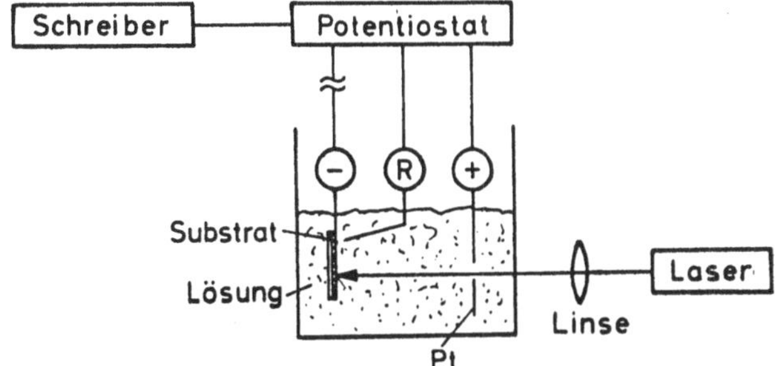

Bild 2. Experimentel-
ler Aufbau

In Bild 2 ist der experimentelle Aufbau zur Untersuchung der lokalen,
Laser erhöhten Abscheidung von Gold auf Kupfer dargestellt. Als Elektrolyt wurde eine Goldcyanidlösung (Aurotron 439 N, Schering) eingesetzt. Bild 3 zeigt die gemessene Goldpunktdicke in Abhängigkeit von
der Außenspannung (ohne Betrieb der Referenzelektrode). Die Laserausgangsleistung war 1 W und die Bestrahlungszeit 180 sec, als Substrat
wurde Phenolharzpapier FR2 benutzt.

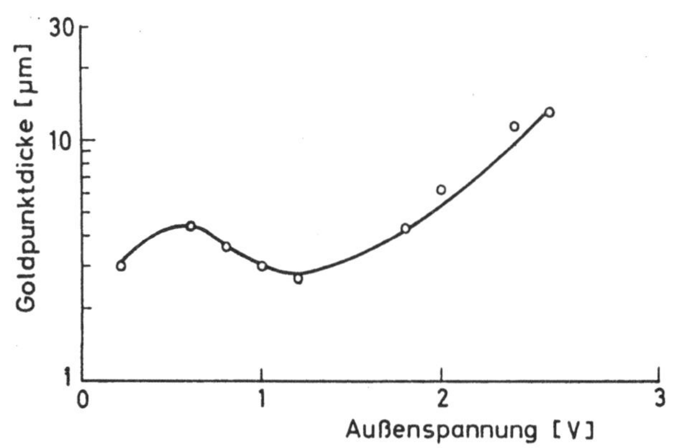

Bild 3. Goldpunktdicke
in Abhängigkeit von
der Außenspannung.

334

In Bild 4 ist die Stromstärke des elektrolytischen Stromes in Abhängigkeit von der Überspannung, gemessen gegen ein Kalomelbezugsystem,
bei verschiedenen Temperaturen des Elektrolyten aufgetragen. Die Kurven zeigen eine über 10^2-fache Stromerhöhung bei Laserbestrahlung des
Substrats. Diese Stromerhöhung bewirkt eine entsprechende erhöhte
Goldabscheidung. Bild 5 zeigt den Erhöhungsfaktor, das ist das Verhältnis der Stromstärke mit Laser zu der ohne Laser, über der Überspannung. Das Maximum dieses Faktors steigt mit steigender Temperatur
der Lösung und verschiebt sich zu kleineren Überspannungen.

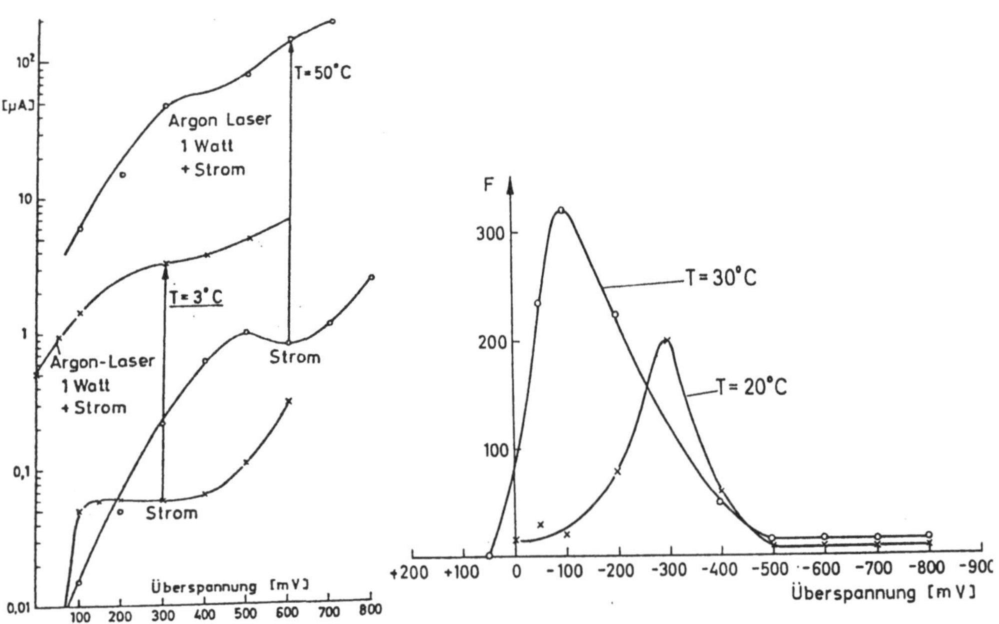

Bild 4. Polarisationskurven
für Au auf Cu/Glasin einer
Goldcanid Lösung (Aurotron
439 N).

Bild 5. Erhöhungsfaktor der Goldabscheidung mit Laser in Abhängigkeit
von der Überspannung.

<u>Literatur</u>

/1/ R. J. von GUTFELD et al, IBM J. Res. Develop. <u>26</u> (1982) 136.

/2/ A. K. AL-SUFI et al, J. Appl. Phys., to be published May 1983.

Laser Melting of Cast Irons

H.W. Bergmann*, B.L. Mordike and T. Bell**

Inst. f. Werkstoffk. und Werkstofft., Techn. Univ. Clausthal, W. Germany
Dept. of Metallurgy and Materials, Univ. Birmingham, UK
*Visiting Scientist, **Wolfson Centre for Surface Engineering

Abstract

Surface hardening of cast irons by laser melting has recently become
of commercial interest.   In this paper the influence of the composition
of the material on the microstructures and the resultant properties is
demonstrated for the major alloying elements.   An attempt is made to
show how the different mechanisms contribute towards the overall hardness.

Introduction and classification of relevant parameters

High power lasers have reached a stage in their development where
they have become widely accepted tools for material processing.   A lot
of data is available in the literature from detailed studies of the
microstructures and hardness.   Although there is a certain merit in
classifying the different microstructures in this manner, little has been
said on their mode of formation.   It would be useful, therefore, to
analyse the relevant physical and metallurgical mechanisms occurring
during laser treatment and the parameters by which they are determined.

Laser surface treatment divides into simple hardening, remelting
and surface alloying.   In all cases the local heat input and structural
changes result in internal stresses in the treated area.   In Table 1
the major parameters, heating rate, quenching rate, solidification rate
and local composition changes are coupled with the active hardening
mechanisms via diffusion time, undercooling and changes in alloy thermo-
dynamics.   For example, high quenching rates lead to higher undercooling
of the melt and result,due to higher nucleation frequency,in grain
refinement.   On the other hand, increased solidification velocities cause
supersaturation, both resulting finally in an increase in hardness.

The parameters which influence the structural changes that occur in
laser surface melting can be subdivided into (a) those depending on laser
variables and (b) those fixed by the material chosen.

Table 1 <u>Mechanisms of surface hardening by laser treatment</u>

<u>Laser hardening</u>	<u>Laser melting</u>	<u>Laser alloying</u>
Local heat input and structural changes always result in internal stresses		
high quenching rates result in: - reduction of diffusion time causing: changes in alloy hardening, transformation hardening and precipitation hardening	high quenching rates result in: - undercooling of the melt causing: changes in grain refinement, formation of metastable phases, transformation kinetics	local changes in composition result in: - changed alloy thermodynamics causing changes in precipitation hardening, grain refinement (e.g., via ternary eutectics) dispersion hardening
high heating rates result in: - incomplete homogenisation causing: formation of metastable phases (e.g., non-stoichiometric carbides, white areas)	high solidification rates result in: - reduction of diffusion time - high supersaturation causing changes in alloy hardening, transformation kinetics - different morphologies of the microstructure	

In Table 2 it can be seen that the parameters, quenching rate, ε, solidification rate, R, and temperature gradient, G, (which are related by $\varepsilon = R \times G$) are functions of mainly one laser variable for the same material and equivalent melting depths in a steady state process.   The choice of the initial composition not only fixes the thermal properties like heat diffusivity, etc., but also the thermodynamic parameters, e.g., the free energies per unit volume in the liquid and solid states which are equal for a solid phase, j, of the temperature $T_o^{\ j}$.

The change in solidification behaviour for a given composition, due to the parameters dependent on laser variables, is demonstrated elsewhere and discussed in the papers of Barton and Fritsch in this volume (1). The correlation between solidification morphology, composition and solidification rate can be seen in Fig. 1a-c from Jones (2).   In eutectic systems the sequence of microstructures formed with increasing solidification rates is different for continuous intersecting or non-intersecting T_o-curves.   In the first case, which may be represented by

Table 2   <u>Parameters infuencing hardening mechanisms in laser melting</u>

Determined by laser variables	Determined by composition
Quenching rate $\varepsilon = \dfrac{\Delta T}{\Delta t}$ Solidification rate $R = \dfrac{\Delta X}{\Delta t}$ Temperature gradient $G = \dfrac{\Delta T}{\Delta x}$ $\boxed{\varepsilon = R \times G}$ for the same $\quad$) $\varepsilon = f(\frac{P}{A})$ composition and) $\qquad\qquad$) $R = f(V)$ melting depth $\quad$) and steady state) $G = f(\frac{P}{A}, V)$ conditions $\qquad$) power density $\quad \dfrac{P}{A}$ traverse speed $\quad V$	Initial composition $\quad C^*$ Temperature of equal enthalpies in solid and liquid states $\quad T_o^{\,j}(C^*)$ Ratio of solidus/ liquidus temperature $\quad \dfrac{T_S}{T_L}(C^*)$ heat diffusivity $h^j(C^*,T) =$ $\dfrac{\lambda^j(C^*,T)}{C_p^{\,j}(C^*,T)\cdot\rho(C^*,T)}$ thermal conductivity $\quad \lambda^j(C^*,T)$ specific heat $\quad C_p^{\,j}(C^*,T)$ density $\quad \rho(C^*,T)$ molar heat of fusion $\quad h_{fm}^{\,j}(C^*,T)$ diffusion coefficient $\quad D_i^{\,j}(C^*,T)$ absorption coefficient $\quad A^j(C^*,T,\lambda_w)$ λ_w wave length of the light i indicating the i-th element j indicating the j-th phase

the system Ag-Cu, finally, a homogeneous phase can be found over the complete range. For a given composition, C^*, the sequence is constitutional homogeneous solidification, eutectic solidification, dendritic formation, and velocity determined homogeneous solidification.

The second type represents the Fe-C system, exhibiting an asymmetric eutectic. In those cases where the field of the γ-Fe is restricted, e.g., in Fe-B or in ternary Fe-C systems containing B, Si, P or strong carbide formers such as Ti, Nb, Mo, W, it is possible to obtain glassy metals above a certain solidification rate. In laser quenched materials it is possible to obtain different solidification rates depending on the temperature profile generated. In Fig. 2 a sequence of different microstructures is shown. Next to the substrate a

homogeneous γ-crystal had solidified epitaxially onto the substrate.
This is followed by a region of eutectic crystallisation. Near the
surface, again, γ-crystals are formed changing from dendritic to planar
solidification. Obviously this causes dramatic variations in
hardness, (3). For further understanding it will be necessary to
produce complete maps of the different solidification morphologies, due
to solidification and quenching rates, in order to obtain optimised and
predictable surface properties.

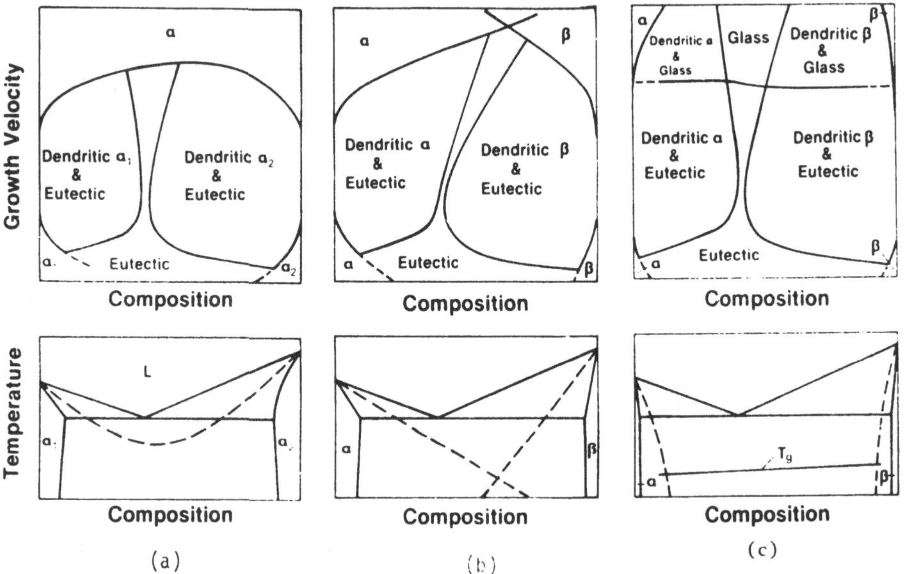

Fig. 1   Change in solidification morphology with compositional and
growth velocity for different T_o-curves in eutectic systems,
from (2).

The influence of the laser variables has been described
elsewhere (4). In this paper the influence of the composition is
demonstrated using interaction times of $\sim$0.01 sec and power densities of
$\sim$8.10^4 $\frac{W}{cm^2}$.

Fig. 2 Microstructure of laser melted cast iron with 3%C, 2%Si, 0.25%Ti, (t_{int} = 1.0 sec, $P/_A$ = 10^6 $w/_{cm^2}$).

Laser melted steels

Laser melting of non-transformable steels, e.g., those with pure ferritic or austenitic structures, does not essentially effect the hardness, see Table 3. It produces homogeneous distribution of segregates and impurities (5). Transformable steels exhibit different microstructures, see Table 4, depending on the carbon content. Although a hardening effect is visible, below 0.4%C laser melting of non-alloyed steels is an ineffective heat treatment process, as δ-ferrite is formed on solidification and retained in the quenched state. The presence of δ-ferrite causes low hardness values which cannot be removed by annealing. Between 0.4 and 0.9%C the hardness variations in the as-quenched state can be balanced by further heat treatments. In

340

Table 3 Microstructural changes in laser melted non-transformable steels

Initial microstructure	Modifications to microstructure	Hardness increase
austenite ferrite	finer distribution of segregates a) " b) in the case of steels containing high melting point precipitates, e.g., TiC, transition to transformable steels is possible	negligible " possible

Table 4 Influence of C-content on laser melted non-alloyed steels

C wt.%	Solidification in the form of	As-quenched microstructure	Hardness/HV	
			As-quenched	After subsequent annealing
0.1<	δ-ferrite	δ-ferrite	no hardening effect	
0.1-0.4	δ-ferrite, coarse columnar crystals containing dendritic segregations	δ-ferrite bainite martensite	200-400 varying	200-900 varying
0.4-0.9	coarse γ-crystals containing dendritic segregations	martensite retained austenite	400-700 varying	900-1000
0.9-1.4	coarse γ-crystals	retained austenite martensite	500-600 varying	900

Table 5 Influence of metallic alloying elements on laser melted transformable steels

γ-extenders	γ-restrictors
Ni, Co, Mn Increase of hardenability Mn increases the tendency of oxide formation	Cr, Mo, Ti, W Cr increases the hardenability and the tendency to segregate Mo, Ti, W cause formation of local ternary carbide eutectics Increase the tendency of oxide formation Tendency of δ-ferrite formation

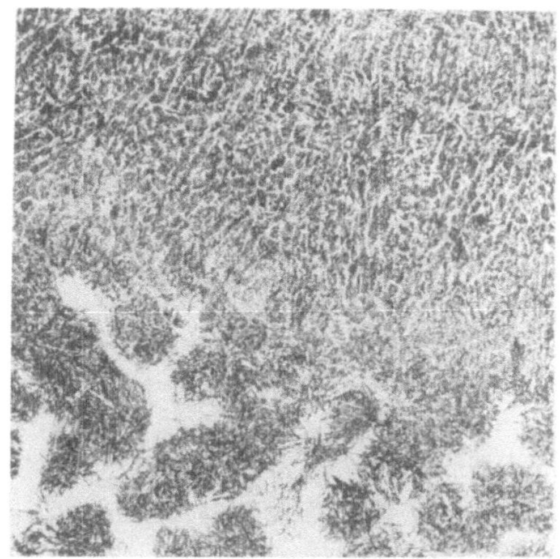

Fig. 3   Transition region of a surface melted low alloyed steel,
         containing 0.74%C.

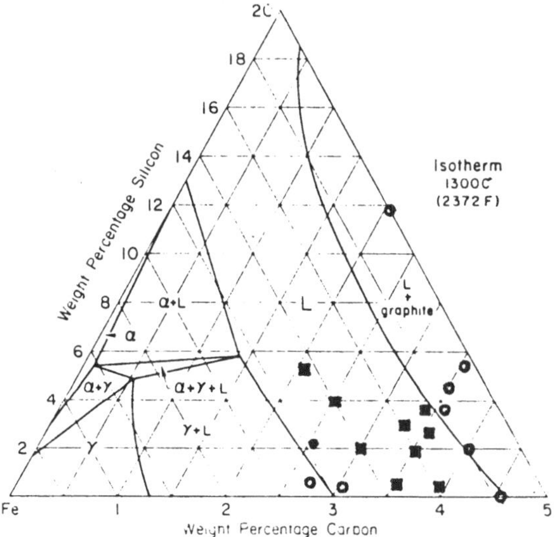

Fig. 4   Isothermal cut of the ternary Fe-Si-C phase diagram.   The
         symbols indicate the compositions investigated.   Retained
         graphite was observed for ■ and not for o.

this concentration range the length of the martensite is limited by the
dendritic type of segregation in the γ-crystals, see Fig. 3.   This form
of segregation was not found at C-contents between 0.9 and 1.4%C and
hence, a coarse martensite is obtained.   Alloying elements in transformable steels can lead to a higher hardenability, e.g., in the case of
Ni, Co, Mn and Cr, see Table 5.   Materials containing strong carbide
formers exhibit local areas of ternary carbide eutectics and are
susceptible to δ-ferrite formation.

Laser melted cast irons

Cast irons are ideal substrates for the laser melting process as
they contain a large amount of carbon in the form of graphite.   In the
concentration range between 2 and 4.4%C epitaxial growth of γ-dendrites
onto the substrate is observed, together with interdendritic eutectic
or Fe_3C monolites.   Further increases in the C-content led to the
formation of both primary cementite and austenite dendrites.   In the
range of 2-3%C and above 4.5%C, graphite was observed in addition to
γ-Fe and Fe_3C, in a finely distributed particle form.   This phenomenon
of retained graphite is often found in Fe-Si-C alloys.   In Fig. 4 the
ternary phase diagram at $1300^{\circ}C$ is shown (6).   The different symbols
indicate whether or not retained graphite was obtained after laser
melting.   The complete solution of carbon correlates well with the
region of low melting point.   A number of samples of microstructures
with different degrees of graphite dissolution are shown in Fig. 5a-c.
The amount of Si has a further influence on the properties of cast irons.
This is particularly relevent in the selection of laser treatment
parameters.   By restricting the γ-field it is possible to obtain
bcc-Fe between room temperature and the melting points, (6), if the
diffusion of carbon is limited by rapid heating.   This effects the
temperature gradient generated in laser surface melting (7).   In
addition, Si is one of the elements which considerably reduces the
thermal conductivity of Fe.   Both influences are demonstrated in Fig. 6.

The reduction in thermal conductivity generally avoids expanded
heat affected zones.   The generation of a steep temperature gradient,
however, may lead to evaporation of the surface.   The effect of Si on
the transformation and melting temperatures shifts the fields of
different thermal conductivity arising from phase transformations.

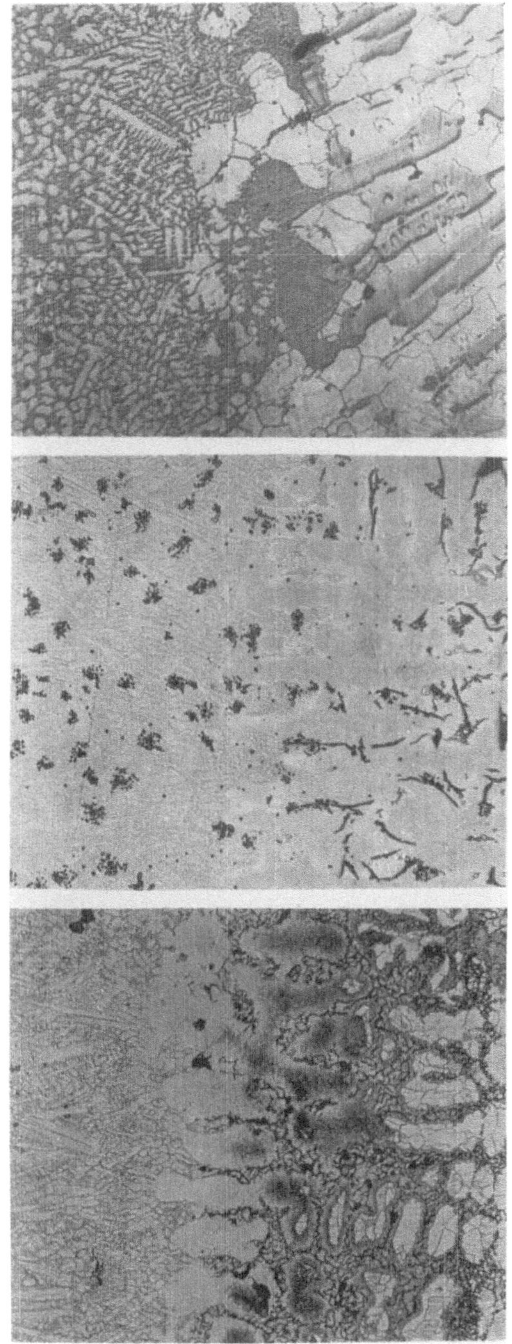

Fig. 5   Laser surface melted cast irons: (a) 2.1%C, 2.0%Si;
(b) 3.0%C, 2.0%Si; (c) 4.0%C 2%Si.

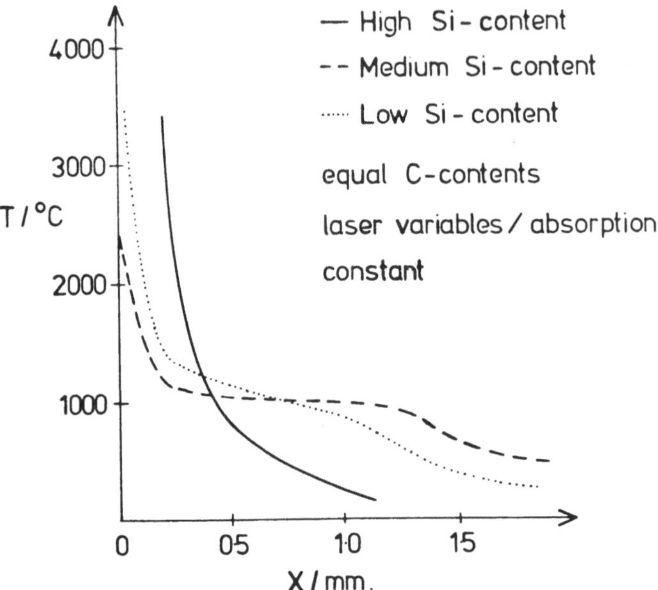

Fig. 6   Influence of Si on the temperature profiles generated in laser melting.

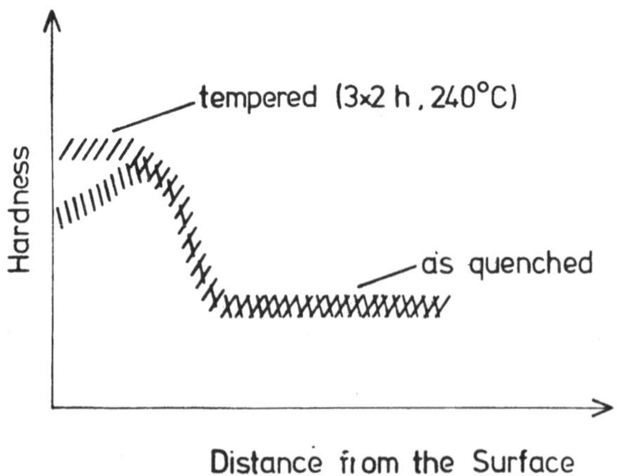

Fig. 7 Typical hardness profiles for laser melted cast irons.

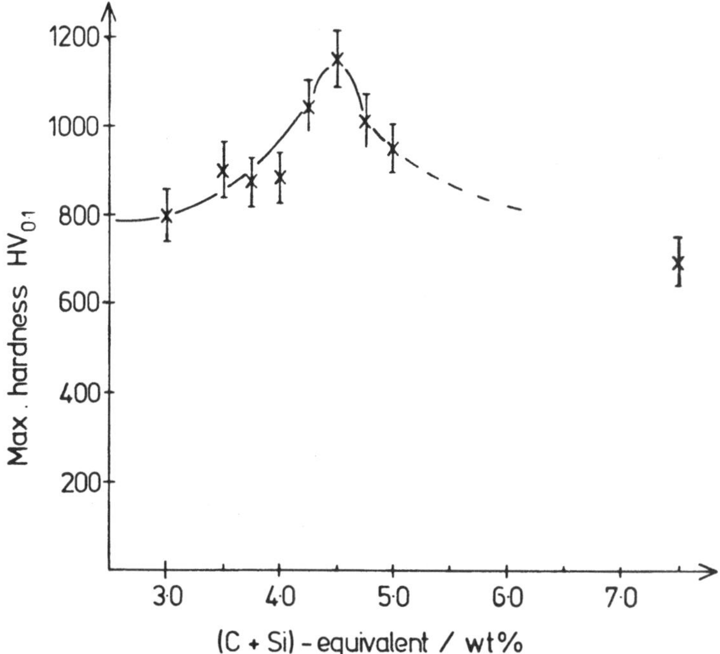

Fig. 8   Optimised hardness values of laser melted and subsequently
annealed cast irons as a function of (C-Si)-equivalent.

The hardness profiles obtained after laser surface melting, although
differing in detail due to the special parameters, generally are of the
shape shown in Fig. 7.   In the as quenched state the hardness drops
from the hardened to the melted area due to the amount of retained
austenite.   This loss in hardness can be recovered by a suitable
annealing treatment.   For the compositions in Fig. 4 the optimum
hardness values after such treatments are correlated in Fig. 8 with the
amount of the carbon-silicon equivalent (C + 0.3 Si).   The higher values
in the eutectic region can be explained by the reduction of grain size
due to undercooling.   Hence, for the chosen laser processing variables
not all carbon was dissolved, the maximum observed might be less
pronounced for other treatment parameters.   The necessity for a post-
heat treatment operation is, to a certain extent, a disadvantage in
commercial applications.   Changing to the quarternary system, Fe-C-Si-P,
one finds that above 0.2%P the hardness values in the as-quenched state
have already reached its maximum, and a continuous profile is obtained,

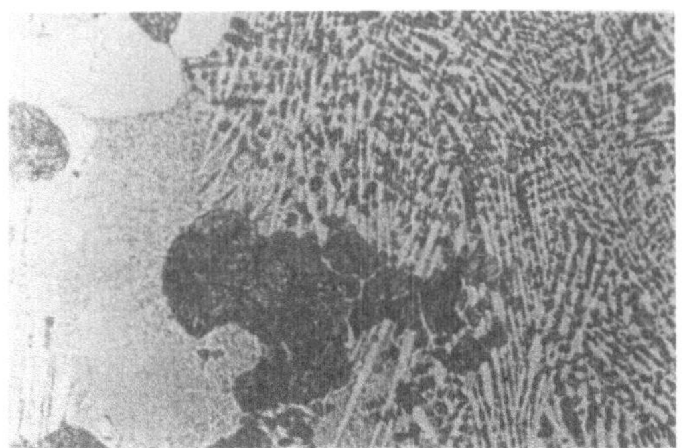

Fig. 9   Effect of increased P-content on the microstructure of laser
          melted cast iron.

Table 6   Hardness values of different cast irons

	Alloy:				Hardness:	
I	3.5 C	2.0 Si	1.0 P		1100-1200	as quenched
	3.5 C	0.5 Si	2.0 P		1100-1200	as quenched
II	2.5 C	1.7 Si	33.0 Ni		450-550	
	3.0 C	2.4 Si	6.0 Mn	12.0 Cr	450-550	
III	2.1 C	5.0 Si			600-650	
	2.7 C	0.3 Si		28.0 Cr	600-650	

see Table 6, (5).   This effect is related to a fine grained microstructure consisting of both primary γ-Fe and Fe_3C dendrites, and the
quarternary eutectic, see Fig. 9.

Alloy cast irons can be hardened by laser melting.   This is shown
in Table 6 for a number of representative compositions, even if they
consist of pure austenitic or ferritic structures.   This hardness
increase results from the number of carbides formed on rapid cooling.
Obviously, it is interesting to understand the contribution of different

mechanisms to the final hardness in rapidly quenched Fe-alloys.
Unfortunately, the different contributions are not additive and the
mechanisms interact with each other.   The following attempt to sub-
divide these different contributions can, therefore, only be a
preliminary approach.   The hardness of laser melted, low carbon steel
components is mainly determined by the relative amounts of δ-ferrite,
martensite and retained austenite present in the microstructure.
An advantage over conventional transformation hardening can be achieved
through the production of a pure martensitic structure free from
ε-carbide precipitates.   Iron alloys containing greater amounts of
carbon are thus much more suited to laser surface melting.   In the case
of plain ferritic and austenitic cast irons the hardness increase is
about 200 ✦ 300 HV.   This is due to carbide formation and the volume
fraction of carbides in the microstructure.   For a fixed volume fraction
the refinement of the carbides, formed on solidification using high
quenching rates, can cause a further increase of about 250 HV through a
reduction in the grain size from 10 μm to 0.1 μm (4).   Optimising both
processes one is able to obtain hardness values of about 800 HV in non-
transformable cast irons.   The actual hardness can then be further
extended if the alloying elements form carbides of higher hardness than
Fe_3C.

The transformation hardening of cast irons and high carbon steels
leads to average hardness values of 800 ✦ 1000 HV.   The formation of
non-stoichiometric carbides may cause local variations in the hardness
values up to 1100 ✦ 1200 HV.   In laser surface melting and subsequent
rapid quenching the grain refinement can modify the transformation
kinetics (4).   The appearance of a fine, twinned martensite is
responsible for an additional gain in hardness up to a final value of
1200 ✦ 1400 HV.

Practical aspects

In practice increasing the hardness itself is not the sole aim of
materials processing, even though this can dramatically improve the
wear resistance.   The abrasive wear properties of cast irons after laser
treatment are compared with untreated material in Fig. 10, taken from the
paper of Bell and co-workers (8).   The advantages of laser treatment
shown are in agreement with the results of other scientists (9).   It can
be seen (10,11) that the increase in hardness of rapidly solidified cast

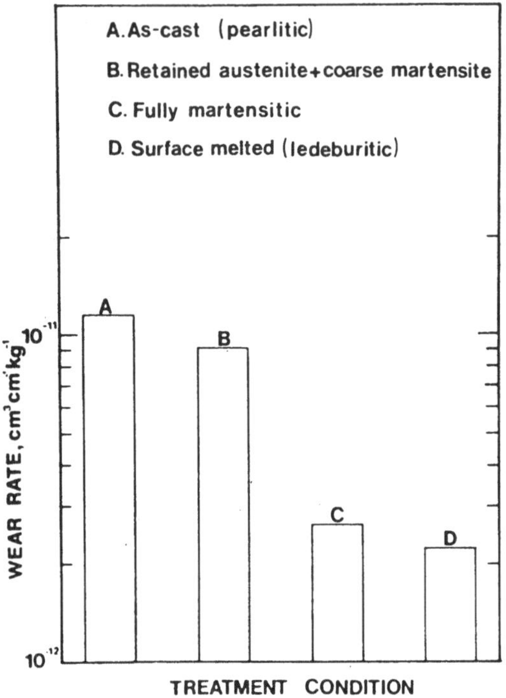

Fig. 10   Abrasive wear resistance of cast iron for different treatments,from (8).

irons does not necessarily mean a decrease in toughness if the increase
is due to grain refinement.   For example, rapidly solidified Fe, 3%C,
1.5%Cr alloys exhibit a tensile strength of about 2700 $\frac{N}{cm^2}$ and an
elongation to fracture of about 25%.   Unfortunately, upto the present
time, no reliable data on the fatigue properties is available.   This
shortage arises from the problem of producing completely crack free
surfaces on laser melting.   On a laboratory scale, it is possible to
avoid cracks by keeping the specimen at a suitably elevated temperature.
It will be the work of industrial firms to offer suitable equipment,
which allows free manipulation of the component at a defined elevated
temperature.

References

(1) H.U. Fritsch and G.J. Barton, this volume

(2) H. Jones, Rapid solidification of metals and alloys, Northway House,
    London, 1982

(3) H.W. Bergmann, B.L. Mordike and H.U. Fritsch, Z. f. Werkstofftechnik, in print

(4) H.W. Bergmann, G.J. Barton and J. Betz, ibid

(5) H.W. Bergmann and B.L. Mordike, ibid

(6) Metals Handbook, $\underline{8}$, A.S.M., Metals Park, Ohio, 1961

(7) H.W. Bergmann and H.U. Fritsch, Z. f. Werkstofftechnik, in print

(8) D.N.H. Trafford, T. Bell, J.H.P.C. Megaw and A.S. Bransden, Proc. Conf. Heat Treatment, 1981, The Metals Society, London

(9) A. Blarasin, S. Corcoruto, A. Belmondo and D. Bacci, 2nd Int. Conf. Heat Treatment, Florence, 1982

(10) G. Frommeyer, H.W. Bergmann and B.L. Mordike, in preparation

(11) B.L. Mordike and H.W. Bergmann, Mat. Res. Soc. Symp. Proc. Vol. 8, "Rapid solidification amorphous and crystalline alloys", Boston, 1982

Zusammenhänge zwischen Eigenschaften von Versuchsparametern beim Laseroberflächenumschmelzen von Eisenwerkstoffen

H.U. FRITSCH und G.J. BARTON
Institut für Werkstoffkunde und Werkstofftechnik,
Technische Universität Clausthal, D-3392 Clausthal-Zellerfeld

Einleitung

In der neueren Literatur findet man häufig Angaben über laserumge-
schmolzene Oberflächen sowie deren Eigenschaften mit unterschiedl-
lichen Gefügen. Meistens findet man bei niedriglegierten Werkstoffen
Angaben über die Ausbildung von Korngefügen, bei höher legierten
Werkstoffen betrifft dies martensitische und zellulare Gefüge und bei
hochlegierten, ledeburitischen Werkstoffen ist dies meist ein Gefüge,
das beim Umschmelzen durch dendritisches Wachstum entstanden ist /1-5/.
Selten jedoch findet man genauere Angaben über die Abschreckbedingungen sowie einen Zusammenhang der erzielten Gefügeparamter mit den
Versuchsbedingungen. Dies betrifft auch den Vergleich zwischen den verschiedenen Versuchsparametern, hauptsächlich Laserleistung und Leistungsdichte. Ziel dieser Arbeit ist es bei unterschiedlichen Um-
schmelzanlagen einen allgemeinen Zusammenhang zwischen den Abschreckbedingungen, dem Gefüge und den Versuchsparametern zu ermitteln.

Ergebnisse und Diskussion

Durch die schnelle Erstarrung beim Oberflächenumschmelzen ergibt sich
im allgemeinen ein feineres Gefüge, das sich meistens auch durch eine
Härtesteigerung auszeichnet. Allgemein wird hierdurch eine Verbesserung der Einsatzeigenschaften wie z.B. Verschleißbeständigkeit und
Korrosionsbeständigkeit erwartet. Beim Laserumschmelzen von ledeburitischen Stählen in einem größerem Bereich von Abschreckbedingungen
bildet sich im allgemeinen ein Gefüge aus γ-Dendriten und einem da-
zwischenliegendem Gefüge aus Cr-Karbiden das in eutektischer Form aus
der Restschmelze erstarrt ist. Im Schliffbild des laserumgeschmolzenen
Stahls X210Cr13 findet man eine typische dendritische Gefügeausbildung.
Man erkennt eine Ausrichtung der γ-Dendriten allgemein in Richtung
des maximalen Wärmeabflusses, wobei die Dendriten leicht in Vorschubrichtung des Laserstrahls gekippt sind. In der Oberfläche stellt

man dagegen eine eher statistische Verteilung der Dendritenorientierung
fest. Diese Dendriten haben sehr gut ausgebildete sekundäre Dendriten-
arme, wodurch sich die Möglichkeit ergibt, durch Messung der sekundären
Dendritenarmabstände λ und Anwendung der von Jones /6/ zuerst für Al-
Legierungen und danach auch auf Fe-Legierungen ausgeweiteten Formel

$$\lambda \, \varepsilon^A = B$$

die Abschreckgeschwindigkeit ε abzuschätzen. Für unsere Untersuchungen wurden die Konstanten A = 0,4 und B = 60 μm benutzt. Mit
diesen Konstanten ergibt sich der im Bild 2 dargestellte Zusammenhang
zwischen sekundärem Dendritenarmabstand λ und Abschreckgeschwindigkeit ε. Die sich ergebenden Abschreckgeschwindigkeiten sind als
untere Grenze zu verstehen, da andere Autoren wie z. B. Suzuki /7/
leicht abweichende Konstanten vorschlugen, wodurch sich die Abschreckgeschwindigkeiten vergrößern können.
Bild 1a, b zeigt zwei REM-Aufnahmen aus der Mitte zweier umgeschmol-
zener Laserraupen für den X210Cr13. Die erkennbaren sekundären Dendritenarmabstände weisen ein Verhältnis von ca. 1 : 2 auf und resultieren aus einer Veränderung der Vorschubgeschwindigkeit beim
Schmelzen von 4 m/min auf 0,5 m/min. Betrachtet man eine Raupe, so

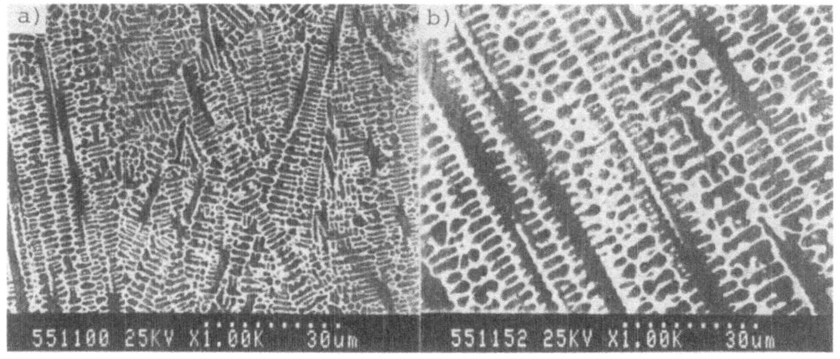

Bild 1. REM-Aufnahmen des X210Cr13, umgeschmolzen mit einem 1,4KW-
Laser bei konstanter Leistungsdichte und einer Vorschubgeschwindigkeit von   a) 4m/min   b) 0,5 m/min

findet man in der Schmelzzone unterschiedliche Abschreckgeschwindigkeiten und zwar die höchsten am Übergang zum Substrat und die niedrigsten an der Raupenoberfläche (Bild 2). Diese höheren Abschreckgeschwindigkeiten am Übergang zum Substrat im Verhältnis zur Raupen-
mitte und -oberfläche führen zu einer wesentlich höheren Übersättigung an Legierungselementen in dieser Zone, was sich auch entsprechend in den Eigenschaften auswirkt. Die an unterschiedlichen Anlagen durchgeführten Umschmelzversuche haben ergeben, daß die Abschreckgeschwindigkeit mit der Vorschubgeschwindigkeit anwächst (Bild 3).

352

Bei gleichen Vorschubgeschwindigkeiten ist ein Ansteigen der Abschreckgeschwindigkeit nicht mit zunehmender Leistung sondern mit zunehmender Leistungsdichte bei gleichzeitiger Verringerung der Belichtungszeit zu beobachten. Bedingt duch die oft gewünschten großen

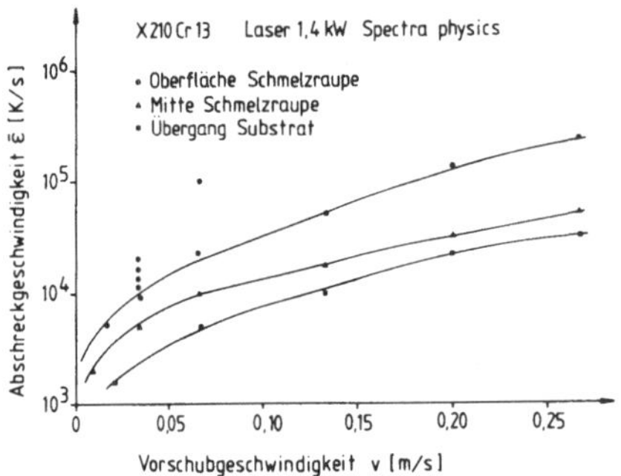

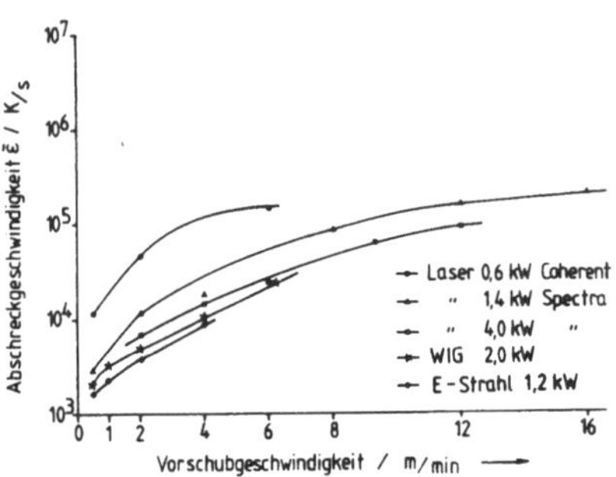

Bild 2. Ermittelte Abschreckgeschwindigkeiten als Funktion der Vorschubgeschwindigkeit in unterschiedlicher Tiefe der Schmelzraupe eines X210Cr13.

Bild 3. Abschreckgeschwindigkeiten, bestimmt aus den sekundären Dendritenarmabständen, als Funktion der Vorschubgeschwindigkeit für unterschiedliche Anlagen.

Einschmelztiefen ist die Erreichung hoher Abschreckgeschwindigkeiten nicht in beliebiger Größe möglich. Weitere Parameter, die die Abschreckgeschwindigkeit beeinflussen können sind der Tiefschweißeffekt, hohe Wärmeleitfähigkeit, α-Gefüge im Ausgangszustand sowie die Belichtungszeit und physikalische Eigenschaften des Werkstoffes wie Schmelzpunkt, spezifische Wärme u.a. Die erzielbare Einschmelztiefe E bei gleicher Vorschubgeschwindigkeit und Strahldurchmesser (Belichtungszeit) kann dagegen durch die Laserleistung bestimmt werden. Einerseits wird dabei der Wert E durch den Beginn der Abdampfung des Werkstoffes begrenzt, andererseits durch die zur Verfügung stehende Laserleistung (Bild 4). Betrachtet man die maximal erreichbaren Einschmelztiefen als Funktion der Vorschubgeschwindigkeit bei verschiedenen Leistungsdichten, so ergibt sich eine annähernd monotone Abhängigkeit beider Größen. Betrachtet man jedoch diese Zusammenhänge aus anderer Sicht, so ergibt sich, daß das pro Zeiteinheit umgeschmolzene Volumen bei kleinen Vorschubgeschwindigkeiten ein Minimum aufweist. Dieser Effekt ist mit der α-γ-Umwandlung und dem Beginn der Abdampfung verbunden. Er findet dann statt, wenn der Wärmetransport durch Wärmeleitung in der Vorschubrichtung größer ist als die Vorschubgeschwindigkeit des Strahls. Die schlechtere Wärmeleitfähigkeit der γ-Phase bewirkt bei verschiedenen Leistungsdichten einen Wärmestau, der zur stärkeren Verdampfung des Materials führt (Bild 5). Dieser Effekt ist auch bei anderen umwandlungsfähigen Stählen beobachtet worden (Bild 6).

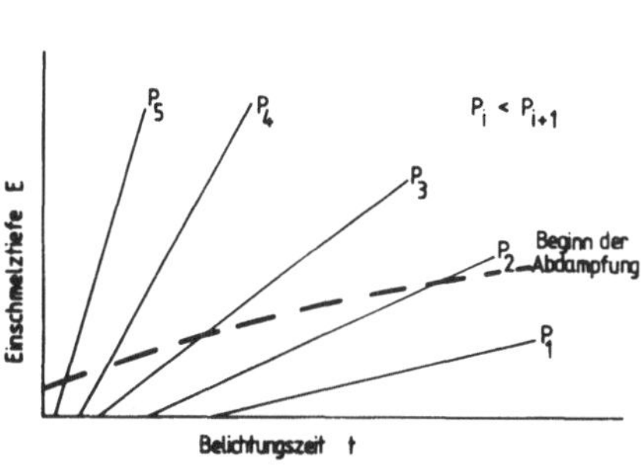

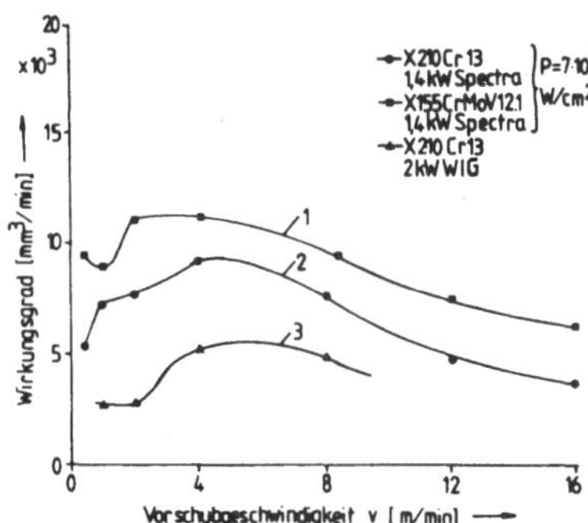

Bild 4. Schematische Darstellung der berechneten Einschmelztiefe E als Funktion der Belichtungszeit t und dem Parameter Leistungsdichte P_i

Bild 5. Erreichter Wirkungsgrad (umgeschmolzenes Volumen pro Zeiteinheit) als Funktion der Vorschubgeschwindigkeit bei verschiedenen Umschmelzverfahren für die Stähle X210Cr13 und X155CrMoV12 1.

In den Bildern 7 und 8 sind die Härteprofile der laserumgeschmolzenen Oberflächenschichten als Funktion der Vorschubgeschwindigkeiten dargestellt. Bei einer Laserleistung von ca. 1,4 kW werden an einem hochlegierten X210Cr13 maximale Einschmelztiefen bis zu 3 mm erreicht. Unabhängig von der Vorschubgeschwindigkeit wird beim Übergang zum Substrat

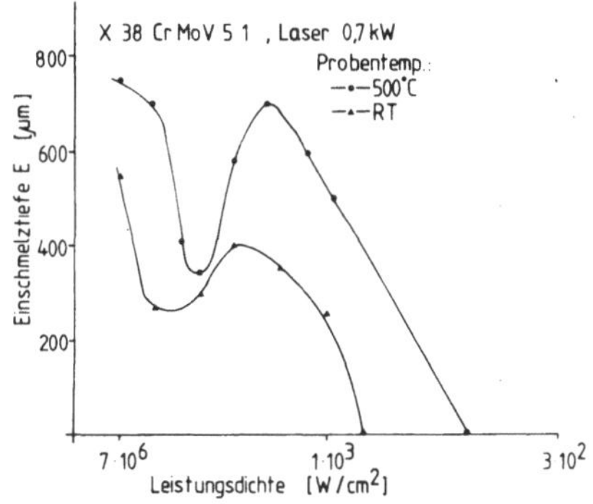

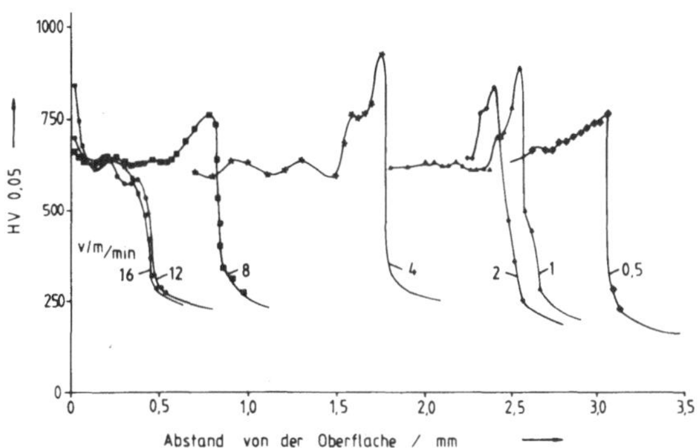

Bild 6. Maximal erreichbare Einschmelztiefe bei einem 0,6KW-Laser als Funktion der Leistungsdichte für unterschiedliche Probentemperaturen (X38CrMoV5 1).

Bild 7. Härteprofile eines X155CrMoV12 1 nach dem Umschmelzen mit einem Laser bei einer Leistungsdichte von 7 10⁵ W/cm².

infolge hoher Übersättigung die maximale Härte von ca. 800 bis 900 HV erreicht, die dann zur Oberfläche einheitlich auf ca. 630 HV abfällt. Die Härtesteigerungsmöglichkeiten des Laserumschmelzens werden erst durch Neuhärtung ausgeschöpft. Das Härtemaximum wird dadurch relativ

354

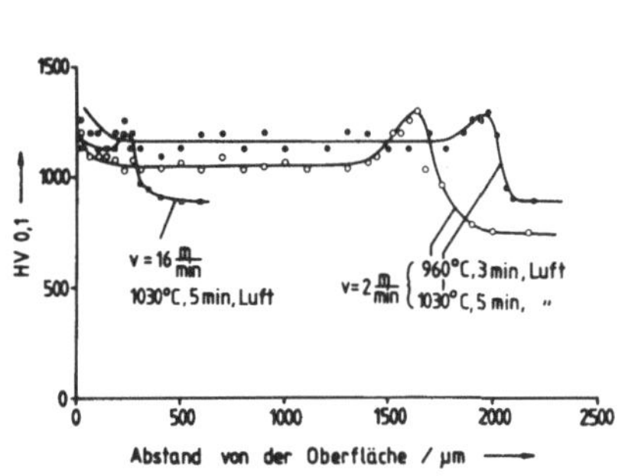

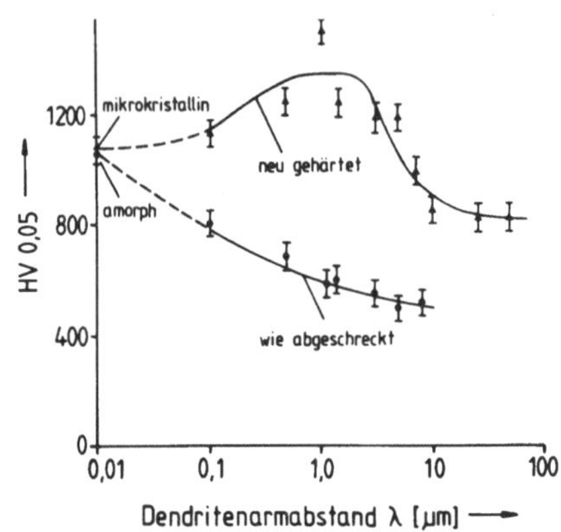

Bild 8. Härteprofile wie in Bild 8 nach einer Neuhärtung.

Bild 9. Erreichbare Härtewerte als Funktion des Dendritenarmabstandes im abgeschreckten und neugehärteten Zustand.

abgebaut und die mittlere Raupenhärte bis auf ca. 1100 HV gesteigert. Ein vollkommener Abbau des Härtemaximums am Übergang zum Substrat ist nur bedingt möglich, da das extrem feine Gefüge stets durch Diffusion eine hohe Löslichkeit an Legierungselementen in den γ-Dendriten zur Folge hat. Die größte Härte wird beim X210Cr13 Stahl nach dem Neuhärten bei einem sekundären Dendritenarmabstand von ca. 1 µm beobachtet. Eine weitere Minderung des Dendritenarmabstandes durch noch größere Abschreckgeschwindigkeiten führt dagegen zu einem Härteabfall (Bild 9). Die Ursachen dafür sind jedoch im Gefüge zu suchen. Bei dem untersuchten X210Cr13 Stahl wird in den so feinen γ-Dendriten eine nur teilweise Umwandlung in Martensit beobachtet. Zusammenfassend läßt sich sagen, daß diese dargestellten Zusammenhänge eine gezielte Einstellung von Eigenschaften ermöglichen, die durch die Abschreckbedingungen beeinflußt werden. Diese Zusammenhänge lassen sich mit geringen Modifikationen auch auf andere ledeburitische, dendritisch erstarrende Stähle übertragen. Inwieweit Aussagen über in nicht dendritischer Form erstarrende Stähle gemacht werden können, wird derzeit geprüft.

Literatur

/1/ H.W. BERGMANN, B.L. MORDIKE, Z. Metallkunde, 71 (1980), 658
/2/ W.M. STEEN, C. COURTNEY, Metals Technology, 1979, 456
/3/ E. RAMOUS et al., Proc. of Conf. on Metallic Glasses, Budapest, 1980, 235
/4/ P.A. MOLIAN, W.E. WOOD, Scripta Met., Vol. 17, 1983, 431
/5/ M. CARBUCICCHIO et al., J. Mat. Sci., 18 (1983), 1543
/6/ H. JONES, Rapid Solidification of Metals and Alloys, Northway House, 1982
/7/ A. SUZUKI et al., Nippon Kinzoku Gakkai-Si, 32, (1968), 1301

Conductivity Measurements on Absorption of Semiconductors under Intense Laser Excitation

E.W. KREUTZ, E. BEYER, H.G. TREUSCH, and W. ZIMMER
Institut für Angewandte Physik, Technische Hochschule Darmstadt
Schloßgartenstraße 7, D 6100 Darmstadt

1. Introduction

Energy can be transferred to solids thermally or electronically through electron
beams or laser beams [1]. The macroscopic melting model [1,2], the plasma annealing-
bond charge weakening model [1,3], and the microstructural double bifurcation model
[4] assumingly describe the different physical mechanisms for the interaction of the
intense radiation with the solid during and after the abrupt energy deposition. Additional phenomena associated with laser materials processing are the intensity-dependent absorption and the plasma ignition by gas breakdown [6 to 8] or target breakdown [9 to 11]. The reflectance of the processing material and the expanding plasma
in front of the surface introduce feedback resulting in stochastic fluctuations of
the temporal and spatial behaviour of the laser beam with its mode properties and
in hydrodynamic instabilities of the dynamics and expulsion of the molten surface
layer.

Time resolved transients of electrical conductivity have been investigated as a function of pressure and composition of the ambient atmosphere during intense laser excitation of single crystal Si and InSb with oxidized surfaces. Magnitude, sign, duration and illumination intensity dependence of conductivity transients give information on fundamental processes whether the changes of electrical conductivity near
threshold for processing contain features which indicate the need for an athermal
model, and what some of the possible athermal mechanisms may be. At very high power
levels, well above the threshold, the present experiments should give clearly evidence for melting and possibly even vaporization to evaluate the dynamics of the
melting and resolidification.

2. Experimental

We used Si (111) single crystalline wafers of (0.003 to 1) Ωcm grown by the
Czochralski method. The InSb (110) wafers are prepared from p-type (excess electron
concentration about 10^{15} cm^{-3}) single crystals grown in the [211] direction by the
Bridgman technique. By standard procedures the substrates are mechanically polished,
chemically etched and subsequently stored in air, leading to the formation of a homo-

geneous natural oxide layer. Further details including sample dimensions and moun-
ting, heating and cooling facilities, and vacuum chamber with gas inlet system have
been described elsewhere |12,14|.

The electrical conductivity σ was measured by suitable dc-technique either in the
constant voltage mode of operation for high-ohmic samples or in the constant current
mode for low-ohmic samples. The fabrication and the arrangement of current contacts
and voltage probes has been reported recently |12|.

Irradiation was provided by a Nd: YAG laser of TEM_{00} mode at a wave-length of
1.06 μm. Square pulses are formed with a Pockels cell outside the resonator giving
with the pulse duration ($25 < \tau_L (\mu s) < 400$) and the laser power ($240 < (P(W) < 300$)
a laser radiation intensity $I < 10^7$ W/cm^2. With a Pockels cell within the resonator
Q-switched laser pulses of nearly Gaussian profile are generated at a pulse length
of about 40 ns full width at half maximum yielding a maximum intensity
$I_{max} = 6 \times 10^9$ W/cm^2. The incident laser energy $h\nu_L$ was varied by neutral density
filters to ensure that the laser pulse shape was independent of energy to avoid any
influence on the melting behaviour |13|. The spot size at the sample surface was mea-
sured by scanning the laser beam with a wedge in combination with a photodiode.

3. Experimental Results

The principal features of all the measured Si and InSb specimens qualitatively are
similar within the range of temperatures (77 to 400 K) and carrier concentrations
(Section 2) investigated. Thus, the drawings show only for Si the photon induced
changes $\Delta\sigma$ of electrical conductivity normalized to the maximum value $\Delta\sigma_{max}$ at I_{max}
correcting for the carrier losses by the sample geometry and arrangement for irradi-
ation |12,14|. Repeated measurements over several runs yielded reproducible results
within the experimental uncertainties considering the changes of electrical con-
ductivity during storage in an ambient atmosphere |15,16|.

The changes $\Delta\sigma$ of electrical conductivity reflect almost the shape of the laser
pulse for low laser radiation ($I < 5 \times 10^7$ W/cm^2) intensities. σ increases steeply
at the onset of the laser pulse and decreases with nearly the same rate after
passing a maximum |12|. $\Delta\sigma/\Delta\sigma_{max}$ is plotted in Fig. 1 versus I indicating different
power laws for the illumination induced $\Delta\sigma$ depending on the intensity regime.

At a critical intensity I_c the conductivity change increases abruptly by an order of
magnitude (Fig. 2). The time dependence of $\Delta\sigma$ |12,14| shows a steep increase of con-
ductivity, which is followed by a nearly saturation region. This enhanced conductivity remained nearly constant until it decays to the initial value. Above I_c, the

duration τ_c of the enhanced conductivity state increased and lasted longer than the laser pulse as the intensity increased |12|. For $I > 5 \times 10^8$ W/cm^2 $\Delta\sigma/\Delta\sigma_{max}$ saturates. I_c depends mainly on the pressure and composition of the ambient (Fig. 2).

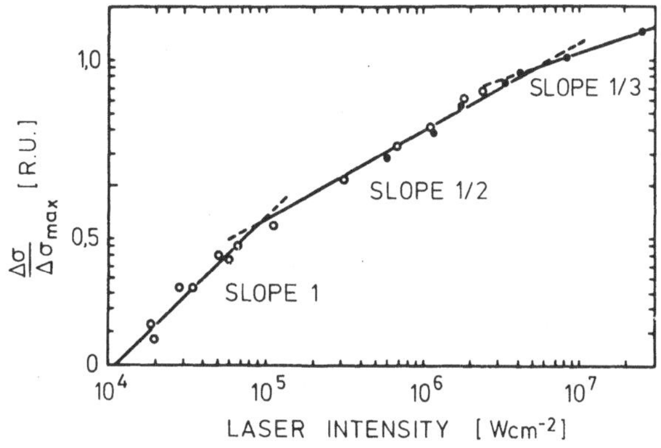

Fig. 1. Normalized electrical conductivity as a function of laser radiation intensity at $p = 10^{-8}$ bar (● this work, o ref. (2)).

I_c becomes higher with increasing pressure in combination with a simultaneous broadening of the threshold (Fig. 2). In oxygen I_c decreases with increasing pressure, in order to show at high pressures the same behaviour as in air (Fig. 2).

The observation of the interacting zone exhibits for $I > I_c$ visually and acustially plasma ignition which is supported by the changes of surface morphology investigated

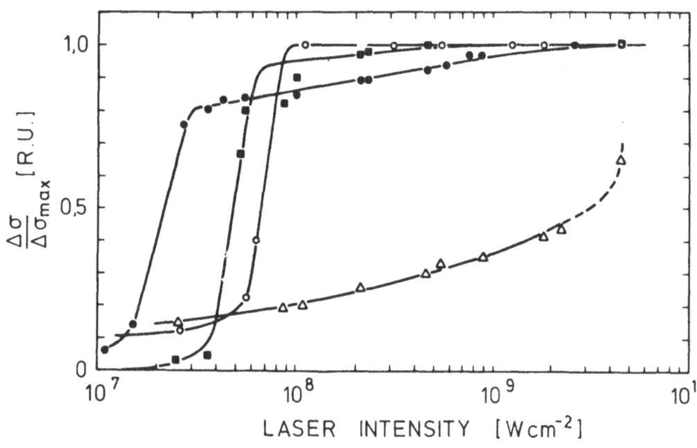

Fig. 2. Normalized electrical conductivity versus laser radiation intensity (o 10^{-8} bar ambient atmosphere, ■ 5×10^{-8}, ● 10^{-6} bar oxygen, Δ 10^0 bar ambient atmosphere, helium or oxygen).

by optical microscopy |12,14|. The resulting $\Delta\sigma_v$ in vacuum with a saturation region (Fig. 2) are much higher than $\Delta\sigma_a$ in air with a smooth increase (Fig. 2).

4. Discussion

The coupling efficiency of a photon beam impinging on a semiconductor is governed by

358

the light absorption length α^{-1}. The smaller this parameter is the greater is the
energy concentrated in the near surface region changing the carrier distribution by
generation and relaxation processes. During high intensity laser irradiation the
rate of carrier creation exceeds the rate of carrier relaxation that an appreciable
density of free carriers builds up. The absorption coefficient α consequently in-
creases with I |17|. Electrons and holes created by excitation across the gap, ioni-
zation of localized impurities within the gap, and transitions within the bands can
acquire enough energy to create additional carriers by impact ionization. As a consequence α increases abruptly from low to high values. The electron concentration
is given by $n = n_0 + \Delta n$, where n_0 is the electron concentration before irradiation
and Δn is the concentration of additionally generated electrons. Similar considera-
tions hold for the holes. Neglecting the effect of carrier diffusion |18|, we can
couple the carrier density to the intensity of the laser radiation inside the sample
by the equation of continuity |14,19|. For homogeneous irradiation the generation
rate of carriers is given by

$$\frac{d(\Delta n)}{dt} = \frac{(1-R)}{h\nu_L} I \, \eta \, \alpha \quad , \tag{1}$$

where η is the quantum yield and R is the reflectance. The recombination rate of excited carriers

$$- \frac{d(\Delta n)}{dt} = r_1 \Delta n + r_2 (\Delta n)^2 + r_3 (\Delta n)^3 \tag{2}$$

can be approximated by a power series. Relating the conductivity changes (Section 3)
only to Δn, since Δn is far higher than possible mobility changes, the solution of
equation (1) and (2) yields different power laws $\Delta \sigma \sim \Delta n \sim I^X$ (0.3 < x < 1.0) depending on the intensity regime (Fig. 1). The first term is associated with phonon
assisted or impurity recombination dominating at low carrier concentrations
($\Delta n \sim I^1$). The second term is associated with inter-or-intra-band recombination
possibly accompanied by photon or phonon emission ($\Delta n \sim I^{1/2}$). The third term is related to the Auger recombination process, which becomes important at high carrier
concentration ($\Delta n \sim I^{1/3}$). The Auger process gives indication for a high carrier
concentration in the surface region probably an electron-hole plasma. Depending on
τ_L and I the coefficients r_i govern for $I < I_c$ by the concentration and lifetime the
magnitude and duration of the observed photoconductivity |12,14|.

Magnitude, sign, duration and illumination intensity dependence of $\Delta \sigma$ indicate sur-
face melting for $I > I_c$ |12,20|. The molten semiconductors investigated are metallic
|21| resulting in an increase of electrical conductivity upon melting. The observed
$\Delta \sigma$'s |20| are composed of three parallel conduction contributions (i) due to photo-
conductivity, (ii) due to the molten surface layer, and (iii) due to thermally generated carriers in the solid. Correspondingly, because the sample resistance decreases with the formation of a metallic layer, the conductivity increases as the thick-

ness of the molten layer increases. There is an enhanced conductivity state at times after the end of the laser pulse |12,14| decaying as the molten layer resolidifies and the temperature decreases. The absorbed energy is stored in the heat capacity of the semiconductor and is mainly distributed over the thickness of the molten surface layer which is governed by diffusion processes via the flow of heat. Detailed numerical calculations |22| were performed to compute the temperature profiles as a function of depth and time in the semiconductors. The calculations, which are based on the flow of heat, represent the kinetics and the intensity dependence of the conductivity changes by heating, melting, and boiling of the irradiated semiconductor supporting the considerations above.

In addition to the recombination processes (equation (2)) the relaxation of the energy stored in the carrier systems during excitation occurs by carrier-carrier collisions and carrier-lattice collisions. The energy transfer from the carriers to the lattice requires phonons. Therefore, the rate of carrier-lattice collisions increases with lattice temperature. A higher temperature favours the formation of oxide layers on the surface. The analysis of the growth rate during scanning simultaneously the surface by a laser beam reveals an enhancement over the normal thermal equilibrium rate of oxidation |23|. The oxidation reaction may be limited by the availibility of free bonds |24|. Since the density of broken bonds is related to the lattice temperature, which increases the number of antibonding states, the presence of photoinduced carriers by band-gap radiation will further enhance this effect. The shift of I_c to lower intensities with increasing oxygen pressure (Fig. 2) might be explained by the heat of reaction ΔH of the oxidation, e.g. ΔH = 69.4 kcal/grammatom oxygen |25| for the forming of a SiO_2 overlayer. The laser intensity in the surface region is enhanced by ΔH becoming free during the interaction of the laser beam with the semiconductor. The heating, melting, and boiling of the irradiated semiconductor obviously start at lower intensities resulting in a decrease of I_c (Fig. 2) for low gas pressures. The lowering of I_c depends on the material, the gas, and the chemical reaction. For high gas pressures the plasma effects predominate (Fig. 2) irrespectively of the type of gas as described in more detail in a recent paper |12|.

Summary

The interaction of high-energy pulsed-laser radiation and single-crystalline Si and InSb is studied by measurements of transients in electrical conductivity giving information on the physical processes in different intensity regimes.

For low intensity laser radiation ($I < 5 \times 10^7$ W/cm^2) the interaction results in a fast heating and quenching process providing data on the energy coupling. The absorption of the incidenting laser radiation via carrier creation and carrier heating

occurs by the generation of electron-hole pairs and the excitation of electron and
hole transitions within the single bands via interband- and free carrier absorption.
Impurity-, band-band- and Auger recombination are the mechanisms for the relaxation
of the electron energy towards equilibrium and for the transfer to the lattice giving indication for an electron-hole plasma.

The interaction yields for high intensity laser radiation ($I > 5 \times 10^8$ W/cm^2) an
ultrarapid heating, melting, evaporation, and quenching. The measurements of transients in electrical conductivity allow to determine the dynamics of melting and
resolidification. The thermal conduction is controlled by the thermal gradients,
which are initially established by the absorption of the laser energy within the
surface region. The thermal properties of the irradiated material, in addition to
the optical properties, play an important role.

References

(1) WHITE, C.W., and PEERCY, P.S. Eds.: Laser and Electron Beam Processing of
     Materials (Academic Press, New York 1980) and references therein
(2) BLINOV, L.M., V.S. VAVILOV, and G.N. GALKIN: Sov.Phys.Semicond. 1 (1967) 1124
(3) VAN VECHTEN, J.A., R. TSU, and F.W. SARIS: Phys. Lett. A 74 (1979) 422
(4) PHILLIPS, J.C.: J. Appl. Phys. 52 (1981) 7397
(5) HERZIGER, G.: Proc. 4th Intern. Symp. Gas Flow and Chemical Lasers (1982)
     to be published
(6) SMITH, D.C.: J. Appl. Phys. 48 (1977) 2217
(7) VEDENOV, A.A., G.G. GLADUSH, and A.N. YAVOKHIN: Sov. J. Quantum Electr. 11
     (1981) 896
(8) MCKAY, J.A., and J.T. SCHRIEMPF: IEEE J. Quantum Electr. QE-17 (1981) 2008
(9) EPSHSTEIN, E.M.: Sov. Phys. Solid State 11 (1970) 2213
(10) ZAKHAROV, S.I.: Sov. Phys. J. Exp. and Theor. Phys. 41 (1975) 1085
(11) WALKER, T.W., A.H. GUENTHER, and P.E. NIELSEN: IEEE J. Quantum Electr. QE-17
     (1981) 2053
(12) KREUTZ, E.W., H.G. TREUSCH, and W. ZIMMER: Proc. Laser-Solid Interactions
     and Transients Thermal Processing of Materials, Material Research Society
     Europe (1983) to be published
(13) WOOD, R.F., and G.E. GILES: Appl Phys. Lett. 38 (1981) 422
(14) ZIMMER, W.: diploma work, TH Darmstadt (1982)
(15) KREUTZ, E.W.: Zeitschrift Angew. Phys. 32 (1971) 280
(16) KREUTZ, E.W.: phys. stat. sol. (a) 40 (1977) 415
(17) BLINOV, M.: Sov. Phys. Solid State 9 (1968) 2537
(18) BOK, J.: Proc. Laser-Solid Interactions and Transients Thermal Processing of
     Materials, Material Research Society Europe (1983) to be published
(19) GRAVE, T., E. SCHÖLL, and H. WURZ: J. Phys. C16 (1983) 1693
(20) GALVIN, G.J., M.O. THOMPSON, J.W. MEYER, P.S. PEERCY, R.B. HAMMOND, and
     N. PAULTER: Phys. Rev. B27 (1983) 1079
(21) GLAZOV, V.M., S.N. CHIZHEVSKAYA, and N.N. GLAGOLEVA: Liquid Semicoductors
     (Plenum Press, New York 1969)
(22) TREUSCH, H.G., L. Bakowsky, and K. WISSENBACH: these proceedings
(23) BOYD; I.W.; Appl. Phys. Lett. 42 (1983) 728
(24) BLANC, J.: Appl. Phys. Lett. 33 (1978) 424
(25) SCHMALZRIED, H., and A. NAVROTSKY: Festkörperthermodynamik (Verlag Chemie,
     Weinheim 1975)

Limitation of Laser Processing Intensity by Laser Induced Gas Breakdown

R.POPRAWE, E.BEYER, L.BAKOWSKY, G.BRUMME and G.HERZIGER
Institut für Angewandte Physik, Technische Hochschule Darmstadt
Schloßgartenstraße 7, D - 6100 Darmstadt

1.Introduction

Laser material processing with intensities below the vapourization threshold I_V such as transformation hardening, alloying or laser annealing, is characterized by the lack of a laserinduced plasma, an absorption not depending on the laser intensity and no material removal.

In the intensity range $I > I_V$ there is evaporization and material removal. This processing mode, used in laser cutting and drilling, is characterized by a hot vapour jet emerging from the target. Because of its temperature and interaction with the laser beam the vapour is partly ionized and therefore will also be refered to as a plasma jet in the following. The plasma causes beam deflection (plasma lensing) and may even absorb the laser beam under certain conditions. Laser processing with intensities above the threshold intensity I_B causes a laserinduced gas breakdown. In this case plasma heating and further ionization increases the absorption of the radiation and therefore increases the processing efficiency. In the limit of even higher intensities the plasma will shield the target completely and the efficiency of material removal decreases drastically. The observed phenomena may be classified in three intensity regions:

a) $I_V < I < I_B$

Vapourized target material formes a plasma jet emerging towards the laser. The plasma is weakly ionized and almost transparent. Besides plasma lensing there is no considerable beam- plasma interaction.

b) $I_B < I < I_D$,   where I_D is the detonation threshold intensity.
The laser beam is absorbed by the target vapour, the surrounding atmosphere or in a mixture of vapour and atmosphere. The laser induced plasma causes the formation of a LSC wave (laser supported combustion /1,2,3/).

c) $I > I_D$
The laser radiation is absorbed completely in a shock wave generated by the

expanding plasma . This limiting case is characterized by the formation of a LSD wave (laser supported detonation /4,5,6/). The LSD wave moves towards the laser with supersonic velocity. A criterion for the onset of a LSD wave is

$$\alpha \, \lambda_c \simeq 1 \tag{1.1}$$

where α (1/cm) is the absorption coefficient and λ_c(cm) is the mean free path of the vapour atoms. The target is shielded completely by a thin , strongly absorbing plasma layer. Material processing in the c)-region does not effect considerable material removal and is used only for very special applications (e.g. shock hardening).

2. Formation of laser induced gas breakdown

The comprehensive literature on laser induced gas breakdown has been summarized by various authors (e.g./7.8/). The threshold intensity for gas breakdown I_B depends on the laser wavelength λ , the pulse duration T_L, the focus radius r_F and the gas properties.

I) First electrons and breakdown criterion

In the case of breakdown formation in the target plasma jet, first electrons are present due to the high temperature of the plasma. In the case of breakdown in the surrounding atmosphere the generation mechanisms of the first electrons are known only qualitatively. Depending on laser- and gas parameters there may be multiphoton processes or field emission involved /9,10/. The first electrons are then heated by inverse bremsstrahlung. Due to inelastic collisions of the electrons with neutral gas atoms (molecules) the electron density increases by avalanche-like multiplication. When the Debye length of the plasma λ_D has gone down to the focus dimensions the electron diffusion will become ambipolar:

$$r_F \simeq \lambda_D = \left(\frac{\varepsilon_0 \, kT_e}{4\pi \, e^2 n_B} \right)^{1/2} \tag{2.1}$$

Here r_F(cm) is the focus radius, T_e(eV) the electron temperature and n_B(1/cm³) the breakdown electron density. In the model this condition is used as a breakdown criterion and equally stands for the upper limit of validity, since only diffusion without the presence of electrical fields is considered.

II) Breakdown volume

The volume of breakdown is determined by the balance of electron generation and electron losses including those due to diffusion. Following Alcock et al./12/ the diffusion length Λ can be approximized by

$$\Lambda^{-2} = \left(\frac{2 \cdot 4}{r_F} \right)^2 + \left(\frac{\pi}{x_F} \right)^2 \tag{2.2}$$

where x_F(cm) is the focus length.

III) Particle- and energy rate equation

The basic equations to describe the processes in the breakdown volume are the electron particle-and energy rate equations:

$$\frac{\partial n_e}{\partial t} = n_e \left(\nu_{ion} - \nu_{loss} \right) \; , \quad \frac{\partial}{\partial t} \left(n_e \frac{m}{2} < v_e^2 > \right) = \alpha I - \sum_i P^i_{loss} \tag{2.3}$$

Here $n_e(1/cm^3)$ is the electron density, $\nu_{ion}(1/s)$ is the ionization frequency, ν_{loss} $(1/s)$ is the diffusion loss frequency, $v_e(cm/s)$ is the electron velocity and $P^i_{loss}(W/cm^3)$ stands for the collisional losses. The absorption coefficient

$$\alpha = \frac{\omega_p^2}{c} \cdot \frac{\nu_c}{\omega^2 + \nu_c^2} \tag{2.4}$$

is an approximation of the general case of inverse bremsstrahlung absorption described by Basov et al./11/ for $\omega_p \ll \omega$, where $\omega_p(1/s)$ is the plasma frequency, $\nu_c(1/s)$ is the collision frequency and ω $(1/s)$ is the laser frequency. In order to give an estimate of the breakdown intensity I_B in terms of the collision frequency , an analytic approximation of the rate equations is obtained by solving the time independent energy equation and integrating the particle rate equation up to the breakdown criterion $n_e=n_B$(see I)):

$$I_B = \frac{\varepsilon_0 m\, c_0}{e^2} \left[\left[\frac{2m}{M} E_0 + \beta E_a \right] v_c^2 + \left[\frac{E_i + E_0}{t_B} \ln \left(\frac{n_B}{n_i} \right) \right] v_c^1 + \frac{kT_e}{m \Lambda^2} \left(E_i + E_0 \right) + \right. $$
$$\left. \left[\frac{2m}{M} E_0 + \beta E_a \right] \omega^2 + \left[\frac{E_i + E_0}{t_B} \omega^2 \ln \left(\frac{n_B}{n_i} \right) \right] v_c^{-1} + \frac{kT_e}{m \Lambda^2} \left(E_i + E_0 \right) \omega^2 v_c^{-2} \right] \tag{2.5}$$

where m/M is the electron/ion mass ratio, E_0 (eV) is the electron energy, $\beta = \nu_c{}'/\nu_c$ is the fraction of inelastic collisions and E_a (eV) is the excitation energy. Fig.1

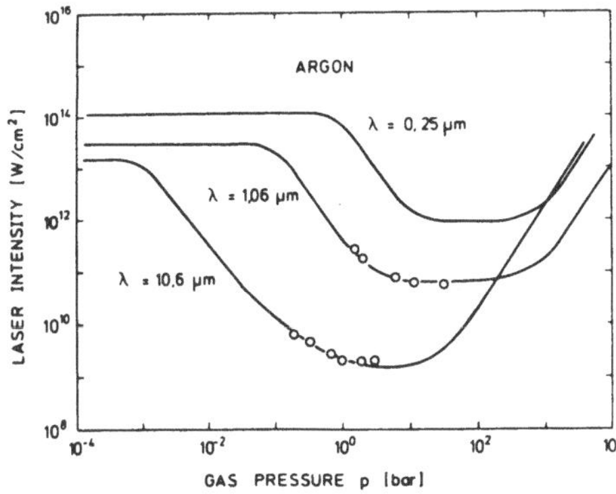

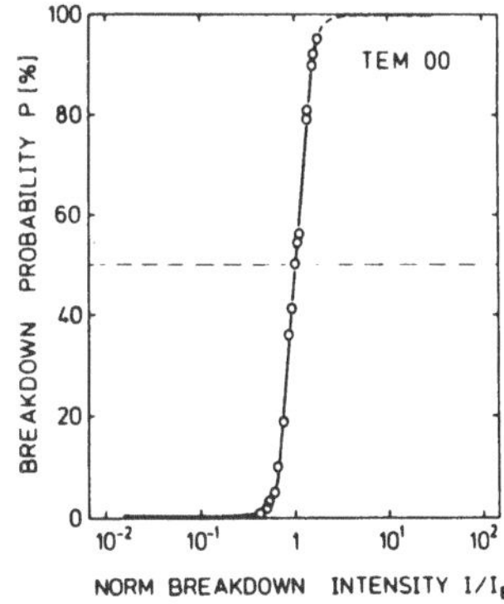

Fig.1 . Calculated estimation of breakdown intensity in argon as a function of pressure P; for comparison experimental data with CO_2 and Nd-Yag lasers are included.

Fig.2 . The breakdown probability changes drastically as a function of the laser intensity.

shows a plot of I_B for various laser wavelengths and argon. Experiments with CO_2-and Nd-Yag lasers in the pressure region of $0.1 < P(bar) < 30$ confirm the estimation of I_B. The appearance of a laser induced gas breakdown depends very sensitive on the laser intensity. Fig.2 shows the experimentally obtained intensity dependence of the breakdown probability P for a single mode Nd-Yag laser. The probability changes from $P \cong 0$ to $P \cong 1$ within a factor of two in intensity change. When multimode lasers are used, the threshold is expected to be less sharp because of statistically destributed intensity peaks.

3) Phenomena observed during material processing with intensities $I > I_B$

As a consequence of the sharp threshold for laser induced gas breakdown material can be removed with laser intensities almost up to the threshold intensity I_B without considerable absorption in the vapour. Exceeding the threshold intensity results in formation of a LSC wave. This can happen either in the surrounding atmosphere or in the plasma jet generated by the laser (Fig.3). The radiation is absorbed partly or completely and the processing mode is not only controlled by the laser any more. At even higher intensity and resulting smaller absorption length $1/\alpha$ the limiting case of a LSD wave will occur. This wave propagates at a velocity

$$v_D = [2(\gamma^2 - 1)\frac{I_o}{\rho_o}]^{1/3} \qquad (3.1)$$

towards the laser /5/, where γ is the adiabatic exponent. Fig.4 shows the spatial and temporal evolution of such a state: after formation of the vapour jet 1 the absorption layer takes off the target (2 and 3). As a consequence of the absorption layer take off the pressure on the target is enhanced strongly (3). In the limit of detonation the pressure P can reach the value behind a detonation wave P_D

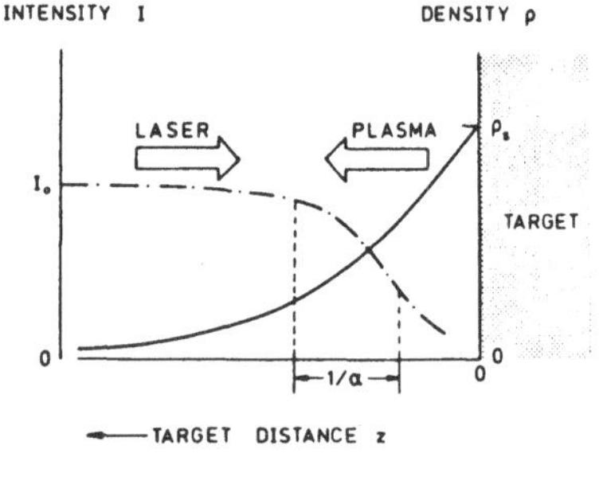

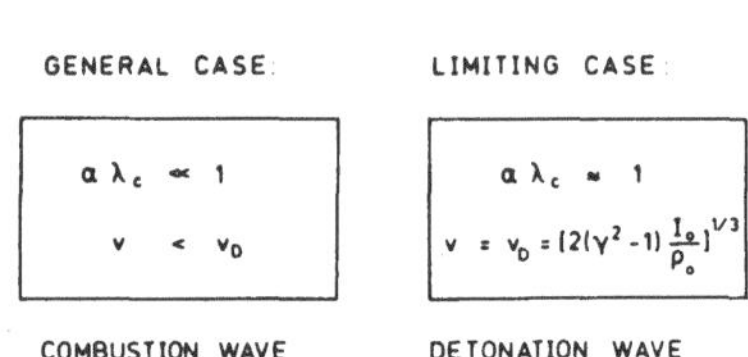

Fig.3. Schematics of laser absorption in the target vapour

$$P \lesssim P_D = [4 \frac{(\gamma - 1)^2}{\gamma + 1} \rho_o]^{1/3} \cdot I^{2/3} \qquad (3.2)$$

Intensities in the region 10^8 W/cm² can lead to pressures of about $10^2 < P(bar) < 10^3$ and result in vigerous expulsion of liquid target material. Because of statistical intensity peaks the use of multimode laser systems can even lead to chaotic processing and unpredictable behavior of melt and processing geometry. To point out the difference between laser- and plasma controlled material removal fig.5 and 6 show the removal efficiency η and the processing result as a function of the laser intensity I. η is defined as

$$\eta = \frac{V \cdot \varepsilon_V}{I \cdot T_L \cdot \pi r_F^2} \qquad (3.3)$$

where V (cm^3) is the volume of removed matter, ε_V (J/cm^3) is the vaporization energy density and T_L is the laser pulse duration. Processing with intensities $I_M < I < I_V$ (I_M (W/cm^2): melting threshold intensity) just melts the target, there is no vaporization (picture 1,fig.6). When processed with the maximum efficiency intensity $I_{\eta\,max}$, the hole diameter matches with the beam diameter. Exceeding $I_{\eta\,max}$ results in hole diameters up to ten times larger than the beam diameter. The processing geometry is determined by plasma expansion.

Similar effects have been oberved in experiments with CO_2- (λ =10.6μm), Nd-Yag- (λ =1.06 μm) and KrF-(λ =0.25 μm) lasers. As a criterion of the processing state the velocity of the emerging absorption waves has been recorded as a function of the laser intensity I (fig.7). For comparison, the velocity of a detonation wave is drawn in according to equation 3.1 . With increasing intensity the absorption wave velocity approaches the detonation limit. The consequence is complete shielding of the target in this limit.

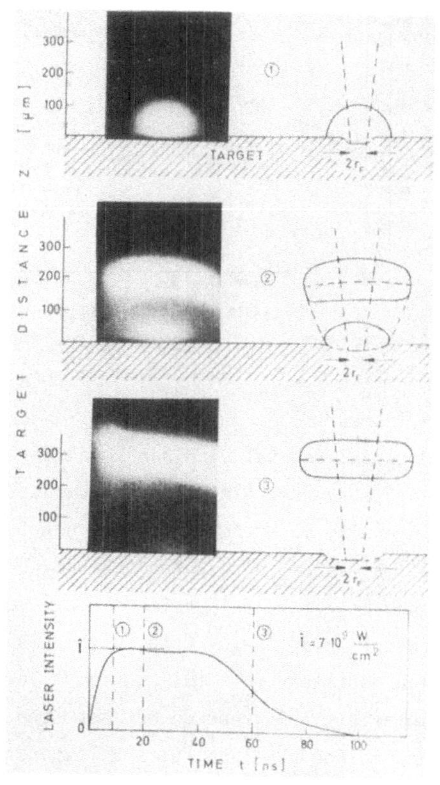

Fig.4.LDS-wave formation;the laser-target interaction is interrupted.

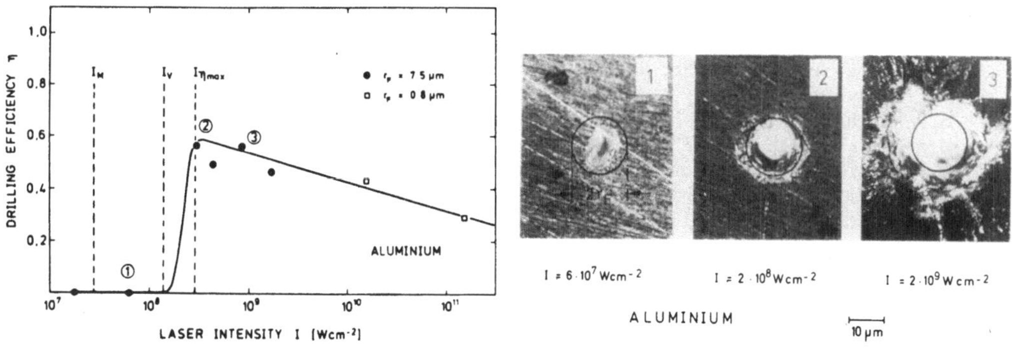

Fig.5,6. Efficiency of material removal η as a function of laser intensity I for KrF-laser (λ =0.25 μm) on aluminium; the pulse length is T_L =50ns. For intensities $I > I_\eta$ max the processing geometry is determined by the expansion of the plasma.

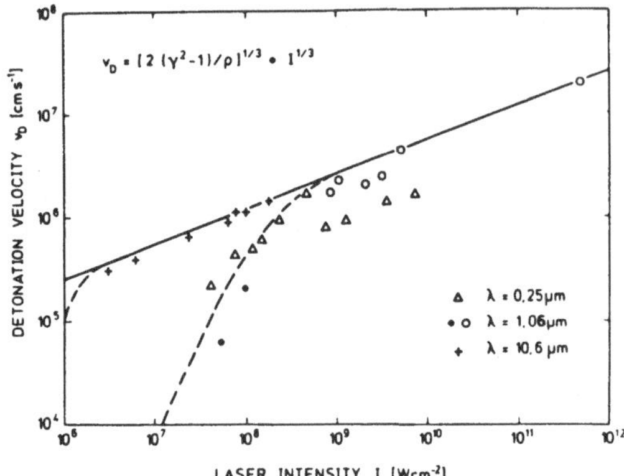

Fig.7. With increasing intensity I the absorption wave velocity approaches the detonation limit. Experiments with CO_2 -, Nd-Yag- and KrF-laser indicate the transition from LSC wave- to LSD wave formation.

4. Summary

Material removal with laser radiation is characterized by the formation of a hot vapour jet emerging from the target towards the laser. Exceeding the threshold intensity I_B, the interaction between laser-and plasma beam leads to the formation of a laser induced gas breakdown. The threshold value depends on laser parameters and material properties. The resulting plasma increases the absorption and the energy coupling. With further increasing intensity the absorption layer takes off and shields the target completely. Due to the associated pressure liquid target material is pressed out of the processing zone and the processing result is determined by the plasma expansion. However, in a small intensity intervall the laser induced plasma acts as an absorption increasing element. In this intensity intervall material removal can be performed with maximum efficiency and accuracy.

5. Literature

/1/ F.Y.Su and A.A.Boni,The Physics of Fluids 19(7), 960(1976)
/2/ F.J.Allen,Ballistic Res. Lab. Aberdeen, Maryland AD-785 602(1974)
/3/ A.N.Pirri, R.G.Root and P.K.S.Wu, AIAA 16(12), 1296(1978)
/4/ P.D.Thomas, AIAA 13(10), 1279(1975)
/5/ Yu.P.Raizer, Sov.Phys.JETP 21(5), 1009(1965)
/6/ A.N.Pirri, The Physics of Fluids 16(9), 1435(1973)
/7/ J.F.Ready, Acad. Press, New York, London 1971
/8/ C.De Michelis, IEEE QE 5(4), 188(1969)
/9/ B.A.Tozer, Phys. Rev. 137(6A), 1665(1965)
/10/ L.V.Keldish, Sov.Phys.JETP 20(5), 1307(1965)
/11/ N.G.Basov and O.N.Krokhin Sov.Phys.JETP 19(123), (1964)
/12/ A.J.Alcock, C.De Michelis and M.C.Richard, IEEE QE 6(10), 622(1970)

Formation and Influence of Laser Induced Plasma during CO_2-Laser Welding

E. BEYER, L. BAKOWSKY, R. POPRAWE, G. HERZIGER
Institut für Angewandte Physik, Technische Hochschule Darmstadt
Schlossgartenstr. 7, D-6100 Darmstadt

1.   Introduction

For CO_2-high power lasers being economical sources of high-energy
radiation the influence of laser induced gasbreakdown (e.g. during welding and cutting) cannot be neglected. During processing of a target with
a laser beam intensity $I > I_c$ a plasma is formed above the processing area.
In welding I_c is the limiting threshold for deep penetration (Fig. 1).
Above the critical intensity I_c the welding depth and the efficiency of
the process increases considerably. Measurements of the reflection coefficient simultaneously show that the increase of efficiency takes
place by increased coupling of the laser energy into the material (Fig.2).
The increase of efficiency always is correlated with a laser induced
plasma.

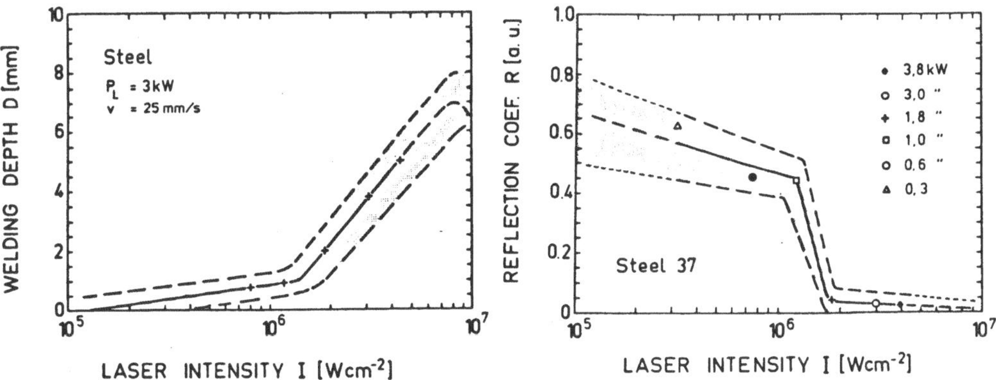

Fig. 1. Welding depth as function
of laser intensity by const.
laser power

Fig. 2. Reflection coefficient as
function of laser intensity
measured with a rotating
wire during the welding
process

2. Threshold intensity for evaporation I_V

The critical intensity I_c for the development of a laser induced plasma
during material processing is some orders of magnitude lower than for
air breakdown in ambient atmosphere (fig. 6 and 7). This is due to the
evaporation of the solid. The ionization energy of a metal atom is considerably lower than the ionization energy of N_2 or inert gas. This originates in plasma development below the intensity I_B of air breakdown.

Further treatment requires the occurence of an atomic metal vapour.
Thus the laser beam first has to heat up a metal target to the evaporation temperature. This means that the rate of energy gain by the absorption of the laser energy in the target has to exceed losses by 3-dimensic
nal heat conduction. Assuming a Gaussian intensity distribution the temperature of the target surface in the beam centre /1/ is given by

$$T_V \ (0,0,0,t) = \alpha I_V r_F \ / \ (2k\sqrt{\tfrac{\pi}{2}}) \cdot \text{arc tg} \sqrt{\frac{8\kappa t}{r_F^2}}$$

with

T_V	=	evaporation temp.
r_F	=	focus radius
t	=	time
K	=	heat conduction
κ	=	temp. conduction
I_V	=	threshold intensity
α	=	absorption coeff.

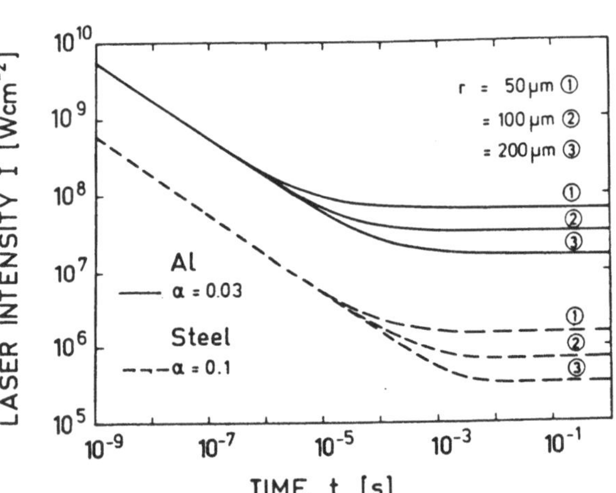

Fig. 3. Laser intensity as function
of time to reach bowling
temperature T_{th} (r=0,z=0,t)
in the beam centre

At short laser pulses the threshold intensity is independent of the
focus radius r_F. With pulse duration longer than 1 ms (cw-working)
I_V increases linearly with r_F (fig. 3). A material specific threshold
intensity results according to the different heat conduction constants.

3. Plasma development and absorption

Laser material processing investigated since 1963 /1,2/ in numerous papers and reviewed by several authors /3-7/ is still a strongly developing field. Nevertheless the mechanism of laser induced gasbreakdown is not understood in all the details. In the laserfield free electrons will be accelerated by inverse bremsstrahlung up to ionization energy. The electron density is increased by means of avalanche ionization. For a single mode beam with a high photon density, laser radiation can be described by a classical electromagnetic wave. Therefore, energy transmission through inverse bremsstrahlung can be approximated by the classical microwave model for plasma generation and heating. With this approximation a simple model describes the starting process of a laser induced plasma.

As the laser radiation interacts substantially with electrons, the influence of plasma is given by the degree of ionization. A plasma with electron densities $n_e < 10^{16} (1/cm^3)$ is transparent for $\lambda \simeq 10 \mu m$.

The following model describes the plasma development as well as the fully devloped plasma (cw-plasma).
Using energy- and particle-rate equations the average electron energy and density as a function of the incident laser radiation can be determined.

a) Particle rate equation

$$\frac{dn_e}{dt} = R_E - R_{Diff.} - R_{REC}$$

R_E = generation rate

R_{Diff} = diffusion rate

R_{REC} = recombination rate

b) Energy rate equation

$$\frac{d\bar{\varepsilon}}{dt} = \frac{1}{n_e}\left(\alpha I - \bar{\varepsilon}\frac{dn_e}{dt} - P_{el} - P_{ion} - P_{rad} - P_{diff}\right)$$

n_e = electron density

$\bar{\varepsilon}$ = average electron energy

$$\alpha \quad = \text{absorption coefficient}$$

$$P_{el} \quad = \text{power losses by electron-atom collisions}$$

$$R_{rad} \quad = \text{power losses by radiation}$$

$$P_{diff} = \text{power losses by diffusion out of interaction volume}$$

$$P_{ion} \quad = \text{power losses by ionization}$$

The single rates are taken from the literature /9 -13/. Assuming isotropic scattering the momentum rate eqation can be neglected. For electron-ion recombination the following relation is assumed:

$$R_{ec} \simeq \text{const } n_e^2 / \sqrt{\epsilon}$$

This rate determines the electron density and the absorption length of a laserinduced plasma. The constant factor has been adjusted by streak photographs of a pulsed TEA CO_2-laser to the absorption length of the calculation.

The produced metal vapour density is a function of absorbed laser intensity, which is considered in the calculations following the formalism by Krokhin /13/.

The solution of the coupled differential equations shows, that for laser intensity $I > 10^6$ W/cm^2 the electron energy grows to a critical value in times $t < 10^{-8}$ sec. The increase of electron energy is a function of laser intensity. At the critical energy the ionization rate is dominant and an avalanche ionization process starts (fig. 4). The process may be compared with a phase transition of first order. At a critical electron density th recombination rate is dominant yielding a constant density n_e (cw-plasma)

Fig. 5 and 6 show the critical intensity I_c for a metal vapour breakdown and the range of free air breakdown I_B. The range of vapour threshold is given by the absorbed intensity $I_v = I_{abs}$ and the incident intensity $I_v = I_{inc}$. The required intensity I_v depends on target reflection coefficient.

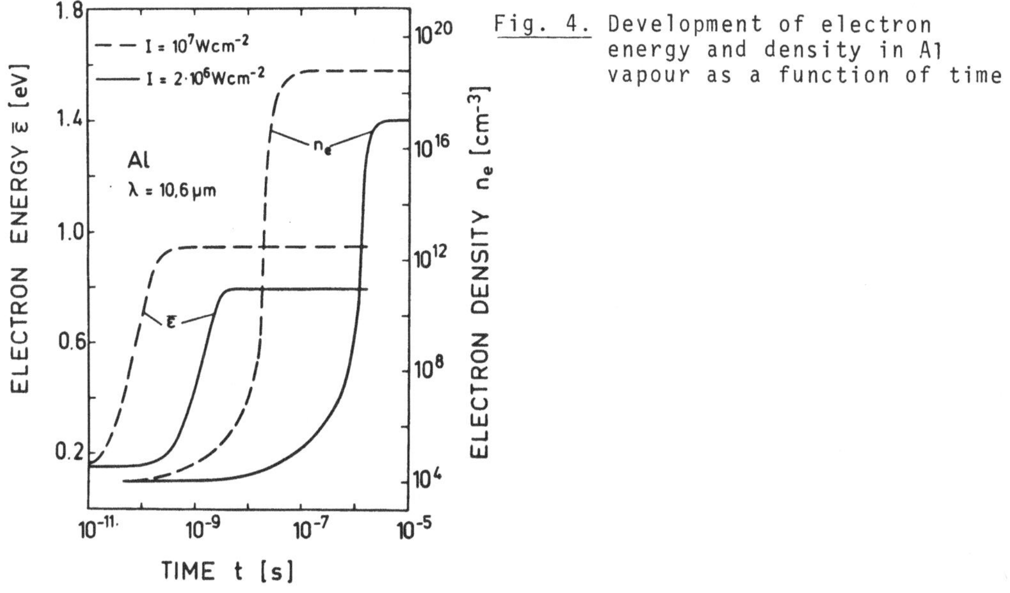

Fig. 4. Development of electron energy and density in Al vapour as a function of time

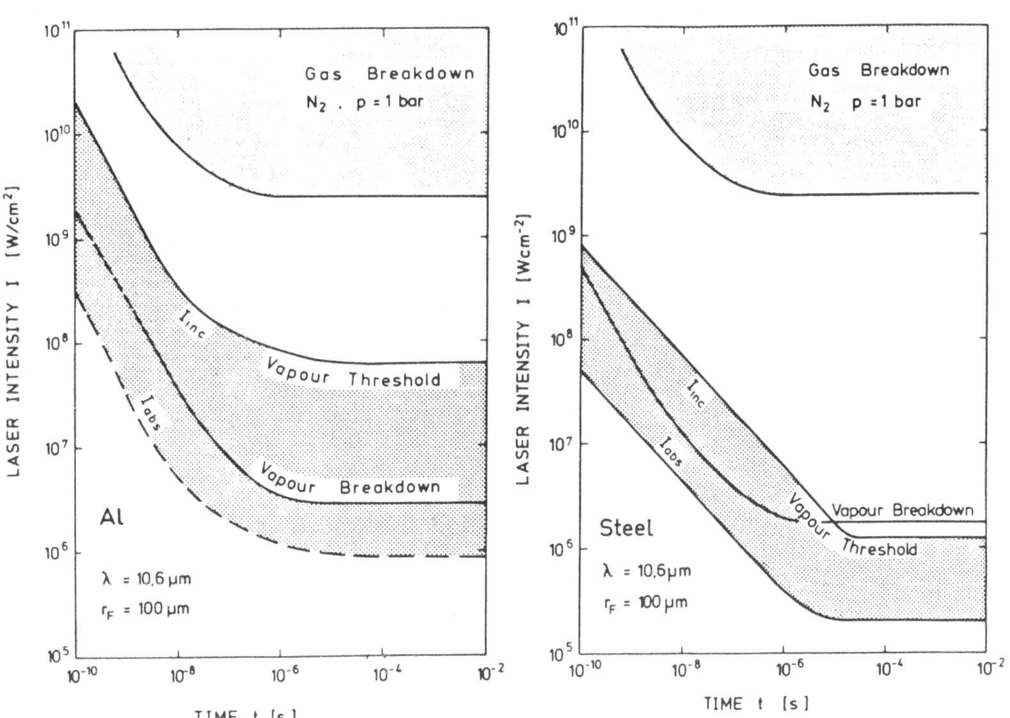

Fig. 5 and 6. Calculated threshold intensity for vapourization, and critical intensity for laser induced metal vapour breakdown and free gas breakdown.

4. Conclusion

During high efficiency material processing a laser induced plasma inevitably changes the processing conditions. Consequently, material processing is no longer controlled by laserbeam, but indirectly by the plasma

In case of low plasma absorption the original beam geometry is changed by lensing effect of plasma. At intensities $> 10^7$ $|W/cm^2|$ the target is screened completely by the plasma. Plasma detaches from the target and moves with detonation velocity towards the laser /14/. The laser target interaction is interrupted until the plasma gets transparent again, based on rarefraction. For these intensities the continous working changes to an undefined pulse processing. Usually the avarage intensities of $CW-CO_2$-lasers are smaller than 10^7 $|W/cm^2|$, but intensity peaks higher than 10^7 $|W/cm^2|$ are possible by mode instabilities or optical feedback /15/.

Literature
(1) CARSLAW, H.S. and JAEGER, J.C.; Conduction of Heat in Solids
    2nd ed University Press (1959)
(2) READY, J.F. Appl.Phys.Lett.3 (1963) 11-13
(3) LINLOR, W.I. Appl.Phys.Lett.3 (1963) 210-211
(4) DE MICHELIS, C. IEEE QE-6 (10) 630-641 (1970)
(5) READY, J.F.; Effects of High Power Laser Radiation
    Academic Press New York (1978)
(6) ARRECHI, F.T. and SCHULZ-DUBOIS, Laser Handbook, North Holland Publ.
    (1972)
(7) DULEY, W.W.; CO_2 Lasers, Effects and Applications,
    Academic Press New York (1976)
(8) NEWSTEIN, M. and SOLIMENE, N. IEEE QE-17 (10) (1981) 2085-2091
(9) PESCHKO, W. Dissertation TH Darmstadt (1981)
(10)BRUMME, G. Dissertation TH Darmstadt (1977)
(11)FABRIKANT, W.A.; Jetp 8, 35 Mitt.d.Akad.d.Wiss.d.UDSSR 25 (1938)
(12)WEYL, G., PIRRI, A. and ROOT, R. AIAA 19, April (1981)
(13)KROKHIN, O.N.: Laser Handbook Vol. 2 North-Holland Publ.Com. (1972)
(14)POPRAWE,R., BEYER,E., BAKOWSKY,L. and HERZIGER,G.: these proceedings
(15)LOOSEN,P., BAKOWSKY,L., HERZIGER,G. and RÜHL,F.: these proceedings

Erzeugung und technische Anwendung laserinduzierter Stoßwellen

J. MUNSCHAU
Bundesanstalt für Materialprüfung
Unter den Eichen 87
D-1000 Berlin 45

Einleitung

Bei Einstrahlung von Laserimpulsen in gasförmige und flüssige Medien wird
in extrem kurzer Zeit Energie umgewandelt, die zur Erzeugung intensiver
Schallwellen bis hin zur Ausbildung starker Stoßwellen führt. Die Erzeugung und technologischen Anwendungsmöglichkeiten derartiger Stoßwellen
werden im folgenden beschrieben.

Grundsätzlich sind zwei Erzeugungsprozesse zu unterscheiden /1/ : Bei
geringen Laserenergien treten durch Strahlungsabsorption bedingte thermische Prozesse auf, d.h. Stoßwellenerzeugung durch "thermische Expansion"
/2,3,4/, dagegen führt die Einstrahlung sehr hoher Laserenergien zu Ioni-
sationsprozessen im Medium und damit zur Erzeugung und explosionsartigen
Ausdehnung heißer Plasmen, d.h. Stoßwellenerzeugung durch "Plasma-Expansion" /5,6,7,8,9/.Da letzteres zu erheblich höheren Stoßwellendrücken
führt, soll im folgenden, insbesondere im Hinblick auf technische Anwendungen, ausschließlich die Stoßwellenerzeugung durch Plasma-Expansion
näher erläutert werden.

Stoßwellenerzeugung durch Plasma-Expansion

Die Erzeugung eines Plasmas im Brennpunkt eines fokussierten Laserimpulses
ist bereits in früheren Arbeiten /5,6,7,8,9/ untersucht worden, jedoch
wurde ein wesentlicher Aspekt zumeist nicht berücksichtigt: Durch das
Expansionsbestreben des heißen Plasmas wächst in extrem kurzer Zeit
(Laserpulsdauer ca. 10^{-8}s) der Druck auf das umgebende Medium außerordent-
lich stark an, vergleichbar mit der Detonationswirkung hochbrisanter Explosivstoffe. Auch die vorhandenen Energiedichten sind durchaus mit denen
von Sprengstoff vergleichbar, können sie sogar erheblich übertreffen. Der
im Plasmabereich entstehende Druck kann daher sehr hohe Werte erreichen.
Nach Berechnungen von Bell u. Landt /6/ und Barnes u. Rieckdorff /7/ werden Drücke von einigen 10^5bar erreicht, was größenordnungsmäßig mit den
eigenen Untersuchungen übereinstimmt.

Die Bestimmung des Drucks im Plasmabereich erfolgt analog zum "Deflagra-
tionsmodell" von Kidder /10/, Ramsden u. Savic /8/ und Raizer /9/. Danach
ergibt sich der Druck p_i an der Bereichsgrenze des Plasmas zu

$$(1) \quad P_i = \varkappa \left(\frac{4\varrho_o}{\varkappa^2 - 1} I^2 \right)^{1/3}$$

mit $\varkappa$ = Adiabatenexponent, ϱ_o = Dichte des Mediums und I = Intensität des Laserstrahls. An der Bereichsgrenze bildet sich so eine Stoßwelle aus, die das umgebende Medium durchläuft.

Die Ausbreitung derartiger Stoßwellen wurde im Zusammenhang mit Unterwasserdetonationen hochbrisanter Explosivstoffe von Kirkwood u. Bethe /11/ untersucht. Die dabei entwickelten Theorien sind als Grundlage für die Ausbreitung laserinduzierter Stoßwellen besonders gut geeignet, da ihre Lösungen im Gegensatz zu den Lösungen der allgemeinen Gleichungssysteme der Hydrodynamik eine relativ einfache Berechnung der Stoßwellenparameter gestatten. Dies ist insbesondere für die praktische Anwendung von Bedeutung. Dieser Lösungsansatz wurde im Hinblick auf die völlig andersartige optische Anregungsenergie modifiziert und führt zu einem System von Gleichungen zur vollständigen Bestimmung des Stoßwellenverlaufs /1/, das ausführlich hier nicht dargelegt werden kann. Aus diesem Gleichungssystem kann jedoch eine Nährungslösung mit hinreichender Genauigkeit für den technischen Anwendungsfall extrahiert werden. Danach ergibt sich der Stoßwellen-Spitzendruck p_m an einem Ort im Abstand R vom Brennpunkt des Laserstrahls zu

$$(2) \quad P_m = \left(2\varkappa\, a_1 t_L \varrho_o^{4/3} c_o^3 \right)^{1/2} \frac{I^{1/3}}{R\left(\ln R/a_1\right)^{1/2}}$$

bzw. in weiter vereinfachter Form

$$(3) \quad P_m = K\, \frac{W_o^{1/3}}{R\left(\ln R/a_1\right)^{1/2}}$$

mit a_1 = Brennpunktradius, t_L = Laserpulsdauer, c_o = Schallgeschwindigkeit im Medium, W_o = eingestrahlte Laserenergie und K = Systemkonstante (Zusammenfassung der Material- und Laserkonstanten des speziellen Anwendungsfalls).

<u>Experimentelle Ergebnisse</u>

Die experimentelle Bestimmung der Stoßwellenparameter erfolgte durch direkte Messung des Druckverlaufs mit Hilfe von piezoelektrischen Druckaufnehmern. Die Impulsdauer der Druckimpulse liegt im Bereich von 10^{-8} - 10^{-5} Sekunden. Druckaufnehmer mit derartig hoher zeitlicher Auflösung sind kommerziell nicht zu erhalten. Es wurde daher ein Druckaufnehmer (Bild 1) nach dem Prinzip des piezoelektrischen Dickenschwingers mit akustisch angepaßtem Druckauslaufstab und integriertem FET-Impedanzwandler entwickelt /12/. Das Signal des Druckwandlers (Bild 2) wird durch einen breitbandigen HF-Verstärker verstärkt. Die maximale Grenzfrequenz dieses Druckmeßsystems liegt bei 200 MHz. Die damit durchgeführten experimentellen Untersuchungen laserinduzierter Stoßwellen ergaben eine befriedigende Übereinstimmung der Meßergebnisse mit den theoretischen Aussagen (Bild 3). Die Darstellung in Bild 3 beinhaltet auch die hier

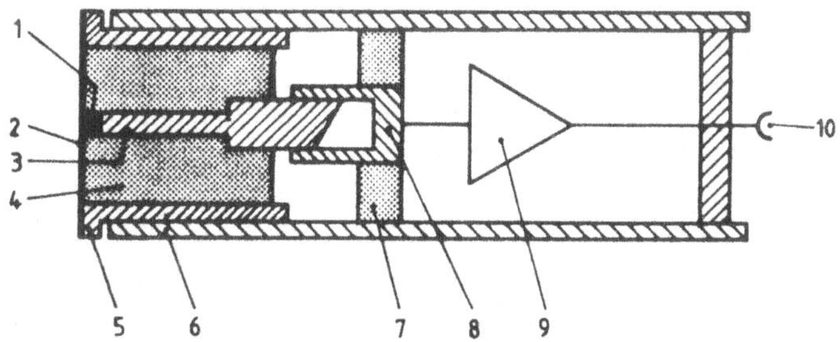

1 - Piezoelement
2 - aufgedampfte Silberschicht
3 - Druckauslaufstab
4 - Polystyrolhalterung
5 - innerer Messingmantel

6 - äußerer Messingmantel
7 - Polystyrolhalterung
8 - Klemmkontakt
9 - Impedanzwandler
10 - Signalausgang

Bild 1. Schematische Darstellung des Druckwandlers

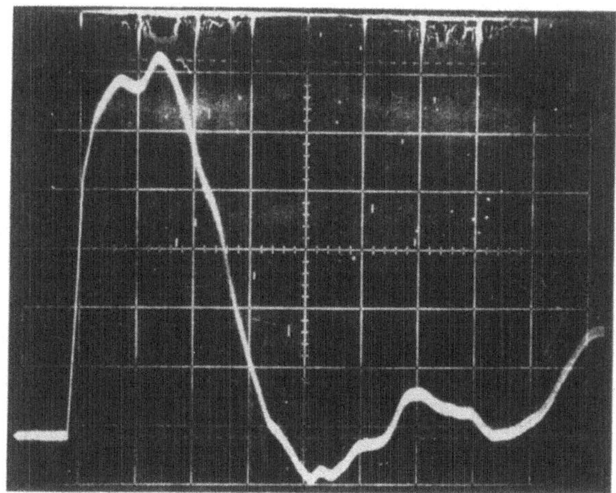

Bild 2. Laserinduzierter Druckimpuls in Wasser
Eingestrahlte Laserenergie: 50mJ
Meßentfernungsverhältnis R/a_1: 100
Horizontale Achse: 200 ns/Div
Vertikale Achse: 2 V/Div bzw. 18,2 bar/Div

nicht näher erläuterten Druckmessungen im Bereich der thermischen Expansion. Bemerkenswert ist der steile Druckanstieg über 2-3 Größenordnungen nach Überschreiten des Breakdown-Schwellwertes, d.h. nach Initialisierung der Plasma-Phase.

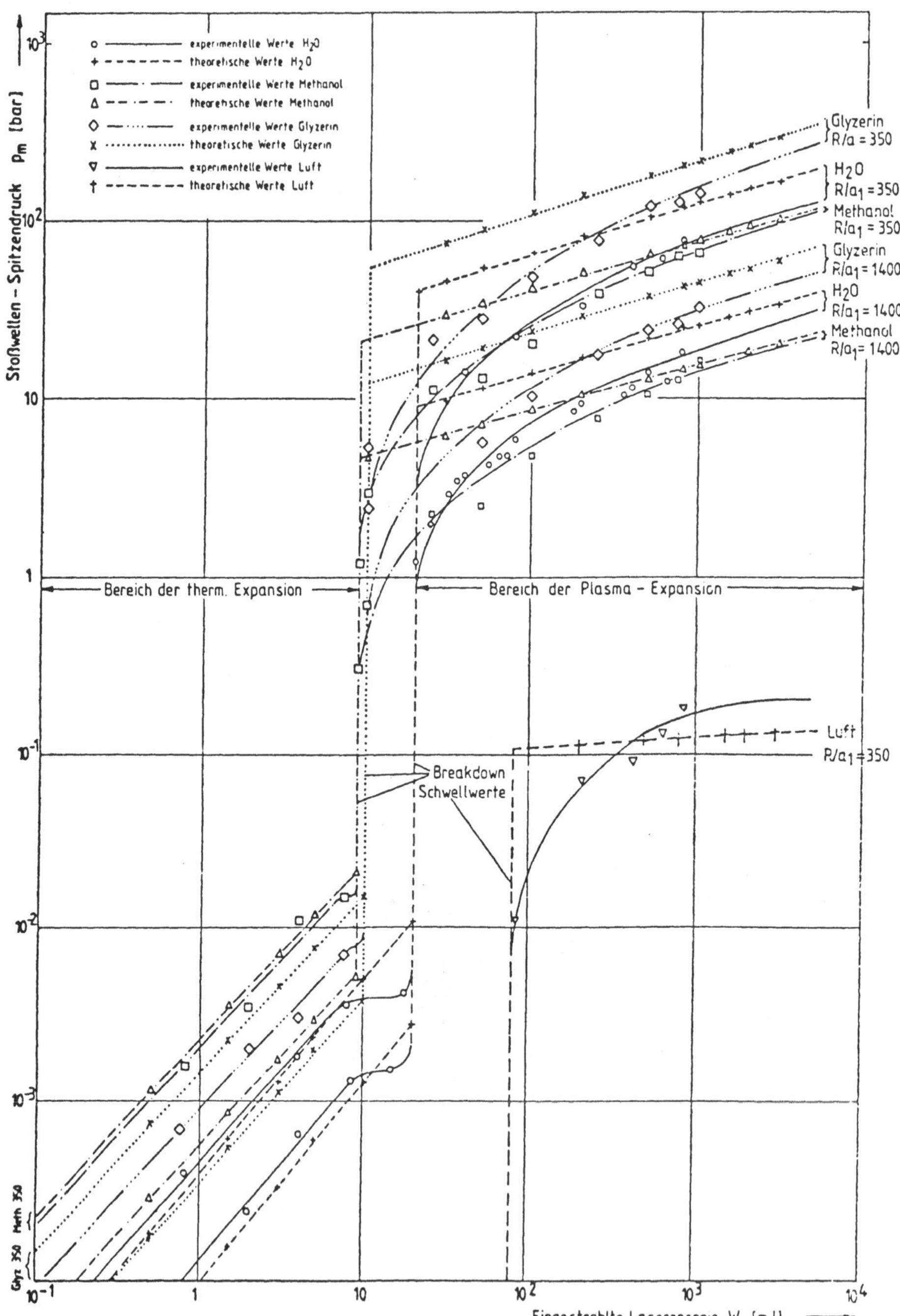

Bild 3. Abhängigkeit des Stoßwellen-Spitzendrucks p_m von der eingestrahlten Laserenergie W_0 bei einigen Flüssigkeiten und Luft
Parameter: Abstandsverhältnis R/a_1

<u>Technologische Anwendungsmöglichkeiten</u>

Die Untersuchung der technischen Anwendungsmöglichkeiten laserinduzierter Stoßwellen war ein Schwerpunkt dieser Arbeit. Es wurden im wesentlichen drei Anwendungsgebiete untersucht /1/:

1. Steuerung fluidischer Logikelemente

Es wurde die Möglichkeit der Steuerung pneumatischer und flüssigkeitsbetriebener Fluidikelemente untersucht. Beide Systeme waren einwandfrei mit Laserpulsen bis zu einer Entfernung von 25m steuerbar. Grundsätzlich sind im freien Raum Steuerungen über Entfernungen bis zu einigen km möglich, ebenso wie die Einleitung des Laserpulses durch Lichtleiter. Der Umschaltvorgang erfolgt sehr kurzzeitig, die Druckanstiegszeit beträgt maximal 0,2 ms. Es ist daher möglich, mit gepulsten Lasern hoher Taktfrequenz die bei fluidischen Schaltelementen derzeit möglichen Schaltfrequenzen voll zu nutzen und wegen der so erreichbaren hohen zeitlichen Signaldichte komplizierte fluidische Steuerungs-und Regelungsvorgänge zu erzielen. Dies ist u.a. auch unter extremen Umweltbedingungen (z.B. starke elektrische / magnetische Störfelder, hohe bzw. tiefe Temperaturen, Strahlung usw.) durchzuführen.

2. Materialumformung durch laserinduzierte Stoßwellen

Bekannt ist die Materialumformung durch Stoßwellen (Hochleistungsumformung) mit Hilfe von Explosivstoffen oder durch Funkenentladung. Trifft eine derartige Stoßwelle auf ein Werkstück, so wird dieses frei verformt oder in ein Gesenk gepreßt. Grundsätzlich in gleicher Weise erfolgt die Stoßwellenumformung durch laserinduzierte Stoßwellen. Die Stoßwelle übernimmt die Funktion des bei konventionellen Verfahren erforderlichen Oberwerkzeugs, z.B. Hammer oder Stempel. Das Umformen durch Stoßwellen hat gegenüber den konventionellen Verfahren eine Reihe von Vorteilen: Schwerverformbare Materialien können damit bearbeitet werden, Werkstoffeigenschaften werden verbessert, komplizierte Formen sind möglich, Werkzeugkosten sind geringer, Beschädigungen der Oberfläche und damit Nachbehandlung werden vermieden usw.. Die bisherigen Verfahren der Stoßwellenumformung weisen jedoch eine Reihe von Nachteilen auf, insbesondere Sicherheitsprobleme, lange Rüstzeiten, geringe Reproduzierbarkeiten. Diese Nachteile werden bei der Anwendung laserinduzierter Stoßwellen vermieden. Die Handhabung eines Lasers ist relativ einfach, die Strahlenergie gut reproduzierbar. Hohe Impulsfolgefrequenzen des Lasers ermöglichen hohe Bearbeitungstaktfrequenzen und in Verbindung mit dem einfachen Aufbau einen hohen Grad der Automatisierung.

3. Laserinduzierte Stoßwellen in der Aero-und Hydrodynamik

Infolge der erzielbaren hohen Drücke von 10^5 bis 10^7bar im Bereich und in der Nähe der Energieumwandlungssphäre ergeben sich völlig neuartige

378

Aspekte zur Untersuchung allgemeiner Probleme der Hydrodynamik und Hoch-
druckphysik, z.B. Ermittlung von Hugoniot-Daten bei Flüssigkeiten und
Gasen usw.. Insbesondere die Anwendung laserinduzierter Stoßwellen in
Stoßrohren ergibt eine Reihe von Vorteilen, z.B. genaue Bestimmbarkeit
und beliebige Einstellbarkeit der Druckamplitude der Stoßwelle; defi-
nierter Auslösezeitpunkt (triggerbar auf einige 10^{-9}s); beliebig kurzzeitig wiederholbare Erzeugung der Stoßwellen; große Anzahl von Stoß-
wellenmedien (gasfömig oder flüssig); keine Fremdstoffe im Stoßrohr
(z.B. Treibgas etc.); keine Beeinträchtigung des Strömungsverlaufs durch
platzende Membranen usw.. Bei Einsatz von Lasern mit hoher Pulswieder-
holungsrate kann praktisch eine Art von kontinuierlichem Stoßwellenbe-
trieb erreicht werden. Damit eröffnen sich völlig neue Möglichkeiten
in der Aero-und Hydrodynamik, der chemischen Analyse-und Verfahrenstechnik (z.B. Einleitung chemischer Prozesse durch intensive Druckimpulse usw
und weiteren Gebieten, in denen Stoßrohr-Verfahren angewandt werden.
Untersuchungsergebnisse zu den oben angeführten Anwendungsbereichen werden im Vortrag dargelegt und sind in der Arbeit des Verfassers /1/
näher erläutert.

(Die Arbeiten wurden als DFG- und ERP-geförderte Forschungsprojekte am
Institut für Luft- und Raumfahrt, TU Berlin, durchgeführt.)

Literatur

/1/ MUNSCHAU, J.: Theoretische und experimentelle Untersuchung zur Er-
    zeugung, Ausbreitung und Anwendung laserinduzierter Stoßwellen,
    Dissertation Berlin 1981, Technische Universität Berlin
/2/ CAROME, E.F., N.A. CLARK und C.E. MOELLER: Appl. Phys. Letters $\underline{4}$
    (1964),95
/3/ HU, C.L.: Acust. Soc. Am. $\underline{46}$ (1969), 728
/4/ HUTCHESON, L., O. ROTH und F.S. BARNES: Record of 11th Symposium
    on Electron, Ion and Laser Technology, IEEE (1971), 413
/5/ CAROME, E.F., C.E. MOELLER und N.A. CLARK: J. Acoust. Soc. Am. $\underline{40}$
    (1966),1462
/6/ BELL, C.E. und J.A. LANDT: Appl. Phys. Lett. $\underline{10}$ (1967), 46
/7/ BARNES, P.A. und K.E. RIECKHOFF: Appl. Phys. Lett. $\underline{13}$ (1968), 282
/8/ RAMSDEN, S.A. und P.SAVIC: Nature $\underline{203}$ (1964), 1217
/9/ RAIZER, Y.P.: Sov. Phys. JETP $\underline{21}$ (1965), 1009
/10/ KIDDER, R.E.: Nucl. Fusion $\underline{8}$ (1968), 3
/11/ KIRKWOOD, J.G. und H.A. BETHE: Office of Scientific Research and
    Development Report, $\underline{OSRD-588}$ (1942)
/12/ MUNSCHAU, J.: ERP-Forschungsbericht $\underline{3534}$ (1980)

Verfahren zur Strukturierung von Materialoberflächen mittels Laserstrahlung

E. HEUMANN, J. KLEINSCHMIDT, K. RÜHLE, K. VOGLER, G. WIEDERHOLD,
W. ZSCHOCKE
Friedrich-Schiller-Universität Jena, Sektion Physik,
Max-Wien-Platz 1, DDR-6900 Jena

Einleitung

In den letzten Jahren hat sich international eine rasche Entwicklung
beim Einsatz von Lasern für die automatische Markierung, Strukturierung, Beschichtung und Beschriftung von Materialien vollzogen /1/.
Bei den meisten kommerziellen Anlagen werden gütegeschaltete cw-Nd-
YAG-Laser als Strahlungsquellen verwendet. Die Strukturierung der
Materialien erfolgt im allgemeinen über thermische Effekte, die infolge einer mehr oder weniger starken Absorption der Laserenergie an
der Oberfläche bzw. im Innern des Materials auftreten. Materialien,
die die Laserstrahlung nicht absorbieren, wie z. B. transparente
Kunststoffe und Glas, können mit dem Nd-YAG-Laser auf diese Weise
nicht bearbeitet werden. Die zur Strukturierung bzw. Beschichtung
solcher Stoffe entwickelten Verfahren verlangen zur Zeit noch einen
hohen technischen Aufwand und sind deshalb bisher nicht über das La-
borstadium hinausgekommen /2/, /3/.
In diesem Beitrag wird ein Verfahren angegeben, das es gestattet,
ohne zusätzlichen Aufwand Materialien mittels Laserstrahlung zu struk-
turieren, die für die Strahlung vollkommen transparent sind.

Charakterisierung des Verfahrens

Als Strahlungsquelle dient ein cw-Nd-YAG-Laser mit einer Ausgangsleistung von 60W im Multimode-Betrieb. Die Güteschaltung erfolgt mit
einem akustooptischen Modulator und liefert Impulse von etwa 200 ns
Dauer bei einer Folgefrequenz von 1 kHz. Die mittlere Leistung im
gütemodulierten Betrieb beträgt 20 W. Über ein programmgesteuertes
Ablenksystem gelangt die Laserstrahlung durch eine Fokussierungsoptik
zur Bearbeitungsstelle (Abbildung 1). Das zu strukturierende Material
(z. B. eine Glasplatte) wird mit der der Strahlung abgewandten Seite
in die unmittelbare Nähe einer Metalloberfläche bzw. mit dieser in
Kontakt gebracht. Die auffallende Laserenergie durchdringt das trans-

parente Material und wird von der Metallunterlage absorbiert. Das auf

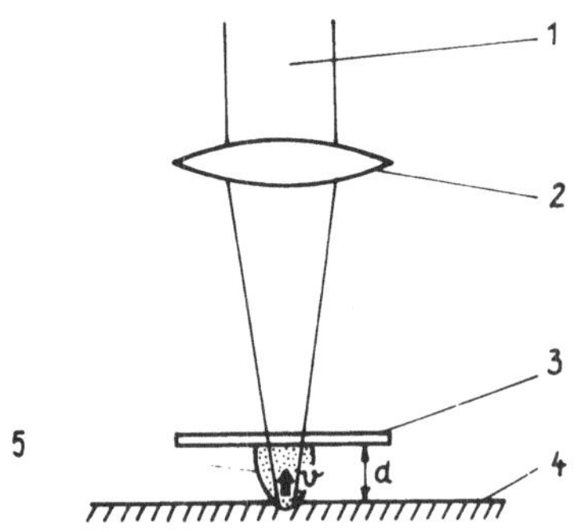

1 laser beam   4 metal surface
2 focusing lense   5 plasma plum
3 glass plate

Abb. 1 Schematische Darstellung des Verfahrens

der Metalloberfläche entstehende Plasma erzeugt eine Schockwelle, die
sich entgegen der Richtung des Laserstrahles ausbreitet und das trans
parente Material auf der dem Metall zugewandten Oberfläche struktu-
riert /4/.
Das Verfahren funktioniert im gesamten Wirkungsbereich der Schockwelle d. h. auch dann, wenn das transparente Medium keinen direkten
Kontakt mit der absorbierenden Unterlage hat. Als Unterlagen können
alle Metalle verwendet werden insbesondere auch diejenigen, die bei
der Wellenlänge des YAG-Lasers nur eine geringe Absorption zeigen
(z. B. Kupfer, Messing, Silber, Gold). Die erzielten Strichbreiten
und Strukturtiefen hängen von den Fokussierungsbedingungen und vom
Abstand zwischen dem zu strukturierenden Material und der Metalloberfläche ab.

Ergebnisse und Diskussion

Abbildung 2 zeigt als Beispiel eine mit dem angegebenen Verfahren
gravierte Glasplatte. Als Metallunterlage diente Stahlblech mit einer
Dicke von 0,5 mm.

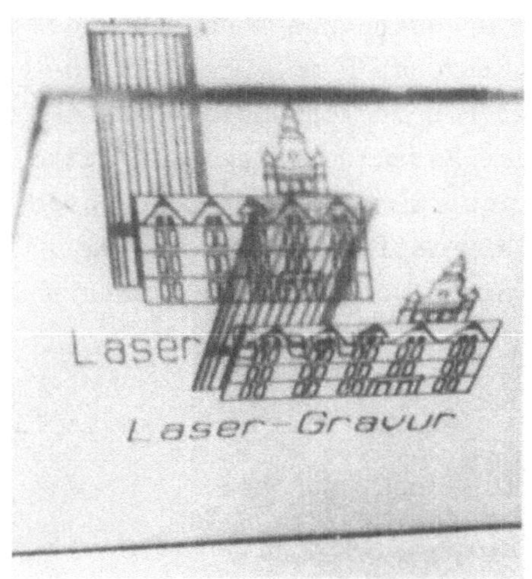

Abb. 2 Gravur auf einer Glasplatte

Mit der gleichen Methode wurden bei Verwendung von Kupfer und Messing
als Unterlagen fest haftende Metallstrukturen auf Glasoberflächen und
auf anderen transparenten Materialien erzeugt.
Der physikalische Hintergrund der Resultate läßt sich wie folgt beschreiben. Die Laserimpulse hoher Intensität ($I \sim 10^8$ W cm^{-2}) bewirken selbst an hochreflektierenden Metalloberflächen mit geringer Rest-
absorption eine extreme Aufheizung des Materials im Fokusbereich der
Strahlung. Dabei können senkrecht zur Materialoberfläche Temperatur-
gradienten bis zu 10^6 grd cm^{-1} auftreten. Das führt zu einer explosionsartigen Verdampfung des Materials /5/, /6/. Die verdampfenden
Partikel expandieren in einer Plasmawolke mit Geschwindigkeiten v
bis zu einigen 10^5 cm s^{-1} und erzeugen in der Metallprobe einen starken Rückstoß, der sich als Schockwelle in diesem Material ausbreitet.
In Abhängigkeit von der Laserleistung und den Materialparametern
können dabei Druckwerte zwischen 10^2 und 10^5 atm auftreten /5/, /6/.
Aus der Messung der Deformation dünner Metallfolien konnten im Vergleich mit statischen Druckmessungen Werte von etwa 300 atm bei dem
hier verwendeten Verfahren ermittelt werden.
Die sich mit dem expandierenden Plasma entgegen der Richtung der
Laserstrahlung ausbreitende Druckwelle führt zu lokalen Zerstörungen
auf der Oberfläche des zu strukturierenden Materials. Gleichzeitig
werden von der Druckwelle Metallpartikel mitgerissen. Je nach Wahl
der Laserparameter (Leistung, Impulsdauer, Folgefrequenz) kann das
zu bearbeitende transparente Material graviert oder mit den Metall-
partikeln der Unterlage bedampft werden.

Die aus der Lichtemission der Plasmawolke abgeschätzte Ausdehnung des Plasmas senkrecht zur Metalloberfläche betrug ca. 5 mm. Die Strukturierung (Gravur oder Beschichtung) erfolgt bis zu Abständen von 2 mm zwischen dem transparenten Material und der Metallunterlage. Im Multi modenbetrieb variiert die erreichbare Strichbreite im Abstandsbereich von 0 bis 2 mm zwischen 0,1 und 0,5 mm. Beim Betrieb des Lasers im transversalen Grundmodus konnten Strichbreiten von 50 bis 100 μm erzielt werden.

<u>Literatur</u>

/1/ Informationsmaterial von Laser-Optronic GmbH, München;
    J. BUCHHOLZ "Laserbeschriftungs- und Markierungsarbeiten"
/2/ D. J. EHRLICH, R. M. OSGOOD, T. F. DEUTSCH, IEEE
    J. of Quant. Electr., QE-16 (1980) 1233
/3/ M. NOWICKI, Vortrag zur Hauptjahrestagung der Physikal.
    Gesellschaft der DDR, Leipzig 1982
/4/ E. HEUMANN, J. KLEINSCHMIDT, K. VOGLER, G. WIEDERHOLD,
    DDR-Patent: WP B44 B/242 261/7
/5/ J. F. READY, Proc. IEEE 70 (1982) 533
/6/ J. F. READY, Effects of high power laser radiation,
    Acad. Press, New York (1971)

Metal Drilling with Lasers

H.-G.TREUSCH, L.BAKOWSKY, K.WISSENBACH, G.HERZIGER
Institut für Angewandte Physik, Technische Hochschule Darmstadt
Schloßgartenstraße 7, D-6100 Darmstadt

1. Introduction

Since 1972 solid state lasers are used as an industrial tool for drilling holes
of precision in ruby and ceramic substances /1,2/.   In achieving the standard of
spark erosive removal of material the laser is superior in metalworking by higher
processing velocities.   Deviation from roundness and cylindricity are below 5 % and
the roughness of the hole surface is less than 5 µm for the erosive process /3,4/.
These tolerances are reached in laser material processing within accurate knowledge
and control of the processing and laser parameters: Intensity I, beam radius r_F and
processing time t_P .

Metalworking differs from those of dielectric substances by losses due to high
reflectance and greater thermal conductivity . The laser parameters have to be
balanced in metal drilling to the material properties like absorption rate and
thermal diffusivity. The drilling process is influenced by melt expulsion,
absorption and refraction of the laser radiation in the expanding vapour. To reach
the aspired hole qualities, the drilling process has to be controlled by a temporal
modulation of the laser power /5/.

For pulsed Nd-YAG laser systems power modulation can be done by pulse cutting
outside and by loss-modulation inside the resonator /6/ . Continous lasers are
modulated for drilling by Q-switching up to high repetition rates (100 kHz).
Precise material removal by laser radiation is only possible in a narrow intensity
range. The process is limited down to smaller intensities by losses due to heat
conduction and up to higher intensities by plasma shielding and lensing of the
radiation /7/.

2. Arrangement of laser processing

A typical arrangement of laser material processing is shown in figure 1.   The
laser radiation of power P is focussed by an optical system to the radius r_F. Laser
systems of TEMoo-mode radiation are commonly used for hole drilling.   These systems
yield a Gaussian intensity distribution at the target surface with an intensity

$$\text{Imax} = 2\ P/\pi r_F^2 \quad ; \quad r_F = w_0\ f/\sqrt{z_R^2+(z-f)^2} \qquad\qquad (1)$$

w_0 = beam waist ($1/e^2$ of intensity)
z_R = Rayleigh length
z = distance from beam waist

in the beam center. The Rayleigh length is given by the distance from beam waist for twofold beam cross section.

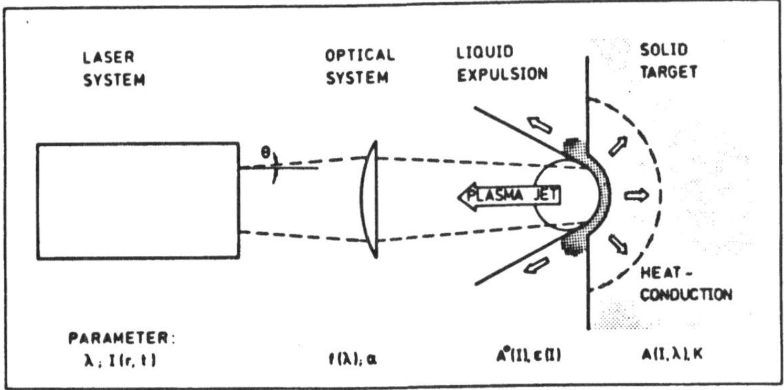

Fig.1. Scheme of laser material processing

Laser radiation is reflected back into the resonator by placing a reflecting target in the focal plane of the system. Target, optical system and laser mirrors represent coupled resonators inducing stochastic fluctuations of temporal and spatial behavior of the laser beam /8/ by the varation of target surface properties. Elimination of optical feedback is possible by inserting Faraday rotators between laser and target. In special cases a combination of linear polarizer and quarter wave plate may be useful.

3. Properties of laser drilling

After heating to the boiling temperature the material removal starts by vapourization and melt expulsion. The ratio of vapourized and liquid material removal depends on laser intensity and changes in favour of the vaporization process for higher intensities. The vapour expanding towards the laser is ionized by the incident radiation. The resulting plasma influences material processing properties not only by refraction and absorption but also by plasma target interaction.

3.1. Target heating

The intensity absorbed by the target surface only is important for material processing. The absorptance A depends on target material and laser wavelength (table 1). The absorbed intensity is converted into heat within the penetration depth δ of laser radiation. The penetration depth of Nd-YAG laser radiation is only a few nanometers for metals. The target heating occurs by thermal conduction out of the absorption volume $V_A = \delta \pi r_F^2$. In addition to the losses of reflection the losses due to heat conduction have to be considered . In case of small penetration depth $(\delta \ll 2 \varkappa t_p)$ the losses caused by heat conduction can be calculated by heat equation including a plane heat source. Therefore the processing parameters of target heating are the absorbed intensity A*Imax, the beam radius r_F and the thermal diffusivity of the material. Based on the condition that

TABLE I			
λ = 1,06μm	STEEL	Al	Cu
A	0.3	0.12	0.06

$$A*Imax \; r_F > \sqrt{8}\, k \; Tv \qquad\qquad (2)$$

k = thermal conductivity

Tv = boiling temperature

the losses can be determined by a one-dimensional calculation.   In case of copper and Imax= 10^8W/cm² the lower limit for the beamradius $r_F <$ 40 μm, where three-dimensional losses can be neclected is given by equation (2).

3.2. Material removal

In exceeding thermal losses by the absorbed laser power the material removal starts by vapourization after a material-dependent heating time. Starting vapourization an abrupt increase of absortance can be observed by on-line reflectance measurements. The laser power is coupled almost totally into the material.

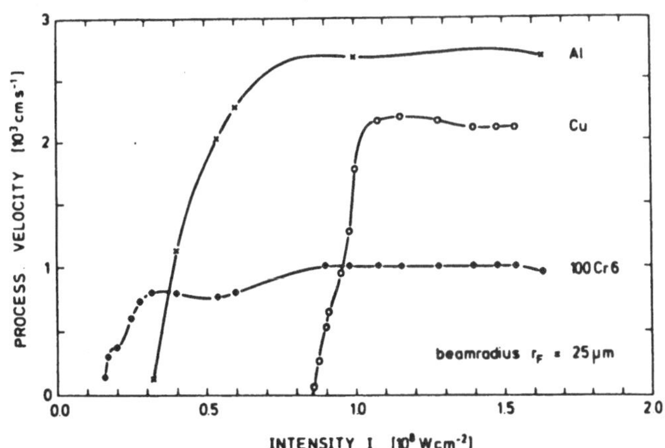

Fig.2. Processing velocity as a function of intensity. Materials: Stainless steel, copper and aluminium. Data from drilling foils of 100 μm thickness

The change in absorptance is reduced to the plasma and the variation of surface geometry (crater formation). The interface vapour-melt moves into the material with an almost constant velocity, driven by vaporization and liquid material removed due to the vapour pressure. The processing velocity v_{PR} shows a characteristic progress as a function of laser intensity in drilling holes of a limited depth (figure 2). Above an intensity threshold given by losses due to reflection and heat conduction the processing velocity unsteadily increases from zero to a material specific value and stays constant with increasing intensity. The saturation of the velocity is reduced to absorption and refraction of the radiation by the expanding vapour plasma and the resulting shielding of the workpiece. The processing velocity

$$v_{PR} = v(r_F, I, z, material)$$

can be described as a function of laser parameters and material properties. The hole depth z (figure 3) increases with $v_{PR}*t$ /9/ and reaches a limiting value $\hat{z}$, which depends on intensity, material properties and beam geometry with respect

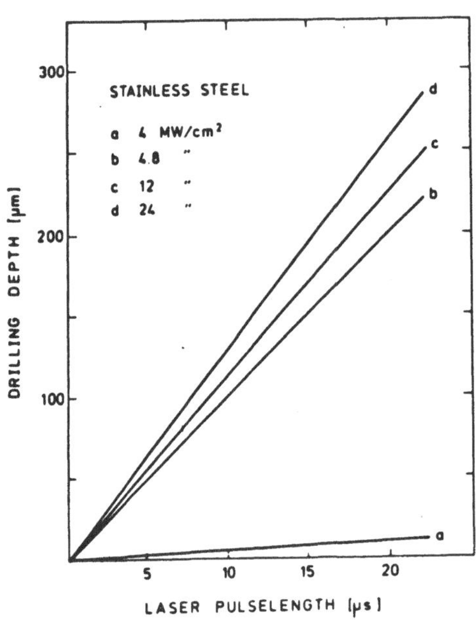

Fig.3. Drilling depth as a function of time in the beginning of material removal for stainless steel.

to the target surface. The geometry of the focussed beam is defined by the Rayleigh length

$$\Delta z_R = f^2 z_R/((z-f)^2 + z_R^2),$$

which is comparable to the focus depth of the optical system. The maximum processing depth $\hat{z}$ and the expected maximum value $z = v_{PR}*t_P$ is shown in fig. 4 as a function of intensity. The deviations from the expected value are due to

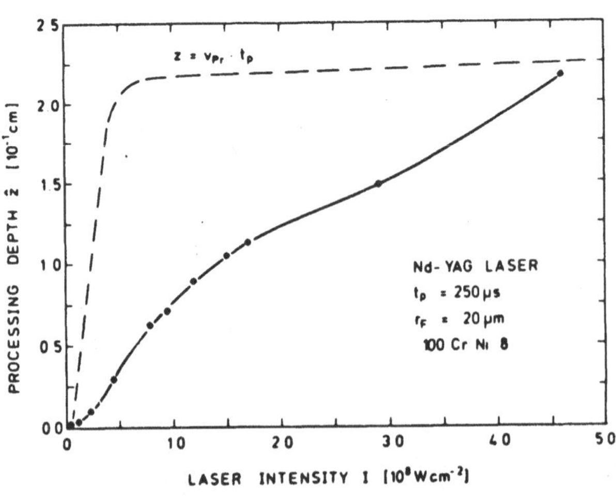

Fig.4. The attainable maximum of processing depth as a function of intensity for stainless steel.

absorption in the vapour plasma and closing of the hole by liquid material.

The main processes in laser material removal are the losses due to reflection and heat conduction, the optical feedback, material removal by vapour and melt expulsion as well as the influence of a laser induced plasma to the material and laser radiation. For hole drilling these processes must be adjusted with respect to the given quality standards. Process adjustment /10/ by simple variation of parameters is impossible due to the gearing and mutual influence of the particular processes (figure 5). Rather a physical description of the processing properties is rather required including their mutual influenes.

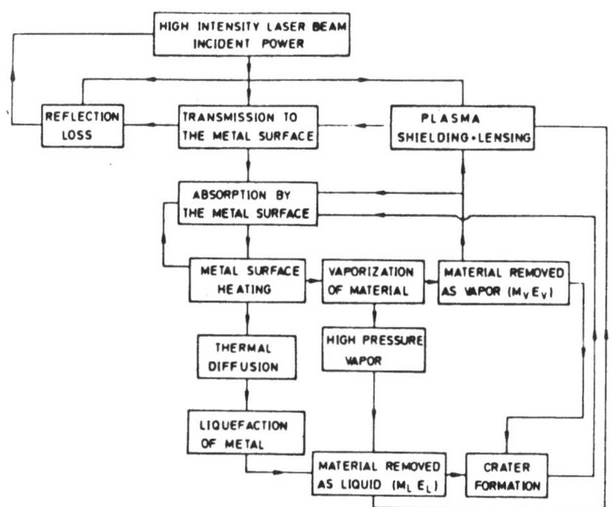

Fig.5. Block diagram for the main processes of material removal and their mutual influences.

4. Parameters of laser drilling

The description of physical properties first of all fixes the limits of laser drilling with respect to intensity and processing time. The given range (figure 6) is valid for stainless steel, an optical system of focussing length f=75mm and focus length Δz =1mm as well as a processing without power modulation. The range of optimum drilling can be expanded by temporal modulation to higher intensities for materials with high thermal diffusivity like copper. For the processing of stainless steel the modulation of laser power is improvement, if the peak power increases up to a value below the threshold for laser induced plasma.

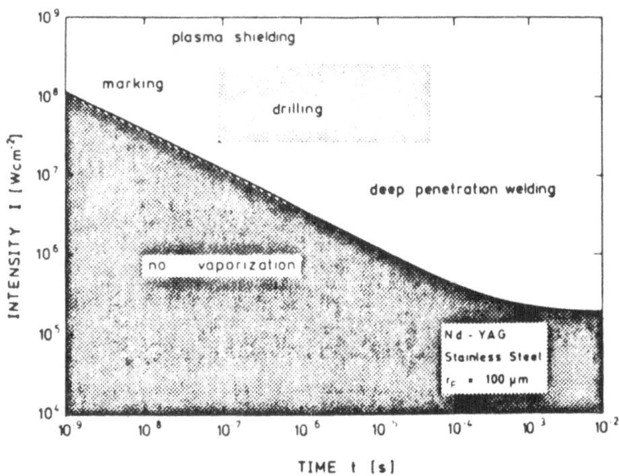

Fig.6. Range of optimum drilling of metals with respect to intensity and processing time.

388

5. Summary

Knowledge and control of the physical mechanisms are important at the current
stage of development in laser material processing, if optimum results should be
reached in respect of quality and economy. Inclusion and control of the optical
feedback have to be considered for reproducible material processing. An accurate
adjustment of absorbed laser power and heat losses is important to control the
process of metal drilling. Assuming the kinetic energy of the molten and vapourized
material to be sufficent to exceed surface tension everywhere in the drilled hole,
no liquid and condensed material is deposited within the range of the produced hole.
The temporal modulation of lasers can improve at the same average power the material
removal and the hole quality.

Literature
1   E.Kocher,L.Tschudi,J.Steffen and G.Herziger
     IEEE J. of Quantum Electronics,Vol, QE-8,No.2,85 (1972)
2   UN-CHUL PEAK and F.P.GAGLIANO
     IEEE J. of Quantum Electronics,Vol. QE-8,No.2,112 (1972)
3   O.Rüdiger and A.Winkelmann, Metall,Jg.12,H.5 (1958)
4   H.Schierholt, Technisches Zentralblatt für prakt. Metallbearbeitung,70 (1960)
5   J.Steffen, Conference on electrical methods of machining Form 283 (1978)
6   G.Herziger, R.Stemme and H.Weber
     IEEE J. of Quantum Electronics,Vol. QE-10,No2,175 (1974)
7   R.Poprawe et al.,these proceedings
8   E.Beyer et al.,these proceedings
9   M.v.Allmen, Thesis, University of Berne (1975)
10   M.K.Chun and K.Rose, J. of Appl. Physics,Vol.11,No.2,612 (1969)

Variable Dämpfung von Laserstrahlung hoher Intensität bei Präzisionsbearbeitungen

J. JUNGHANS, E. HUBER
GRETAG Aktiengesellschaft, Althardstrasse 70, CH 8105 Regensdorf

Zur Leistungsvariation (Dämpfung) von Laserstrahlung sind die verschiedensten Methoden und Systeme bekannt, die jedoch mit diversen Nachteilen behaftet sind. Dämpfung kann erreicht werden, indem direkt in den Strahlungserzeugungsprozess im Laserresonator eingegriffen wird – interne Dämpfung – oder die Strahlung erst ausserhalb des Resonators geschwächt wird – externe Dämpfung. Als gebräuchlichste Verfahren werden verwendet:

extern	intern
- Absorptionsfilter	- Variation über die Anregung
- Polarisator und Analysator (elektro- bzw. magnetooptisch)	- Polarisator und Analysator (elektro- bzw. magnetooptisch)
- akustooptischer Kristall	- akustooptischer Kristall
- reflektierende Oberflächen	- Auskoppelmodulation
- Prismen mit "frustrierter Totalreflexion" /1/	
- Zerhackerscheiben	
- statistische Lochscheiben	

Bei präzisen Lasermaterialbearbeitungen, optischen Messungen, Empfängerkalibrierungen, etc. werden Abschwächer benötigt, welche den nachfolgenden Anforderungen genügen sollten:

- kontinuierliche Variation des Dämpfungsgrades;
- Verwendung hoher Strahlungsintensitäten;
- keine Beeinflussung der Laseremission (keine Veränderung der "thermischen Linse" der optischen Komponenten);
- keine Strahlversetzung und Richtungsänderung;
- keine Erwärmung und dadurch Veränderung des Dämpfungsverhaltens;
- keine Veränderung der Strahldivergenz;
- keine Veränderung der geometrischen Intensitätsverteilung des Strahles;
- keine wesentliche Aenderung des Polarisationszustandes der Strahlung (Eignung sowohl für polarisiertes wie auch unpolarisiertes Licht).

Die von uns vorzustellende Anordnung verwendet planparallele Platten mit dielektrischer Beschichtung. Die Dämpfung wird durch Neigen der Platten eingestellt. Fällt ein Lichtstrahl auf eine drehbare Glasoberfläche, so ändert sich das transmittierte Licht in Abhängigkeit des Einfallswinkels, je nach vorhandener Polarisationsrichtung, gemäss

390

der bekannten Formeln zu:

$$n_i \sin\theta_i = n_t \sin\theta_t \qquad R_s = \frac{\sin^2(\theta_i - \theta_t)}{\sin^2(\theta_i + \theta_t)}$$

$$R_p = \frac{\tan^2(\theta_i - \theta_t)}{\tan^2(\theta_i + \theta_t)} \qquad T_p = 1 - R_p,\ T_s = 1 - R_s$$

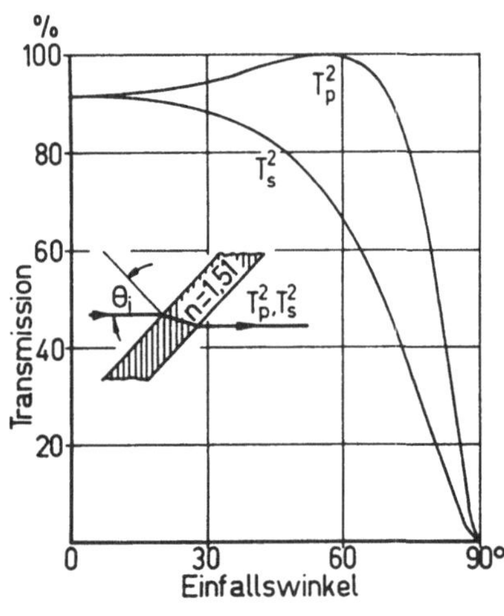

Abb. 1. Transmission eines Lichtstrahles durch eine planparallele Platte BK7 ohne Berücksichtigung von Mehrfachreflexionen

Eine kontinuierlich neigbare Platte ist die einfachste Form eines variablen Dämpfungsgliedes. Da die Dämpfung über veränderliche Reflexion erfolgt, kann diese Anordnung auch bei sehr hohen Intensitäten verwendet werden; begrenzt nur durch die Zerstörungsschwelle des Plattenmaterials.

Die endliche Plattendicke ergibt leider eine Strahlversetzung der einfallenden Strahlung. Diese Strahlversetzung lässt sich durch 2 mit gleichem Winkel sich gegeneinander neigende Platten beheben. Die Transmissionskurve ergibt allerdings noch unterschiedliche Werte für senkrecht- bzw. parallelpolarisiertes Licht: Polarisationszustand von ein- und austretender Strahlung unterscheiden sich.

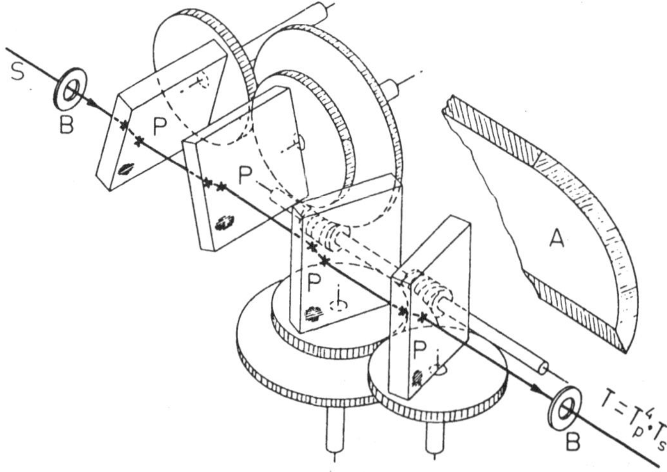

Abb. 2. Anordnung von 2 um 90° verdrehten Plattenpaaren zur Dämpfung von Strahlung: keine Strahlversetzung und keine Veränderung des Polarisationszustandes

S = Lichtstrahl
B = Blende
P = drehbare Platten
A = Strahlauffänger

Werden 2 dieser Plattenpaare in der optischen Achse gegeneinander um 90° verdreht, so wird die s-(p-) Komponente der einfallenden Strahlung der ersten Anordnung mit der p-(s-) Komponente der zweiten Anordnung

gefaltet; Strahlversetzung und Variation des Polarisationszustandes
sind nun behoben. Die Einstellung der Dämpfung erfolgt, wie Abb. 2
zeigt, durch konformes Neigen der Platten.

Mit ihren 8 reflektierenden Oberflächen (4 Platten) beträgt die Grunddämpfung bei senkrechtem Lichteinfall bereits 29 %. Wir haben deshalb
die Platten mit einer dielektrischen Antireflexionsbeschichtung belegt.
Hierdurch wird die Grunddämpfung auf vernachlässigbare Werte verringert. Die Dämpfungskurve ist jetzt - Vor- oder Nachteil - für die
berechneten Wellenlängen (meistens eine) festgelegt.
Wie aus Abb. 1 ersichtlich, steigt die Dämpfungskurve, ohne Verwendung
dielektrischer Beschichtung, erst bei hohen Neigungswinkeln stark an.
Auch dieser Verlauf lässt sich durch geeignete Auswahl der Beschichtung
beeinflussen.

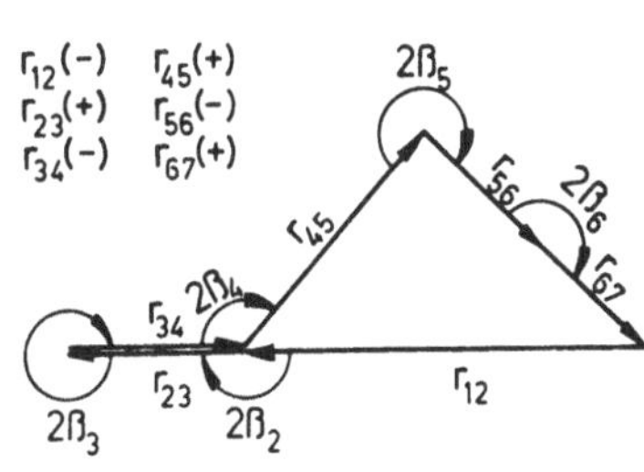

Die Berechnung der Beschichtung wird mit
5 dielektrische Schichten dargelegt /2/.
Das aus den Teilreflexionen

$$r_{12} = \frac{n_1 - n_2}{n_1 + n_2} \; ; \; r_{23} = \frac{n_2 - n_3}{n_2 + n_3} \; ; \; \ldots$$

gebildete Vieleck muss sich bei Antireflek-
tionsbeschichtung schliessen. Die Winkel
an den Ecken ergeben nach der Beziehung

$$\beta_i = \frac{2\pi n_i d_i}{\lambda_0} \qquad d_i = \frac{\beta_i \lambda_0}{2\pi n_i}$$

Abb. 3. Prinzip eines
Reflexionsvieleckes für
Antireflexbeschichtung

die dielektrische Schichtdicke. Eine Abschätzung des Dämpfungsverlaufes in Abhängigkeit vom Neigungswinkel ergibt die Vor-
stellung des sich öffnenden Vieleckes;
die Veränderung der Teilreflexionen wird hierbei vernachlässigt. Die Berechnung des Dämpfungsverlaufes führt vom Einschichtsystem zum Viel-
schichtsystem /4/,

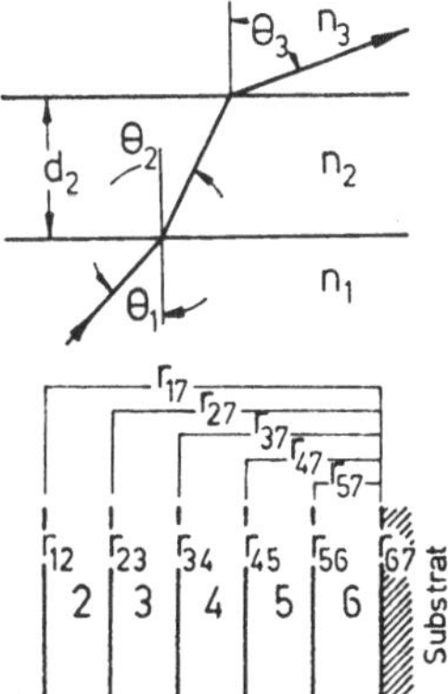

$$r_{p12} = \frac{n_1 \cos\theta_1 - n_2 \cos\theta_2}{n_1 \cos\theta_1 + n_2 \cos\theta_2}$$

$$r_{s12} = \frac{(\cos\theta_1/n_1) - (\cos\theta_2/n_2)}{(\cos\theta_1/n_1) + (\cos\theta_2/n_2)}$$

$$r_{p23}, r_{s23}, \ldots \quad \text{analog}$$

$$r_{57} = \frac{r_{67} + r_{56} e^{2i\beta_6}}{1 + r_{67} r_{56} e^{2i\beta_6}} \quad r_{p47}, r_{s47} \ldots r_{p17}, r_{s17}$$

$$r_{p57} = f(r_{p56}, r_{p67}, \beta_6)$$
$$r_{p47} = f(r_{p45}, r_{p57}, \beta_5)$$
$$r_{p17} = f(r_{p12}, r_{p27}, \beta_2) \qquad R = |r_{17}|^2$$
$$\beta_2 = 2\pi n_2 (d_2 \cos\theta_2)/\lambda_0$$

Abb. 4. Transmission eines Strahles durch ein
homogenes, geschichtetes Medium

wobei n_i der betreffende Berechnungsindex, Θ_i der Strahlwinkel, d_i die Schichtdecke, λ_0 die gewünschte Vakuumwellenlänge und r_{ij} die Teilreflexion ist. Für die ganze Anlage ergibt sich die totale Transmission zu

$$T = T_p^{\,4} \cdot T_s^{\,4}$$

Durch die von uns verwendete Antireflexionsbeschichtung werden Veränderungen der Intensitätsverteilung des Strahles durch innere Mehrfachreflexionen vermieden. Werden grosse Strahlquerschnitte (grösser 4 mm) verwendet, kann die Beschichtung derart bestimmt werden, dass das Transmissionsmaximum bei höheren Werten (z.B. 10°) liegt. Die dann bei zunehmendem Neigungswinkel beginnenden Mehrfachreflexionen überlagern sich nicht mehr dem ursprünglichen Strahl.

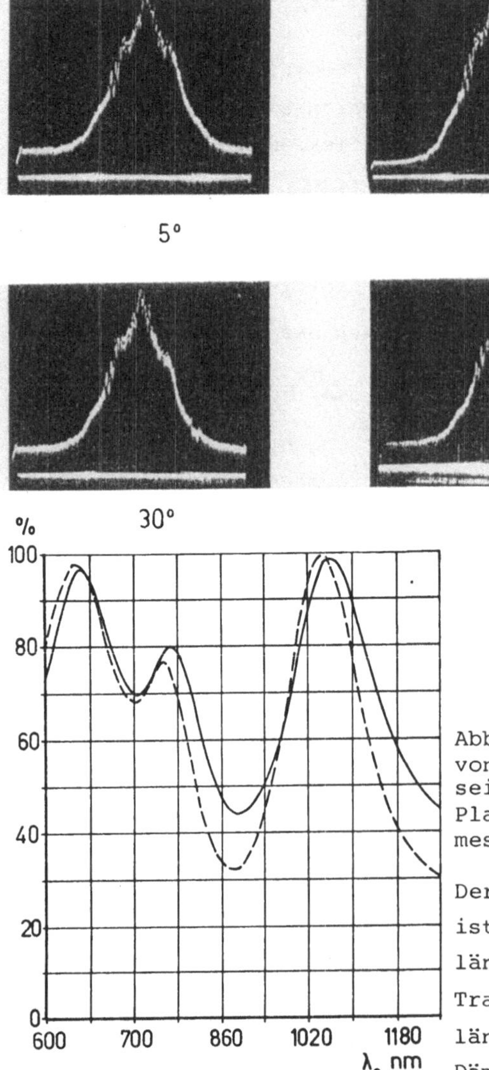

Abb. 5. Relative Intensitätsverteilung für verschiedene Neigungswinkel des transmittierten Strahles durch die in Abb. 2 gezeigte Anordnung.

Die Messung der relativen Intensitätsverteilung (Abb. 5) wurde mit einem Diodenarray (1024 Dioden) durchgeführt. Die der Intensitätsverteilung überlagerten Interferenzerscheinungen sind auf allen Messungen gleich und rühren von Reflexionen der Glasschutzschicht des Diodenarrayes her.

Abb. 6. Transmission in Abhängigkeit von der Wellenlänge durch eine beidseitig beschichtete, planparallele Platte --- berechnete Werte, —— gemessene Werte

Der Reflexions- bzw. Dämpfungsverlauf ist aufgrund der Beschichtung wellenlängenabhängig (Abb. 6). Je steiler der Transmissionsanstieg zu längeren Wellenlängen hin abfällt, desto höher ist die Dämpfung bei steigendem Neigungswinkel.

Die gesamte Anlage besitzt bei senkrechtstehenden Platten eine minimale
Dämpfung von nur 4 % und erreicht bei einem Plattenneigungswinkel von
75° einen Abschwächungsfaktor grösser als 1000 (Abb. 7). Die an den
Plattenoberflächen reflektierte Leistung fällt auf einen Auffänger und
wird über das Gehäuse als Wärme abgeführt.

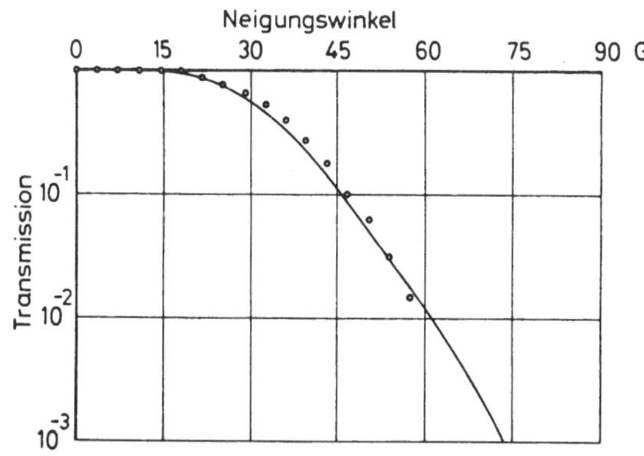

Abb. 7. Transmission der gesamten Anlage (Abb. 2) bei
1.06 um. --- berechnete Werte, o Messwerte

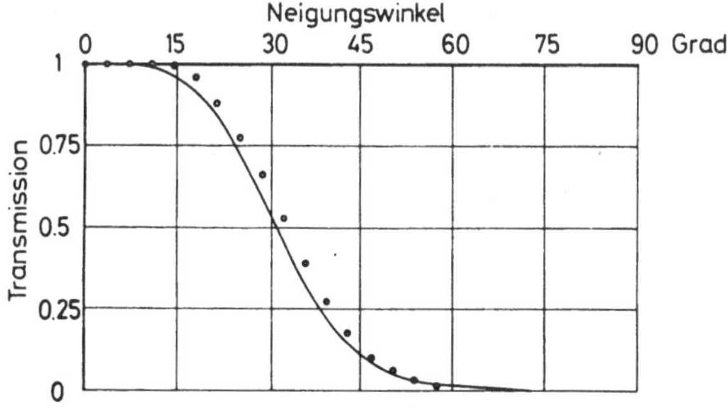

Zusammenfassung

Wir haben gezeigt, dass es möglich ist, ein Gerät mit kontinuierlicher
Abschwächung mit hoher optischer Qualität, beginnend bei einer
vernachlässigbaren Dämpfung bis zu extrem hohen Werten (Faktor grösser
1000) herzustellen. Werden Dämpfungsverläufe höherer oder flacherer
Steilheit in Abhängigkeit vom Neigungswinkel benötigt, so ist dieses
durch geeignete Wahl der Beschichtungen möglich.

Literatur

/1/ G.T. SINCERBOX, J.G. GORDEN, "Modulating light by attenuated total
    reflection", Laser Focus, Nov. 1981, S. 55 - 58
/2/ H.A. MACLEAD, "Thin-Film Optical Filters", London, Ad. Hilger 1969
/3/ K. BENNET, R.L. BYER, "Variable Laser Attenuation - Old and New"
    Laser Focus, April 1983, S. 55 - 62
/4/ M. BORN, E. WOLF, "Prinziples of Optics", Pergamon Press, Oxford
    1975, Kapitel 1.5

Industrielle Einsatzmöglichkeiten von Lasermaterialbearbeitungsmaschinen mit periodisch gepulsten Lasern

J. JUNGHANS
GRETAG Aktiengesellschaft,
Althardstrasse 70, CH 8105 Regensdorf

Der industrielle Benützer heutiger Lasermaterialbearbeitungsmaschinen
wünscht ein Gerät mit ausgereifter Technik und hohem Bedienungskomfort.

Natürliches, d. h. inkohärentes Licht lässt sich aufgrund der Gesetze
der geometrischen Optik nie über die Leistungsdichte der erzeugenden
Quelle erhöhen. Im Punkt der fokussierten Strahlung kann maximal die
Temperatur der Quelle errreicht werden. Beim Laser ist die Fokussierbarkeit jedoch nur durch den Divergenzwinkel der Strahlung begrenzt.

Beim Auftreffen von Strahlung auf eine Materialoberfläche wirkt nur die
Absorption durch thermische Beeinflussung auf das zu bearbeitenden
Material. Da die Eindringtiefe in stark absorbierenden Materialien
gering ist, wächst die Oberflächentemperatur im Zentrum der
fokussierten Strahlung sehr rasch an, und die thermische Beeinflussung
des umliegenden Materials wird somit auf ein Minimum beschränkt. Das
Absorptionsverhalten ändert sich stark mit der Oberflächentemperatur
des zu bearbeitenden Metalles /1/.

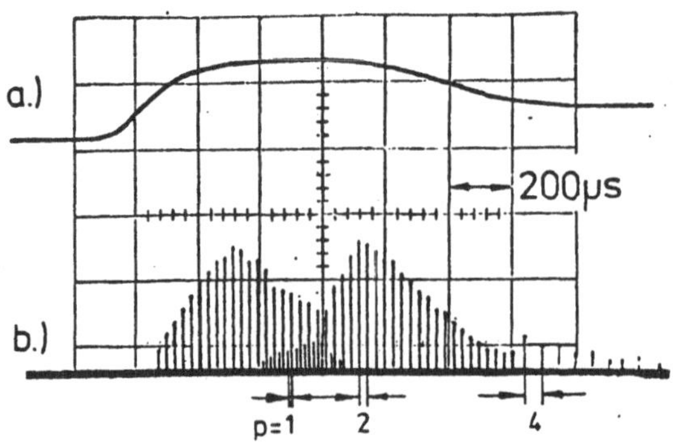

Abb. 1. Beispiel eines Pulszuges periodischer Q-switch Pulse (b). Die Repetitionsfrequenz ändert sich bei konstant gehaltener Modulationsfrequenz (75 kHz) in Abhängigkeit von der Pumpintensität (a) /2/.

Um die für die Bearbeitung hohen Temperaturen erreichen zu können, verwenden wir einen periodisch gepulsten Nd:YAG-Laser. Die emittierte Strahlung besteht aus Pulszügen je nach verwendeter Repetitionsfrequenz mit Pulsbreiten von einigen 10 ns bis zu ungefähr einer Mikrosekunde. Die Spitzenintensität liegt bei einigen kW/cm^2. Der steile Pulsanstieg bringt uns augenblicklich in das Gebiet hoher Absorption. Die emittierte Strahlung folgt exakt der Modulationsfrequenz nur bei hohen Pumpleistungen oder niedrigen Modulationsfrequenzen läuft /2/.

Für eine gute Oberflächenbearbeitung (Beschriften, Gravieren, Verfärben) arbeitet man bei Metallen mit Modulationsfrequenzen um 5 kHz und Intensitäten von 20 MW/mm^2. Bei Nichtmetallen (hauptsächlich Kunststoffen) muss durch Versuche abgeklärt werden, ob

- das Grundmaterial
- oder das beigemischte Farbpigment thermisch verfärbt werden
- oder thermisches Aufschäumen erzielt werden soll.

Zu hohe Intensitätsdichten führen hier zu einer feinen Brandspur. Auch diese schwarze Brandspur lässt sich für Beschriftungen verwenden. Werden jedoch Beschriftungen mit Farbeffekt benötigt, müssen Strahlintensität, Modulationsfrequenz und Verfahrgeschwindigkeit genau bestimmt und eingehalten werden. Die optimalen Daten werden in der Regel durch einen kurzen Versuch ermittelt. Die notwendigen Intensitäten sind bei Beschriftungen von Kunststoffen normalerweise kleiner als bei Metallen und die Modulationsfrequenz höher (grösser 10 kHZ).

Beim Beschriften ist nicht unbedingt die Gesamtleistung des Laserstrahles ausschlaggebend, sondern die Intensität (Leistungsflussdichte) auf bzw. im zu bearbeitenden Material. In der Regel wird mit Gaussstrahlen quasi nullter Ordnung gearbeitet. Werden breitere Beschriftungen verlangt, arbeitet man mit einer elektronsichen Strichverbreiterung.

Die Strahlablenkung erfolgt über 2 Galvanometerspiegel mit nachfolgender Fokussierlinse. Eine Konstanthaltung der Fokussierebene mittels eines verstellbaren Strahlaufweiters gewährleistet die Einhaltung des Beschriftungsmassstabes.

Für die verwendete Laserstrahlung transparente Materialien lassen sich durch sog. "Lasersputtering" beschriften. Hierzu liegt dieses transparente Material auf einer gut absorbierenden und verdampfbaren Unterlage (in der Regel Metall). Der Laserstrahl wird durch das Werkstück hindurch auf das absorbierende Material fokussiert. Dieses Material verdampft und schlägt sich in transparentem Material gut sichtbar

nieder. War der Laserstrahl exakt auf die Unterlage fokussiert, und
wird nun das transparente Material mit dem Brechungsindex n_2 und der
Dicke d auf die Unterlage gelegt, so erhöht sich der Fokusabstand um
die Länge d.$(1 - n_1/n_2)$; n_1 ist in der Regel 1 (Luft). Der Durchmesser
des fokussierten Laserstrahles ändert sich nicht.

Eine Lasermaterialbearbeitungsmaschine besteht aus:

- dem eigentlichen Laser;
- dem Handling;
- der Steuerelektronik;
- der Strahlführung;
- der Dateneingabe
- und der Software.

Laser, Elektronik, Strahlführung, grosse Teile der allgemeinen Maschi-
nensoftware und die Art der Dateneingabe bilden das Grundkonzept der
Maschine, auf dem aufbauend das kundenspezifische Handling und die
hierfür adaptierte Software speziell angepasst werden. Diese Anpas-
sungen kann der Kunde selbst mit geringstem Aufwand vornehmen. Auch das
Handling wird über unseren zentralen Rechner angesteuert. Die Programmierung erfolgt, da es sich hauptsächlich um die Anwendung von Relais,
Ventilen und Abfrage von Messdaten handelt, analog zu den bekannten
speicherprogrammierbaren Steuerungen und ist damit leicht an die
geänderten Gegebenheiten anpassbar. Neben digitalen Daten können auch
analoge Werte verarbeitet werden.

Die Programmerstellung soll ohne vorherigen Softwarekurs durchführbar
sein. Bei der Erstellung des Benutzerprogrammes haben wir deshalb von
einer freien Programmierung, wie sie z. B. bei CNC-Maschinen üblich
ist, Abstand genommen. Eine freie Programmierung ermöglicht eine Vielzahl von Variationsmöglichkeiten, welche allerdings erst nach
gründlicher Ausbildung des Programmierers verwendbar sind.
Wir haben einen anderen Weg beschritten und uns für die Dialogprogrammierung entschieden. Dialogprogrammierung bedeutet, die Maschine stellt
dem Benützer Fragen, die er beantwortet bzw. in einer auf dem Bildschirm erscheinenden Tabelle einträgt. Hierdurch wird eine übersicht-
liche Darstellung mit leichter Ueberprüfbarkeit etwaiger Fehler
erreicht. Der Benützer setzt sich an die Maschine und erstellt das
Programm (fast) ohne vorherige Unterweisung.

Um dem Benützer einen guten Ueberblick über die verschiedenen Programm-
teile zu geben, werden die Programme in einzelne selbständige Programm-
teile untergliedert, die zusammengelinkt werden.

Der Programmierer erstellt im Dialogverfahren

- den Beschriftungstext,
- das Programm für Firmenlogos und Zeichen
- das Teach-in-Programm
- bestimmt die Art der Strichverbreiterung
- die fortlaufende Numerierung
- sowie das Programm fürs Handling.

Liegen fertig vermasst Werkstattzeichnungen vor, so werden die Daten
via Tastatur in die Tabelle eingegeben (Dialogprogrammierung). Sollen
aber Figuren, Firmensignete und andere graphische Darstellung graviert
werden, so würde die Erstellung einer genauen Massskizze einen grossen
Arbeits- und damit Lohnaufwand erfordern. Zur Zeitreduktion lässt sich
das Lernprogramm (Tech-in) verwenden. Das zu programmierende Zeichen
wird auf den Bearbeitungsplatz unter die Laseroptik gelegt und die
Konturen via TV-Bildschirm und Steuerkugel abgefahren (Digitalisierung). Auf diese Art und Weise lassen sich nicht nur Geraden, sondern
auch Kreise eingeben. Die Steuerkugel, mit welcher der Ort des Laserstrahles direkt eingestellt wird, wurde statt dem seit langem wohlbe-
kannten Steuerknüppel (Joy-stick), mit dem die Verfahrgeschwindigkeit
des Strahles in Abhängigkeit von der Auslenkung eingestellt wird, ausge-
wählt. Die Programmierzeiten konnten, wie Versuche ergaben, um die
Hälfte verkürzt werden.

Während der Laserbearbeitung wird der Arbeitsplatz auf dem Bildschirm
des Computerterminals mit den notwendigen Daten (Programmieranweisun-
gen) abgebildet. Bei der Durchführung des Lernprogrammes erscheint dann
automatisch der Bildausschnitt mit Fadenkreuz, wobei die augenblick-
lichen Koordinaten am oberen Bildrand eingeblendet sind.

Bei einer Lasermaterialbearbeitungsmaschine handelt es sich um eine
Maschine, je nach Ausstattung, im Hundertkilo-DM-Bereich. Das bedeutet,
eine derartige Maschine muss produzieren, um die Amortisationskosten
einzubringen. Programmier- und Einrichtzeiten sollten somit auf ein
Minimum beschränkt werden. Aus diesem Grunde ist die Anlage mit einer
separaten Programmierstation versehen. Hier können die Maschinen-
programme, einschliesslich Lernprogramm, welches dann über Plotter und
Steuerkugel abläuft, erstellt werden. Die Floppy-Disk mit diesem neuen
Beschriftungsprogramm, welches schon auf dem Plotter getestet wurde,
wird in die Laserbearbeitungsmaschine gesteckt, und das Programm ist
sofort benützbar. Die Maschine ist auf Multiuserbetrieb vorbereitet und
kann als DNC-Maschine betrieben werden.

398

<u>Zusammenfassung</u>

Wir sind auf die Probleme bei der Laserbeschriftung eingegangen und
haben eine Maschine vorgestellt, die es gestattet, die anstehenden
Probleme optimal zu lösen.

<u>Literatur</u>

/1/ W.R. Englisch, "Materialbearbeitung mit dem Laser", Battelle
    Information, Mai 1977, S. 7 - 20
/2/ J. Junghans, H. Weber, "Geração dos pulsos periódicos de Q-switch",
    Revista do Circúlo de Engenharia Militar, ano XL, 1978, No. 80,
    S. 1 - 7

YAG-Laser-Bearbeitungszentrum

H.P. SCHWOB und U. GRUETZNER
LASAG AG
Steffisburgstrasse 1, CH-3600 Thun

1. Einführung

Das hier beschriebene YAG-Laser-Bearbeitungszentrum dient zum Bohren
von Kühllöchern in die wärmebeanspruchten Teile von Flugzeugturbinen.
Durch diese Löcher wird dann beim laufenden Triebwerk kühle Luft ein-
geblasen, welche als Luftfilm die Metalloberflächen vor den heissen
Verbrennungsgasen abschirmt. Mit der damit möglichen Erhöhung der
Verbrennungstemperatur wird ein besserer Wirkungsgrad und eine geringere Schadstoffemission erreicht.

Die wichtigsten Bearbeitungsanforderungen sind:

Material	hochwarmfeste Verbindungen auf der Basis von Eisen, Nickel oder Kobalt.
Materialdicke	0.8 mm - mehrere mm
Lochdurchmesser	0.2 mm - grösser 1 mm
Lochneigung	90° - 30° (z.T. - 15°) gegenüber Oberfläche
Werkstückdimension	0.1 - 1 m
Zahl der Löcher	bis über 100'000 pro Werkstück
Bohrzeit	0.1 - 1 s

Für das Laser-Bohren dieser Löcher werden vor allem zwei Verfahren
benützt. Das Bohren mit Impulsserie, für welche die beschriebene Anlage optimiert ist, zeichnet sich im wesentlichen durch seine Produk-
tivität aus (bis zu 10 Löcher/s). Dem gegenüber wird das Bohren im
Trepanierverfahren für die grösseren Lochdimensionen und bei strenge-
ren Qualitätsanforderungen eingesetzt. Zur Beurteilung der Qualität
dienen die Toleranzen der Lochdimensionen, der Zustand der Oberfläche
im Loch und die Zahl und Grösse von Auswurf und Spritzern.

Im folgenden werden zuerst diese beiden Bohrverfahren und sodann
die gesamte Anlage vorgestellt.

2. Bohrverfahren

2.1 Bohren mit Impulsserie

Bei diesem Bohrverfahren werden die Löcher durch eine Serie von Laserimpulsen gebohrt, wobei sich die Laserstrahlachse relativ zum Werkstück nicht bewegt. Die benötigten Laserdaten hängen von den geforderten Lochdimensionen ab. Interessant sind vor allem die hohen Produktionsraten.

Allgemeine Daten:

Lochdurchmesser	0.1 - 1.0 mm	Laserimpulsenergie	2 - 30 J
Tiefe zu Durchmesser	2:1 - 15:1	Laserimpulsdauer	0.2-1.0 ms
Max. Lochtiefe	6 mm	Laserimpulsleistung	10 - 30 kW
Lochneigung	15 - 90°	Laserimpulsfrequenz	5 - 50 Hz

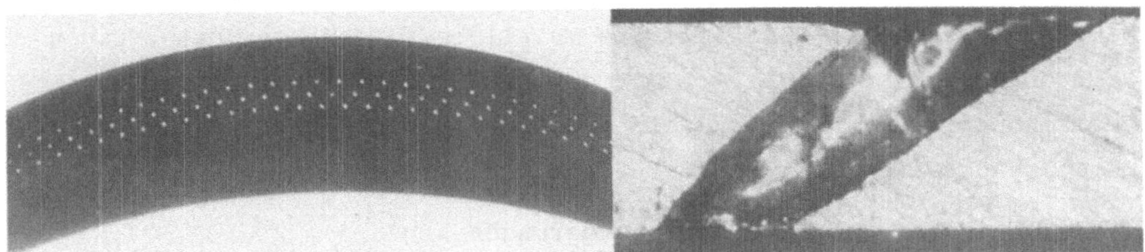

Beispiel 1: <u>Mit Impulsserie gebohrte Löcher in Zylinder</u>

Lochdurchmesser	0.5 mm	Bohrwinkel	30°
Materialdicke	1.2 mm	Produktion	5 Löcher/s

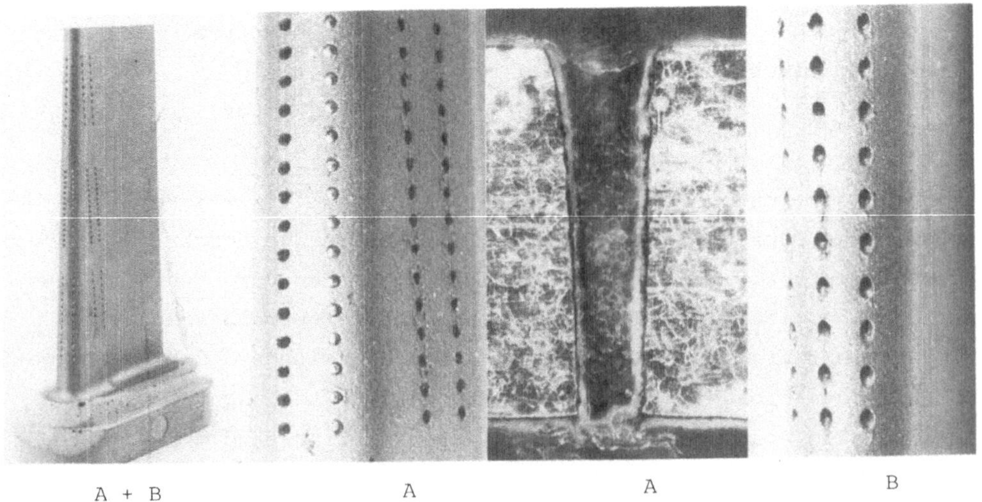

Beispiel 2: <u>Mit Impulsserie gebohrte Löcher in Turbinenschaufel</u>

Lochdurchmesser	0.3 mm	Materialdicke	A: 1.5 mm	B: 1.2 mm
Produktion	10 Löcher/s	Lochneigung	A: 90°	B: 45°

2.2 Bohren im Trepanierverfahren

Beim Trepanierverfahren werden die Löcher durch den gepulsten Laser
ausgeschnitten, d.h. die Laserstrahlachse rotiert mehrmals entlang der
Lochwand. Die Rotation geschieht meist durch kreisförmiges Drehen der
Fokussieroptik. Damit werden Löcher grösseren Durchmessers und verbes-
serter Qualität der Lochwand möglich. Bei den folgenden Beispielen
sind insbesondere die Parallelität und Gleichmässigkeit der Lochwand,
die scharfen Ein- und Austrittskanten, sowie die kleinen Lochabstände
zu beachten.

Allgemeine Daten:

Lochdurchmesser	0.4 - 5 mm	Laserimpulsenergie	2 - 5 J
Max. Lochtiefe	10 mm	Laserimpulsdauer	0.2 - 0.5 ms
Lochtoleranz	$\pm$ 0.02 mm	Laserimpulsleistung	ca. 10 kW
Lochneigung	15 - 90°	Laserimpulsfrequenz	40 - 100 Hz

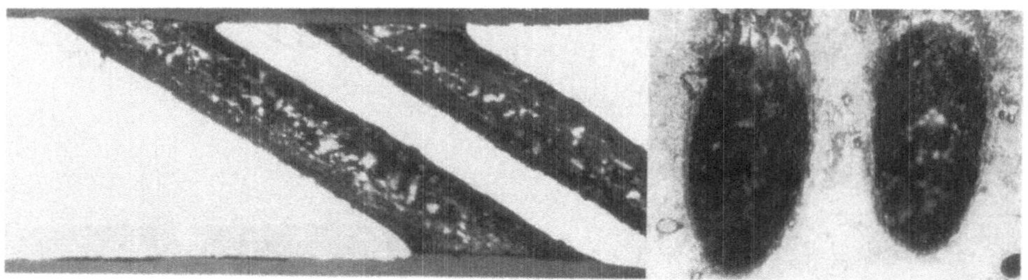

Beispiel 3: Trepanieren

Lochdurchmesser	0.7 mm
Materialdicke	2.4 mm
Lochneigung	30°
Bohrzeit	1.5 s

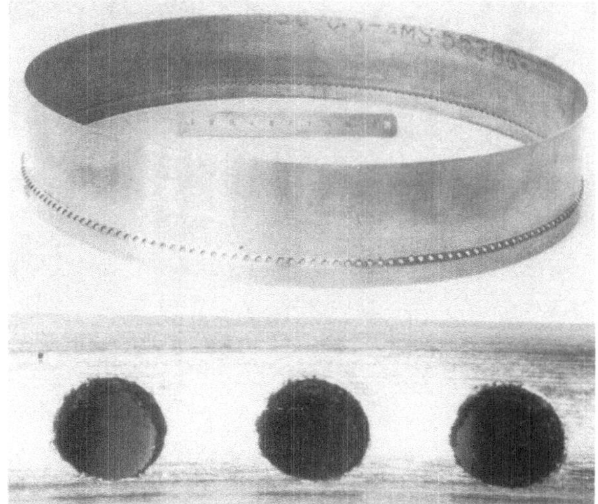

Beispiel 4: Trepanieren

Lochdurchmesser	2.0 mm
Materialdicke	0.8 mm
Lochneigung	90°
Bohrzeit	2 s

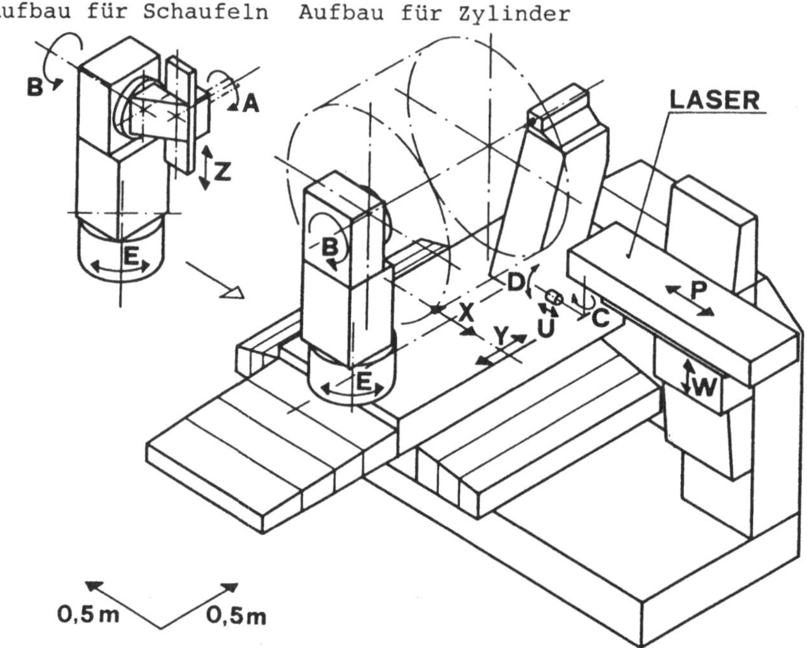

Figur 1: Schematische Darstellung der Anlage

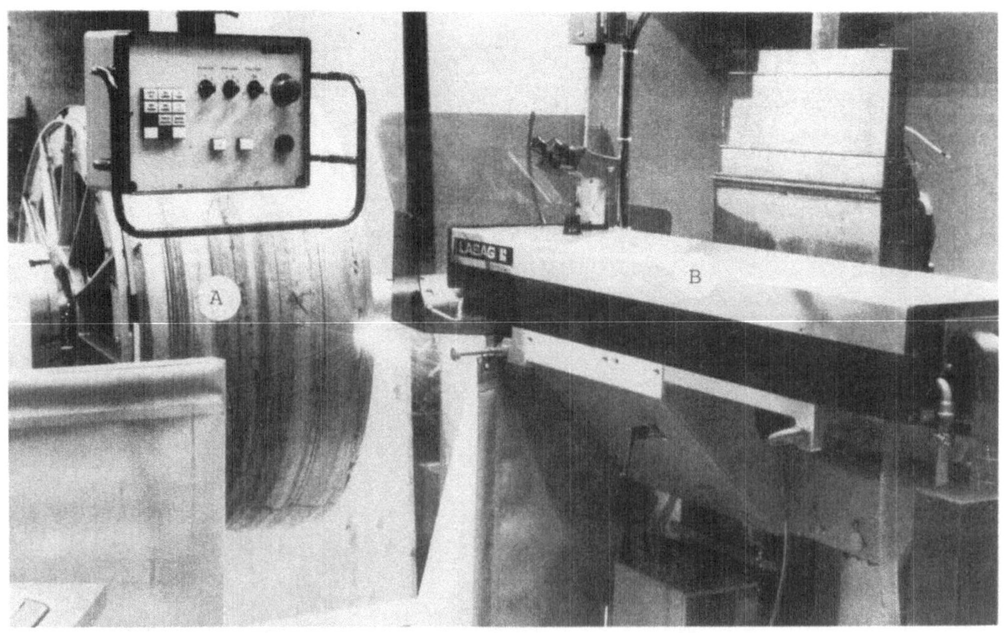

Figur 2: Teilansicht der Anlage während der Bearbeitung

Bohren mit Impulsserie bei gleichförmig drehendem Zylinder
A: gewellter Zylinder des Nachbrenners (Durchmesser 1 m)   B: Laser

3. Anlage

3.1 Laser

Das Lasersystem besteht aus zwei Laserköpfen in einer Oszillator-
Verstärker-Anordnung, um die erwünschte Laserimpulsleistung bei
kleiner Divergenz zu erreichen.

Laser-Daten

Lasertyp	Nd:YAG, gepulst
Strahldivergenz	2 mrad (halber Winkel)
Impulsdauer	0.2 - 10 ms, stufenlos einstellbar
Impulsfrequenz	0.1 - 100 Hz, stufenlos einstellbar

Maximale Laserimpulsenergie und entsprechende Parameter

Impulsdauer	0.2	0.5	1.0	ms
Max. Impulsenergie	5	12.5	25	J
Entspr. Impulsfrequenz	40	16	8	Hz
Max. Impulsleistung	25	25	25	kW

3.2 Mechanischer Aufbau und Steuerung (siehe Figur 1)

Verfahren der Zylinder:	X, Y, B (gesteuert)
Verfahren der Schaufeln:	X,Y,Z,A,B (gesteuert); E (von Hand)
Verfahren des Lasers:	C,D,U (gesteuert); P, W (von Hand)

3.3 Bohren mit Impulsserie bei gleichförmig drehendem Werkstück

Die schweren Werkstücke können nicht mit der nötigen Geschwindigkeit
bewegt und gestopt werden. Deshalb verfahren sie mit konstanter
Geschwindigkeit, und ein schnelles Strahlablenksystem lässt den Laserstrahl während des Bohrens eines Loches (typisch 3-6 Laserimpulse) dem
Werkstück folgen. Damit entfällt bei regelmässigen Lochanordnungen die
Positionierzeit und es werden Produktionsraten bis zu 10 Löchern/s
erreicht.

4. Zusammenfassung

Dieses Bearbeitungszentrum zeigt die hervorragende Eignung des Lasers
für Bohrungen hoher Qualität und Kadenz bei variablen Lochdimensionen.
Seine Steuerbarkeit und die tiefen Betriebskosten machen ihn zu einem
idealen Werkzeug der Produktion.

Festkörper-Lasersysteme mit annähernd rechteckförmigem Intensitätsprofil durch kombinierte Anwendung von nichtlinearen Absorptionsprozessen und Phasenkonjugation

E. HEUMANN, S. I. SCHASTAK

Friedrich-Schiller-Universität, Sektion Physik

Max-Wien-Platz 1, DDR-6900 Jena

Einleitung

Viele wichtige Anwendungen von Hochleistungslasern auf ausgewählten
Gebieten der Materialbearbeitung, z.B. Laser-Lithografie, Laser-Aus-
heilung, Laser-Glättung, -Härtung und -Legierung von Oberflächen,
erfordern ein homogenes Intensitätsprofil der Strahlung. Eine Reihe
von direkten oder indirekten Methoden werden verwendet, um solche
Profile zu erzeugen oder zu formen:

- räumliche Filterung /1/,
- Verwendung speziell dotierter Laser-Verstärkermaterialien, bei dene
  das Zentrum weniger stark angeregt wird als die äußeren Bereiche /2
- Phasenkonjugation /3/

Diese Methoden erfordern einerseits einen vergleichsweise hohen technischen Aufwand. Andererseits kann damit zwar eine Homogenisierung de
Intensitätsverteilung über den Strahlquerschnitt, nicht aber die Formung rechteckförmiger Profile erreicht werden.

In diesem Beitrag wird über eine einfache Methode zur Erzeugung und
Formung nahezu rechteckförmiger Strahlprofile von Hochleistungslaser-
Impulsen innerhalb und außerhalb des Resonators gütegeschalteter
Festkörperlaser (Rubin-, Nd-Glas-, Nd-YAG-Laser) unter Ausnutzung
nichtlinearer Absorptionsprozesse berichtet.

Experimentelle Anordnung und Ergebnisse

Die experimentelle Anordnung ist in Abbildung 1 schematisch dargestellt. Der Laseroszillator 1 arbeitet im transversalen Grundmodus,
die Impulsdauer beträgt 25 ns und die Impulsenergie 30 mJ. Die Laserleistung entspricht bei dem Strahlquerschnitt von 0,3 mm^2 einer Pho-
tonenflußdichte (N_L) von etwa 10^{27} $cm^{-2}s^{-1}$. Im Resonator befindet
sich zusätzlich zum Güteschalter (sättigbarer Einphotonen-Absorber)
ein Stufenabsorber 3. Unter einem Stufenabsorber soll ein Medium verstanden werden, das eine Einphotonen-Absorption aus dem durch einen
Mehrphotonenprozeß besetzten Zustand zeigt /4/,/5/. Für den Rubinlaser eignet sich eine Lösung von Anthranilsäure, und für Nd-Glas-

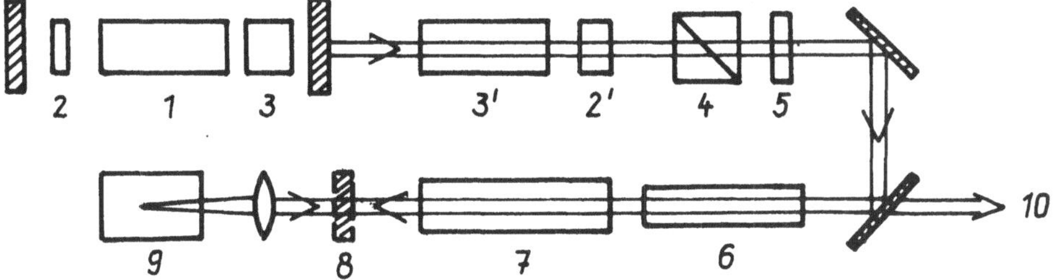

Abb. 1. Schema der experimentellen Anordnung

oder Nd-YAG-Laser eignen sich spezielle Farbzentrenkristalle /6/ als
Stufenabsorber. Diese Absorber beeinflussen den Generationsprozeß im
gütegeschalteten Laser zunächst nicht, da sie bei anfangs kleinen
Intensitäten im Resonator keine Verluste zeigen. Erst nachdem der
Gütemodulator gesättigt und damit der Weg für die Herausbildung des
Riesenimpulses freigegeben ist, erzeugt die zunehmende Intensität
eine merkliche Besetzung, z. B. des ersten angeregten Elektronenzu-
standes im Stufenabsorber durch Zweiphotonen-Absorption. Von diesem
Zustand findet dann eine effektive Einphotonen-Absorption in einen
höheren Anregungszustand statt. Der Stufenabsorber bewirkt somit Ver-
luste im Resonator, die mit steigender Intensität anwachsen. Das
führt zu einer Begrenzung der Maximalintensität und somit zu Impulsen
mit abgeflachtem räumlichen Intensitätsprofil (Abb. 2a).
Eine zusätzliche Profilformung kann in analoger Weise außerhalb des
Resonators durchgeführt werden. Dazu passiert der Impuls zur weiteren
Abflachung des Maximums nochmals einen Stufenabsorber (3') und einen
sättigbaren Einphotonen-Absorber (2') zur Aufsteilung der Flanken
(Abbn. 2b und 2c).
Nach dem Durchgang durch die Absorberkombination ist die Impuls-
energie nur noch etwa ein Drittel von der des Eingangsimpulses, die
Energiestabilität von Impuls zu Impuls ist besser als 10 % und das
Intensitätsprofil ist nahezu rechteckförmig (Abb. 2c).
Der Impuls gelangt über einen Polarisator 4 und eine λ/4 Platte in
ein Verstärkersystem 6, 7 und wird zunächst durch einen dielektrischen
Spiegel 8 zurückgekoppelt. Aufgrund der Inhomogenitäten in den Ver-
stärkermaterialien wird das Profil des Impulses zerstört. Am Ausgang 10 erhält man die inAbb. 3a gezeigte Intensitätsverteilung.

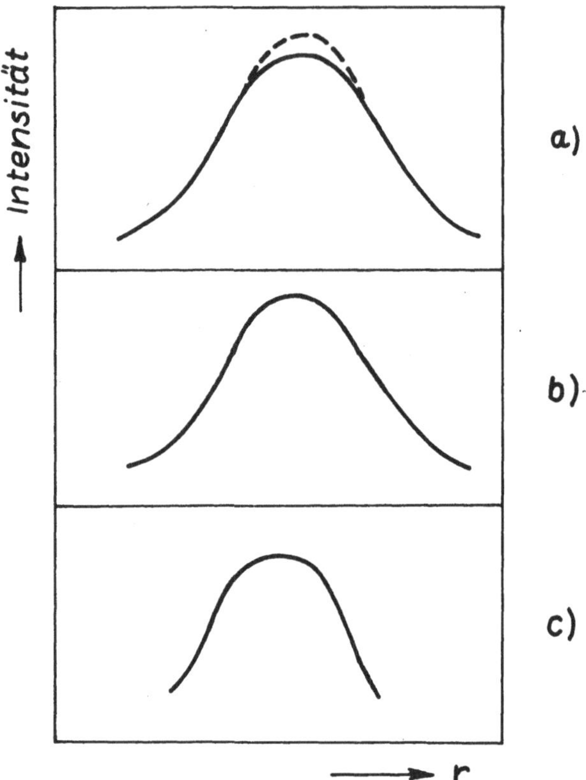

Abb. 2. Registrierte Intensitätsprofile
a) Oszillator Impuls, ----- ohne Stufenabsorber,
————— mit Stufenabsorber
b) nach dem Durchgang durch den zusätzlichen Stufenabsorber außerhalb des Resonators
c) nach der Absorberkombination

Um diesen Nachteil zu beseitigen, wird zur Rückkopplung ein phasen-konjugierender Reflektor 9, der auf der Basis der stimulierten Brillouin-Streuung (SBS) arbeitet, verwendet. Dazu wird der Strahl in eine 30 cm lange Küvette mit Azeton fokussiert. Die Reflektivität des "SBS-Spiegels" beträgt ca. 30 %. In diesem Fall hat der Impuls am Ausgang 10 des Verstärkersystems eine Energie von 300 mJ. Das rechteckförmige Intensitätsprofil des Eingangsimpulses reproduziert sich bei der Phasenkonjugation durch SBS mit hoher Genauigkeit (Abb. 3b).

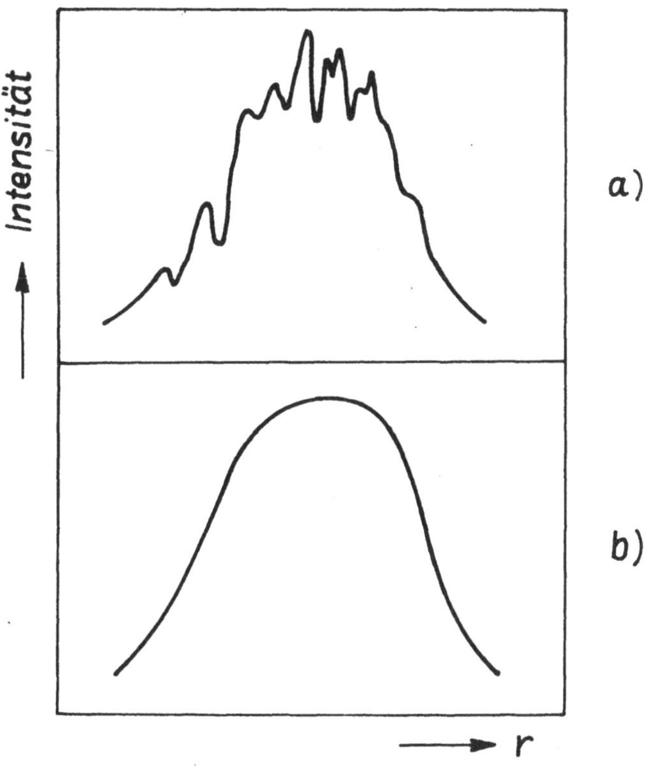

Abb. 3. Intensitätsprofile des verstärkten Impulses
    a) Rückkopplung mit dielektrischem Spiegel
    b) Rückkopplung durch phasenkonjugierenden Reflektor

Diskussion

Zur qualitativen Interpretation der Resultate betrachten wir den Stufenabsorber als 3-Niveausystem. Viele organische Moleküle zeigen unter Einwirkung intensiver Laserimpulse neben einer Zweiphotonen-Absorption $S_0 \longrightarrow S_1$ eine Einphotonen-Absorption vom ersten angeregten Singlett-Zustand S_1 zu einem höheren Singlett-Zustand S_x (Abb.4). Solche Stufenabsorptionsprozesse wurden z. B. in Lösungen von Stilben Anthranilsäure, Diphenylanthrazen unter Einwirkung der Rubinlaserstrahlung /4/,/5/ und in Farbzentren-Kristallen unter Einwirkung der Strahlung von Nd-Glas- und Nd-YAG-Lasern beobachtet /6/. Eine merkbare Besetzung des S_1-Niveaus durch Zweiphotonen-Absorption und somit eine effektive Einphotonen-Absorption $S_1 \longrightarrow S_x$ erfolgt nur im Bereich des Maximums des räumlichen Intensitätsprofils.
Unter der Annahme $T_{10} > T_L$ (Laser-Impulsdauer) kann man den Stufen-

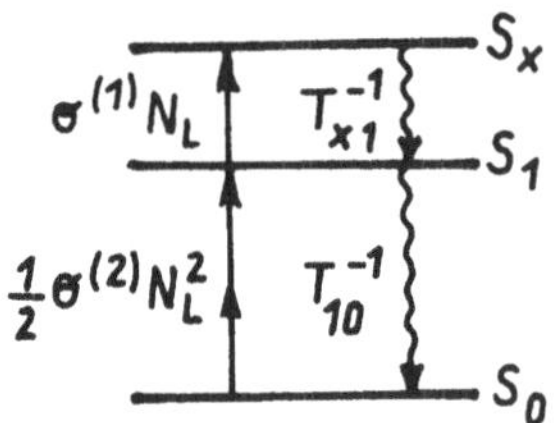

Abb. 4. Schema des Stufenabsorbers

$\frac{1}{2}\,\sigma^{(2)}N_L^2$; $\sigma^{(1)}N_L$ - Übergangswahrscheinlichkeiten für die Zweiphotonenabsorption $S_0 \longrightarrow S_1$ und die Einphotonen-Absorption $S_1 \longrightarrow S_x$

T_{x1}^{-1}, T_{10}^{-1} entsprechende Relaxationsprozesse

absorptionsprozeß außerhalb des Laserresonators durch folgendes System von Bilanzgleichungen beschreiben

$$\frac{\partial n_1}{\partial t} = \frac{1}{2}\,\sigma^{(2)}N_L^2 n$$

$$\frac{\partial N_L}{\partial t} = -\,\sigma^{(1)}n_1 N_L \tag{1}$$

n, n_1 - Teilchenzahldichten im Grundzustand und im S_1-Zustand; $\sigma^{(2)}$, $\sigma^{(1)}$ - Absorptionsquerschnitte für den Zweiphotonen- und den Einphotonen-Prozeß.

Die Lösung dieses Systems /7/ zeigt die impulsformende Wirkung des Stufenabsorbers

$$N_L(r,l,t) = \frac{N_L(r,o,t)}{1 + \frac{1}{2}\sigma^{(2)}\sigma^{(1)}nl \int_{-\infty}^{t} N_L^2(r,o,t')dt'} \tag{2}$$

Wenn $T_{10} \lesssim T_L$ ist, muß der Relaxationsprozeß $S_1 \longrightarrow S_0$ berücksichtigt werden und der Einfluß auf das Intensitätsprofil wird durch die Differentialgleichung

$$\frac{\partial^2}{\partial t \partial z}\ln N_L^2 + \frac{1}{T_{10}}\frac{\partial}{\partial z}\ln N_L^2 + \sigma^{(2)}\sigma^{(1)}N_L^2 n = 0 \tag{3}$$

beschrieben.

Zur Wechselwirkung des durch den Stufenabsorber verformten Impulses $N_L(r,l,t)$ mit dem sättigbaren Einphotonenabsorber wird ein Zwei-Niveausystem betrachtet. Unter den gegebenen experimentellen Bedin-

ungen ist die Relaxationszeit des verwendeten Farbstoffes T'_{10} sehr
lein im Vergleich zur Dauer des Laserimpulses. Damit kann die Wech-
elwirkung als stationär durch folgende Gleichung beschrieben werden

$$\frac{dN_L}{dz} = - \frac{n_o}{2T'_{10}} \frac{N_L}{N_L + \dfrac{1}{2\sigma T'_{10}}} \tag{4}$$

σ – Absorptionsquerschnitt; n_o – Teilchenzahldichte im Grundzustand

m Ausgang der Küvette $z = 1'$ ergibt sich die Form des Intensitäts-
rofiles aus der Lösung von Gl.(4) und unter Verwendung von Gl.(2)

$$N_L(r,1',t) = N_L(r,1,t)\, \exp(-\sigma n_o 1')$$

$$\cdot\, \exp\left\{ 2\sigma T'_{10} N_L(r,1,t) \left[1 - \frac{N_L(r,1',t)}{N_L(r,1,t)} \right] \right\} \tag{5}$$

.iteratur

'1/ N.B. BARANOVA, u.a., Trudy FIAN, <u>103</u> (1978), 103
'2/ Recent developments in CILAS glass lasers, Werbematerial der
     Firma CILAS 1980
'3/ T.G. KARR, H.J. HOFFMANN, Lasers' 81 (New Orleans 1981),
     Vortrag FF7
'4/ J. KLEINSCHMIDT, S. RENTSCH, W. TOTTLEBEN, B. WILHELMI,
     Chem. Phys. Lett. <u>24</u> (1974), 133
'5/ E. HEUMANN, W. TRIEBEL, G. SLOBIN, Exp. Tech. Phys. <u>22</u>,3(1975),253
'6/ T. DAMM, E. HEUMANN, F. NOACK, K. VOGLER, wird veröffentlicht
'7/ J. HERRMANN, J. WIENECKE, B. WILHELMI, Optical and Quant.
     Electr. <u>7</u>, 5 (1975), 337

Qualitätsaspekte der Laserbearbeitung von Dünnschichten

Z.Drozd, L.Boruc
Technische Universität Warszawa
Institut für Elektronischen und Feingerätebau
ul.Chodkiewicza 8, PL 02-525 Warszawa

1.Einleitung

Aus Gründen der Erhöhung der Präzision und Zuverlässigkeit den elektronischen Mikroschaltungen erfolgt die Notwendigkeit den stetigen Verbesserungen des in Herstellung /besonders in Widerstandsabgleich/ angewandten Laserbearbeitungsprozesses. In diesem Vortrag sind einige Methoden und Ergebnisse der Qualitätsuntersuchungen der Bearbeitung von
dünnen Widerstandsschichten dargestellt. Diese Untersuchungen wurden
in folgenden Richtungen durchgeführt:
- Theoretische Analyse des Bearbeitungsprozesses,
- Auswahl der Leistungsdichte und Schnittbreite,
- Mikroskopische Untersuchungen des Bearbeitungsgebietes,
- Zuverlässigkeitsanalyse.

2.Untersuchte Erscheinungen

Für die Bearbeitung der Dünnschichten werden meistens die Q-switched
Nd-YAG Laser verwendet. Während der Wirkung der fokussierten Licht -
impulse auf die auf Glas-, Keramik-, oder Siliziumsubstrat aufgetra-
gene Metallschicht erfolgt das Schmelzen und Dampfen des Schicht -
stoffes $[1,2,3]$. Bei jedem Impuls wird die Schicht im Gebiet mit
Diameter von 5 bis zum 50 µm im Zeitdauer von einigen ns bis zum etwa
100 ns entfernt. In diesem Zeitdauer können folgende Prozesse /Bild 1a
beobachtet werden:

- Schmelzen und Dampfen des Schichtsstoffes unter der Wirkung der
  Lichtsleistungsdichte q,
- Vertreibung den Teilchen des flussigen Schichtsstoffes,
- Verschiebung den Teilchen des flussigen Schichtstoffes auf der Boden-
  fläche des Schnittes /B/ mit der Geschwindigkeit V unter der Wirkung
  des entstehenden Druckes P, den Gasbewegungen und molekulären Kräfte
- Erhitzung der Schicht und Substrat mit hohem Temperaturgradient im
  Abstand bis zum einigen µm von der Bearbeitungszone.

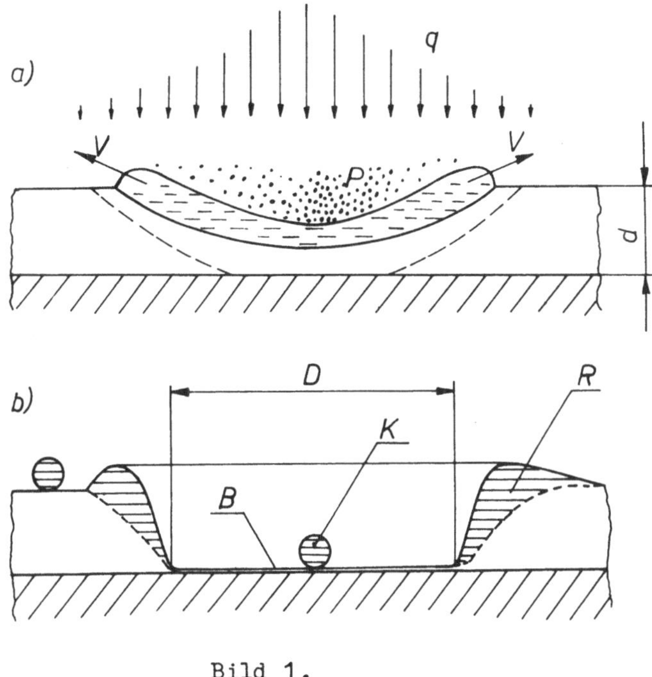

Bild 1.

Diese Prozesse entscheiden über die Eigenschaften des erzeugten Schnittes /Bild 1b/. Die Durchgeführten Untersuchungen [4] zeigen, dass die Qualität des Widerstandes wird von der Struktur den Randgebiete /R/, der Struktur von Bodenfläche des Schnittes /B/ und den während der Bearbeitung entstehenden Verunreinigungen des Schichtes bedingt.

3.Theoretische Analyse des Bearbeitungsprozesses

Die theoretische Analyse des Bearbeitungsdurchlaufes wurde mit der Methode des thermischen Bilanzes [5] durchgeführt. Für die Bestimmung des Profils der Schichtsstoffabnahme wurde der Durchschnitt des Schichtes in der Stelle der Kraterbildung in diskrete Elemente $r_i z_j$ /Bild 2/ untergeteilt.

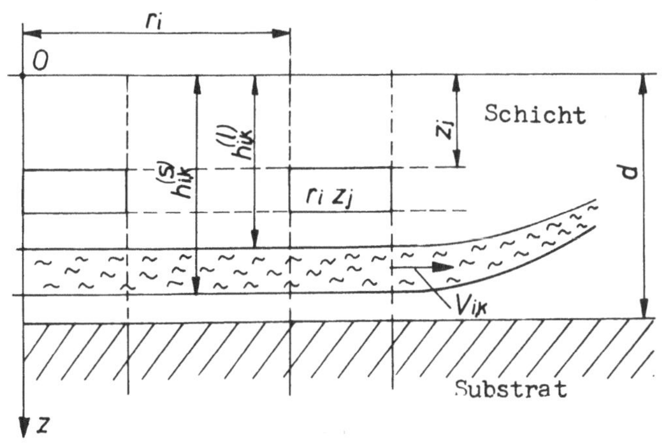

Bild 2.

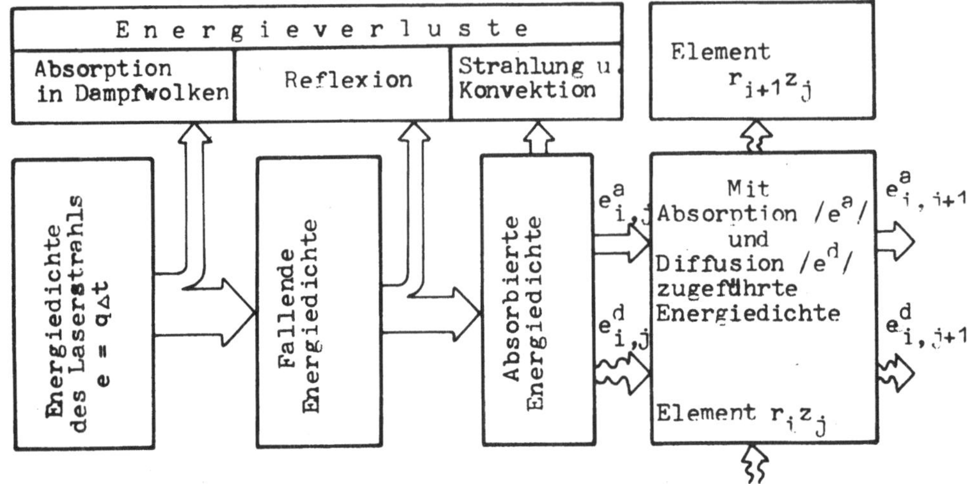

Bild 3.

Das Schema des mit dem Rechner analysierten Energiedurchflusses wird im Bild 3 dargestellt. Als Ergebniss dieser Analyse wurde der Spiegel der flussigen / $h^{/l/}_{i,k}$ / und soliden / $h^{/s/}_{i,k}$/ Phase /Bild 2/ für die einzelne Zeitperiode Δt_k bestimmt. Die mit dieser Methode bestimmten Durchlaufe diesen Spiegeln sind im Bild 4 dargestellt.

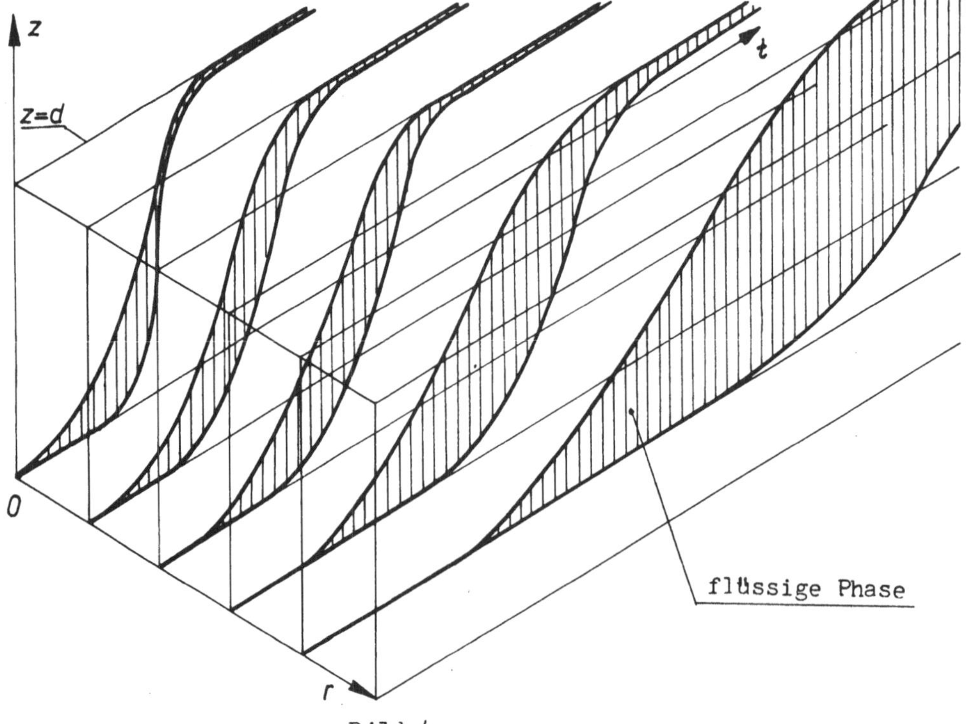

Bild 4.

4. Schnittbreite und Leistungsdichte

Bei der Vorprüfung des Einflusses der Schnittbreite und Leistungs -
dichte auf die Qualität des Schnittes ist es günstig einige Prüfschnitte in den mit Winkel α = 2...5° zum Arbeitstisch befestigten Musterstücken /Bild 5a/ mit verschiederen Frequenzen, Geschwindigkeiten und
Lichtleistungen auzuführen. In erzeugten Schnitten /Bild 5b/ können

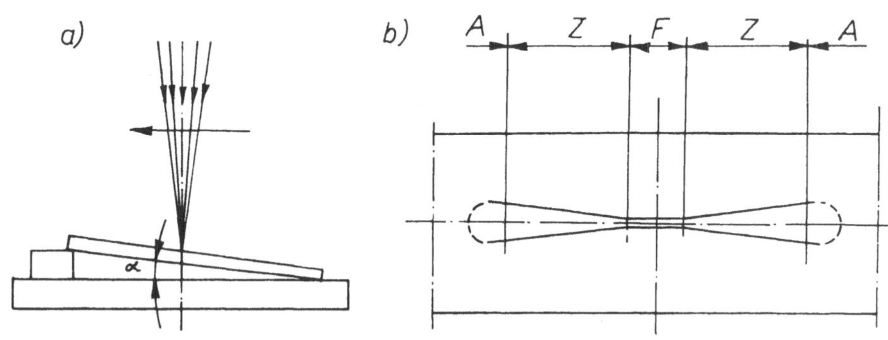

Bild 5.

drei Gebiete festgestellt werden [6]:

    F - volle Fokusierung der Laserstrahl,

    Z - veränderlicher Querschnitt der Laserstrahl,

    A - Anfang der Bearbeitungsgebietes.

Diese Methode erlaubt die mikroskopische Beobachtung und Auswahl der
optimalen Schnittbreite und Leistungsdichte für die bearbeitete Stoffe.

5.Mikroskopische Untersuchungen des Bearbeitungsgebietes

Die erzeugten Schnitte können mit Hilfe des biologischen Mikroskops
mit der Auf-,Durch-, und Mischbeleuchtung untersucht werden. Für manche
Fälle ist die Verwendung von Grün-, Rot-, Blau- oder Gelbfiltern nötig.
Bei Durchbeleuchtung untersucht man die Struktur der Bodenfläche des
Schnittes und die Schichtdefekte. Für die Untersuchungen und Messungen
der Koaleszenzinseln und Struktur des Randgebietes wird die Mischbe -
leuchtung angewandt. Mit Aufbeleuchtung werden die Verunreinigungen
auf der Schichtoberfläche und Überhitzungen beobachtet.

6. Zuverlässigkeitsuntersuchungen

Für die Zuverlässigkeitsanalyse kann das theoretische Modell des Schnittes in Form einer morphologischen kontinuierlichen Struktur [7] angewandt werden. Die notwendigen Daten erzeugt man von beschleunigten

414

Prüfung einer Menge von Prüfschnitte.

Die in Bodenfläche und Randgebieten des Schnittes auftretenden Ausfälle
können in Form eines Koeffizientes δ_λ der linearen Ausfallrateverteilung bestimmt werden:

$$\delta_\lambda = \frac{\lambda_L}{L}$$

mit: λ_L - Ausfallrate der in Bodenfläche und Randgebieten auftretenden
Ausfälle für die gesamte Länge $L = \sum_{i=1}^{n} l_i$ der geprüften Schnitte.

Die in der Schichtfläche auftretende Ausfälle sind mit dem Koeffizient
δ_λ der Flächenausfallverteilung bestimmt:

$$\delta_\lambda = \frac{\lambda_\lambda}{S}$$

mit λ_λ - Ausfallrate der in der Schichtfläche auftretenden Ausfälle
für die gesamte Fläche S der Prüflinge.

Für die auf den Schnittenden auftretenden Ausfälle kann die Ausfallrate λ_e bestimmt werden:

$$\lambda_e = \frac{m}{\sum_{i=1}^{m} t_i + (n-m)T}$$

mit: n - Anzahl der geprüften Schnitte, m - Anzahl den auf den Schnittenden auftretenden Ausfälle, T - Zeit der Prüfung.

Obengenannte Koeffizienten erlauben die Bestimmung des Richtwertes
der Ausfallrate eines Widerstandes:

$$\lambda_w = k_B \left(l_w \delta_\lambda + s_w \delta_\lambda + \lambda_e \right)$$

mit: $0 < k_B < 1$ - Koeffizient der Belastung, l_w - Schnittlänge,
s_w - Widerstandsfläche.

Literatur
[1] Ready,J.F: Material Processing-An Overview.Proceedings of the IEEE,
    Vol.70 NO 6, 1982.
[2] Cohen,M.G.,Kaplan,R.A.,Arthurs,E.G: Micro-Materials Processing.
    Proceedings of the IEEE, Vol.70, NO 6, 1982.
[3] Drozd,Z.,Lemanowicz,J: Obróbka laserowa mikroukładów elektronicz-
    nych.EM-82.ATR Bydgoszcz, 1982.
[4] Drozd,Z.,Boruc,L.,Lemanowicz,J: Korekcja laserowa rezystorów
    warstwowych.Mikronika 82.SIMP, Warszawa, 1982.
[5] Boruc,L: Obróbka laserowa elementów warstwowych.EM-82. ATR Bydgoszcz
    1982.
[6] Drozd,Z.,Lemanowicz,J:The Effect of Laser Machining Parameters on
    the Properties of integrated Circuits.ISEM 7.IFS Ltd u.North-
    Holland Publ.Co.Birmingham,1983.
[7] Drozd,Z: Einfluss der Technologie auf die Zuverlässigkeit der Bau-
    elemente.VDI Berichte 395.VDI, Düsseldorf 1981.

Laserfeinbearbeitung in der Mikroelektronik zur Bearbeitung von Masken und integrierten Schaltungen

J. BUCHHOLZ

Rofin-Sinar Laser GmbH, D-8063 Odelzhausen und 2 Hamburg 74

1. Einleitung

Der Trend der Mikroelektronik zu immer weitergehender Integration beinhaltet Masken zunehmender Abmessungen bei feineren Geometrien. Es stehen heute Masken bis 6 x 6 " Größe - in der Entwicklung sogar bis 8 x 8" Größe - zur Verfügung, wobei die Bahnbreiten 1 µm und weniger betragen können. Hier gewinnt die Reparatur der Masken zunehmende Bedeutung, da damit sowohl die Kosten für Reticles, Mutter- und Arbeitsmasken als auch die Ausschußrate bei der IC - Fertigung gesenkt werden können. Ferner ist die direkte Feinbearbeitung von Halbleitern durch eine Art "Mikrochirurgie an Leitungsbahnen" mit fein fokussierbaren Laserstrahlen von zunehmendem Interesse, wobei die verschiedensten Technologien (MOS, Bipolar, LSI, Speicher und monolytische Schaltkreise) dieses Verfahren zumindest in Entwicklungslabors einsetzen.

Das Wort Laserfeinbearbeitung wurde hier gewählt, um die zwei wichtigsten Eigenschaften der hier zu diskutierenden Laserstrahlung zu kennzeichnen:
a. Die Fokussierung des Laserstrahls auf geringe Durchmesser von weniger als 10 µm oder 1 µm.
b. Die exakte Begrenzung des Randes der Strahlung.

2. Bearbeitungsprinzip

Der für die Laserfeinbearbeitung heute am häufigsten eingesetzte Laser ist der YAG-Laser mit bzw. ohne SHG (Frquenzverdoppler). Dieser Lasertyp ermöglicht gute Strahlqualität bei kurzen, hohen Impulsen. Es werden bis etwa 1 mJ in weniger als 30 ns an der Bearbeitungsstelle appliziert. Durch optische Komponenten höchster Qualität, sorgfältiges Design und gutes Know How lassen sich heute im industriellen Maßstab Fokusdurchmesser von nur dem 1,6-fachen der Wellenlänge des Lasers realisieren. Besonders die Strahlformungsoptik (Bild 1 + 2) ist hier für die Laserfeinbearbeitung von Bedeutung, da sie u.a. eine scharfe Begrenzung des Laserstrahls und eine variable Strahlgeometrie über eine einstellbare

Rechteckblende ermöglicht.

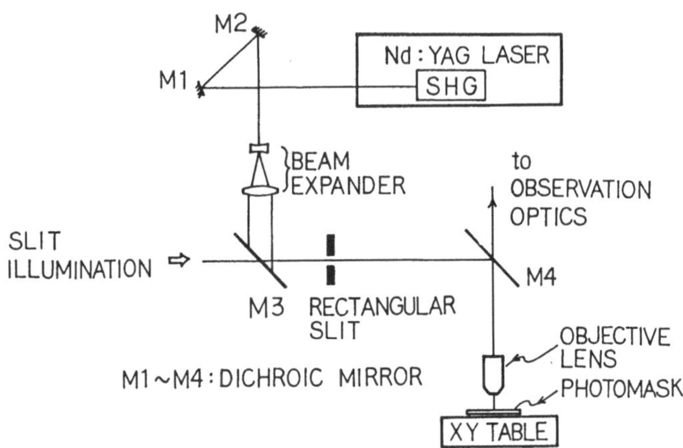

Bild 1. Typischer Strahlengang eines Laserfeinbearbeitungssystems

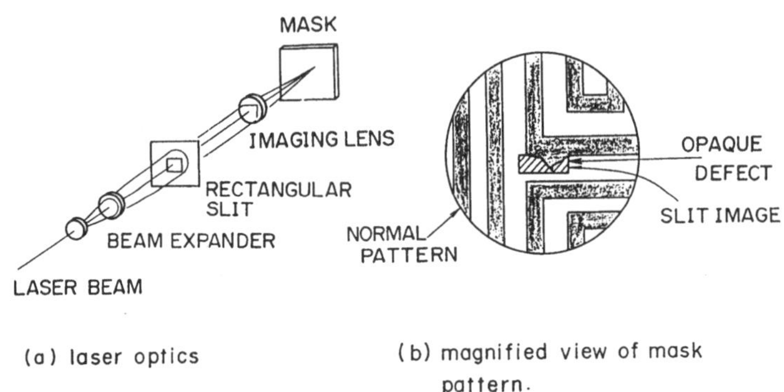

Bild 2. Strahlformungsoptik mit Rechteckblende

Die resultierende Intensitätsverteilung über den Strahlquerschnitt ist nicht gaußähnlich, sondern rechteckähnlich, was den Anforderungen an eine präzise begrenzte Bearbeitung entgegenkommt. Die Anstiegsflanken können dabei nur ca. 0.1 µm betragen. Auf dem Plateau sind Interferenzstrukturen erkennbar, die bei der Bearbeitung von Dünnschichtmaterial wiederzufinden sind, wenn die Laserimpulsenergie nahe der Verdampfungsschwelle liegt, siehe Bild 3. Bei hinreichend hoher Impulsenergie verschwindet die gitterähnliche Struktur und eine saubere, rückstandfreie Bearbeitung erfolgt.

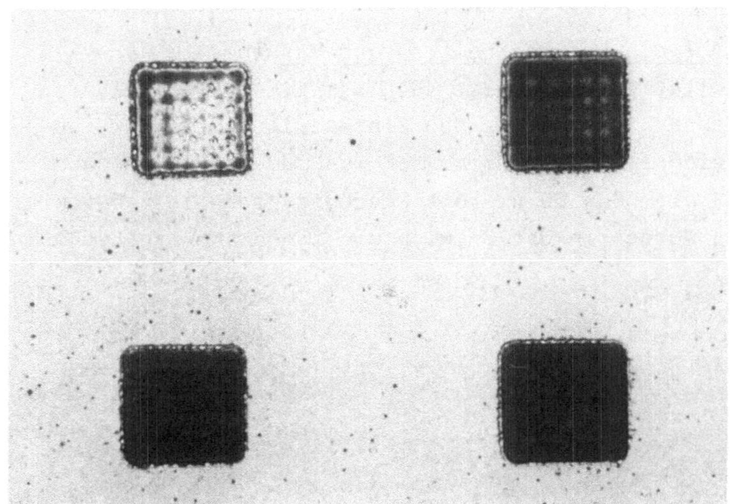

Bild 3. Durch YAG-Laserstrahlen erzeugte Quadrate von ca. 10 µm auf
Eisenoxidmaske bei verschiedenen Laserimpulsenergien

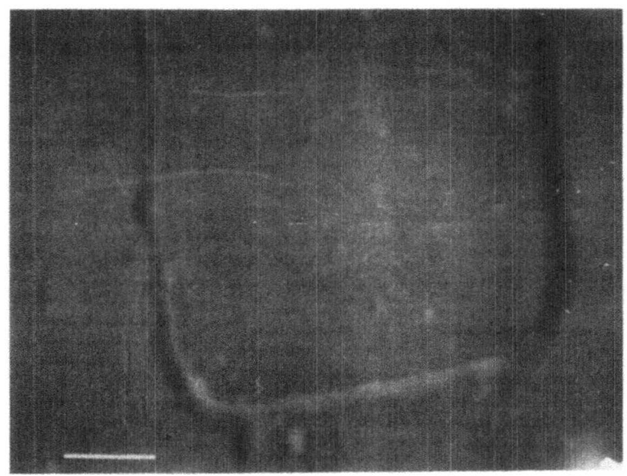

Bild 4. REM-Aufnahme einer Chrom-Maske, die durch YAG-Laser mit 1,06 µm
Wellenlänge bearbeitet wurde

Die Bilder belegen, daß es möglich ist, mit Laserstrahlen von 1,06 µm
Wellenlänge Übergangszonen von nur einem Bruchteil der Wellenlänge
(hier ca. 0,2 µm) zu erzielen. Sowohl Wärmeeinflußzone als auch Schmelzzone und Aufwurfhöhe sind extrem gering. Auch das Substratmaterial
(hier: Glas) wird nicht negativ beeinflußt. Entsprechend gut sind die

Ergebnisse beim Projizieren reparierter Masken aus Photoresist.

3. Beispiele für die Feinbearbeitung von Masken

Neben der Qualität der Kante ist auch das minimal erzielbare Strahlrechteck von Interesse. Systeme für den Einsatz in der Produktion decken heute den Bereich 1,5 x 1,5 µm bis 60 x 60 µm (ohne Frequenzverdoppler) bzw. 0,8 x 0,8 bis 30 x 30 µm (mit Frequenzverdoppler) ab. Mit Frequenzvervierfachung wurden in der Entwicklung schon Bahnbreiten 0,7 µm und weniger erzielt.

Bild 5. Laserfeinbearbeitung auf CrO_2-Maske bei 532nm Wellenlänge, Schlitzbreite 0,8 bzw. 2,4 µm (Auflicht)

Eine weitere, neue Anwendung ist das Beschriften von Masken im Mikrometerbereich, indem einzelne Rechtecke mittels Computersteuerung (z.B. BASIC-Programm) zu Buchstaben aneinandergereiht werden. Die Feinbeschriftung kann auf kleinstem Raum erfolgen und somit vereinzelt zusätzliche Informationen auf Maske und Wafer übertragen (Buchstaben-Größe z.B. 80 x 60 µm).

Die bei dieser Bearbeitung präzise bearbeitbaren Werkstoffe sind u.a.:

Cr	Schichtdicke: Dünnfilmtechnologie
CrO_2	Substratmaterial: Glas, Quarz, Saphir
FeO_3	
Al	Schichtdicke bis einige µm
Poly-Si	Substratmaterial: Si-Wafer
Ni Fe	Oberfläche darf auch Glas-passiviert sein
Ti-Leg.	

4. Beispiele für die Bearbeitung von integrierten Schaltungen

Die zuletzt genannte Gruppe von genannte Gruppe von Werkstoffen findet sich i.a. auf Wafern, wo z.Zt. die folgenden Laserfeinbearbeitungen sinnvoll sind:

4.1 <u>Unterbrechen von Leiterbahnen</u> z. B. Bipolar-Technik) zur Fehleranalyse von Schaltungen und für andere Entwicklungs- und Testarbeiten.
Bild 6 zeigt als Skizze den Bearbeitungsvorgang. Wichtig ist die exakte Adaption der Rechteckgröße auf die zu entfernende Al-Brücke. Das Laserintensitäts-Rechteckprofil muß mit dem "Werkstoffrechteck" übereinstimmen, um eine Bearbeitung bzw. Beschädigung des Si unter und unmittelbar neben der Al-Brücke zu verhindern.

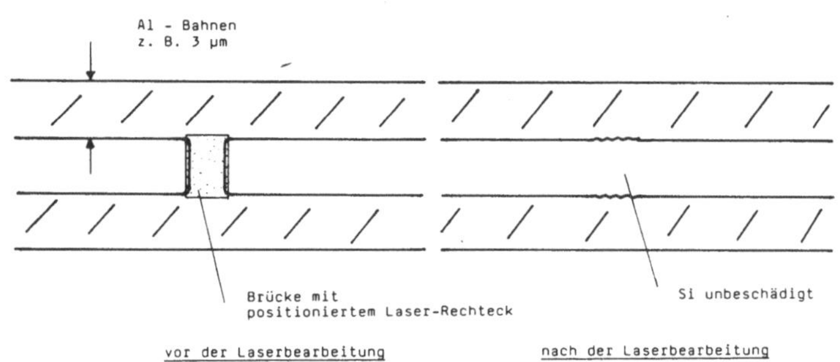

<u>Bild 6</u>. Schema der Entfernung von Brücken aus Al auf Si-Wafer

Bei dieser Art der Laserfeinbearbeitung wird besonders deutlich, daß gaußähnliche Intensitätsverteilungen des Lasers ungeeignet wären. Aber auch die eingestellte Länge der Brücke ist von entscheidender Bedeutung für die Qualität der Bearbeitung. Werden Rechteckbreite und -länge sowie Impulsenergie innerhalb eines Erfahrungsbereichs richtig vorgewählt, so läßt sich die genannte Unterbrechung einer Brücke ohne Zerstörung des darunterliegenden Si durchführen, siehe Bild 7.

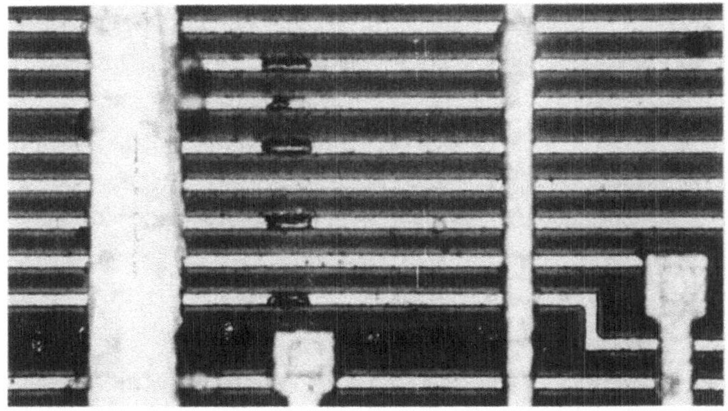

<u>Bild 7</u>. Unterbrechung von 3 µm Al-Bahnen auf Si mit YAG-Laser

4.2 Trennen von benachbarten Al-Bahnen (Entfernung von unbeabsichtigten Brücken) zur Rettung einzelner IC in der MOS-Technologie. Bearbeitung erfolgt wie 5.1. Evtl. in Verbindung mit IC-Tester.

4.3 Markieren von IC-Teilen ähnlich wie beim Maskenfeinbeschriften.

4.4 Trimmen monolytischer Schaltkreise. Für das Trimmen werden kontinuierlich gepumpte anstatt durch Blitzlampen gepumpte YAG-Laser bevorzugt. Auch dort werden Strahlformungsoptiken eingesetzt.

4.5 Redundanzabgleich zur Rettung von Speicher-IC in Verbindung mit Memory-Tester-System, ähnlich wie 4.1.

5. Beispiel Lasersystem für die Laserfeinbearbeitung

NEC hat vor kurzer Zeit neben einem bewährten System, das bis zur 2 µm - Technologie einsetzbar ist, ein neues System für 1 µm - Technologie auf den Markt gebracht. Die wesentlichen Baugruppen dieses Submicron-Systems sind: frequenzverdoppelter YAG-Laser, Strahlformungsoptik, Fokussieroptik mit 10 mm Arbeitsabstand, Auto-Fokus, 8 x 8 " Koordinaten-

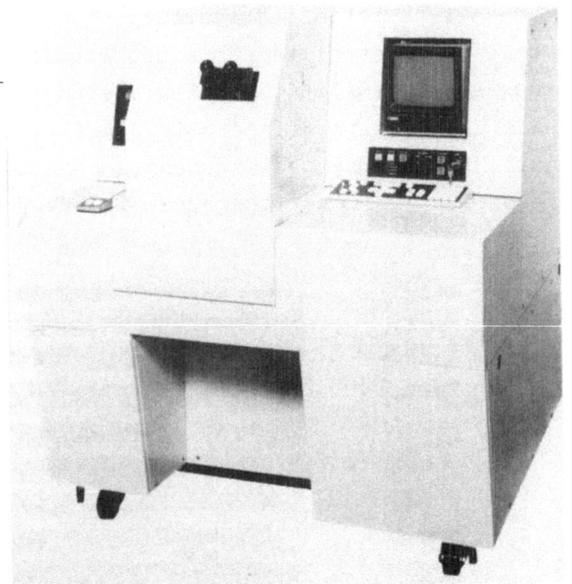

Bild 8. Komplettes Laserfeinbearbeitungssystem für die Mikroelektronik (Typ YL 452 A für Submicron-Bearbeitung bei 532 nm Wellenlänge)

tisch mit 0,02 µm Feinauflösung, Multi-BUS Computer sowie Interfacevor-
richtungen für automatische Datenübertragung.

6. Zusammenfassung und Ausblick

Merkmale des Verfahrens "Laserbearbeitung in der Mikroelektronik"

- o Berührungslose Bearbeitung
- o exakte Positionskontrolle mittels
  Pilotstrahl und Mikroskop
- o Scharfe Strahlbegrenzung durch
  rechteckähnliche Intensitätsverteilung
  über den Strahlquerschnitt und kurze Impulsdauer
- o Variable Laserstrahlgeometrie (z.B. Rechtecke, Quadrate) ermöglicht optimal angepaßtes Bearbeitungsareal
- o Abtragungstiefe stufenweise wählbar bei
  Schichtmaterial

<u>Nutzen der Laserfeinbearbeitung für die Mikroelektronik</u>

Reduzierung der Maskenkosten um bis 30 % und mehr (erhebliche Reduzierung der
Defektdichte, Kostenreduzierung besonders hoch bei Printverfahren)

Erhöhung der Ausbeute bei der IC-Herstellung aufgrund der geringeren Defektdichte der Masken (Ausbeute um 5 % oder mehr erhöhbar wegen:

$$\text{IC - Ausbeute} \approx \left(\frac{1}{1 + \text{Defektdichte} \times \text{IC-Fläche}} \right)^{\text{Anzahl von Prozessen}}$$

Schnelle und gezielte Analyse von IC in der Entwicklung
Erhebliche Beschleunigung der Zeitspanne Design bis Vorserienmusterlieferung

Erhöhung der Ausbeute bei der IC - Fertigung durch Rettung von IC (z.B. Redundanzabgleich bei Speicher-IC, Entfernung von Brücken bei MOS-IC etc.)

Neben den hier diskutierten Laserfeinbearbeitungen gibt es in der Mikroelektronik noch einige weitere, etwas gröbere Bearbeitungen, die hier
kurz erwähnt sein sollen: das Ritzen von Substratmaterial, das Beschriften
von Wafern und IC-Gehäusen sowie das Reflow-Löten von IC-Anschlüssen.
Diese Auflistung zeigt mit dem vorangegangenem, daß Laserverfahren heute schon eine erhebliche Bedeutung innerhalb der Halbleiterfertigung
haben. In der Forschung und Entwicklung geht der Trend zu noch höherer
Präzision der Laserbearbeitung z.B. durch Einsatz von Excimer-Laser-
strahlung. Es werden aber auch weitere, neue Wege beschritten, wie z.B.
bei der Entwicklung des Laser-Pluggers, der Material definiert mit Laser-
unterstützung deponieren soll.

Optoelektronische Signalübertragung

Optoelectronic Signal Transmission

Die zukünftige Entwicklung der optischen Nachrichtentechnik

M. Börner
Lehrstuhl für Technische Elektrophysik, Technische Universität München,
8000 München 2, Arcisstr. 21

1. Einführung

Die prinzipiellen Möglichkeiten, die durch die Existenz von modulierbaren Halbleiterlasern, von Photo-Empfangsdioden und der Glasfaser als Übertragungsmedium gegeben
sind, wurden in der optischen Übertragungstechnik genutzt. Bild 1, Bild 2 zeigt ein
Faserkabel mit einer Vielzahl von Einzelfasern. Die Nutzung erfolgte nicht wegen der
Neuheit dieser Technik, sondern wegen der Vorzüge gegenüber herkömmlichen Übertragungsverfahren, wobei insbesondere das Koaxialkabel und die zweidrähtigen Teilnehmer-
anschlußleitungen als Übertragungsmedium zu nennen sind. Bild 3 zeigt die prinzipielle Überlegenheit der Glasfaser gegenüber dem Kupferkabel. Es gibt aber noch eine
Reihe weiterer technischer Möglichkeiten der Glasfasertechnik, die man sich klar
machen muß, um Anhaltspunkte über ihre zukünftige Entwicklung zu gewinnen.

Zu den technischen Vorzügen kommen Kostenvorteile. Über dieses Gebiet will ich hier
nur einige abschließende Bemerkungen machen. Es ist wohl durch Feldversuche inzwi-
schen hinlänglich gesichert, daß die optische Übertragungstechnik, soweit sie vorhandene Technik substituiert, auch preislich vorteilhafte Lösungen bieten kann. Eine
Abschätzung zeigt Bild 4. Man kann allerdings noch keine endgültigen Angaben machen,
da abgewartet werden muß, wie der Stückzahleffekt auf die Preise durchschlägt. Über
die Nutzung in der traditionellen Fernmeldetechnik hinaus bietet die optische Nachrichtentechnik völlig neue Anwendungsmöglichkeiten, wie beispielsweise die ungestörte Nachrichtenübertragung in mit Strahlungsfeldern verseuchter Umgebung oder den
breitbandigen Teilnehmeranschluß im öffentlichen Telefonnetz, bei denen nur zu fragen ist, ob der technische Vorteil auch preislich akzeptiert werden kann. Das gleiche gilt generell für breitbandige Verteilnetze oder vollvermittelte Netze, in
denen die niedrigere Dämpfung verglichen mit dem Koaxialkabel vorteilhaft ist und
auch die Laufzeitverzerrungen besser beherrscht werden können.

Auf diesen von technischen Neuerungen bestimmten Gebieten darf man aber wohl ebenfalls sicher sein, daß die optische Übertragungstechnik auch wirtschaftliche Lösungen anbieten w d. Ja, man kann sagen, daß die optische Übertragungs- und Verarbei-
tungstechnik überhaupt erst die Grundlage für neue Anwendungen liefert. Im folgenden
wollen wir uns mit diesen neuen Lösungen nachrichtentechnischer Probleme, die die

Glasfaser anbietet, beschäftigen. Wir gehen aus vom heutigen Stand und leiten daraus
die neuen Aufgaben ab.

2. Der Stand der Übertragungstechnik

Zwar benutzen heute noch viele Übertragungsverfahren eine analoge Darstellung der
Signale, aber die Zukunft gehört eindeutig den digitalen Signalverarbeitungs- und
Übertragungsverfahren. Wenn wir den Stand der optischen Übertragungstechnik kurz
darstellen wollen im Hinblick auf eine zukünftige Entwicklung, so können wir uns
also beschränken auf die heutige digitale Übertragungstechnik. Sie wird realisiert
im öffentlichen Fernsprechnetz bei der Übertragung von Bündeln von 64 kbit/s-Kanä-
len. Der Grundkanal von 64 kbit/s überträgt einen Sprachkanal und kann prinzipiell
natürlich auch als allgemeiner Datenkanal genutzt werden. Die Übertragung eines solchen Kanales über die Glasfaser ist kein Problem. Die Reichweite der Übertragung ist
begrenzt nur durch die Faserdämpfung von weniger als 1db/km. Bei einer maximalen
Anschlußlänge des Teilnehmers an das öffentliche Netz von 15 km entsteht kein Zwang,
 Zwischenverstärker zu benutzen. Im Netz selber werden durch Multiplextechnik viele
64 kbit/s-Kanäle gebündelt. Um gleich zu den höchsten Bündeln zu kommen, betrachten
wir eine Fernstrecke. Hier hat man mit maximal 5.10^4 Kanälen je in den beiden Richtungen zu rechnen. Mit heutiger Analogtechnik realisiert man das im sogenannten
V 10800-System. Man benötigt im genannten Beispiel 5 solcher Systeme. Die Verstärker-
feldlänge der mit Koaxialkabeln ausgerüsteten Anlagen beträgt 1,55 km! In digitaler
Technik wird man in Zukunft dafür Systeme mit einer Bitrate von etwa 1 Gbit/s benutzen. Für $5 \cdot 10^4$ Sprachkänale in zwei Richtungen wären dann 10 Glasfasern nötig.
Bild 5 zeigt, daß mit der Monomodefaser hier Verstärkerfeldlängen von 20 ... 30 km
möglich sind, bei einer Wellenlänge von 1,3 μm. Der Fortschritt, der bei dieser Art
von Aufgabenstellung zu erwarten ist, ist der Fortschritt, der in vielen kleinen
Schritten Qualität und Technologie in heute schon absehbaren Schritten verbessert.
Die wesentliche Grundlagen sind in den Laboratorien schon erarbeitet. Zusätzlich
muß man beachten, daß der Bedarf der reinen Fernsprechtechnik an mehr Kanälen in der
Orts- und den Fernebenen durchaus begrenzt ist. Die neue optische Technik wird hier
also nur langsam sich durchsetzen und nur langfristig im Rahmen des Ersatzbedarfes
die Koaxtechnik verdrängen. Immerhin: Dieser Prozeß ist nicht aufzuhalten. Aber:
Die Auswechselzykluszeit in der Fernmeldetechnik beträgt 20-30 Jahre!

3. Neue Fernmeldetechnik

Im gleichen Augenblick nun, in dem die Techniker das Problem einer kostengünstigen
und leistungsfähigen Übertragungstechnik für das Fernsprechen gelöst haben, entsteht
darüber hinaus neuerdings der Wunsch nach gänzlich neuen Fernmeldediensten: Neben
der Sprache sollen Daten und Fernsehen voll vermittelt, also zwischen beliebigen
Teilnehmern weltweit ausgetauscht werden. Dieser Wunsch besteht zwar im Prinzip
schon lange. Aber erst heute ist eine große wirtschaftliche Bedeutung dieser Dienste

zu erkennen. Und mit der Glasfaserübertragungstechnik verfügt man auch über ein
Medium, das prinzipiell die zu erwartenden hohen Datenströme aus den Bildtelefonen
zu übertragen gestattet. Gegenüber diesen Datenströmen sind alle in der eigentlichen
Datenverarbeitung entstehenden Datenflüsse so gering, daß wir von nun an nur noch
von der Aufgabe sprechen wollen, wie ein weltweites, vollvermitteltes Bildfernsprechen technisch realisiert werden kann. Im Gegensatz zum Fernsprechen sind hier
die technischen Probleme durchaus noch nicht gelöst. Auch sind viele Fragen im Zusammenhang mit dem zu erwartenden Benutzerverhalten und den Benutzerwünschen offen.
Diese Fragen, zusammen mit technischen Realisierungsmöglichkeiten, sollen in einem
Großversuch der Bundespost, Bigfon genannt (Breitbandiges integriertes Glasfaser-
Fernmelde Ortsnetz), in der nächsten Zeit geklärt werden. In der übrigen Welt sind
ähnliche Bestrebungen im Gang, so daß man sicher sein kann, daß auf die Techniker
neue große Aufgaben zukommen werden. Neben der Vermittlung von ungeheuren Datenströmen wird deren weltweite Übertragung eine neue Herausforderung sein. Auch die
optische Übertragungstechnik wird Neuland betreten müssen.

4. Neue optische Übertragungstechnik: Wellenlängenmultiplex

Im Zusammenhang mit den Bigfon-Versuchen der Deutschen Bundespost gibt es Vorschläge,
mehrere Trägerfrequenzen über die Glasfaser zu übertragen. Bild 6 Man macht dabei von der
Tatsache Gebrauch, daß im optischen Bereich eine fast unermeßliche Bandbreite prinzipiell verfügbar ist. Allein zwischen den Wellenlängen von 1µm ($3 \cdot 10^{14}$ Hz) und
1,5 µm ($2 \cdot 10^{14}$ Hz) hat man eine Bandbreite von 10^5 GHz verfügbar. Der Teilnehmer
könnte beispielsweise auf der Wellenlänge λ_1=1,3 µm Fernsprechen und auf der Wellenlänge λ_3=1,5 µm Bildfernsprechen. Er empfängt entsprechend bei den Wellenlängen
λ_2=1,4 µm bzw. λ_4=1,6 µm. Die Glasfaser selbst verträgt das mühelos, solange die
Leistungsdichte begrenzt bleibt. Bei der sogenannten Multimode-Faser mit großem
Kerndurchmesser gibt es hier keine Schwierigkeiten. Das eigentliche Problem ist das
Ein- und Ausfädeln der verschiedenen Farben λ_1, λ_2, λ_3, λ_4. Bild 7 zeigt, wie mit
Hilfe eines Beugungsgitters das ankommende Licht, bestehend aus den vier Farben,
in 4 getrennte Wellenfelder aufgespalten wird. Die vier getrennten Farben werden
in vier verschiedenen Fasern weitergeleitet. Da der Aufbau der Beugungs- und Fokus-
sierungsanordnung reziprok ist, kann die Strahlrichtung beliebig umgekehrt werden;
beispielsweise können die Farben λ_1 und λ_3 der abgehenden Richtung, λ_2 und λ_4 der
ankommenden Richtung zugeordnet werden. Bild 8 zeigt den praktischen Aufbau der
Anordnung mit Mitteln der Mikrooptik. Bild 9 zeigt die Durchlaßkurven für einen
Demultiplexer mit 8 Kanälen.

5. Die zukünftige optische Übertragungstechnik: Die kohärente optische Signalverar-
arbeitung und -übertragung

Mit dem eben geschilderten Verfahren des Wellenlängenmultiplex kann die Glasfaser
mehrfach genutzt werden: Die Grenzen dieses Verfahrens sind jedoch abzusehen. Viel

mehr als 10 Kanäle wird man aus Gründen der Genauigkeit der mikrooptischen Anordnung nicht zusammenfügen können. Schließlich kommt auch noch das Problem auf, daß die Leistungsdichte auch in der Multimodefaser wegen des Auftretens nichtlinearer Streuprozesse nicht beliebig gesteigert werden kann. Immerhin benötigt man etwa eine Leistung von einigen mW pro Kanal.

Die verfügbare Frequenzbandbreite der Glasfaser ist von der Kanalbelegung hingegen in keiner Weise ausgeschöpft. Zunächst wollen wir uns klar machen, daß auf den Fernstrecken des öffentlichen Netzes langfristig ein wirklicher Bedarf an dieser noch verfügbaren Bandbreite besteht.

Wir gehen dabei davon aus, daß wieder langfristig die gleiche Zahl von Breitbandkanälen mit vermittelten Fernsehtelefongesprächen verfügbar sein muß wie heute für die Ferngespräche. Der Kanal (in einer Richtung) hat dann eine Bitrate von etwa 250 Mbit/sec zu übertragen. Bandbreitesparende, analoge Modulationsverfahren scheiden bei der Fernübertragung aus. Im schon genannten Beispiel mit $5 \cdot 10^4$ Kanälen in jeder Richtung könnte man ein Übertragungssystem in bisheriger optischer Technik bei Benutzung von 1Gbit/s auf jeder Faser errichten, also mit $1/4 \cdot 5 \cdot 10^4$ Glasfasern in jeder Richtung, insgesamt also mit $2,5 \cdot 10^4$ Fasern, untergebracht in mehreren Kabeln.

Schon die Spleiss- und Steckertechnik stellt uns vor große Schwierigkeiten, vor allem deshalb, weil man sicherlich Monomode-Fasern mit wenigen µm Kerndurchmesser verwenden muß, um große Verstärkerfeldlängen von 20-30 km zu erreichen. Die folgende kurze Abschätzung über die Verlustleistung in den Zwischenregeneratoren zeigt aber, daß ein solches System praktisch nicht zu verwirklichen ist: Pro Faser benötigt man im Zwischenregenerator bei einer Bitrate von 1 Gbit/s auch in überschaubarer Zukunft wenigstens 20 W Leistung, die als Wärme abgeführt werden muß, bei $2,5 \cdot 10^4$ Fasern sind das insgesamt 500 kW.

Um diese Schwierigkeit wird man nur herumkommen, wenn man die Leistung pro Kanal drastisch senkt. Der Weg dahin ist deutlich vorgezeichnet, und die Forschungen in dieser Richtung haben schon begonnen: Forschungen für kohärent-optische Signalverarbeitung und integrierte Optik.

Das Ziel ist die Realisierung einer aus der Hochfrequenztechnik bekannten echten Frequenzmultiplextechnik unter Benutzung von Homodyn- oder Heterodyn-Frequenzumsetzungsverfahren. Mit solchen Verfahren lassen sich im Prinzip um etwa 20 db niedrigere Leistungspegel im Übertragungskanal erreichen, bei gleicher Bitfehlerrate im Detektor wie beim bisherigen Verfahren der direkten Gleichrichtung. Doch das allein ist noch nicht ausreichend. Ein entscheidender Fortschritt wird es sein, die Kanäle in der Frequenz dichter zu packen und dann viele Kanäle gemeinsam zu verstärken, in einfachen linearen, relativ schmalbandigen Laser-Verstärkern.   Bild 10 zeigt die Dämpfung in einer Glasfaser in Abhängigkeit von der Frequenz. Bei einer Wellenlänge von 1,1 µm (272,540 THz) ist mit einem Strich ein Band der relativen Breite von 1,4 ‰ eingetragen. Es entspricht einer absoluten Breite von 500 GHz. Wie man in diesem Band lagemäßig 100 je mit 1 Gbit/s modulierte Kanäle unterbringen kann, zeigt Bild 11. Die gesamte Bandbreite ließe sich in einem linearen Halbleiter-laser-Verstärker verstärken, wobei also für 4 x 100 Kanäle (4 Kanäle à 250 Mbit/s in einem 1 Gbit/s-Kanal) wieder mit etwa 20 W Verlustleistung im Verstärker zu rechnen wäre. Statt 500 kW hätten wir dann in einem nichtregenerativen Zwischen-verstärker nur eine Verlustleistung von 5 kW abzuführen. Das erscheint schon eher möglich. Aber mit welchen Mitteln wäre eine solche Technik realisierbar? Sie ist nur denkbar mit Hilfe von monolithisch integrierten elektronischen und optischen Schal-tungen. Zunächst zeigt uns das Bild 12, was im Signal-Empfänger prinzipiell gelei-stet werden muß. Das in einer Faser ankommende Signal hat die Frequenzen $f_1 \pm f_{m1}$ ... $f_{100} \pm f_{m100}$. $f_1 \ldots f_{100}$ sind die Träger, $f_{m1} \ldots f_{m100}$ sind die Seitenbänder. Zur De-modulation (Abbereitung bis ins Basisband 0...1Gbit/s) müssen die Träger $f_1 \ldots f_{100}$ mit hoher Leistung und phasenstarr wieder zugesetzt werden. Bei der hohen Frequenz von etwa 270 THz gelingt das nur, wenn in dem gesamten Übertragungssystem nur ein einziger Mutteroszillator (Laser) vorhanden ist, der in einer getrennten Faser mitge-führt und im Empfänger entsprechend linear in einem Laserverstärker verstärkt wird. Das Prinzip des eigentlichen frequenzselektiven Demodulators zeigt das Bild 13. In einem Interferenz-Resonator mit Bragg-Reflektoren kommt die Frequenz f_1 zur Resonanz. Wenn innerhalb und längs des Resonators ein p-n-Übergang einer Halbleiter-Diode liegt, werden vorzugsweise die Resonanzfrequenz und entsprechend der Bandbreite des Resonators auch seine Seitenbänder durch Ladungsträgererzeugung absorbiert, die anderen Frequenzen werden durchgelassen. Setzt man mit sehr hoher Amplitude z.B. den entsprechenden Träger f_1 zu, so entsteht im Diodenstrom das Basisbandsignal f_{1m}.

Viele Probleme zur Realisierung eines solchen Vielkanalhomodyn-Frequenzmultiplex-Empfängers und natürlich auch des entsprechenden Senders als optisch-elektronischer Großschaltkreis sind noch offen. Die Probleme reichen vom Substrat-Material, das sowohl optisch als auch elektronisch für die Integration geeignet sein muß, bis zur Strukturierung beispielsweise der Resonatoren mit hoher Güte. Die Wahl der rich-tigen Frequenz für das Übertragungsband ist dagegen fast belanglos, da es einen weiten Bereich mit niederer Dämpfung und niederer Dispersion der Glasfaser gibt, der prinzipiell brauchbar ist. Die Dispersion kann man übrigens zum Ausgleich des Lauf-zeitverhaltens der einzelnen Spektralanteile mit Hilfe der Kompensation der Wirkung der Materialdispersion und der Wellenleiterdispersion im Monomode-Wellenleiter in ausreichend schmalen Frequenzbereichen praktisch vollständig beseitigen. Die Länge der Übertragungsstrecke ist dann wieder nur Dämpfungsbegrenzt, während heute noch die Laufzeitverzerrungen die Länge der Strecke sehr stark bestimmen. Wie erste optisch-elektronisch integrierte Schaltungen aussehen, zeigt Bild 14.

Es ist noch ein weiter Weg bis zur Integration so komplexer Schaltungen, wie wir sie eben besprochen haben. Es ist aber ein Weg, an dessen Ende die Möglichkeit stehen wird, weltweit und vollvermittelt, Fernsehgespräche höchster Vollkommenheit zu führen.

Eine Reihe von Bildern stammen aus dem Siemens-Telcom-Report 6. Jahrg. April 83, Beiheft Nachrichtenübertragung mit Licht, sowie aus dem AEG-Forschungsinstitut Ulm. Dafür sei gedankt.

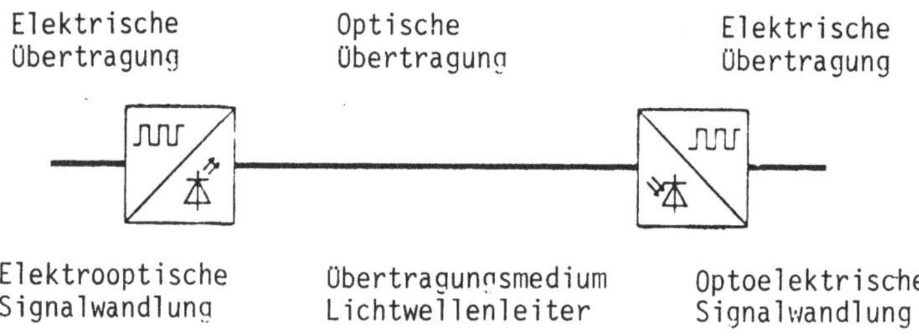

Bild 1. Prinzip der optischen Nachrichtenübertragung

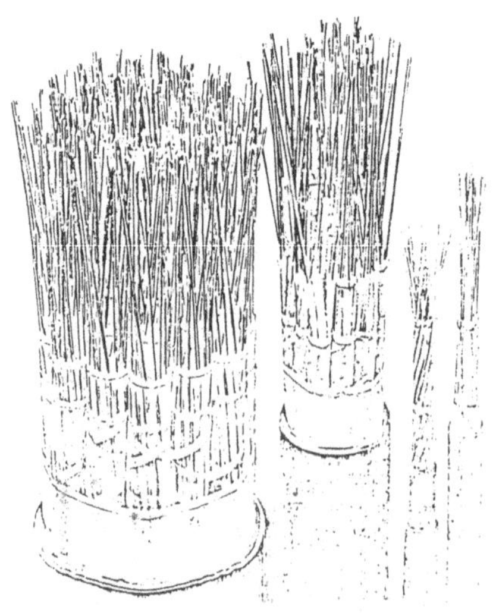

Bild 2. Faser-Kabel

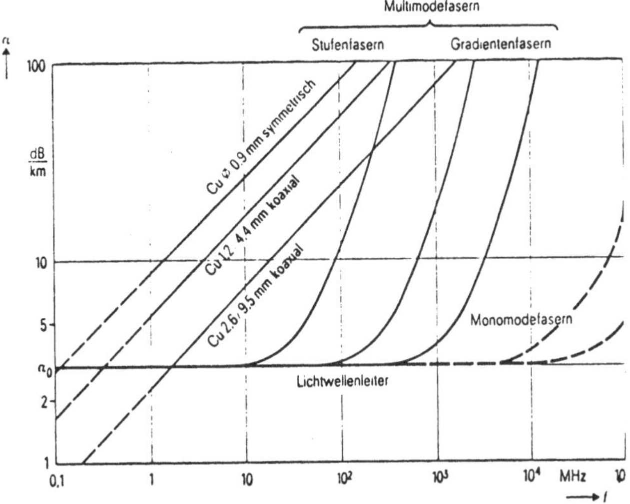

α₀   Grunddämpfung der Fasern, abhängig von der Lichtwellenlänge λ
(bei λ = 850 nm liegt sie bei rund 3 dB/km, bei λ = 1300 nm unter 1 dB/km)

Bild 3. Übertragungsverhalten der Lichtwellenleiter im Vergleich mit Kupferleitern (Cu) am Beispiel der kilometrischen Dämpfung α und der Frequenz f

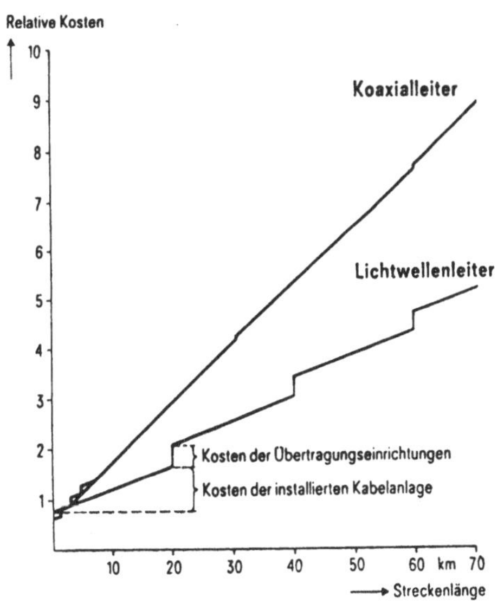

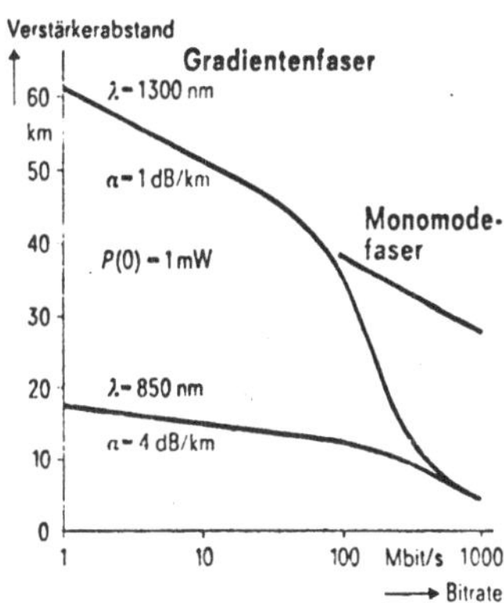

α Dämpfungskoeffizient (Faser, Spleiße, Stecker)
λ Lichtwellenlänge
$P(0)$ Lichtleistung des Senders (Laserdiode)

Bild 4. Wirtschaftlichkeitsvergleich für eine Übertragungsrate von 565/s über Koaxial- und Lichtwellenleiter

Bild 5. Theoretisch mögliche Verstärkerabstände und Bitraten in Lichtwellenleiter- Übertragungssystemen

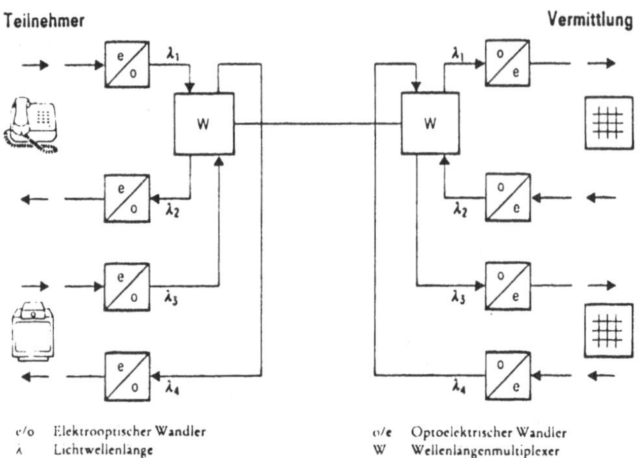

Bild 6. Übertragung von Breitbandsignalen mit Hilfe der Lichtwellenleitertechnik

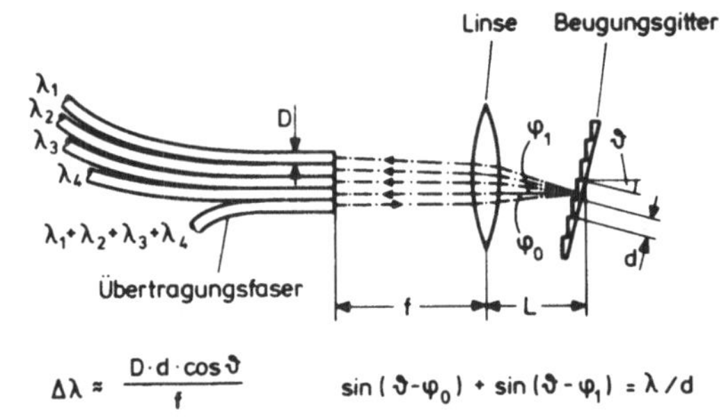

$$\Delta\lambda \approx \frac{D \cdot d \cdot \cos\vartheta}{f} \qquad \sin(\vartheta - \varphi_0) + \sin(\vartheta - \varphi_1) = \lambda/d$$

Bild 7. Gitter-Demultiplexer (Littrow-Aufbau)

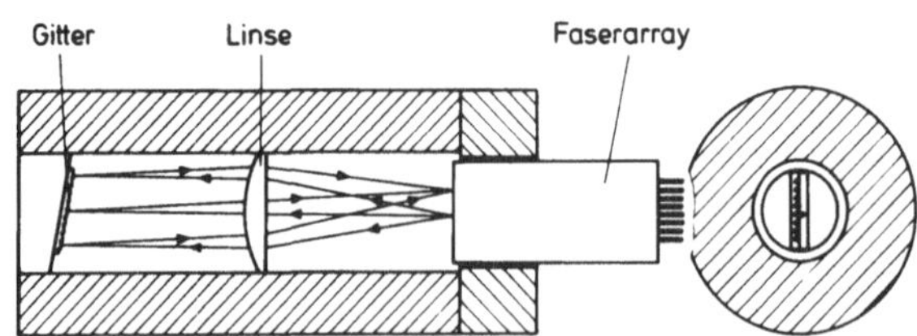

Bild 8. Aufbaukonzept für Gitter-Multi-/Demultiplexer

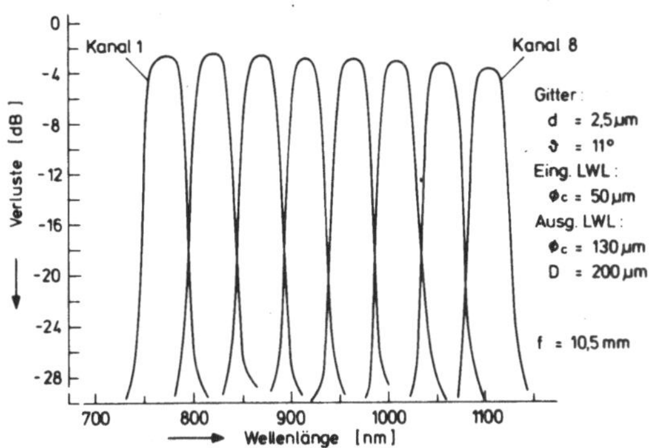

Bild 9. Gitter-Demultiplexer: Durchlaßkurven

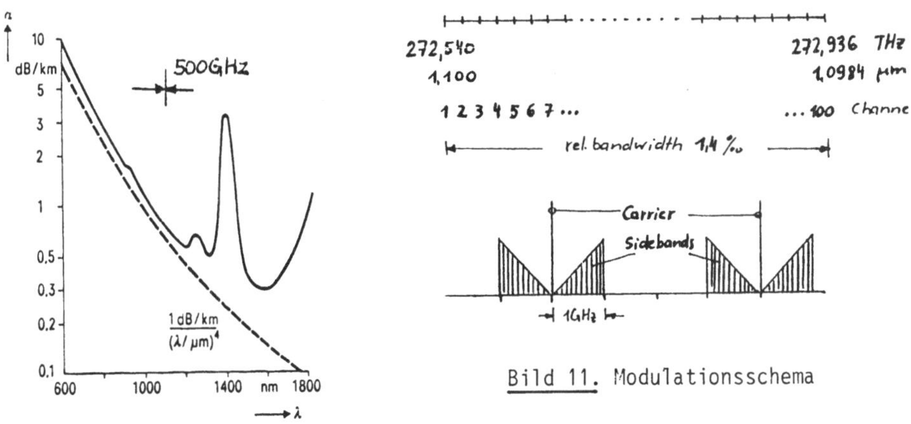

Bild 11. Modulationsschema

Bild 10. Dämpfungskoeffizient α einer Quarzglasfaser in Abhängigkeit von der Lichtwellenlänge λ

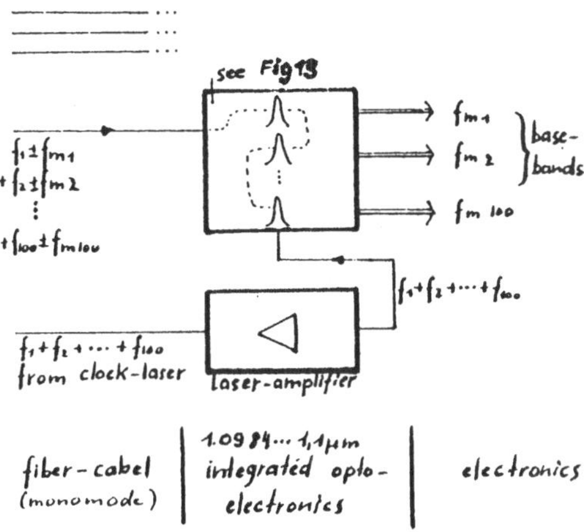

Bild 12. Homodyndemodulation

434

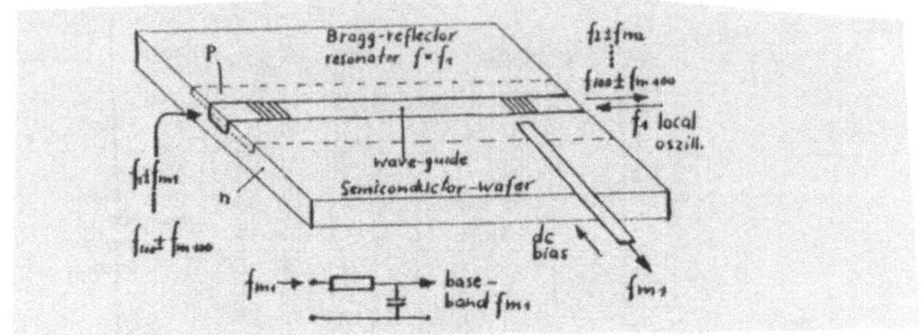

Bild 13. Selektive Resonator-Struktur

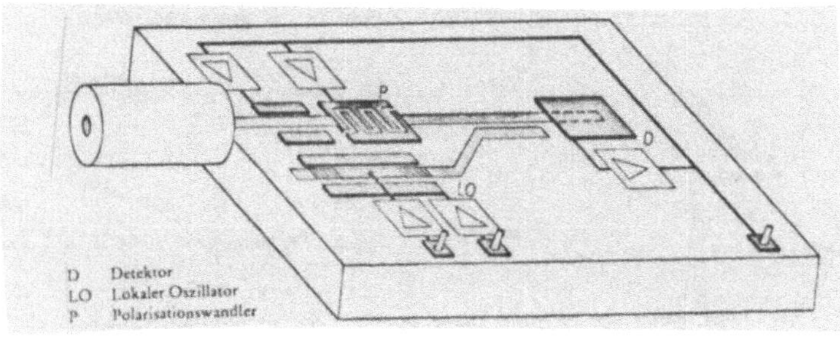

Bild 14. Integrierter optischer Heterodynempfänger

Einsatz der Glasfasertechnik im Netz der deutschen Bundespost mit Schwerpunkt im Ortsnetz

A. NAAB
Bundesministerium für das Post- und Fernmeldewesen
5300 Bonn

1   Zielsetzungen

Schon seit mehreren Jahren bereitet die Deutsche Bundespost (DBP) durch
Forschungsvorhaben die Einführung der optischen Nachrichtentechnik in
ihr Fernmeldenetz technisch-wissenschaftlich vor. Hierbei verfolgt die
DBP die Zielsetzung, etwa ab 1985/86 mit dem Aufbau eines Integrierten
Breitband-Fernmeldenetzes (IBFN) zu beginnen. Die optische Nachrichtentechnik kann bereits gegenwärtig im Weitverkehrsnetz unter wirtschaftlichen Gesichtspunkten eingesetzt werden. Im Bereich des Ortsnetzes,
d.h. im Teilnehmeranschlußbereich, ist der Beginnzeitpunkt hingegen noch
an die Erfüllung technischer und wirtschaftlicher Voraussetzungen ge-
knüpft.

Um diese Voraussetzungen schaffen zu können, führt die DBP den Systemversuch BIGFON (Breitbandiges Integriertes Glasfaser-Fernmelde-Orts-Netz)
durch.

2   Systemversuche

Bereits 1980 wurde die deutsche Fernmeldeindustrie gebeten, firmenindividuelle Konzepte für die Erstellung und Erprobung von Prototyp-Systemen
vorzuschlagen. Die am Systemversuch beteiligten Firmen sollen die technische Realisierung ihrer Projekte nach eigenen Vorstellungen entwerfen.
Die DBP hat lediglich einen Rahmen vorgegeben, der die gemeinsame Ziel-
richtung festlegen soll. Dieser Rahmen beinhaltet insbesondere die Forderung, daß dem Versuchsteilnehmer die gleichzeitige Übertragung von mehreren Fernsprech-, Daten-, Text- und Faksimilekanälen, 2 bis 4 verteilver-
mittelten Fernsehkanälen (die Zahl der angebotenen Programme richtet sich
nach den örtlichen Gegebenheiten), 24 Stereotonkanälen sowie einem Bild-
fernsprechkanal mit Farbfernsehqualität angeboten werden soll.

Das Vorhaben hat in seinem bisherigen Verlauf an der technisch-technolo-
gischen Realisierbarkeit einer Integration aller Fernmeldedienste in
einer optischen Teilnehmer-Anschlußleitung keine Bedenken aufkommen
lassen. In den zurückliegenden 3 Jahren, in denen die Entwicklung der

erforderlichen Systemtechniken vorgenommen wurde, traten keine grundsätzlichen Probleme auf, die diesbezügliches in Frage stellten; es kann vielmehr vorausgesetzt werden, daß die Inbetriebnahme der BIGFON-Projekte ab
Dezember 1983 planmäßig stattfinden wird.

Bereits im Herbst 1982 wurden die ersten Glasfaser-Kabel in den BIGFON-
Anschlußbereichen verlegt. Der Aufbau der zentralen technischen Einheiten wird im Sommer 1983 und die Einrichtung der Teilnehmer-Einheiten
voraussichtlich ab Oktober 1983 erfolgen. Anschließend werden 6 unterschiedliche, firmenindividuelle BIGFON-Konzeptionen in die Testphase ein-
treten.

Das Spektrum der von den Firmen angebotenen Konzepte eines Breitbandigen-
Integrierten-Glasfaser-Fernmeldeortsnetz umfaßt sämtliche mit z.Z. verfü-
barer Technologie erstellbaren Alternativen, von der Anwendung optischer
Multiplexverfahren, analoger Breitbandübertragungs- und Vermittlungstech-
nik bis hin zur hochbitratigen Digitalübertragungs- bzw. Vermittlungs-
technik.

Der Systemversuch BIGFON sieht die firmenspezifischen Erstellungen der
Prototyp-Systeme in den Städten Berlin (Krone, Siemens, SEL), Hamburg
(TEKADE/F&G), Hannover (AEG-Telefunken, fuba), Düsseldorf (AEG-Telefunken), Stuttgart (SEL), Nürnberg (TEKADE/F&G) und München (Siemens) bis
Ende des Jahres 1983 und die anschließende Erprobung bis 1986 vor. Die
Anzahl der angeschlossenen Teilnehmer ist den Bedingungen einer Proto-
typ-Erprobung angepaßt und liegt demnach bei 320 Teilnehmern. Die Glas-
faser-Versuchsnetze sind integraler Bestandteil des öffentlichen Fern-
meldenetzes, d.h. über sie wird normaler Fernmeldebetrieb abgewickelt
werden. 68 Teilnehmern der BIGFON-Teilnehmer wird die Möglichkeit ge-
boten werden, das Bildfernsprechen zu testen.

Während des 3jährigen BIGFON-Systemversuchs plant die DBP, die BIGFON-
Aufbauorte für den überregionalen Bildfernsprechverkehr untereinander
zu verbinden. Dies wird erstmals im Rahmen des Systemversuchs BIGFern
geschehen, mit dem u.a. der Einsatz der Glasfaser auch im überregionalen
Fernnetz erprobt werden soll. Zu diesem Zweck ist beabsichtigt, zwischen
Hamburg und Hannover 140 Mbit/s-Glasfaser-Übertragungssysteme zu errichten. Die geplante Errichtung der optischen Weitverkehrs-Strecke dient
- der heutigen technologischen Entwicklungsphase entsprechend - der Erprobung digitaler 140 Mbit/s-Prototypsysteme auf Gradientenprofilfasern.

Da zu erwarten ist, daß bis 1985/86 die serienmäßige Einsatzreife derar-

tiger Systeme erreicht werden kann, bilden sie die übertragungstechnische
Grundkonzeption der DBP für die in der zweiten Hälfte der 80er Jahre mögliche Einführung eines neuen Dienstes "Breitband-Individualkommunikation"
wie z.B. Bildfernsprechen.

Die Festlegung der System-Bitrate auf 140 Mbit/s ergibt sich u.a. aufgrund der technischen Entwicklung in der digitalen Fernsehstudiotechnik
und der damit einhergehenden internationalen Normung. Aus ihr läßt sich
ableiten, daß für die Übertragung hochwertiger digitaler Fernsehsignale
die in der PCM-Hierarchie vorhandene Übertragungsgeschwindigkeit von
140 Mbit/s zur Standardgeschwindigkeit werden wird. Nicht nur technische,
sondern auch wirtschaftliche und dienstspezifische Gründe sprechen nun
dafür, auch für das Bildfernsprechen, wie es im BIGFON-Versuch konzipiert ist, dieselbe Bitrate sogar bis hin zum Teilnehmer zu verwenden.

Ebenso entsprechen die weiteren Systemkennwerte der Versuchsstrecke den
gegenwärtig erkennbaren technischen und technologischen Entwicklungen:
Das Kabel wird 60 Gradientenprofilfasern enthalten, von denen zunächst
30 Fasern mit 140 Mbit/s-Systemen beschaltet werden. Da zunächst die
Fasern nur im Simplex-Verfahren betrieben werden, steht dann eine Übertragungskapazität für 15 simultane Bildfernsprechverbindungen zur Verfügung.

Allerdings ist die Spezifikation der Fasern so gewählt, daß bei später
verfügbaren Wellenlängenvielfach-Technologien pro Faser mindestens zwei
140 Mbit/s-Kanäle übertragen werden können. Somit bietet die Prototyp-
Anlage eine Gesamtübertragungskapazität von 60 Duplex-Breitbandkanälen.

Die optischen Systeme werden im 1 300 nm Wellenlängenbereich arbeiten:
der Abstand der elektro-optischen Zwischenregeneratoren wird fast ausnahmslos 18 km betragen. Die Realisierbarkeit dieser übertragungstechni-
schen Parameter konnte in einem Forschungsvorhaben, genannt Berlin III,
bereits im Frühjahr 1983 bestätigt werden. In diesem Vorhaben, für das
6 Firmen der deutschen Fernmeldeindustrie den Auftrag für jeweils eine
Versuchsanlage erhielten, wurde gezeigt, daß der Einsatz von 140 Mbit/s-
Systemen auf Gradientenprofilfasern tatsächlich eine günstige Lösung für
ein zukünftiges Glasfaser-Breitband-Netz darstellen.

Die von den Auftragnehmern eingesetzten Kabel enthielten Fasern unterschiedlicher Herstellungsverfahren. So waren z.B. auch Fasern aus japa-
nischer und französischer Fertigung vertreten.

Als vorläufiges Ergebnis dieses Vorhabens kann festgestellt werden, daß die gegenwärtig produzierten Gradientenprofilfasern es gestatten, prinzipiell eine Nutzbitrate von 140 Mbit/s über eine Entfernung von bis zu 36 km zu übertragen, ohne die geforderte Bitfehlerrate von 10^{-9} zu überschreiten.

Weiterhin zeigte sich, daß Spezifikationsparameter, die bisher üblicherweise auf die Längeneinheit bezogen werden (z.B. Bandbreite in MHz · km) nicht das tatsächliche übertragungstechnische Verhalten einer optischen Weitverkehrs-Strecke wiedergeben können, d.h. Theorie und Praxis weichen hier noch erheblich voneinander ab.

3 Systemkonzeption der DBP

Über die Systemkonzeption eines Integrierten Breitband- Fernmelde-Netzes lassen sich gegenwärtig folgende Aussagen treffen: Es wird drei Dienstleistungskategorien umfassen:
- schmalbandige Individualkommunikation (Fernsprechen, Daten und Texte),
- breitbandige Individualkommunikation (Bildfernsprechen) und
- breitbandige Verteilkommunikation (Fernseh- und Tonrundfunk).

Für die technische Konzeption eines Ortssystems gelten folgende Überlegungen:
Die Teilnehmeranschlußleitung kann aus einer Faser bestehen, da die Trennung von Hin- und Rückrichtung bei der Signalübertragung durch optische Weichen leicht realisierbar ist. Die Übertragung der einzelnen Dienstleistungskategorien sollte zweckmäßigerweise so erfolgen, daß eine Zuordnung optischer Signalströme zu dienstspezifischen Vermittlungseinrichtungen möglich ist, um einen modularen Systemaufbau (bedarfs orientiert) zu erleichtern.

Kapazitätsarme PIN-Photodioden für 1,1 bis 1,55 μm

A.-E. SCHURR und R. DEUFEL
AEG-TELEFUNKEN, Forschungsinstitut Ulm
Sedanstraße 10, D-7900 Ulm

1. Einleitung

In diesem Beitrag werden InGaAs-PIN-Photodioden vorgestellt. Diese
Dioden sind als sehr schnelle und empfindliche Empfänger für das nahe
Infrarot von 0,95 bis 1,65 μm Welleniänge einsetzbar /1/. Sie werden insbesondere in Empfangseinheiten von optischen Nachrichtensystemen bei
1,3 μm verwendet. Von solchen Detektoren werden niedrige Sperrschicht-
kapazitäten, kapazitätsarmer Aufbau, geringe Dunkelströme, hohe Schnel-
ligkeit und möglichst hoher Quantenwirkungsgrad verlangt. Ferner soll
eine hybride Integration mit einem FET /2/ und eine einfache Ankopp-
lung einer Lichtleitfaser an die Photodiode möglich sein. Speziell
im Hinblick auf eine Integration ist besonders auf eine möglichst
kleine Zusatzkapazität durch den Diodenaufbau zu achten.

2. Struktur und Aufbau der InGaAs - PIN-Photodioden

In Fig. 1 ist die Struktur dieser Dioden dargestellt.

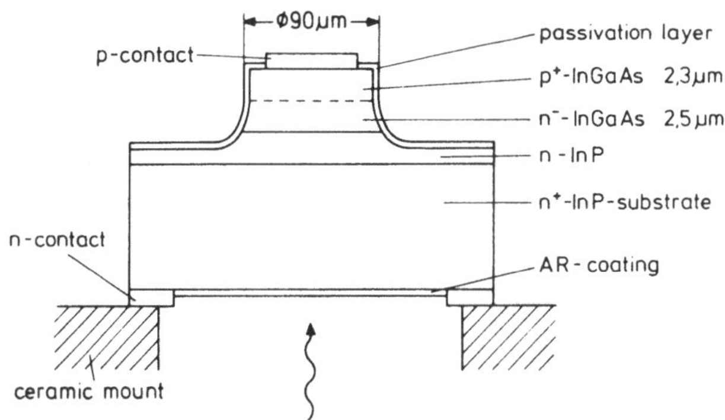

Fig. 1. Struktur der InGaAs - PIN-Photodiode

Durch den Mesadurchmesser von nur 90 μm und die niedrige Dotierung der
n⁻-InGaAs-Absorptionsschicht von einigen 10^{-15} cm^{-3} wurde eine niedrige
Sperrschichtkapazität und ein geringer Dunkelstrom erreicht. Allerdings

muß dadurch substratseitig eingestrahlt werden. Die Passivierungs- und Entspiegelungsschichten sind durch pyrolytisch aufgebrachtes SiO_2 hergestellt.
Die Dioden werden substratseitig so auf eine Keramik-Subsenke montiert, daß das Einstrahlfenster über einem Schlitz in der Subsenke zu liegen kommt. Durch diesen Schlitz erfolgt das Ankoppeln der Lichtleitfaser. Die Subsenken sind mit Leiterbahnen versehen, so daß die Dioden einfach kontaktiert und mit FET's integriert werden können. Die Streukapazitäten der Subsenken liegen bei 73 fF.

3. Eigenschaften der InGaAs - PIN-Photodioden
Dunkelstrom

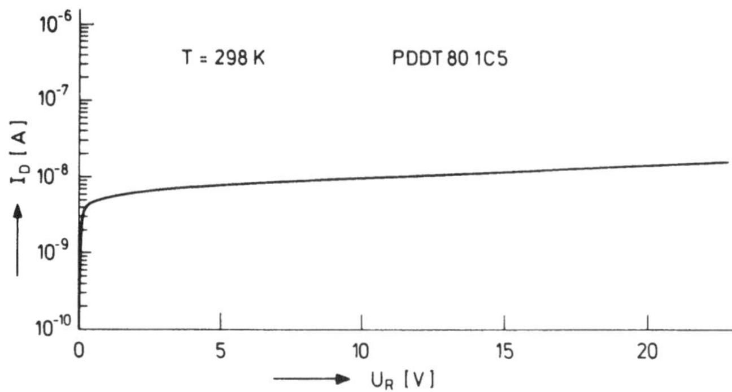

Fig. 2. Dunkelstromkennlinie einer InGaAs-PIN-Photodiode

Wie aus Fig. 2 ersichtlich ist, beträgt der Dunkelstrom bei einem Arbeitspunkt von -10 V etwa 10 nA. Pro 10 Grad Temperaturerhöhung verdoppelt sich der Dunkelstrom.

Empfindlichkeit

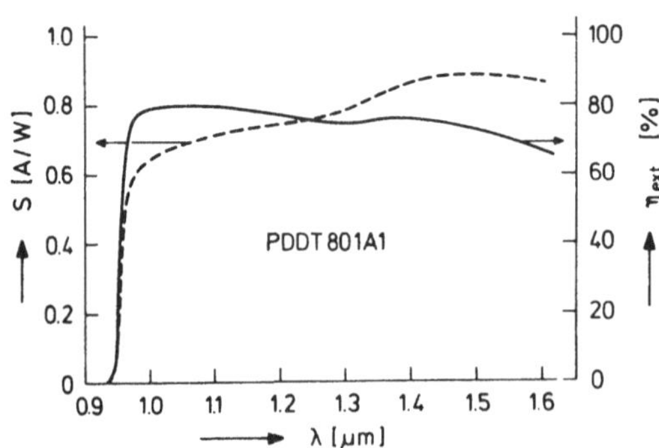

Fig. 3. Spektrale Empfindlichkeit S und Quantenwirkungsgrad η_{ext} einer InGaAs-PIN-Photodiode

Der Verlauf der Empfindlichkeit und des Quantenwirkungsgrades einer dieser Dioden ist in Fig. 3 aufgetragen. Der externe Quantenwirkungsgrad ist im Wellenlängenbereich zwischen 1 und 1,55 µm größer 70 %. Die Dioden können auch für die Detektion von Licht der Wellenlänge von 1,55 µm, also im Bereich minimaler Faserdämpfung, eingesetzt werden. Der Empfindlichkeitsbereich der Dioden wird auf der kurzwelligen Seite durch die Absorption im Substratmaterial und auf der langwelligen Seite durch die Bandkante des ternären Materials begrenzt. Durch weitere Optimierung der Entspiegelung und durch Verringerung der freien Ladungsträgerabsorption im Substratmaterial sollte ein externer Quantenwirkungsgrad von ca. 90 % möglich sein.

<u>Kapazität</u>

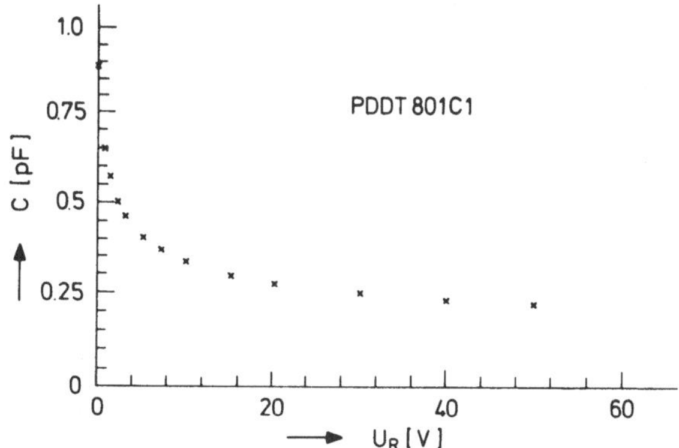

Fig. 4. C-V-Kennlinie einer InGaAs - PIN-Photodiode

Die Dioden haben bei 10 V Sperrspannung Sperrschicht-Kapazitäten unter 0,4 pF. Die Auswertung der C(V)-Messung ergibt, daß die n^--InGaAs-Absorptionsschicht bei dieser Spannung bis hin zur Heterogrenze ausgeräumt ist, wodurch ein langsamer Diffusionsanteil zum Photostrom vermieden wird (Fig. 5).

<u>Schnelligkeit</u>

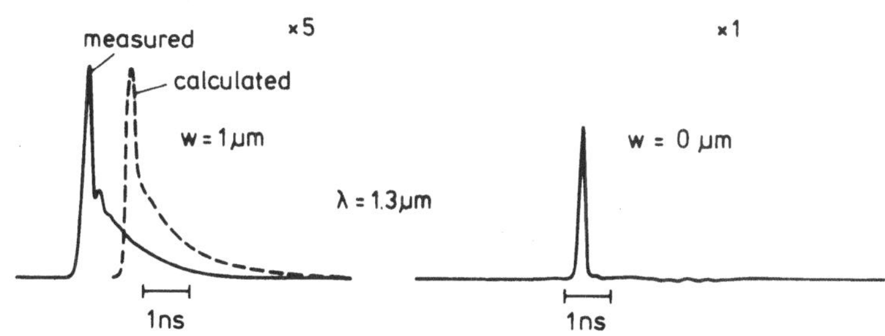

Fig. 5. Impulsverhalten der InGaAs - PIN-Photodioden·

Im rechten Teil von Fig. 5 ist die Impulsantwort einer dieser Photodioden bei 10 V Sperrspannung dargestellt. Die Lichtimpulse der Wellenlänge von 1,3 µm lieferte ein mit "mode-locking" betriebener Ramanlaser.
Die Auswertung ergibt, daß die Pulshalbwertsbreite der Photodioden kleiner 130 psec ist.
Im linken Teil des Bildes ist die experimentelle und berechnete Impulsantwort einer Photodiode mit einer nichtausgeräumten n^--InGaAs-
Absorptionsschicht zwischen Heterogrenze und Sperrschichtrand mit einer
Dicke von 1 µm aufgetragen. Der Diffusionsanteil zum Photostrom aus dieser Schicht führt zu einem Ausläufer des Photostromsignals von einigen
nsec Dauer. Die maximale Pulsamplitude sinkt gleichzeitig auf etwa 25 %.
Schnelle Photodioden können also nur realisiert werden, wenn die Absorptionsschicht im Arbeitspunkt vollständig ausgeräumt ist.

Die diesem Artikel zugrunde liegenden Arbeiten wurden mit Mitteln des
Bundesministerium für Forschung und Technologie gefördert. Die Verantwortung für den Inhalt liegt allein bei den Autoren.

Literatur
/1/ C. A. BURRUS, A. G. DENTAI und T. P. LEE
    Electron. Lett. 15, 655 - 657, (1979)
/2/ D. R. SMITH, R. C. HOOPER, K. AHMAD, D. JENKINS, A. W. MABBITT
    und R. NICKLIN
    Electron. Lett. 16, 69 - 70 , (1980)

1,3 μm V-Nutlaser als Sender für optische Nachrichtenübertragungssysteme

G. ARNOLD, P. MARSCHALL und E. SCHLOSSER
AEG-TELEFUNKEN, Forschungsinstitut Ulm
Sedanstraße 10, D-7900 Ulm

1. Einleitung

Für Glasfaser-Nachrichtenübertragungssysteme im langwelligen Spektralbereich sind Halbleiterlaser, die bei 1,3 μm emittieren, von besonderem
Interesse. Beim Ankoppeln der Laser an die Faser kann es zu Rückwir-
kungseffekten kommen, die zu einem starken Zusatzrauschen führen /1/.
Besonders rückwirkungsempfindlich sind spektral schmale Laser wie z. B.
die indexgeführten Streifenstrukturlaser. Gaingeführte Streifenlaser,
wie der in diesem Beitrag beschriebene 1,3 μm V-Nutlaser, können dagegen grundsätzlich spektral breiter und damit rückwirkungsunempfindli-
cher sein. Diese Laser zeichnen sich zudem aus durch einfache Herstellbarkeit, hohe Zuverlässigkeit, Stabilität der Emissionseigenschaften,
Modulierbarkeit bis zu hohen Bitraten /5/ und sind damit geeignet als
Sender für optische Nachrichtenübertragungssysteme.

2. Laserstruktur

Die Laserstruktur ist in Fig. 1 wiedergegeben /2/. Sie basiert auf der
GaInAsP/InP Doppelheterostruktur mit 2 zusätzlichen Schichten von
n - GaInAsP und n - InP. Die Stromeinengung geschieht durch eine Zn-
Diffusion, deren Verlauf durch die Oberflächenstruktur vorgegeben ist.
Unterhalb der V-Nut erzeugt die Diffusion einen schmalen, streifenför-
migen p-leitenden Kanal von der Oberfläche zur p - InP Schicht. Die La-
sereigenschaften hängen von der Breite dieses Kanals und der Diffusions-
tiefe ab. Damit die Laser transversal im Grundmodus stabil emittieren,
muß die Streifenbreite hinreichend schmal sein (Durchdringweite $\lesssim 8$ μm).
Für die Lage der Diffusionsfront sind in Fig. 1 zwei extreme Beispiele,
gekennzeichnet als Typ 1 bzw. Typ 2, angedeutet. Die Eigenschaften beider Lasertypen werden im folgenden dargestellt.

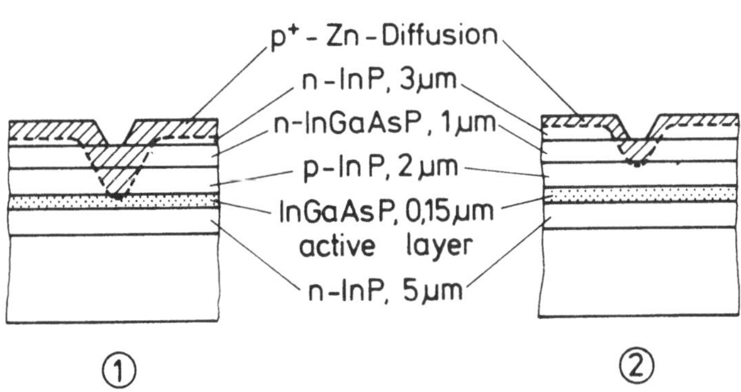

Fig. 1. Struktur des 1,3 μm V-Nutlasers

3. Lasereigenschaften

3.1 Licht/Stromkennlinie

Eine typische Licht/Stromkennlinie ist in Fig. 2 wiedergegeben. Bei einer Durchdringweite unter 8 µm sind die Kennlinien innerhalb des Meßbereiches bis 35 mW "kink"-frei. Der Schwellstrom liegt bei 200-300 µm langen Lasern zwischen 80 und 100 mA für Typ 1 und zwischen 150 und 200 mA für Typ 2. Der Wirkungsgrad beträgt ca. 0,2 W/A. Die Laser können bis 70°C im Dauerstrich betrieben werden.

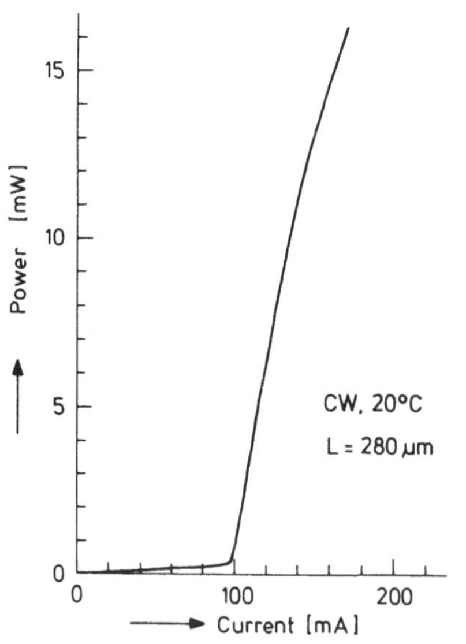

Fig. 2. Licht/Stromkennlinie

3.2 Abstrahlcharakteristik

Die Fernfeldverteilungen parallel und senkrecht zur aktiven Schicht sind in Fig. 3 zusammengestellt. Das Fernfeld senkrecht zum pn-Übergang hat eine Halbwertsbreite von 40 - 50°. Parallel zur aktiven Zone ergeben sich für Typ 1 - Laser schmale Fernfelder von ca. 10° Halbwertsbreite, die sich auf etwa 30° bei Typ 2 - Lasern verbreitern.

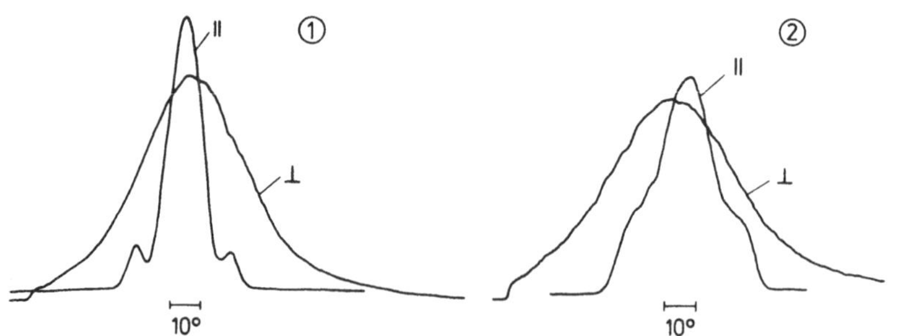

Fig. 3. Fernfeldverteilung parallel und senkrecht zum pn - Übergang

3.3 Laserspektrum

Die Laserspektren, dargestellt in Fig. 4, sind multimodig, wobei die Hüllkurve bei Lasern von Typ 2 im Vergleich zu Typ 1 wesentlich breiter ist. Bei höheren Leistungen werden die Spektren schmaler, so daß Typ 1-Laser nahezu monomodig werden. Die Abhängigkeit der spektralen Breite von der Ausgangsleistung ist in Fig. 5 aufgetragen.

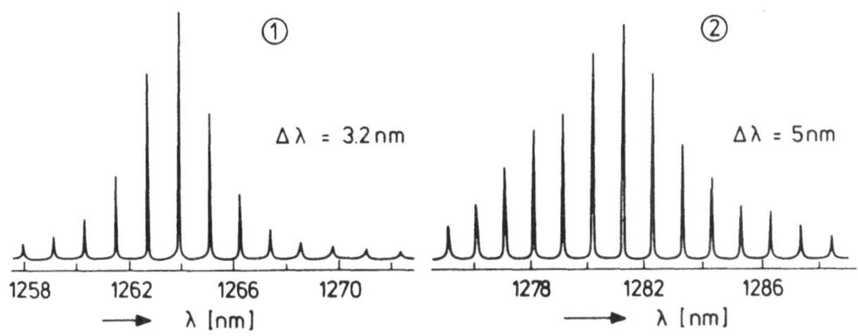

Fig. 4. Laserspektren bei 2 mW Ausgangsleistung

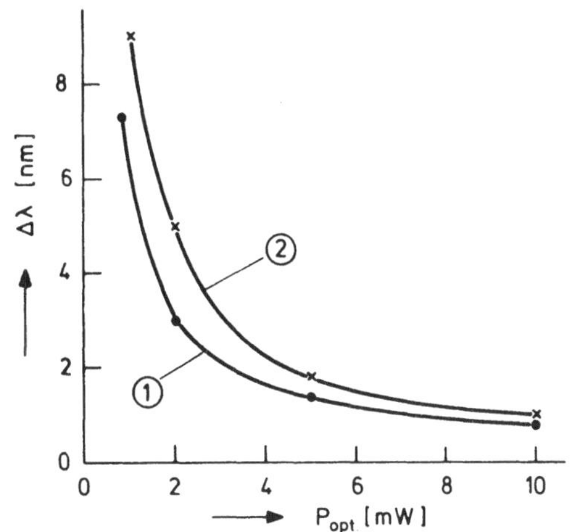

Fig. 5. Spektrale Breite als Funktion der Ausgangsleistung

3.4 Rückwirkungsempfindlichkeit

In Fig. 6 ist das beim Ankoppeln einer Gradientenfaser hervorgerufene Zusatzrauschen in Abhängigkeit von der spektralen Breite für verschiedene Laser aufgetragen /3/. Es zeigt sich, daß spektral breite Laser, so auch der Typ 2 - V - Nutlaser kaum auf Rückwirkungen reagieren. Andererseits können spektral schmale Laser durch Aufbringen einer Entspiegelungsschicht nachträglich spektral verbreitert und damit unempfindlicher gemacht werden /4/.

3.5 Zuverlässigkeit

Acht Testlaser laufen bereits mehr als 13000 h bei 40°C und 5 mW Lichtleistung pro Spiegel im Dauerstrichbetrieb. In diesem Zeitraum ist der erforderliche Ansteuerstrom um 10 - 20 % angestiegen, während sich die übrigen Eigenschaften wie Quantenwirkungsgrad, Abstrahlcharakteristik, Spektrum und Rauscheigenschaften praktisch nicht verändert haben.

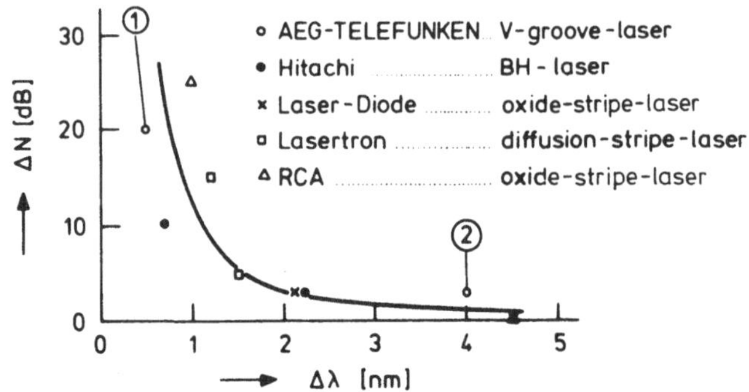

Fig. 6. Reflexionsbedingtes Zusatzrauschen; P = 5 mW

Die diesem Artikel zugrunde liegenden Arbeiten wurden mit Mitteln des Bundesministeriums für Forschung und Technologie gefördert. Die Verantwortung für den Inhalt liegt allein bei den Autoren.

Literatur.
/1/ G. ARNOLD
    Proc. 7th ECOC, Copenhagen, paper 10.4, (1981)
/2/ P. MARSCHALL, W. PFISTER, E. SCHLOSSER
    Proc. 8th ECOC, Cannes, 179 - 182, (1982)
/3/ H. STORM
    Priv. Mitteilung
/4/ G. ARNOLD, K. PETERMANN und E. SCHLOSSER
    To be published in IEEE J. Quant. Electron. (1983)
/5/ G. WENKE, B. ENNING
    J. OPT. Commun. 3, 122 - 128, (1982)

Optical Feedback Noise Characteristics of Semiconductor Lasers

J.W.M. Biesterbos   Philips Research   Eindhoven / The Netherlands

G.A. Acket   Philips Research   Eindhoven / The Netherlands

Introduction

In most applications of semiconductor lasers the presence of reflecting surfaces in front of the laser is unavoidable. Hence, emitted light is partly fed-
back into the laser and participates again in the process of stimulated emission. When this contribution to the stimulated emission varies in amplitude or
phase, the noise level in the laser light increases, sometimes by orders of
magnitude. This paper deals with detailed experiments on this feedback induced
optical noise.

Experimental

The experiments are performed in the set-up of fig. 1, where the focussing of
mirror M and the length of the external cavity (L_{ext}) can be accurately controlled by means of piezo electrical actuators. The optical noise is detected
by means of a photodiode and determined using a selective voltmeter (HP 3591 A)
in the frequency band 0.03-0.6 MHz.
The noise is expressed as RIN (relative intensity noise)$=\dfrac{\langle \Delta I^2 \rangle}{\langle I \rangle^2}$, which is defined as the ratio of the mean squared value of the fluctuation spectrum and
the squared value of the optical signal intensity.

Results

Figure 2 shows the typical excess noise behaviour, due to optical feedback, for
a single longitudinal mode CSP laser. The various longitudinal mode patterns
appearing under feedback are shown in the insets. The (intrinsic) noise level
of the free running diode is 2.10^{-13} Hz^{-1}, thus sometimes an increase of the
RIN by orders of magnitude is observed. The excess noise, which appears as
noise bursts during modulation of L_{ext} over a few wavelengths, is synchronized
with certain types of feedback-induced mode-hopping as shown in fig. 3. Between
the excess noise bursts the laser is single mode and the noise has its intrinsic level.

448

Discussion

In single longitudinal mode lasers optical feedback influences effectively the
reflection of the laser mirror reflectivity R_{eff} according to
$R_{eff}=\left\{\sqrt{R_o}+(1-R_o)\sqrt{r}\cos\theta\right\}^2$. R_o is the reflectivity of the laser facet without
feedback, r that of the external cavity and θ the phase of the reflected light.
R_{eff} influences the gain of the various modes by an amount $\Delta g=\frac{1}{2}L^{-1}\ln R_{eff}/R_o$.
(L is the length of the laser diode).

As a result of this feedback induced gain variation mode hopping between longitudinal modes occurs. During this process the balance between the various modes
is disturbed and the very high optical noise per longitudinal mode predominates.
From fig. 3 it is clear that also mode hopping transitions exist, which, in
the frequency range investigated, are not accompanied by excess noise. These
are transitions with hysteresis /1/ due to the well known gain suppression of
the non-lasing modes /2,3/.

The various possibilities to overcome the optical noise bursts due to feedback
induced mode hopping will be discussed. Although the experiments reported here
were performed at frequencies of interest for reading of information from disks,
similar problems are expected in telecommunication if phase variations in reflected light occur.

References

/1/ J.W.M. Biesterbos et.al., IEEE J.Quantum Electron, vol.QE-19, June 1983.
/2/ M. Nakamura et. al. J. Appl. Phys. 49 (1978), 4644.
/3/ M. Yamada et. al. IEEE J.Quantum Electron, QE-15 (1979), 743.

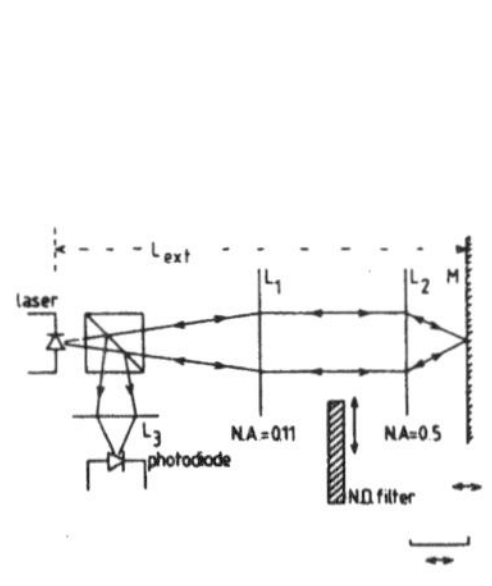

fig.1 experimental set-up for measuring feedback
induced optical laser noise

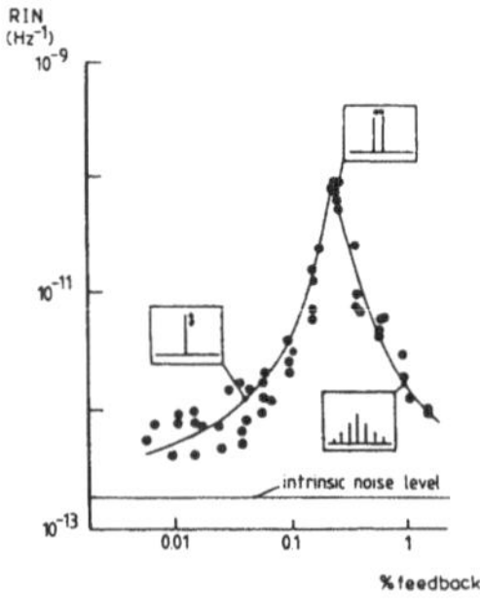

fig.2 maximum feedback induced RIN versus
the relative feedback intensity

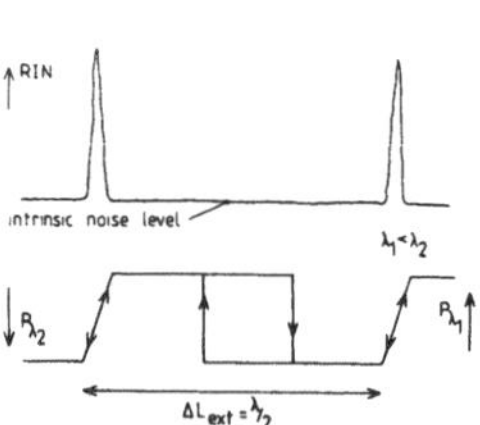

fig.3 feedback induced mode-hopping between λ₁
and λ₂ and the resulting optical noise bursts

Wellenlängenmultiplexer und -demultiplexer mit Beugungsgittern

B. HILLERICH und M. RODE
AEG-TELEFUNKEN, Forschungsinstitut Ulm
Sedanstraße 10, D-7900 Ulm

Einleitung

Der Wellenlängen-Multiplexbetrieb ermöglicht eine Vervielfachung der
Übertragungskapazität von Lichtwellenleiter-Übertragungsstrecken. Bei
Strecken mit mehr als 3-4 Übertragungskanälen werden im Multiplexer
bzw. Demultiplexer als wellenlängenselektive Elemente vorzugsweise
Beugungsgitter eingesetzt. Sie bewirken durch ihre Winkeldispersion
in Verbindung mit einer Abbildungsoptik die gewünschte räumliche Wel-
lenlängentrennung.

Dimensionierungsfragen

Der schematische Aufbau eines Demultiplexers ist in Bild 1 gezeigt.
Das aus der Übertragungsfaser austretende Licht besteht aus dem Wellen-
längengemisch $\lambda_1+\lambda_2+$ ... Es wird durch die Linse kollimiert und gemäß
der Beziehung

$$\sin(\vartheta+\varphi_0) + \sin(\vartheta-\varphi_1) = \lambda/d \qquad (1)$$

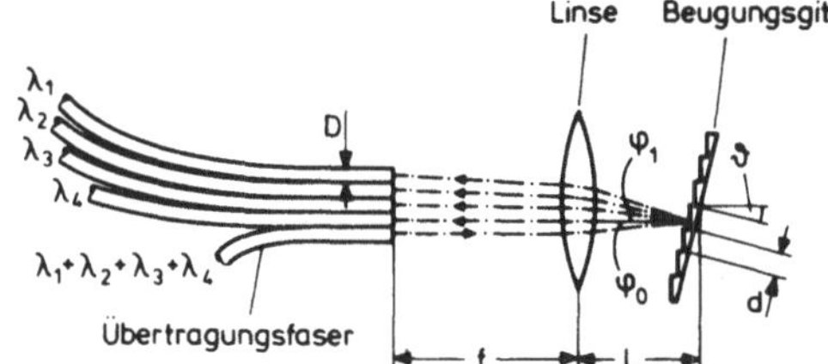

Fig. 1. Prinzip des Gitter-
Demultiplexers

für verschiedene Wellenlängen λ_1, λ_2, ... mit unterschiedlichen Winkeln
φ_1 abgebeugt. Die Linse dient ebenfalls zur Fokussierung des abgebeugten
Lichtes auf die verschiedenen Ausgangsfasern. Der spektrale Abstand
zwischen benachbarten Übertragungskanälen $\Delta\lambda$ ist gegeben durch

$$\Delta\lambda \approx \frac{D \cdot d \cdot \cos\vartheta}{f} \qquad (2)$$

Eine wichtige Größe bei Wellenlängen-Multiplexsystemen ist die Paßbandbreite in den einzelnen Übertragungskanälen. Sie sollte möglichst
groß sein, da durch sie die zulässige Toleranz der Senderwellenlängen
bestimmt wird. Durch Rechnung läßt sich zeigen, daß die Paßbandbreite
im wesentlichen vom Verhältnis zwischen Kerndurchmesser und Faserab-
stand D abhängig ist. Profil und numerische Apertur der Eingangs- bzw.
Ausgangsfasern spielen eine untergeordnete Rolle. In Bild 2 sind berechnete Durchlaßkurven für zwei Extremfälle gezeigt: Kurve a für eine
Ausgangsfaser, deren Kerndurchmesser von 50 µm identisch ist mit dem
Kerndurchmesser der Eingangsfaser. Die 1 dB-Paßbandbreite beträgt nur
18 % des Kanalabstandes. Wählt man eine Ausgangsfaser mit einem Kerndurchmesser von 130 µm, so steigt die Paßbandbreite auf 70 % (Kurve b).
Linsenaberrationen und Verluste durch das Beugungsgitter wurden bei
diesen Rechnungen vernachlässigt.

Bei Multiplexern ist die Situation ungünstiger (siehe Bild 3), da die
Ausgangsfaser mit einem Kern-Durchmesser von 50 μm festgelegt ist.
Werden identische Fasern für Eingang und Ausgang verwendet, so beträgt
die 1 dB-Paßbandbreite nur 9 % des Kanalabstandes (Kurve a). Durch Re-
duzierung des Kerndurchmessers der Eingangsfasern (Kurve b) läßt sich
die Paßbandbreite um den Faktor 2 ... 3 verbessern, wobei man in der
Regel erhöhte Koppelverluste zwischen Lichtquelle (Laserdiode oder
LED) und Faser in Kauf nehmen muß. Eine Paßbandbreite, die vergleich-
bar ist mit der von Demultiplexern, ist nur bei Verwendung sehr dünner
Eingangsfasern (Kurve c) erreichbar.

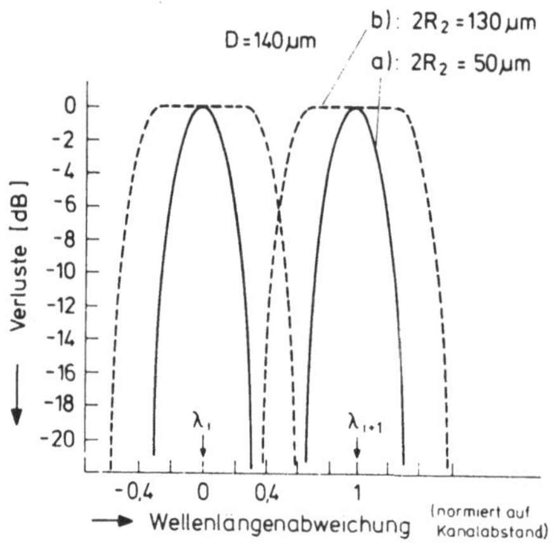

Fig. 2. Berechnete Durchlaßkurven
für Demultiplexer (Eingangsfaser:
Kern-Ø: 50 μm, parabol. Profil,
Ausgangsfaser: Stufenprofil)

Fig. 3. Berechnete Durchlaßkurven für Multiplexer
(Ausgangsfaser: Kern-Ø: 50 μm)

Gitterherstellung
Beugungsgitter hoher Qualität können durch anisotropes Ätzen von
Silizium-Wafern hergestellt werden /1 - 2/. Auf Wafern, die um einen
Winkel ϑ = 11° bzg. der (111)-Kristallebene fehlorientiert sind, wird
in Photolithotechnik eine Oxidstreifenmaske aufgebracht. Der Streifen-
abstand beträgt 2,5 μm bzw. 3,5 μm, je nachdem,ob die Gitter für den
Spektralbereich um 0,85 μm oder 1,3 μm eingesetzt werden sollen.

In der anschließend erfolgenden chemischen Ätzung werden Vertiefungen
geätzt, deren Seitenwände (111)-Ebenen entsprechen. Dadurch ergibt
sich das für ein "blazed" Gitter erforderliche asymmetrische Dreiecks-
profil (Bild 4). Zur Erhöhung der Reflektivität wird eine Cr-Au-
Schicht aufgedampft.
Derartige Gitter weisen einen hohen Beugungswirkungsgrad über einen
großen Spektralbereich auf (Bild 5). Der maximale Beugungswirkungsgrad
liegt über 85 %; 80 % sämtlicher Gitter eines Wafers weisen einen Wirkungsgrad über 70 % auf. Im Vergleich zur traditionellen Herstellung
von Gittern durch Replikation /3/ ermöglicht dieses Verfahren eine
kostengünstige Fertigung mit hoher Ausbeute.

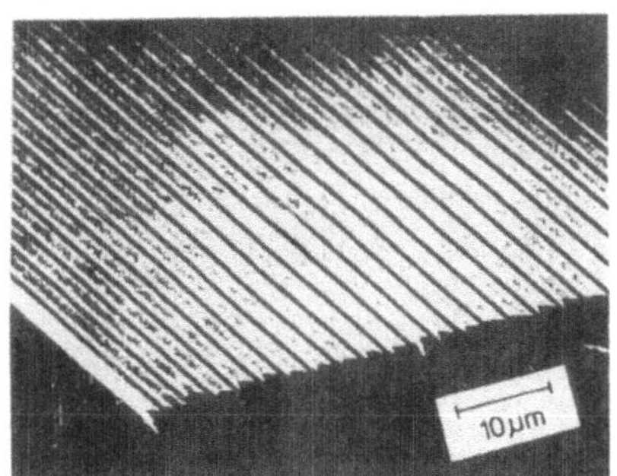

Fig. 4. REM-Foto der Gitter-
struktur

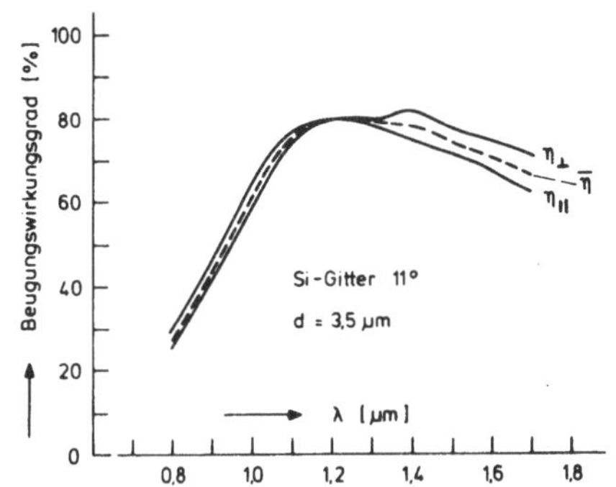

Fig. 5. Typischer Spektralverlauf
des Beugungswirkungsgrades

Protoypen von Demultiplexern

Das in Bild 6 dargestellte Gehäusekonzept für Gittermultiplexer und
-demultiplexer ermöglicht einen stabilen, kompakten Aufbau und eine
einfache Justage.
Die in V-Nut-Arrays eingelegten Eingangs- und Ausgangslichtwellenlei-
ter befinden sich in einer zylindrischen Hülse, die in 3 Achsen verschiebbar und um die Längsachse drehbar ist. Als kollimierendes bzw.
fokussierendes Element dient eine einfache Linse (Plankonvex- oder
Kugellinse oder ein Achromat). Bild 7 zeigt einen 10-Kanal-Demulti-
plexer im Gehäuse.

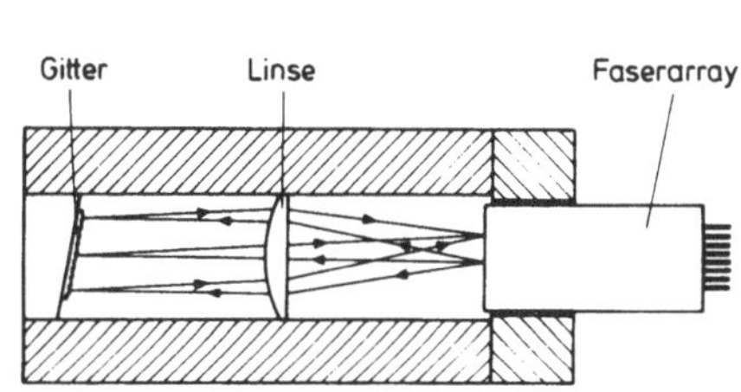

Fig. 6. Aufbaukonzept für
Gitter-Multi-/Demultiplexer

Fig. 7. Gitter-Demultiplexer

Prototypen von Demultiplexern mit maximal 10 Kanälen im Wellenlängenbereich um 0,85 µm und 1,3 µm wiesen Einfügungsdämpfungswerte zwischen
2,2 und 3,4 dB auf (Bild 8). Die erhöhten Verluste bei den äußeren
Kanälen sind auf Abbildungsfehler der Linse (Achromat) zurückzuführen.
Der Kanalabstand liegt bei ca. 50 nm, die Übersprechdämpfung zum Nach-
barkanal ist besser als 27 dB. Durch die Verwendung von Ausgangs-Licht-
wellenleitern mit großem Kerndurchmesser wird eine relativ große 1 dB-
Paßbandbreite von 25 nm erzielt. Die Verluste lassen sich durch geeignete Antireflexschichten auf Linsen und Faserenden um ca. 1 dB reduzieren, so daß dann die Verluste der Demultiplexer im wesentlichen
durch die Qualität des Beugungsgitters bestimmt sind.

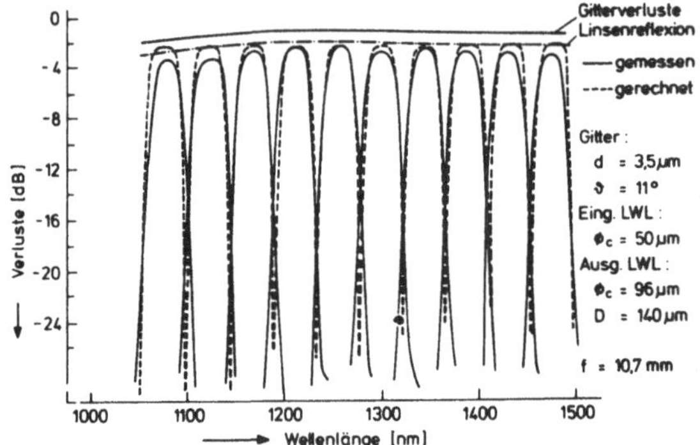

Fig. 8. Durchlaßkurven eines 10-Kanal-Demultiplexers

Die diesem Bericht zugrunde liegenden Arbeiten wurden mit Mitteln des Bundesministeriums für Forschung und Technologie unterstützt. Die Verantwortung für den Inhalt liegt bei den Autoren.

Literatur

/1/ J. MÜLLER et al.
    SPIE Vol. 192 Interferometry (1979), 244 - 250
/2/ Y. FUJI et al.
    IEEE Journ. Quant. Electr., QE-16 (1980), 165 - 169
/3/ R. T. JARREL, G. W. STROKE
    Appl. Opt. 3 (1964), 1251 - 1261

Mid Infrared Waveguides

C. Le Sergent
C.G.E. Research Center - MARCOUSSIS 91460-F

There is an increasing interest for waveguides operating beyond
the 1.55 µm transmission window of silica optical fibers and
specially within the 4 and 10 µm ranges.
Possible applications include communications, remote detection
imaging, radiometry and power transmission for laser surgery,
tooling, etc...

This paper presents a material oriented, "state of the art"
review of the field.
Following topics will be considered :

1)   The requirements regarding,
     - losses
     - thermal and mechanical properties : temperature range,
       flexibility
     - power levels
     - chemical resistance
     - toxicity.

2)   The solutions which are being studied or developed in
     various laboratories,
     - fluoride or chalcogenide glass fibers
     - polycrystalline fibers (AgCl, KRS_5)
     - monocrystalline fibers (CsI, CsBr)
     - hollow waveguides (metallic light pipes or fibers with
       hollow core).

Some of these conductors are already available mostly for
short range applications.

Losses down to 0.3 dB/m and c.w. power acceptances higher than 50 W have been reported at 10.6 μm for $\sim$ 1 m long crystalline fibers (KRS_5-AgCl).
Present glass fibers can exhibit either medium losses (0.5-10 dB/m) over a wide range e.g., 3-11 μm (heavy chalcogenides) or low-losses, down to 0.01 dB/m, within a very narrow window around 2.5 μm (ZrF_4 based compositions).

A considerable amount of work has still to be done to get the practical transmission closer to the intrinsic values and to achieve optimized trade-off with secondary characteristics.

Laser und Optoelektronik
in der Weltraumtechnik

Lasers and Optoelectronics
in Space Technology

Lasers and Advanced Optics in Space

E. David Hinkley
Jet Propulsion Laboratory
California Institute of Technology
Pasadena, California 91109/USA

1. Introduction

Spaceborne observations of extraterrestrial objects as well as Earth have resulted
in major advances in optics and remote sensing technology. Improved optical systems
have been developed to meet the need for higher spatial and spectral resolution, greater spectral coverage, reduced focal-plane aberrations, adaptability to sophisticated
image processing, and autonomous operation under a variety of conditions. These systems have enabled planetary missions of the Voyager type, employing sophisticated vidicon cameras, which have greatly enhanced our understanding of Jupiter and Saturn.
Galileo, a more advanced mission to Jupiter, will utilize virtual-phase CCD-type array
detectors for increased sensitivity and spectral coverage. This paper describes the
unique optical features of Voyager, Galileo, and several other planetary, astronomy,
and astrophysics missions such as Space Telescope, Infrared Astronomical Satellite,
Solar Optical Telescope, Advanced X-Ray Astronomical Facility, Large Deployable Reflector, and Imaging Spectrometer.

Laser techniques offer capabilities beyond those of passive remote sensing. For
example, LIDAR (laser radar) can provide accurate range information as well as remote,
range-resolved measurements of chemical species and wind velocity. The NASA Shuttle
LIDAR Working Group prepared a Technical Report (Browell, 1979) which outlined 26 experiments which could be performed with laser systems onboard an orbiting spacecraft.
Measurement scenarios are described along with the expected accuracy and coverage. Although lasers have yet to be used for global remote-sensing applications, progress is
being made toward the development of advanced LIDAR instrumentation for space. Another
future application of laser technology in space is free-space communication where a
highly coherent and directional laser beam can transmit information at rates much
greater than is currently possible.

2. Advanced Optics

The definition of "advanced optics" used in this paper is broad and encompasses
imaging, spectrometry, detectors, optical information processing, cryogenics, large

optical structures, and pointing -- all of which relate to remote sensing. Advanced optical systems are being developed for a variety of space experiments: Earth observations, planetary studies, solar physics, astrophysics, and astronomy. Table 1 lists several space missions requiring advanced optics -- from IRAS, launched in 1983, to LDR, which is still in the early planning stage.

Table 1. Space Missions using Advanced Optics

Mission	Launch	Duration	Special Feature
Infrared Astronomical Satellite (IRAS)	1983	1 yr	Superfluid helium (2K)
Atmospheric Trace Molecule Spectroscopy (ATMOS)	1984	1 wk	Fast FTIR scan (1 sec)
Galileo	1986	2 yr	CCD imaging detector
Space Telescope (ST)	1987	5 yr	Precise pointing
Shuttle Imaging Spectrometer (SIS)	1988	1 wk	Triple Schmidt reflection optics
Upper Atmospheric Research Satellite (UARS)	1989	2 yr	Cryo-cooled spectrometer
Shuttle IR Telescope Facility (SIRTF)	1989	1 yr	Cryo-cooled optics
Solar Optical Telescope (SOT)	1990	2 yr	High heat rejection
Shuttle LIDAR	1991	3 yr	Precise pointing
Large Deployable Reflector (LDR)	1995	6 yr	Accurate receiver surface (1 μm)

For Earth resource observations, recent Landsat satellites have much better spatial resolution and spectral coverage than the earlier ones; and future Earth-observing systems will involve imaging spectrometers and laser systems. Planetary science made giant strides as a result of the observations of Jupiter and Saturn (and their moons and rings) by the two Voyager spacecraft with their vidicon cameras. Galileo, a successor to Voyager, will utilize an 800x800 CCD array to provide better spatial resolution and infrared (IR) coverage, along with an ultraviolet (UV) and near-IR mapping spectrometer. In the astronomy and astrophysics area, the Infrared Astronomical Satellite has provided outstanding new discoveries, including a new comet and possible formation of solar system around the star, Vega. Solar, stellar, and astrophysical missions of the future will utilize such advanced instrumentation as the Space Telescope, Solar Optical Telescope, and Large Deployable Reflector.

2.1 Imaging

Historically, the most common use of advanced optics for spaceborne missions has been in imaging or photography with restricted or extended spectral coverage. The main performance criteria for an imaging system are:

- Spatial resolution
- Spectral coverage
- Speed/sensitivity

The tradeoffs among these are dictated by the mission objectives. For a discussion of a model for the Space Telescope low-light camera, for example, see Breckinridge _et al_ (1982).

2.2 Spectrometry

Another major application of advanced spaceborne optics is spectrometry, which provides spectral (wavelength) information in addition to spatial. Historically, spectrometry was first performed in space using simple spectral-bandpass filters. These are still used in missions where the spectral signatures are relatively broad. Very high resolution instruments based upon grating or Fourier-transform infrared (FTIR) spectrometers can greatly improve our ability to identify gaseous species in planetary atmospheres. As an example of advances made in remote spectrometry, Table 2 shows the operational capabilities of three FTIR instruments developed for airborne and Shuttle-borne measurements of trace atmospheric species, represented in order of increased performance as Mark I, Mark II, and ATMOS. (See papers by Farmer and Raper [in preparation] and Schindler [1970] for details.) The Mark I has been flown on aircraft and balloon platforms, the Mark II on balloon platforms for stratospheric measurements, and ATMOS is scheduled for Spacelab 3 in November 1984. Note, in particular, the short scan time (1 sec) of ATMOS -- needed to obtain 1 km vertical resolution for species measurements from Shuttle orbit.

Table 2. Advanced FTIR Spectrometers for Remote Sensing

	Coverage (μm)	Resolution (cm^{-1})	Scan time (sec)
Mark I	2-6	0.1	60
Mark II	2-16	0.02	60
ATMOS	2-16	0.02	1

2.3 Other Examples of Advanced Optics

Detectors: New detectors are needed to extend wavelength coverage and improve sensitivity and speed of response (e.g., heterodyne detection).

Information Processing: With the very high data rates from imaging and other instruments, the speed offered by optical information processing will be needed for some missions of the future.

Cryogenics: Certain advanced detectors and cooled optical systems require cryogenic temperatures. The requirements for planetary missions (high heat loads) are often different from those for astrophysics and astronomy (low background radiation), and appropriate refrigerators must be developed to meet these disparate needs.

Large Optics: Optical structures such as the Space Telescope and Large Deployable Reflector, which must maintain critical alignment in spite of varying thermal conditions, will represent a major challenge.

Pointing: The ability to precisely point a receiver (e.g. Space Telescope) or transmitter (laser) from a spaceborne platform is another major challenge of optics. It will require collective disciplines of structures, materials techology, mechanical engineering and physics, as well as optics.

3.   Spaceborne Remote-Sensing Missions

The three main types of spaceborne remote-sensing missions requiring advanced optics are Astronomy and Astrophysics, Planetary, and Terrestrial.

3.1 Astronomy and Astrophysics

Most astronomy and astrophysics missions are directed toward measuring very weak signals; hence large optics, sensitive detectors, and telescopes and detectors cooled to cryogenic temperatures are often required. The important optics-related capabilities of several of these missions are indicated below (see Table 1 for definitions):

     IRAS   - Detectors with 10 to 120 μm wavelength coverage
               - Superfluid liquid helium cooling (2K)
               - 60-cm IR telescope
               - 900 km orbit (polar)

ST   - CCD camera
         - 1600-1600 pixels for Widefield Planetary Camera

SOT   - 1000x1000 CCD
         - High heat rejection

AXAF   - Solid-state imager for x-rays
         - Large-glancing-angle spectrometer for 0.5-10 keV

LDR   - Large size (10-20 m diameter)
         - High surface perfection (1 μm)
         - May use adaptive optics to maintain surface figure

SIRTF   - Cryo-cooled detector and optics

3.2 Planetary Exploration

The features of three spacecraft to Jupiter demonstrate progress made in long-term autonomous operaton and imaging capabilities:

    PIONEER - Spacecraft spins for stabilization (3 rpm)
            - Imaging system (Pioneer Imaging Photoplanimeter) uses the
              spinning motion to scan objects; employs two discrete detectors
              and visible-wavelength filters
            - Encountered Jupiter in 1973 (Pioneer-10) and 1974 (Pioneer-11)

    VOYAGER - 3-axis-stabilized spacecraft
            - Vidicon imaging system
            - Wavelength coverage: 0.35-0.6 μm
            - Encountered Jupiter in 1979 (Voyagers I and II)

    GALILEO - Dual-spin spacecraft, imaging system despun
            - 800x800 virtual-phase CCD
            - Wavelength coverage: 0.4-1.1 μm
            - Will reach Jupiter in October 1988, after 1986 launch

The purpose of Galileo is to perform a close-range study of Jupiter, its satellites, and its magnetosphere for almost 2 years. It is scheduled to be launched in May 1986 by Shuttle. In August 1988 (150 days prior to its arrival at Jupiter), the Galileo probe will be separated from the Orbiter, which will then fly within 1000 km of Io.

During its 20-month mission the Orbiter, carrying 11 instruments, will transmit data at rates up to 134 kbits/sec. It will make close flybys of Io, Europa, Ganymede, and Callisto at various times. Its camera section, in the stabilized de-spun portion of the spacecraft, consists of an identical telescope system to that of Voyager, but with more sensitivity (over an order of magnitude) and broader spectral coverage made possible with the 800x800 virtual-phase CCD. Galileo's spatial resolution (determined by spacecraft velocity and closeness to the planetary surface as limited by detector speed) will be comparable to that of Landsat D (which is better than 30 m), vs. 1000 m for Voyager.

In addition to those missions mentioned above, there are three separate spacecraft being planned (from Europe, Japan, and USSR) to intercept Halley's comet in 1986. The optical features of these are:

- High-resolution photography and electronic imaging: nucleus data (rotation rate, surface structure, general activity);

- Spectroscopy and spectrometry: data on physical composition of nucleus, coma, and tail;

- IR spectroscopy and radiometry: temperature, size, and composition of dust particles.

3.3 Terrestrial Missions

The optical demands of terrestrial missions have primarily been in the area of increased spatial resolution and, more recently, combining spatial and spectral information. The following types of Earth-observing instruments are either in orbit or being planned:

Landsat D Multispectral Scanner - called Thematic Mapper, it has 30 m spatial resolution in 6 spectral bands from the visible to the near IR. In the thermal IR, the spatial resolution is 120 m. Scanning perpendicular to the ground is implemented with a mechanically- scanned mirror.

Multispectral Linear Array - a Landsat follow-on instrument (circa 1986) in which the scanning mirror will be replaced by multiple solid-state detector linear arrays sensitive to several narrow spectral bands.

<u>Shuttle Imaging Spectrometer</u> - advanced land-remote-sensing instrument which generates a 128-channel high-resolution spectrum between 0.5 and 2.5 μm for each spatial resolution element. It is implemented with a unique triple Schmidt optical design (Breckinridge <u>et al</u>, 1983) to disperse the image onto a mosaic of area-array detectors.

<u>ATMOS</u> - to be launched on Spacelab 3 in November 1984, it has 0.02 cm^{-1} spectral resolution with 2-16 μm wavelength coverage which it can scan in 1 sec.

Lasers are also being proposed for terrestrial observations from space to perform remote sensing measurements not possible with passive instruments. Several of these potential applications of spaceborne laser technology are described in the Secton 4.

4.   <u>Laser Systems in Space</u>

Laser systems will eventually be used in space to measure global distributions of key atmospheric species and meteorological parameters and detect tiny relative movements of the Earth's surface. Table 3 summarizes several proposed applications of spaceborne lasers.

Table 3. Spaceborne Laser Applications and Techniques

<u>Application</u>	<u>Technique</u>
Wind Field Measurements	Doppler
Atmospheric Species	Differential Absorption
Upper Atmospheric Species	Differential Fluorescence
Earth Resources	Differential Reflectometry
Earthquake Detection	Ranging
High-Data-Rate Communications	Free-Space Transmission

Experiments toward some of these objectives are being pursued by NASA. They involve the use of a UV-LIDAR (laser radar) system onboard the NASA advanced ER-2 high-flying aircraft, using a tunable dye laser pumped by a Nd:YAG laser. The results of this series of experiments will yield important information as to the potential utility of spaceborne laser remote sensing. Another application being considered is global measurements of tropospheric winds using Doppler LIDAR. Because of the strong relationship between the atmospheric backscatter coefficient (at the laser wavelength) and the pulse energy needed (which has a direct bearing on the spacecraft power demand), world-wide technical cooperation is being sought in order to obtain representative values of these coefficients under a variety of climatic conditions. Spaceborne LIDAR is also being proposed for precise distance-measuring to detect small movements of the Earth's crust, possibly leading to an ability to predict earthquakes and volcanic eruptions. In another type of application, lasers are being proposed for long-distance communication in space, during which they would provide high-data-rate transmission to an Earth orbiter even from the most distant regions of the solar system.

4.1 Wind-Measuring LIDAR

In 1978 Huffaker proposed that a CO_2 infrared laser heterodyne system could measure tropospheric winds from an orbiting spacecraft -- either Shuttle at an altitude of 250-300 km, or an operational satellite at 800 km altitude. An airborne CO_2 laser system using the same principle of detecting the Doppler shift of backscattered laser radiation has been operated by NASA Marshall Space Flight Center scientists for several years (Bilbro, 1980); thus, the technique has been demonstrated on a small scale by the MSFC team and, more recently, by Woodfield and Vaughan (1983). For particles with an average velocity component v along the direction of the laser beam, the Doppler shift of the backscattered radiation is $2v_o v/c$, where v_o is the unshifted laser frequency and c the speed of light.

A spaceborne CO_2 LIDAR system to measure tropospheric winds would need 10-joule pulses at a repetition rate of 20 Hz, with a frequency stability during each pulse of 200 kHz or better. A major consideration is the large power need of several kilowatts for the laser itself; which may be available on a 1-2 week Shuttle mission, but for those of longer duration, will require advances in energy storage or production onboard the spacecraft and/or improvements in laser operating efficiency.

Abreu (1979) has studied an alternative technique to measure winds from a satellite platform using a LIDAR system operating in and near the visible portion of the spectrum, and incoherent (rather than heterodyne) detection. Because of the enhanced backscatter coefficient at shorter wavelengths, such measurements can extend to altitudes well into the stratosphere; whereas LIDAR systems operating in the 10-μm region will generally be limited to the troposphere, but have the advantage of being eye-safe and having relatively high efficiency.

Because of the greatly varying meteorological, aerosol, molecular, and pressure conditions with altitude, more than one technique will probably have to be used from space to provide global wind information for the entire atmosphere.

4.2 Species-Measuring LIDAR

Global measurements of several key tropospheric gases are expected to be possible
using active laser techniques. Airborne experiments based on the principles of resonance fluorescence (Browell et al, 1981) and differential absorption (Menzies and Shumate, 1978; Weiseman et al, 1978) have shown potential for future spaceborne applications. The simplest approach for this purpose is one based upon differential absorption of laser radiation backscattered from the Earth's surface.

Laser Absorption Spectrometer. The Laser Absorption Spectrometer (LAS) is a particular laser instrument (Menzies and Shumate, 1978) for the detection of gaseous species in the atmosphere by differential absorption of laser radiation backscattered from
the Earth's surface. Using the LAS, airborne measurements of tropospheric ozone have
been made over the past few years. The system uses two small CO_2 lasers, and airplane-
to-ground transmission for a laser line which is absorbed by ozone is compared with
that for a second laser line which is beyond the ozone-absorbing region. By ratioing
the two return signals, effects of turbulence and changing reflectivity of the Earth's
surface are largely eliminated. The present LAS system utilizes two 0.3-watt, continuous (cw) CO_2 lasers and infrared heterodyne detection for high signal-to-noise ratio.
Ozone measurements have been made on the west coast of the United States (Grant and
Shumate, 1980), east coast (Shumate et al, 1981), and midwest (Shumate, 1980). Since
range-gating is not possible with a cw system, only the path-averaged ozone concentration below the airplane is measured. Measurements cannot be made with high vertical
resolution unless the airplane flies at different altitudes, although some altitude information can be obtained (resolution of approximately 5 km) using the pressure dependence of the spectral line shapes. Extrapolations to Shuttle indicate that two 10-watt
lasers will be able to perform measurements of trace atmospheric species. Those which
have spectral overlaps with CO_2 laser lines include O_3, NH_3, C_2H_4, CCl_4, C_2H_3Cl, and
HNO_3.

Differential Absorption LIDAR (DIAL). A pulsed, tunable infrared laser can provide range information lacking with a cw system, and has the potential of providing approximately 1 km vertical resolution. For a Spacelab orbit, a laser with the required
wavelength coverage, energy (25 J/pulse), pulse repetition frequency (15 Hz), stability, and overall efficiency is not yet available, but development is continuing. Recent
airborne measurements of water vapor and ozone with a visible-UV DIAL LIDAR system have
been successful, however, and potential spaceborne systems are further along than in
the infrared. Experimental flights of the UV DIAL system onboard the NASA ER-2 aircraft at the Ames Research Center are expected to take place within the next two or
three years. The results of these experiments will be important in determining when
Shuttle LIDAR will become a reality.

466

4.3 Aerosol-Measuring LIDAR

Global measurements of atmospheric aerosols can lead to an understanding of the formation and evolution of haze layers in the troposphere as well as enable study of stratospheric heterogeneous reactions relating to ozone depletion. In order to delineate lower atmospheric hazes more clearly, a special infrared channel has been installed on Landsat D. Clusters of sulfuric acid molecules (from sulfur compounds emitted at ground level) are thought to act as condensation nuclei leading to the growth of stratospheric aerosols which are the principal components of the Junge layer. The sink for sulfur occurs when the heavier aerosols settle out of the stratosphere into the troposphere, forming a dilute "acid rain." Consequently, in addition to measuring the concentration and size distributions of aerosols in the stratosphere, it is important to determine their chemical composition. Russell et al (1981) have published a comprehensive report on the potential measurements of aerosols from Shuttle.

A High Spectral Resolution LIDAR (HSRL) has been developed at the University of Wisconsin by Shipley et al (1979) to measure the spatial distributions of the atmospheric aerosol optical extinction coefficient on both regional and global scales. It uses a nitrogen UV laser to optically pump a high-spectral-resolution dye laser, the output of which is directed into the atmosphere. The HSRL measures the aerosol optical extinction coefficient by distinguishing light which is backscattered by the aerosol from that backscattered by air molecules. Quantities such as the aerosol optical extinction coefficient, backscattering phase function, aerosol-to-molecular scattering ratio, and visibility can be derived from this information. A 35-cm-diameter receiver telescope collects the light backscattered by the aerosol particles and air molecules, and the return signal is analyzed using a high-spectral-resolution, two-channel Fabry-Perot spectrometer.

Identification of the chemical composition of stratospheric aerosol particles may be possible using a multiwavelength (infrared) backscattering technique called DISC (Wright et al, 1975). A preliminary study of this approach indicates that adequate LIDAR sensitivity can be obtained at 220-km orbital altitude with a 10-J CO_2 laser and a 1-m-diameter collector. The measurement principle is based on the fact that the aerosol particle backscatter coefficient shows a dependence on wavelength that is characteristic of its composition. This could also be a powerful tool for tropospheric aerosol analysis if the backscatter signatures of different aerosols are distinctive enough.

Laser remote sensing of aerosol particles is extremely active at the present time, and a number of research laboratories are looking for ways to improve detection of the size distribution and shape as well as density and chemical composition.

4.4 Earthquake Forecasting by Spaceborne LIDAR

The use of a spaceborne LIDAR to detect movement of the Earth's crust was proposed by Siry in 1973 and analyzed more recently by Kahn _et al_ (1979). The concept is based on time-of-flight measurements of very short laser pulses to a set of ground-based retroreflectors and back, yielding a precision of 1-2 cm for the measured distances between them. By repeating the measurements at intervals of several months (possibly years), data essential to determining strains near earthquake zones, ground uplifting prior to volcanic eruptions, and land subsidence can be obtained.

The proposed laser system consists of a Nd:YAG laser with 0.2 nanosecond pulse width and a repetition rate of 10 Hz. The receiver system employs a 15-cm-diameter collecting telescope and a photomultiplier. Because of the rather modest system requirements of a LIDAR system of this type, and the fact that the laser is solid-state and already space-qualifiable, this application of a spaceborne laser may occur before any of the others described above. Moreover, the necessary precision has already been demonstrated by range measurements made between a ground-based LIDAR and an orbiting satellite containing a set of retroreflectors.

5. Summary and Conclusion

The recent successes of such space missions as Voyager, IRAS, several Landsat orbiters, and Shuttle have demonstrated the benefits of spaceborne remote sensing. Advanced optics and laser technology represent broad and very challenging fields to meet some of these objectives, from high-resolution passive imaging and spectrometry to active probing of atmospheres and surfaces. Intrumentation based on advanced optics will undoubtedly play an important role in nearly every future space mission based on remote sensing.

ACKNOWLEDGMENTS

Appreciation is expressed to many colleagues for providing material for this paper; in particular, James B. Breckinridge, James A. Cutts, Torrence V. Johnson, Robert T. Menzies, Ray L. Newburn, and Fred Vescelus at JPL, and John J. Degnan at NASA Goddard Space Flight Center. Work performed at the Jet Propulsion Laboratory, California Institute of Technology, under contract with the National Aeronautics and Space Administration.

REFERENCES

Abreu, V. J. (1979): "Wind Measurements from an Orbital Platform using a LIDAR System with Incoherent Detection: An Analysis," Applied Optics 18, 2992.

Bilbro, J. (1979): "Atmospheric Laser Doppler Velocimetry: An Overview," Optical Engineering 19, 533.

Breckinridge, J. B., T. G. Kuper, and R. V. Schack (1982): "Space Telescope Low Scattered Light Camera -- A Model," SPIE Instrumentation in Astronomy 331, 395.

Breckinridge, J. B. , N. A. Page, R. R. Shannon, and J. M. Rogers (1983): "Reflecting Schmidt Imaging Spectrometers," Applied Optics 22, 1175.

Browell, E. V. (1979): Shuttle Atmospheric LIDAR Research Program, NASA SP-433.

Browell, E. V., A. F. Carter, and S. T. Shipley (1981): "Measurements of Ozone and Aerosol Profiles with the NASA Langley Airborne DIAL LIDAR System," Paper V13, IEEE Conference on Lasers and Electro-optics, Washington, D.C.

Farmer, C. B. and O. F. Raper (in preparation): "High Resolution Infrared Spectroscopy of the Upper Atmosphere: A Description of the ATMOS Experiment."

Grant, W. B., and M. S. Shumate (1980): Remote Measurements of Ozone Burdens in the L. A. Basin on September 20 and 24, 1979 using the JPL Airborne Laser Absorption Spectrometer. Final report to the NASA Office of Technology Utilization.

Huffaker, R. M. (1978): Feasibility Study of Satellite-Borne Lidar Global Wind-Monitoring System. NOAA TM-ERL WPL-37.

Kahn, W. D., F. O. Vonbun, D. E. Smith, T. S. Englar, and B. P. Gibbs (1979): Performance Analysis of the Spaceborne Laser Ranging System, Goddard Space Flight Center, NASA Technical Memorandum 80330.

Menzies, R. T., and M. S. Shumate (1978): "Tropospheric Ozone Distributions Measured with an Airborne Laser Absorption Spectrometer," J. Geophysical Research 83, 4039.

Russell, P. B., B. M. Morley, J. M. Livingston, and G. W. Grams (1981): Improved Simulation of Aerosol, Cloud, and Density Measurements by Shuttle LIDAR. Final report of SRI International to NASA, Contract NAS1-16052.

Schindler, R. A. (1970): "A Small, High-Speed Interferometer for Aircraft, Balloon, and Spacecraft Applications," Applied Optics 9, 301.

Shipley, S. T., E. W. Eloranta, and D. H. Tracy (1979): "Measurements of the Optical Extinction Coefficient of Atmospheric Aerosols by Means of a High Spectral Resolution LIDAR." Paper 2-10 presented at the 9th International Laser Radar Conference.

Shumate, M. S. (1980): Participation of the JPL Laser Absorption Spectrometer in the 1980 PEPE/NEROS Program in Columbus, Ohio. JPL Report, 715-84 (internal document).

Shumate, M. S., R. T. Menzies, W. B. Grant, and D. S. McDougal (1981): "Laser Absorption Spectrometer: Remote Measurement of Tropospheric Ozone." Applied Optics 20, 545.

Siry, J. W. (1973): Crustal Motion Measurement by Means of Satellite Techniques, GSFC X-590-73-273.

Weiseman, W., R. Beck, W. Englisch, and K. Gürs (1978): "In-flight Test of a Continuous Laser Remote Sensing System," Applied Physics $\underline{15}$, 257.

Woodfield, A. A. and J. M. Vaughan (1983): "Airspeed and Wind Shear Measurements with an Airborne CO_2 CW Laser," AGARDograph $\underline{AG.272}$, <u>Advancements in Sensors and their Integration into Aircraft Guidance and Control Systems</u>, edited by J. L. Hollington (to be published).

Wright, M. L., J. Pollock, and D. S. Colburn (1975): "DISC: A Technique for Remote Analysis of Aerosols by Differential Scatter," 7th International Laser Radar Conference, Menlo Park, California.

CO_2 Laser Transceiver System with Gbit/s Capability for Space-to-Space Communications – State of Technology

W. ENGLISCH, M. ENDEMANN, W. DIEHL AND W. WIESEMANN
Battelle-Institut E.V.
Am Römerhof 35, D 6000 Frankfurt am Main 90

1. Introduction

For high-rate data transmission in space the CO_2 laser is best suited
to meet the very demanding requirements, i.e. two-way communications
between satellites with data rates $>$ 1 Gbit/s at a bit error rate
$\leq 10^{-8}$, reliable operation over a period of up to 6 years, compatibi-
lity with the general requirements for space flight hardware. The first
system concept presented at the Laser 81 (1) has been further developed
and refined leading to the detailed design (2) of a CO_2 laser transceiver package with improved performance. In the following description
emphasis is given to the improved technical solutions not included in
the previous concept.

2. System Description

Fig. 1 presents a somewhat simplified exploded view of the transceiver
package. Fig. 2 shows a block diagram indicating the functional arrange
ment of the various subassemblies and components and their mutual inter
action.

The transmitter subsystem is optimized for minimum prime power consump
tion. The conventional type cw CO_2 transmitter laser (3) provides a
nominal output power of 6.2 W (unmodulated) and is designed for excellent passive frequency stability. The external CdTe light pipe modulato
(4) with aspect ratio L/d = 400 is of travelling wave type and generate
an average modulated sideband power of 1.3 W. The CdTe crystals are cut
for maximum phase modulation in order to establish the preferred modulation format, i.e. binary phase shift keying (PSK) with a strong resi-
dual carrier. The modulator driver provides a bipolar drive signal in
NRZ code. The bandwidth of the transmitter subsystem can be as high as
5 GHz.

modulator driver
transmitter laser power supply
aquisition flip mirror
transmitter laser
detector cooler
mixer element
modulator
upper base plate
point ahead ass.
nutator
focusing lens
λ/4 plate
LO beam combiner
LO - laser
fine pointing assembly
transmit/receive diplexer
internal alignment assembly
lower base plate
electronics package
LASER TRANSCEIVER PACKAGE
coarse pointing assembly (> 2π coverage)
LO - laser power supply

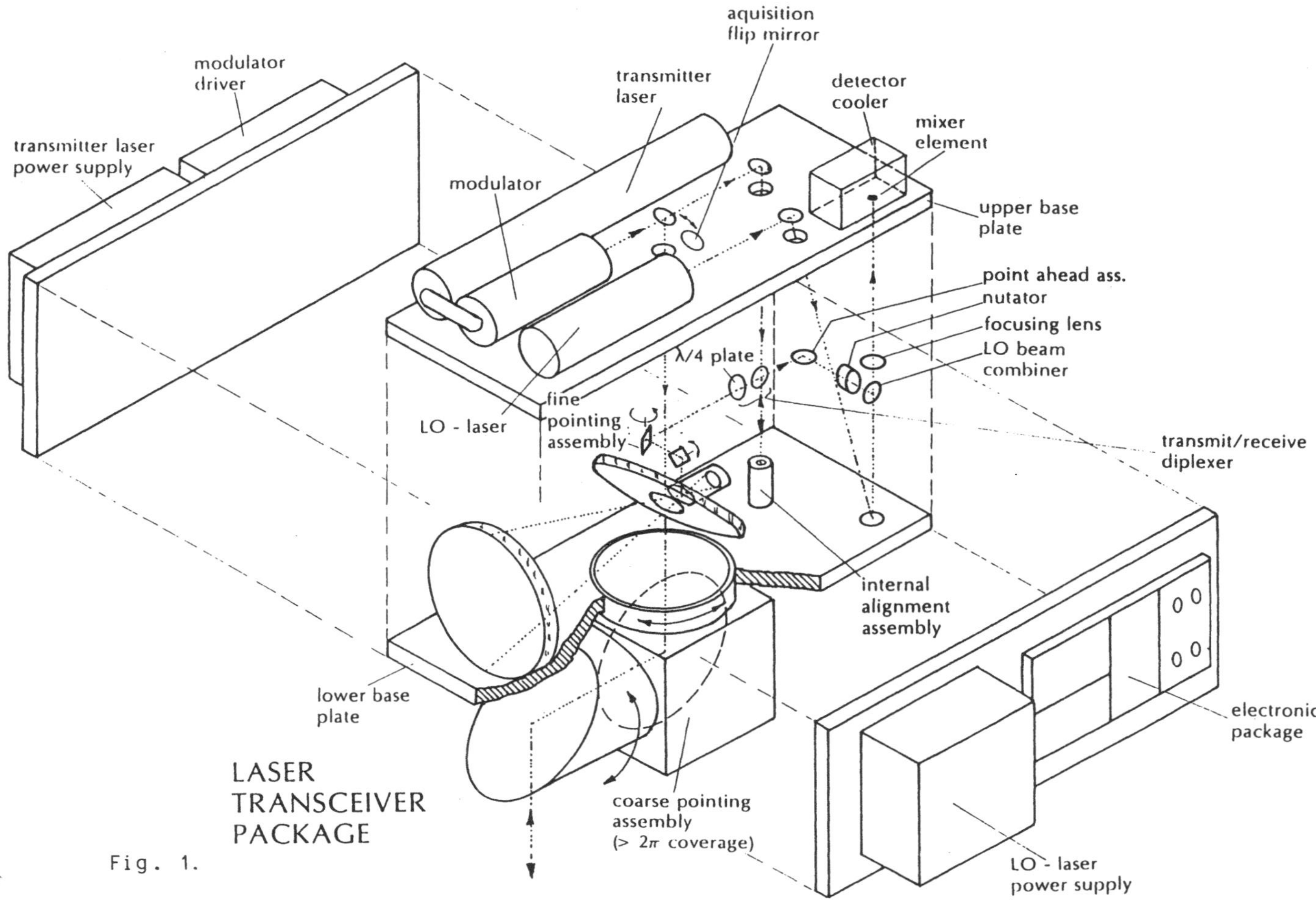

Fig. 1.

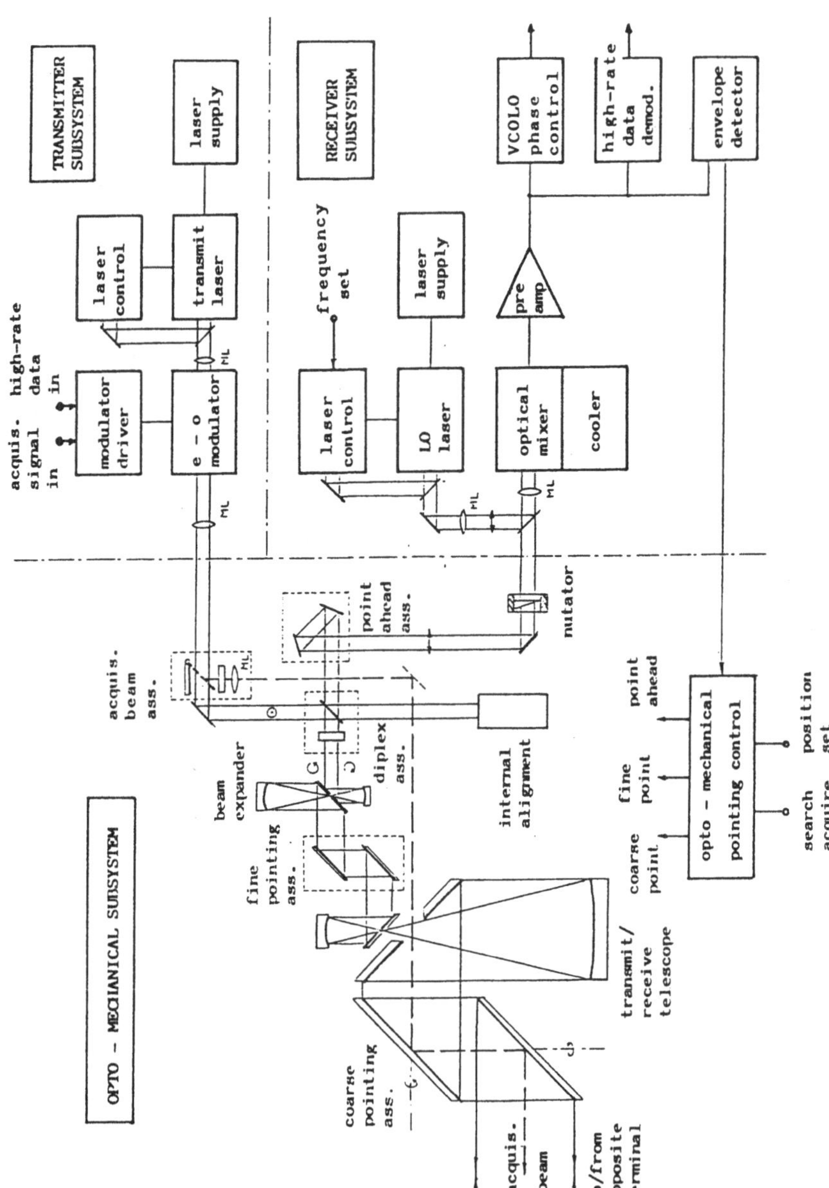

Fig. 2. Functional block diagram of CO_2 laser transceiver package

The <u>receiver subsystem</u> is designed for coherent optical homodyne detection. The homodyne receiver consists of a phase sensitive detector (PV HgCdTe photomixer) and a voltage controlled optical local oscillator (VCOLO) whose phase is synchronized to the instantaneous phase of the incoming signal wave. The VCOLO is realized by a CO_2 waveguide laser (5) with wide frequency tuning range phase control is accomplished within two stages: large but slow frequency (and hence phase) deviations are generated by a PZT which changes the resonator length, while fast variations are realized by an external acousto-optical modulator. The feasibility of the optical homodyne detection concept has been demonstrated experimentally (6). It offers significant advantages over the conventional heterodyne detection scheme; in particular, it removes all the constraints imposed by intermediate frequency requirements. Thus full use can be made of the frequency response of available HgCdTe diodes resulting in a useful bandwidth of the receiver subsystem of about 2 GHz. This bandwidth can be further increased by multiplexing different CO_2 laser lines.

The <u>optomechanical subsystem</u> comprises the main optics, the moving optical parts of the pointing acquisition and tracking (PAT) subsystem, and the box-like mechanical structure. Both the telescope (D = 25 cm) and the beam expander are of confocal Gregorian design and operate in a diffraction limited beam expanding mode with overall magnification factor m = 51. The final communication transmit beam width is $\Theta_{TC} \simeq$ 36 μrad. The main optics are common to the transmit and receive channels. Transmit/receive diplexing is accomplished by means of a $\lambda/4$ Fresnel rhomb plus a wire-grid polarizer which act as an efficient polarization switch. The internal alignment assembly allows for in-flight boresighting of the transmit and receive axes by means of the point ahead assembly.

Accurate pointing of the very narrow beamwidths is performed in two stages: The coarse pointing assembly located in object space of the telescope provides a more than hemispherical coverage at comparatively low speed. The fine pointing assembly is located close to the exit pupil of the telescope and provides a high speed fine pointing capability with accuracy of $\pm$ 6 μrad (as measured in object space). The control signals for continuous spatial tracking are provided via the conical scan tracking technique (7) where the optical nutator generates a slight spatial modulation of the signal focused onto the detector element. The S/N within the tracking loop is high enough to allow for sub-arc second tracking accuracy. The point ahead assembly accounts for the aberration angle between the transmit and receive optical axes which is caused by the relative velocity of the communicating terminals transverse to the line of sight.

The acquisition beam assembly provides the diverged beacon transmit bea (Θ_{TA} = 3.7 mrad) required for the first step in the acquisition procedure. In the receiving mode during acquisition the receiver must search for the opposite illuminating terminal. Its narrow instantaneous FOV ($\Theta_{RC} \simeq$ 20 µrad) is scanned over the entire acquisition uncertainty field ($\Phi_A \simeq \pm$ 0.15°). The spiral of Archimedes scan pattern is produce by means of the fine pointing elements. Due to the rather high S/N rati in the acquisition channel the mean time for mutual acquisition of terminals is < 0.5 s with a probability for correct acquisition of > 99,9 and a false alarm probability < 10^{-6}.

3. System Performance

Table 1 presents the calculated nominal S/N ratios for the communication, the acquisition, and the tracking channels. As can be realized the various technical solutions and improvements incorporated in the transceiver design result in a largely improved performance as compared to the previous concept.

Table 1. Summary of S/N ratios   (LEO-GEO link, R   43000 km

	Comm. channel (1 Gbit/s, BER < 10^{-8})	Acqu. channel (widened beacon)	Track. channel (pointing error < 6 µrad)
S/N calc.	22.0 dB	26.5 dB	51.4 dB
S/N requ.	13 dB	16 dB	10 dB
Margin	9.0 dB	10.5 dB	41.4 dB

4. State of Technology

The development of key technologies and components required to realize the transceiver package has been initiated by ESA and the German Space Authority.

All critical areas of development are covered and the results obtained
up to now look very promising. Practical application of laser communication in space appears to be feasible around 1990.

<u>Literature</u>

(1) ENGLISCH, W., DIEHL, W., Proc. Laser 81, p. 312-316, Springer Verlag,
    Heidelberg 1982
(2) ENGLISCH, W.: Design of a CO_2 Laser Transceiver Package for Space-
    to-Space Optical Communication. Final report ESTEC contract no. 3488/
    79 (1983)
(3) DIEHL, W. et al.; Proc. Laser 83
(4) SCHOLZ, A., LEEB, W., BONEK, E.; IEEE J. QE-18 (1982) 1021
(5) ENDEMANN, M. et al.; Proc. Laser 81, p. 308-311, Springer Verlag,
    Heidelberg 1982
(6) PHILIPP H.K. et al.; Proc. Laser 83
(7) ENGLISCH, W., ENDEMANN, M.; Proc. Laser 83
(8) HARTL, PH., WITTIG, M.: Evaluation of Detection Schemes for CO_2 Laser
    Applications Final Report ESTEC contract no. 4640/81 (1982)

Raumflugtauglicher CO_2-Laser

W. Wiesemann, W. Diehl, M. Rother, M. Endemann, W. Englisch
Battelle-Institut e.V.
Am Römerhof 35, D 6000 Frankfurt am Main 90

Kontinuierliche CO_2-Laser werden für verschiedene Anwendungen im Weltraum benötigt, wie z.B. Breitbandkommunikation zwischen Satelliten (1)
oder Erdbeobachtung mit kohärentem cw-Lidar (2).

Die Anforderungen an einen Laser für diese Anwendungen sind:
- Ausgangsleistung ca. 10-20 W
- Emission linear polarisiert im TEM_{00}-Modenprofil
- kurzzeitige Frequenzschwankungen $<$ 1 MHz/μs
- abstimmbar auf verschiedene Linien
- abgeschlossenes Gassystem
- Betrieb mit isotopischen Gasgemischen möglich
- Lebensdauer einer Gasfüllung $>$ 1500 hr
- Lebensdauer mit Austausch der Gasfüllung bis 60000 hr
- elektrischer Wirkungsgrad $>$ 5 %
- Kühlung unter 0-g Bedingungen möglich
- mechanische Belastbarkeit entsprechend WR-Anforderungen
- Masse $<$ 20 kg
- Volumen $<$ 30 l

Am Battelle-Institut e.V. wird zur Zeit der Prototyp eines Lasers entwickelt, der diesen Anforderungen genügen soll. Eine Analyse der Ent-
wurfsparameter zeigt, daß ein konventioneller Niederdrucklaser mit Hoch-
spannungsanregung die größten Erfolgsaussichten bietet.

Die aufgeführten Bedingungen ergaben folgende Konzeption

o Modulare Metall-Keramik-Konstruktion:
   Wegen der beim Start auftretenden mechanischen Belastung wird für
   die Plasmaröhren und die Elektrodenbecher an Stelle der sonst üblichen Gläser ein höher belastbarer Keramikstoff verwendet, der außerdem elektrisch isolierend und gut wärmeleitend ist. Die An-

kopplung an den Kühlkreislauf und das Gasreservoir erfolgt mit metallischen Werkstoffen.

o Abgeschlossenes Gasvolumen mit Gasreservoir und Möglichkeit zum Gasaustausch:
Eine ausreichende Lebensdauer wird durch den Einsatz geeigneter Elektroden und Fügetechnologien erreicht. Zur zusätzlichen Erhöhung der Betriebszeit wird ein Lasergasgemisch mitgeführt, das für mehrere Nachfüllungen ausreicht.

o Passive Kühlung durch Wärmeleitung zur Wärmesenke:
Da die konventionellen flüssigen oder gasförmigen Kühlmittel für Raumfluganwendungen ungeeignet sind, erfolgt die Kühlung passiv mittels einer Heat-Pipe.

o Passive Frequenzstabilisierung:
Die üblichen aktiven Frequenzstabilisierungen benutzen eine permanente Modulation der Laserfrequenz, die für Kommunikationsanwendungen störend ist. Es wird daher angestrebt, die gewünschte Frequenzstabilität weitgehend passiv durch konstruktive Maßnahmen und Verwendung geeigneter Materialien zu erreichen.

<u>Konstruktionsbeschreibung des Lasers</u>

Die Ausgangsleistung von über 10 Watt bedingt eine aktive Entladungsstrecke von ca. 1,30 m Länge. Um zu kompakten Abmessungen zu gelangen, werden vier getrennte Entladungsstrecken von je 34 cm Länge verwendet, die optisch hintereinandergeschaltet sind (siehe Bild 1). Jede dieser Entladungsstrecken ist mit Brewster-Fenstern abgeschlossen. Zur Verbesserung der Wärmeabfuhr sowie der höheren mechanischen Belastbarkeit sind die Entladungsröhren aus BeO-Keramik hergestellt. Die vier Entladungsröhren sind um einen Aluminium-Kern angeordnet, der die Verlustwärme an eine Heat-Pipe überträgt, die sie zu einer geeigneten Wärmesenke weiterleitet. Als Voraussetzung für die Erzielung einer langen Lebensdauer wurden geeignete Hard-Seal-Techniken (Glaslot, Au-In-Lötung) entwickelt und erprobt.

Die Entladungsrohre sind mit einem gemeinsamen Leichtmetall-Gasvorratsgefäß verbunden, das als doppelwandiger Zylinder die Entladungsröhren umschließt. In dem Vorratsvolumen (1,5 l) befindet sich ein Druckgefäß

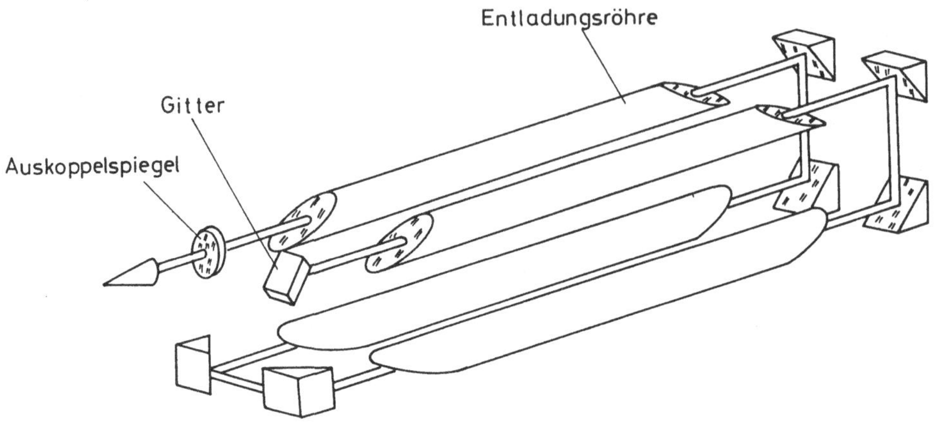

Bild 1. Anordnung der optischen Bauteile

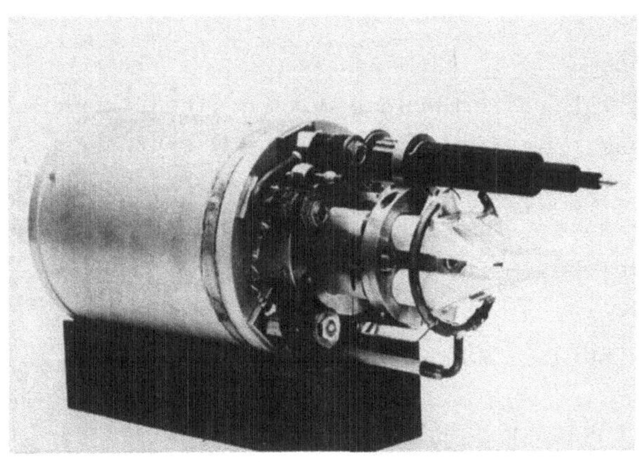

Bild 2. Laserkern

mit dem Gasvorrat für spätere Nachfüllungen. Ein elektromechanisches
Dosierventil erlaubt eine ferngesteuerte Abfüllung des Gasgemisches.
Gasvorratsgefäß und Entladungsröhren mit Kühlkörper bilden den aktiven
Laserkern (siehe Bild 2).

Die tragende Struktur des Lasers besteht aus einem zylindrischen Rohr
aus kohlefaserverstärktem Kunststoff, das an den Enden zwei Edelstahl-
Montageplatten für die optischen Elemente des Resonators trägt (siehe
Bild 3). Die CFK-Struktur gewährleistet trotz geringstem Gewicht eine
hervorragende mechanische Festigkeit. Sie ist für extrem geringe thermische Längenausdehnung ausgelegt, so daß eine gute passive Frequenzstabilität garantiert ist. Struktur und Laserkern sind thermisch weitgehend voneinander entkoppelt.

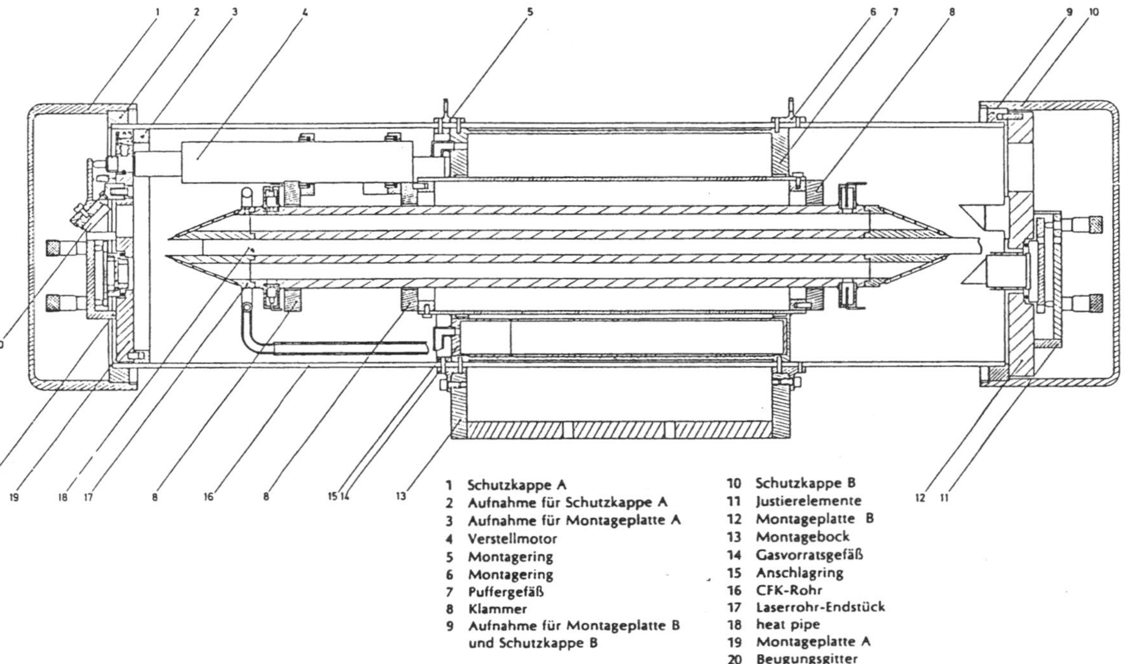

Bild 3. Entwicklungsmodell des raumflugtauglichen CO_2-Lasers, M 1:3,5 (Justierschrauben werden vor den Tests entfernt)

Der Resonator besteht aus einer ZnSe-Auskoppelplatte, einem Gitter
(150 l/mm) zur Linienselektion und den Umlenkspiegeln zur Strahlfaltung
Alle optischen Komponenten sind zunächst justierbar, können aber mit
einfachen Mitteln nach der Einstellung fixiert werden.

References

(1) ENGLISCH, W.: "Study on Intersatellite Laser Communication Links",
    Final Report ESTEC Contract No. 3555/78/NL/PP (SC) (1979)

(2) ENGLISCH, W., WIESEMANN, W.: "Remote sensing of atmospheric trace
    gases by differential absorption spectroscopy".
    Proc. OST Conf. 6.-11.3.1978, Toulouse; ESA-SP-134, p. 465-473

Optimierung eines Mach-Zehnder-Wellenleitermodulators in LiNbO$_3$

B. FURCH und E. BRATENGEYER
Technische Universität Wien, Institut für Nachrichtentechnik
Gußhausstraße 25, A-1040 Wien, Austria

1. Einleitung

Wegen des kleinen Platzbedarfes, des geringen Gewichtes, des niedrigen
Leistungsverbrauchs und der elektromagnetischen Verträglichkeit kommt
dem Einsatz optischer Systeme (Glasfasern, Halbleiterlaser und inte-
griert-optische Bauelemente) zur Verarbeitung und Übertragung hoher Da-
tenraten in Satelliten und Raumfahrzeugen große Bedeutung zu /1/. Ein
grundlegendes Element der integrierten Optik ist der Mach-Zehnder-Wel-
lenleitermodulator /2/, dessen prinzipielle Wirkungsweise an folgende
Einsatzgebiete denken läßt.

- in der Telekommunikation als externer Modulator für Halbleiterlaser
  für extrem hohe Bitraten, und als Modulator in Heterodynesystemen, wo
  durch direkte Strommodulation der Monomodebetrieb gestört wird
- in der "on-board"-Vorverarbeitung breitbandiger Signale als Analog-
  Digital-Wandler, optischer Abtaster und Multiplexer

Wir beschreiben die Herstellung von Wellenleitermodulatoren in Y- und
Z-geschnittenem Lithiumniobat (LiNbO$_3$), untersuchen die Optimierungsmög-
lichkeiten zur Erzielung eines hohen Auslöschungsvermögens bei niedriger Modulationsspannung, einer niedrigen Einfügungsdämpfung und einer
hohen Bandbreite, und berichten über Modulationsexperimente bei
$\lambda_o = 0.63$ µm.

2. Aufbau und Wirkungsweise

Der untersuchte Modulatortyp ist eine planare Version des Mach-Zehnder-
Interferometers, das eine Phasenmodulation des Lichtes in eine Inten-
sitätsmodulation überführt. Der Eingangswellenleiter verzweigt sich,
wie in Abb.1 gezeigt, Y-förmig in zwei parallele Wellenleiter und teilt
eine einfallende Lichtwelle in zwei gleich große Komponenten. Der relativen Phasenlage der beiden Teilwellen entsprechend wird am Vereini-
gungspunkt der Grundmodus oder, durch Interferenz der beiden antisymme-
trischen Felder, der Modus zweiter Ordnung angeregt. Führt nun der Ausgangswellenleiter nur den Grundmodus, strahlt das Licht des höheren Modus ins Substrat ab, wodurch die Intensität im Ausgangswellenleiter auf

ein Minimum sinkt. In elektrooptischen Kristallen läßt sich eine Phasen
verschiebung durch elektrische Felder erzielen, die den Brechungsindex
im Wellenleiter ändern. $LiNbO_3$, ein für die integrierte Optik bevorzug-
tes Material, zeichnet sich einerseits durch gute elektrooptische Eigen
schaften aus, andererseits lassen sich dämpfungsarme Wellenleiter durch
Eindiffusion von Titan erzielen, wobei die Streifenstruktur durch photo
lithographische Abhebetechnik hergestellt wird /3/. Das elektrische
Steuersignal wird an planare Elektroden angelegt, die ebenfalls in Abhebetechnik auf dem Kristall aufgebaut werden. Sowohl im Entwurf als
auch in der Herstellung besteht nun die Möglichkeit Modulationsspannung
Auslöschungsvermögen (= Modulationstiefe), Einfügungsdämpfung und Bandbreite zu optimieren.

3. Optimierung

Eine <u>niedrige Halbwellenspannung</u> V_π (= Modulationsspannung für Ein/Aus-
Modulation) kann nur erzielt werden, wenn die Polarisation des Lichtes
und die Lage der Elektroden so gewählt werden, daß das elektrische Feld
über den stärksten Koeffizienten des elektrooptischen Tensors, nämlich
r_{33}, mit dem Licht in Wechselwirkung tritt. Für Y-geschnittene Kristall
und TE-polarisiertes Licht ist demnach die optimale Elektrodenposition
neben den Wellenleitern (Abb.1a), für Z-geschnittene Kristalle und TM-
polarisiertes Licht aber über den Wellenleitern (Abb.1b). Die verwendete 3-Elektrodenstruktur halbiert durch die Gegentaktwirkung die erforderliche Spannung. Die effektive elektrische Feldstärke wird durch
den Elektrodenabstand d und das Überlappungsintegral Γ von elektrischem
und optischem Feld bestimmt, sodaß sich für vollständige Ein/Aus-Modu-
lation ein Spannung-Länge Produkt von $V_\pi L = (\lambda_o d)/(2n_e^3 r_{33}\Gamma)$ ergibt. n_e
ist der außerordentliche Brechungsindex.

Für ein <u>hohes Auslöschungsvermögen</u> (AV) ist die richtige Dimensionierung der Wellenleiter maßgeblich. Es muß jene Kombination von Streifenbreite w, Titandicke τ, Diffusionstemperatur und -zeit gefunden werden,
die einen Wellenleiter erzeugt, der den Grundmodus gut führt, einen
höheren Modus aber kräftig dämpft. Bei dieser Parameterwahl muß auf die
verschiedenen Diffusionseigenschaften des jeweiligen Kristallschnittes
Rücksicht genommen werden. Dazu gehört auch die Unterdrückung der durch
Ausdiffusion von Li_2O entstehenden Ausbildung eines planaren Wellenlei-
ters.

Die <u>Einfügungsdämpfung</u> kann klein gehalten werden durch sorgfältiges
Polieren der Wellenleiterstirnflächen. Die Größe des Verzweigungswinkel

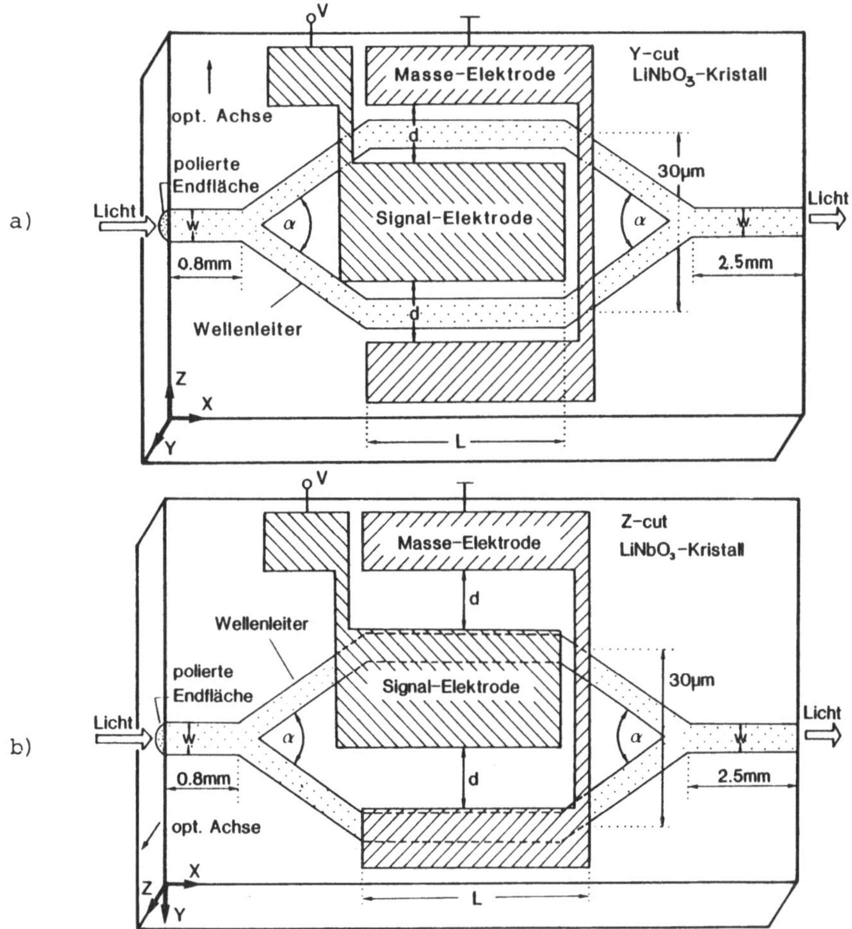

Abb.1. Geometrie und Abmessungen der hergestellten Wellenleitermodulatoren:
a) in Y-geschnittenem LiNbO$_3$ für TE-polarisiertes Licht
b) in Z-geschnittenem LiNbO$_3$ für TM-polarisiertes Licht
Ti-Streifenbreite w = 2.8 bis 6.0 µm, Elektrodenabstand d = 3 bis 8 µm,
elektrooptische Wechselwirklänge 3 mm, 4.5 mm und 6 mm. Der Verzweigungswinkel α beträgt 1.14°.

α bestimmt die Strahlungsverluste an den Diskontinuitäten der Y-Gabeln
und der Wellenleiterbiegungen. Beim Z-geschnittenen Modulator muß die
durch die über den Wellenleitern liegende Metallelektrode auftretende
Dämpfung durch eine dielektrische Trennschicht reduziert werden.

Die <u>Bandbreite</u> wird durch die RC-Zeitkonstante der Elektrodenstruktur
bestimmt. Für eine große Bandbreite ist daher eine kleine Kapazität an-
zustreben. Bei gegebener Modulationsspannung ist die größte Bandbreite
mit einer kurzen Elektrode und möglichst kleinem Elektrodenabstand zu
erzielen /4/.

484

4. Experimentelle Untersuchungen

Zur Erfassung der genannten Einflußgrößen haben wir eine Reihe von Wellenleitermodulatoren in Y- und Z-geschnittenem $LiNbO_3$ aufgebaut. Dimensionen und Geometrie sind Abb.1 zu entnehmen. Die 15-29 nm dicken und 2.8-6.0 µm breiten Titanstreifen für die Wellenleiter wurden 5.2 Stunden bei 980°C in Ar-Atmosphäre eindiffundiert. Während der Abkühlphase wurde O_2 durch das Rohr geblasen. Beide Gase wurden vor dem Durchströmen des Diffusionsrohres in einer Blubberflasche mit Wasserdampf angereichert /5/. Durch Einstellung der Wassertemperatur konnte exakte Kompensation der Li_2O-Ausdiffusion erzielt werden. Nach der Diffusion wurden die Kristallflächen nach einer eigens entwickelten Methode /6/ kantenscharf poliert. In Abhebetechnik wurden 150 nm dicke Goldelektroden hergestellt, wobei 3 nm Cr als Haftschicht dienten. Der Meßaufbau zur Messung von Lichtführung und Modulationsverhalten ist in Abb.2 dargestellt. Linear polarisiertes Licht eines HeNe-Lasers bei λ_O = 0.63 µm wurde durch ein 40X Mikroskopobjektiv auf die polierte Stirnfläche des Eingangswellenleiters fokussiert. Ein zweites 40X Mikroskopobjektiv bildete die Stirnfläche des Ausgangswellenleiters stark vergrößert gleichzeitig auf eine Photodiode sowie ein Vidikon für die Darstellung auf einem Monitor ab.

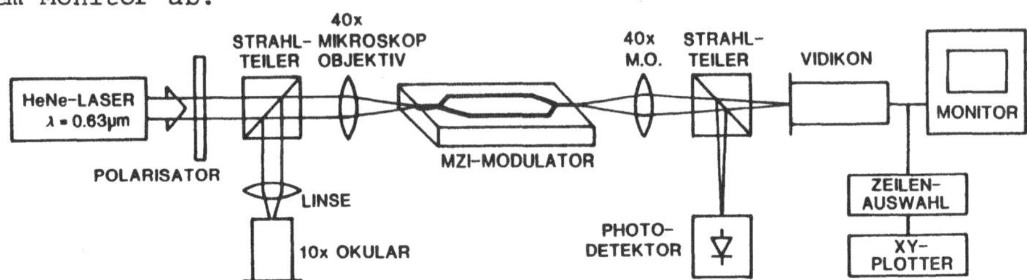

Abb.2. Aufbau zur Messung der Lichtführung und des Modulationsverhaltens. Das $LiNbO_3$-Substrat kann mit Hilfe von Mikromanipulatoren in drei Richtungen und um zwei Achsen auf ca. 1 µm genau positioniert werden. Der erste Strahlteiler wird zur Einstellung der exakten Fokussierung des Lichtes auf die Wellenleiterstirnfläche benutzt.

Für einige Modulatoren ist das gemessene Spannung-Länge Produkt zusammen mit dem Auslöschungsvermögen, der Einfügungsdämpfung und der 1/e-Modenbreite in Tab.1 zusammengefaßt. Nach dem Zerschneiden eines Modulators wurden die Verluste eines gut geführten Modus an einer Gabel mit <2 dB, an einer Knickstelle mit <1 dB gemessen. Bei ungenügender Lichtführung steigen die Verluste an den Verzweigungen und Biegungen stark an. Die Abweichung in der Symmetrie der Lichtaufteilung lag unter 5%. Die erzielte hohe Symmetrie ist mitverantwortlich für das hohe Auslöschungsvermögen. Die Kapazität einer 6 mm langen Modulatorelektrode wurde mit <4 pF gemessen, was einer Bandbreite von 1.6 GHz entspricht.

Tabelle 1. Wellenleiterbreite w, 1/e-Modenbreite $\omega_{1/e}$, Einfügungsdämpfung ED, Auslöschungsvermögen AV, Spannung – Länge Produkt $V_\pi L$ und Überlappungsintegral Γ für zwei Y-geschnittene Modulatoren (MY1, $\tau = 27$ nm; MY2, $\tau = 29$ nm) und einen Z-geschnittenen Modulator (MZ1, $\tau = 18$ nm).

	$w[\mu m]$	$\omega_{1/e}[\mu m]$	$d[\mu m]$	ED[dB]	AV[dB]	$V_\pi L[Vcm]$	Γ
MY1	5.8	4.8	6.9	14.7	21.5	1.02	0.65
		4.9	5.8	15.0	21.1	0.89	0.63
		5.0	5.0	14.9	23.2	0.73	0.65
$\tau = 27$ nm	6.0	5.0	5.8	14.0	22.3	0.90	0.62
		5.0	4.8	14.3	21.9	0.81	0.57
MY2	4.3	4.7	6.0	18.7	24.8	0.87	0.66
		4.5	4.9	19.3	23.2	0.81	0.59
		4.7	3.6	19.9	24.0	0.72	0.48
$\tau = 29$ nm	4.7	4.3	6.6	14.7	28.2	0.96	0.66
		4.1	5.1	15.0	24.6	0.87	0.56
MZ1	4.9	4.7	7.9	31.0	10.0	0.95	0.81
		4.7	5.5	32.0	8.4	0.70	0.76
$\tau = 18$ nm	4.6	5.0	7.6	26.0	13.5	0.96	0.76
		5.0	6.9	25.5	12.0	0.87	0.77
		4.9	5.2	24.0	18.1	0.87	0.58

5. Schlußfolgerung

Unser Vergleich zeigte auf, daß Modulatoren auf Y-geschnittenem $LiNbO_3$ hinsichtlich Einfügungsdämpfung und Auslöschungsvermögen jenen auf Z-geschnittenem $LiNbO_3$ überlegen sind, bei denen zwar eine höhere Feldüberlappung ($\Gamma = 0.8$) erreicht werden kann, sich aber eine Elektrodenfehljustierung viel stärker auswirkt. Das Aufbringen einer Si_3N_4-Trennschicht zwischen Wellenleiter und Elektroden konnte zwar die Einfügungsdämpfung um ≈ 10 dB reduzieren, erfordert aber einen zusätzlichen technologischen Aufwand und verursachte DC-Driftprobleme. Bei Y-geschnittenen Modulatoren ist die Feldüberlappung zwar etwas geringer ($\Gamma = 0.65$ für $d > \omega_{1/e}$, /7/), jedoch konnte stets ein Auslöschungsvermögen von 21-28 dB erreicht werden, das über ein Vielfaches der Halbwellenspannung konstant blieb.

6. Danksagung

Diese Arbeit wurde vom Fonds zur Förderung der wissenschaftlichen Forschung im Projekt S-22/01 gefördert.

7. Literatur

/1/ LUTZ,H., Laser '83, München, Proc. (1981) 289
/2/ AURACHER,F., KEIL,R., Wave Electr. 4 (1980) 129
/3/ SCHMIDT,R.V., KAMINOW,I.P., APL 25 (1974) 458
/4/ ALFERNESS,R.C., IEEE MTT-30 (1982) 1121
/5/ JACKEL,J.L. et al., APL 38 (1981) 509
/6/ FURCH,B. et al., z.Veröff.angen. in J.Opt.Comm (1983)
/7/ MARCUSE,D., IEEE QE-18 (1982) 393

Conical Scan-Tracking-Technik mit kohärentem optischen Empfang bei 10 μm

W. ENGLISCH, M. ENDEMANN
Battelle-Institut e.V.
Am Römerhof 35, D 6000 Frankfurt am Main 90

1. Introduction

The exceedingly small beam widths involved in space-to-space optical communications require cooperative efforts by both laser terminals to actively track one another with high accuracy. There are two tracking techniques known from Radar technology which may be applied to the CO_2 laser communication system (1) considered here: The monopulse technique requires a separate quadrant detector while the conical scan technique utilizes the same single detector element as used for data detection. Evaluation of the latter one for the special case of coherent optical detection (i.e. heterodyning or homodyning) at λ = 10 μm is the subjec of this paper.

2. Coherent Detection Receiver Gain Curve

It is well known that the sensitivity of an optical heterodyne or homo- dyne receiver depends strongly on the spot sizes and the field distri- butions of the focused signal and local oscillator (LO) waves. Conside- ring an Airy signal spot and a Gaussian LO we find from ref. (2) the combination R_D/R_A = 0.86, R_D/W_0 = 2.5 to be near optimum. Here R_D = detection radius, R_A = Airy spot radius (maximum to first zero), and W_0 = Gaussian spot radius of LO. For tracking purposes the variation of sensitivity with lateral displacement ϱ of the signal spot, i.e. the receiver gain curve, must be known. It has been calculated by numerical integration of the superimposed field distributions using the para- meters given above. Fig. 1 shows this gain curve which is normalized to the ideal alignment case where detector, signal spot, and LO spot ar centered exactly. For further calculation it is convenient to approxi- mate the gain curve by the analytical expression.

$$g(\varrho) = \exp\left[-2K(\varrho/R_D)^2\right] \qquad (1)$$

With K = 1.452 an excellent fit results.

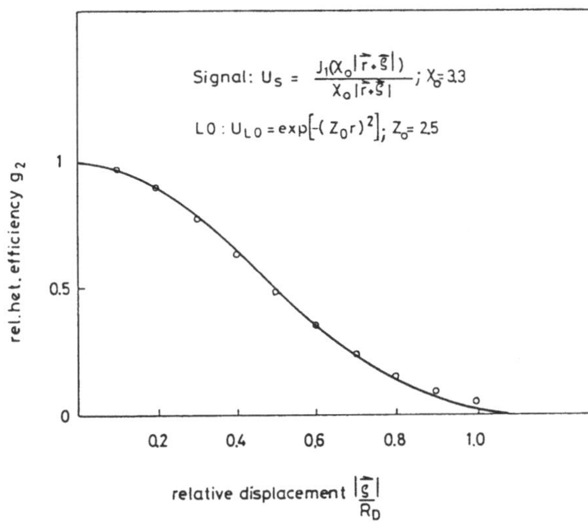

Fig. 1. Degradation of relative heterodyne efficiency by lateral displacement of signal spot (Airy signal, Gaussian LO). Circles indicate Gaussian approximation

3. Conical Scan Tracking

With the conical scan technique an optical nutator (e.g. a rotating wedge) generates a periodical spatial modulation of the received beam which is focused onto the detector. The geometry is depicted in fig. 2. The nutator produces a small conical motion (bias angle δ, radian frequency ω) of the incoming beam around the nominal optical axis which is centered at the detector. Under perfect pointing conditions the center of the sig-

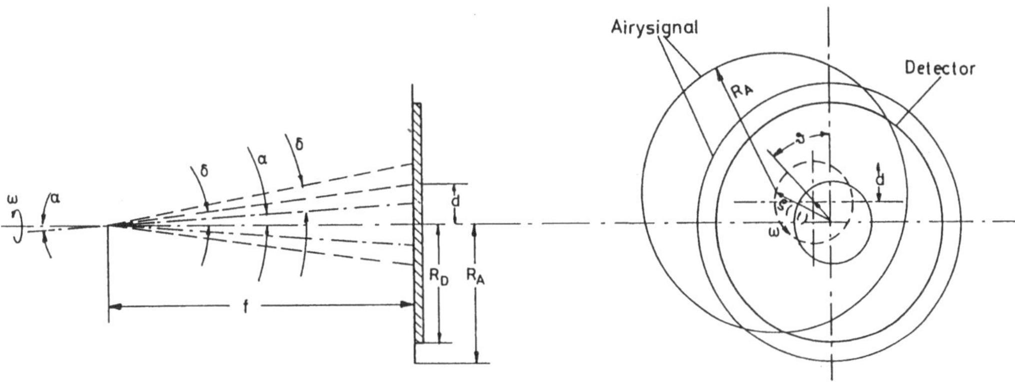

Fig. 2. Conical scan tracking technique; geometrical arrangement

nal Airy spot describes a circle with radius d = fδ around the detector center. f is the distance between detector and focusing element or nutator (whatever is smaller). The net effect is a slight decrease in receiver gain but no modulation of the signal occurs. If a pointing error angle α is superimposed to the bias angle δ the center of the scan circle is displaced from the detector center by a distance a = fα while the direction of displacement is given by ν. Since the signal Airy spot moves along the scan circle the distance ϱ of the Airy spot

center to the detector center varies with time and causes a small
amplitude modulation of the detector output signal. ϱ is given by

$$\varrho(t) = \left[a^2 + d^2 + 2ad\cos(\omega t - \nu) \right]^{1/2} \tag{2}$$

Normalizing a and d to the detector radius R_D by $d = KR_D$, $a = \varrho R_D$
and using equ. (1) yields the tracking receiver gain function

$$g(t) = \exp\left[-2K\left\{ \varrho^2 + K^2 + 2K\varrho \cos(\omega t - \nu) \right\} \right] \tag{3}$$

4. Tracking Error Signal

Taking the square root of g(t), making a series expansion, and taking
only the term with the fundamental frequency yields the relevant detector output voltage:

$$U(t) = U_0 \exp\left[-K(\varrho^2 + K^2) \right] +$$
$$+ U_0 \exp\left[-K(\varrho^2 + K^2) \right] \cdot \left[2K\varrho \cos(\omega t - \nu) \right] \tag{4}$$

The first term represents the signal in the data channel. The exponential factor accounts for degradation of the nominal signal (K_0) due to
conical scanning. The second term with periodical modulation represents
the tracking error signal. As can be seen the modulation amplitude
depends on the magnitude of the pointing error, while the modulation
phase indicates the direction of incorrect pointing. Fig. 3 illustrates

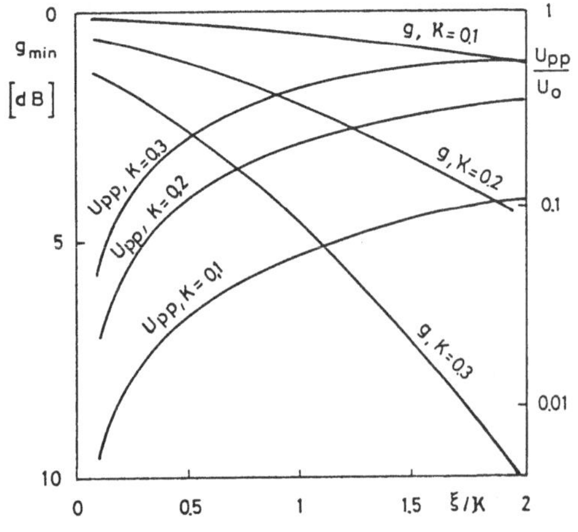

Fig. 3. Variation of minimum receiver gain
and peak-to-peak amplitude of
tracking error signal with conical
scan bias   and tracking error

the variation of the peak-to-
peak amplitude U_{pp} of the
tracking error signal as well
as the variation of minimum
receiver gain g_{min} as function
of the normalized pointing
error and the conical scan bias
angle. There is clearly a trade
off between conflicting require
ments:

- Keep the receiver gain as high
  as possible in order to minimize losses in the data channel
- Make U_{pp} as large as possible
  in order to quarantee a good
  tracking performance.

5. Illustrative Example

For the CO_2 laser communication package described in ref. (1) (however
with heterodyne detection assumed) a conical scan tracking system has
been designed and analyzed in detail (3). The optimized values of the
relevant parameters are: detector radius R_D = 100 µm, conical scan bias
angle δ = 0.11 mrad yielding K = 0.05, F-number of focusing system F
= 9.5, tracking loop bandwidth B_P = 200 Hz. The calculated performance
is shown in fig. 4. As can be seen the specified pointing accuracy of
$|\Delta\theta| \leqslant 6$ µrad is easily met by the tracking system designed and the
nominal signal loss in the communication channel is negligible for this
value. The rather high signal-to-noise ratio $(S/N)_P$ within the tracking
loop would, in principle, allow for sub-microradian tracking accuracy.
However the actual achievable pointing accuracy is determined by mechanical limitations of the moving optical parts.

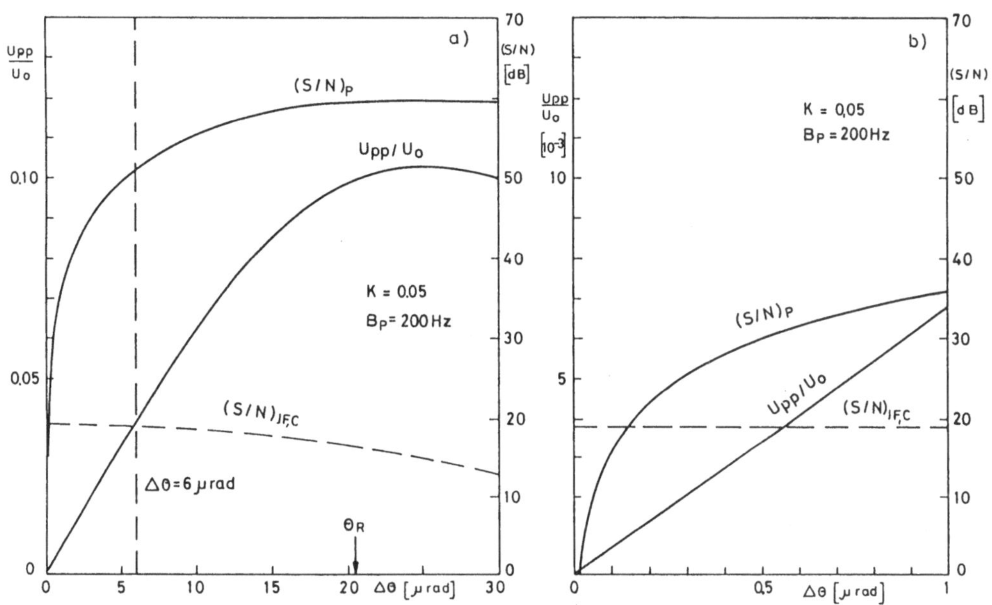

Fig. 4. Variation of signal voltage an $(S/N)_P$ in the tracking loop with
pointing error as measured in object space. b) is an enlarged
portion of a) at small pointing errors. The mean $(S/N)_{IF,C}$ in the
communication channel is indicated too

Literature

(1) ENGLISCH, W. et al.; Proc. Laser 83
(2) COHEN, S.; Appl. Optics 14, p. 1953 (1975)
(3) ENGLISCH, W.: Design of a CO_2 Laser Transceiver Package for Space-
    to-Space Optical Communications. Final report ESTEC contract
    no. 3488/79

Homodynempfang von amplituden- und phasenmodulierten Infrarotsignalen

H.K. PHILIPP, E. BONEK, A.L. SCHOLTZ und W.R. LEEB
Technische Universität Wien, Institut für Nachrichtentechnik
Gußhausstraße 25, A-1040 Wien, Austria

1. Einführung

Für schnelle Datenübertragung und für Entfernungsmessung im Weltraum,
sowie für Echolotung in der Atmosphäre untersucht die Europäische Weltraumbehörde (ESA) Empfänger für CO_2-Laserstrahlung. Im Zuge der Arbeiten stellte sich das Homodyn-Verfahren - das ist ein kohärenter Überlagerungsempfang mit einer Zwischenfrequenz gleich Null - als sehr erfolgversprechend dar.

Das Prinzip eines optischen Homodyn-Empfängers zeigt Fig.1. Ein optisches Eingangssignal der Frequenz f_o und der Phase Θ_i wird in einem
optischen Phasendetektor mit dem Signal des lokalen Oszillators gemisch
Wird nun der lokale Oszillator nach dem Prinzip eines Phasenregelkreise
(PLL) dem optischen Eingangssignal in Frequenz und Phase nachgeführt, s
ergibt sich im synchronisierten Zustand der PLL am Ausgang des Phasendetektors ein elektrisches Mischsignal proportional dem Phasenfehler
$\Theta_e = \Theta_i - \Theta_o$. Dieses wird über das Schleifenfilter dem lokalen Oszillator
zur Aufrechterhaltung der Synchronisation zugeführt. In dieser Konfigu-
ration eignet sich der Empfänger unmittelbar zur Demodulation von

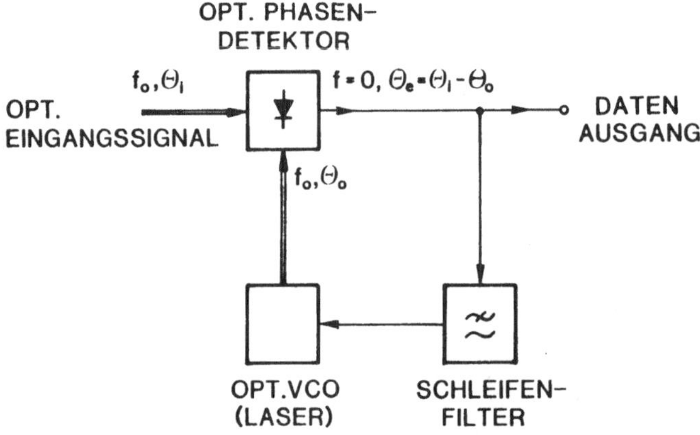

Fig.1. Prinzip eines optischen Homodyn-Empfängers

phasenmodulierten Eingangssignalen, da Phasenänderungen mit Frequenzen höher als die Eigenfrequenz der PLL nicht mehr ausgeregelt werden und daher am Ausgang des Phasendetektors ausgekoppelt werden können.

Der große Vorteil des Homodyn-Verfahrens gegenüber herkömmlichem Überlagerungsempfang liegt nun einerseits im besseren Signal-zu-Geräusch Verhältnis, andererseits in einer Halbierung der Detektorbandbreite, die gerade in dem Wellenlängenbereich des CO_2-Lasers von $\lambda = 10$ µm der Erhöhung der Datenrate derzeit eine technologische Grenze setzt. Weiters erlaubt das Verfahren eine sehr einfache selbsttätige Verfolgung Doppler-verschobener Eingangssignale.

Der spannungsgesteuerte optische lokale Oszillator (VCO) muß langsame Frequenzänderungen über einen weiten Abstimmbereich (z.B. Kompensation einer Doppler-Verschiebung), sowie schnelle Frequenzänderungen (Kompensation der Kurzzeitfrequenzschwankungen der Laser) ermöglichen. Die schnelle Nachführung von Frequenz und Phase wurde auf zwei Arten demonstriert, nämlich mit einem internen elektrooptischen Phasenmodulator (EOM) bzw. mit einem externen akustooptischen Modulator (AOM). Beide Nachführmechanismen wurden ergänzt durch eine langsame Frequenznachführung über eine piezoelektrische Längenänderung des Laserresonators.

Das Anbringen eines elektrooptischen Modulators innerhalb des optischen Resonators (siehe Fig.2) ermöglicht durch elektronische Änderung der optischen Resonatorlänge eine schnelle Frequenzänderung des VCO-Lasers. Die Einfügungsdämpfung des Modulators führt allerdings zu einer Reduktion des Abstimmbereichs des VCO insgesamt, sodaß die im folgenden beschriebene Methode zur Frequenznachführung für die ins Auge gefaßten Anwendungen zielführender erscheint.

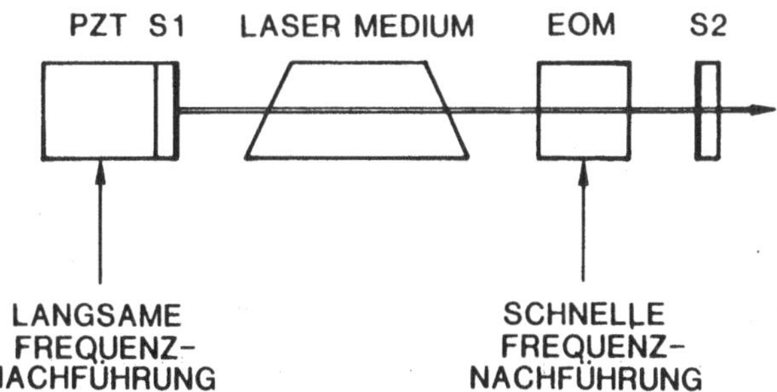

Fig.2. Spannungsgesteuerter opt. VCO mit internem elektrooptischen Phasenmodulator (EOM); S_1,S_2...Resonator-Spiegel

492

Fig.3 zeigt die Ausführung des optischen VCO mit einem externen akustooptischen Modulator, der im Bragg-Bereich arbeitet. Durch Wechselwirkung des Laserstrahls der Frequenz f_1 mit einer akustischen Welle der
Frequenz f_a entsteht ein in der Frequenz um f_a versetzter Laserstrahl,
der um zweimal den Bragg'schen Winkel Θ_B räumlich versetzt aus dem Modulator austritt. Durch Frequenzverstimmen des elektrischen VCO, der
die akustische Welle der Frequenz f_a erzeugt, ergibt sich somit indirekt
eine Frequenzvariation des optischen VCO um die Mittenfrequenz f_1+f_a.

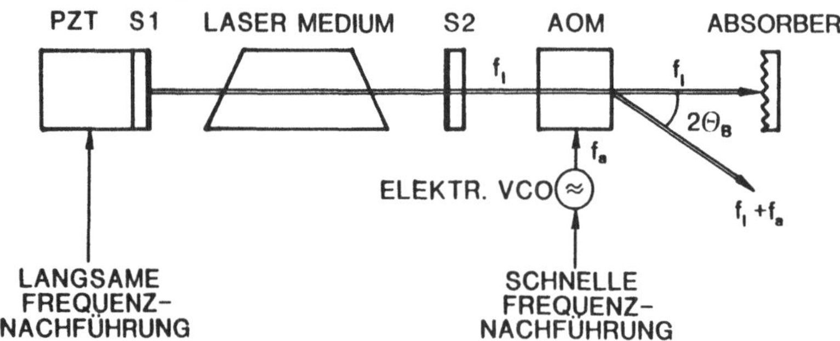

Fig.3. Spannungsgesteuerter opt. VCO mit externem akustooptischen Fre-
quenzmodulator (AOM); S_1,S_2...Resonator-Spiegel, Θ_B...Bragg-Winkel

2. Homodyn-Empfänger für PM-Signale

In Fig.4 ist das Blockschaltbild eines Homodyn-Empfängers für PM-Signale
zu sehen. Als optischer Phasendetektor (PD) wurde eine auf 77 K gekühlte
HgCdTe-Photodiode verwendet. Ist die Schleife eingerastet, so ist das
Ausgangssignal des PD proportional der Phasendifferenz Θ_e zwischen Eingangssignal (i) und VCO-Signal (o). Da der Mischprozeß in der Photodiod‹
nicht einer Multiplikation entspricht, enthält das PD-Ausgangssignal zusätzlich einen störenden, den auftreffenden Laserleistungen proportiona-
len Gleichterm, der kompensiert werden muß. Das so gewonnene Regelsignal
steuert, gefiltert durch einen aktiven Tiefpaß 1.Ordnung, Frequenz und
Phase des optischen VCO derart, daß der Phasenfehler zwischen Eingangs-
und VCO-Signal ein Minimum wird. Die Zeitkonstante des aktiven Filters
und damit die Eigenfrequenz und der Dämpfungsfaktor der PLL bestimmen
sich aus den relativen Frequenzschwankungen der optischen Signalquellen,
zwei konventionell aufgebauten CO_2-Laser. In dem typischen Laboraufbau
und ohne besondere Vorkehrungen gegen thermische und akustische Stör-
einflüsse ergeben sich Frequenzschwankungen von einigen Megahertz pro
Sekunde. Daraus läßt sich ableiten, daß gute Ergebnisse im Regelverhal-
ten der PLL - d.h. ein Phasenfehler Θ_e in Größenordnung einiger Grad -
für Eigenfrequenzen von einigen zehn Kilohertz und einer Dämpfung $\xi \approx 1$
zu erwarten sind.

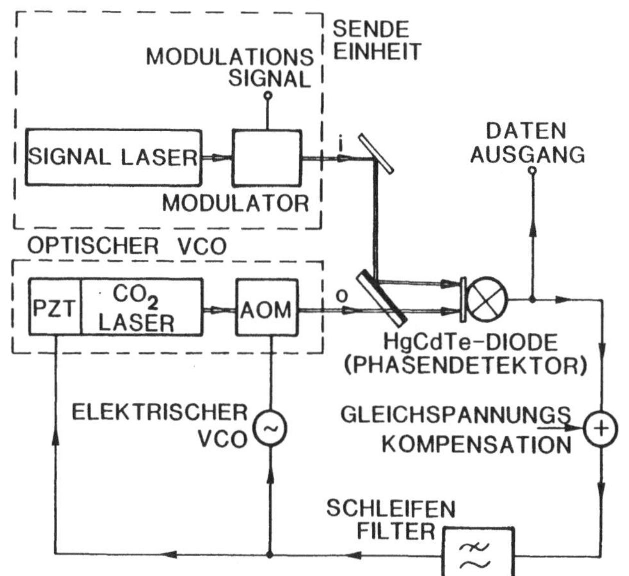

Fig.4. Homodyn-Empfänger für PM-Signale

Zur Demonstration der demodulierenden Eigenschaft der PLL wurden geeignete Eingangssignale simuliert. Die Sendeeinheit, in der das Eingangssignal mit dem gewünschten Modulationsformat erzeugt wird, besteht aus einem CO_2-Laser, dessen Strahl, durch Linsen geeignet geformt, einen Wanderwellenmodulator /1/ durchläuft. In diesem wird dem Laserstrahl das von einem 16-bit Wortgenerator im NRZ-Code und mit einer Datenrate von 25 Mbit/s bis 100 Mbit/s gelieferte elektrische Testsignal aufgeprägt. Mit der Sendeeinheit wurde ein digitales PM-Signal mit einem Phasenhub von 8°, entsprechend einem Leistungsverhältnis von Träger zu Seitenband von 21 dB, erzeugt und dem Homodyn-Empfänger zugeführt. Wurde nun die PLL durch Wobbeln des VCO zum Einrasten gebracht, so konnte das demodulierte Datensignal am Datenausgang des Empfängers beobachtet werden. Bei einer auf den Phasendetektor auftreffenden Leistung $P_o \approx$ 100 µW des VCO-Lasers konnte die Funktion der optischen PLL bei Eingangsleistungen von einigen hundert Mikrowatt bis zu einigen zehn Nanowatt erfolgreich demonstriert werden. Auch wurde ein Schwund der Eingangsleistung bis zu 8 dB simuliert und damit die Eignung dieses Empfängers für eine typische Anwendung (Datenverbindung geostationärer - niedrig fliegender Satellit) gezeigt.

3. Homodyn-Empfänger für AM-Signale

Zur Demodulation von AM-Signalen benötigt der oben beschriebene Empfänger einen zusätzlichen Phasendetektor (siehe Fig.5). In einem optischen Hybrid /2/ werden Eingangssignal (i) und VCO-Signal (o) derart überlagert, daß an seinem Ausgang i- und o-Signal in Phase (i+o), bzw. i- und

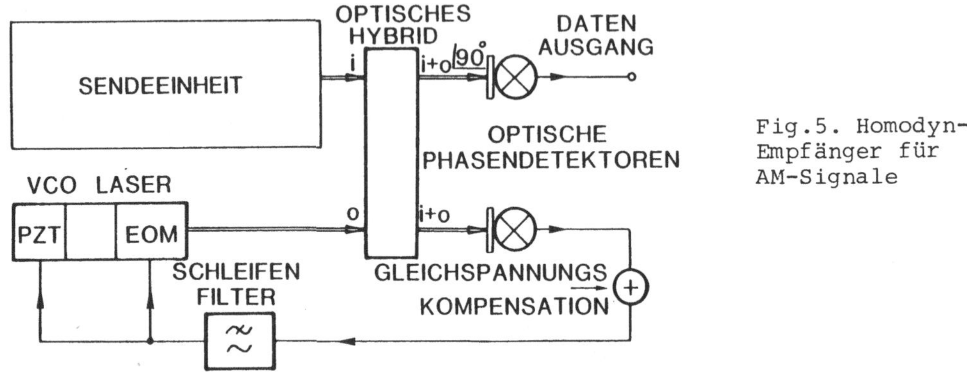

Fig.5. Homodyn-Empfänger für AM-Signale

das um 90° phasenverschobene o-Signal (i+o $\lfloor 90^\circ$) vorhanden sind. Das Mischprodukt aus dem Eingangssignal und dem VCO-Signal liefert nun wieder das Regelsignal für die PLL, während am Ausgang des Quadratur-Detektors bei eingerasteter PLL das demodulierte Datensignal beobachtet wird. Mit der bereits beschriebenen Sendeeinheit wurden digitale AM-Eingangssignale mit einem Modulationsindex zwischen 0,3 und 1 simuliert und bei einer VCO-Leistung von einigen hundert Mikrowatt und einer Eingangsleistung im Bereich von einigen Mikrowatt demoduliert. Fig.6 zeigt ein typisches Schirmbild eines Augendiagramms vom demodulierten Datensignal bei einer Bitfolgefrequenz von 100 Mbit/s, einem Modulationsindex von 0,4 und einer Eingangsleistung von $P_i \approx 0,2$ µW.

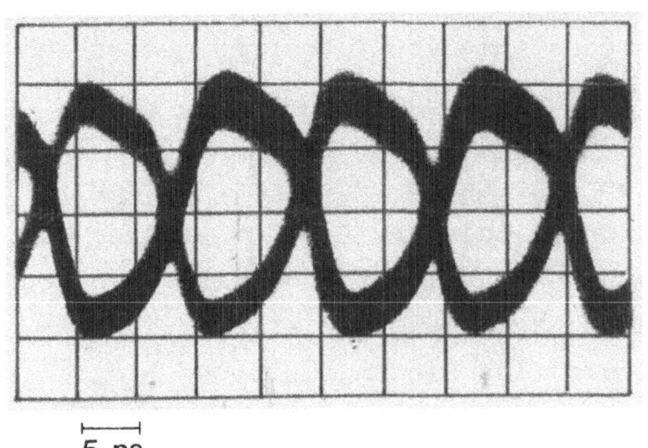

Fig.6. Schirmbild eines Augendiagrammes der demodulierten Daten; Datenrate 100 Mbit/s NRZ, Modulationsindex $m \approx 0,4$, Eingangsleistung $P_i \approx 0,2$ µW

Diese Arbeit wurde im Auftrag der Europäischen Weltraumbehörde (ESA) durchgeführt.

Literatur

/1/ SCHOLTZ, A.L., LEEB, W.R. und BONEK, E.: IEEE J. Quantum Electron., vol. QE-18 (1982), 1021
/2/ LEEB, W.R.: Arch. Elek. Übertragung, vol. 37 (1983), 203

Testergebnisse eines CO_2-Laser-Heterodyn-Entfernungsmessers

F. MALOTA, H. HERRMANN und K. KAIN
DFVLR, Institut für Optoelektronik
Münchener Straße 20, D 8031 Oberpfaffenhofen

Der verstärkte Einsatz von CO_2-Lasern bei der Entfernungsmessung kann
auf die folgenden Punkte zurückgeführt werden:

- geringere Störungen durch die atmosphärische Turbulenz bei 10,6 µm
  Wellenlänge gegenüber dem Sichtbaren oder dem nahen IR,
- größere Augensicherheit,
- hoher Laser-Wirkungsgrad.

Demgegenüber steht aber die in der Größenordnung von 100 liegende geringere Empfindlichkeit der Detektoren bei $\lambda = 10,6$ µm. Für Entfernungs-
messer bei 10,6 µm kann man als Ausgleich dafür den Heterodynempfang
einsetzen, der darüberhinaus noch zwei weitere Vorteile mit sich bringt:

Einmal ist aufgrund der Empfangstechnik eine hochauflösende Richtungs-
selektion möglich, zum anderen kann man, wenn zur E-Messung Impulse verwendet werden, gleichzeitig Entfernung und Geschwindigkeit aufgrund des
Dopplereffekts eines Targets messen.

Zur Entfernung von festen Objekten wurde ein Heterodynsystem aufgebaut
(siehe Fig. 1). Es besteht aus einem Sender, welcher Impulse von 300
bis 500 ns Dauer emittiert. Die Strahlungsleistung beträgt etwa 30 kW
und die Impulsfolgefrequenz ist zwischen 7 und 12 Hz einstellbar. Der
Lasersender emittiert bei der P(20) Linie. Der Lokal Oszillator (LO)

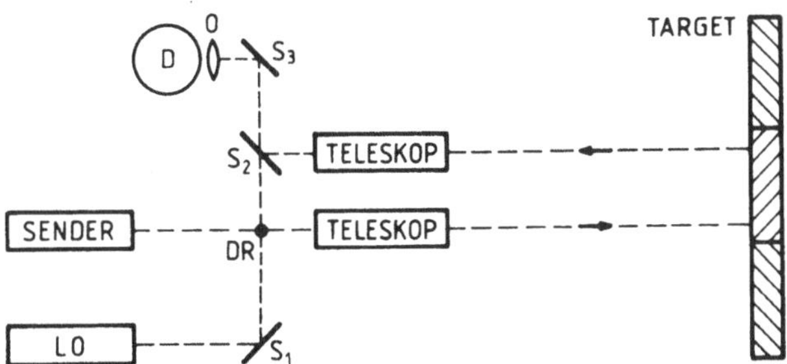

Fig. 1. Schemazeichnung des Heterodyn-E-Messers

emittiert cw-Strahlung und liegt ebenfalls auf der P(20) Linie. Der Detektor D (Mischer), mit einer Optik O, ist ein Hg Cd Te Detektor, der mit flüssigem Stickstoff gekühlt werden muß. Weiterhin kommen zu der Anlage noch die folgenden elektronischen Komponenten: Verstärker, Filter, Biomation-Recorder, Oszillographen, Zähler.

Da die beiden Laser, Sender und LO, passiv stabilisiert sind, ist es erforderlich, daß das ausgehende Signal auch in Heterodyntechnik empfangen wird. Das geschieht dadurch, daß von der ausgehenden Impulsleistung ein Teil durch einen Draht DR auf den Detektor D gebracht wird, wobei auch gleichzeitig die Strahlung des LO eintritt. Der vom Target reflektierte Impuls wird über den teildurchlässigen Spiegel S_2 und den Umlenkspiegel S_3 auf den Detektor D gebracht. Auch hier ist gleichzeitig die cw-Strahlung des LO vorhanden, und es kann sich ein Überlagerungssignal ausbilden. Die Zwischenfrequenz ν_{if} kann durch Betätigen von Piezokristallen, auf denen ein Gitter sitzt, sowohl am Sender als auch am LO, eingestellt werden. Der Bereich, der bei der vorhandenen Anlage auszunutzen ist, liegt zwischen etwa 1 MHz und 30 MHz.

Mit der angegebenen Apparatur wurden verschiedene Messungen durchgeführt, welche sich auf die Qualität und die Größe des Heterodynsignals im Vergleich zum direkten Empfang bezogen, wie z.B. Einfluß der Spiegeleinstellungen und der unterschiedlichen Polarisationsrichtungen, Einfluß der Depolarisation durch das Target, Einstellung des Detektors, etc. Eine Untersuchung bezog sich auf die Vergrößerung des Heterodynsignals als Funktion der Leistung P_{LO} des LO.

Nach der Theorie steht der Heterodynstrom I_h mit der Leistung des LO im folgenden Zusammenhang:

$$I_h \approx \sqrt{P_s} \cdot \sqrt{P_{LO}} \quad .$$

In der Fig. 2 ist dieser Zusammenhang experimentell untersucht worden, wobei die Leistung des Senders P_s konstant gehalten wurde. Zur E-Messung wurde eine Teststrecke mit einem Reflektor in einer Entfernung von 148 m aufgebaut. Fig. 3a zeigt den ausgehenden und den reflektierten Impuls, aufgenommen mit direktem Empfang. In Fig. 3b sind beide Impulse mit dem Heterodynverfahren detektiert, wobei die Zwischenfrequenz in diesem Fall 23 MHz betrug. Fig. 3c zeigt ebenfalls eine E-Messung, wobei das Target bewegt wurde.

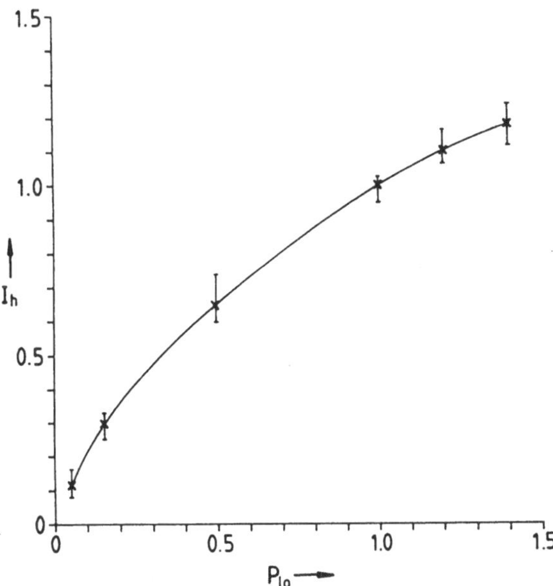

Fig. 2. Heterodynstrom I_h als Funktion der Strahlungs-leistung P_{LO} des LO

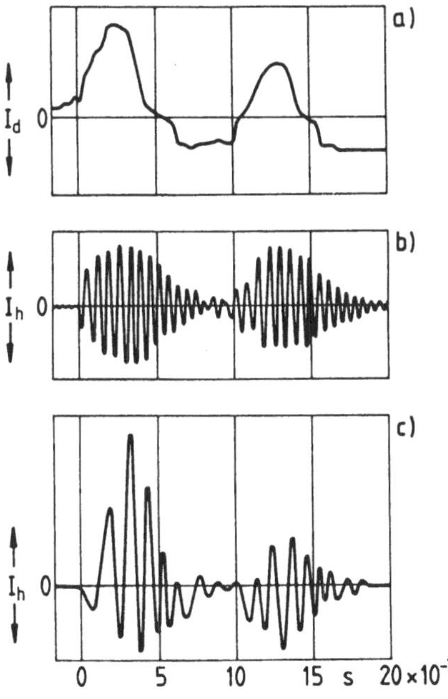

Fig. 3.

a) Zwei Pulse mit direktem Empfang detektiert.

b) Zwei Pulse mit Heterodynempfang detektiert.

c) Zwei Pulse mit Heterodynempfang detektiert. Bewegtes Target

Bei der Laufzeitmessung muß nun der zeitliche Abstand zwischen den zwei
Impulsen gemessen werden, was im vorliegenden Fall einige Schwierigkeiten bereitet. Es wurden drei Methoden gestestet, um die Laufzeit zu bestimmen, wenn beide Signale in Form von Heterodynsignalen vorliegen.
a) Spitze-Spitze-Messung; b) Schwellwertmessung; c) Nulldurchgangsmessun

In Fig. 4 sind die Ergebnisse aus je 95 Messungen aufgetragen. Es zeigt
sich, daß die Nulldurchgangsmessung am besten abschneidet. Weiterhin
ist aber anzumerken, daß die Schwellwertmessung technisch am einfachsten
von den drei Methoden auszuführen ist. Der Entfernungsfehler hängt von
der Bandbreite und dem SNR ab.

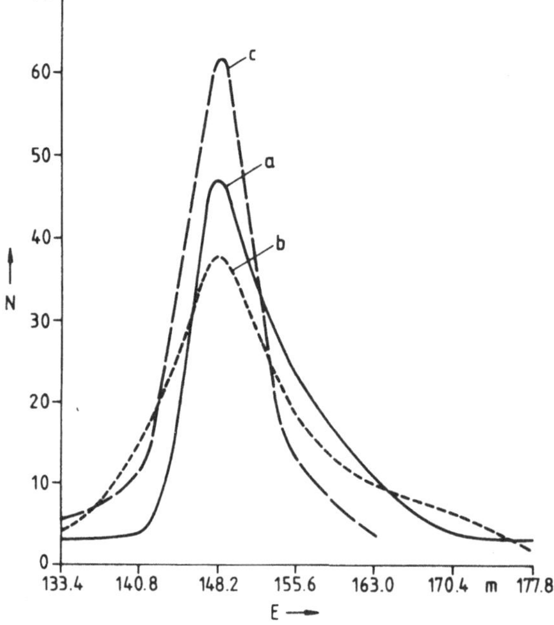

Fig. 4.

E-Messungen auf
3 Arten.

a) Spitze-Spitze-
   Messung.

b) Schwellwertmessung.

c) Nulldurchgangs-
   messung

In der Fig. 5 ist der Meßfehler δR als Funktion der Bandbreite aufgetragen, mit SNR als Parameter. Weiterhin sind bei vier Zwischenfrequen-
zen ν_{if} die Fehler δR aus je 60 Messungen eingetragen, wobei ν_{if} mit
der Bandbreite B identifiziert wurde und das SNR in etwa 100 betrug.

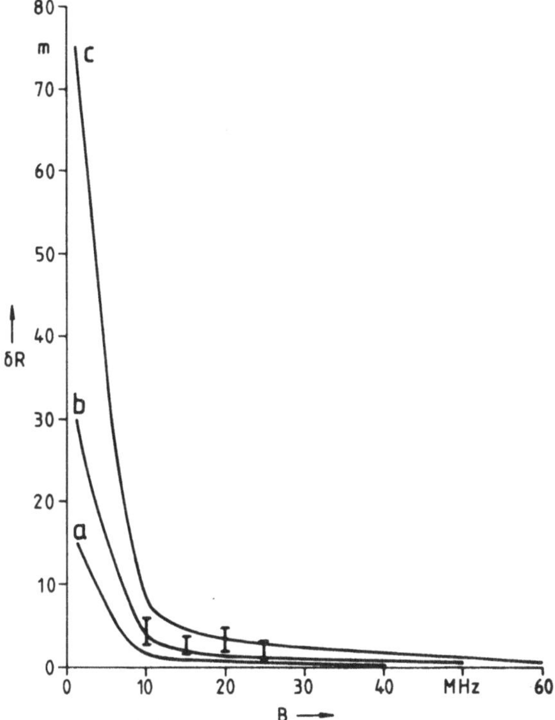

Fig. 5. E-Messfehler als Funktion der Bandbreite
für 3 verschiedene SNR.

a) 100,   b) 25,   c) 4.

I Meßdaten SNR ≈ 100

Experimentelle Untersuchungen an Halbleiterlasern für optische Nachrichtensysteme im Weltraum

A. Popescu, W.R. Leeb, B. Furch, A.L. Scholtz
Institut für Nachrichtentechnik, Technische Universität Wien
Gusshausstrasse 25, A-1040 Wien, Österreich

Aufgrund des hohen Entwicklungsstandes von optischen Sende- und Empfangs-
elementen stellen optische Nachrichtensysteme nunmehr auch im Weltraum
eine Alternative zu Mikrowellensystemen dar. Optische Systeme bieten
neben einer hohen potentiellen Bandbreite kleinere Antennenabmessungen,
Unempfindlichkeit gegenüber elektromagnetischen Störungen und eine
kleinere Abhörwahrscheinlichkeit. Als Nachteile der optischen Nachrichtensysteme gelten zur Zeit der kleinere Spielraum bei der Auswahl
der Modulationsart, eine relativ kleine Ausgangsleistung bei akzepta-
bler Lebensdauer der Sendeelemente, sowie erschwertes Ausrichten und
Nachführen der Antennen zufolge der kleinen Strahldivergenz.

Für den künftigen Einsatz zur Nachrichtenübertragung über kurze bis
mittlere Distanzen (< 1000 km) im Weltraum hat die Europäische Raum-
fahrtbehörde (ESA) GaAlAs-Halbleiterlaser vorgesehen. Unsere experimentellen Untersuchungen hatten zum Ziel, mehrere kommerziell erhältliche
Laser hinsichtlich ihrer Eignung als Sendeelemente für digitale optische Nachrichtensysteme im Weltraum zu prüfen. Die getesteten Lasertypen und die von den Herstellern angegebenen Kenndaten sind in Tabelle
1 zusammengefaßt. Die Wellenlänge, bei der diese Laser emittieren, bewegt sich um $\lambda = 850$ nm. Unter Berücksichtigung der systemspezifischen
Anforderungen an die Laserdioden, wie hohe optische Pulsleistung, gute
Bündelbarkeit, schmales Spektrum und große Lebensdauer, wurden Messungen
durchgeführt und Kriterien für die Auswahl der Halbleiterlaser und der
Ansteuerungsparameter vorgeschlagen.

Aus der klassischen Formel für die Freiraum-Übertragungsdämpfung geht
hervor, daß die Anzahl n_D der Photonen, die bei Aussenden eines Licht-
impulses den Empfänger erreichen, proportional zur optischen Laser-
spitzenleistung P und zur Pulsdauer τ ist. Es gilt

$$n_D = KP\tau\eta_S\eta_D\eta_P \ , \qquad\qquad\qquad [1]$$

wobei $\eta_S\eta_D\eta_P$ Verluste in der Sender- und Empfängeroptik (η_S, η_D) sowie Nachführungsverluste (η_P) charakterisiert, und K einen von den Antennengewinnen und der Entfernung Sender-Empfänger abhängigen Proportionalitätsfaktor bedeutet.

Tabelle 1. Kenndaten der untersuchten Halbleiterlaser (Firmenangaben).

TYP, HERSTELLER	SCHWELLSTROM (mA)	ANSTIEGS- UND ABFALLZEIT (ns)	MAXIMALE DAUERLEISTUNG (mW)	TRANSVERSALER MONOMODUS
HLP3400 HITACHI	20	< 0,5	11	JA
NDL3205S NEC	55	< 0,5	3	JA
CV294P AEG	70	< 1,5	10	NEIN
LS7749 ITT	125	—	7	—
C86030E RCA	125	< 1	20	JA

Die maximal erzielbare optische Spitzenleistung ist von der Pulsdauer, der Pulsfolgefrequenz, von der Pulsamplitude I_L des Laserstroms, und von der Lasertemperatur abhängig. Gemessene Pulskennlinien sind in Fig.1 dargestellt. Dabei wurden die Pulsfolgefrequenz f = 10 MHz, die Pulsdauer τ = 20 ns, und die Lasertemperatur T = 15°C bei allen Laserdioden konstant gehalten. Die Pulskennlinien wurden bis zum Einsetzen von Sättigungserscheinungen der optischen Leistung gemessen. Die im gewählten Arbeitspunkt gemessenen Kurven zeigen, daß die im Pulsbetrieb erreichbaren optischen Spitzenleistungen ein Vielfaches der maximalen Dauerleistung darstellen. Obwohl die Konstruktionstypen der fünf untersuchten Laserdioden verschieden waren, (buried heterostructure (HLP3400), V-Nut (CV294P), double heterostructure (LS7749 und NDL3205S), constricted double heterostructure large optical cavity (C86030E)), haben die Sättigungserscheinungen in jedem Fall bei einem Laserstrom von $I_L \approx$ (7-9)I_{th} eingesetzt (I_{th}...Schwellstrom). Der Quantenwirkungsgrad

η lag zwischen η = 0,35 und η = 0,45 je nach Lasertyp und Arbeitspunkt.
Daraus folgt daß das Produkt Schwellstrom x Quantenwirkungsgrad ein
Maß für die erreichbare optische Spitzenleistung ist.

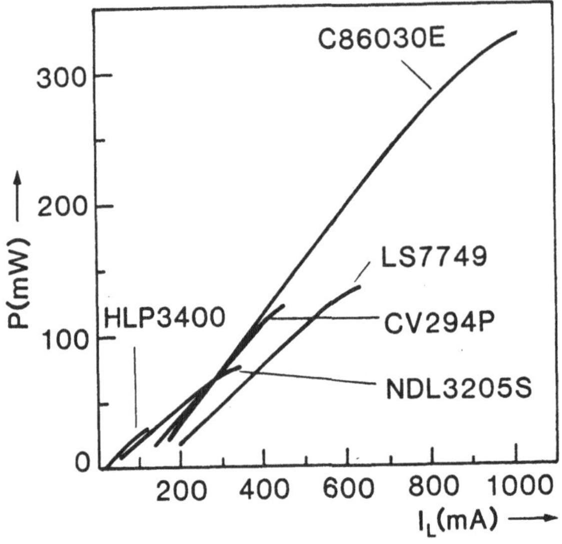

Fig.1. Pulskennlinien der getesteten Lasertypen. Pulsfolgefrequenz 10 MHz, Pulsdauer 20 ns, Temperatur 15°C.

Die Abhängigkeit der Pulskennlinie von der Pulsfolgefrequenz und der
Pulsdauer ergab, wie erwartet, daß bei Verkleinerung des Tastverhältnisses höhere maximale Pulsleistungen erzielt werden können. Verringert
man z.B. das Tastverhältnis d beim Laser C86030E von d = 0,2 auf d = 0,04
so steigt die maximale optische Spitzenleistung um ca. 20%. Das Tastverhältnis wurde sowohl über die Pulsfolgefrequenz als auch über die Pulsdauer verringert. Beim Laser C86030E ergab sich in beiden Fällen dieselbe Wirkung , während beim Laser HLP3400 die Einflüsse der Pulsfolgefrequenz und der Pulsdauer quantitativ verschieden waren. Der Einfluß
des Tastverhältnisses auf die maximale Spitzenleistung stellte sich als
relativ gering heraus. Dies bedeutet, daß Kriterien wie Lebensdauer,
Modulationsformat und Systembandbreite für die Wahl des Tastverhältnisses entscheidender sein werden.

Die Lasertemperatur hatte im Bereich T = 5°C bis T = 40°C einen starken
Einfluß auf den Quantenwirkungsgrad und dadurch auch auf die maximale
optische Spitzenleistung. Je nach Tastverhältnis war die Leistung bei
T = 5°C um 50% bis 100% höher als bei T = 40°C.

Das Verhalten des Nahfeldes mit steigender optischer Spitzenleistung ist sowohl für die Lebensdauer der Laserdioden als auch für die strahlformende Optik wesentlich. In Fig.2 wird die Abhängigkeit des Intensitätsmaximums I am Laserspiegel von der optischen Spitzenleistung dargestellt. Der auf dem "large optical cavity"-Prinzip aufgebaute Laser C86030E weist zum Unterschied von den anderen getesteten Halbleiterlasern eine mit der Lichtleistung proportional zunehmende Nahfeldfläche auf. Dies bewirkt, daß auch bei hohen Spitzenleistungen (P = 300 mW) die maximale Strahlintensität $I \approx 1,5$ MW/cm^2 nicht überschreitet. Hingegen erreichen die anderen untersuchten Laser diese Strahlintensität schon bei Leistungen von $P \approx 100$ mW. Ein großflächiges Nahfeld und dessen Ausweitung mit zunehmender optischer Spitzenleistung ist daher ein wesentliches Auswahlkriterium für die vorliegende Anwendung.

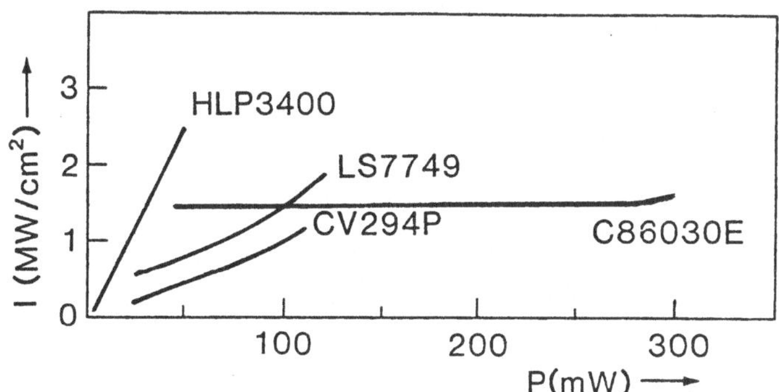

Fig.2. Maximale optische Intensität I am Laserspiegel als Funktion der Pulsspitzenleistung P.

Die Untersuchung des Zusammenhanges zwischen Nah- und Fernfeld ergab, daß das von den Lasern emittierte Licht nicht perfekt kohärent ist und kein reiner Grundmodus abgestrahlt wird. Der Gewinn der Sendeantenne ist dann kleiner als jener, der aus einfachen theoretischen Überlegungen folgt.

Der mögliche Einfluß von störender Hintergrundstrahlung läßt sich durch ein optisches Filter im Empfänger verringern. Die unterste Grenze für die Filterbandbreite ist durch die spektrale Breite der Laseremission gegeben. Bei den untersuchten Laserdioden <u>wurden im Impulsbetrieb spektrale Breiten von 5 nm bis 10 nm gemessen</u>. Selbst jene Laser, die im kontinuierlichen Betrieb in einem einzigen Longitudinalmodus arbeiteten, zeigten eine derartige Verbreiterung des Spektrums.

504

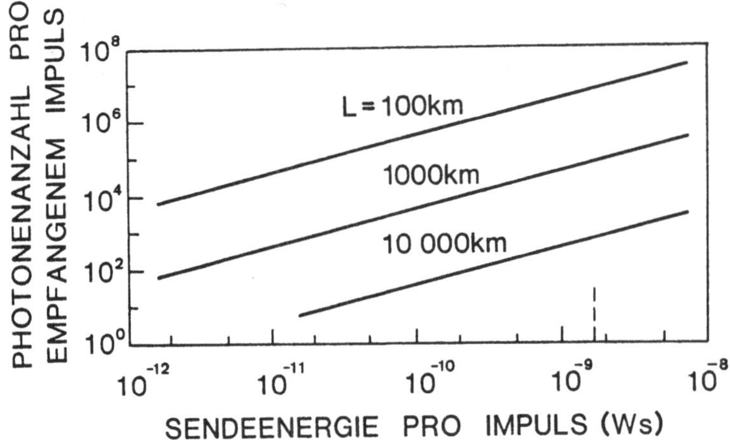

Fig.3. Anzahl der empfangenen Photonen als Funktion der Sendeenergie (jeweils pro Impuls) für verschiedene Sender-Empfänger Distanzen L. Weitere Systemparameter siehe Text.

Von den untersuchten Lasern erscheint die Type C86030E für ein optisches Nachrichtensystem im Weltraum am besten geeignet. Zur Bestimmung typischer Systemparameter wurde von einer Spitzenleistung von $P = 200$ mW einer Pulsdauer von 10 ns, einer Pulsfolgefrequenz von 10 MHz, Antennen durchmessern von jeweils 0,1 m, einer Wellenlänge von 830 nm und einen Transmissionsgrad $\eta_S\eta_D\eta_P = 0,12$ ausgegangen. Mit der Sendeenergie von 2.10^{-9} Ws/Impuls ergibt sich aus Fig.3 für die Zahl der auf den Detekto auffallenden Photonen/Impuls $n_D \approx 600$, wenn die Entfernung L zwischen Sender und Empfänger 10.000 km beträgt. Damit läßt sich eine Bitfehlerrate von weniger als 10^{-9} erreichen.

Diese Arbeit wurde im Auftrag der Europäischen Weltraumbehörde (ESA) durchgeführt.

Spaceborne Lidars for Global Metereological Measurements

M. ENDEMANN, W. ENGLISCH
Battelle-Institut e.V.
Am Römerhof 35, D 6000 Frankfurt am Main 90

Design and performance predicitions of a preliminary lidar sensor concept for future operational meteorological observation satellites have
been examined in a recent ESA study (1). In its final development stage,
this sensor will have the capability to determine the following basic
atmospheric parameters with a height-resolution ranging from 150 m to
1.5 km:
- cloud-top height and aerosol distribution
- humidity profiles
- temperature and pressure profiles

In terms of system development needs, the cloud-top height and aerosol
distribution sensor is least demanding and thus will be the first
development stage. Cloud-top height can be used in combination with
data from passive sensors to determine rough profiles of wind velocity,
temperature, and humidity. The aerosol distribution in troposphere and
stratosphere is an important climatological parameter that can also be
determined with this lidar sensor.

Addition of a second transmitter laser allows to perform differential
absorption lidar (DIAL) measurements. Humidity as well as temperature
and pressure profiles can be determined using absorption lines of H_2O
and O_2 between 760 and 780 nm (2,3).

The ESA study has lead to the conceptual design for a lidar sensor that
can perform these measurements from an observation satellite in an 800 km
orbit. Its main features are shown in Fig. 1.

New tunable solid-state lasers operating between 700 and 800 nm are
ideally suited for the proposed lidar sensor. Of particular interest
for this transmitter laser are new Cr^{3+}-doped crystals like Alexandrite,
Emerald, and GGG. The preliminary performance data of these laser
materials show that a transmitter laser with one joule pulse energy,

506

10 Hz pulse repetition frequency, and a 2 % wall-plug efficiency
(prime-power consumption 500 W) can be engineered.

The cloud-top and aerosol sensor has only one transmitter laser that
is mounted on the side of the receiving telescope. For DIAL measurements a second laser (probe laser) is placed above the first one. It
emits the probe pulse that is tuned onto a selected absorption line of

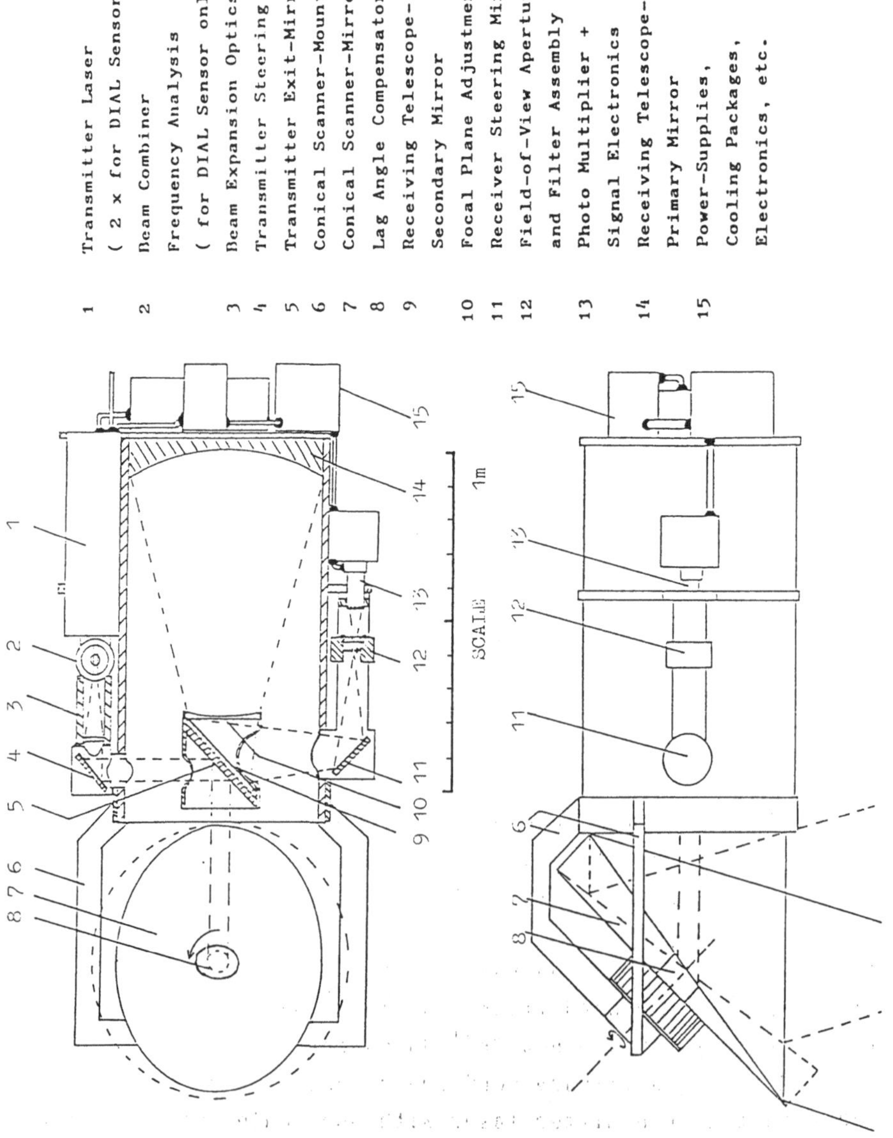

Fig. 1: Layout of the Conceptual Lidar Sensor

H_2O or O_2. The probe laser fires with a short time delay of about 100 µs after the reference laser. Thus the same backscattering volume is probed and errors due to the high temporal and spatial variation of the back-scattering volume are eliminated.

The probe laser must fulfill rather stringent requirements with respect to line width and frequency accuracy. These parameters are observed by an on-line frequency monitor mounted in front of the probe laser. The frequency setting can be observed in an opto-acoustic cell, while line-width parameters can be monitored in an etalon.

The transmitter pulses are expanded in a simple telescope unit to their final beam divergence of 0.07 mrad (half-cone angle). Two steering mirrors direct the beam coaxial to the receiving telescope.

The backscattered light is collected by a Newtonian telescope with 60 cm diameter (F = 2). Two steering mirrors direct the light to the photo multiplier tube that allows essentially shot-noise or background limited detection. The quantum efficiency can be above 10 % at 780 nm when a GaAs-cathode is used.

The field of view is restricted to 0.1 mrad (half cone angle) by an aperture in the focal plane to reduce the background signal. Further reduction is achieved with a filter of 0.5 nm bandwidth. An overall optical transmission of 50 % is assumed in the performance estimates.

A broad measurement track is essential for an operational sensor. Thus the lidar will be scanned across the satellite track using the conical scan technique with lag-angle compensation. A 45° steering mirror in front of the (horizontally mounted) transmitter/receiver assembly per-forms the scan. The mirror is tilted with an angle θ to one side and rotates around the 45°-axis. Thus a conical scan pattern is produced with a scan angle of 2θ.

A track width of 500 km is selected. This track width allows to collect each day data from a large portion of the earth surface. For an 800 km orbit the scan angle then is 17.3^O. The measurement distance is 840 km which is only slightly more than the height of the orbit. The scan mirror rotated with a period of about 6.5 s to produce a rather even spacing of the measurement-foot points of 24 km.

The signal electronics has a bandwidth of about 1 MHz, so that the
vertical resolution is 150 m. The digitized data are stored for each
measurement and used to determine the cloud-top height. Later processing allows a noise reduction by averaging over vertical resolution
elements as well as over adjacent pixel elements. Thus the spatial
resolution can be adapted to the specific measurement needs.

Performance estimates of the lidar sensor have been carried out. For
a lidar measurement over 840 km, the dependence of the S/N on the
varying range over the last 15 km is less than 4 % and thus negligible.
If the atmospheric transmission is assumed to be near unity (as is the
case for the window-frequency in cloud-free conditions), the S/N of
this lidar sensor depends only on the backscattering coefficient $\beta(z)$
and the strength of the background radiation.

Fig. 2 shows the dependence of the S/N on $\beta(z)$. Values for day and night
time measurements are shown. During daytime the S/N is reduced due to
background signal. A background spectral radiance of 100 W/m^2
µm is assumed. A value of $\beta(z)$ = 4.10^{-6} m^{-1} is "normal" for
the lower troposphere and can be expected for most measurements. It is marked by a verti-cal line.

As can be seen from Fig. 2, the S/N of nighttime measure-ments is in all cases better than unity. The addition of a background signal reduces the S/N significantly for backscattering coefficients below 10^{-5} m^{-1}. For "normal" scattering from the low troposphere the S/N of a

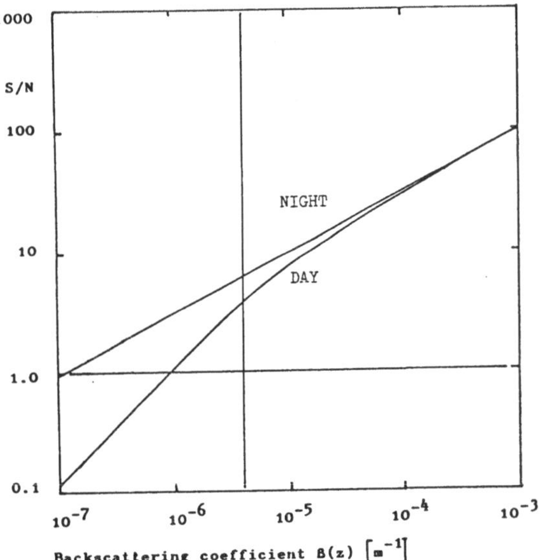

Fig. 2: S/N as function of the
backscattering coefficient
$\beta(z)$ for daytime and
nighttime measurements

single measurement still is about 4 with full background, and increases
to 6 at night.

For cloud-top measurements and strong aerosol plumes $\beta(z)$ can be several orders of magnitude larger. A S/N of nearly 100 is to be expected for $\beta(z) = 10^{-3}$ m^{-1}. Cloud-top and aerosol distribution measurements can be performed with a single pulse. The horizontal resolution is given by the spacing of the pixels along the conical scan pattern of 31 km.

For DIAL measurements the S/N must be improved by averaging over a number of statistically independent samples. For an estimate of the spatial resolution of the DIAL sensor we assume that a 2 % uncertainty level is desired. For a measurement at night time with "normal" back-scattering, data from 70 independent 150 m range intervals must then be averaged. When the required height resolution is 1.5 km (= 10 resolution elements), data from 7 adjacent pixels are used, giving a horizontal resolution near 80 km. Under daylight conditions about 175 independent range intervals must be averaged to obtain the same uncertainety level. This corresponds to a horizontal resolution of about 120 km.

A more detailed performance analysis of cloud-top, aerosol, and the different DIAL measurements is currently being performed. The results of these computer simulation will be used to assess the utility of the proposed lidar sensor on board of operational observation satellites.

Ref.:
(1) ENDEMANN, M.: "Applications of Lasers for Climatology and Atmo-
    spheric Research", Final Report for ESTEC Contract No. 4868/81/
    NL/HP(SC) (1983)
(2) BROWELL, E.V., WILKERSON, T.D., MCILLRATH, T.J.: Appl. Opt. 18,
    3474 (1979)
(3) KORB, C.L., SCHWEMMER, G.K., DOMBROWSKI, M: Chapt. 3.1 in "Optical
    and Laser Remote Sensing", Springer Series in Opt. Sci. Vol. 39,
    Ed.: D.K. Killinger, A. Mooradian, Springer Verlag (1983)

Laser in der Umweltmeßtechnik

Lasers in Environmental Measuring Techniques

Monitoring of the Stratospheric Ozone Layer Using a XeCl Excimer Laser

J. WERNER, K.W. ROTHE, H. WALTHER[+]
Sektion Physik der Universität München
Am Coulombwall 1, D-8046 Garching

The use of lasers for remote sensing of atmospheric trace gases was
already well established in the late sixties /1/. This technique, based
on the Radar principle but operating with light instead of radio waves,
is therefore called Lidar.

First field experiments were performed with pulsed fixed-frequency lasers (Ruby, Nd:Yag). Later on, a more extended range of applications
was opened using the wavelength tunability of dye lasers and the numerous lines of fixed-frequency IR lasers coinciding with vibronic transitions of atmospheric constituents of interest.

However, most electronic transitions lie in the UV. Here high power
laser sources became available with the invention of excimer lasers.
The already well established technique of stimulated Raman scattering
allows the shift of the fixed laser frequencies to shorter and longer
wavelengths with high conversion efficiencies.

One of the most important trace gases in the atmosphere is ozone. It
plays a dominant role in the energy balance of the troposphere and
stratosphere, since it determines the temperature distribution being
of importance for the weather. In addition the ozone acts as a shield
against the solar UV-radiation, which otherwise could affect life on
earth.

Theoretical models /2/ predict the impact of man made chemical species
on the ozone layer. Photolysis of chlorofluoromethanes at altitudes
around 40 km releases free chlorine, which acts as a catalyst for the

---

+) Also Max-Planck Institut für Quantenoptik, Garching

ozone destruction at this height. It is argued that the ozone content
at 40 km will be reduced by 20 % in the next 100 years. At the same
time an increase is expected at lower altitudes, reaching nearly the
same percentage. This is due to catalytic nitrogen oxide cycles and to
a temperature shift induced by increasing carbon dioxide. For the experimental proof of these predictions precise measurements over a long
period are required.

Among various Lidar techniques the differential absorption Lidar (DIAL)
is the most appropriate one for monitoring of the stratospheric ozone
layer up to heights of 50 km.

This method utilizes the strong wavelength dependence of absorption by
gaseous molecules in combination with the backscattering by Rayleigh and
Mie scattering, which show much weaker wavelength dependences. Measuring
simultaneously within and out of a strong absorption line or band allows
the elimination of the unknown backscatter behaviour of the atmosphere.

Our system /3/ uses a commercial 100 mJ XeCl excimer laser operating
at repetition rates of up to 100 Hz. An unstable resonator reduces the
beam divergence to 1 mrad. The laser radiation at 308 nm is strongly
absorbed by ozone.

Radiation in a reference line at 338 or 352 nm, insignificantly absorbed by ozone, is generated by stimulated Raman scattering in a high
pressure gas cell. Methan (338 nm) or hydrogen (352 nm) is used as the
Raman medium. To obtain comparable backscatter intensities from the
upper stratosphere at both wavelengths a conversion rate of about 15 %
is sufficient. This set-up allows simultaneous emission of both wavelengths, thus improving the precision of measurement.

The backscattered signal is received by a 60 cm diameter mirror. The
two lines are then separated by a dichroic filter and focused onto two
photomultipliers used in the photon counting mode. A PDP-11 computer
processes the signals and stores the data on floppy-disks. The whole
apparatus is installed in a container, now being operated from the top
of the Zugspitze (2964 m) in the Alps to avoid the signal reduction
by aerosol absorption in the lower atmospheric layers.

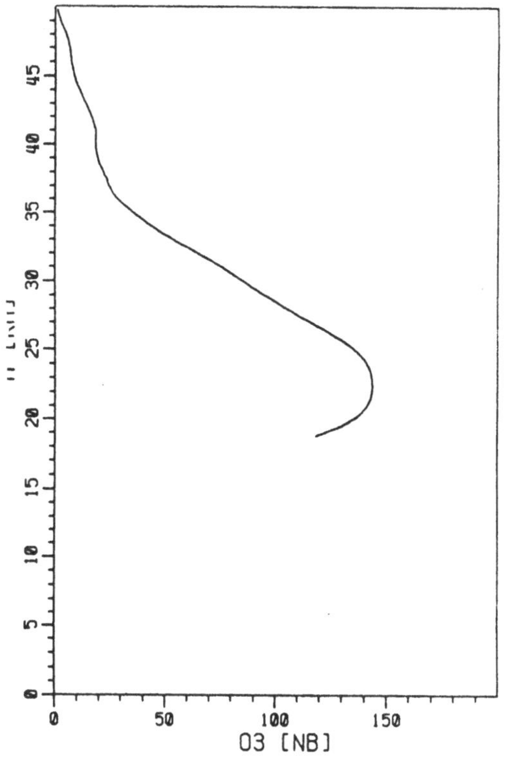

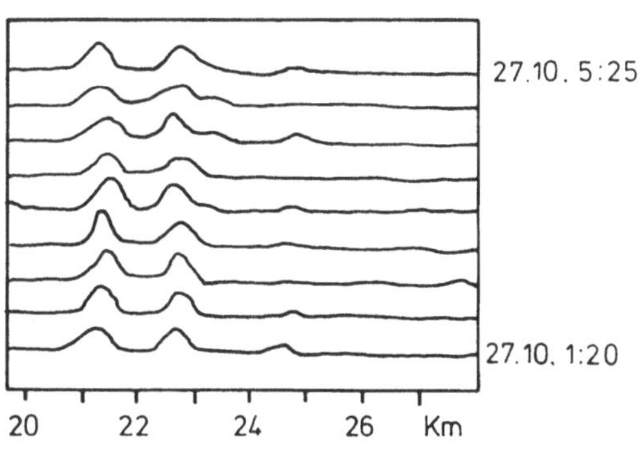

Fig. 1. Monthly average of ozone partial pressure (nanobar) in November 1982.

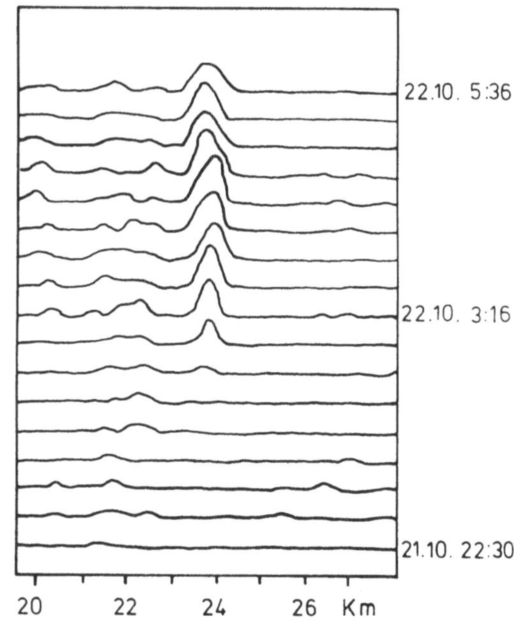

Fig. 2. Rapid change in Mie component of backscattering (arbirary units) within a few days. The strong enhancement is due to the eruption of the volcano El Chichon (Mexico).

Ozone profiles are obtained from 10 to 50 km altitude. An accuracy of a few percent with a resolution of better than one kilometer requires about 15 minutes integration time at 20 km, but the whole night at 50 km. Fig. 1 shows the monthly average for November 1982, indicating the usual high winter time ozone concentration.

The Mie component of the backscattered intensity also enables us to obtain information on the stratospheric aerosol concentration. The temporal behaviour of the aerosol cloud from the El Chichon (Mexico) volcano eruption in April 1982 is of special interest. Fig. 2 demonstrates the high variability of this aerosol layer within a few days.

References

/1/ E.D. HINKLEY (Ed.), Laser Monitoring of the Atmosphere, Springer Verlag, Berlin 1972

/2/ D.J. WUEBBLES et al., J. Geophys. Res. (to be published)

/3/ K.W. ROTHE, H. WALTHER and J. WERNER, Differential-Absorption Measurements With Fixed-Frequency IR and UV Lasers, in: D.K. KILLINGER and A. MOORADIAN (Ed.), Optical and Laser Remote Sensing, Springer Verlag, Berlin 1983

LIDAR Investigation of Oil Films on Natural Waters

G.CECCHI, P.MAZZINGHI
Istituto di Elettronica Quantistica - CNR, Via Panciatichi 56/30,
Firenze (Italy)

L.PANTANI, C.SUSINI
Istituto per la Ricerca sulle Onde Elettromagnetiche - CNR, via Panciatichi 64, Firenze (Italy)

INTRODUCTION

Oils are between the most common pollutants at the surface of natural
water, therefore the identification of oil films is one of the relevant problems in the remote sensing of the sea surface.
The knowledge of the thickness and thickness gradients of an oil film
on the sea surface gives useful informations both for the measurement
of the amount of oil in a spill and for the understanding of how oil
moves on the sea surface or mixes into the water column.
In this paper a laboratory study is presented, showing the possibility
of oil class identification by applying cross-correlation techniques
to the fluorescence spectra. The possibility of oil thickness measurement by fluorescence intensity or water Raman scattering is also investigated.

OIL IDENTIFICATION

Cross-correlation techniques are very popular in signal shape identification and are particularly suitable for automatic identification procedures. Since the fluorescence spectrum depends on the chemical composition of the oil and can be used as a "fingerprint" it seems suitable to investigate the potential of fluorescence-spectra cross correlation as a technique for oil film identification.
The cross correlation between two spectra $S(\lambda)$ and $G(\lambda)$ is given by:

$$C(L) = \int_{-\infty}^{+\infty} S(\lambda) \cdot G(\lambda - \lambda_0)\, d\lambda \quad . \tag{1.1}$$

When $S(\lambda) = G(\lambda)$ equation (1.1) gives the autocorrelation of $S(\lambda)$
which is symmetrical and with the principal maximum at $L = 0$:

$$C_a(0) = \int_{-\infty}^{+\infty} S^2(\lambda)\, d\lambda \quad . \tag{1.2}$$

If the spectra are normalized in order to give $C_a(0) = 1$ the value of $C_L(0)$ is always lower than 1. The position of the maximum and the value for $L = 0$ may allow the identification of an unknown spectrum when it is cross correlated with known reference spectra.

Fluorescence spectra were recorder at different excitation wavelength and, after the normalization, cross and autocorrelations were computed[1] for the spectra obtained at the same excitation wavelength. The auto and cross correlation functions were then collected in terms of position of the principal peak and value for $L = 0$. As it can be seen from the example of Fig.1 the cross correlation between the spectra of two crude oil is very similar to an autocorrelation while the cross correlation between the spectra of a crude oil and a diesel light oil shows a strong shift of the maximum.

TELEDETECTION OF THE THICKNESS OF OIL FILMS

Two optical methods were proposed for the teledetection of the thickness of oil films. The first uses the fluorescence of the oil. The film thickness is then given by[2]:

$$d = - \frac{1}{k + k_f} \, \ln\left[1 - \frac{F(d)}{F(\infty)}\right] \tag{2.1}$$

where k and k_f are the absorption coefficients of oil at the excitation and the measured fluorescence wavelength. $F(d)$ is the measured fluorescence intensity, and $F(\infty)$ is the intensity from a thick film, which completely absorbs the laser light. When such a thick film is not available, in real LIDAR operation, $F(\infty)$ can be evaluated with an excitation wavelength in the ultraviolet, because of the high extinction coefficient of oils in this region of the spectrum (Fig.2). $F(d)$ is then measured with a visible excitation, thus permitting an higher penetration of the light in the oil film. The two measurements should be corrected to the same excitation power level, measuring distance, and detector sensitivity. The disadvantage in this case consi-

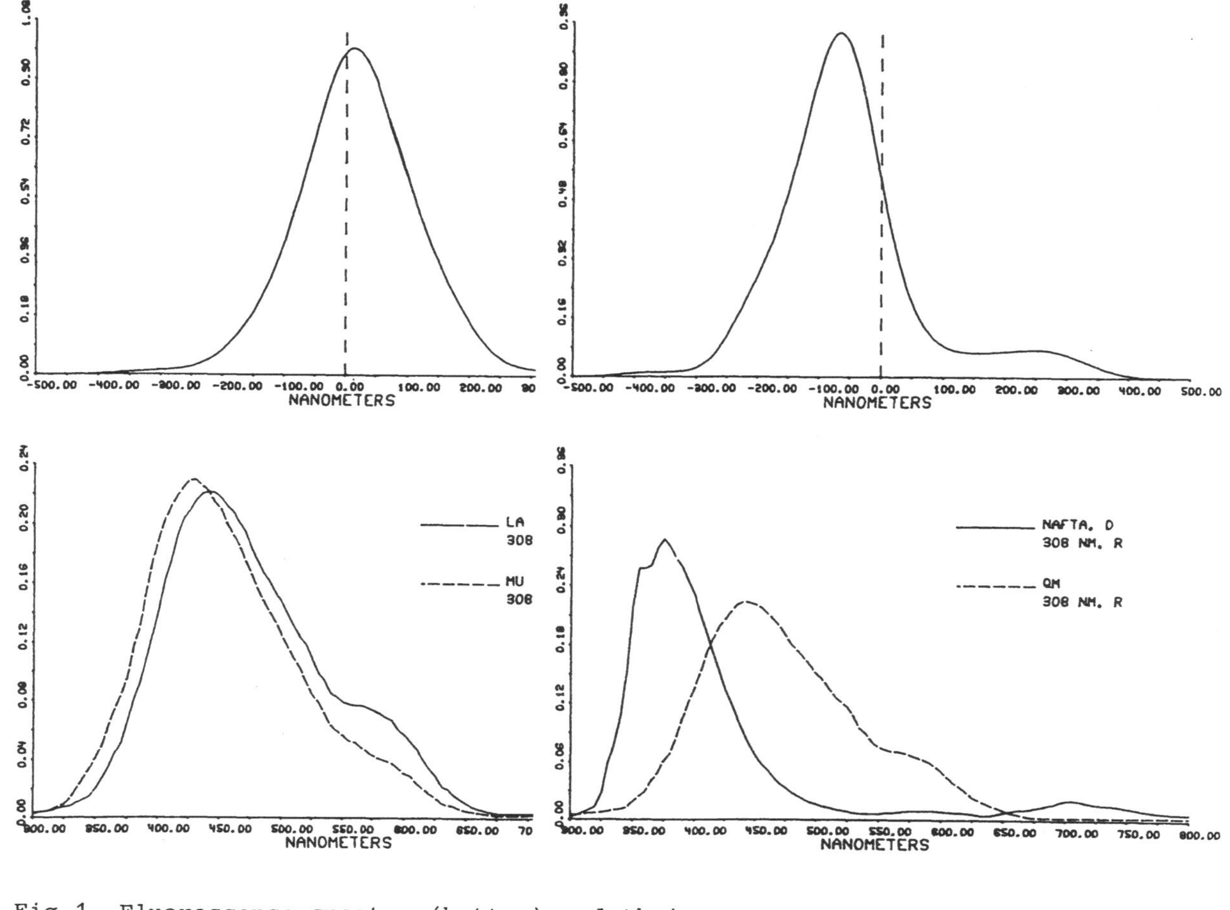

Fig.1. Fluorescence spectra (bottom) and their corss correlation
(top. Left: Two crude oils; right: Crude oil and diesel gasoline

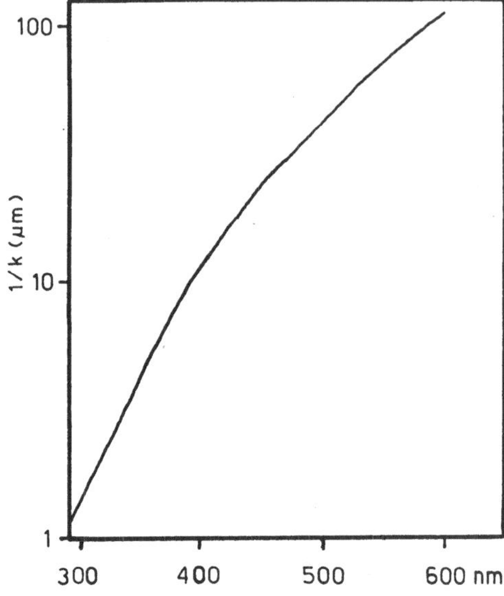

Fig.2.

Absorption path

versus wavelength

of a crude oil sample

sts in the necessity to use at least two laser sources (better results
are obtained with three[2]) and in the difficulty of measuring $F(\infty)$
for thickness below $\backsim 1\,\mu$m, where also an UV excitation is partially
transmitted.

The other method uses the laser induced Raman backscatter of the
3400 cm^{-1} OH stretch band of the sea water beneath the oil slick. In
this case the film thickness is given by[3]

$$d = - \frac{1}{K + K_R} \, \ell n \left[\frac{R(d)}{R(0)} \right]$$

(2.2)

where k and k_R are the absorption coefficients of oil at the excitation and the Raman wavelength. R(d) and R(0) are the Raman signal intensity over and outside the slick, respectively. This method has the
limitation that the film thickness can be measured only if the laser
beam is not completely absorbed by the oil. This limits the maximum
measurable film thickness to $0 \backsim 5$ μm with UV excitation nad to 500
μm with a deep red ($\backsim 700$ nm) one[3].

In order to evaluate the useful measuring range of these methods we
have used the data from the fluorescence oil detection experiment described elsewhere[4] for film thickness determination by means of (2.1).
A new experiment was then performed for the evaluation of the water
Raman method, using essentially the same experimental equipment, ex-

cept for a higher resolution grating in the monochromator (bandwidth
∿ 0.2 nm/channel). Typical spectra are shown in Fig.3, for several
oil film thicknesses, and with a 308 nm excitation. The Raman signal
integrated intensity was then measured after the subtraction of the
interpolated fluorescence intensity (dashed line in Fig.3), and used
in eq.(2.2).

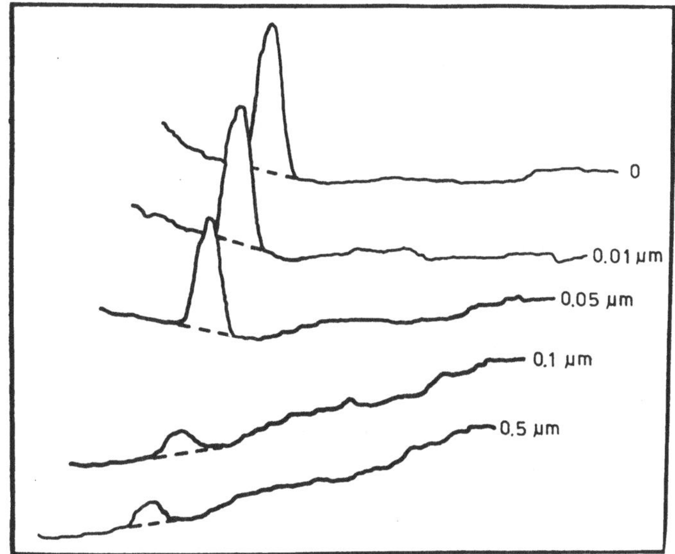

Fig.3.   Water Raman spectra with oil films of different thickness with
         a 308 nm excitation. The peak of Raman band is at 344 nm

Results of calculation with the two methods reported in Tab.1 show
that these measurements can only give the order of magnitude of the
film thickness, infact the accuracy is difficult to estimate.   The
foundamental source of error is the determination, with good precision,
of the absorption spectra which, for highly absorbing liquids such
oils, needs special equipment[5]. It is also to be considered that
there is some uncertainty in the preparation of films of calibrated
thickness on the relatively small area of the water sample used[4].

Tab.1

d_{oil} (μm)	d_{fluor} (μm)	d_{Raman} (μm)
0.01	0.05	0.016
0.05	–	0.15
0.1	0.14	0.97
0.5	–	0.99
1	0.53	0.38
10	3.8	–

CONCLUSIONS

The LIDAR detection and measurements of thin oil films have been demonstrated using a laboratory LIDAR simulator. The fluorescence spectra of oil excited by UV pulsed lasers can be used with cross correlation techniques for the identification of the oil class.
The fluorescence intensity can be then used to measure the film thickness. Thinner films (< 1 μm) can be better measured with a water Raman scattering technique.

REFERENCES

1) V.CAPPELLINI, A.G.COSTANTINIDES, P.L.EMILIANI : "Digital filters
   and their applications". Academic Press (1978)

2) H.VISSER: Appl.Opt. **18**, 1746 (1979)

3) F.E.HOGE, R.N.SWIFT: Appl.Opt. **19**, 3269 (1980)

4) P.BURLAMACCHI, G.CECCHI, P.MAZZINGHI, L.PANTANI: Appl.Opt. **22**, 48,
   (1983)

5) F.E.HOGE, J.S.KINCAID: Appl. Opt. **19**, 1143 (1980)

A Laser Transmissometer with Variable Receiver's Field of View for Measurements of Forward Scattered Power in Dense Media

P.Bruscaglioni, G.Del Fante, A.Ismaelli, L.Lo Porto, G.Zaccanti

Istituto di Fisica Superiore - Università di Firenze, Italy

1 - Introduction

A transmissometer whose receiver has a variable angular field of view has been employed in a campaign of measurements of fog extinction coefficient. The variation of the
received power with the angle α of the receiver FOV showed that an appreciable amount
of forward scattered power is added to the direct transmitted power. An analysis of
the received scattered power in terms of a simple model of the medium scattering function allowed us to follow the variations of the medium scattering properties continuously.

2 - The variable FOV transmissometer

Fig.1 shows the block diagram of the transmissometer with the control and acquisition
data system. The transmitter is a 10 mW HeNe laser. A beam splitter takes out a portion of the emitted power that is used as a reference to take into account the source
fluctuations. The transmitted beam is mechanically choppered at 200 Hz. The optical
section of the receiver is an optical condenser with 105 mm focal length, 100 mm
diameter and a maximum semiaperture angle of 3°. A set of six diaphragms on a rotating circular corona in the focal plane of the optical system, permits to vary its
semiaperture angle α from $.5^{\circ}$ to 3° by steps of $.5^{\circ}$. Four sets of 6 diaphragms are
present on the corona, and are operated for each complete round.
The receiver detector is a silicon PIN photodiode, with a usable surface of 1 cm^{2},
put behind the diaphragm. Two filters (a colour filter and an infrared suppressor)
are used in order to reduce the background radiation which arrives on the photodiode.

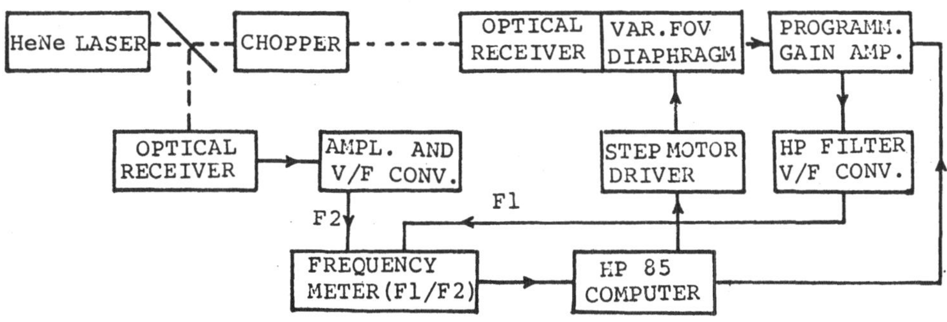

Fig.1 - Block diagram of the transmissometer.

This is to avoid the saturation of the receiver preamplifier.

The current supplied by the photodiode is sent to an amplifier with a programmable gain that allows us operate with an optical depth ranging from 0 to 12. The choppered output voltage of the amplifier is filtered with a high pass filter in order to suppress the remaining background contribution and is then rectified. The resulting signal is sent to a voltage to frequency converter and successively to the frequency meter. The voltage to frequency conversion permits to transmit the signal through long line with a best noise immunity.

Also the reference signal, after a voltage to frequency conversion is sent to the frequency meter. This is set up to measure the ratio between the two frequencies. Such a ratio is directly related to the beam attenuation due to the media. In such a way the power fluctuations of the source are taken automatically into account.

The whole apparatus is controlled by a personal computer mod.HP 85. The computer accomplishes the following operations: 1) setting the diaphragms by a step motor which moves the corona containing the diaphragms; 2) controlling the amplifier gain; 3) acquiring the frequency meter data and recording on a cassette.

The apparatus usually operates by the continuous scanning of the diaphragms. The measurement rate was of one measurement per second, i.e. 30 seconds for each complete round of the corona.

3 - Measurements of the extinction coefficient

In the presentation of the measurements results the following symbols will be used: P_n, will represent the received power when the receiver's angular field of view is equal to α_n . n ranges from 1 to 6, while α_n ranges from $.5^o$ to 3°, by steps of $.5^o$. P_e will represent the emitted beam power and P_o the power that would be received if no forward scattered power would reach the receiver.

In february 1983 the apparatus was used in a campaign of measurements at the CNR basis of San Pietro Capofiume, in the Po valley. Fig.2 shows an example of measured attenuation over a transmission link of 50 m. The figure shows the results obtained by using the largest optical receiver aperture: $\alpha = 3^o$. In the first half of the mea surement interval, lasting about 3 hours, the fog optical depth stayed almost con-

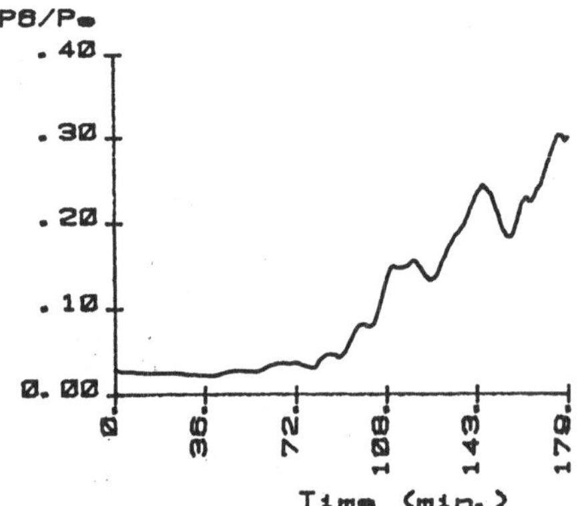

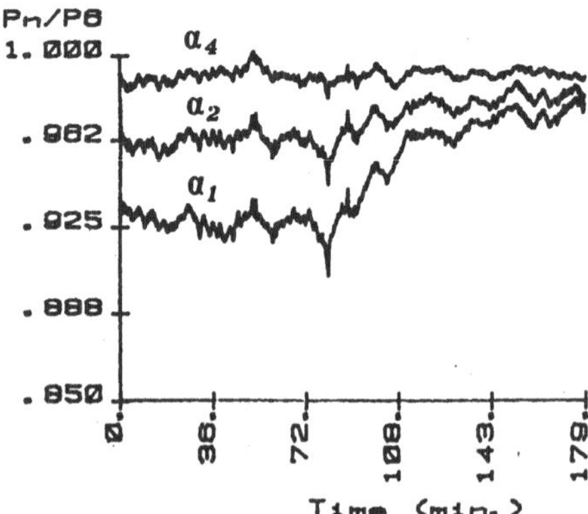

Fig.2. An example of measured attenuation $\alpha = \alpha_6 = 3^o$.

Fig.3. Ratio P_n/P_6 for the same time interval as Fig.2. n = 1,2,4. Results averaged over 60 sec.

stant, and was slowly decreasing in the second half. For the same time interval
Fig.3 shows the ratios P_n/P_6, with $n = 1,2,4$ ($\alpha_1 = .5^0$, $\alpha_2 = 1^0$, $\alpha_4 = 2^0$, $\alpha_6 = 3^0$).
One sees that the relative differences between the received powers is evident in the
case of the dense fog, and tends to disappear as the fog becomes lighter.
A confirmation that the differences of received power for different angles are not
due to errors introduced by the apparatus, comes from Fig.4, which shows the observed ratios $(P_n - P_1)/P_1$ versus α_n in a laboratory test with no appreciable link attenuation. As seen in the figure, where curves belonging to 8 successive sets of 6
angles are presented, errors are limited to less than 0.5%. What is shown by Fig.3
therefore was ascribed to a forward scattering effect.

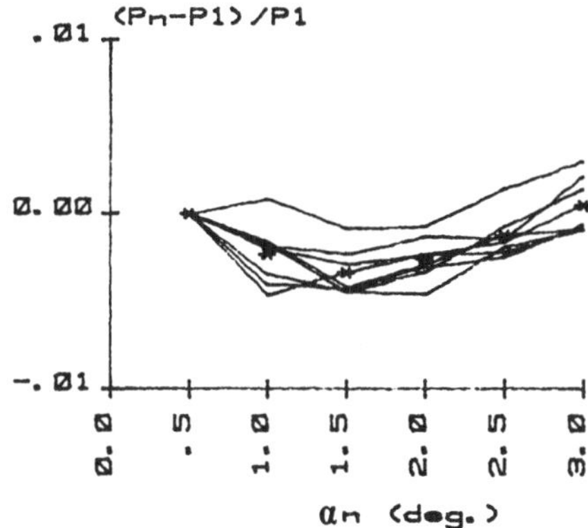

Fig.4. Measured ratio $P_n/P_1 - 1$ with no
practical attenuation for 8 successive
sets of FOV angles.

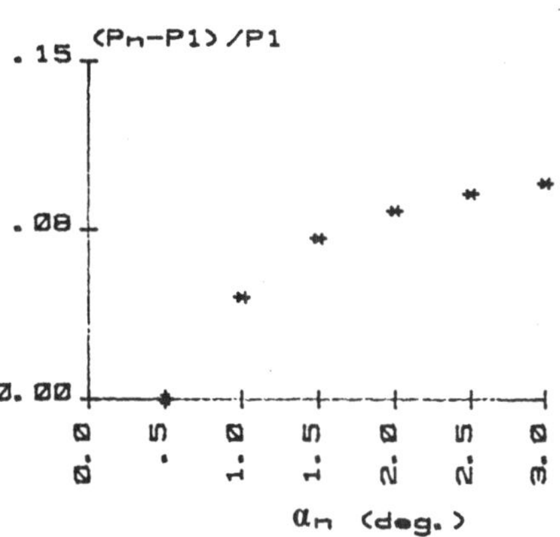

Fig.5. Measured ratio $P_n/P_1 - 1$. Results
averaged over 60 sec.

Fig.5 shows a case of measured ratio $(P_n-P_1)/P_1$ versus α_n, in condition of heavy fog.
In order to smooth the effects of fast fog fluctuations the presented values of P_n
are the result of an average over an interval of time, as indicated in the figure
caption. The apparent downward curvature of the curve of Fig.5 is related to the
scattering properties of the medium (see sect.4). Figs 3 and 5 show clearly that
the amount of the received forward scattered power encreases as α encreases.
Let us indicate as P_o the amount of power that would be received if forward scattering was absent. An estimation of P_o could be obtained by an extrapolation to $\alpha = 0$,
of the experimental results. Since only six values of α are available, we used a
least square fitting procedure of the experimental points with a simple curve:
$\alpha = A + B P^2$, with A and B constants. Fig.6 shows some examples of the estimated ratio P_o/P_e.
As a conclusion of this section one can deduce that the apparatus presented in Sect.1,
can be continuously employed in measurements of the attenuation coefficient of fogs.
The amount of forward scattered power affecting the attenuation measurements is easily evidenced, so that an estimation of P_o, directly related to σ is possible. Small
and relatively fast changes in the fog depth can also be detected, as shown by Fig.3,
where one can notice that apparent oscillations with period of the order of the minute occur together for the three curves pertaining to different values of α_n, so
that one can ascribe them to changes of fog conditions.

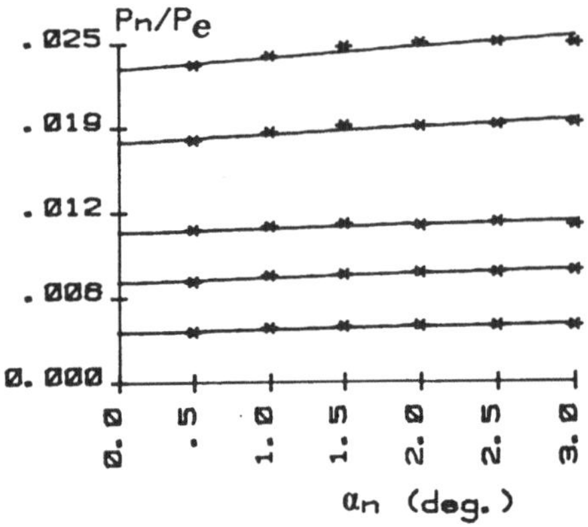

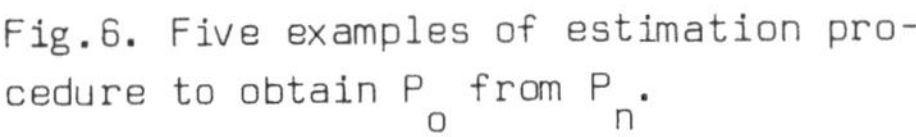

Fig.6. Five examples of estimation procedure to obtain P_o from P_n.

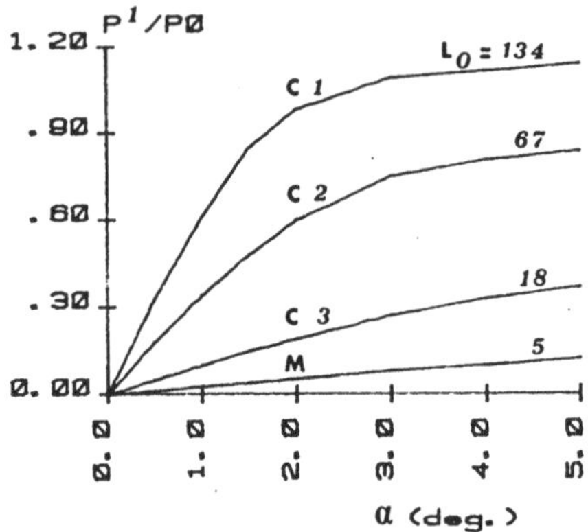

Fig.7. Ratio P^1/P_o calculated by using four Deirmendjian models.

4 - Analysis of the scattered power

Our purpose has been to test the capability of the apparatus to detect continuously the changes in the scattering properties of the medium, which possibly occurs together with changes of the extinction coefficient. Given some model of the scattering function $L(\theta)$, one important parameter was chosen to be representative for the medium situation: this was the value of $L(\theta)$ for $\theta = 0$, indicated in the sequel as L_o. In order to have a guide for the analysis some calculations were performed, for our measurement geometry, of the contribution of forward single scattering on the received power. The calculations were made in the hypothesis of a homogeneous medium. The details of the calculations can be found in (1). If $P^1(D,R,\sigma,\alpha)$ represents the singly scattered received power, where D is the distance transmitter-receiver, R is the receiver aperture radius, Fig.7 gives the results of the calculated ratio P^1/P_o, when several Deirmendjian's scattering functions have been used (2). In our case, and for th figure, D = 50 m, R = 5 cm. As P^1/P_o is proportional to σ, only the results obtained for $\sigma = 1$ m^{-1} are presented. Fig.8 presents the analogous results, obtained by using th Henyey-Greenstein scattering function.

Two features are apparent from Figs 7-8. The ratio P^1/P_o encreases as L_o encreases. Also as L_o encreases the dependence of P^1/P_o on α changes from a rather rectilinear shape t a more and more curved one. This second feature has been used in our first analysis, in order to search for a fitting between our results and the ones prevised by a model describing the medium.

From Figs 7-8 the following ratio R_L has been calculated:

$$R_L(n) = (P^1(\alpha_n) - P^1(\alpha_1)) / (P^1(\alpha_6) - P^1(\alpha_1))$$

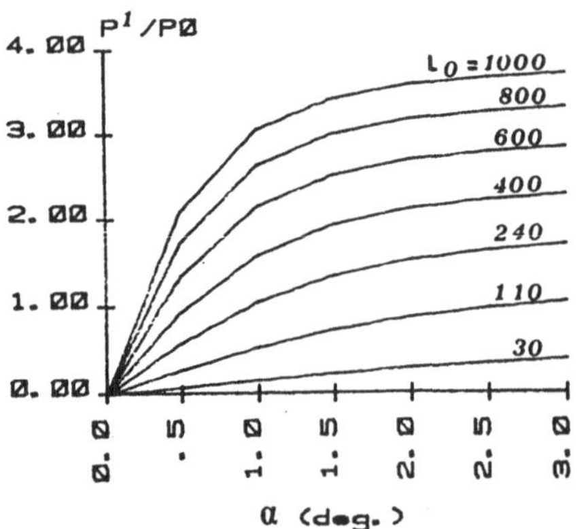

Fig.8. Ratio P^1/P_o calculated by using the Henyey-Greenstein function with several values of the asymmetry parameter, g.
$$L(\theta) = (1/4\pi)(1-g^2)/(1+g^2-2g\cos\theta)^{1.5}.$$

If one assume that in our measurement conditions the received forward scattered power is due to single scattering, one has, according to our symbols: $P^1(\alpha_n) = P_n - P_o$. The calculated ratio R_L was compaired with the experimentally determined ratio R_n:

$$R_n(n) = (P_n - P_1)/(P_6 - P_1)$$

Figs 9 a,b show the parameter L_o obtained, by a least square difference procedure, from a comparison of $R_n(n)$ with $R_L(n)$ calculated by using the data of Fig.8. The same time intervals have been considered for Fig.9 and Fig.2. One can notice that L_o tends to decrease as the fog optical depth decreases. This is to be related to a general decrease of the drops dimensions.

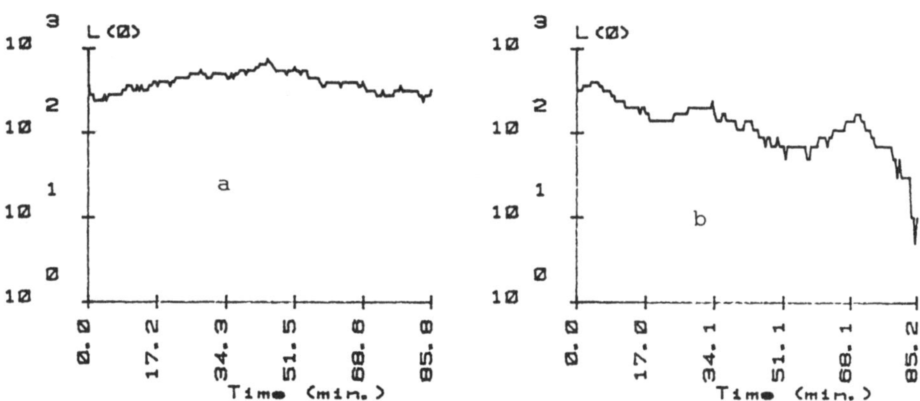

Fig.9. a: Estimated values of L_o - first half of time interval of Fig.2.
        b: Same as a - second half of time interval of Fig.2.

Bibliography

1. P.Bruscaglioni et al. A transmissometer for measurements of atmospheric exstinction coefficient. Rept. IFS, Università di Firenze, June 1982 Progetto Finalizzato ATC of CNR. Contract 81.00198-81.
2. D.Deirmendjian Electromagnetic scattering on spherical polydispersions. Elsevier N.Y. 1969.

Erprobung eines augensicheren Laser-Schrägsichtmeßgerätes auf dem Flughafen München-Riem

Ch.Werner
Institut für Optoelektronik,DFVLR, D-8031 Wessling

F.Bachstein
FB Elektronik,Wesendonkstr.22, D-8000 München 81

E.Gelbke
Impulsphysik GmbH, Sülldorfer Landstr.400, D-2000 Hamburg 56

Über die Notwendigkeit der Messung der Schrägsicht an Flughäfen wurde mehrfach berichtet [1,2,3]. In Europa kommt es jedes Jahr,speziell im Herbst,späten Winter und frühem Frühjahr,vor,daß Flugzeuge wegen zu geringer Sichtweite am Zielflughafen nicht landen können. Kosten und Probleme entstehen daraus für Fluggesellschaften und Passagiere.

Die Erlaubnis zur Landung bei verringerter Landebahnsicht hängt neben der Eignung des Fluggeräts und Qualifikation der Besatzung davon ab,daß die mit Transmissometern an der Landebahn bestimmte Sichtweite mindestens behördlich genehmigten Grenzwerten entspricht. Die eigentliche Entscheidung für die Landung fällt der Flugzeugführer während des letzten Teils des Anflugs in Abhängigkeit von der tatsächlichen Sicht auf die Landebahn, im Zweifelsfall muß ein Durchstartverfahren eingeleitet werden. Die ermittelte Sichtweite gilt streng genommen nur für Aufstellungsort und Meßhöhe des benutzten Transmissometers. Die Bestimmung der Höhenabhängigkeit der Nebelschicht bis zur Entscheidungshöhe bei Betriebsstufe II (30 m) ist das Ziel. Eine Art der Annäherung besteht darin,ein Nebelmodell zur Berechnung heranzuziehen. Eine andere, hier engewendete Art besteht in der Messung der Nebelschichtung ,also des aktuellen Nebelprofils,bis in eine Höhe von 30 m mit Hilfe eines Laser-Wolkenhöhenmessers.

Das Augensicherheitsproblem wurde durch die Benutzung eines GaAs-Laser-Dioden-Arrays niedriger Leistung gelöst. Die erwünschte Kosteneinschränkung wurde durch Verwendung eines handelsüblichen Laser-Wolkenhöhenmessers (mit dem augensicheren Laser) in Kombination mit einem Lidar Empfänger erreicht. Weiterhin wurde die Lidar-Signalverarbeitung auf das Wesentliche begrenzt.

Das Meßgerät sollte mit Blickrichtung auf ein anfliegendes Flugzeug etwa 175 m vor der Landebahnschwelle 25 in Verlängerung der Landelichter aufgestellt werden. Das Problem der möglichen Rückwirkung auf den Landekurssender 07 wurde anläßlich einer Flugvermessung als gegenstandslos erkannt.

Führt man analog zur Bestimmung der Normsichtweite V_N eine Schrägsichtweite $V_S(\alpha)$ ein, so wird für einen Erhebungswinkel α:

$$V_S(\alpha) = \frac{3{,}9 \cdot R_\alpha}{\ln(1/\tau_r(\alpha))}$$

Man kann die Transmission auf dem schrägen Sichtpfad $\tau_r(\alpha)$ durch Aufstellung eines Transmissometers im Anflugbereich nicht messen. Deshalb bietet sich der Einsatz von Fernmeßverfahren und hierbei das Lidar-Verfahren an.

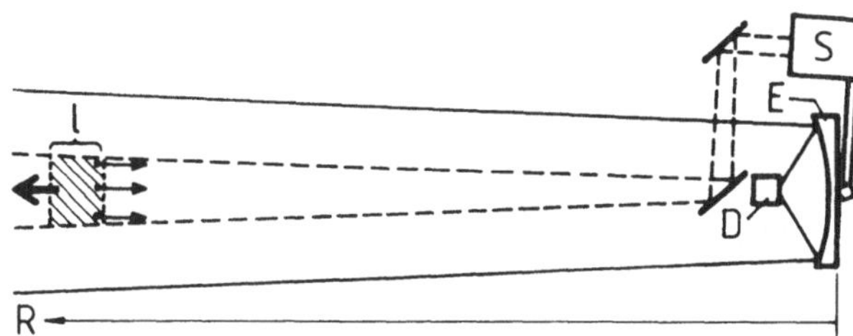

Abbildung 1. Schema des Lidar Schrägsichtmessgeräts

Ein gepulster Laser wird als Sender verwandt (S). Der Laser sendet einen Lichtimpuls der Dauer Δt und der Leistung F_o bei der Wellenlänge λ in die Atmosphäre.Das Licht wird auf dem Weg durch die Atmosphäre extingiert. Ein Teil des Lichts wird von den Luftmolekülen und den Aerosol- bzw. Nebelteilchen in Rückwärtsrichtung gestreut. Das zurückgestreute Licht kann durch einen Empfänger (E) mit Detektor (D) empfangen werden. Die Voraussetzung der Übertragung der mit dem Lidar gewonnenen Extinktion auf die Sichtweite ist die Wellenlängenunabhängigkeit. Nebel ist weiß, so daß man die mit einem Laser ($\lambda < 3\,\mu$m) gemessene Extinktion auf die für das Auge gültige Wellenlänge bei der Bestimmung der Normsicht (grünes Licht, $\lambda = 0.55\,\mu$m) übertragen kann. Augensicher sind die verwendeten Laser dann,wenn ihre Ausgangsenergiedichte einen gewissen Schwellwert nicht überschreitet.
Die Lidar Signatur, das entfernungskorrigierte Signal S(R) ist definiert als

$$S(R_i) = \frac{U(R_i) \cdot R_i^2}{K} = \beta(R_i) \cdot \tau_i^2$$

mit $U(R_i)$ als Signal am Detektor D aus einer Entfernung R_i ,
K als Systemkonstante beinhaltend die Laserleistung F_o und die Detektorcharakteristika
$\beta(R_i)$ ist der wellenlängenabhängige Rückstreukoeffizient
τ_i ist die Transmission bis zur Entfernung R_i mit

$$\tau_i = \exp\left(-\int_0^{R_i} \sigma(R_i)\,dr\right)$$

wobei $\sigma(R_i)$ der Extinktionskoeffizient ist.
Zur Bestimmung der für Sichtweiten notwendigen Extinktion muß der Verlauf der Sig-

530

natur S bestimmt werden. Bei Annahme einer homogenen Nebelschichtung mit der Entfernung R ($\beta(R_i)$ = const.) ist

$$\frac{1}{S} \frac{dS}{dR} = -2 \; \sigma$$

und daraus

$$V_S(\alpha) = \frac{-7,8 \cdot \Delta R}{\Delta \ln S(R)} \; .$$

Aus dem Verlauf des Signals läßt sich für bestimmte Entfernungsbereiche ΔR die Sichtweite $V_S(\alpha)$ berechnen. Dies ist die Slope-Methode [1] .

In einer vereinfachten Methode analog zur Slope-Methode kann bei zwei Entfernungen R_B und R_R die zugehörige Spannung herausgegriffen werden und man erhält folgende Beziehung:

$$V_S(\alpha) = \frac{7,82 \cdot (R_R - R_B)}{\ln\left(\dfrac{U(R_B) \cdot R_B^2}{U(R_R) \cdot R_R^2} \right)} \; .$$

Dieses 2-Punkt-Verfahren [3] kommt mit wesentlich geringerem elektronischen Aufwand aus . Es enthält nur noch 4 Meßgrößen, eine Entfernung R_R bei einer bekannten Spannung $U(R_R)$ und eine Spannung $U(R_B)$ bei einer festen Entfernung R_B nahe am Gerät.

Bei einer inhomogenen Nebellage ($\beta(R_B) \neq \beta(R_R)$) muß mit dem 2-Punkt-Verfahren unter mehreren Erhebungswinkeln gemessen werden um anschließend per Computer die aktuelle Nebelschichtung herausrechnen zu können.

Abbildung 2 zeigt eine schematische Darstellung der Landebahn,eine sich im Anflug befindliche Boeing 737 mit eingezeichneten Winkeln und die Darstellung der SVR-Meßrichtungen

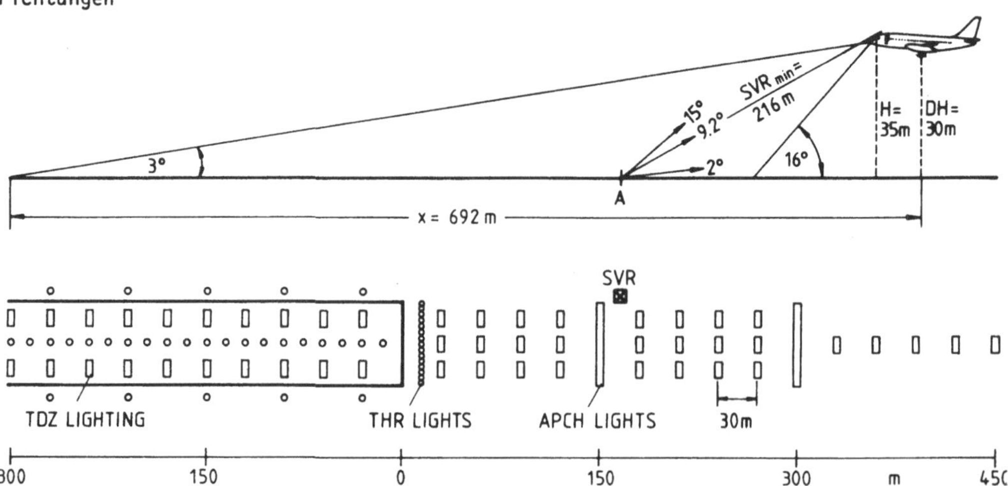

Abbildung 2. Schematische Darstellung der Landebahn (unten) und der Seitenansicht des Landeanflugs einer Boeing 737 mit Winkel- und Höhenangaben

Der Aufstellungsort A für das SVR-Meßgerät wurde als Mittelwert für die verschiedenen Flugzeugtypen gewählt. Die schematische Darstellung des SVR-Lidars ist in Abbildung 1 zu sehen. Es besteht aus einem Impulsphysik-Laser-Ceilograph und einem Lidar Empfänger. Tabelle 1 zeigt die System-Parameter.

Tabelle 1: System Parameter des SVR-Lidar-Prototyps

Laser Sender: GaAs-Laser-Dioden-Array im Impulsphysik-Laser-Ceilograph	
Wellenlänge	0.906 μm
Puls-Energie	1.6 μWs
Puls-Dauer	80 ns
Puls-Wiederholfrequenz	2.4 kHz
mittl.Ausgangsleistung	20 W
Empfänger	
Spiegel-Durchmesser	40 cm
PIN Photodiode	

Abbildung 3 zeigt das System am Aufstellungsort.

Abbildung 3. SVR-Lidar Prototyp,Seitenansicht des Gerätes aufgebaut in München-Riem

Der Wolkenhöhenmesser ist justierbar zum Lidar-Empfänger aufgebaut. Das System ist im Schwerpunkt gelagert und kann über einen Motor mit Regelkreis in der Elevation gesteuert werden. Der elektronische Teil wird vom Rechner CBM 3032 kontrolliert,die Programmiersprachen sind BASIC und ASSEMBLER. Der Analog-Digital-Wandler arbeitet mit einem Speicherintervall von 50 ns. Das empfangene Lidar Signal ist stark verrauscht,

bei 2.4 kHz Pulsfolgefrequenz kann eine Mittelbildung über 500 Einzelsignale innerhalb von 0.2 sec. erfolgen. Der verwendete Wolkenhöhenmesser ist im Sendeteil so
modifiziert, daß einmal der Laser ständig pulst (ohne Sparschaltung) und daß andererseits der Trigger sauber vor dem eigentlichen Laserpuls geliefert wird,um die Kabel-
verzögerungen (250 m) elektronisch ausschalten zu können.

Die Auswahl von 2 Meßwerten (U_B und R_R) findet in einem Zeitabstand von 416 us statt,
zusätzlich hat die Aufsummierung stattzufinden. Diese Art der Operation wird vom
ASSEMBLER Programm bewerkstelligt.

Das Programm läuft völlig automatisch ab. Nach dem Start tastet das SVR-Lidar 3 vor-
gegebene Elevationswinkel 2^0, 9^0 und 15^0 ab und hält bei jeder Stellung an,um die
Lidar Signale zu verarbeiten. Dabei wird ein spezielles Programm gestartet, um nach
dem Anfangswert $U(R_B)$ bei R_B die Rauschdistanz R_R zu bestimmen. Diese Bestimmung
wird von einem Wert R (z.B.100 m) aus gestartet, in Schritten von 7.5 m (entsprechend
der möglichen Auflösung von 50 ns), bis der Signalpegel unter einem vorher eingegebe-
nen Rauschpegel liegt. Dieser Wert wird im Buffer für die spätere Berechnung gespeichert. Nach dieser Routine wird der nächste Elevationswinkel eingestellt. Bei 9^0 und
15^0 wird die Rauschentfernungs-Suchroutine abgekürzt,da stets von dem beim vorhergehenden Winkel bestimmten Wert R_R ausgegangen wird. Die Richtung der 7.5-m-Schritte
liegt dann daran,ob der Signalwert bereits unter dem Rauschwert liegt oder nicht.
Nach Beendigung des Meßzyklus bei der Elevation 15^0 wird die Schrägsichtweite berechnet und über ein Interface dem Wetterdienstrechner ASDUV als Spannungswert
(entsprechend einer zusätzlichen Transmissometerangabe) zugeleitet.

Der Wetterdienstrechner ermittelt die Schrägsicht SVR mit den zusätzlichen Angaben
wie Lampenhelligkeit und Hintergrundlicht. Das SVR-Lidar liefert alle 30 s neue Daten
die Geschwindigkeit kann erhöht werden.

Für die für die Messung zur Verfügung stehende Zeitspanne vom 27.Januar 1983 bis zum
14.April 1983 erwarteten wir einige Nebeltage. Leider kamen in der Meßzeit nur 2 Nebeltage und einige Hochnebellagen vor. Bei Hochnebel konnte die Erwartung bestätigt
werden,daß die SVR kleiner ist als die RVR.

Die ersten und einzigen Tage,an denen das SVR-Lidar für die Bestimmung der Schrägsicht im Nebel eingesetzt und getestet werden konnte,waren der 3. und 4. März 1983.
Aus den Messungen unter verschiedem Winkel war zu erkennen,daß die Sichtweite mit
der Höhe besser wurde. Trotz starkem Nebels am Boden mit Sichtweiten um 200 m
konnte man mit dem Auge in Zenitrichtung den blauen Himmel sehen. Abbildung 4 zeigt
die gemessenen Werte SVR und RVR (vom Transmissometer A) vom 4.März 1983.

Die gestrichelte Kurve gibt den Verlauf der RVR an,die Zahlen sind verbunden mit
Meldungen der Lufthansa-Piloten,die an dem Test teilnahmen. Ihre Aussage,ob die
RVR oder die SVR die angetroffene Nebelsituation besser trafen,wurde bei der Landung
mit einem Fragebogen erfaßt. Die Angaben sind mit in der Abbildung enthalten, +
steht für SVR und o für RVR.

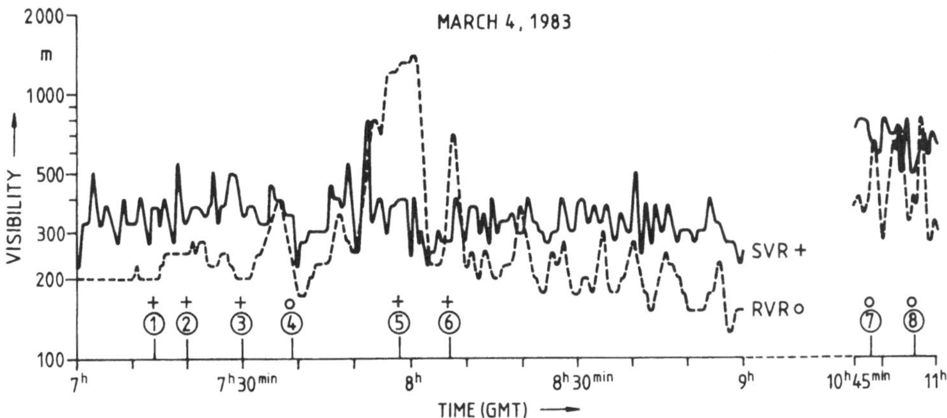

Abbildung 4. Gespeicherte Werte der Sichtweiten RVR(A) und SVR von der Kassette des Deutschen Wetterdienstes im zeitlichen Verlauf für den 4.3.83. Zusatzinformationen sind Pilotenmeldungen bei der Landung

Von der Deutschen Lufthansa wurde für die Piloten,die im Rahmen der flugbetrieblichen Vereinbarung an dem SVR-Test teilnahmen,ein Fragebogen entwickelt. Die Piloten waren nach der Landung aufgefordert,eine persönliche Einschätzung zu geben,ob RVR-oder SVR-Angaben die Situation zur Landezeit besser trafen. Außerdem sollten sie auf einer Skizze den Lampenbereich ankreuzen,den sie aus einer Höhe von 30 m erkennen konnten. Die Auswertung dieser Fragebögen sollte einen Hinweis auf die Brauchbarkeit von SVR-Werten liefern. Von fast allen während der Beobachtungszeit angeflogenen LH-Flugzeugen liegen ausgefüllte Fragebogen vor.
Durch die geringe Zahl von Nebeltagen und die damit begrenzte Einsatzzeit des SVR-Lidar stehen jedoch nur 14 Pilotenangaben zur Verfügung. Aus den 7 Fällen für SVR und 7 Fällen für RVR kann allein noch keine gesicherte Aussage hinsichtlich des Einsatzes des Schrägsichtlidars getroffen werden.

Literatur.
[1] Viezee,W.;Oblanas,G. and Collis R.T.H.: AFCRL-RT-73-0708,Final Report (1973)
[2] Ruppersberg,G.H. und Schellhase,R.: Optica Acta 26 (1979) 699-709
[3] Werner,Ch.:Optics and Laser Technology (1981) 27-36

Analysis of Two Different Techniques for Lidar Visibility Measurements

L.Pantani, I.Pippi and B.Radicati

Istituto di Ricerca sulle Onde Elettromagnetiche, C.N.R., Via Panciatichi 64

I 50127 Firenze (Italy)

Introduction

The measurement of the atmosphere extinction coefficient is the primary element for the calculation of atmospheric visibility. Different techniques were suggested in the past for its extraction from lidar returns ([1, 2, 3, 4]). It is often useful to have a quick-look technique which allows a first exam of lidar returns for test and classification purposes before the processing.

Two techniques are here discussed which are based on the assumption of a constant value of the extinction coefficient along the propagation path, namely on the same assumption which is the base of the popular "slope method" for lidar visibility-measurements. The first technique is based on the connection between the extinction coefficient and the position of the lidar-return maximum while the second one is based on the fitting of the far-field return with a theoretical exponential return.

The return maximum and the extinction coefficient

The lidar return $P(R)$ can be expressed by the well-known "lidar equation":

$$P(R) = (K/R^2)\,\beta_\pi(R)\,\eta(R)\,\exp\left[-2\int_0^R \sigma(r)dr\right] \tag{1}$$

where R is the distance from the lidar along the propagation path, $\beta_\pi(R)$ and $\sigma(R)$ the atmospheric backscattering and extinction coefficients, $\eta(R)$ the "geometrical form factor" ([5, 6]) and K a constant depending from the lidar structure. Under the assumption of a constant value of $\beta_\pi(R)$ and $\sigma(R)$ along the propagation path the range-corrected return becomes:

$$RP(R) = K\,\beta_\pi\,\eta(R)\,\exp(-2\,\sigma R) \tag{2}$$

The first derivative of equation (2) is zero in correspondence to the position R_m of the RP(R) maximum:

$$K \, \beta_\pi \exp(-2\sigma R_m) \left\{ \left[\frac{\partial \eta(R)}{\partial (R)} \right]_{R=R_m} - 2\sigma \eta(R_m) \right\} = 0 \qquad (3)$$

and the extinction coefficient :

$$\sigma = \frac{1}{2\,\eta(R_m)} \left[\frac{\partial \eta(R)}{\partial R} \right]_{R=R_m} \qquad (4)$$

can be calculated if $\eta(R)$ and its first derivative are known analytically ([5, 6]) or experimentally ([7]).

In order to check the accuracy of this technique a Fortran program was written for the calculation of equation (1) with preselected values of the lidar parameters and atmospheric models. The program was operated with different behaviour of $\sigma(R)$, σ was then computed through equation (4) and compared with the original value. An example of the errors done with this procedure when σ is not constant is shown on fig.1 for a σ model:

$$\sigma(R) = \sigma_0 + GR \qquad (5)$$

with positive and negative values of G. From these simulation the technique seems more suitable for standard visibilities below 2 km than for high visibilities.

Exponential signal approximation

Before some limited situations there is always a distance R_0 after which $\eta(R)$ can be assumed as a constant ([5, 6]) and equation (2) is a decreasing exponential:

$$RP(R) = K \, \beta_\pi \exp\left(- \frac{7.82 \, R}{V_s}\right) \qquad V_s = \frac{3.91}{\sigma} \qquad (6)$$

where V_s is the "standard visibility". A first check of the standard visibility value in the far-field can therefore be obtained comparing on a video terminal the lidar signal with a sequence of exponential courves all starting from a point on the far-field return itself (see fig.2).

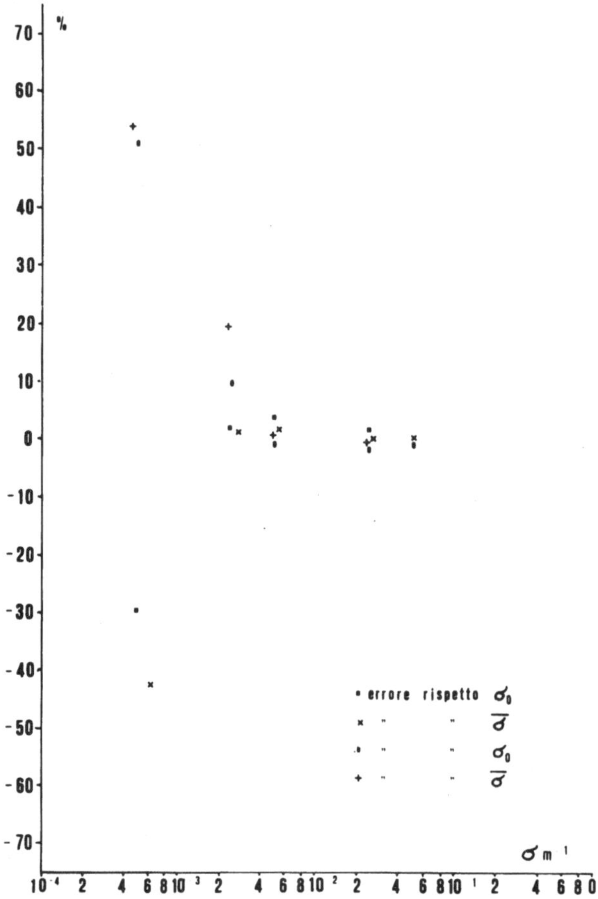

Figure 1. Error done on the extimation of σ on the basisis of the maximum of the lidar return when $\sigma(R)=\sigma_0+GR$ and $\overline{\sigma}=\sigma_0+(GR_m/2)$. A) Error respect to σ_0 (squares) and to $\overline{\sigma}$ (X) when $G=\sigma_0/9\cdot10^2$. B) Error respect to σ_0 (points) and to $\overline{\sigma}$ (crosses) when $G=-\sigma_0/18\cdot10^2$.

A program was prepared in order to apply the abovementioned procedure having as input the recorded data of the lidar returns. The program was tested on lidar returns obtained by means of the lidar IROE IV ([8]) during a field experiment campaign which took place in the Po Valley in March 1982 at the C.N.R. meteorological base of S.Pietro Capofiume. The tests showed that this technique is suitable for a quick-look selection and classification of lidar returns in a medium visibility range. In fact when the visibility is too high there is a small difference in the far-field return between signals corresponding to large changements of visibility, when the visibility is very low the return rapidly falls below the noise level and the applicability range of equation (5) is too small for a good fitting.

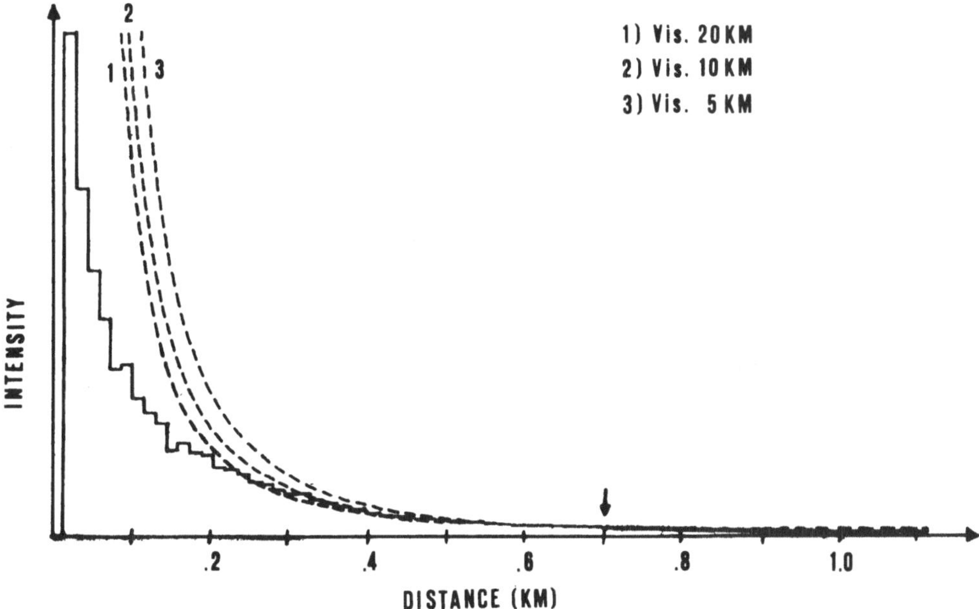

Figure 2 . Comparation between lidar returns (solid line) and theoretical returns, without geometrical form factor correction, for a given standard visibility (dashed lines). The marker shows the overlapping point.

Conclusions

Two techniques were shown for a quick-look analysis of lidar returns from a visibility point of view. The first technique is suitable for near-field applications with standard visibilities below 2 km, the second one is suitable for far field applications with standard visibilities between 1 and 20 km.

References

[1]) COLLIS R.T. et Alii:"Visibility measurements for aircraft landing operations" Final Report FAA No.DoT-FA70 WAI - 178, Stanford Research Institute, Sept.1970

[2]) VIEZEE W., OBLANAS J. and COLLIS R.T.H.:"Lidar evaluation of fog dissipation techniques" AFCRL-TR-73-0052, Stanford Research Institute, Febr.1973

[3]) HERRMANN H., PANTANI L., STEFANUTTI L. and WERNER C.:"Lidar measurements of atmospheric visibility" Alta Frequenza, 43, 732 (1974)

[4]) WERNER C.:"Slant range visibility determination from lidar signatures by the two point method" Optics and Laser Technology, 27, Feb.1981

[5]) HALLDORSON T. and LANGERHOLC J.:"Geometrical factors for the lidar function" App. Opt., 17, 240 (1978)

[6]) PANTANI L.:"Geometrical factors in lidar signals" Atti della Fondazione Giorgio Ronchi, 36, 292 (1981)

[7]) SASANO Y.et Alii:"Experimental determination of the geometrical form factor in the laser radar equation" 9th Int. Laser Radar Conf., Munchen July 1979

[8]) CASTAGNOLI F., MORANDI M., PIPPI I. and RADICATI B.:"Lidar system for visibility monitoring" Opt. Quant. Electr.,15, 261 (1983)

Remote Measurement of Wind Profiles Using the DFVLR Laser Doppler Anemometer

F. KÖPP, R.L. SCHWIESOW and CH. WERNER
DFVLR - Institut für Optoelektronik
D-8031 Oberpfaffenhofen

Application

Knowledge of wind profiles in the atmospheric boundary layer is important for a number of environmental concerns. In meteorology, the profile is related to momentum transfer and the surface roughness scale. Wind-wave interaction at the sea surface depends on the profile near wave tops. The trapping and dispersion of air pollutants is largely controlled by the wind field. For safe aircraft operations it is important to detect vertical shear of the horizontal wind and the strength and persistence of wingtip vortices. The stress on, and productivity of, large wind turbines is determined by the wind profile. For these and other reasons, we have developed a laser-based, remote-sensing anemometer and tested it against more conventional instrumentation. Some advantages of lidar (light detecting and ranging) measurements over balloon sondes and tower-mounted anemometers are (a) nearly continuous-in-time monitoring of the profiles, (b) spatial averaging for more representative values, and (c) measurements to altitudes of at least 750 m without towers.

Principle

The principle of operation is shown in Fig. 1. Natural particles with radii near 1 μm in the atmospheric aerosol move with the air parcels and scatter incident laser light (at optical frequency f_o) back to the lidar system. If the particles are mov-

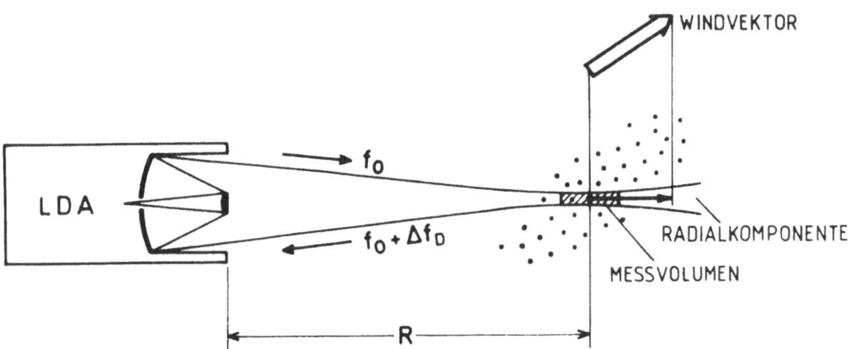

Fig. 1. Operating principle of the laser Doppler anemometer, which measures the radial component of the velocity of natural particles at a remote distance R that are drifting with the wind.

ing with respect to the lidar, then the scattered light is Doppler-shifted by a frequency Δf_D. In the case of backscatter, only the component of velocity along the lidar line of sight gives rise to a Doppler shift. For our system, which operates at 10.6-μm wavelength, the frequency shift is approximately 189 kHz for a 1 m/s radial velocity component. The spatial volume where we sense the wind is determined by the focal volume of the telescope; it may be moved anywhere in the atmosphere over a full hemisphere by azimuth and elevation steering mirrors. When we measure the Doppler shift, we determine the radial component of the wind with an accuracy that is limited only by our knowledge of the laser wavelength. The Doppler lidar is an inherently self-calibrated instrument.

Laser Hardware

Our laser Doppler anemometer uses two lasers, one as transmitter and one as local oscillator, operating together. Figure 2 gives a schematic diagram of the system, whic is described more fully by Köpp (1). The transmitter CO_2 laser is DC-excited and

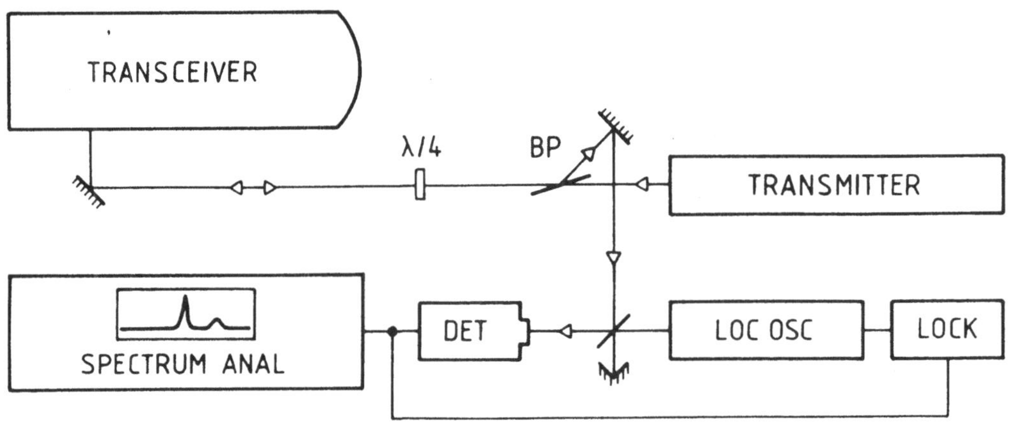

Fig. 2. Schematic diagram of the DFVLR laser Doppler anemometer.

operates at 4 W continuous-wave on the P-20 line. Its output is TEM_{00} in a single longitudinal mode, and it is stabilized to the peak of the P-20 gain curve by control of the cavity length. The local oscillator waveguide laser is RF-excited and also operates in a single cavity mode. We utilize the temporal coherence of the lasers to give a line width of less than 5 kHz, far less than typical Doppler shifts, and we use the spatial coherence to produce a small focal volume and provide a defined remote-sensing sample volume. In addition, single-mode operation is required to produce plane-parallel wave fronts that permit optical heterodyning between the signal and local oscillator. Optical heterodyning allows us to measure Δf_D with very high spectral resolution. In order to achieve good range resolution, both transmitted anc received beams use a common 30-cm, diffraction-limited telescope as a transceiver

element. With this arrangement, both beams automatically overlap in the focal region.
We can scan in range from 40 to 1000 m and have a range resolution of 10 m at 100 m
range. Because the lasers are linearly polarized, a germanium plate at Brewster's
angle and a quarter-wave plate serve as a transmit-receive beam switch.

System Operation

The frequency spectrum of the detector output contains two main peaks. The larger is
the RF beat between the transmitter at f_o and the local oscillator at f_{lo}. This difference is usually locked to a constant 6 MHz by a servo circuit that controls the
cavity length of the local oscillator laser. A smaller RF beat is the difference
between the local oscillator at f_{lo} and the backscattered signal at $f_o + \Delta f_D$. This
signal beat ranges between 3 MHz, corresponding to a velocity of approximately 16 m/s
away from the lidar, through 6 MHz, corresponding to zero radial component, to 9 MHz,
corresponding to 16 m/s toward the lidar. Figure 3 shows the two heterodyne peaks
from the DFVLR laser Doppler anemometer when the target was a rotating wheel with a
radial velocity component moving away from the lidar. It is also possible to measure

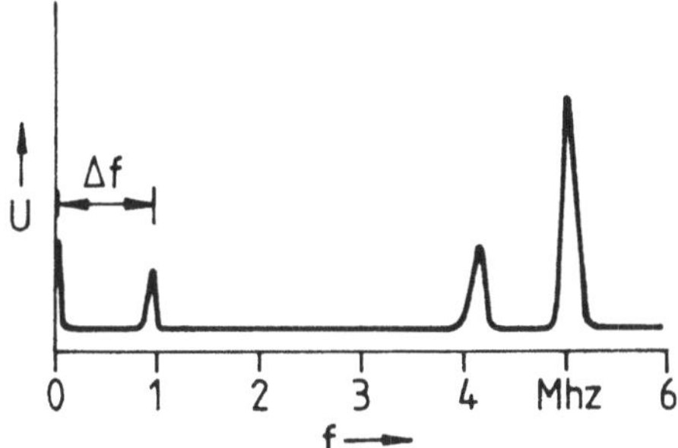

Fig. 3. Frequency spectrum showing the
transmitter-to-local-oscillator beat
frequency (higher amplitude peak at
5 MHz) and signal-to-local-oscillator
beat frequency (lower amplitude peak
at 4 MHz). The homodyne beat (signal-
to-transmitter) is at 1 MHz, outside
the heterodyne range of interest.

the absolute value of the radial velocity component when $f_{lo} = f_o$ (homodyne operation),
but information on the sense (toward or away from the lidar) is lost. The homodyne
peak can be seen in Fig. 3. Typical atmospheric frequency (velocity) spectra are
wider than the signal peak in Fig. 3 because the wind is different in different parts
of the focal volume as a result of turbulence and horizontal inhomogeneity.

In the lidar system, the scanning, on-line data reduction, and data recording are
controlled by a microcomputer (2). The average radial velocity of scatterers in the
sensing volume is approximated by the velocity at the peak of the velocity spectrum,
which is a plot of the backscattered intensity as a function of velocity. 1024
spectra are digitally averaged to produce a velocity estimate every 52 ms. The output of the overall program that processes the radial velocity values consists of a
wind magnitude and direction at selected altitudes. A method for obtaining the horizontal wind vector from radial wind values taken around a circle centered at the
altitude of interest (i.e. a conical scan) is discussed by Köpp et al. (3). The

542

entire system is installed in a standard 6.1-m container, which allows the lidar to
be easily moved to various experimental sites and quickly made ready for operation.

<u>Performance</u>

We demonstrated the ability of the DFVLR laser Doppler anemometer to measure accurately the radial component of the wind by comparing the lidar wind values with measurements from a standard sonic anemometer, which was at a range of 200 m from the
lidar. One-minute averages from the two instruments show a standard deviation from
perfect identity of 0.12 m/s and a correlation coefficient of 0.98; 48-min averages
differ by only 0.05 m/s, which is an estimate of the maximum systematic uncertainty
in the lidar. Twenty-nine vertical profiles of the horizontal wind up to 750-m
altitude made by the lidar and by a conventional balloon sonde are similar to each
other with a standard deviation of 1.3 m/s and a correlation coefficient of 0.83 for
magnitude, and 12^{0} and 0.91 for direction. We attribute the differences in profiles
to horizontal inhomogeneity in the wind field, the different spatial sampling and
averaging inherent in the lidar conical scan and in the drifting balloon, and to the
uncertainty in tracking balloon sondes. The comparison profiles were taken in weather conditions ranging from fog (visibility 0.1 to 1 km) through haze (1 to 10 km)
to clear atmosphere (10 to 40 km visibility), which shows the capability of the lidar
for operation in a wide variety of circumstances.

Figure 4 gives an example of the ability of the lidar to produce a time series of
wind profiles. Such profiles can be taken as rapidly as every 5 minutes if rapidly-

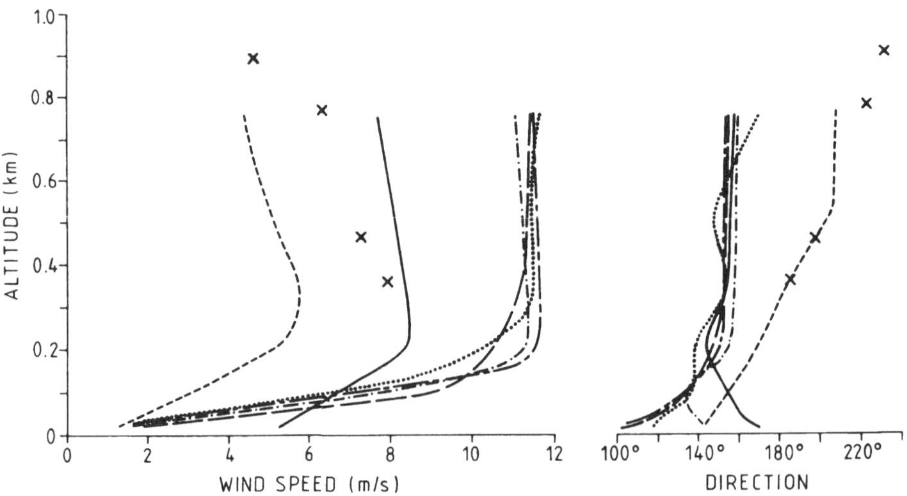

Fig. 4. Composite of vertical wind profiles, 29-30 Sept 82: (times in MEZ)
——————— 1621   ———— ——— 2108   ——— ·—— ·—— 2137   — ·—— ·—— ·— 2226
················· 2308   — — — — — — — 0820. The crosses are balloon sonde data at 0600.

changing conditions make a high data rate worth while. This ability to produce
nearly continuous-in-time profile data is an advantage of the lidar over such instruments as balloon sondes or aircraft probes. The profiles in Fig. 4 show a large increase in wind at approximately 200-m altitude and a concurrent decrease in surface
wind after sunset. A maximum in the wind aloft occurred between 2130 and 2220, after
which time the wind decreased to the morning profile. The morning profile still
showed a relative wind maximum (low level jet) near 300-m altitude. The balloon-
sonde data at 0600 provide time continuity between lidar data at 3208 and 0820 and
verify the decrease in wind magnitude above 300 m.

Conclusions

The data show that the DFVLR laser Doppler anemometer can measure wind profiles to
altitudes of at least 750 m whenever the visual range is a few hundred meters or
greater. The lidar system utilizes the temporal and spatial coherence properties of
CO_2 lasers to achieve its performance objectives, and it involves frequency locking
of one laser to another to give a direct heterodyne velocity spectrum. Because the
conically-scanned lidar gives a spatially-averaged wind value, lidar wind profiles
are probably more representative of wind fields in the atmospheric boundary layer
than are profiles to 750-m altitude that are measured by other types of instruments.

Doppler lidar has shown its measurement accuracy and some results of environmental
importance (3-8), but there are many environmental-measurement applications that
remain to be explored and improvements in the lidar instrumentation, scan methods,
and analysis techniques that can be tested. We expect that the field of usefulness
of this particular laser application will expand significantly in the future.

Literature.
(1) KÜPP,F.,: A direct laser Doppler anemometer for remote sensing of atmospheric
flows, Internat. Symposium on Applications of Laser-Doppler-Anemometry to Fluid
Mechanics, Lisbon (1982) 12.1.1-12.1.10
(2) KÜPP,F.,H.HERRMANN,R.SCHWIESOW,CH.WERNER und F.BACHSTEIN: Erstellung und Erprobung des Laser-Doppler-Anemometer zur Fernmessung des Windes, DFVLR-FB 83-11
(1983) 101 pp
(3) KÜPP,F.,R.L.SCHWIESOW and CH.WERNER: Remote measurements of boundary-layer wind
profiles using a CW Doppler lidar, submitted to J. Appl. Meteor. (1983)
(4) BROWN,A.,E.L.THOMAS,R.FOORD and J.M.VAUGHAN: Measurements on a distant smoke
plume with a CO_2 laser velocimeter, J. Phys. D: Appl. Phys. 11 (1978) 137
(5) DIMARZIO,C.,C.HARRIS,J.W.BILBRO,E.A.WEAVER,D.C.BURNHAM and J.N.HALLOCK: Pulsed
laser Doppler measurements of wind shear, Bull. Amer. Meteor. Soc. 60 (1979) 1061
(6) VAUGHAN,J.M.: Remote wind measurements in the atmosphere using laser Doppler
methods, Proceedings of LASER 79, München (1979) IPC Sci. & Tech. Press Ltd, England
(7) SCHWIESOW,R.L.,R.E.CUPP,P.C.SINCLAIR and R.F.ABBEY,Jr: Waterspout velocity
measurements by airborne Doppler lidar, J. Appl. Meteor. 20 (1981) 341
(8) SCHWIESOW,R.L., and R.S.LAWRENCE: Effects of a change of terrain height and
roughness on a wind profile, Boundary Layer Meteor. 22 (1982) 109

Optoelektronische Displays

Optoelectronic Displays

Opto-electronic Display Technologies, State-of-the-Art and Trends

B. Kazan
Xerox Palo Alto Research Center
Palo Alto, CA 94304, U.S.A.

I. Introduction

For many years the cathode-ray tube completely dominated the display field, satisfying the needs of television, radar and oscilloscope applications. During the past 10 years, however, the requirements of new applications for compact, low-power displays has stimulated the development of many new types of displays. These displays, each with its own combination of special advantages and disadvantages, are based on an unusually wide variety of different materials and physical phenomena. This makes it difficult to discuss the subject in a cohesive manner and also make it difficult to compare one display with another on a common basis.

Most display devices fall into one of the following broad categories:

(a) Cathode-Ray Tubes	(e) Liquid-Crystal Displays
(b) Gas-Discharge Displays	(f) Electrophoretic Displays
(c) Electroluminescent Layer Devices	(g) Electrochromic Displays
(d) Light-Emitting-Diode Displays	

Displays in the first five categories are, to varying degrees, commercially available. Displays in the last two categories, despite much development effort, are still in the laboratory stage. Since each category itself contains a number of displays which may differ from each other in their principle of operation, it is not possible to more than touch upon the various displays in this brief survey. Before starting on the discussion, however, it is well to note that cathode-ray tubes today still account for about 80% of the display market. In keeping with this, such displays are stressed to some extent below.

II. Cathode-Ray Tubes

A.   Monochrome Tubes

In view of the very satisfactory performance of monochrome tubes in TV applications, much of the emphasis in the past has been on cost reduction. Today, because of the growing emphasis on displaying computer-generated information, there is a strong incentive to improve their quality. Making use of computer modeling to improve the design of electron guns and deflection yokes, tubes operating at 18kV can now display up to about 1500 x 2000 spots[1] with acceptable brightness. Using "dispenser" cathodes with higher emission current density, and operating at higher anode voltages, this resolution may possibly be increased by 50% or more during the next few years.

Since many phosphor screens presently have efficiencies[2] as high as 20%, major improvements in this respect are not expected. However, there is a continuing development of new phosphors, particularly of the rare-earth type, for individual applications to satisfy special requirements of color, persistence and power handling capability. These usually have an emission concentrated in one or more narrow bands. When viewed through an appropriate narrow-band filter, most of the ambient reflected light may be absorbed, allowing images to be easily viewed in an airplane cockpit even with direct sunlight falling on the screen.

In the case of oscilloscope tubes, traveling-wave deflection systems are frequently used to allow deflection at frequencies up to several GHz or higher. To obtain adequate brightness from a single trace of such a high-frequency transient, channel-plate secondary-emission multipliers with a current gain of more than 1000 are now being incorporated adjacent to the phosphor screen.[3] These greatly increase the brightness, making it possible to photograph single transients having a rise time less than 1/3 nanosecond.

B. Color Cathode-Ray Tubes

The most well-known color tube[4] is the shadow-mask type used in television. In such tubes the three guns employed are arranged either in a delta or in-line configuration in combination with an appropriate shadow mask and phosphor screen structure. For high-resolution applications, monitor tubes are now being made with masks whose hole pitch has been reduced to half or a third of the usual 0.6 mm spacing. These allow color images with more than 1000 scan lines to be produced. In all shadow-mask tubes, however, 80% or more of the beam current is intercepted by the mask. To avoid this, several laboratories are investigating new types of mask structures that enable an electron transmission of about 50% to be obtained. In one arrangement,[5] the mask consists of two sets of mutually perpendicular conductors across which about 1 kV is maintained. The resultant local fields then cause the electrons passing through each hole to be compressed to a narrow stripe, confining them to a single color phosphor element.

For some years attempts have been made to develop beam-index tubes[4] which eliminate the shadow-mask entirely. In these tubes, a single beam scans successive vertical strips of red, green and blue phosphor. By switching the beam on at the correct moment the desired color or color mixture can be produced. To indicate the exact beam position, vertical UV-emitting phosphor strips are deposited at regular intervals on the rear of the screen. The resultant UV pulses pulses are then picked up by an external photocell mounted at the rear of the tube at one side. Although such tubes are capable of producing TV images with acceptable resolution, their brightness is somewhat limited because of the need to use a beam whose diameter at all times is less than the width of a phosphor strip, thus limiting the beam current. However, because of their ruggedness and freedom from color shifts due to external magnetic fields tubes of this type are potentially useful in military and other special applications.

For applications such as radar where a limited color gamut is acceptable, penetration tubes[4] are frequently used. These employ a dual-layer continuous screen. At an accelerating potential of 8 kV, for example, the beam excites the red phosphor. If the anode voltage is raised to 16 kV the beam penetrates to the second phosphor, producing light which is predominantly green. A major problem, however, results from the power consumed in switching the anode voltage. For example, if the color is switched 10^4 times a second, of the order of 100 watts may be dissipated. In a new color switching scheme recently developed,[6] a single-layer phosphor screen is used which emits both red and blue light. By using suitable dichroic polarizers and a large area liquid-crystal layer to rotate the polarization angle of the transmitted light, color switching from red to green can be accomplished in about a msec by applying only 20 volts to the liquid crystal cell.

C. Storage Tubes

In one class of storage tubes, known as the bistable type,[7] the writing beam is used to establish a potential pattern on the surface of the viewing phosphor screen. A low-voltage flood beam landing on the phosphor then shifts the elements to one of two equilibrium potentials, at the same time producing a. stored luminescent image corresponding to the potential difference maintained between the on and off elements. Used in oscilloscopes, small tubes are able to capture a single transient signal containing frequencies up to about 10–20 MHz. Although simple and low in cost, they are likely to be replaced in the coming years by oscilloscopes containing a conventional cathode-ray tube with circuits which can sample and hold a transient signal for subsequent display. Oscilloscopes of the latter type now available, for example, can sample a transient signal at a rate up to about 100 MHz.

In the second or transmission-mesh type of storage tube,[7] an insulator-coated metal mesh is mounted adjacent to the phosphor screen. Initially the insulator is uniformly charged to a slightly negative potential. This prevents low-velocity electrons from a flood beam from landing on the insulator or passing though the mesh holes. The action of the writing beam is to shift the insulator potential less negative, allowing the flood beam to pass through the mesh holes at selected areas. By maintaining a voltage of 5 kV or more on the phosphor screen images with a brightness of several thousand cd/m^2 can then be produced. The additional ability to electronically control the rate of decay[8] of the image has made such tubes useful for airborne radar applications. Because of their very high writing speed, which may be more than 10^9 cm/sec, such tubes have also found major use in storage oscilloscopes, allowing single transients with frequencies up to about 400 MHz to be captured. Because of their very high frequency capability, it seems unlikely that tubes of this type will be replaced for some years to come by oscilloscopes with sampling circuits.

D. Projection Tubes

For producing pictures with a diagonal size greater than about 75 cm, projection systems must be used.[9] To obtain color pictures the output images from three tubes with different color phosphors are projected onto a common screen. Because of the limited light collection of ordinary refractive lenses, tubes containing a Schmidt reflective system within the vacuum envelope are now being made by a number of manufacturers. In these, the phosphor is deposited on a small metal plate facing the scanning beam. The light from this is then collected and directed forward by a spherical mirror on the rear wall of the envelope. Such systems can collect more than 50% of the light from the phosphor, making them several times more efficient than conventional optical systems. In all projection tubes, however, the power of the electron beam must be limited to avoid damage to the phosphor or saturation of the light output. Using improved phosphors as well as directional viewing screens, color images of about 1 meter size can now be produced whose brightness is comparable with direct-view color tubes.

For very large screen applications, however, it is necessary to resort to electron-beam-controlled light-valve systems.[7] The most successful example of this is the Eidophor system. In this, the light from a 5kW xenon arc is directed onto a spherical mirror coated with a thin film of oil. If the oil film is undisturbed, this light is reflected back to the source. However, if the oil is locally deformed by electric charges deposited by a scanning electron beam, scattered light from these areas can pass through the system, and reach the viewing screen. This system allows bright pictures with 1000-line resolution to be projected onto a screen as large as 10 meters square, making it very useful for large theaters.

A more recent system, makes use of a thin insulating crystal of deuterated potassium dihydrogen phosphate as the light valve.[10] When charged by the electron beam local changes in birefringence are produced which rotate the polatization angle of light passing through the crystal. Using a suitable optical system, the amount of light which can pass through the system and reach the screen can then be controlled. The image produced in such devices may be retained for hours, or new images may be written at TV frame rates.

E. Flat-Panel Cathode-Ray Tubes

Although small TV display panels based on the newer technologies have been demonstrated (as discussed in the following sections), the achievement of large displays having a combination of high resolution, wide color gamut, high brightness and good gray scale, appears to be a much more difficult task than initially anticipated. This realization has stimulated the investigation of several new ideas for flat panel displays based on electron-beam technology.

In one approach[11] being pursued by the Philips Co., a horizontally-scanned electron beam is injected from the upper edge of the panel parallel to the phosphor screen. The beam is then sharply bent toward the phosphor at a vertical position which can be continuously varied by successively reducing the potentials of about 10 horizontal electrodes strips located behind the beam. An important key to this device is the recent development of a large-area multiplier structure which is mounted adjacent to the phosphor. This provides a current gain of about 10^3 and enables a high-resolution, low-current scanning beam operating at only 300 volts to be used. Experimental devices have already been made, capable of producing a picture whose brightness and resolution are comparable with conventional TV. Considerably larger pictures are expected in the future using structures less than 8 cm thick.

In another approach,[12] being followed by the Siemens Co., an X – Y electrode structure is used to control, in a line-by-line fashion, the flow of electrons from the rear of the panel to a 4 kV phosphor screen. To avoid the need for a large-area heated cathode as a source of electrons, which would consume high power, a low-pressure gas discharge is used as a source of electrons. With such structures, about 6 cm thick, 35 cm diagonal pictures have been produced containing 448 x 720 elements. The total power consumption, including driver stages, is only 20 watts. Further work in this area may lead to high-resolution color displays.

A third approach,[13] being followed by RCA, is intended for displays of 1 meter size or larger. Here an array of electron beams injected from the lower edge of the panel parallel to the phosphor screen, are kept from spreading by means of the periodic focusing action of an extended mesh structure. By successively switching each of the horizontal electrodes on the back side of the panel from $+330$ to -100 V, the array of beams (which are individually current modulated) is bent sharply into the phosphor screen, producing a horizontal row of bright spots at an arbitrary height. Experimental devices of this type with a viewing area of 13 x 25 cm^2 have already been built. The final aim of this work is to produce a panel 75 cm x 100 cm in size with a thickness less than 10 cm, capable of producing high resolution color TV pictures.

Summarizing the situation with respect to cathode ray tubes in general, it should be noted that the beam current density and spot size are, in some cases, approaching the theoretical limits. However, improvements in areas such as deflection yokes, electron emitters and storage techniques will probably result in gradual but significant improvements in overall tube performance. Aside from this, large-area, flat panels now under development based on new concepts may result in wall-type, high-resolution color television displays.

III. Gas-Discharge Displays

One of the primary advantages of gas-discharge cells[14] is their suitability for use in X–Y addressed displays containing a large number of elements. This arises from the fact that current flow and light emission occur at a cell only if the voltage applied across it exceeds a threshold, e.g., about 250 volts. In such displays, for example, the cathodes of successive rows of elements are connected to a set of X-conductors and the anodes connnected to a set of Y conductors. In practice the elements of a row are addressed by applying a voltage pulse $-V_x$, slightly below the threshold, to one of the cathode strips and simultaneously applying voltages $+V_y$ to the desired anode strips. This process is then repeated in sequence for all the rows. Although satisfactory brightness can be obtained for panels with up to about 100 rows, the limited peak brightness which can be obtained without shortening the life of the cells makes it difficult to obtain sufficient average brightness in panels with more than about 200 rows.

This brightness limitation is avoided in displays which are ac operated and which have an internal memory.[14] In such displays, the X and Y electrodes are coated with a film of insulator. In operation about 100 volts ac at about 50 kHz is maintained between the two sets of electrodes. This voltage is slightly below the level required to produce a discharge. If during some half cycle, an additional voltage is applied across a pair of X and Y electrodes, a discharge is initiated at their intersection. This is rapidly quenched by the build-up of charges on the insulating walls of the cell. The resultant wall potential, added to the potential applied during the next half cycle, then causes the cell to continue firing indefinitely. Such storage panels are being manufactured in various sizes. The largest of these is about 80 cm square and contains more than 10^6 elements.

Although small panels are being used in some bank terminals, because of the high cost of the semiconductor drivers the larger panels are used primarily for military applications. However, semiconductor chips with 32 drivers per chip are now commercially available. If manufacturers' claims can be believed, it is expected that the cost of drive circuits in a few years will be sufficiently low to allow larger gas panels to compete with some cathode-ray tube systems.

To eliminate most of the of drivers along one axis, self-scan displays[14] were developed. In these, an auxiliary discharge, hidden from view is made to propagate from the back side of one cathode strip (or Y) conductor to the next conductor (contained in a common envelope) by applying suitable three-phase dc voltage pulses. This causes the breakdown potential of the display elements on the viewing side of the perforated cathodes to be lowered in sequence. In operation, time-varying voltage pulses are applied by means of individual drivers to the viewing anodes (X conductors) oriented perpendicular to the cathodes. In effect, irrespective of the length of the display, only three drivers are required for all the Y conductors. Commercial devices of this type have been manufactured in various sizes, the largest of which can display 8 rows of 40 characters each. Recently self-scan devices of this type have been developed in which the electrodes of the viewing side are provided with an insulating coating and operated with ac. Because of the internal memory of the viewing section,[15] the need for continuously refreshing the panel is thus avoided.

All of the above panels employ neon as the primary gas constituent, resulting in their typical orange-red emission color. For producing large, color-television images, special panels are being investigated which contain a helium-xenon mixture whose emission is rich in ultra-violet. This is used to excite photoluminescent phosphors of different color deposited on the cell walls. The largest experimental panel of this type has a diagonal of 40 cm and contains 240 x 320 cells. Although pictures with good gray scale and color gamut have been produced, the efficiency of such panels remains disappointingly low, being more than two order of magnitude below that of color cathode-ray tubes. In view of the fundamental problems that must be overcome to improve the efficiency (aside from the costs of panel fabrication) it is not likely that such panels will become practical in the next few years.

IV. Electroluminescent Phosphor Displays

Although there are several types of electroluminescent phosphor displays,[16] in almost all cases ZnS is used as the primary material. In the oldest type, a thin layer of ZnS powder in an insulating binder is used with ac voltage applied to produce light output. Because of its limited peak light output and lack of a sharp voltage threshold, this type is now used only for displaying fixed messages. A second type of powder-binder layer, operated with dc voltage has a much higher light output and a sharp threshold. However, it is receiving only limited attention today because of the longer life and other advantages of the thin-film devices discussed below.

Thin-film electroluminescent structures consist of a 0.5 micrometer evaporated film of light-emitting ZnS (activated with Mn) sandwiched between thin films (each about 0.2

micrometer thick) of an insulator such as Y_2O_3. If an ac voltage of 5 kHz is applied the light output rises very steeply when the voltage exceeds about 200 volts. Because of this sharp voltage threshold, devices based upon such layers can be addressed in the same manner as gas panels.

At present, X–Y addressed panels using such films are commercially available in sizes up to 8 x 12 cm^2 with 240 x 320 elements. These are not only useful for displaying alpha-numeric information but because of their good gray scale capability, they can also be used to display TV pictures. With their drive circuits these panels consume about 8 – 14 watts, producing images with about 75 – 100 cd/m^2 brightness, about 1/4 that of conventional TV. In brightness and efficiency such panels are roughly comparable to neon-filled gas panels. As in the case of gas panels the cost of the large number of high-voltage drivers is a deterrent to their widespread use. Since only yellow light can be produced with acceptable efficiency, such panels are presently of limited interest for TV. However, because of their compactness, light weight and high resolution capability, they are expected to find important future applications in portable computer terminals.

V. Light-Emitting Diode Displays

Another important display technology is based on light-emitting diodes[17] or LED's. These are essentially p-n junctions in a semiconductor crystal which emit light when they are forward biased. Unlike electroluminescent displays, which require hundreds of volts and operate at high fields, LED's require only 2 – 3 volts. To produce visible radiation the semiconductor crystal used must have a band-gap greater than 1.8 eV, thus eliminating silicon and many other semiconductors. The material chosen must also be one in which it is possible to fabricate p-n junctions.

These conditions are well satisfied by $GaAs_{1-x}P_x$ which is used for all present commercial diodes. One of the most common is the red-emitting diode which contains 40% phosphorus and 60% arsenic. By increasing the phosphorous (which increases the bandgap) and also adding nitrogen (to increase the efficiency), diodes emitting in the orange, yellow and green are obtained. In general the efficiency of commercial LED's is of the order of a few tenths of a lumen/watt, making them comparable with gas panels and electroluminescent layers in this respect. (Although the highest efficiency is obtained from red diodes produced by adding Zn and O to GaP, such diodes are of limited use because of the saturation of their light output at high currents.)

For many applications a small GaAsP chip 1–2 mm in size is used on which a pattern of diode elements is fabricated. By applying voltage to selected diodes a character or numeral can then be displayed. Because of their small size, however, such chips are usually viewed through a magnifying lens. For larger displays an array of individual diodes that are X–Y connected is sometimes used. In operation, the diodes are then addressed in lin-by-line-fashion. For displaying more than a row of characters the cost of assembling the large number of diodes may become prohibitive. Despite this problem, this approach has been experimentally pushed to a somewhat extreme degree by workers of the Sanyo Co. who assembled an array of 240 x 320 green-emitting diodes[18] covering an area of 12 x 16 cm^2. Using this structure, TV pictures with good gray scale and with a brightness of about 125 cd/m^2 (perhaps a third of conventional TV) were produced. The total power consumption for the display and its driver circuits was 60 watts.

Despite competition from other displays, light-emitting diodes are expected to remain important because of their low-voltage operation, making them compatible with low-voltage semiconductor circuits. Also they have a very long operating life, frequently exceeding 10,000 hours, and are capable of producing very high brightness under pulsed conditions. One of the major challenges to workers in this field is the unavailability of a blue-emitting diode having satisfactory efficiency and capable of operating at low voltage. Although limited success has been achieved with SiC, for example, the quantum efficiency so far obtained with diodes of this material is about two orders of magnitude less than that of GaAsP.

VI. Liquid-Crystal Displays

Most liquid-crystal displays in present use are of the twisted-nematic type.[19] In these the elongated liquid-crystal molecules (oriented parallel to the glass plates confining the material) assume a 90° twist in direction as a result of the surface treatment of the plates. Polarized light entering the cell then follows the twist of the molecules and is able to pass through the crossed polarizer at the output. However, if a potential of about 2 – 3 volts is applied across the layer, the molecules tend to align perpendicular to the plates because of their dielectric anisotropy. The polarized light is now no longer twisted and it is blocked by the output polarizer. Usually a reflector is placed below the cell allowing it to be viewed from the same side as is illuminated.

For watches and calculators groups of electrode segments are provided to which voltages are applied in different combinations to display numerals or characters. For displaying more complex information the glass plates confining the liquid crystal are provided with X and Y electrodes on their respective surfaces. However, unlike gas-discharge displays, for example, which produce light only when the threshold voltage is exceeded, when liquid-crystal cells are X–Y addressed they respond to the time-integrated effect of the rms voltage applied over a period corresponding to their response time which, at room temperature is about 0.1 second. The integrated effect on "off" elements during a complete frame scan is thus almost as great as the integrated voltage applied to the "on" elements . This situation becomes progressively worse as the number of rows is increased. Today, panels are being sold, for example, which have up to 64 rows of elements. However, since the on elements have only about 13% higher effective voltage than the off elements the contrast is rather low. The incomplete orientation of the "on" molecules by the addressing voltage pulses (whose magnitude must be sufficiently low to avoid exciting "off" elements) also results in a very limited viewing angle.

To obtain higher contrast and better viewing angle, displays are now being developed which incorporate a thin-film transistor in series with each picture element. In this arrangement the gate electrode of each row of transistors is connected to a separate X conductor. By applying + 10 volts, for example, to a selected X conductor, all the transistors of this row are made conducting. This allows the signal voltages on the Y conductors (connected to the source electrodes of the columns of transistors) to appear on the liquid-crystal elements of this row. Since the transistors of all other rows are assumed to be nonconducting, the signal voltages on the Y leads are prevented from appearing across any other liquid-crystal elements. Although a number of laboratory groups have been exploring transistor arrays using CdSe and amorphous silicon, perhaps the best results so far have been obtained by the Seiko Co. using polycrystalline silicon.[20] With this material small displays for use in pocket TV have been developed which are 4.3 x 3.2 cm^2 in size and have 240 x 240 elements. By incorporating a mosaic of red, green and blue filters full color pictures have been obtained. In operation the display is illuminated from the rear by a miniature fluorescent light which consumes about 1/2 watt, another 1/2 watt being consumed by the remaining circuitry of the pocket TV set.

Using other liquid crystals of the smectic type, displays are also being developed which have an internal memory. These depend on local heating for the writing process. In one arrangement a scanning laser beam is used to heat selected elements of a small cell.[21] Upon cooling, the heated elements (initially transparent) become optically scattering, remaining in this state for days. Using an appropriate optical system the resultant pattern can be projected as a black-and-white image onto a large screen. When desired, the image can be erased by applying about 200 volts ac across the liquid-cyrstal cell for a fraction of a second. Using, for example a liquid-crystal cell, about 10 cm square, which is mechanically scanned by an array of 32 GaAlAs diodes, a stored image with 64 x 10^6 elements can be produced in about one minute.

Another arrangement for thermal writing[22] makes use of a form of X–Y addressing. Here each row of elements is heated in sequence by applying a 25-volt, 10 msec. pulse across the opposite ends of successive X conductors (which act as resistive heaters). By applying signal voltages on selected vertical conductors during the cooling of each row, the desired elements of successive rows are prevented from shifting to the scattering state upon cooling. Using this principle, panels 10 cm square with 250 x 250 elements have been developed. At present, at least one company is planning to manufacture such displays.

Summarizing the situation with regard to liquid-crystal displays, the very low power consumption of conventional twisted-nematic displays (about 1 microwatt/cm^2) makes them ideally suited for battery-operated applications. Although X–Y addressed twisted-nematic displays with 16 or more rows of characters are now commercially available, because of their viewing characteristics they are likely to be eventually replaced by displays which incorporate an array of thin-film transistor elements, enabling high-contrast images with much higher information content to be produced. Although the cost of fabricating such transistor arrays is presently a serious obstacle, the large potential market for pocket TV may help to accelerate the development of this technology. For special applications, such as computer-aided design, where very complex images with long storage are required, it is expected that laser-addressed smectic liquid crystals will find important specialized uses.

VII. Electrophoretic Displays

Electrophoretic displays[23] generally consist of a dark insulating fluid in which pigment particles of a contrasting color are suspended. Because of the charge acquired by the particles when mixed with the fluid they can be attracted to the front or back transparent electrode of a thin cell, depending on the polarity of the applied voltage. If yellow particles in a black fluid are used, for example, at areas where the particles are drawn to the viewing side their yellow color can be seen. Where they are drawn to the rear they are hidden and only the black fluid can be seen.

With about 50–100 volts applied, switching occurs in about 10 msec. requiring a current flow of only about 1 $\mu A/cm^2$. Of interest is the fact that the particles can remain on the electrodes for hours or days after the switching voltage is removed. Aside from problems of particle agglomeration and settling which may occur, such displays are not suitable for X–Y addressing because of the lack of a threshold voltage. Because of these factors such displays have not come into commercial use. However, successful X–Y addressing has been reported by the Phliips Company using new devices which contain an additional control electrode[23] within the cell. If successful, this work may lead to compact large-area devices with unusually low power consumption.

VIII. Electrochromic Displays

In electrochromic displays a reversible electrochemical effect[24] is used to change the color or light transmission of a material. One material which has been extensively studied is WO_3. If a transparent film of WO_3 is placed in contact with an electrolyte containing hydogen ions, and a positive potential applied to the transparent substrate of the WO_3, hydrogen ions are drawn into the tungsten oxide forming a layer with a deep blue coloration. This process, which takes a few tenths of a second, can be reversed by applying a potential of opposite polarity. Although only 1 or 2 volts are required for switching, relatively high currents, of the order of 10 mA/cm^2, must be used. Despite the attractive appearance of such displays however, there is a tendency for them to degrade after many switching cycles. Two problems have so far prevented X–Y addressing. Because of the high conductivity of the electrolyte the current flow spreads to elements beyond the crossover of a selected pair of X and Y electrodes. Also, cells which are colored develop a back emf. This voltage can then cause current to flow through other elements of the array, causing them to become partially colored as well. Although many research groups are still investigating electrochromic displays, their future use remains uncertain.

X. Conclusion

It is hoped that the foregoing discussion has served to indicate the complexity of the display field and the directions in which it is moving. It is of interest to note that the market for each of the newer technologies seems to be growing at a rate of 10–15% per year. At the same time, the market for cathode-ray displays is also growing at a rate of about 6% per year. It is thus not likely that any single technology will dominate the display field in the near future. The user, presently faced with difficult choices among different display technologies, is thus not likely to find his choices easier in the next few years. Although the various display technologies have somewhat different short-term application objectives, a major long-term objective of many of them is to replace the present bulky cathode-ray tube. However, in view of the numerous difficult technical problems which must yet be overcome, this is not likely to occur in the next few years but rather in a gradual and evolutionary manner over a more extended period.

References

1. C. Infante, D. Denham and B. McKibben, A 230 MHz Bandwidth High-Resolution Monitor, SID 1983 Digest, pp 124-125.
2. S. Larach and A.E. Hardy, Cathode-Ray-Tube Phosphors, Principles and Applications, Proc. of IEEE, Vol 61, pp 915-926 (1973).
3. B. Janko, A New High-Speed CRT, Proc. of SID, Vol. 20/2, pp 55-60 (1979).
4. A.M. Morrell et al., "Color Television Picture Tubes," Academic Press, New York, 1974.
5. M. van Alphen and J. van den Berg, Quadrupole Post-Focusing Shadow-Mask CRT, SID 1980 Digest, pp 46-47.
6. R. Vatne et al., An LC/CRT Field-Sequential Color Display, SID 1983 Digest, pp 28-29.
7. B. Kazan and M. Knoll, "Electronic Image Storage," Academic Press, New York, 1968.
8. K. Zeppenfeld, Wideband, Variable Persistence Oscilloscope Tube, Electronic Components and Applications, Vol. 3, pp 138-141 (1981).
9. A. Ohkoshi et al., Projection TV Systems for Consumer Use, Advances in Image Pickup and Display, Vol. 7 (to be published), Academic Press, New York.
10. G. Marie et al., Pockels Effect Imaging Devices and Their Applications, Advances in Image Pickup and Display, Vol. I (1974), Academic Press, New York.
11. D. Lamport et al., The Channel Electron Multiplier CRT: Flat Deflection System, SID 1982 Digest, pp 210-211.
12. A. Schauer, Plasma Lights Up 14 in. Flat-Panel Display, Electronics, Dec. 15, 1982, pp 128-130.
13. T. L. Credelle, Large-Screen Flat-Panel Television: A Guided-Beam Display, Electro-Optical Systems Design, Jan. 1982, pp 31-42.
14. G. F. Weston, "Glow-Discharge Displays," Mills and Boon, Ltd. London, 1972.
15. G. Holz, et al., A 2000 Character Self-Scan Memory Plasma Display, SID 1983 Digest, pp 130-131.
16. W. E. Howard, Electroluminescent Display Technologies and Their Characteristics, Proc. of SID, Vol. 22/1, pp 47-56 (1981).
17. C. Weyrich, Light-Emitting Diodes for the Visible Spectrum, Festkoerperprobleme, Vol. XVIII, J. Treusch, Ed., Vieweg, Braunschweig, 1978.
18. T. Niina et al., A High-Brightness GaP Green LED Flat-Panel Device for Character and TV Display, IEEE Trans. on Elect. Dev., Vol. ED-26, pp 1182-1186 (1979).
19. E. Kaneko, Liquid-Crystal Matrix Displays, Advances in Image Pickup and Display, Vol. 4, Academic Press, New York, 1981.
20. S. Morozumi et al., B/W and Color LC Video Displays Addressed by Poly-Si TFT's, SID 1983 Digest, pp 156-157.
21. A. G. Dewey et al., A 64 Million Pel Liquid Crystal Projection Display, SID 1983 Digest, pp 36-37.
22. S. LeBerre et al., A Flat Smectic Liquid-Crystal Display, SID 1982 Digest, pp 252-253.
23. B. Singer and A. L. Dalisa, An X-Y Addressable Electrophoretic Display, Proc. of SID, Vol. 18/3-4, pp 255-266 (1977).
24. B. W. Faughnan and R. S. Crandall, Electrochromic Displays Based on WO_3, Topics in Applied Physics, J. Pankove, Ed., Vol. 40, Springer, 1980.

Laser-Chemie

Laser-Chemistry

Preparative Organic Photochemistry with Lasers

F. P. SCHÄFER
Max-Planck-Institut für biophysikalische Chemie, Abteilung Laserphysik,
Postfach 28 41, D-3400 Göttingen

Early ideas on the use of lasers in chemistry were concentrating on
tunable infrared lasers to superexcite selectively an infrared-active
normal vibration of a molecule in order to break the weakest of the
bonds participating in this normal vibration. We now understand that
these hopes could not materialize because of the strong inharmonic
coupling of modes that leads to a very fast (tenths to tens of pico-
seconds) random distribution of the absorbed energy over many normal
modes. The result is in a first approximation that of an unselective
thermolysis.

Classical photochemistry with lamps makes use of electronic excitation
of molecules. For this mostly ultraviolet or short-wavelength visible
radiation is needed. Looking for lasers in this wavelength region,
until a few years ago there were mainly weak nitrogen lasers at the
single wavelength of 337 nm or complex nonlinear optics schemes. Now-
adays commercially available excimer lasers offer quite a number of
wavelengths with considerable output power. Their reliability and ease
of handling is certainly adequate for applications in an organic
chemistry laboratory. A really sensible application would not, however, look at a laser as a mere replacement of a mercury lamp, but
rather try to make clever use of those properties of the laser for
which a laser is far superior to a lamp (vide infra).

But even considering only of the photon flux at a specified wavelength
(which determines the turnover in conventional photochemistry) lasers
compare well with typical lamps. This can be seen in Fig. 1 presenting
the spectrum of a 700 W medium-pressure mercury arc lamp (TQ 718,
Heraeus, Hanau, Fed. Rep. Germany) and the lines obtainable from a
typical excimer laser (EMG 201, Lambda Physik, Göttingen, Fed. Rep.
Germany) and some indication of the average output available today
from the many different types of dye lasers. The lamp emits all lines
simultaneously and its output must be filtered accordingly to reduce

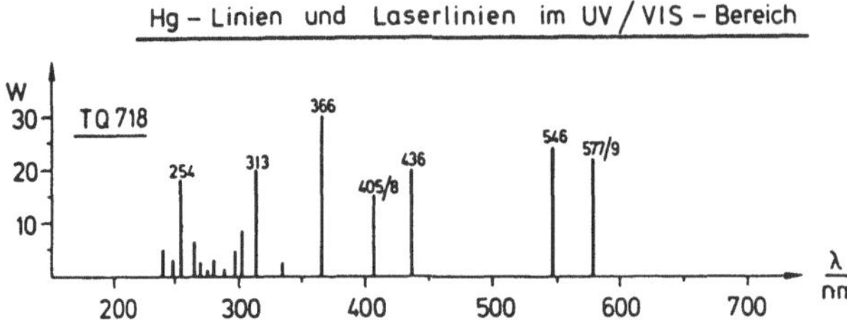

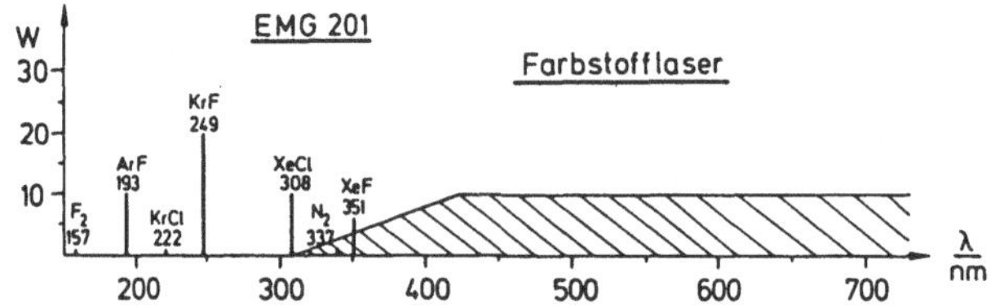

Fig. 1

the output at unwanted wavelengths, which normally also reduces the output at the specified wavelength. Excimer lasers on the other hand, emit only one of the lines shown (determined by the gas filling) and dye lasers can be tuned to any specified single wavelength in the hatched region.

A comparison of the total cost of 1 mole of photons (including capital investment, maintenance, replacement parts, energy, cooling water, gas fills etc.) at present gives about 40 DM for the lamp and 500 to 600 DM for the excimer laser shown in Fig. 1. One must bear in mind, however, that lamp technology has more or less come to its limits, so that one cannot expect any future decrease in lamp cost, whereas excimer lasers are still at their infancy and have a high potential for further developments. A reasonable extrapolation based on developmental trends gives an estimate of total cost of one mole of photons from an excimer laser in five to six years from now below that mentioned above for the lamp.

Figure 2 is a comparison of lamps and lasers considering not only photon flux but also those properties for which the lasers excel. Besides the laser and lamp type discussed above it also includes the largest commercial lamp used in photochemistry (production of nylon 6) and a high-pressure short-arc mercury lamp, much used in the laboratory, dye lasers, and CO_2-lasers. Although the latter are infrared lasers, they are nevertheless included in the comparison, since the absorption of about 20 infrared photons per molecule can put as much energy into

		Lampe		Laser		
	60 kW Hg-Tl bei 535 nm	700 W Hg Mitteldruck (Original Hanau TQ 718) bei 254 nm	HBO 200 + Kondensor + Filter bei 366 nm	KrF-Laser (Lambda-Physik EMG 201) bei 248,5 nm	Farbstofflaser 10 W bei 590 nm	CO_2-Laser 10 kW bei 10,6 µm
Quantenfluss $\left(\frac{Einstein}{h}\right)$	211	0,14	$<5\cdot10^{-4}$	0,15	0,18	$3,2\cdot10^{3}$ (162)
Intensität $\left(\frac{W}{cm^2}\right)$	3,5	0,01	$2,5\cdot10^{-3}$	$6\cdot10^{6}$	140	10^{4}
Spektrale Halbwerts-breite [nm]	ca. 20 + Linien	ca. 2 + Linien	ca. 2 + Untergrund	1	$<10^{-3}$	$<0,1$
Divergenz-winkel [°]	360	360	2,5	0,1×0,2	0,1	$<0,1$
Pulsdauer [s]	CW	CW	CW bis $>10^{-3}$	$\sim10^{-8}$	CW bis $<10^{-12}$	CW bis $<10^{-10}$
Polarisation	unpolarisiert			Polarisationsgrad >100:1 (mit Brewsterfenster)		

Fig. 2

the molecule as one ultraviolet photon. The listing is mostly self-ex-
planatory. One might additionally point out that intensities given for
lasers are those of the unfocussed laser beams and can be manipulated
by orders of magnitude due to the good collimation of laser beams by
telescopes, which is hardly possible with extended area light sources,
like lamps.

In the following, a few examples are given, which demonstrate clearly,
how some of the five most important properties of lasers, namely intensity, monochromaticity (including tunability), pulse width, collimation, and polarization, can be exploited for novel types of photochemical reactions. Thinking up new applications that make full use of
these laser-specific properties, singly or combined, is a real challenge for the creative imagination of a photochemist.

An example of preparative organic photochemistry where the monochromaticity and wavelength of lasers are important, is the production of
vitamin D. In this process the starting material (7-dehydrocholesterole
or ergosterole) is irradiated to form previtamin D (which is then
thermally transforming into the endproduct, vitamin D). Since the
spectra of educt and product overlap to a large extent, the backreac-
tion is also possible, leading to a photostationary mixture of educt
and product. At the same time several photochemical side-reactions
occur leading to several unwanted by-products, which are either physio-
logically inactive or even toxic. In the industrial production care-

ful filtering of the mercury lamp radiation and exact time-control is necessary to minimize the by-products and maximize the content of previtamin D, which is then isolated from the mixture. Fuss and Braun of the Max-Planck-Institut für Quantenoptik, Munich, used an antistokes-raman-shifted dye laser for irradiation. Scanning the wavelength region of absorption of the educt, they found that 296 nm is the optimum wavelength, resulting in an 80 % turnover with a 90 % previtamin-D content [1]. For higher efficiency one could use a raman-shifted excimer laser.

Another approach was used by Malatesta et al. [2], who first irradiated the educt with a KrF-excimer laser at 248 nm, resulting in a mixture of 25.8 % product and 71.3 % of the by-product tachysterole. In a second step they irradiated this mixture with a nitrogen laser at 337 nm where practically only tachysterole absorbs, resulting in a mixture of 80 % previtamin-D content.

Another interesting example of preparative organic laser chemistry is the production of vinylchloride from dichloroethane, as described by Wolfrum and coworkers [3]. In this radical chain reaction the production of the starting radicals is the rate-determining step, necessitating a temperature of 450° to $550\ ^{\circ}C$ in the industrial process. Using an excimer laser to produce these starting radicals not only allows the temperature to be lowered to $300\ ^{\circ}C$, but also results in a purer product by suppressing several thermal side-reactions. Here one needs the high intensity of an excimer laser to produce the necessary concentration of starting radicals, as well as the right wavelengths, to obtain the necessary penetration depth of several meters under the normal process conditions. In addition, the high collimation of the laser beam is necessary, in order to be able to enter the laser beam through a small quartzglass window at one end of a several meter long section of a tube furnace and efficiently illuminate the whole volume of it. These three conditions could never be fulfilled by any lamp, only by a laser.

The high intensities of lasers open up a whole new field of reactions from higher excited states, triplets as well as singlets, reached by the stepwise absorption of two photons. This is only possible with lasers, as one can immediately see, realizing that it needs an intensity of about 10 MW/cm^2 to keep half of the molecules in the first excited singlet state, assuming a singlet lifetime of 1 ns. Only such high photostationary excited-state concentrations result in a sufficiently high probability of the absorption of a second photon by the excited molecule. The highly excited state which is thus reached is, of course, very short-lived and only very fast reactions can compete with the fast radiationless deactivation rate. Some possibilities are intramolecular rearrangements, ring-opening reactions, four-center eliminations, and reactions with solvent molecules.

We have investigated several examples of these processes in our laboratory. A first example is the rearrangement of carvone (C) to give carvone camphor (K) and, in a sequential reaction with the ethanol solvent, a partial conversion to an ester (E). This reaction with mercury lamp irradiation has only a quantum yield of less than 3×10^{-3} and results in many polymeric by-products. Using a XeCl-excimer laser at 308 nm or a XeF-excimer laser at 351 nm, the quantum yield increases with intensity to 7 % (>30 MW/cm^2). Since carvone camphor also absorbs at 308 nm, a XeCl-laser will produce some ester, while a XeF-laser at 351 nm, where the product has a very low absorption, will give a very pure product. No polymeric by-products are seen at all. Figure 3 gives a comparison of the results with lamp and lasers [4]. Recently, a similar study, confirming our results, was published [5].

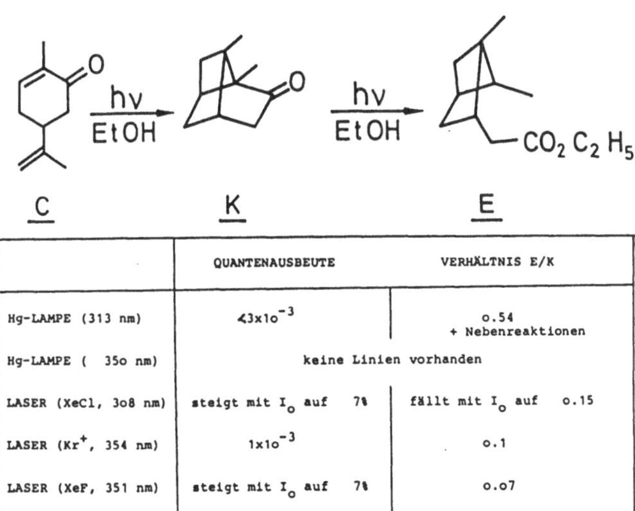

	QUANTENAUSBEUTE	VERHÄLTNIS E/K
Hg-LAMPE (313 nm)	$<3\times10^{-3}$	0.54 + Nebenreaktionen
Hg-LAMPE (350 nm)	keine Linien vorhanden	
LASER (XeCl, 3o8 nm)	steigt mit I_o auf 7%	fällt mit I_o auf 0.15
LASER (Kr$^+$, 354 nm)	1×10^{-3}	0.1
LASER (XeF, 351 nm)	steigt mit I_o auf 7%	0.07

Fig. 3

A second example are some electrocyclic reversion reactions, studied by D. Plaas in our laboratory [6]. The reaction shown in Fig. 4 was studied earlier by several groups in an organic glass matrix at 77 K [7]. In liquid solution at room temperature, no trace of the product

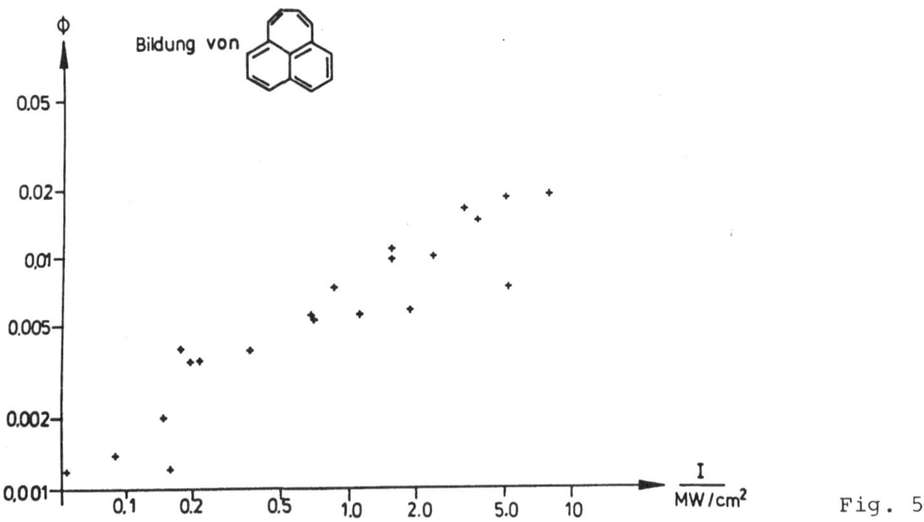

is formed with lamp irradiation. Using a XeCl-laser at 308 nm, the
quantum yield for the production of pleiadiene is seen to rise with
intensity in Fig. 5, reaching 2 % at 10 MW/cm^2. The reactions shown

in Fig. 6 give a similar functional dependence of quantum yield on
laser intensity. One of the reasons why the theoretical limit of 50 %
for the quantum yield is not reached, is the absorption by the product
formed during the irradiation. If the product is not photoreactive itself it just acts as a photon sink, dissipating the absorbed energy as

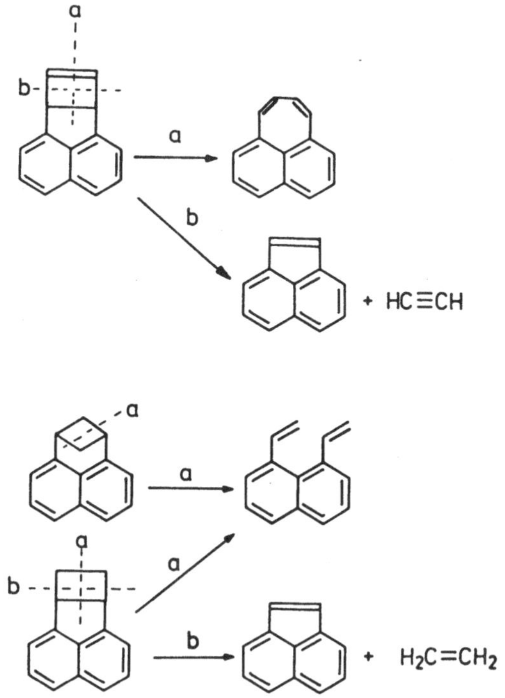

Fig. 6

heat. An especially instructive example is given in Fig. 7, which shows
the apparent extinction coefficient of a solution of the educt, shown
in the insert, time-averaged over the 10 ns pulse of the XeCl laser,
as a function of the laser intensity.

This effect should be avoidable to some extent, if the pulse width
would be sufficiently reduced, so that the pulse would be terminated
before a significant fraction of the excited molecules has reacted.
Most probably this would entrail the use of picosecond pulses.

Up to now very little work has been done on the use of picosecond
pulses in preparative organic photochemistry. One most interesting
example was recently reported by Matveetz and Letokhov [8]. An overview
of their results is shown in Fig. 8, adapted from reference [8]. Irradiation of maleic acid in aqueous solution with a lamp or low intensity
laser light only leads to isomerisation to fumaric acid (reaction I).
Using high intensity irradiation with nanosecond pulses from a KrF-
laser (248 nm) mostly leads to a dimerisation forming a cyclobutane
ring (reaction II). This is rationalized as a reaction from a highly
excited triplet state. If picosecond pulses of very high intensity from

ε × 10⁻³
l/mol×cm
308nm
ε(308)=1300
ε(308)=10200
5
4
3
2
1,25
2   5   10   20   50   100   200
I
MW/cm²

Fig. 7

I
HOOC   COOH
C=C
H   H
hν
H   COOH
C=C
HOOC   H
Maleinsäure
Fumarsäure

II
HOOC   COOH
C=C
H   H
H   H
C=C
HOOC   COOH
2hν
H H
HOOC-C-C-COOH
HOOC-C-C-COOH
H H

III
HOO   COOH
C=C
H   H
2hν
H₂O
H   H
HOOC-C-C-COOH
H   OH
Äpfelsäure

II
I (x50)
III (x10)
QUANTUM YIELD
3
2
1
0
10¹   10⁷   10⁷   10⁸   10⁹   10¹⁰
LASER INTENSITY, W·CM⁻²

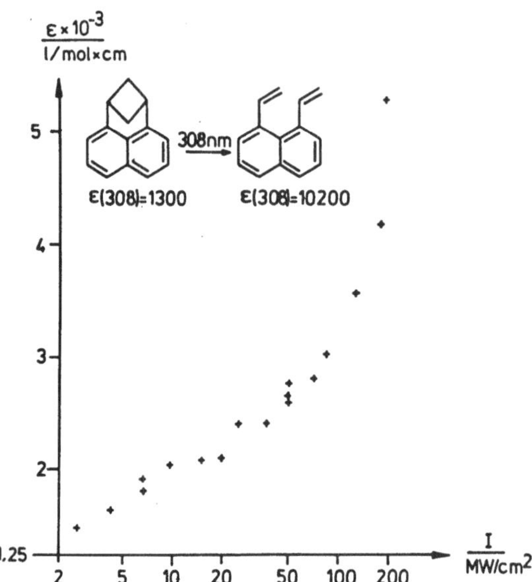

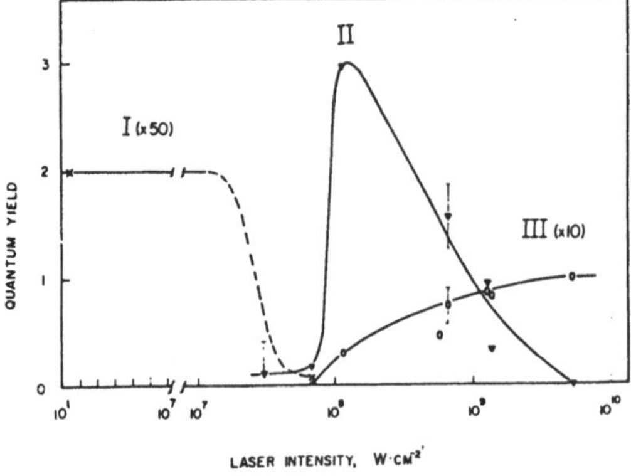

Fig. 8

a frequency-quadrupled (266 nm) Nd-YAG laser are used, most of the
molecules do not have sufficient time for an intersystem crossing before they absorb a second photon, exciting them to a higher singlet
state. A very fast addition of a water molecule to the double bond
leads to malic acid (reaction III).

This indicates the potential of very short pulses for photochemical
applications. A summary of some advantages of short pulses for photochemical work is given in Fig. 9.

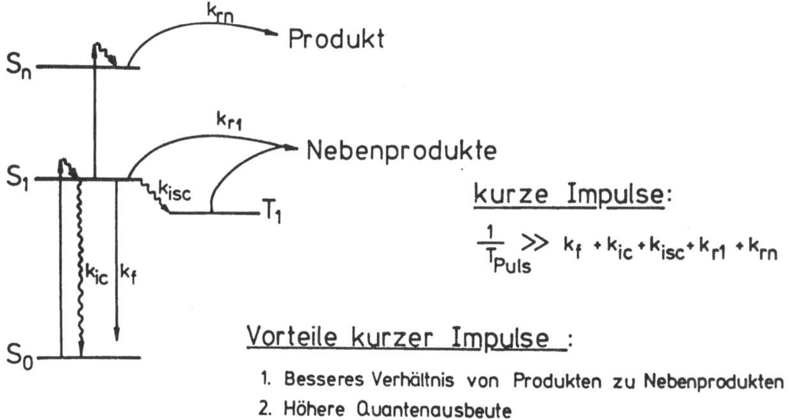

$$\frac{1}{T_{Puls}} \gg k_f + k_{ic} + k_{isc} + k_{r1} + k_{rn}$$

$$(k_{rn} \sim I^2)$$

Fig. 9

The most important obstacle to the application of picosecond pulses for
preparative organic photochemistry is the expense, complexity, ineffi-
ciency, and low reliability of state-of-the art picosecond pulse lasers,
which only give ultraviolet pulses using nonlinear optical methods. At
present we are developing an efficient, relatively inexpensive, and
reliable excimer laser source of ultraviolet picosecond pulses, using
a scheme indicated in Fig. 10. A small excimer laser is used to pump
via a small dye laser a special dye laser ("distributed-feedback dye
laser" [9]) that produces single pulses of some ten ps duration. After
frequency-doubling the weak ultraviolet pulse is fed into a second,
larger excimer laser, used as a regenerative amplifier, which results
in very high intensity picosecond pulses at the high repetition rate
possible with excimer lasers [10].

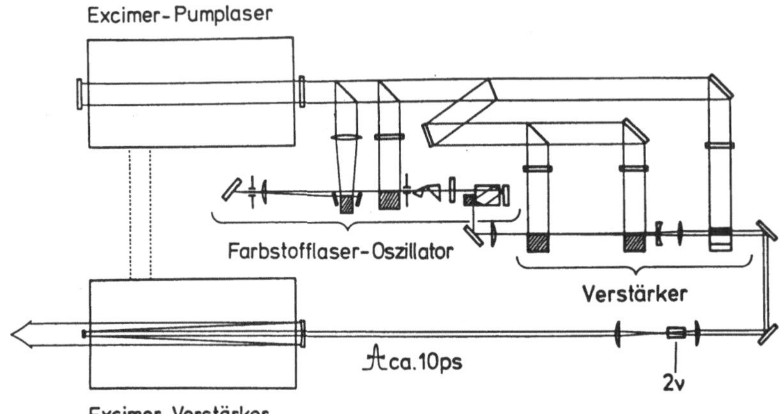

Fig. 10

In conclusion, one can expect that the rapid development of lasers
suitable for preparative organic photochemistry will open up a new
era in photochemistry in the near future for which a few early examples
were mentioned above.

References

[1] FUSS, W.: private communication
[2] MALATESTA, V., C. WILLIS and P.A. HACKETT: J. Am. Chem. Soc. 103
    (1981) 6781
[3] SCHNEIDER, M. and J. WOLFRUM: these Proceedings
[4] BRACKMANN, U. and F.P. SCHÄFER; Chem. Phys. Lett. 87 (1982) 579
[5] MALATESTA, V., C. WILLIS and P.A. HACKETT: J. Org. Chem. 47
    (1982) 3117
[6] PLAAS, D.: unpublished
[7] MEINWALD, J., G.E. SAMUELSON, M. IKEDA, R.M. PAGNI and
    J.A. BUTCHER, jun.: J. Am. Chem. Soc. 92 (1970) 7604;
    TURRO, U.J. and V. RAMAMURTHY: J. Org. Chem. 42 (1977) 92;
    CASTELLAN, A., J. KOLC and J. MICHL: J. Am. Chem. Soc. 100
    (1978) 6687
[8] KHOROSHILOVA, E.V., N.P. KUZMINA, V.S. LETOKHOV and
    Yu.A. MATVEETZ: Appl. Phys. B 31 (1983) 145
[9] BOR, Zs., ALEXANDER MÜLLER, B. RACZ and F.P. SCHÄFER:
    Appl. Phys. B 27 (1982) 9;
    BOR, Zs., ALEXANDER MÜLLER, B. RACZ and F.P. SCHÄFER:
    Appl. Phys. B 27 (1982) 77
[10] SZATMARI, S. and F.P. SCHÄFER: unpublished

Multi-Photon Ionization and Laser Mass Spectrometry

H.J.Neusser
Institut für Physikalische Chemie, Technische Universität München, Lichtenbergstr.4,
8046 Garching

I.   Introduction

Today mass spectrometry is a very common analytical method for sensitive detection in industrial research, physics and chemistry.

The principle of operation in mass spectrometry is based on the ionization of the neutral particles and the mass-selective detection of the ions due to their different e/m ratio. Therefore, one of the important parts of a mass spectrometer is the ionization region. In most commercial mass spectrometers ionization is achieved by electron impact. This is certainly a broadly applicable technique but as many other ionization methods it suffers from a lack in selectivity. For efficient ionization an electron energy as high as 70 eV is necessary which is much larger than the ionization energy of the species under consideration. A not defined fraction of this high energy is transferred to the molecules which are not only ionized but also dissociated to a large amount. As a consequence of this non selective energy transfer many competing reactions (e.g. dissociations) take place within the ionization region which are often dependent even on the type of the instrument.

It is therefore of interest to use a laser for ionization which is already known to lead to very selective excitation of atoms and molecules. The simplest way to obtain ionization by laser light is to heat the material within the laser focus. Then at sufficient power levels evaporization and ionization occurs due to the high temperature of the heated material. This type of ionization, however, is not selective in energy since as in every sample at thermal equilibrium there exists a broad energy distribution.

Another possibility which is going to be discussed in this work is to induce resonant absorption of the material which leads to ionization without thermal heating.

This is usually achieved for low particle densities, i.e. in the gas phase. Unfortunately, the wavelength necessary for a resonant one-photon ionization process is in the vacuum ultraviolet region of the spectrum ($h\nu > 7eV$), where laser sources are hardly available or operate with very low efficiency. This type of ionization, the conventional photoionization, is shown in Fig.1a for the typical organic molecule benzene.

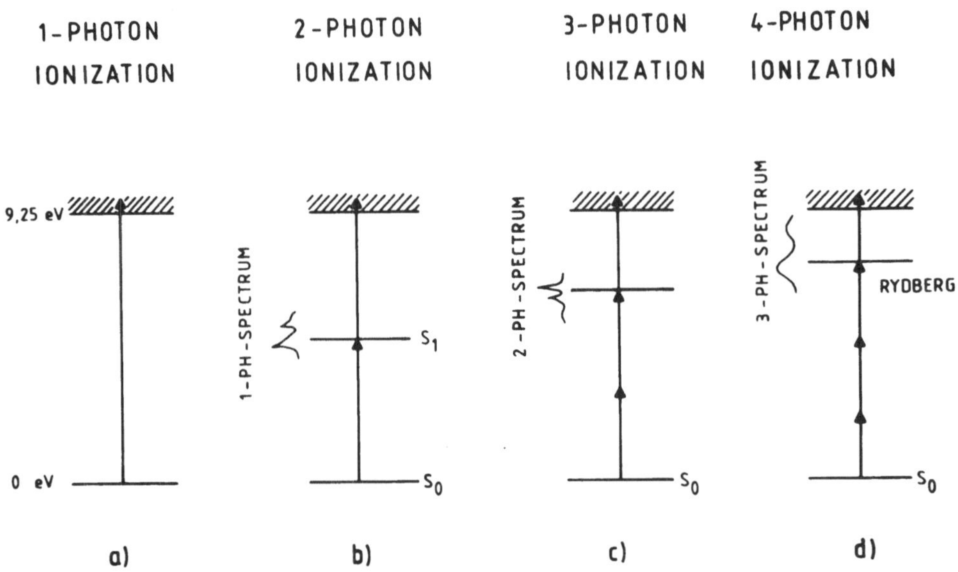

Fig.1.　Several possible paths for multi-photon ionization with resonant intermediate states at different energy levels.

II. Characteristics of Multi-Photon-Ionization in a Mass Spectrometer

An alternative method is the ionization of the molecule by multi-photon absorption, i.e. the absorption of several UV or visible photons by a single molecule within a short time. The efficiency of this multi-photon ionization is drastically increased if real states of the molecules are in resonance with the photon energy (resonance enhancement)/1/. Several possibilities with real states at the energy of the first, the second or the third photon are shown in Fig.1b, c, d,. For example, resonance enhanced two-photon ionization (Fig.1b) is obtained if the laser wavelength is tuned to a one-photon absorption band of the molecule under investigation. Then, even for moderate laser intensities of some 10^7 W/cm^2 which can be obtained with a focused N_2 laser pumped dye laser beam a very efficient ionization of nearly all molecules within the focus volume is achieved.

In addition to the high efficiency, multi-photon ionization via a resonant intermediate state is very species selective due to the resonance condition in the intermediate state. Particularly, in a mixture of molecules that component is preferentially ionized whose absorption band is in resonance with the laser wavelength. Recently, we have demonstrated that the spectral selectivity is sufficient to distinguish between the isotopes of polyatomic molecules e.g. mono$-^{13}$C$-$benzene and ^{12}C$_6$H$_6$ /2/. This then led to a strong optical enhancement of the ^{13}C^{12}C$_5$H$_6$ signal within the mass spectrum of benzene. Generally, a trace material within a mixture can be detected with increased sensitivity if the laser wavelength is tuned to one of the numerous peaks within the absorption spectrum of this trace material and if the absorption coefficient of the various components differs at the chosen wavelength. The coupling of a mass spectrometer with a laser adds the wavelength as a second dimension to the mass and makes mass spectrometry two dimensional /3/.

It has been found in our early experiments that resonance-enhanced two-photon ionization can be a very "soft" ionization /4/. This means, that the molecules are ionized without destruction even at those intensity levels where ionization is already very efficient ($\sim 10^6$ W/cm^2) (see Fig.2a). A "soft" ionization is important for detection of trace amounts since then no fragment ions appear which may produce an interfering background. Also, for molecular weight determination the appearance of fragment ions would complicate the identification of the parent ion.

On the other hand, if the laser intensity is further increased to some 10^8 W/cm^2 ionization is no longer soft but in addition to the parent ion many fragment ions appear (hard ionization)/5/. This is shown for C_6H_6 in Fig.2b. One of the strongest ion peaks is now the C^+ fragment ion. If intensity is further increased to 10^9 W/cm^2 the C^+ fragment ion is found to be the by far strongest peak in the mass spectrum /6/. This means that benzene ions have now absorbed such an amount of energy that they burst into an atomic C^+ ion fragment and a neutral rest. Simple thermodynamic calculations readily show that at least an energy as high as 35 eV is necessary to produce C^+ from neutral C_6H_6. Hence, it is clear that the absorption of at least 8 UV photons is necessary to produce C^+ fragments from neutral C_6H_6. The mechanism of this "superexcitation" has been investigated in our recent

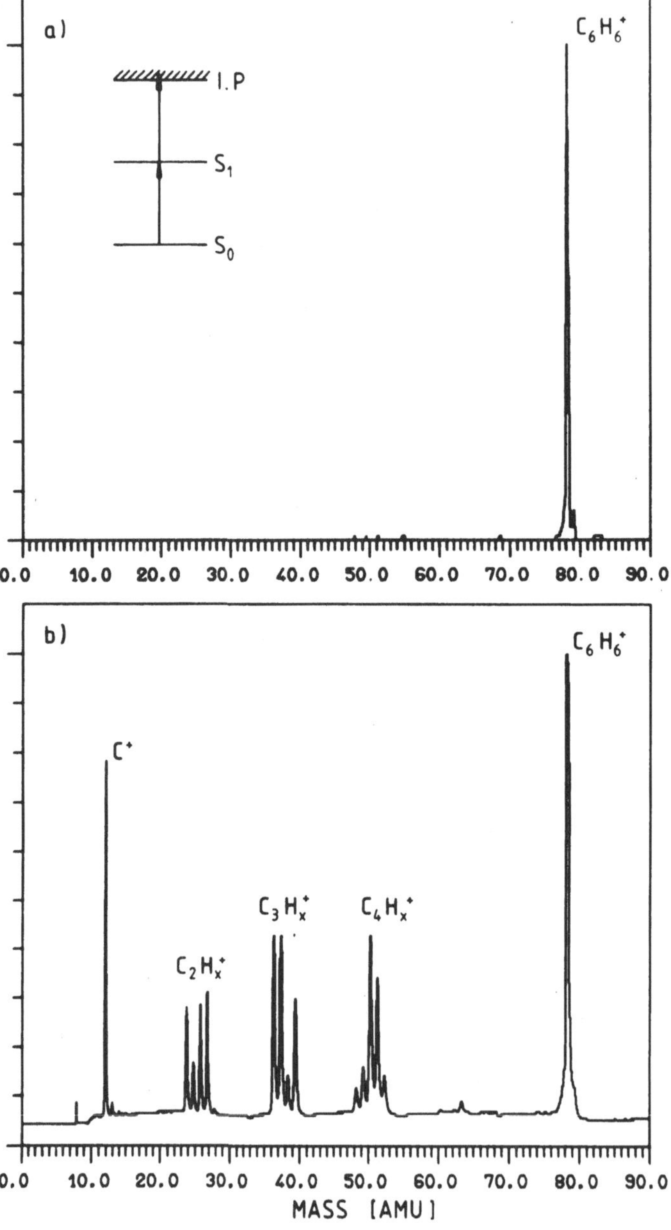

Fig.2. Multi-photon mass spectra of C_6H_6 for different laser intensities
a) low intensity ($10^6 W/cm^2$): soft ionization
b) high intensity ($10^8 W/cm^2$): fragmentation

work both experimentally /6/ and theoretically /7/. It has been found that photon absorption switches from the parent ion after its dissociation to the smaller fragment ions which then in turn again absorb and dissociates until ultimately C^+ is produced (ladder switching process). It is important that this excitation mechanism is different from that in electron impact ionization and therefore leads to a different and more specific energy release within the molecular ions. This specific energy release is important for ion kinetic studies and for a control of fragmentation in the mass spectrum.

III. <u>Mass Spectrometer</u>

In the following section some of the basic properties of this new type of ion source and the resulting mass spectrometer will be discussed.

Since the laser beam can be focused down to about 50 μm all ions are produced within a very small volume of about 10^{-5} cm^3. Furthermore in this point like ion source all ions are produced within a very short time due to the short duration (5 nsec) of the laser pulse. This high spatial and temporal resolution constitute an ideal source for a time-of-flight mass spectrometer.

Due to these characteristics of the ion source a time-of-flight mass spectrometer is established in a straight forward manner by adding an acceleration region with an acceleration field and a field-free drift tube with an ion detector at the end of it /8/. This set up is simpler than in conventional time-of-flight instruments since the ion source is already pulsed and therefore no pulsed acceleration voltage is needed. The mass resolution of this prototype laser time-of-flight mass spectrometer was found to be M/Δ M $\sim$ 250. This mass resolution is sufficient to resolve single mass differences for molecules up to amu 250.

It has been shown for a number of molecules that in resonance-enhanced two-photon ionization already for moderate light intensities of less than 10^7 W/cm^2 a complete saturation of both absorption steps (see Fig.1b) is achieved and nearly all molecules within the focus volume are ionized. After a single laser pulse about 10^6 ions appear at the detector yielding a strong ion signal which is sufficient to obtain a complete mass spectrum with good signal to noise ratio. Therefore, laser multi-photon ionization is a very fast and sensitive detection method which allows e.g. for the fast and selective detection of short-lived reaction products. With a 10 Hz laser a complete mass spectrum can be sampled every 100 msec. If the gas inlet into the mass spectrometer is pulsed and synchronized with the laser pulse only very small amounts of material are necessary ($< 10^{-12}$g) for the measurement of a complete mass spectrum with good signal to noise ratio.

Another benefit of the laser multi-photon ionization is the small kinetic energy difference of the ions produced within the pointlike ion source. In conventional ion sources the kinetic energy distribution is determined by the extension of the ionization region as well as by the spread of velocities of the ions and ion fragments upon preparation. If the ions originate from different places within the acceleration region they are accelerated in a somewhat different potential giving rise to different kinetic energies. In the case of laser multi-photon ionization all ions are produced within a small focus of 100 μm diameter and the kinetic energy spread is only about 1% of the final kinetic energy. Even though this is a narrow kinetic energy distribution it is the dominating factor for a restriction of the mass resolution to about 250.

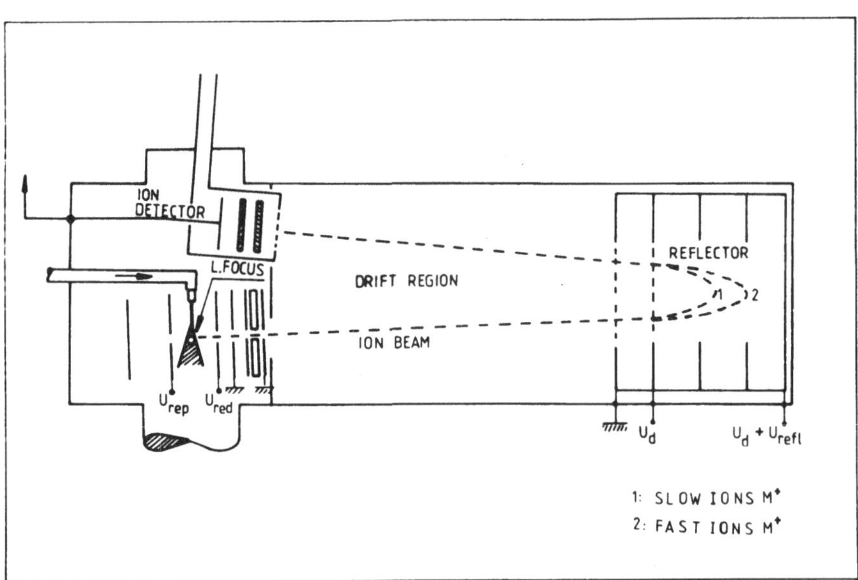

Fig.3.   Schematic illustration of the reflecting field high resolution time-of-flight mass spectrometer.

We have increased the resolution of the time-of-flight mass spectrometer by using a reflecting field set up /9/ which compensates for kinetic energy differences of the ions. A scheme of the reflectron type time-of-flight mass spectrometer is shown in Fig.3. The molecular ions are produced in an effusive molecular beam in front of the nozzle by resonantly enhanced two-photon ionization (see Fig.1b). The ions produced in the laser focus are accelerated by an attracting electrode. After having passed the field-free drift region and decelerated to low velocities the ions enter the reflecting field region and are finally reflected back towards the detector. Due to a long residence time within the reflecting field differences in time-of-flight for ions of the same mass can be compensated. The difference in time-of-flight is corrected since faster ions with higher kinetic energy penetrate more deeply into the reflecting field region and therefore remain for a longer time as opposed to ions of the same mass but with lower kinetic energy. (See for illustration ion pathways 1 and 2 in Fig. 3). Such differences in time-of-flight are due to the differences of the accelerating potential across the laser focus. Since in multi-photon ionization the ionization region is small compared to other ionization methods the resulting kinetic energy differences of some percent can be fully corrected in the reflecting field mass spectrometer. Therefore, the combination of laser multi-photon ionization with the reflecting field set up yields a promising technique for obtaining high resolution mass spectra with high sensitivity.

This is demonstrated in Fig.4a,b where a small part of the multi-photon mass spectrum of C_6H_6 is shown on an extended scale. It represents the mass peaks around the parent ion mass peak $C_6H_6^+$. For illustration the high mass resolution obtained with the reflecting field set up (Fig.4a) is compared with the mass spectrum obtained with the time-of-flight mass spectrometer without reflecting field (Fig.4b). The mass resolution in Fig.4b is $M/\Delta M \sim 250$ whereas the mass resolution has been increased by use of the reflecting field by more than one order of magnitude to $M/\Delta M \sim 3900$ (Fig.4a). At present time the mass resolution is limited mainly by the length of the laser pulse, and to some extent also by space charge effects but it is conceivable that it can be increased to about 10000, so that unit mass resolution seems to be possible even for biological macromolecules.

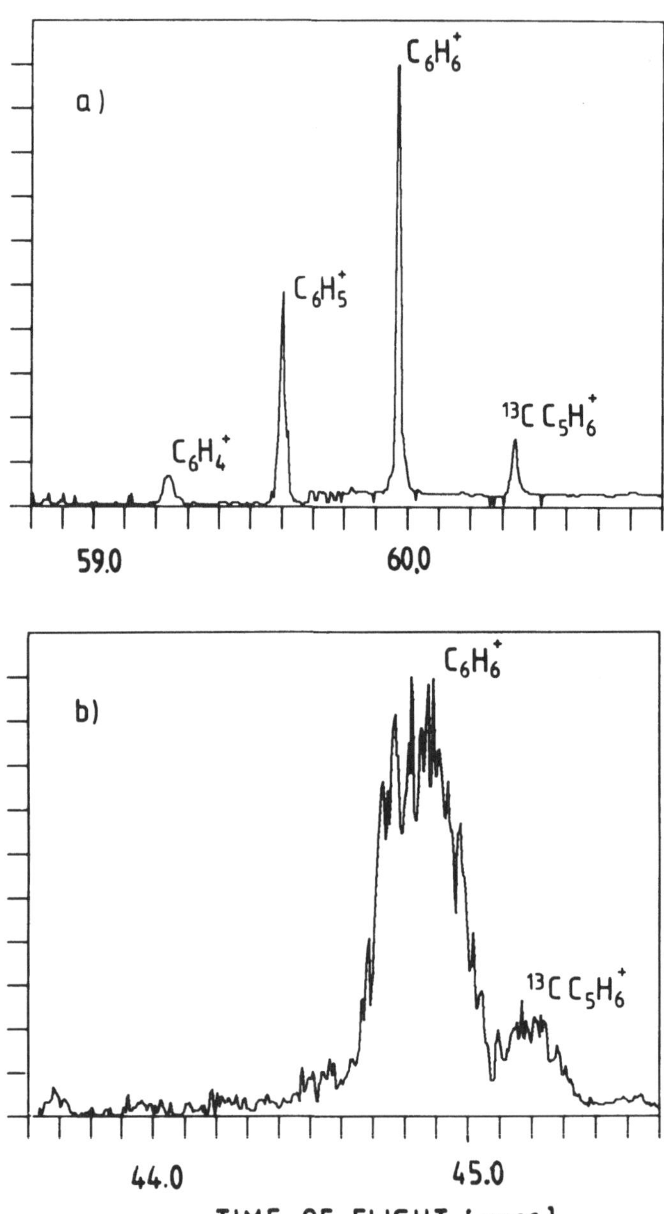

Fig.4.   Part of the multi-photon mass spectrum of C_6H_6 recorded

a) with the high resolution reflectron type time-of-flight mass spectrometer

b) with the same time-of-flight mass spectrometer operating without reflecting field.

IV. Structure Determination and Metastable Ions

One of the important methods to reveal the unknown structure of molecules from their mass spectral pattern is to study the decay channels of resulting molecular ions. These are a typical finger-print for molecular groups and are used in order to identify the unknown structure of a molecule under investigation. For the identification of ion decay channels metastable ions historically play an important role, since for metastable decay channels the precursor ion as well as the product ions can be determined in a reliable manner. This is possible, since metastable ions decay on the way from the ion source to the detector resulting in an energy shift of the product ions which is typical for the mass of the product ions and the mass of the precursor ion.

Multi-photon ionization mass spectrometry brings about a new impetus for metastable ion detection due to three reasons. Firstly, the energy distribution of the ions (particularly of the parent ions) is sharp, i.e. the internal energy of the parent ions is rather well defined for a given wavelength of the laser light. As a result it can be found which decay channels are active at a given internal energy of the ions /10/ and these decay channels are stronger than in a conventional experiment with broad energy distribution resulting in many different competing decay channels at different energy levels.

Secondly, the sensitivity for metastable ion decay detection is considerably increased in a reflecting field time-of-flight mass spectrometer with laser multi-photon ionization. This can readily be seen from Fig.4 a, b. In the reflecting field mass spectrum of Fig.4a two additional strong mass peaks ($C_6H_4^+$, $C_6H_5^+$) appear which are not seen in Fig.4b. These mass peaks are due to a metastable decay of excited $C_6H_6^+$ in the field-free drift region of the mass spectrometer according to the following reactions

$$C_6H_6^{+^*} \longrightarrow C_6H_5^+ + H$$

$$C_6H_6^{+^*} \longrightarrow C_6H_4^+ + H_2$$

This metastable decay, however, cannot be observed in the conventional time-of-flight mass spectrometer.

Often it is difficult to detect ions resulting from a metastable decay since the corresponding peaks in the mass spectrum are blended by the much stronger stable ions. A third advantage of this technique is that the signal of stable ions can be suppressed by suitable choice of the reflecting voltage due to their different kinetic energy. In this way the sensitivity for detection of metastable decay channels is increased substantially in multi-photon mass spectrometry.

V. Summary

It has been shown that multi-photon ionization within the small focus of an intense pulsed laser beam can lead to an efficient ionization of the absorbing molecules. The efficiency of the multi-photon ionization process is drastically increased if the laser wavelength is tuned to the maximum of the molecular absorption spectrum (resonance-enhanced multi-photon ionization). Particularly, in a mixture of molecules that component is preferentially ionized whose absorption band is in resonance with the laser wavelength.

Combining multi-photon ionization with a mass spectrometer leads to a highly selective and sensitive detection of a particular species within a mixture (trace analysis). The suitability of multi-photon ionization for trace analysis is further improved at low light intensities (10^6 W/cm^2) because under these experimental conditions soft ionization without fragmentation of the parent molecular ion takes place.

Due to the properties of this new ion source a time-of-flight mass spectrometer is the instrument of choice in multi-photon mass spectrometry. We have increased the limited mass resolution of a time-of-flight mass spectrometer to $M/\Delta M \sim 4000$ using a reflecting field set up which compensates for kinetic energy differences of the ions.

Furthermore, it has been shown that metastable slow decay processes of the ions after the laser pulse can be observed with particularly high sensitivity in a laser mass spectrometer. These metastable decay processes are known to be important for the determination of the structure of molecules.

In conclusion, the use of lasers brings about a series of benefits in mass spectrometry due to the selective ionization of the species under investigation.

References

/1/   P.M.JOHNSON, J.Chem.Phys. $\underline{62}$ (1975) 4562

/2/   U.BOESL, H.J.NEUSSER, E.W.SCHLAG, J.Am.Chem.Soc. $\underline{103}$ (1981) 5058

/3/   V.S.LETOKHOV, Phys.Today. May 1977, p.23

/4/   U.BOESL, H.J.NEUSSER, E.W.SCHLAG, Z.Naturforsch. $\underline{A33}$ (1978) 1546

/5/   L.ZANDEE, R.B.BERNSTEIN, J.Chem.Phys. $\underline{71}$ (1979) 1359
J.P.REILLY, K.L.KOMPA, J.Chem.Phys. $\underline{73}$ (1980) 5468

/6/   U.BOESL, H.J.NEUSSER, E.W.SCHLAG, J.Chem.Phys. $\underline{72}$ (1980) 4327

/7/   W.DIETZ, H.J.NEUSSER, U.BOESL, E.W.SCHLAG, S.H.LIN, Chem.Phys. $\underline{66}$ (1982) 105

/8/   U.BOESL, H.J.NEUSSER, E.W.SCHLAG, Chem.Phys.Letter $\underline{87}$ (1982) 1

/9/   U.BOESL, H.J.NEUSSER, R.WEINKAUF, E.W.SCHLAG, J.Phys.Chem. $\underline{86}$ (1982) 4857

/10/   H.KÜHLEWIND, H.J.NEUSSER, E.W.SCHLAG, Int.J.Mass Spectrom.Ion Phys., in press

Fluoreszenzspektroskopie hoher Zeitauflösung und hoher Empfindlichkeit

W. BECKER, S. DÄHNE, K. JUNGE und E. KLOSE
Zentralinstitut für Optik und Spektroskopie
Akademie der Wissenschaften der DDR
Rudower Chaussee 6, DDR-1199 Berlin

Die Erzeugung ultrakurzer Laserlichtimpulse mit Impulsdauern im Pico-
sekundengebiet (10^{-12} s) und neuerdings auch im Femtosekundenbereich
bis herab zu 4 . 10^{-14} s hat die Möglichkeit geschaffen, diesen Zeitbereich für viele Zweige von Wissenschaft und Technik zu erschließen
(1,2). Eine häufig verwendete Methode zur Signalerfassung derartig
kurzer Impulse ist die zeitkorrelierte Einzelphotonenzählung (TCSPC).
Die Methode hat sich insbesondere in der Fluoreszenzspektroskopie
bewährt. Der Grund dafür liegt in der relativ hohen Zeitauflösung
und der sehr hohen Empfindlichkeit, die lediglich durch die Quanten-
struktur des Lichtes begrenzt ist.

Das Signal-Rausch-Verhältnis des Meßsignals (SNR) ist bei einem
idealen Empfänger durch die Statistik der Photonenregistrierung gegeben durch

$$SNR = \sqrt{\overline{I}_m \cdot \eta \cdot K \cdot T \cdot \frac{\Delta t}{\tau}}$$

In dieser Gleichung bedeuten: $\overline{I}_m$ – gemittelter Photonenstrom am
Photoempfänger; η – Quantenausbeute des Empfängers; K – Verhältnis
der Zahl der vom Meßverfahren verarbeiteten Photonen zur Zahl der
registrierten Photonen; T – Meßzeit; Δt – Zeitauflösung des Meßsystems; τ – charakteristische Zeit des Meßlichtes (z. B. Fluores-
zenzlebensdauer). Ein hohes SNR ist bei vorgegebenem Δt und T
durch hohe Werte von $\overline{I}_m$ und K zu erreichen. Solange die untersuchten Vorgänge linear sind, ist $\overline{I}_m$ proportional zur Intensität
$\overline{I}_a$ des Anregungslichtes. Damit kann das Produkt S = K . $\overline{I}_a$ als
Maß für die Empfindlichkeit eines Meßverfahrens angesehen werden.
Die TCSPC zeichnet sich durch einen Photonenausnutzungsfaktor
K = 1 aus. Bei Kombination dieser Methode mit modensynchronisierten kontinuierlich laufenden Lasern mit hohem $\overline{I}_a$ sind Empfindlich-
keiten zu erwarten, die um mehrere Größenordnungen über den mit
herkömmlichen Meßverfahren erreichten Werten liegen (z. B. 4 - 6

Größenordnungen im Vergleich zu TCSPC mit Blitzlampenanregung oder
Boxcarintegrator mit Stickstofflaseranregung).

Bei der TCSPC werden die registrierten einzelnen Photonen der
Fluoreszenz einer impulsförmig angeregten Probe nach ihrer zeitlichen Lage in bezug auf die Anregungsimpulsfolge sortiert und in einem
Digitalspeicher summiert. Unter der Bedingung, daß die Wahrschein-
lichkeit des Eintreffens mehrerer Photonen pro Anregungsimpuls ausreichend gering ist, erhält man im Speicher die zeitliche Dichte-
funktion der Photonen und damit den gesuchten Zeitverlauf des Lumi-
neszenzlichtes (3). Während die TCSPC im Zusammenwirken mit gepulsten Lichtquellen mit Folgefrequenzen bis in den kHz-Bereich seit
längerer Zeit technisch beherrscht wird, ergeben sich in Verbindung
mit modensynchronisierten Lasern mit einer Impulsfolge um 100 MHz
erhebliche Probleme. Das Prinzip des Systems (Fig. 1) zeichnet sich
durch folgende Vorteile aus:

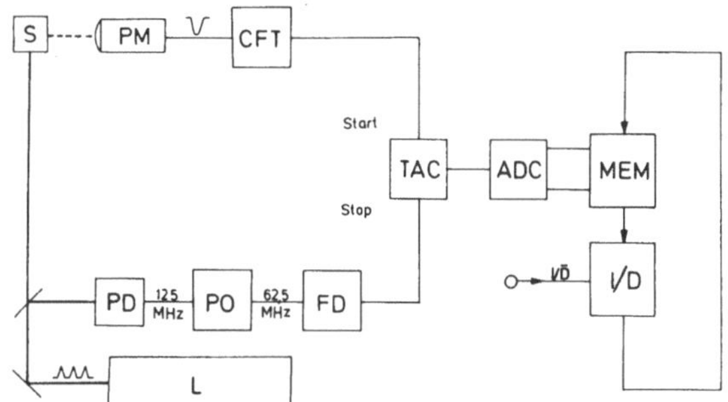

Fig. 1. Prinzipschaltbild der entwickelten TCSPC-Elektronik

- Verarbeitung der vollen Impulsfolgefrequenz des verwendeten
  Lasers von 125 MHz.
- Realisierung einer Photonenzählrate bis 500 kHz, wodurch
  kurze Meßzeiten und hohe Meßgenauigkeit erreicht werden.
- Möglichkeit zur Unterdrückung von Dunkelimpulsen des SEV,
  von Streulicht und von störenden Fluoreszenzkomponenten mit
  Hilfe einer Wechsellichtanordnung mit Meß- und Vergleichs-
  küvette.

584

- Exakte, von Leistungsschwankungen des Lasers weitgehend unabhän-
  gige Synchronisation der Messung mit der Anregungsimpulsfolge
  durch einen parametrischen Oszillator sowie Möglichkeit zur Darstellung von mehr als einer Impulsperiode.
- Numerische Kurvenentfaltung durch eingebauten Mikrorechner, wodurch die Abklingzeit der Fluoreszenz im Bereich von 20 ps bis
  40 ns bestimmt werden kann.
- Durch Rechnersteuerung des Meßablaufs wahlweise Aufnahme von
  Fluoreszenzabklingkurven und von zeitaufgelösten oder statischen
  Fluoreszenzspektren.

Der Meßablauf vollzieht sich in folgender Weise:
Die Probe (S) wird durch die Lichtimpulse des Lasers (L) zur Fluoreszenz angeregt. Die einzelnen Photonen erzeugen am Ausgang des Photo-
nen-Vervielfachers (PM) Impulse, die im anschließenden Constant
Fraction Trigger (CFT) zu definierten Triggerimpulsen aufbereitet
werden. Parallel dazu wird über eine Photodiode (PD) ein Signal erzeugt, das zur Synchronisation des Meßablaufs mit der 125 MHz-Im-
pulsfolge des Lasers dient. Das Signal wird selektiv verstärkt und
begrenzt und steuert einen parametrischen Oszillator (PO), an dessen
Ausgang ein Signal von 62,5 MHz zur Verfügung steht. Dieses Signal
ist weitgehend vom Rauschen befreit und phasenstarr mit der Laser-
impulsfolge gekoppelt. Durch den anschließenden Frequenzteiler (FD)
kann die Frequenz dieses Synchronisationssignals weiter reduziert
werden. Beide aufbereiteten Signale werden dem Zeit-Amplituden-Kon-
verter (TAC) zugeführt, an dessen Ausgang für jedes Photon eine
Spannung entsteht, die linear mit dem Zeitpunkt seiner Registrierung
in bezug auf das Synchronisationssignal verknüpft ist. Im hier angewandten System sind gegenüber herkömmlichen Systemen die Start-Stop-
Impulse vertauscht. Das Photon des Meßlichtes gibt das Start-Signal
und der darauf folgende Impuls des Anregungslichtes das Stop-Signal.
Das ist deshalb möglich, weil bei Verwendung von modensynchronisierten Lasern der Impulsabstand von 8 ns mit einer Genauigkeit von besser als 10^5 konstant ist. Zum anderen hat diese Variante den Vorteil,
daß das elektronische System nur dann ausgelöst wird, wenn ein Photon
des Meßlichtes vorhanden ist.   Der im TAC erzeugte Spannungswert
wird im Analog-Digital-Wandler (ADC) in eine Binärzahl umgewandelt,
die im Meßwertspeicher (MEM) eine bestimmte Adresse einstellt. Das
auf dem adressierten Speicherplatz vorhandene Datenwort wird durch
eine Inkrement/Dekrement-Schaltung (I/D) inkrementiert, d. h. im
Speicher wird eine 1 addiert bzw. eine 1 subtrahiert.

Für die Unterdrückung von Streulicht, der Eigenfluoreszenz von Lösungsmitteln oder von Wirtsgittern sowie der Dunkelimpulse des PM wird im optischen Teil (Probenraum S) eine Wechsellichtanordnung mit periodischer Umschaltung zwischen der Probe und einer Vergleichsprobe verwendet. Mit der Lichtumschaltung wird gleichzeitig die I/D-Schaltung zwischen Inkrementieren und Dekrementieren umgeschaltet. Dadurch wird erreicht, daß die Photonen aus der Probe (Lumineszenz- und Streulicht einschließlich Dunkelimpulse) addiert, die Photonen aus der Vergleichsprobe subtrahiert werden.

Der Ablauf der gesamten Messung wird durch ein Mikrorechnersystem gesteuert. Es sorgt auch für die Datenausgabe aus dem Meßspeicher über Display, Schreiber, Drucker oder Lochstreifenstanzer, für numerische Kurvenentfaltung, für die Steuerung des optischen Systems sowie für die Umsortierung der Meßwerte bei der Aufzeichnung von statischen Fluoreszenzspektren.

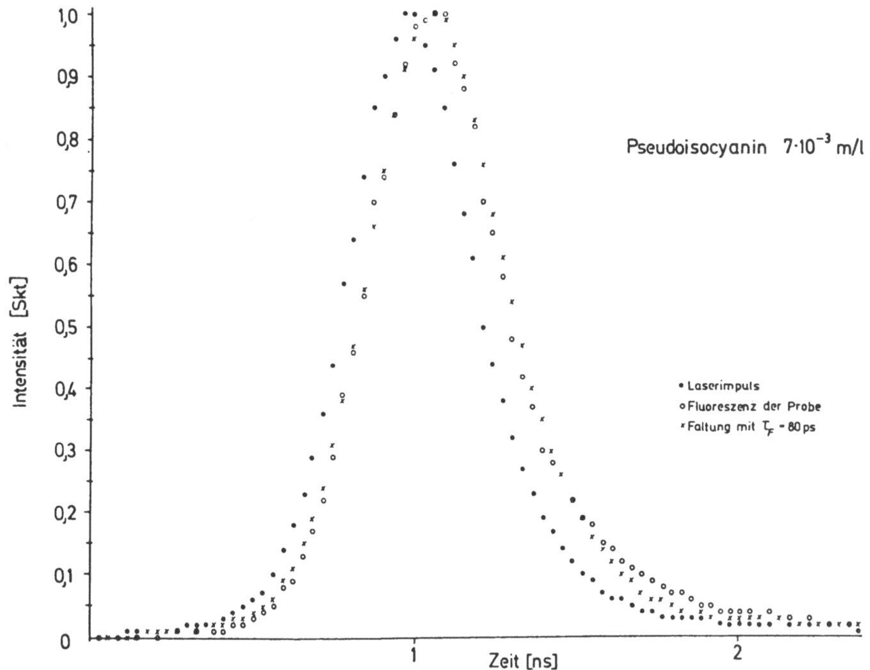

Fig. 2. Fluoreszenzabklingfunktionen
Kurve ●: modensynchronisierter Argonionenlaser ILA 120
       (125 MHz Folgefrequenz) bei 514 nm
Kurve ○: 7 . 10⁻³ molare, wäßrige Lösung von Pseudo-
       isocyaninchlorid bei 579 nm

In dem Meßbeispiel (Fig. 2) wurde die Fluoreszenzabklingzeit einer wäßrigen Lösung von Pseudoisocyanin untersucht. Der Farbstoff besitzt Interesse für die photographische Sensibilisierung von Silberhologenid-Emulsionen. Er bildet J-Aggregate, deren photophysikalisches Verhalten es noch aufzuklären gilt (4). Kurve ● ist die Impulsantwort des Meßsystems mit einer Halbwertsbreite von 425 ps. Kurve o zeigt das zeitliche Fluoreszenzverhalten der Meßprobe. Die numerische Entfaltung beider Kurven ergibt eine tatsächliche Fluoreszenzabklingzeit – unter der Annahme eines einfach exponentiellen Fluoreszenzverlaufes – von 80 ps, wobei der mittlere Fehler nur $\pm$ 5 ps beträgt.

Durch Erweiterung des Wellenlängenbereiches des Anregungslichtes sowie durch Kombination dieser Nachweismethode mit Methoden der nichtlinearen Optik zum Zwecke der Erzielung einer höheren Zeitauflösung (Bildung von Kreuz-Korrelations-Funktionen zwischen Anregungslicht und Meßlicht (2)) ist diese Methode hervorragend für Untersuchungen photophysikalischer und photochemischer Primärprozesse und Elementarreaktionen geeignet.

Die Autoren danken den Herren H. Dürr und H. Stiel für die Bereitstellung von Meßergebnissen.

Literatur
(1) Picosecound Phenomena I, II, III Springer, Berlin (West),
    1978, 1980, 1982
    Ultrafast Phenomena in Spectroscopy   I Tallinn (USSR) 1978
                                          II Reinhardsbrunn (GDR) 1980
(2) W. BECKER, K. BERNDT, E. KLOSE: Ultrashort Phenomena in
    Spectroscopy UPS-83 to be published
(3) A. MUENTER: J. Phys. Chem. 80 (1976) 2178
(4) F. FINK, E. KLOSE, K. TEUCHNER und S. DÄHNE: Chem. Phys. Lett.
    45 (1977) 548
    S.K. RENTSCH, R.V. DANELIUS, R.A. GADONAS, A. PISKARSKAS:
    Chem. Phys. Lett. 84 (1981) 446
    B. KOPAINSKY, W. KAISER: Chem. Phys. Lett. 88 (1982) 357.

Einfluß der Jodzellenreinheit auf die Frequenz I_2-stabilisierter Laser

F. SPIEWECK
Physikalisch-Technische Bundesanstalt
Bundesallee 100, D-3300 Braunschweig

Die international empfohlenen Frequenzen bzw. Wellenlängen I_2-stabilisierter He-Ne- und Ar^+-Laser (ν_i = 473 612 214,8 MHz, λ_i = 632 991 398,1 fm; ν_o = 489 880 355,1 MHz, λ_o = 611 970 769,8 fm; ν_{a3} = 582 490 603,6 MHz, λ_{a3} = 514 673 466,2 fm) /1/ gelten für "reine" Jodzellen. Enthalten jedoch die Zellen außer Jod noch Luft /2/ oder unbekannte Verunreinigungen, so weisen die stabilisierten Frequenzen der Laser im allgemeinen eine Rotverschiebung auf /2-4/. Ziel der vorliegenden Arbeit war es, abzuschätzen, welche Rotverschiebung bei zur Stabilisierung benutzten Jodzellen auftreten kann. Hierzu eignen sich Untersuchungen der I_2-Fluoreszenz bei Anregung mit einer Laserlinie (bei 502 nm) als Funktion des I_2-Dampf(partial)drucks /4-6/; denn sowohl die Breite der mit Hilfe des Hanle-Effekts (Depolarisation der Fluoreszenzstrahlung durch ein Magnetfeld) gemessenen Kurven /4,5/ als auch die Intensität der Fluoreszenzstrahlung selbst /6/ werden durch Verunreinigungen stark verändert.

Es wurde eine Apparatur zur Beobachtung des Hanle-Effekts aufgebaut /4,5/. Zur Anregung der I_2-Fluoreszenz wird ein Ar^+-Mehrfrequenzlaser bei λ = 502 nm benutzt, da bei dieser Wellenlänge ausreichend starke Magnetfelder noch mit Permanentmagneten erzielbar sind. Der Strahl passiert nacheinander eine rotierende Sektorscheibe (Zerhackerfrequenz: 393 Hz) und die zu untersuchende I_2-Zelle. Die Laserleistung beträgt 5 mW. Die Kühlfinger der Jodzellen wurden nicht mit einem Peltierkühler versehen, da der Peltierstrom ein zusätzliches Magnetfeld verursachen würde, sondern entweder in Eiswasser oder in Wasser mit einer Temperatur knapp unterhalb der Raumtemperatur getaucht. Die Fluoreszenz wird durch ein Kantenfilter (Schott, KV-550) und eine doppelte Polarisationsfolie mit einem rauscharmen Photodetektor (EG & G, HAD 1000 A) beobachtet. Die mit Hilfe zweier Synchrondetektoren und eines Quotientenbildners /4/ auf die Laser-Intensität bezogene Fluoreszenz gelangt nach einer Mittelungszeit von 10 s (Solartron/Schlumberger: Time Domain Analyser JM 1860) zur Anzeige. Unter dem Einfluß des starken Feldes eines Permanentmagneten (vier Samarium-Kobalt-Scheiben /4/ ergeben eine magnetische Flußdichte bis zu 0,2 T) beträgt die Signaländerung im Falle von $^{127}I_2$ 5 % und im Falle von $^{129}I_2$ 4 %. Dabei ist die Intensität der Fluoreszenzstrahlung von $^{127}I_2$ etwa doppelt so groß wie die von $^{129}I_2$. Die Meßanordnung mit dem Permanentmagneten wurde etwa 1,2 m vom Laser entfernt aufgestellt, um eine Beeinflussung der Laserintensität durch das Magnetfeld zu verhindern.

Untersucht wurden insgesamt 23 $^{127}I_2$-Zellen (gefüllt: 1973 (2 x), 1975, 1977 (3 x), 1978, 1979 (2 x), 1980 (4 x), 1981, 1982 (2 x) und 1983 (7 x)) und vier $^{129}I_2$-Zellen (gefüllt: 1973, 1976, 1982 und 1983). Einige der $^{127}I_2$-Zellen aus dem Jahre 1983 waren absichtlich verunreinigt worden (z.B.: Luftzusatz von $\leqq$ 5 Pa, $\leqq$ 17 Pa und $\leqq$ 53 Pa).

Bild 1 zeigt die Ergebnisse der Messungen mit Hilfe des Hanle-Effekts; eingetragen sind zusätzlich die Frequenzverschiebungen gegenüber "reinen" $^{127}I_2$-Zellen bei gleichem Joddampfpartialdruck für die a_3-Komponente der Rotationslinie P(13), bei 582 THz, gemessen als Schwebungsfrequenzen zweier I_2-stabilisierter Ar^+-Laser /2/ (gleiche Verschiebungen zeigen auch die Komponenten a_1, a_2 und a_4). Selbst bei den ältesten "reinen" $^{127}I_2$-Zellen blieb die Halbwertsbreite unterhalb von 5 mT, d.h. auf Grund der mit Hilfe des Hanle-Effekts vorgenommenen Messungen sollte die maximale Rotverschiebung im Falle "reiner" Zellen für ν_{a3} unterhalb von 5 kHz bleiben ($\Delta\nu/\nu \leqq 8,5\cdot10^{-12}$). Die (in Bild 1 nicht enthaltenen) Messungen an einer 1973 gefüllten Quarzzelle, die (ohne Übergangsstück) bei einer höheren Temperatur der Brennerflamme abgeschmolzen wurde, ergaben bei einer Kühlfingertemperatur von 0 ^oC ($\hat{=}$ 4,12 Pa) eine Halbwertsbreite von 5,5 mT; tatsächlich wurde dann eine Frequenzverschiebung der a_3-Komponente von $-(10 \pm 5)$ kHz gemessen. Meßpunkte für verunreinigte Jodzellen liegen in Bild 1 deutlich oberhalb derjenigen für "reine" Zellen.

Die in Bild 1 eingetragene maximale Rotverschiebung von $-$ 100 kHz stellt praktisch die maximal feststellbare Frequenzverschiebung der a_3-Komponente dar, da bei noch größerer Verunreinigung die Fluoreszenz so schwach wird, daß sich die entsprechenden Signale der gesättigten Absorption nicht mehr zur Stabilisierung der Laserfrequenz benutzen lassen.

Bei den Messungen wurde noch ein weiteres Kriterium für die Zellenreinheit gefunden. Es stellte sich nämlich heraus, daß die Intensität der Fluoreszenz selbst bereits stark von der Zellenreinheit abhängt. In Bild 2 wurde daher die Frequenzverschiebung als Funktion der reziproken Fluoreszenzintensität dargestellt, wobei die Kühlfingertemperatur immer 0 ^oC betrug. Außer den Frequenzverschiebungen für Hyperfeinstrukturkomponenten der $^{127}I_2$-Rotationslinie P(13) der (43-0)-Bande bei ν = 582 THz wurden hier auch die - größeren /4/ - Frequenzverschiebungen für Hyperfeinstrukturkomponenten der $^{127}I_2$-Rotationslinie R(127) der (11-5)-Bande bei ν = 474 THz eingetragen, die mit Hilfe I_2-stabilisierter He-Ne-Laser /7/ gewonnen wurden. Da bei "reinen" Zellen die Werte der reziproken Fluoreszenz um maximal 17 % über dem Wert für die "reinste" Zelle liegen, sollte die obere Schranke einer Rotverschiebung bei "reinen" $^{127}I_2$-Zellen - entsprechend den ausgezogenen Geraden in Bild 2 - für die a_3-Komponente der Rotationslinie P(13), 43-0   5 kHz ($\Delta\nu/\nu$ = $8,5\cdot10^{-12}$) und für die i-Komponente der Rotationslinie R(127), 11-5   12,8 kHz ($\Delta\nu/\nu$ = $2,7\cdot10^{-11}$) betragen. Bei der erwähnten - 1973 gefüllten - Quarzzelle liegt der Wert der reziproken Fluoreszenz um 19,5 % über dem Wert für die "reinste" Zelle; hiernach ist die Frequenz der a_3-Komponente um -6 kHz verschoben.

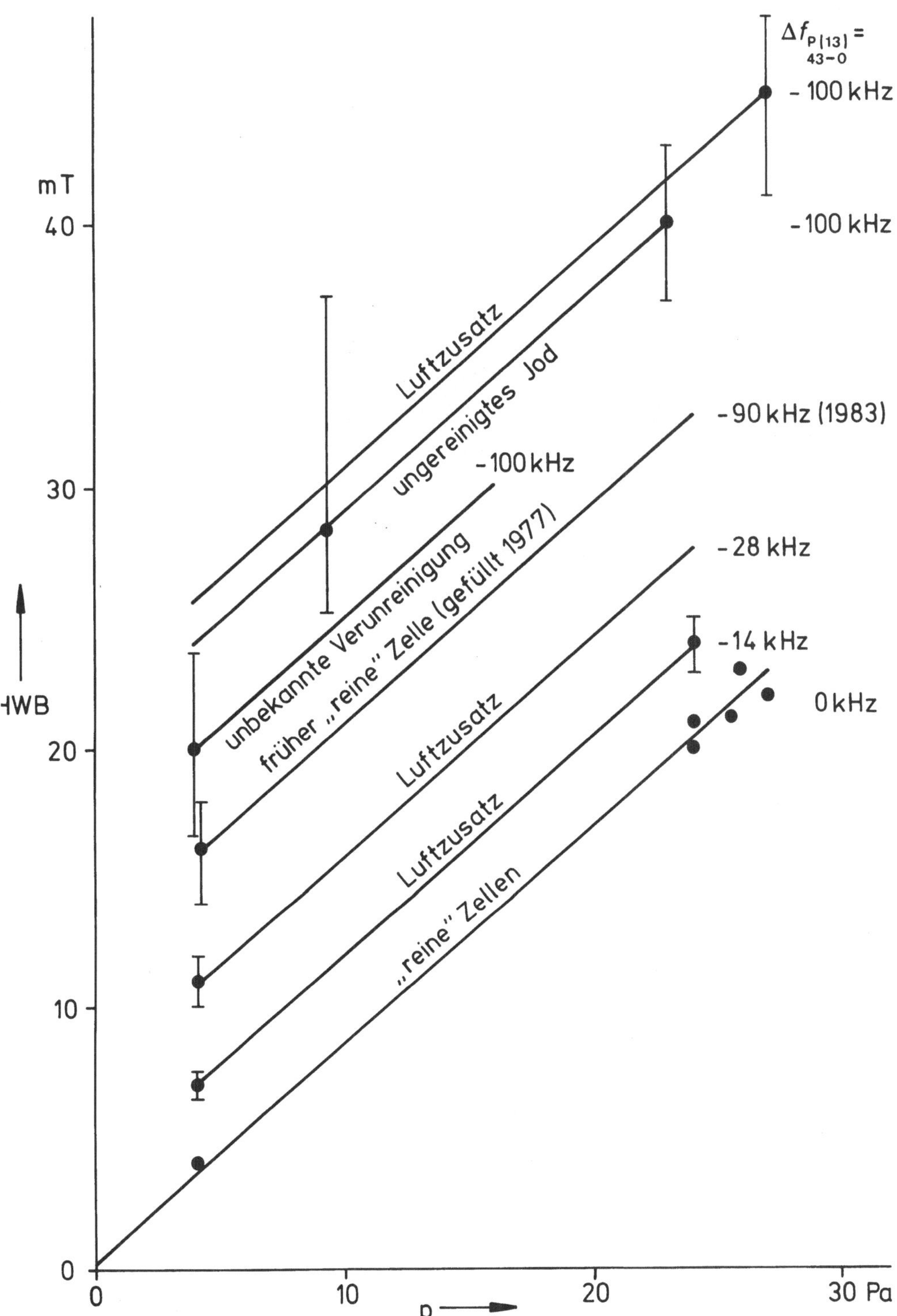

Bild 1. Halbwertsbreite der magnetischen Depolarisation (HWB) als Funktion des Jod-dampfdrucks p; Δf : zugehörige Rotverschiebung; Unsicherheitsbalken: (2σ)-Werte

Die Fluoreszenz nimmt mit sinkendem Joddampf(partial)druck umso stärker ab, je stärker die Jodzelle verunreinigt ist. Wurde der Joddampf(partial)druck von 25 Pa auf 4,12 Pa erniedrigt, so sank die Fluoreszenzintensität bei "reinen" $^{127}I_2$-Zellen

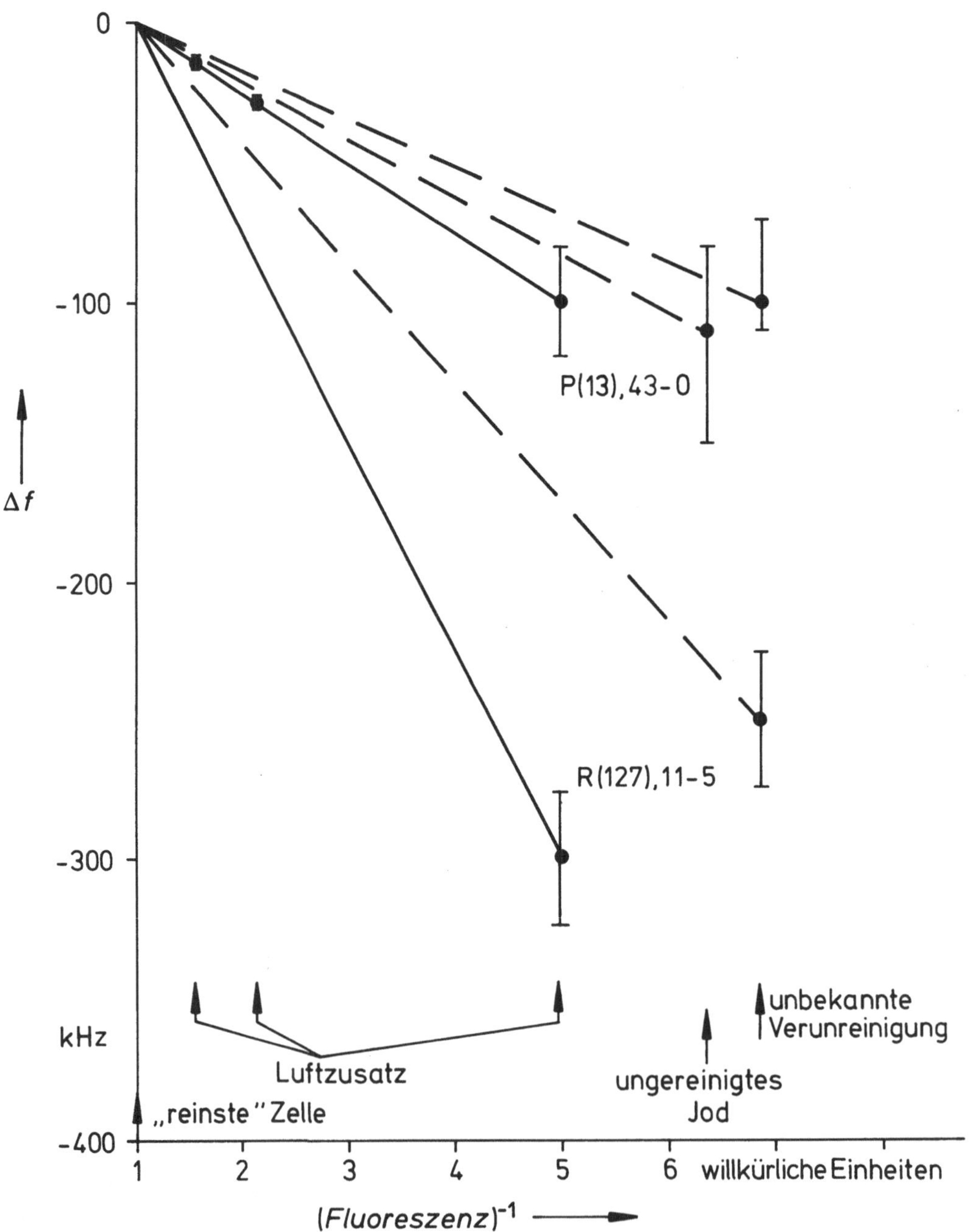

Bild 2. Frequenzverschiebung Δf als Funktion der reziproken Fluoreszenz

auf etwa 85 %, bei der erwähnten Quarzzelle auf etwa 70 % und bei den am stärksten
verunreinigten Zellen auf Werte um 30 %. Daher sollte sich bereits aus dem Inten-
sitätsverhältnis der Fluoreszenz für unterschiedliche Joddampf(partial)drucke in
eindeutiger Weise auf das Vorhandensein einer Verunreinigung - und damit einer
Rotverschiebung - schließen lassen /6/.
Die größere Rotverschiebung für Hyperfeinstrukturkomponenten der Rotationslinie
R(127), 11-5 ließ vermuten, daß die Rotverschiebung mit der Rotationsquantenzahl
zunimmt. Deshalb wurde zusätzlich die Frequenzverschiebung von Hyperfeinstrukturkomponenten der $^{127}I_2$-Rotationslinie R(98), 58-1 bei ν = 582 THz /8/ gemessen.

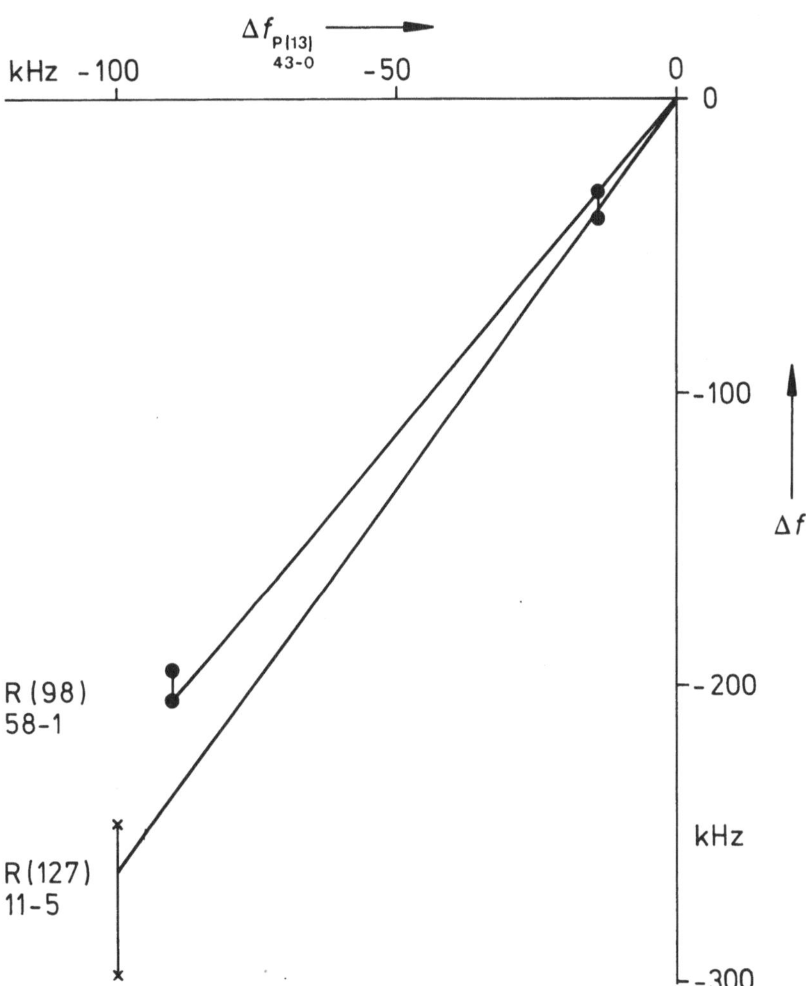

Bild 3. Frequenzverschiebungen von Hyperfeinstrukturkomponenten der
$^{127}I_2$-Rotationslinien R(98), 58-1 und R(127), 11-5 als Funktion der Frequenzverschiebung von Hyperfeinstrukturkomponenten der $^{127}I_2$-Rotationslinie P(13), 43-0

Die in Bild 3 eingetragenen Meßwerte, die im Fall stark verunreinigter Jodzellen zumeist bei Kühlfingertemperaturen von 12 ^{o}C gewonnen wurden, deuten auf eine Bestätigung dieser Vermutung. Daher ist anzunehmen, daß die Rotverschiebung der o-Komponente der Rotationslinie R(47), 9-2 bei ν_o = 490 THz kleiner als die Rotverschiebung von Hyperfeinstrukturkomponenten der Rotationslinie R(98), 58-1 (sowie von Hyperfeinstrukturkomponenten der Rotationslinie R(127), 11-5), aber größer als die Rotverschiebung von Hyperfeinstrukturkomponenten der Rotationslinie P(13), 43-0 sein wird.

Zusammenfassend läßt sich feststellen, daß die zur Stabilisierung von Laserlinien in der PTB hergestellten und bisher für "rein" gehaltenen Jodzellen nur eine sehr geringe Rotverschiebung der stabilisierten Laserfrequenzen bzw. -Wellenlängen verursachen. Verunreinigungen, die zu Rotverschiebungen führen, lassen sich sowohl mit Hilfe des Hanle-Effekts als auch durch Messung der Fluoreszenzintensität bei verschiedenen Joddampf(partial)drucken /6/ nachweisen.

Herrn Dipl.-Ing. H. Darnedde danke ich für die Messung der Frequenzverschiebungen von Hyperfeinstrukturkomponenten der $^{127}I_2$-Rotationslinie R(127), 11-5.

Literatur

/1/ List of Recommended Radiations, Recommendation M2 (1982) of the CCDM
/2/ F. Spieweck, IEEE Transact. Instrum. Measurem. IM-29 (1980), 361
/3/ Com.Int.Poids Mes., P.-V. 50 (1982), Rapport du Directeur du BIPM
/4/ W.R.C. Rowley and B.R. Marx, Metrologia 17 (1981), 65
/5/ G. Flory, M. Broyer, J. Vigué et J.C. Lehmann,
    Rev.Phys.Appliquée 12 (1977), 901
/6/ F. Spieweck, Veröffentlichung in Vorbereitung
/7/ J.-M. Chartier, J. Helmcke, and A.J. Wallard, IEEE Transact.
    Instrum.Measurem. IM-25 (1976), 450
/8/ H.-J. Foth and F. Spieweck, Chem.Phys.Lett. 65 (1979), 347

A New Method for Designing a Paraboloidal Reflector

Cao Wen Zhuang, Wu Bao Ye

Kunming Institute of Physics

Kunming Yunnan Provice / China

Abstract: In this paper a new method of designing a paraboloidal reflector (abbreviated as PR) has been developed and its calculative formula, the block diagram of the calculative program as well as a practical example have been presented.

I. The Preliminary analysis of the PR

A principal function of the PR is to focus (or spread) luminous energy emitted from a luminous body into a desired angle (i.e. scattered-angle). However, it is impossible that all luminous energy can be collected. The reason is as follow:

1. The PR, not allowed to be made up a closed form, must have an opening with a diameter of $2r_{max}$, as shown in Fig. 1, in order for the converged light rays to come out through it.

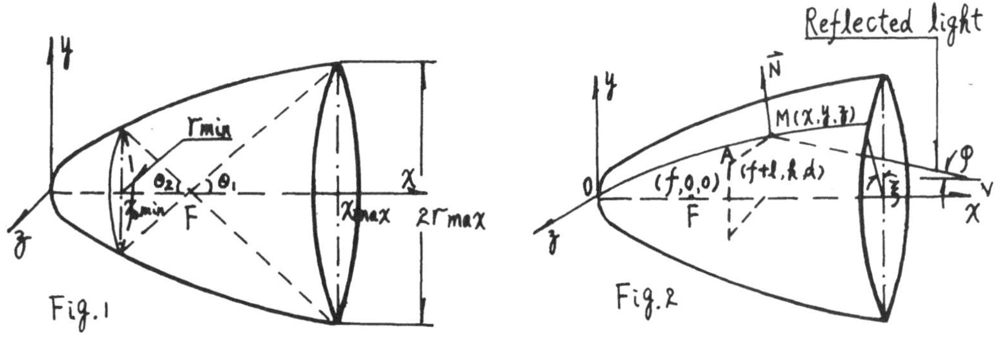

2. For some instruments, an opening has been made on the top of the PR

594

because a light source must be put in the cavity and fixed on the PR from its top. Suppose that the opening diameter is $2r_{min}$ (see Fig.1)or $r_{min}=0$ in the case the top of the PR is not opened. The light rays emitted by the luminous body from the solid angles W_1 and W_2 formed with point F subtended by the openings can not be converged if they do not strike the PR.

II. The efficiency of both reflection and convergence of a PR

The design of the PR is to find out the optimum focal distance f_{op} with which a maximum converging efficiency η_{max} can be obtained according to the given shape and its parameters of the luminous body.

Let us assume that P_t is the total luminous energy emitted from a luminous body and P_r is the luminous energy reflected from the inside surfac of the PR. Obviously, P_r is only a part of P_t since there is one or two openings on the PR. After the reflection of the energy striking the PR takes place, the luminous energy being focused into a desired scattering angle 2σ is only a part of P_r, denoted by P_σ ; then the ratio

$$K = P_\sigma / P_r \qquad \dots\dots(1)$$

is called as the reflective efficiency of the PR. thus, the convergency efficiency η should be

$$\eta = P_\sigma / P_t = P_r \cdot K / P_t = B \cdot K \qquad \dots\dots(2)$$

where $B = P_r / P_t$ is called as the applicable coefficient of the luminous energy. Evidently, $\eta < 1$, $B < 1$, $K < 1$. Therefore, in order to get the maximum η , two of the maximum value, K and B, have to be found out. Generally speaking, the distribution of the value P_t is irregular within 4π steradian, but most of them possess the property of an axial or rotational symmetry. The value of P_t depends on the two solid angles, W_1 and W_2 as well as the position of the luminous body within the PR. The value of P_t and P_r can be obtained by using the method of either Rousseau /1/ or angular coefficient.

III. The calculation of reflective angle φ

In Fig.2, suppose a luminous point A is placed in a PR with a focal distance f and the coordinate of the point A is (f+1, h, d), we may imagine that when a light ray $\overline{AM}$ emitted from the luminous point A strikes any point M (X,Y,Z) on the PR, it will be reflected so that a reflective light ray $\overline{MV}$ occures and comes out through the PR. The reflective angle φ between the reflective light ray $\overline{MV}$ and X-axis is what we are much interested in, for it is the covergent angle of the luminous energy. If the meridian plane containing the point M, as shown in Fig.2, has been turned through an angle ξ in respect to the XOY- plane about the X-axis; thus we have:

$$\cos\varphi = \frac{(X-f)(X-f-l)+2\sqrt{fX}(2\sqrt{fX}\cdot\cos\xi-h)\cos\xi+(2\sqrt{fX}\,\mathrm{Sin}\,\xi-d)\cdot2\sqrt{fX}\,\mathrm{Sin}\,\xi}{(f+X)\sqrt{(X-f-l)^2+(2\sqrt{fX}\cos\xi-h)^2+(2\sqrt{fX}\,\mathrm{Sin}\,\xi-d)^2}} \quad \dots\dots(3)$$

where l is a distance between a projecting point of point A on the X-axis and the focal point F. If the projecting point is located at the right side of the point F, l takes a positive sign, otherwise a negative. Suppose the distance between any two of the adjacent points among those incident ones on the inside surface of the PR is given as a constant ΔS and ΔS is very small, then we may write:

$$\Delta X_i = \Delta S \Big/ \sqrt{1+f/X_i} \quad \dots\dots(4) \qquad\qquad X_{i+1} = X_i + \Delta X_i \quad \dots\dots(5)$$

The increments $\Delta\xi_j$ (by degree) of the turning angle for every value X_i should be:
$$\Delta\xi_j = 90\cdot\Delta S \Big/ (\pi\sqrt{fX_i}) \quad \dots\dots(6)$$

IV. The calculation of K, the efficiency of reflection of PR

In respect of a point source with a given focal distance, let us further assume that the number of all the incident points been calculated on the PR is S. Among them, the number of the points whose reflective angles satisfy the condition $\varphi \leqslant \sigma$ is T. Obviously, the value of S and T are proportionate respectively to P_r and P_σ. Thus, according to equation (1), the efficiency of the reflection of a point source should be K=T/S.

For a given luminous body with any shape, we can divide it into a number of point sources as m, the coordinates of which are $(f+l_1, h_1, d_1)$, $(f+l_2, h_2, d_2)$ ......$(f+l_m, h_m, d_m)$. By summing up the calculation of each point source, then we get: $T= \sum_i^m T_i$, $S= \sum_i^m S_i$. The efficiency of reflection for the whole luminous bodies can be calculated by $K=T/S$ and the value of P_r, B and η obtained. The calculative block diagram is at the back.

V. Designing example

There is an instrument whose luminous body is of a solenoid shape. Its diameter is ϕ1mm and its length is 1.5mm. We want to design a PR. whose scattering-angle is $4°(2\delta=4°)$. The diameters of the two openings of the PR are $2r_{max}=100$mm and $2r_{min}=22$mm.

1. Under a standard working state, we measured the luminous intensity of the luminous body and by Rousseau's method in accordance with the measured values we get $P_t=475.8150889$ (relative unit).

2. calculation of the value of P_r, K, B and η

We practised this task by using a computer model TQ-16. In the calculation we took $\triangle S=0.5$mm, $\triangle l=0.1$mm as the step-size and -90 to +90 as the range of the turning angle. The final result is that the center of the luminous body is located at the F and its axis is coincide with that of PR. The focal distance we get is f=15mm, the abscissa and the opening angle corresponding to its opening diameter ϕ22mm and ϕ100mm are $X_{min}=2.017$mm, $X_{max}=41.67$mm and $\theta_2=40.27°$, $\theta_1=61.93°$ respectively. Thus, by the use of Rousseau's method, we may get $P_r=335.9319369$ (relative unit). By further calculation we can get: B=70.6%, S=378480, T=369165, K=97.54% and $\eta=68.86\%$.

It is clear that to improve η, the key problem is to decrease the solid angles W_1 and W_2 as much as we can, but this attempt is limited by the pre-determined diameters of the openings. Therefore, the problem to be solved in the designing is to take measures to utilize those parts of light source which passes through ω_1 and ω_2 but does not strike the PR.

The Block Diagram of Calculation Program

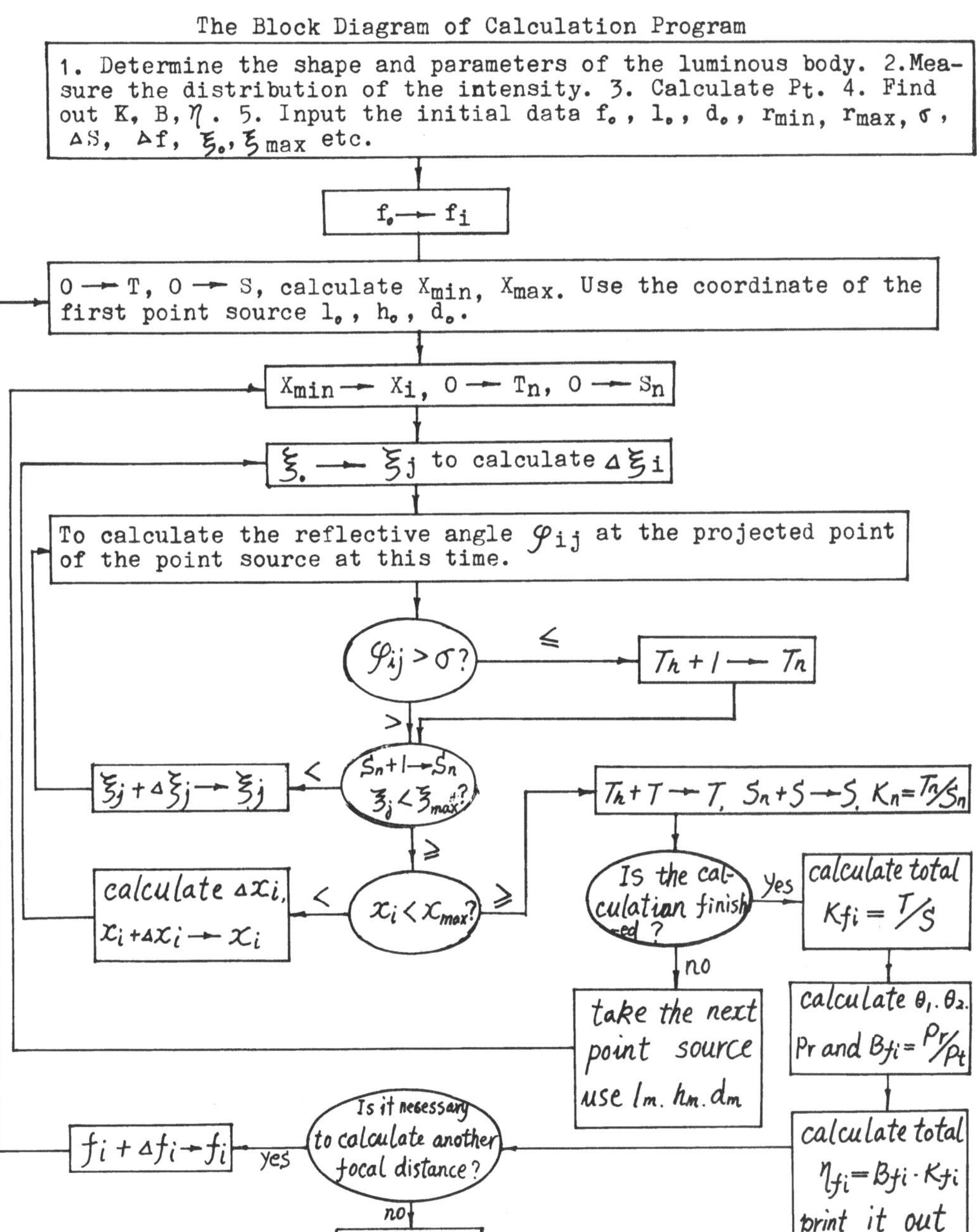

Reference.

/1/ H.A.E.Keitz, "Light Calculations And Measurements", ChapterIV,
Philip's Technical Library, Holland, March 1955.

UV-Laserinduzierte Chlorwasserstoffabspaltung aus Alkylhalogeniden

M. SCHNEIDER, J. WOLFRUM
Physikalisch-Chemisches Institut der Universität
Im Neuenheimer Feld 253, D 6900 Heidelberg

Einleitung

Die thermische Eliminierung von Chlorwasserstoff aus 1,2-Dichlorethan ist heute
die wichtigste technische Quelle der PVC-Monomeren Vinylchlorid /1/. Das Produkt wird über einen Radikalkettenmechanismus gebildet:

$$(1) \quad C_2H_4Cl_2 \quad \longrightarrow \quad C_2H_4Cl + Cl \qquad \Delta H = 80 \text{ Kcal/mol}$$

$$(2) \quad Cl + C_2H_4Cl_2 \quad \longrightarrow \quad HCl + C_2H_3Cl_2 \qquad \Delta H = -6 \text{ Kcal/mol}$$

$$(3) \quad C_2H_3Cl_2 + M \quad \longrightarrow \quad C_2H_3Cl + Cl + M \qquad \Delta H = 23 \text{ Kcal/mol}$$

Wegen der hohen Aktivierungsenergien für den Kettenstart (1) sind Temperaturen
von 450 bis 550 $^{\circ}$C notwendig, um vertretbare Umsatzgeschwindigkeiten zu erreichen. Leitet man dagegen die Reaktion photochemisch ein, so wird Reaktion (3),
die eine wesentlich geringere Energiezufuhr benötigt, geschwindigkeitsbestimmend,
so daß eine deutliche Senkung der Temperatur möglich ist /2/. Als qualifizierte
Lichtquelle steht seit einigen Jahren der Excimer-Laser zur Verfügung, dessen
räumliche Kohärenz auch anspruchsvollen geometrischen Anordnungen wie Rohr-
reaktoren genügt. Außerdem erlauben die kurzen Pulse von einigen ns Dauer die
Untersuchung der Kinetik der Kettenreaktion getrennt vom Kettenstart /3/. Durch
Auswahl geeigneter Wellenlängen und Variation der Pulsenergie läßt sich die Radi-
kalkonzentration über weite Bereiche variieren.

Experiment

Zur Bestimmung der Quantenausbeuten der laserinduzierten Reaktion wurden zunächst die Absorptionsquerschnitte δ von 1,2-Dichlorethan in einer geheizten
Quarzzelle in Abhängigkeit von Temperatur (T = 380 - 770 K) und Wellenlänge
(λ = 190 - 250 nm) gemessen. In diesem Bereich fällt die Absorption zu größeren
Wellenlängen steil ab - man erhält einen linearen Verlauf der Funktion $\log \delta / \lambda$ -
und ist am langwelligen Ende stark temperaturabhängig (Zunahme um drei Größenordnungen im betrachteten Temperaturintervall).

Die Absorptionszelle wurde auch für die Umsatzmessungen eingesetzt. Bestrahlt wurde mit einem Excimer-Laser (Fa. Lambda Physik, Typ EMG 501) bei 193 nm (ArF), 222 nm (KrCl), 248 nm (KrF) und 308 nm (XeCl) und Pulsenergien zwischen 1 und 100 mJ/cm^2. Die Produktanalyse erfolgte gaschromatographisch bzw. infrarotspektroskopisch nach Expansion des Gases in eine IR-Küvette.

Für die zeitaufgelösten Messungen des Reaktionsablaufs wurde die in Abb. 1 gezeigte Anordnung verwendet. Als Nachweismethode diente die UV-Absorption des

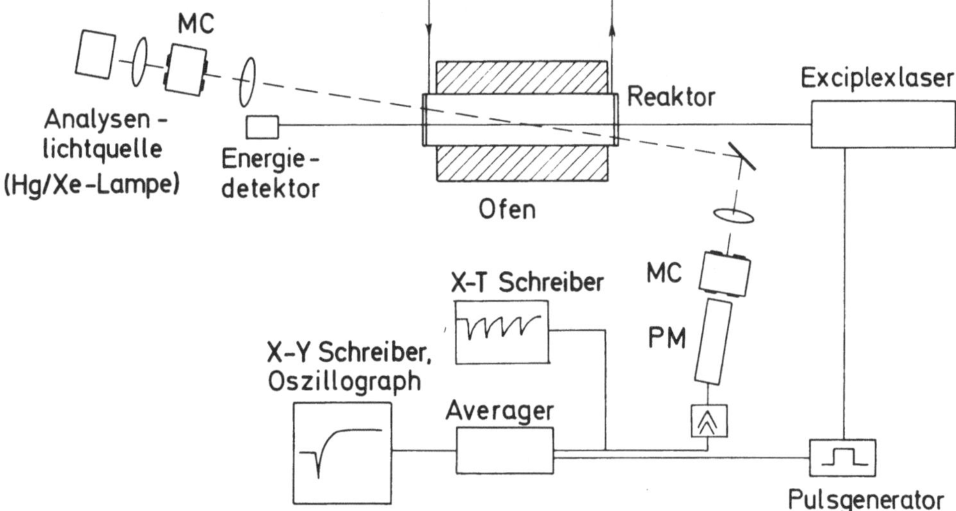

Abb. 1. Experimentelle Anordnung zur Untersuchung laserinduzierter Radikalkettenreaktionen.

Vinylchlorids, das bei 215 nm um einen Faktor 20 stärker absorbiert als Dichlorethan. Zur Rauschverminderung war es notwendig, über mehrere Signale zu mitteln. Ein typisches Meßsignal ist in Abb. 2 dargestellt.

Zusätzlich zur schnellen Signalaufzeichnung wurde über einen Schreiber die Stabilität der Analyselichtquelle und der vollständige Gasaustausch in der Zelle nach jedem Laserpuls protokolliert. Die Aufenthaltszeit des Reaktanden in der Zelle lag bei 500 ms, während der Laser mit einer Folgefrequenz von 1 Hz arbeitete. Als günstigste Photolysewellenlänge erwies sich die des KrF-Lasers, da Dichlorethan bei 248 nm nur schwach absorbiert und damit eine räumlich homogene Radikalerzeugung längs des Reaktors gewährleistet war.

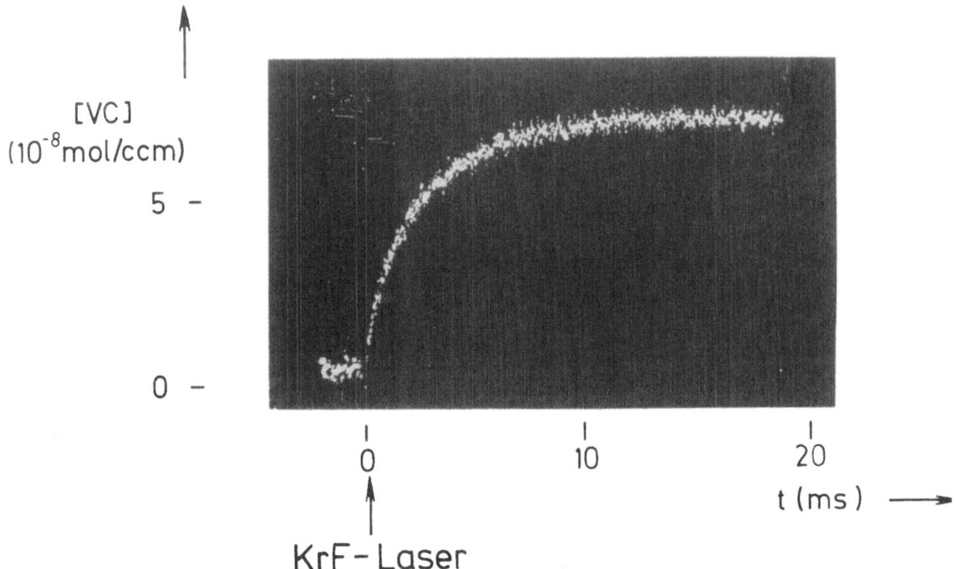

Abb. 2. Bildung von Vinylchlorid nach einem Laserpuls

Ergebnisse und Diskussion

Aus den experimentellen Daten wurden Informationen über die Abhängigkeit der Quantenausbeute und der Geschwindigkeit der Zerfallsreaktion (3) von Parametern wie Druck, Temperatur, Laserwellenlänge und Pulsenergie gewonnen. Es zeigte sich, daß die Quantenausbeute, die im Bereich von 10 bis 10^4 liegt, zu kürzeren Wellenlängen und höheren Pulsenergien abnimmt und bei Steigerung der Temperatur mit einer Aktivierungsenergie von 8 - 10 Kcal zunimmt. Beide Resultate werden über den Reaktionsmechanismus in quantitativer Übereinstimmung mit Rechnersimulationen erklärt: Höhere Startradikalkonzentrationen $[Cl]_o$ führen zu vermehrtem bimolekularen Kettenabbruch und damit zu einer Verringerung der Quantenausbeute ϕ (Abb. 3):

$$\phi \sim [Cl]_o^{-0,5}$$

Der Temperatureinfluß läßt sich unmittelbar aus der Endothermizität von Reaktion (3) ableiten, die auch zu einer Begrenzung des Umsatzes pro Puls durch adiabatische Abkühlung führt. Mit steigender Konversion wird außerdem die Autoinhibition durch Vinylchlorid wirksam.

Für die Geschwindigkeitskonstante k_3, deren Kenntnis für die Simulation des Systems notwendig ist, ergab sich aus unseren Messungen der Ausdruck

$$\log k_3 = 13,8 - 19800/2,303\ RT$$

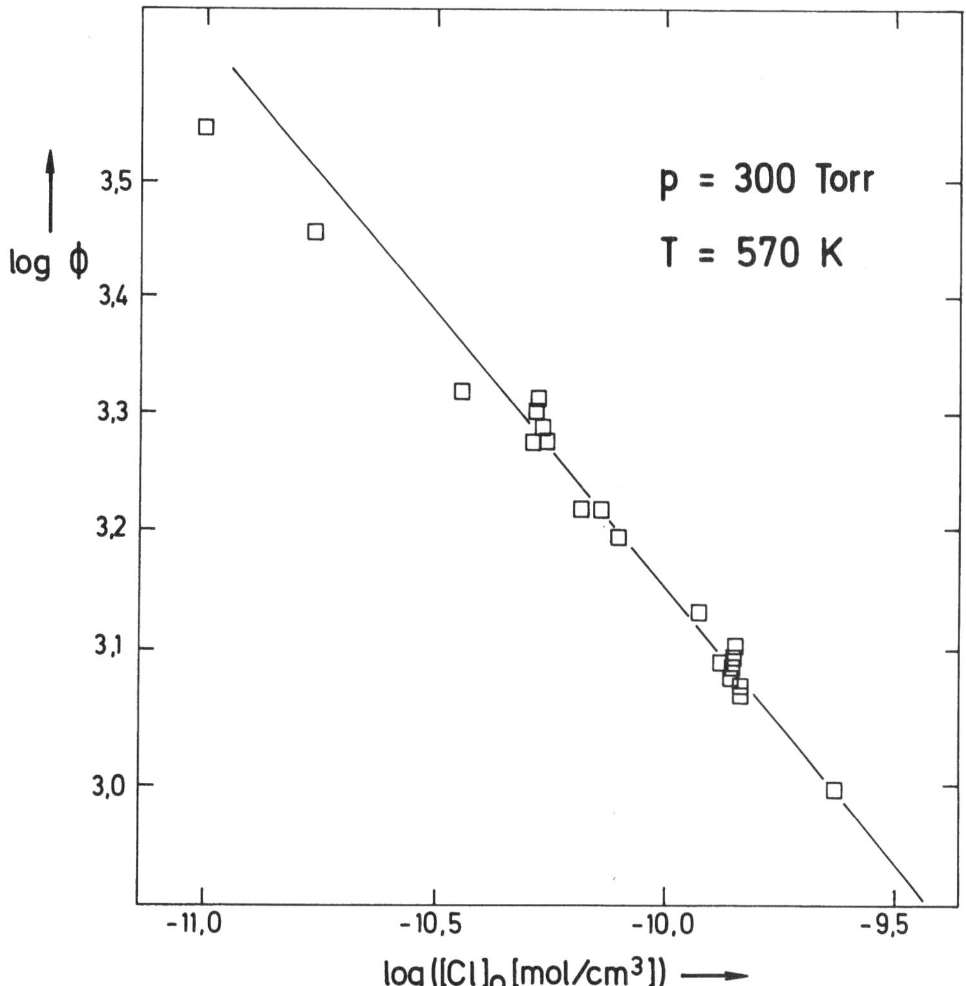

Abb. 3. Quantenausbeute als Funktion der Startradikalkonzentration

Die Konstante ist im Bereich von 60 bis 600 Torr deutlich druckabhängig (Abb. 4). Dieses Ergebnis wird durch neuere Untersuchungen einer anderen Arbeitsgruppe /4/ bestätigt.

Ein analoges System wie die Dichlorethanphotolyse stellt die photochemische Synthese von Vinylidenfluorid aus 1,1,1-Chlordifluorethan dar, das wir momentan untersuchen. Es stellt sich jedoch heraus, daß die Kettenlängen hier wesentlich kürzer sind, verursacht durch die selbstinhibierende Wirkung des primär gebildeten Radikals $C_2H_3F_2$. Bessere Ergebnisse erzielt man daher durch Photolyse von zugesetztem CCl_4.

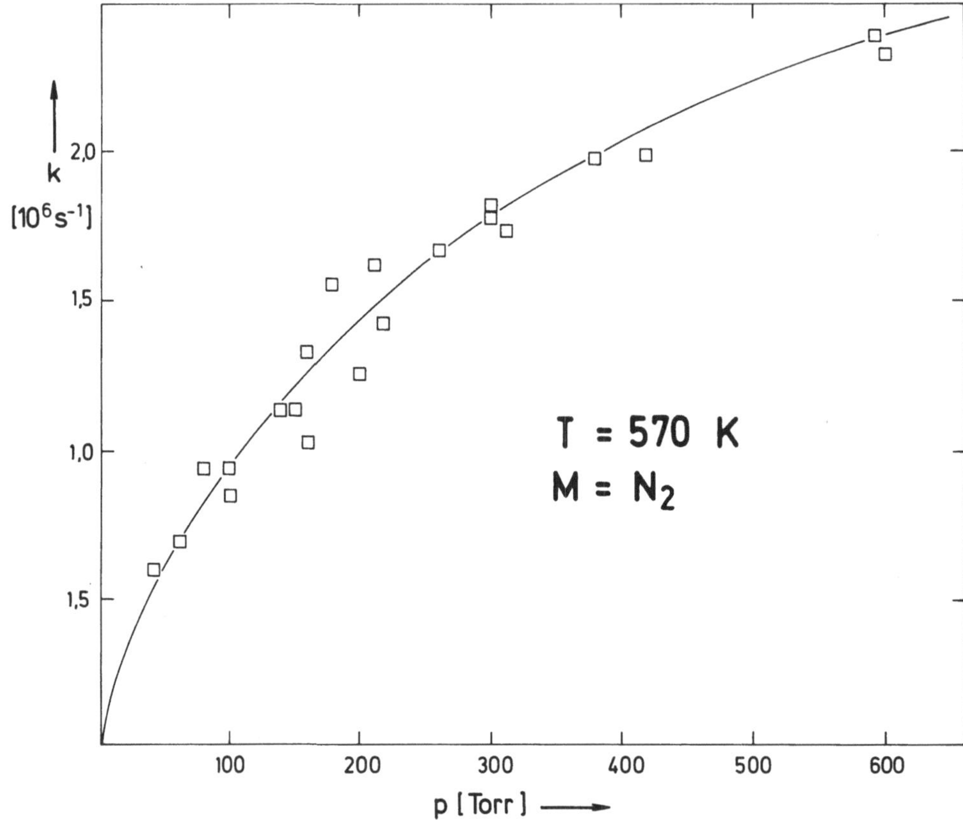

Abb. 4. Druckabhängigkeit der Geschwindigkeit des Zerfalls von $C_2H_3Cl_2$

Abschließend sei darauf hingewiesen, daß einer Anwendung des Excimer-Lasers in der technischen Photochemie zur Zeit noch sein hoher Preis entgegensteht. Allerdings zeichnet sich hier eine ähnliche Entwicklung ab wie im infraroten Spektralbereich beim CO_2-Laser. Vom Standpunkt der laserinduzierten Radikalkettenreaktion ist eine hohe Repetitionsrate oder sogar ein cw-Betrieb des Lasers anzustreber

Literatur
/1/ MCPHERSON, R. W. , C. M. STARKS, G. J. FRYAR: Hydrocarbon Processing,
    März 1979, S. 75
/2/ SCHNEIDER, M. : Bericht 21/1981 des MPI für Strömungsforschung, Göttingen;
    WOLFRUM, J. , M. KNEBA, P. CLOUGH, M. SCHNEIDER: EP 0027554 A 1
/3/ NESBITT, D. J. , S. LEONE: J. C. P. 72, 1722 (1980)
/4/ ASHMORE, P. G. , A. J. OWEN: Faraday Trans. I, 78 (1982) 657

Laser Sicherheit

Laser Safety

Status of Laser Safety Requirements

Robert Weiner
Weiner Associates 544-23rd Street Manhattan Beach, CA 90266 USA

Introduction

In recent years there has been considerable activity in the field of laser safety standards and regulations. This paper briefly discusses the various documents and presents a status listing of requirements in most industrialized nations.

Types of Requirements

There are several approaches to categorizing these documents: manufacturer requirements vs user requirements; legal regulations vs voluntary standards. Manufacturer's requirements specify product classification, engineering features, labels, and user manuals. User requirements specify installation constraints, operator training, and exposure limitations.

Laws or regulations may be published which must be met by the manufacturer before a product can be sold or imported. Likewise, user regulations are laws which must be satisfied before the product can be installed or used.

In many countries where the requirements are in the form of a voluntary standard, it is used by courts in evaluating liability in the event of a lawsuit.

Laser Radiation Standards

In most countries, requirements are being developed based on the standard provided by the Technical Committee No. 76 of the International Electrotechnical Commission (Reference 33). This IEC standard, approved in February 1983 and due to be formally published in late 1983, specifies requirements for both manufacturers and users.

In the United States, regulations for manufacturers are published by the Office of Devices and Radiological Health (formerly Bureau of Radiological Health), under the Food and Drug Administration. These regulations (Reference 1) are in the process of being amended, with the majority of the changes to ease requirements. User requirements in the U.S. are the responsibility of the individual states, but only a few states currently have such laws.

The differences between the U.S. and IEC requirements are primarily in the area of labelling, Class I limits for some products, interlocks, class level for required features, and collateral (ancillary) radiation.

Other Standards

The IEC TC-76 Committee has also prepared a draft electrical safety standard (Reference 38). This is quite similar to the German VDE 0836 and CENELEC HD 194 documents. It affects only products which are not subject to another IEC electrical safety standard.

The American National Standards Institute (ANSI) Committee for the Safe Use of Lasers has developed a draft standard on fiber optics for communications (Reference 34). This covers both laser diodes and LED sources. The .IEC TC-76 committee will be reviewing this draft, and a comparable IEC document will probably be available in the next few years.

Another topic under study by the ANSI committee is the use of lasers in medical applications. A preliminary draft standard should be available in 1984. France is preparing a document covering medical lasers for use in hospitals.

Status Summary

The following Table lists laser safety requirements by country or agency. It indicates whether the document applies to manufacturers or users, and whether it is in the form of a legal requirement or a voluntary standard. Following the Table is a list of requirements as they are known to the author in June 1983. Mailing addresses are shown for use in obtaining copies of most of the documents.

The status of the requirements will be updated as information becomes available. These updates can be requested (at no charge) from the author (See Reference 41). Any information on changes which can be provided to the author will be greatly appreciated.

SUMMARY - LASER SAFETY REQUIREMENTS

Country/ Agency	Document Number	Requirement Mfgr	Requirement User	Application Law	Application Std	Comment	Ref #
USA	21CFR1040	C,M		X		Amend in 84-5	1
ANSI	Z136.1-1980		C		X		2
USA(Texas)			C	X			3
"(NY)	Ind.Code 50		C	X		Amend in '84-5	4
"	SSRL		'84+	X		Optional by State	5
UK	BS 4803	C	C		X	IEC	6
W. Germany	VDE 0837	'84	'84		X	IEC	7
"	VBG 93		C,G	X		Amend in '84	8
"	DIN 56 912		C		X		9
"	DIN 58 126T6		C		X		10
France		'84	'84	X		IEC	11
"		'84		X		Medical Lasers	12

REQUIREMENTS (continued)

Country/ Agency	Document Number	Requirement Mfgr	User	Application Law	Std	Comment	Ref #
Sweden	SSI FS 1980:2	C,G		X		Approval Req'd	13
"	SSI FS 1983:3		C	X			14
"	AFS 1981:9		C	X			15
"		'84-5		X		Video disc sys.	16
Norway		'84-5	'84-5	X		IEC	17
Denmark		'84-5	'84	?	X	IEC	18
Switzerland			'85	X		IEC	19
Japan			'86		X,G	IEC	20
Australia	AS 2211-1981	C	C		X	IEC	21
"	AS 2397-1980		C		X	Const. lasers	22
"(W.A.)			C		X		23
"(Tas)			C		X		24
Canada	Chapter 34	C		X		Enabling law	25
"	PC 1977-3127	C		X		Education lasers	26
"	PC 1977-3123	C		X		Scanners	27
"		'85		X		All products	28
"	I.L.574		C		X	Light shows	29
"(Alberta)	45/73		C	X			30
Holland	1978/6		C		X		31
China (PR)		'86	'86	X			32
IEC	TC-76	'83	'83		X	Radiation	33
ANSI	Z136.2-?		'84		X	Fiber Optics	34
ANSI	Z136.3-?		'85-6		X	Medical lasers	35
WHO	E.H.C.#2		'83		X		36
ACGIH			C		X		37
IEC	TC-76	'84	'84		X	Electrical	38
W. Germany	VDE 0836	C	C	X		Electrical	39
CENELEC	HD 194	C	C		X	Electrical	40
Weiner	Safety Update					Newsletter	41

Notes:
C   = current requirement   M   = manufacturer's test required
'84   = estimated date of publication   G   = government test required
IEC   = based on IEC TC-76 standard

REFERENCES

1. 21CFR 1040, Performance Standards for Laser Products.

    Source: U.S. Office of Devices and Radiological Health (HFX-430), 5600 Fishers Lane, Rockville, MD 20857, USA

2. ANSI Z-136.1-1980, ANSI Standard for the Safe Use of Lasers

    Source: American National Standards Institute, 1430 Broadway, New York City, NY 10018, USA

3. Texas Regulations for the Control of Laser Radiation Hazards

    Source: Nonionizing Radiation Program, Division of Occupational Health, Texas State Dept. of Health, 1100 W. 49th St., Austin, TX 78756, USA

4. Industrial Code Rule 50, Lasers

    Source: Radiological Health Unit, 6th Floor, New York State Division of Safety and Health, Two World Trade Center, New York City, NY 10047, USA

5. Suggested State Regulations for Lasers

    Source: Same as Reference 1

6. BS4803, Radiation Safety of Lasers Products and Systems (3 parts)

    Source: British Standards Institution, 2 Park Street, London W1A 2BS, England

7. VDE 0837, (Draft: Identical to IEC 76(CO)6 draft)

    Source: VDE-Verlag GmbH, Bismarckstr. 33, D 1000 Berlin 12, W. Germany

8. Unfallverhütungsvorschrift VBG 93 "Laserstrahlen" (Laser rays)

    Source: Carl Heymanns Verlag KG, Gereonstr. 18-32, D 5000 Köln 1, W. Germany

9. DIN 56 912, Sicherheitstechnische Anforderungen für Bühnenlaser und Bühnenlaseranlagen

    Source: Beuth Verlag GmbH, Burggrafenstr. 4-10, D 1000 Berlin 30, W. Germany

10. DIN 58 126 T 6, Sicherheitstechnische Anforderungen für Lehr-, Lern-, und Ausbildungsmittel, Laser.

    Source: Same as reference 9

11. (Version of IEC TC-76, title unknown)

    Source: Comite' Electrotechnique Francais, 12 Place des Etats-Unis, 75783 Paris Cedex 16, France

12. (Title unknown)

    Source: Ministire de la Sante'

13. SSI FS 1980:2, Statens strålskyddsinstituts föreskrifter m.m. om lasrar (The regulations of the Swedish National Institute of Radiation Protection concerning lasers).

    Source: National Institute of Radiation Protection, Box 60204, S-10401 Stockholm, Sweden

14. SSI FS 1983:3, Statens strålskyddsinstituts föreskrifter om verksamhet med laser (The regulations of the Swedish National Institute of Radiation Protection for use of Class 3B and 4 lasers)

    Source: Same as reference 13

15. AFS 1981:9, Arbetarskyddsstyrelsens kungörelse om laser med kommentarer (Regulations of the Board of Occupational Safety and Health concerning work with lasers)

    Source: Liber Förlag, 16289 Vällingby, Stockholm, Sweden

16. (Concerning tests of audio and video recorders/players)

    Source: Same as reference 13

17. Bestemmelser om Laser (Requirements to Lasers)

    Source: State Institute of Radiation Hygiene, Box 55, N-1345, Osteras, Norway

18. (Version of IEC TC-76, title unknown)

    Source: Arbejdstilsynet, Rosenvaengets Alle 16-18, DK-2100, Koebenhavn, Denmark

19. Guidelines for the Safe Use of Lasers

    Source: SUVA, Sektion Physik, Postfach, CH-6002 Luzern, Switzerland

20. Japanese Industrial Standard

    Source: Japanese Standards Association, 1-24, Akasaka 4 Chome, Minato-ku, Tokyo, Japan 107

21. AS 2211-1981, Laser Safety

    Source: Standards Association of Australia, Standards House, 80 Arthur St., North Sydney, N.S.W. 2060, Australia

22. AS 2397-1980, Guide to the Safe Use of Lasers in the Construction Industry

    Source: Same as reference 21

23. Guidelines for Users of Lasers in Western Australia

    Source: Radiological Council, Dept. of Public Health, Verdun Street, Nedlands, Western Australia 6009, Australia

24. Radiation Control Action, 1977

    Source: Dept. of Public Health, Public Buildings, Davey Street, Hobart 7000, Tasmania, Australia

25. Chapter 34, Radiation Emitting Devices Act

    Source: Radiation Protection Bureau, Environmental Health Centre, Room 233, Tunney's Pasture, Ottawa, Ontario K1A 0L2, Canada

26. P.C. 1977-3127, Radiation Emitting Devices Regulations, amendment (Demonstration lasers)

    Source: Same as Reference 25

27. P.C. 1977-3123, Radiation Emitting Devices Regulations, amendment (laser scanners)

    Source: Same as Reference 25

28. (Requirements covering all laser products, similar to U.S. regulations, title unknown)

    Source: Same as Reference 25

29. I.L. No. 574, Information Letter - Safe Use and Operational Guidelines for Laser Light Shows

    Source: Same as Reference 25

30. 45/73, Regulations Respecting the Protection of Persons from the Hazards of Laser Operation

   Source: Alberta Dept. of Health, Radiation Health Branch, Oxbridge Place, 9820-106 Street, Edmonton, Alberta T5K 2J6, Canada

31. 1978/6, Acceptable Levels of Micrometre Radiation

   Source: Ministry of Health and Environmental Protection, Postbox 439-2260 AK, Leidschendam, The Netherlands

32. Laser Safety Standard

   Source: Beijing Institute of Opto-Electronic Technology, P.O. Box 648, Beijing, P.R. China

33. IEC TC-76, Radiation Safety of Laser Products, Equipment Classification, Requirements and User's Guide (Draft 76(CO)8, July 1982; Approved version due late 1983)

   Source: International Electrotechnical Commission, 1-3 Rue de Varembe, 1211 Geneva 20/, Switzerland

34. ANSI Z-136.2-198?, American National Standard for the Safe Use of Optical Fiber Communications Systems Utilizing Laser Diode and LED Sources

   Source: Same as Reference 2

35. ANSI Z-136.3-198?, (Standard for safe use of medical lasers)

   Source: Same as reference 2

36. Environmental Health Criteria #2, Lasers and Optical Radiation

   Source: World Health Organization, European Office, Copenhagen, Denmark

37. A Guide for Control of Laser Hazards, 1976

   Source: American Conference of Governmental Industrial Hygienists, P.O. Box 1937, Cincinnati, Ohio 45201, USA

38. IEC TC-76, Electrical Safety of Laser Equipment and Installations (Draft: 76(CO)9, September 1982; Approved version due 1984)

   Source: Same as Reference 33

39. VDE 0836, VDE-Bestimmung für die elektrische Sicherheit von Lasergeräten und -anlagen (VDE specification for the electrical safety of laser equipment and installations)

   Source: Same as Reference 7

40. HD 194, Requirements concerning the electrical safety of laser apparatus and installations.

   Source: European Committee for Electrotechnical Standardization

41. Laser Safety Update (newsletter providing status update on regulations - no charge)

   Source: Weiner Associates, 544-23rd St., Manhattan Beach, CA 90266 USA

Leached AR Surfaces for High Energy Optics:
Scale-Up and Performance

Lee M. Cook, Karl-Heinz Mader
(Schott Optical Glass Inc.)

N.J. Brown and G. Richard Wirtenson
(Lawrence Livermore National Laboratory)

Introduction

Laser damage to components is a major factor limiting output power of lasers.
Surface damage usually occurs at lower levels than bulk damage, and conventional
antireflective (AR) coatings typically damage at lower levels still. However,
leached AR surfaces have damage thresholds approaching those of bare surfaces.
A leaching process developed by Schott[1-3] is being used under license by Lawrence
Livermore National Laboratory (LLNL) for the AR treatment of non-active BK-7
elements of the NOVETTE laser, currently the world's most powerful system. This
paper will review the design constraints imposed by the process in scale-up, and
the performance of these optics.

The Treatment Process

The glass is leached by immersion in a fluid which attacks the surface to produce a porous low-index layer, yielding the desired reflectance characteristics.

The process sequence used in large-scale treatment is discussed in detail in
Reference 4. The treatment system is shown in Figure 1. In the scale-up to 90 cm
diameter optics, the process constraints in decreasing order of importance were;
(a) maintenance of constant surface volume (S/V) ratio, (b) control of activator
ion contamination, (c) maintenance of uniform glass surface chemistry, and
(d) fluid flow requirements. The fragility of the leached surfaces and the severe
cleanliness specifications were also significant factors in design and operation.

Design Constraints on a Treatment System

Layer growth rate is directly proportional to the S/V ratio. However, the maximum ratios do not produce the minimum reflectivity and risk potential differences
in reaction rates sufficient to produce figure irregularities that can accumulate
through a large number of surfaces. Too low a rate can result in undesirably long
processing periods. Fixing S/V ratio at a convenient level led to a thin tank
placed vertically to avoid gas accumulation beneath or precipitates above the surfaces. Free liquid volume was set by the largest elements to be treated, and
smaller sizes were accomodated by plastic inserts to reduce free volume and by

Fig. 1. Operation of the treatment system. An 80 cm diameter crystal-array apodizer
is being loaded for treatment. (From ref. 4, photo courtesy LLNL).

varying amounts of ground make-up glass. Total reliance on make-up glass to control S/V ratio is precluded by differences in leach rate between polished and
ground surfaces. Vertical positioning of the tanks also minimizes free surfaces and
evaporation, reducing make-up water requirements.

A number of chemicals mimic the effects of high S/V ratios, increasing rates
to undesirable levels and increasing reflectance minima. The major chemical increasing the rate of film formation is silica, precluding the use of silicone
rubber and unnecessary glass components. Contamination by metal ions such as
Fe^{3+}, Zn^{2+}, Al^{3+}, Cr^{3+}, etc., also accelerates leaching undesirably. This forced

the design of an all plastic system limited to polypropylene and teflon. Most
other plastics contain metal catalysts which were found to leach out during exposure. Polypropylene is used above normal temperature limits, leading to per-
sistent problems of oxidative degradation and non-isotropic shrinkage of components as described in Reference 4.

Uniform surface chemistry is also required for reproducible treatment.
Trends of note are; (a) a strong effect of annealing rate on treatment rate,
(b) prior exposure to acids decrease rates, (c) lower pressure polishes yield
rates closer to "virgin" glass surfaces than high pressure, and (d) contact with
metal ions prior to treatment accelerates rates. These factors are controlled by
specifying fabrication conditions. While variations in rate have been observed,
they have not been observed across a single surface. On one occasion, four pairs
of lenses of different sizes were processed in the same tank. Each set processed
at a slightly different rate, but each pair was identical and uniform. During
processing rates are monitored by checking small witness samples processed with
the large optics. When only one type of optic is processed the leaching
rate of the witness samples appears to be dominated by that of the larger optic,
provided that excessive amounts of differently processed make-up glass are not
used. In-situ reflectance measurements have been attempted on a small scale, but
have not been successful to date due to low reflectance levels and scattering
from solution cloudiness.

There was initial concern over flow uniformity in the narrow flow passages
available at the selected S/V ratio, but a closed-loop circulation system distributed flow across the surface metered by ports in the plastic fixtures. No problems with irregularity have been observed and larger parts are more uniformly
coated than small lab samples.

The entire circulation system is of teflon including the circulation pump.
All process liquids are filtered and the process takes place inside portable HEPA
filtered work stations. Cleanliness requirements are met by maintaining surfaces
in a wet condition, transported to a clean room in water-filled tanks, where
filtered freon rinses and vapor drying complete the cleaning process (Figure 2).
Special handling and storage fixtures permit only edge contact with the fragile
surfaces.

Performance of Treated Optics

The process has been used to treat all BK-7 transmissive optics and diagnostics
in the NOVETTE system, ranging from 10 cm diameter spatial filter lenses to 90 cm
windows. Following treatment an automated measurement system evaluates transmittance and reflectance. Typical reflectance and scattering losses through two surfaces at two wavelengths (1.06 and 0.53 μm) range from 0.5 to 0.8% (figure 3) quite

Fig. 2. Removal of 80 cm diameter f/20 spatial filter lens from final freon cleaning in class 10,000 NOVETTE assembly area (from ref. 4, photo courtesy LLNL).

comparable to results from conventional processes. Damage thresholds of system components have actually been somewhat higher than those of research samples (15 j/cm^2 vs. 12 j/cm^2) and well above the 5 j/cm^2 of conventional AR coatings. No laser damage to system components has been observed.

A potentially serious problem arose with the appearance of slow, long term changes in reflectance characteristics. In both air and vacuum environments changes in reflectance ranging from a few tenths to three or four percent occurred in times ranging from several weeks to several months. The degradation under vacuum usually occurred more rapidly and appeared more severe. Original reflectance levels are easily restored with solvent washes (alcohol or acetone) while water has little effect.

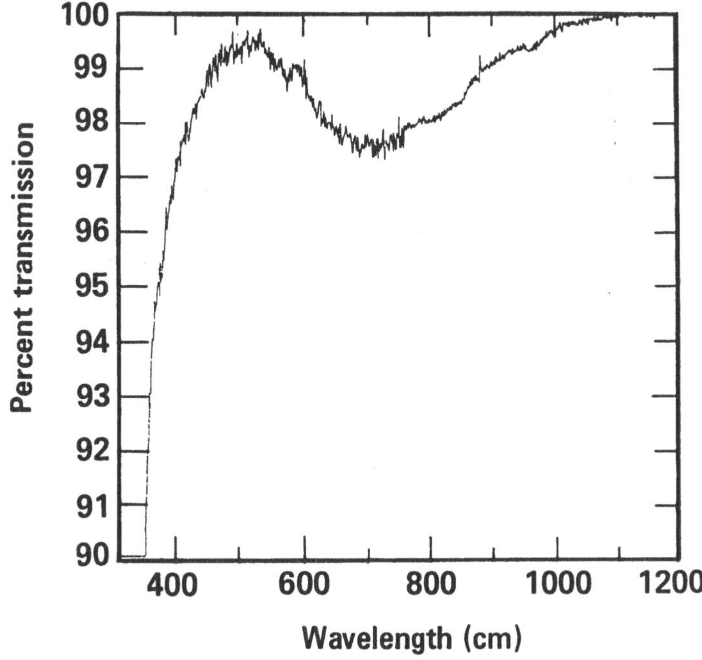

Fig. 3. Typical transmission characteristics of a treated optic for simultaneous operation at 1.06 and 0.53 μm (thickness 1.0 cm).

Degraded samples were rinsed with Methylene Chloride at LLNL and the washings were analyzed by Gas Chromatography/Mass Spectrometry (GC/MS) and Fourier Transform Infrared Spectroscopy (FTIR). The major contaminants in all cases were found to be a mixture of phthalate esters with lesser amounts of caprolactams, and in the case of surfaces exposed to vacuum, trace quantities of silicones. Despite the fact that the degradation appeared more severe in a vacuum environment, significantly lower levels of contaminants were found on the vacuum exposed surfaces. Similar hydrocarbon contamination effects degrading performance have been reported for sol-gel AR coatings.[5]

Possible sources of contamination have been identified as HEPA filters regularly tested by Dioctyl phthalate (DOP) smoke tests, plastic storage fixtures containing DOP as a plasticizer, floor tiling containing DOP, nylon fixture components containing caprolactams, and silicone oils and vacuum greases.

Primary concern has centered on the vacuum behavior as these surfaces are not amenable in-situ to simple solvent rinsing. A number of tests have been made in a carefully cleaned and pre-baked NOVETTE spatial filter vacuum assembly on which a mass spectrometer was installed to monitor contaminants. Eight samples exposed to 1×10^{-6} torr for 168 hours at 100°C showed no degradation and a 0.5% decrease in reflectance (2 surfaces) presumably due to desorption of water. A subsequent test of seven samples in the same system at 22°C for 23 days again showed no degradation. A 16 hour exposure to vacuum showed the same slight reflectivity decrease accompanied by a 0.5 ± 0.1 mg. weight decrease. This behavior is consistent with changes in the surface index profile due to water desorption.[3]

These tests are continuing in order to pinpoint the source of contamination and elucidate the mechanism. Contamination of the vacuum system itself, earlier use of contaminated solvents, and air filtered through contaminated filters is suspected. A suspected mechanism for the accelerated "aging" observed under vacuum is the desorption of trace water in surface porosity, allowing more rapid penetration of pre-existing surface contamination into the pore structure.

Tests to date indicate degradation effects are solely caused by hydrocarbon contamination rather than structural changes in the AR layer itself. Long term testing is continuing and cleaning procedures are being reviewed to reflect current understanding of the problem, which appears correctable.

Conclusions

The transition of the neutral solution process from laboratory to production-scale operation has been smooth. This has occurred in large part because the system was designed around the special chemical requirements of the process. Few engineering problems were encountered and process yields have been very high. Most system malfunctions have occurred because construction materials were used above their normal temperature limits. Treated optics installed in the NOVETTE system have resulted in significant improvements in the output of the laser. The primary problem encountered in use has been long-term increases in reflectance caused by adsorption of hydrocarbon contaminants. Procedures for eliminating this problem in the future are now being evaluated. Especially for vacuum operation, careful attention to clean technique is necessary to avoid degradation effects.

References

1.) L.M. Cook, K.-H. Mader, Proceedings Electro-Optic/Laser '81, Nov. 17-19, 1981, Anaheim, CA., p. 211-20.

2.) L.M. Cook, S. Ciolek, K.-H. Mader, J. Amer. Ceram. Soc. 65 (9), pp. C152-5 (1982).

3.) L.M. Cook, W.H. Lowdermilk, D. Milam, J. Swain, Appl. Opt. 21, 1482-5 (1982).

4.) G.R. Wirtenson, N.J. Brown, L.M. Cook, Proc. Fourteenth Annual Symposium on Optical Materials for High-Power Lasers, Boulder, Colo., Nov. 15-17, 1982 (to be published).

5.) C.T. Solaga, Appl. Opt. 20 (20), 3464 (1981).

Photovoltaische Solartechnik

Photovoltaic Solar Technology

Perspektiven der Photovoltaik

O. HINTRINGER
Siemens AG, München, BRD

Zusammenfassung

Basismaterial für die Photovoltaik in Energieanwendungen wird voraussichtlich bis in die 90iger Jahre kristallines Silizium bleiben.Modul-
wirkungsgrade von 15 % und Kosten unter DM 10,--/Wp[1] für Module scheinen
bis dahin erreichbar. Die nächste Generation könnte eine Dünnschichttech-
nik mit amorphem Silizium sein,vorausgesetzt es gelingt, das Lebensdauer-
problem zu lösen. Das Marktpotential der Photovoltaik liegt in absehbarer
Zukunft bei dezentralen Stromversorgungen entlegener Verbraucher. Aussagen, wann und in welchem Ausmaß von ihr ein Beitrag zur Weltenergieversorgung geleistet wird, basieren heute weitgehend auf Spekulation.

Einleitung

Die auf die Erdoberfläche auftreffende Sonnenenergie entspricht etwa dem
10.000fachen, heutigen Weltenergiebedarf. Sie ist eine regenerative Energie, dezentral verfügbar und kann zur direkten und indirekten Energie-

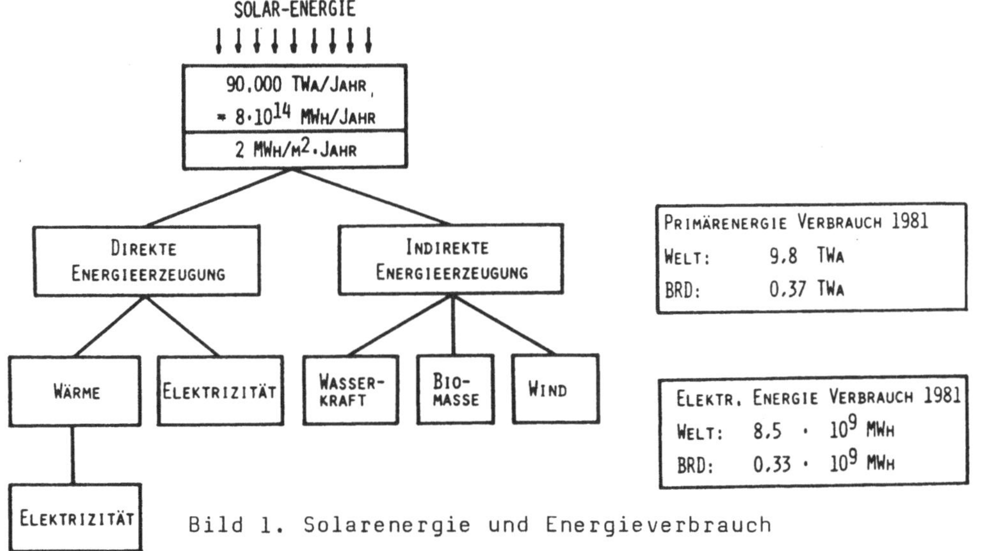

Bild 1. Solarenergie und Energieverbrauch

620

erzeugung benutzt werden, s. Bild 1. Die direkte Umwandlung in elektrische Energie basiert auf Halbleitereffekten und wird als Photovoltaik bezeichnet. Sie taucht in zunehmendem Maße bei der Diskussion mög-
licher zukünftiger Energiestrategien auf. Hierbei sind jedoch einige
Randbedingungen zu beachten.

Die Sonnenenergie steht nur mit starken zeitlichen, klimatischen und
geographischen Schwankungen zur Verfügung. Bei der photovoltaischen
Nutzung spielt daher die ökonomische, raum- und gewichtsparende Speicherung der elektrischen Energie eine entscheidende Rolle. Eine befriedigende Lösung konnte hierfür bis heute nicht gefunden werden.

Ferner ist zu beachten, daß die eingestrahlte Sonnenenergie pro Flächen-
einheit gering und der Wirkungsgrad für die Umwandlung in elektrische
Energie klein ist, wodurch große Flächen erforderlich sind.

Eine weitere, nicht minder wichtige Voraussetzung für eine zunehmende
Anwendung der Photovoltaik zur elektrischen Energieerzeugung ist die
erforderliche drastische Senkung der Herstellkosten für Solargeneratoren.

Bei der Analyse des Welt-Energieproblemes wurde entdeckt, daß die Durch-
setzung neuer Energietechnologien eigenständigen Gesetzen unterliegt
/1,2/, die als Basis für ein logistisches Ersetzungsmodell geeignet
erscheinen, s.Bild 2. Es wurde eine charakteristische, ziemlich gleich-

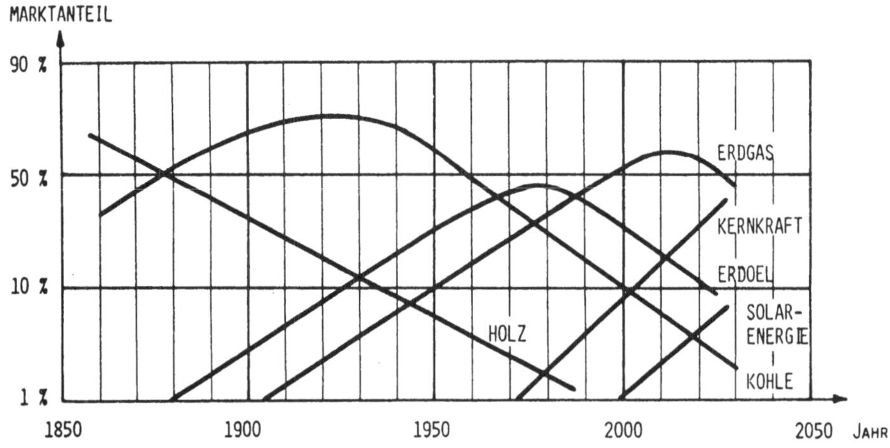

Bild 2. Ablösung von Energietechnologien /1/

bleibende Ablösezeit erkannt, nämlich die Zeit, in der die jeweils neue
Energietechnologie einen Marktanteil von 50 % gewinnt. Für die Solar-

energie wird nach diesem Modell prognostiziert, daß sie um das Jahr
2000 am Primärenergiemarkt Fuß zu fassen beginnt und ihr Anteil in
50 Jahren, d.h. um das Jahr 2030 knapp 10 % betragen könnte. Wenn man
annimmt, daß davon der photovoltaische Energieanteil 6 % beträgt, so
wären das selbst nach der niedrigsten Energiefortschreibung (16 TWa
Szenario /2/) immerhin 1 Mill. GWh/Jahr. Dies entspräche 2 % des pro-
gnostizierten elektrischen Energiebedarfes im Jahre 2030.

Für die Bedeutung der photovoltaischen Solarenergie ab der 2. Hälfte des
nächsten Jahrhunderts wird vermutlich wesentlich sein, ob damit in Kon-
kurrenz zur Kernenergie, Wasserstoff als Sekundärenergieträger und Ersatz fossiler Brennstoffe erzeugt werden kann /3,4/. Nur mit einer Was-
serstoffwirtschaft ließe sich voraussichtlich das Problem der Speicherung photovoltaischer Solarenergie in großem Stil lösen. Im Gegensatz
zu anderen Energieoptionen wie etwa der Kernfusion steht die technische
Machbarkeit der Photovoltaik außer Frage und auch die politische Durch-
setzbarkeit wird aus heutiger Sicht möglich sein. Wann und in welchem
Ausmaß die Photovoltaik einen Beitrag zur Weltenergieversorgung leisten
wird, kann heute niemand mit Sicherheit voraussagen.

In absehbarer Zukunft werden die Marktchancen der terrestrischen Photovoltaik bei dezentralen Stromversorgungen entlegener Verbraucher liegen,
insbesondere wo technische Gründe wie Wartungsfreiheit und Zuverlässigkeit vor ökonomischen stehen. D.h. sie wird dort wirtschaftlichen und
sozialen Fortschritt bringen, wo die Infrastruktur der herkömmlichen
Elektrizitätsversorgung fehlt. Beispiele sind die Bewässerung von Agrar-
land und die Trinkwasserversorgung in Ländern des Sonnengürtels, die
Stromversorgung abgelegener Häuser und Siedlungen, Nachrichten- und Meß-
stationen sowie der kathodische Korrosionsschutz von Pipelines. Es sei
jedoch darauf hingewiesen, daß in Ländern der Dritten Welt, in denen ein
großer Bedarf für dezentrale Stromversorgungen besteht und viel Sonnenenergie vorhanden ist, die Finanzierung von Solarprojekten sehr schwierig und nur mit Entwicklungshilfe-Organisationen möglich ist.

Im folgenden wird nach Erläuterung der prinzipiellen Wirkungsweise von
Solarzellen und der Materialauswahl hierfür untersucht, welche Kosten-
senkungen möglich erscheinen.

<u>Wirkungsweise und Materialalternativen</u>

Basis für die photovoltaische Energieumwandlung ist die Absorption von
Photonen durch einen Halbleiter. Dadurch werden paarweise Ladungsträger
(Elektronen und Löcher) erzeugt, wenn die von der Lichtwellenlänge ab-

hängige Photonenenergie ausreicht, die materialspezifische Valenzbindung
(Bindungsenergie oder Bandabstand E_g) aufzubrechen. Überschüssige Photonenenergie geht als Wärme verloren. Werden nun die lichtzu- und abgewandten Halbleiterbereiche unterschiedlich dotiert (pn-Übergang), so
entsteht ein in seiner Stärke vom Bandabstand abhängiges elektrisches
Feld, das Elektronen und Löcher trennt. Über Metallkontakte kann die
entstehende elektrische Leistung abgegriffen werden, s. Bild 3.In Bild 4

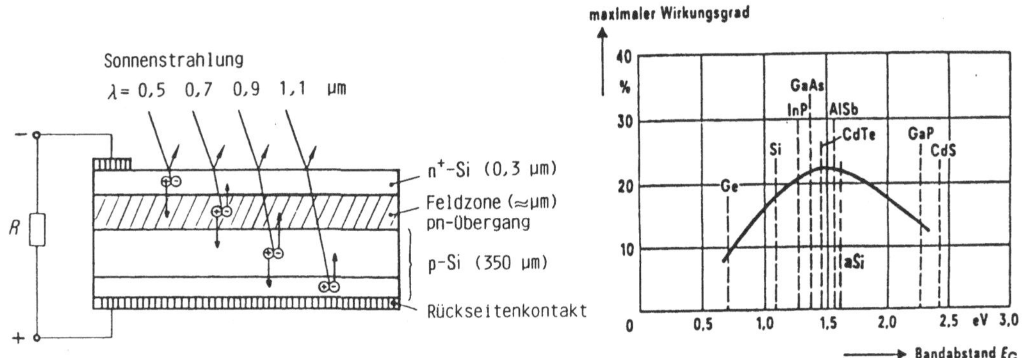

Bild 3. Prinzip der Solarzelle

Bild 4. Vergleich von Umwand-
lungsalternativen

ist der erreichbare Umwandlungs-Wirkungsgrad in Abhängigkeit vom Bandabstand E_g dargestellt. Man erkennt, daß ein Bandabstand von etwa 1,5 eV
optimal wäre und daß eine ganze Reihe von Materialien diesem Wert sehr
nahe kommt, insbesondere Galliumarsenid und amorphes Silizium /5,6/.

Aus mehreren Gründen wird heute jedoch kristallines Silizium bevorzugt:
Silizium ist praktisch unbegrenzt verfügbar (27,7 % der Erdkruste bestehen aus Si), es ist ein äußerst stabiles Material und der Si-Halb-
leiterprozeß wird großtechnisch sehr gut beherrscht. Substanzen mit hö-
herem Wirkungsgrad als Si enthalten entweder seltene und teure Elemente
wie Gallium und Indium oder zusätzlich giftige wie Cadmium und Arsen
oder sind instabil wie Aluminium-Antimon. Das überwiegende Ziel der Untersuchung dieser III-V und II-VI Verbindungshalbleiter liegt in ihrer
möglichen Verwendung als Dünnschicht-Solarzellen. Die Ergebnisse für
Wirkungsgrad und zeitlicher Stabilität (Degradation) sind jedoch bis
heute enttäuschend. Nur kristallines GaAs spielt als Sonderanwendung
in Konzentratorsystemen eine gewisse Rolle.

Eine Ausnahmestellung kann amorphes Silizium (a-Si) beanspruchen. Seit
der ersten Solarzelle aus amorphem Silizium im Jahr 1976 /7/ werden welt-

eit große Anstrengungen unternommen, dieses Material zu erforschen und
en Wirkungsgrad der daraus hergestellten Solarzellen zu erhöhen /8/.
s bietet die Möglichkeit, wahrscheinlich besonders kostengünstige Dünnschicht-Solarzellen herzustellen, weil es eine hohe Photonen-Absorption
nd einen idealen Bandabstand von 1,6 eV besitzt, so daß Schichtdicken
on nur 0,5 /um ausreichen. Es kann in einer Silan-Glimmentladung auf
eliebigen Substraten, vorzugsweise Stahlblech, abgeschieden werden,
obei durch einfaches Hinzugeben von Dotiergasen zum Trägergas während
es Abscheideprozesses gleich der pn-Übergang entsteht. Auch eine Ab-
cheidung auf flexiblen Folien aus Metall oder Kunststoff ist möglich,
as kontinuierlichen und damit kostengünstigen Herstellverfahren sehr
ntgegenkommt. Durch die Verwendung von Silan (SiH4) werden bei der Ab-
cheidung auch Wasserstoffatome in das a-Si eingebaut, die die freien
alenzen im amorphen Netzwerk absättigen, s.Bild 5. Dadurch wird ein

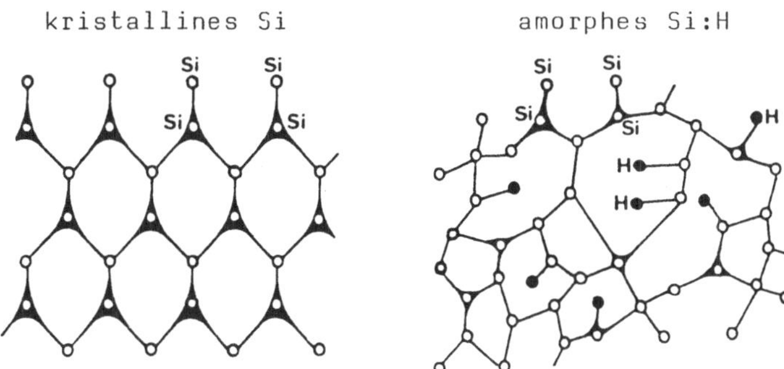

ild 5. Vergleich von kristallinem und amorphem Silizium

roßer Teil der Defekte elektrisch neutralisiert, so daß diese keine
chädlichen Rekombinationszentren für Ladungsträger mehr darstellen und
ine Dotierung möglich ist. Das so hergestellte a-Si enthält bis zu
0 % Wasserstoff und wird als a-Si:H (hydrogenisiertes a-Si) bezeichnet.

en auf den ersten Blick ins Auge springenden Vorteilen von a-Si:H-
olarzellen stehen einige ernstzunehmende Nachteile gegenüber: die grund-
egenden physikalischen Eigenschaften werden noch nicht verstanden und
ie Forschungs- und Entwicklungsarbeiten sind auf empirische Verfahren
ngewiesen. So kann auch der maximal erreichbare Wirkungsgrad nur empi-
isch abgeschätzt werden. Er steht z. Zt. für den einfachen pn-Übergang
ei 12,5 % /9/ ·und für a-Si-Solarzellen mit mehrfachen pn-Übergängen
Tandem- oder Kaskadenzellen) bei 19 % /10/. Die Langzeitstabilität ist
ragwürdig; Degradation wird sowohl im Dunklen (chemische Reaktionen
wischen Metallkontakten und a-Si:H) als auch unter Beleuchtung (Ladungs-

träger-Rekombinationen) beobachtet. Die Abscheideraten sind klein, so
daß der Herstellprozess für eine Solarzelle (bzw. für eine Charge) ca.
1 Stunde beansprucht.

Der höchste bisher berichtete Wirkungsgrad für eine relativ kleine Labor-
zelle beträgt 10,1 % /11/.

Obwohl bereits in Japan hauptsächlich für Taschenrechner sehr große
Stückzahlen mit allerdings kleinen Wirkungsgraden um 3 bis 4 % und kleinen Flächen produziert werden, befindet sich die a-Si-Solarzelle noch
in einer frühen Entwicklungsphase. In Energieanwendungen wird sie nicht
vor den 90er Jahren zum Einsatz kommen und es bleibt, wie bereits gesagt,
vorerst hierfür bei Solarzellen aus kristallinem Silizium.

Die Energieverluste einer einkristallinen Si-Solarzelle /6/ zeigt
Bild 6: 3 % werden trotz Antireflexbelag reflektiert, 23 % der Einstrahlung haben keine ausreichende Energie, um Elektron-Lochpaare zu

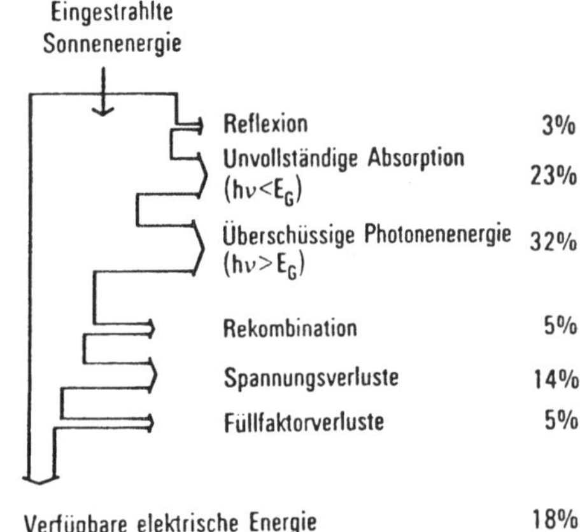

Bild 6. Energieverluste in
einer Solarzelle aus
einkristallinem Si

erzeugen, 32 % gehen verloren, weil Photonen mit höherer Energie als
der Bandabstand trotzdem nur ein Ladungsträgerpaar erzeugen. Nicht alle
Ladungsträger erreichen den pn-Übergang, das ergibt Rekombinationsver-
luste von 5 %. 14 % Verlust entstehen durch Spannungsabfall innerhalb
der Diode und 5 % dadurch, daß es nicht gelingt, eine Solarzelle mit
idealer, rechteckiger Diodenkennlinie zu erzeugen (Kennlinien-Füllfaktor)
Maximal stehen somit 18 % der eingestrahlten Sonnenenergie als elektrische Energie zur Verfügung.

raktisch hat man im Labor einen Wirkungsgrad von 17,5 % unter terre-
trischen Einstrahlungsbedingungen erreicht. Bei Konzentration des Son-
enlichtes ergeben sich theoretisch als auch praktisch höhere Wirkungs-
rade von 25 % bzw. 20,5 % z.B. bei 70 Sonnen. In der Massenfertigung
im Jahre 1982 weltweit ca. 5 Mill. Stück mit 100 mm Durchmesser) er-
ielt man Wirkungsgrade von 12 bis 13 %. Die Schwerpunkte bei der Weiter-
ntwicklung liegen darin, kostengünstige, massenfertigungsgerechte Halb-
eiterprozesse für Diffusion und Metallisierung (z.B. Lack- und Sieb-
ruckverfahren) zu entwickeln und hierbei auch den Wirkungsgrad hochzu-
reiben. Bei Verwendung von polykristallinem Si gewinnen Verfahren zur
orngrenzenpassivierung zunehmend an Bedeutung.

ür die Anwendung in Solargeneratoren müssen die Solarzellen vor Umwelt-
inflüssen wie Feuchte, Sandsturm, Hagelschlag, Winddruck, Salzwasser
.ä. geschützt werden. Hierzu werden bis zu 144 Zellen elektrisch ver-
unden und in einem sandwichförmigen Aufbau aus Glas, Kunststoff- und
etallfolien eingebettet, s. Bild 7. Diese Einheit wird als Solarmodul
der Solarpanel bezeichnet. Durch die Einbettung entstehen unvermeid-
iche Absorptions- und Reflexionsverluste. Ferner kann die Modulfläche
icht zu 100 % mit Solarzellen - auch nicht mit rechteckigen - belegt
erden, da Verbindungstechnik und Rahmung einen gewissen Flächenbedarf
rfordern. Dies führt heute mit Solarzellen von 12 % Wirkungsgrad zu
inem Modulwirkungsgrad von ca. 9 %, der hauptsächlich durch Verbesse-
ung der Solarzellen auf maximal 15 % gesteigert werden kann.

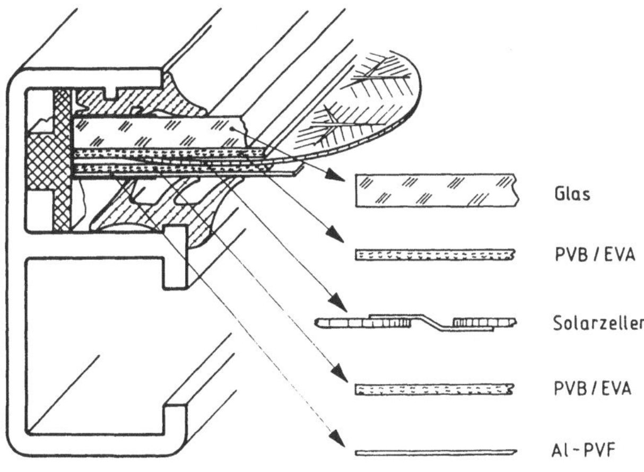

Bild 7. Modulaufbau

Silizium-Herstellverfahren

Die Analyse der Kosten von Solarmodulen ergibt, daß ca. 40 % auf die
Siliziumscheiben, dem Ausgangsprodukt für die Solarzelle, entfallen
und die heute überwiegend aus Halbleiter-Silizium hergestellt werden.
Hierbei wird mit dem carbothermischen Reduktionsprozeß (Lichtbogenofen)
aus Quarz (SiO_2) technisches Silizium gewonnen, daraus Trichlorsilan
($SiHCl_3$) hergestellt, das destilliert wird und unter Wasserstoffein-
wirkung bei 1000 ^{o} C im Siemens-Reaktor zu polykristallinen Stäben auf-
wächst. Daraus werden durch Tiegel- oder Zonenziehen einkristalline
Stäbe hergestellt, die in Scheiben (Dicke ca. 350 μm) zersägt werden.

Die erste Maßnahme zur Kostensenkung ist der Ersatz der teuren Kristall-
ziehverfahren durch Gießen /12/. Hierbei entstehen Blöcke oder Stäbe,
aus denen Scheiben gesägt werden, die zwar polykristallin sind, aber
relativ große kristalline Bereiche enthalten und gute Solarzellen ergeben. Eine weitere wesentliche Kostensenkung ist durch die Entwicklung
neuer Herstellprozesse für Polysilizium möglich /5,12/. Ein besonders
aussichtsreicher beruht darauf, den ökonomischen, carbothermischen Reduktionsprozeß beizubehalten und die nachfolgenden aufwendigen Reini-
gungsstufen (Umwandlung in Trichlorsilan, Destillation, Abscheidung im
Siemens-Reaktor) durch eine kostengünstige Vorreinigung der Ausgangs-
materialien Quarz und Kohlenstoff zu ersetzen /13/. Eine Schlüsselrolle
nimmt hierbei die Entwicklung eines Verfahrens zur Herstellung von hoch-
reinem Quarz ein.

Als Ausgangsmaterial dient billiger Quarzsand; er enthält vor allem
Metall-Verunreinigungen wie Al, Fe und Ti, die in die regelmäßige Struktur des Quarzes eingebaut sind. Auf Grund der hohen Bindungsenergie
zwischen den SiO_4-Tetraedern ist Quarz sehr widerstandsfähig gegen chemische Einflüsse und es ist nahezu unmöglich, daraus Verunreinigungen
zu entfernen. Wenn jedoch der Quarzsand mit glasformenden Oxiden wie
Natriumoxid und Calciumoxid verschmolzen wird, brechen die Silizium-
Sauerstoffbindungen auf und der Schmelzpunkt wird von 1750 ^{o} C auf ca.
1000 ^{o} C herabgesetzt. Das Ergebnis ist eine lockere, offene Struktur,
s. Bild 8, aus der die Verunreinigungen durch Säuren ausgelaugt werden
können, so daß hochreines SiO_2 übrig bleibt. Der Säureauslaugprozeß
wird sehr erleichtert, wenn das Glas in Form von Glaswolle vorliegt,
da so eine große Oberfläche der Säure ausgesetzt wird. Zur anschlies-
senden carbothermischen Reduktion von SiO_2 im Lichtbogen-Reduktions-
ofen dient mit Säure gereinigter Ruß. Das erzeugte Si kann wie bei einem
Hochofen abgestochen und zu Blöcken oder Stäben gegossen werden.

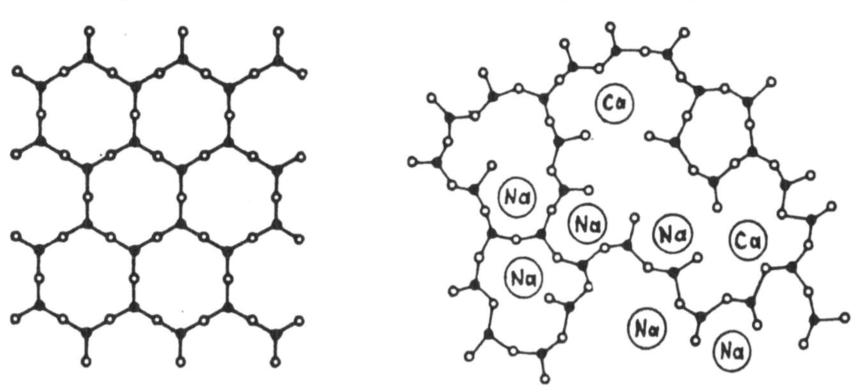

Bild 8. Schematische Darstellung der Struktur von Quarz und eines
Multikomponenten-Glases

Es wird erwartet, daß durch diesen neuen Prozeß zur Herstellung von
polykristallinem Solar-Silizium die heutigen Kosten für Polysilizium
von ca. DM 140,--/kg auf etwa DM 20,-- bis 30,--/kg für Solarsilizium
gesenkt werden können.

Um die hohen Schnittverluste (ca. 50 %)und die Sägekosten zu ersparen,
arbeiten einige Forschungsgruppen daran, Silizium in Flächenform zu
kristallisieren, d.h. Bänder anstatt Stäbe aus der Silizium-Schmelze
zu ziehen /12/. Am weitesten entwickelt sind hierfür das EFG-Verfahren
("Edge Defined Film Fed Growth") /14/ und das "Dendritic Web Growth"-
Verfahren /15/. Beim ersten wird das Si-Band durch eine Graphit-Zieh-
düse geformt. Die Si-Schmelze steigt hierbei durch die Kapillarkräfte
in der Ziehdüse hoch, s.Bild 9. Mit Hilfe eines Kristallkeimes kann
nun ein Band mit guten Kristalleigenschaften hochgezogen werden, wobei
die Ziehgeschwindigkeit durch die Kristallisationsgeschwindigkeit begrenzt wird.

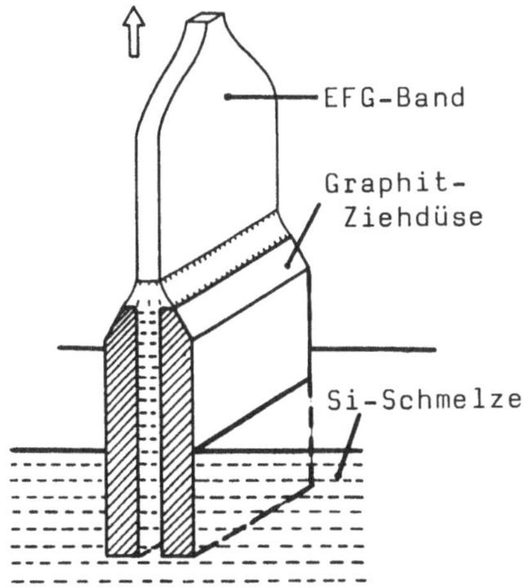

Bild 9. EFG-Verfahren
(edge-defined
film-fed growth)

628

Das Dendritic Web-Verfahren benötigt keine Ziehdüse. Hierbei wird
ein Nadelkristall (Dendrit) als Kristallkeim auf die Si-Schmelze auf-
gesetzt und beim Hochziehen verzweigt. Dazwischen bildet sich ein Si-
Film (Web), der zu einem einkristallinen Band kristallisiert und hochgezogen wird, s. Bild 10. Mit Si-Bändern aus beiden Verfahren konnten
sehr gute Solarzellen (Wirkungsgrad 12 bis 15 %) hergestellt werden.

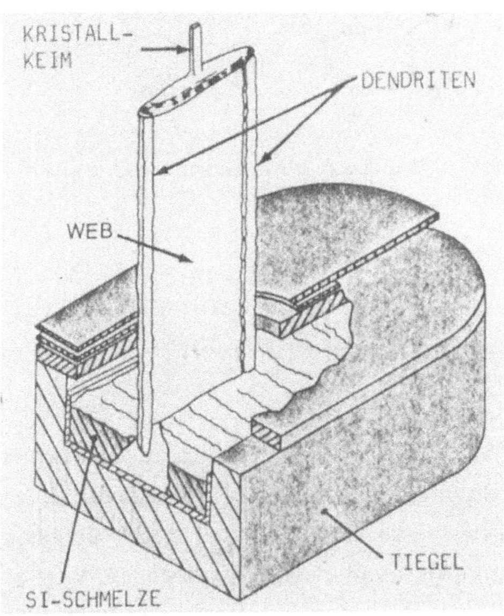

Bild 10. Prinzip des
Dendritic Web
Verfahrens

Beide Verfahren lassen nur langsame Ziehgeschwindigkeiten von 6 bis
8 cm^2/min zu und liegen damit gleich mit der Herstellgeschwindigkeit
bei konventionellem Tiegelziehen und anschließendem Sägen. Es wird daher an zahlreichen Varianten gearbeitet mit dem Ziel, die Ziehgeschwindigkeit zu erhöhen. Eine dieser Varianten ist das S-Web-Verfahren
("Supported-Web-Technik") /16/, s. Bild 11: in einem Tiegel mit flüssigem Si befindet sich ein Schacht, durch den ein Graphitnetz mit Geschwindigkeiten bis 1 m/min gezogen werden kann. Flüssiges Si benetzt
Graphit sehr gut und hat eine hohe Oberflächenspannung, daher bilden
sich in den Maschen Siliziumhäute, die in einer Nachheizzone kristallisieren können. Auf diese Weise kann die Ziehgeschwindigkeit von der
Kristallisationsgeschwindigkeit entkoppelt werden. Durch leichte Schräg
stellung des Netzes wird es nur auf einer Seite beschichtet und es entsteht eine ziemlich plane Oberfläche mit großen einkristallinen Bereichen, die sich über die Maschen hinweg erstrecken.

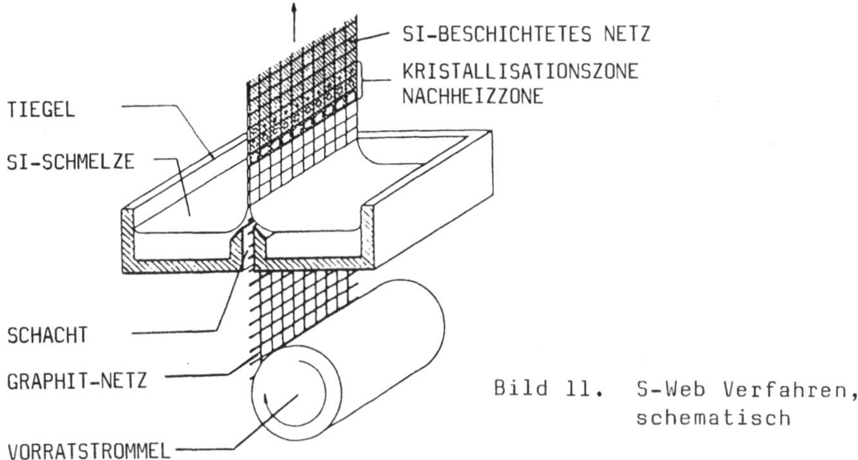

Bild 11. S-Web Verfahren, schematisch

Im Bild 12 sind die beschriebenen Möglichkeiten zur Kostenreduzierung zusammenfassend dargestellt. Man erkennt drei Generationen von Solar- modulen. Eine erste, die auf der konventionellen Kristallzieh- und Säge- technik beruht und bei der Siebdruck, Sprühtechnik sowie neues Solar- silizium zur Kostenreduktion eingesetzt wird. Damit kommt man zu Modul- kosten von ca. DM 10,--/Wp[1] bis Ende der achtziger Jahre. Eine zweite Generation könnte auf Bandziehverfahren und vermehrten Einsatz von Kunststoff bei der Modulherstellung beruhen und bis Mitte der neunziger Jahre zu Modulkosten von 2 bis 3 DM/Wp führen. Eine dritte Generation werden wahrscheinlich Dünnschicht-Solarzellen, vielleicht aus amorphem Silizium, sein, womit ein Kostenziel von 1 DM/Wp möglich erscheint.

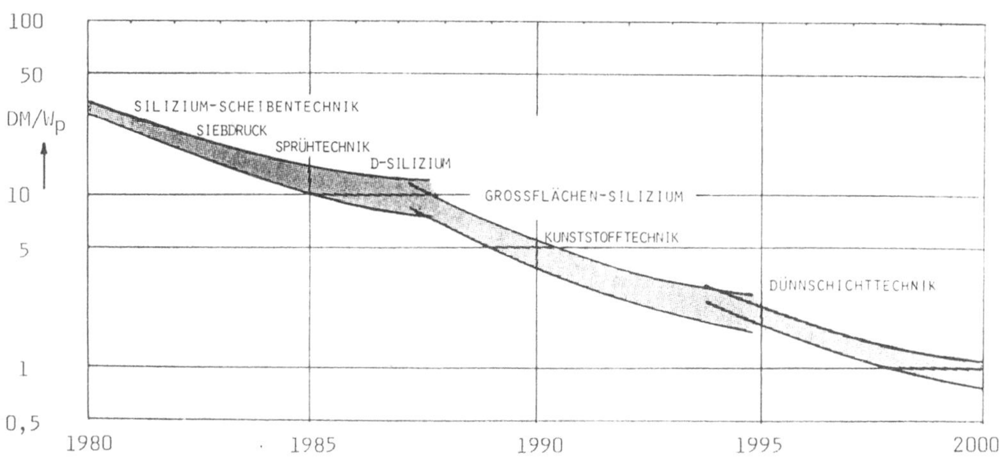

Bild 12. Kostenziele für Solarmodule

Die genannten Kosten beziehen sich nur auf Solarmodule. Um daraus photc voltaische Stromgeneratoren herzustellen, kommen Kosten für die System-technik (Land und Bauwerke, Tragekonstruktion, Installation, Umrichter, Laderegler, Speicher, Kontroll- und Sicherheits-Einrichtungen, Vertei-ler etc.) hinzu, die je nach Größe und Anforderung an den Solarstrom-generator ein Mehrfaches der genannten Kosten betragen können.

LITERATUR

1 IIASA-Analyse: Energy in a Finite World.Ballinger Publ.Comp.,Cambridge,Mass.,USA

2 GERWIN,R.: Die Welt-Energie-Perspektive,Deutsche Verlags-Anstalt 1981, 152

3 BOCKRIS,J. O'M., JUSTI,E.W.: Wasserstoff die Energie für alle Zeiten.
   Udo Pfriemer Verlag, München 1980

4 DFVLR-Nachrichten: Wasserstoff als Sekundärenergieträger.Heft 34 (Nov.1981)

5 VAN OVERSTRAETEN,R., MERTENS,R., NIJS,J.: Progress in photovoltaic energy
   conversion. Rep. Prog. Phys. 45 (1982) 1041-1111

6 WINSTEL,G.H.: Elektrische Energie aus Solarzellen. Siemens Energietechnik, 2
   (1980), Heft 7, 266-269

7 CARLSON,D.E., WRONSKI,C.R.: Amorphous Silicon Solar Cell.Appl.Phys.Lett.28
   (1976) 671-673

8 PLÄTTNER,R.D., KRÜHLER,W.W., MÖLLER,M.: Properties of Amorphous Silicon Solar
   Cells. Siemens F.u.E. Ber. 11 (1982) 284-290

9 KUWANO,Y., TSUDA,SH., OHNISHI,M.: Energy Conversion Process of pin Amorphous
   Si Solar Cells. Jap. J. Appl. Phys. 21 (1982) 235

10 NAKAMURA,G., SATO,K., KONDO,H., YUKIMOTO,Y., SHIRAHATA,K.: High-Efficiency
   Tandem Type Solar Cells Consisting of a-Si:H and a-Si Ge:H. Proc. 4th E.C.
   Photovoltaic Solar Energy Conf., Stresa 1982, Reidel Publ. Co., 616-620

11 CARLSON,D.E.: US-Japan Joint Seminar on "Technological Applications of Tetra-
   hedral Amorphous Silicon" Palo Alto, Ca., July 1982

12 DIETL,J., HELMREICH,D., SIRTL,E.: Solar Silicon. Crystals Bd. 5, Silicon,
   Editor: Grabmaier J., Springer-Verlag 1981, 43-107

13 AULICH,H.A., DIETZE,W., EISENRITH,K.-H., SCHÄFER,J., SCHULZE,F.W., URBACH,H.P.:
   Production of Solar Grade Silicon in an Arc Furnace Using High Purity Starting
   Materials. Proc. 4th E.C. Photovoltaic Solar Energy Conf., Stresa 1982,
   Reidel Publ. Co., 868-873

14 WALD,F.V.: Crystal Growth of Silicon Ribbons. Crystals Bd. 5, Silicon,
   Editor: Grabmaier J., Springer-Verlag 1981, 147-198

15 SEIDENSTICKER,R.G.: Dendritic Web Silicon for Solar Cell Application.
   J. of Crystal Growth, Vol. 39 (1977) 17

16 GRABMAIER,J.G., FÖLL,H., FREIENSTEIN,B., GEIM,K.: Fast Silicon-sheet Growth
   with the supported-Web Method. Proc. 4th E.C. Photovoltaic Solar Energy Conf.,
   Stresa 1982, Reidel Publ. Co., 976-979

---

1) W_p (Watt peak) ist die maximale photovoltaisch erzeugte elektrische Leistung
   bei AM1-Bedingung, d.h. Sonne im Zenit und klare Atmosphäre.

Hocheffiziente Dünnschichtsolarzellen-Systeme

F. PFISTERER und H.W. SCHOCK
Institut für Physikalische Elektronik, Universität Stuttgart,
Pfaffenwaldring 47, D-7000 Stuttgart 80

Wirkungsgrade von Solargeneratoren von ca. 15 % erscheinen mit Tandemstrukturen realisierbar. Ein Aufbau bestehend aus einem beidseitig beschichteten Substratglas - auf der einen Seite eine ZnTe-ZnCdS- und auf der anderen Seite eine Cu_2S-ZnCdS-Dünnschichtsolarzelle - wird vorgestellt. Die Produktionskosten solcher Strukturen dürften unterhalb der von ein- und multikristallinen Si-Solarzellen liegen. Es werden erste Ergebnisse über die Entwicklung von ZnTe-ZnCdS Solarzellen berichtet.

Einleitung

Dünnschichtsolarzellen bleiben im Wirkungsgrad bisher meist unter 10 %. Dieser Nachteil gegenüber ein- und multikristallinen Si-Solarzellen erscheint überwindbar mit Dünnschicht-Tandemstrukturen. Die theoretische Situation /1/ ist in Abb. 1 dargestellt: bei Aufteilung des Spektrums in mehrere Bereiche und Zuführung dieser Bereiche zu speziell abgestimmten Solarzellen gewinnt man gegenüber der Einzelsolarzelle mit E_{g1} zusätzliche elektrische Energie aus dem roten Spektralbereich (für $h\nu<E_{g1}$) und nutzt die höhere Quantenenergie im kurzwelligen Spektralbereich in einer Solarzelle mit hohem Bandabstand und dementsprechend höherer Ausgangsspannung.

Grundsätzlich sind zwei Realisierungsmöglichkeiten für Mehrfachsolarzellen-Systeme zu erwähnen:
1. Aufteilung des Spektrums mit optischen Elementen;
2. Anordnungen von Zellen mit transparenten Kontakten in Kaskaden, wobei die Zelle mit dem größten Bandabstand der Sonne zugewandt ist. Kaskadenanordnungen können
   a) galvanisch gekoppelt (als Beispiel sei die monolithische AlGaAs-GaAs-Kaskade erwähnt /2/) oder
   b) galvanisch getrennt sein.

Materialauswahl für Dünnschicht-Tandemsysteme

Zunächst haben die in Frage kommenden Halbleitermaterialien die bekannten Kriterien zu erfüllen, die schon für einfache Dünnschichtsolarzellen gelten:
- hoher Absorptionskoeffizient;

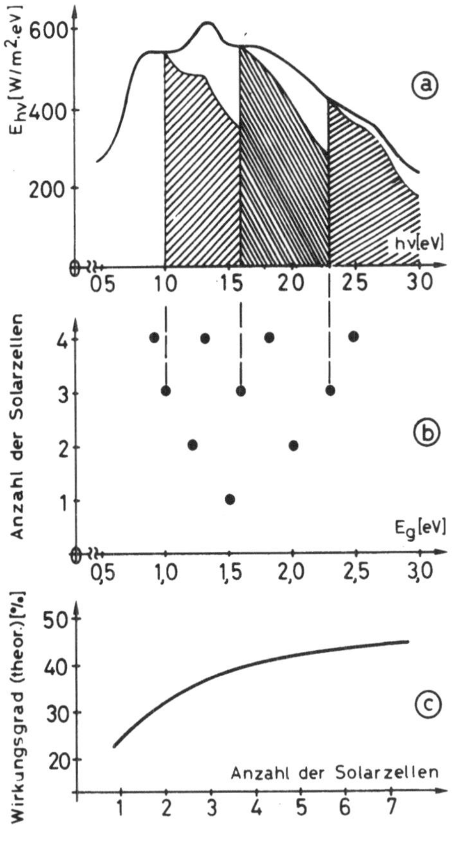

Abb. 1. Tandemsysteme

a) mit 3er-Tandem nutzbarer Anteil der
Sonnenstrahlung (schraff. Fläche);

b) optimale Bandabstände (nach /1/);

c) theoretisch erreichbare
Wirkungsgrade (nach /1/).

- gutes Schichtwachstum bei Substrattemperaturen unterhalb 400° C; dann ist Glas als
kostengünstiges Substratmaterial möglich;
- Anpassung der Kristall- bzw. Halbleiterparameter bei Heteroübergängen: Gitterkonstante, Elektronenaffinität, thermischer Ausdehnungskoeffizient.
Eine Übersicht über als geeignet bekannte Halbleiter gibt Tabelle I. In der Tabelle
fällt auf, daß für Solarzellen mit hohem Bandabstand als p-Leiter neben amorphem SiC
lediglich ZnTe existiert. Als n-Material bietet sich das Mischkristallsystem
$Zn_xCd_{1-x}S$ an. Durch die Variation von x läßt sich die Elektronenaffinität χ an die
des ZnTe anpassen, wie Abb. 2 zeigt. Gleichzeitig verschlechtert sich jedoch die Anpassung der Gitterkonstanten im Heteroübergang ZnTe-ZnCdS, wie Abb. 3 zeigt. Wollte
man beide Parameter, χ und a, anpassen, so müßte man auf quarternäre Materialien
übergehen.

Aufbau einer Dünnschicht-Tandem-Struktur

Der geplante Aufbau eines Dünnschicht-Tandems ist in Abb. 4 zu sehen. Statt ZnTe-
$Zn_xCd_{1-x}S$ ist hier die Kombination ZnTe-ZnS dargestellt. Das Substratglas wird zu-

Tabelle I. Halbleitermaterialien für Dünnschichtsolarzellen und
Dünnschichttandemstrukturen

p-Halbleiter	Eg/eV	n-Halbleiter	Eg/eV
Cu_2S	1,2	CdS	2,4
$CuInS_2$	1,53	$CuInS_2$	1,53
$CuInSe_2$	1,0	$CuInSe_2$	1,0
InP	1,35	ZnS	3,66
ZnTe	2,26	$Zn_xCd_{1-x}S$	2,4 - 3,66
a-Si:H	1,7 - 1,8	a-Si:H	1,7
$a\text{-}Si_{1-x}Ge_x$:H	1,1 - 1,7	$a\text{-}Si_{1-x}Ge_x$:H	1,1 - 1,7
$a\text{-}Si_{1-x}C_x$:H	1,7 - 3.0	$a\text{-}Si_{1-x}C_x$:H	1,7 - 3,0

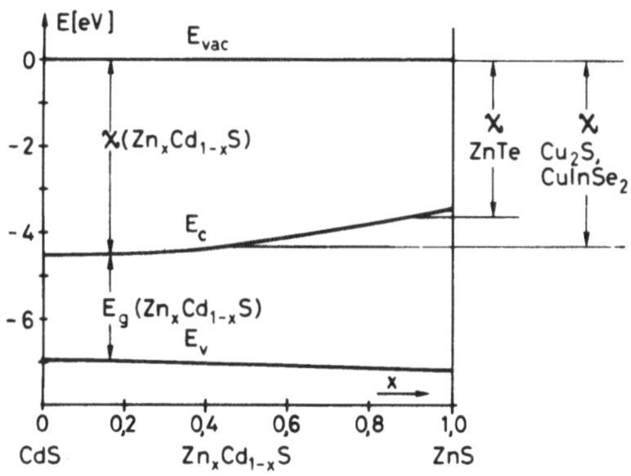

Abb. 2. Bandstruktur des Systems $Zn_xCd_{1-x}S$ als Funktion von x.

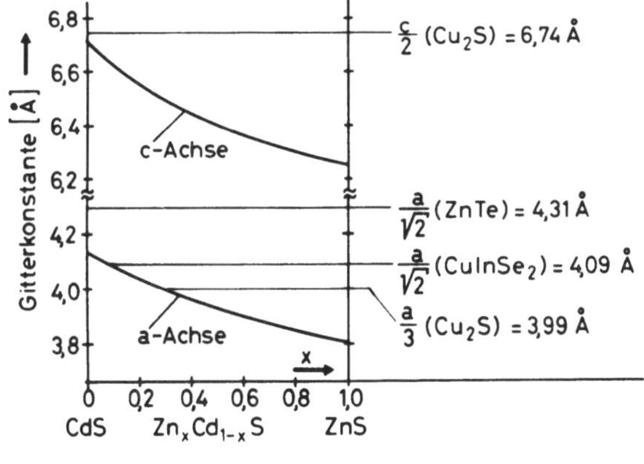

Abb. 3. Gitterkonstante des Systems $Zn_xCd_{1-x}S$ nach /3/ und Vergleich mit div. p-Halbleitern.

nächst beidseitig mit transparentem Kontaktmaterial (ITO = indium tin oxide) und
ZnCdS bzw. ZnS beschichtet mittels Sputtering bzw. Aufdampfung. Die ZnTe-Schicht
wird ebenfalls aufgedampft, die Cu_2S-Schicht entsteht in einem Tauchbad, das Cu^+-

Ionen enthält /4/. Die Frontkontaktierung und Versiegelung gegen Witterungseinflüsse kann in analoger Weise wie bei Cu_2S-CdS Dünnschichtsolarzellen erfolgen /4/.

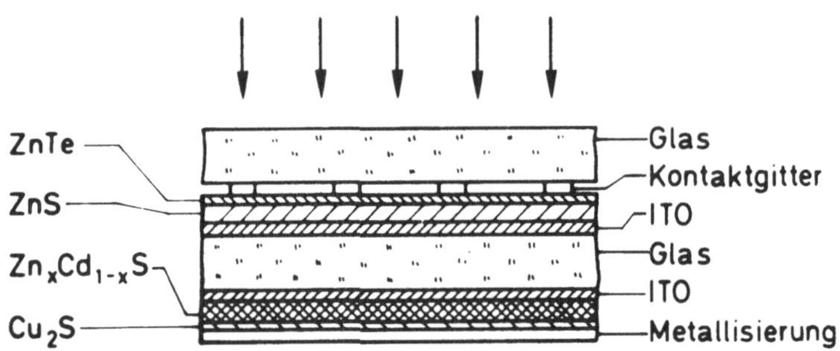

Abb. 4. Schematischer Aufbau einer ZnTe-ZnS/$Zn_xCd_{1-x}S$-Cu_2S-Tandemzelle.

Bisherige Ergebnisse mit ZnTe-ZnCdS Heteroübergängen

Der spektralen Quantenausbeute, Abb. 5, ist zu entnehmen, daß der größte Teil des Photostromes bei den bisher hergestellten Proben im CdS erzeugt wird (schraffierte Fläche). EBIC (electron beam induced current) Messungen deuten auf eine kleine Diffusionslänge im ZnTe und auf hohe Oberflächenrekombination hin.

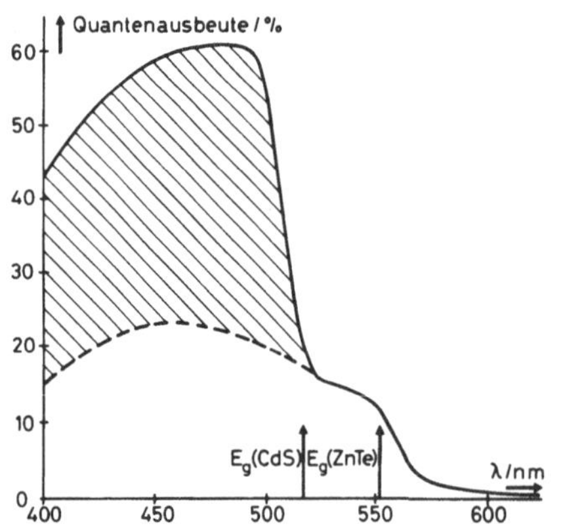

Abb. 5. Quantenausbeute eines ZnTe-ZnCdS Heteroüberganges.

Die Kennlinie einer ZnTe-ZnCdS Zelle ist in Abb. 6 zu sehen. Der Knick in ihrem Verlauf tritt besonders bei Zellen mit hohem Kurzschlußstrom auf. Dies und ein häufig zu beobachtender langwelliger Ausläufer bei der Quantenausbeute deuten darauf hin, daß parallel zum Übergang ZnTe-ZnCdS ein weiterer existiert, gebildet entweder durch Cu_2S oder durch Cu_2Te. Letztere Materialien können beim Dotierungsprozeß der ZnTe-Schicht mit Kupfer entstehen. Ein Indiz dafür ist u.a. die beobachtete Abhängigkeit der Leerlaufspannung von der Cu-Konzentration: statt wie zu erwarten mit der Cu-Kon-

zentration anzusteigen, sinkt die Leerlaufspannung mit zunehmendem Cu-Gehalt. Bei
geringer Cu-Dotierung wurden Leerlaufspannungen bis 930 mV gemessen.

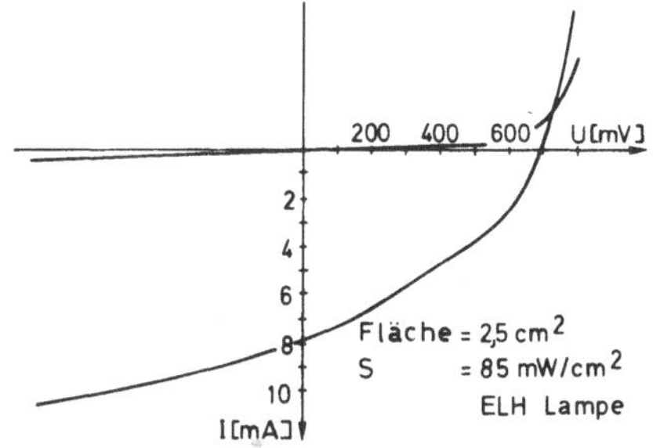

Abb. 6. Kennlinie einer ZnTe-ZnCdS Zelle.

Perspektiven von ZnTe-ZnCdS/Cu$_2$S-ZnCdS Tandemstrukturen

Abschätzungen, die auf gemessenen und auf für realisierbar erachteten Daten für die
einzelnen Zellen beruhen, lassen für die beschriebene Tandemstruktur Wirkungsgrade
von ca. 16 % erwarten. In Tabelle II sind die Einzeldaten aufgelistet.

	Leerlauf-spannung	Kurzschluß-strom bei AM 1,5	Füll-faktor	P_{max} bez. auf AM 1,5	mit Sammlungs- und optischen Verlusten
ZnTe-ZnCdS	1,5 V	6,5 mA/cm²	0,75	7,3 mW/cm²	6 mW/cm²
Cu$_2$S-ZnCdS	0,7 V	20 mA/cm²	0,70	9,8 mW/cm²	7 mW/cm²
insgesamt bezogen auf 100 mW/cm² Einstrahlung					13 mW/cm² 15,8 mW/cm2.

Zusammenfassung

Die Untersuchungen zeigen, daß über Tandemanordnungen auf Dünnschichtbasis Wirkungsgrade erzielt werden können, die bei realistischer Einschätzung diejenigen von besten
einkristallinen Si-Solarzellen erreichen oder gar übertreffen können.

Literatur

/1/ N.A. GOKCEN a.J.J. LOFERSKI, Solar Energy Mater. 1, 271 (1979).
/2/ S.M. BEDAIR u.a., Proc. 3rd EC Photov. Solar Energy Conf., Cannes, 1980; S. 408.
/3/ O.P. AGNIHOTRI u.a., Proc. 13th IEEE Photov. Spec. Conf., Washington, 1978;S.195.
/4/ F. PFISTERER und W.H. BLOSS, Proc. "LASER 81" - Optoelektronik, München, 1981.

Lichtblitzgerät für Solarzellen-Kennlinien-Messungen

Andreas Wagner
Institut für Theorie der Elektrotechnik, Universität Stuttgart
Breitscheidstr. 3, D-7000 Stuttgart 1

Reproduzierbare Messungen unter natürlichem Licht sind bei unseren
meteorologischen Bedingungen meist nur an wenigen Tagen des Jahres
möglich. Kommerziell erhältliche Lichtquellen, die eine gute Strah-
lungsverteilung und ein dem Sonnenlicht angepaßtes Spektrum besitzen,
sind sehr teuer. Deshalb wurde eine Bestrahlungsquelle mit einer
Blitzlampe aufgebaut (Bild 1).

Bild 1. Lichtblitzgerät zur Erzeugung hoher Bestrahlungsstärken

Mit diesem Lichtblitzgerät können die Kennlinien von Solarzellen und
Solarzellenmodulen gemessen werden.

Technische Daten des Lichtblitzgerätes:

Blitzlampe FX-132 von EG&G
Maximale Blitzenergie: 200 J
Blitzkondensator: 280 μF (7 x 40 μF MP, 2.5 kV, Bosch)
Maximale Bestrahlungsstärke ohne Reflektor:
        2 kW/m^2 auf einer Fläche von 60 x 60 cm^2
        Homogenität der Bestrahlung: Anstieg um 2.5 %
        von der Mitte zum Rand.
Maximale Bestrahlungsstärke mit Reflektor:
        bis zu 100 kW/m^2 auf einer Fläche von 7 x 7 cm^2.
Zeitlicher Abstand der Blitzfolge:
        0.86 ... 2 sec, abhängig von der Blitzenergie.

Das Lichtblitzgerät besteht aus einem Leistungsteil (mit dem Entladestromkreis der Blitzlampe) und einem Steuerteil für die Meßdatener-
fassung. Zur Festlegung des zeitlichen Verlaufs des Lichtblitzes wird
der Blitzkondensator (C = 280 μF) über eine Spule mit der Selbstinduk-
tivität L = 22,6 μH entladen (Bild 2).

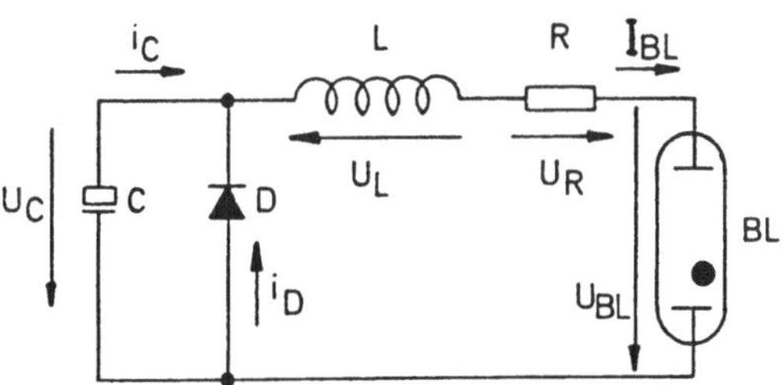

Bild 2. Entladestromkreis der Blitzlampe

Dadurch steigt der Strom I_{BL} nach der Zündung der Blitzlampe BL in
T/4 = 125 μs auf seinen Maximalwert. Nach diesem Zeitpunkt übernimmt
die Diode D den Strom I_{BL}, wodurch ein Vorzeichenwechsel der Spannung
am Kondensator weitgehend vermieden wird (vgl. Bild 3). Dies ist wün-
schenswert, da eine häufige Umpolarisierung des Dielektrikums die
Lebensdauer der Kondensatoren herabsetzt. Das kleine Überschwingen
der Kondensatorspannung wird von der parasitären Induktivität der
Diode und deren Zuleitung verursacht, die deshalb den Strom I_{BL}

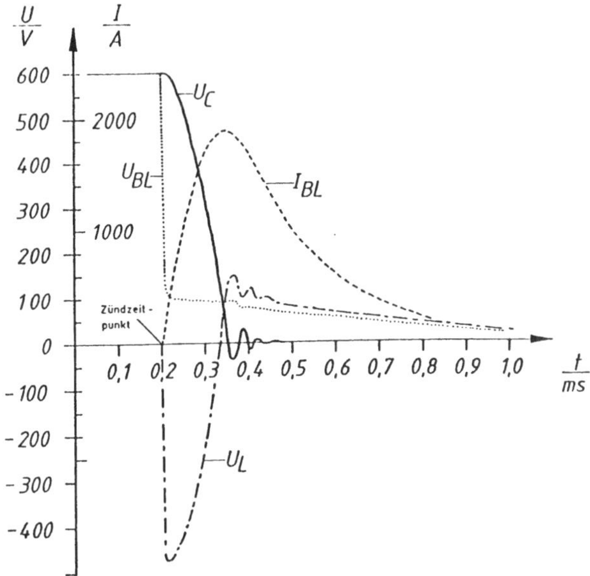

Bild 3. Zeitlicher Verlauf der elektrischen Größen im Entladestrom-
         kreis der Blitzlampe

638

nicht schlagartig übernehmen kann. Zur Veranschaulichung: Bei einer
Kondensatorspannung von 1200 V beträgt der von der Diode zu überneh-
mende Strom ca. 4000 A. Dabei stellt sich ein negativer Überschwinger
der Kondensatorspannung U_C = 50 V ein. Nachdem die Diode den Strom I_{BL}
übernommen hat, klingt dieser Strom exponentiell ab. Der von der
Blitzlampe erzeugte Lichtfluß ist näherungsweise proportional zur um-
gesetzten Leistung $U_{BL} \cdot I_{BL}$.
Der Einfluß des Spektrums von künstlichen Strahlungsquellen wurde in
/1/ und /2/ untersucht. Für gepulste Xenon-Lampen ergeben sich gute
Vergleichswerte zum natürlichen Sonnenlicht unter AM O und AM 1,5.
Bezüglich der Erwärmung der Solarzelle bietet die gepulste Strahlungsquelle erhebliche Vorteile. Damit die Kennlinienpunkte bei gut repro-
duzierbaren Bestrahlungsstärken gemessen werden können, muß allerdings
der Lichtimpuls in seinem zeitlichen Intensitätsverlauf bekannt sein.
Eine Solarzellenkennlinie wird aus mindestens 125 Einzelmessungen zusammengesetzt. Die Messung eines Kennlinienpunktes wird durch einen
Komparator getriggert, der die aktuelle Bestrahlungsstärke im abfallenden Ast des Lichtpulses mit einem vorgegebenen Sollwert vergleicht.
Die Meßwerte von Strom und Spannung werden an einen Kleinrechner
(CBM 8032) übertragen, mit dem anschließend eine automatische Auswertung durchgeführt werden kann. Die Anordnung zur Messung von Strom
und Spannung zeigt Bild 4 (Die Steuerspannung U_{St} wird für jede Ein-
zelmessung schrittweise erhöht).

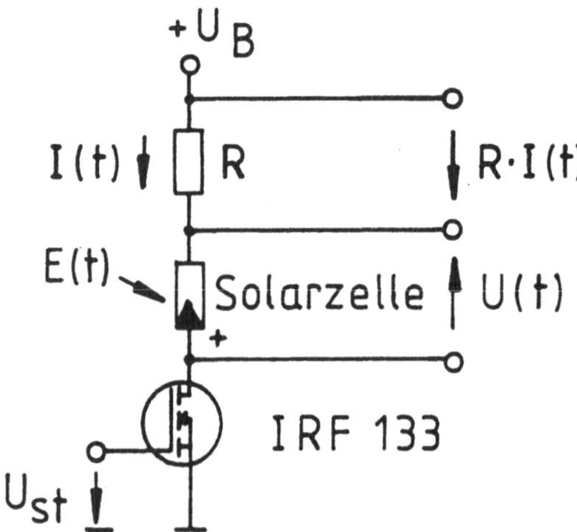

Bild 4. Anordnung zur Messung von Strom und Spannung

Für die Beurteilung der Qualität der gemessenen Kennlinien muß der
Einfluß der Sperrschichtkapazität der Solarzelle im gepulsten Betrieb

auf die Kennlinie, sowie die Meßgenauigkeit und die Reproduzierbarkeit
bekannt sein.

Würden sich bei einer momentanen Bestrahlungsstärke E(t) die gleichen
Strom-Spannungswerte am Solarzellenmodul wie im stationären Fall ein-
stellen, so müßte U (t) im Falle E(t) = O und I(t) = O ebenfalls Null
sein. Dies ist im gepulsten Betrieb nicht so.

Bei sehr kleinen Strömen (Größenordnung einige μA) und im Leerlauf
verursacht die Sperrschichtkapazität ein langsames Abklingen der Spannung, nachdem der Blitz vorüber ist. Dieser Speichereffekt wirkt sich
auf den aufgezeichneten Kennlinienverlauf kaum aus, da die Meßpunkte
der Kennlinie üblicherweise im abfallenden Ast des Lichtpulses im
Bereich zwischen 50 % und 90 % der maximalen Bestrahlungsstärke aufgenommen werden.

Lediglich bei sehr hohen Bestrahlungsstärken (bis 100 kW/m^2) tritt je
nach Zellentyp ein mehr oder weniger ausgeprägter Knick in der Kennlinie in der Nähe des Leerlaufpunktes auf (Bild 5). Dieser Bereich der
Kennlinie kann für den stationären Zustand der Solarzelle nachträglich
durch Extrapolation ermittelt werden.

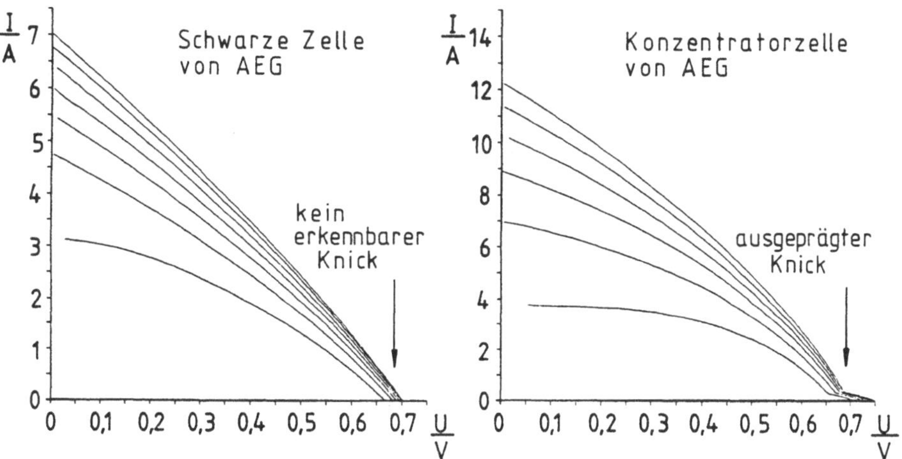

Bild 5. Verzerrung der Solarzellen-Kennlinie im gepulsten Betrieb
         bei hohen Bestrahlungsstärken

Der Vergleich von Solarzellen-Kennlinien, die sowohl unter Dauerlicht
als auch mit dem Lichtblitzgerät aufgenommen wurden, zeigten im Rahmen
der Meßgenauigkeit (1,5 % vom Meßbereichsendwert) gute Übereinstimmung.
Die Reproduzierbarkeit der Kennlinienmessung selbst liegt innerhalb
der Strichstärke der Kennlinienplots (Bild 6).

640

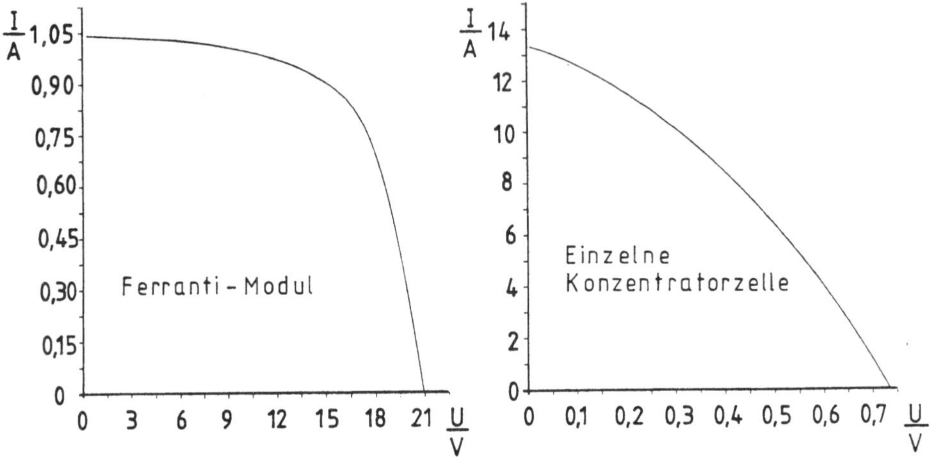

Bild 6.   Reproduzierbarkeit der Kennlinienmessung (In jedem Diagramm
sind zwei Kennlinien eingetragen)

Über die Problematik bei der Messung von Konzentratorzellen unter
künstlichen Lichtquellen (z. B. die Bestimmung der Bestrahlungsstärke)
wird ausführlich in /3/ berichtet.

Diese Arbeit wurde vom Bundesministerium für Forschung und Technologie
(BMFT) im Rahmen des Projektes ET 4045 B (Technische Nutzung solarer
Energie)gefördert.

Literatur

/1/   SEAMAN C. H. et al: The spectral irradiance of some solar simu-
lators and its effect on cell measurements. - Proceedings 14 th
Photovoltaic Specialists'Conference, San Diego 1980.

/2/   CURTIS H. B.: Global calibration of terrestrial reference cells
and errors involved in using different irradiance monitoring
techniques. - Proceedings 14 th Photovoltaic Specialists'Conference, San Diego 1980.

/3/   NASBY R. D., SANDERSON P. W.: Performance measurement techniques
for concentrator photovoltaic cells. - Solar Cells, Vol. 6, No.
1982.

/4/   BMFT-Forschungsbericht 03E-4045 B: Technische Nutzung von Sonnenenergie. - Stuttgart 1982.

Messung der spektralen und integralen Empfindlichkeit von photovoltaischen Solarelementen

H. KAASE und J.METZDORF
Physikalisch-Technische Bundesanstalt
Bundesallee 100, D 3300 Braunschweig

1. Einleitung

Die kritischste Größe bei der Kalibrierung von Solarelementen ist die Kurzschlußstromdichte $i_{sc,AMn}$. Deren Messung unter definierten Bedingungen (AM1.5 oder AMO
bei $25^{o}C$ /1/) und damit die Bestimmung der integralen Empfindlichkeit

$$s_{AMn} = i_{sc,AMn}/E_{AMn}$$

erfordert insbesondere die Einhaltung der beiden Strahlungsgrößen Gesamtbestrahlungsstärke E_{AMn} und die dazugehörige, durch den Index AMn charakterisierte spektrale Verteilung.

Die übrigen Größen, Leerlaufspannung V_{oc} und Füllfaktor FF, zur Berechnung des
Wirkungsgrades

$$\eta_{AMn} = (i_{sc} \, V_{oc} \, FF)_{AMn}/E_{AMn}$$

sind durch die elektrischen Parameter und die Werte des Ersatzschaltbildes des Solarelements festgelegt und werden bei gegebener Kurzschlußstromdichte nicht von der
Zusammensetzung des eingestrahlten Spektrums beeinflußt.

Für die Empfindlichkeit von Strahlungsempfängern ist in der PTB eine hinreichend
gesicherte Skala realisiert, die durch Kalibrierungen von Strahlungsempfängern bei
spektralen Strahlungsleistungen von $\frac{\partial E}{\partial \lambda} \Delta\lambda \approx 1$ µW an interessierte Stellen in Forschung und Industrie weitergegeben wird. Die entsprechenden monochromatischen Be-
strahlungsstärken $\frac{\partial E}{\partial \lambda} \Delta\lambda$ liegen bei homogener Ausleuchtung eines Solarelementes
i.allg. unter $100nW/cm^2$. Es ist jedoch nicht ohne weiteres zulässig, die Gültig-
keit dieser Kalibrierung ohne zusätzliche Prüfung in den um 6 Größenordnungen höheren Bereich der Solarkonstanten zu übertragen. Deshalb werden im Folgenden verschiedene Methoden zur Bestimmung der AMn-Empfindlichkeit diskutiert und verglichen.

2. Faltung der spektralen Empfindlichkeit

Es wird die absolute differentielle spektrale Empfindlichkeit $\Delta s(\lambda,E_L)$ von Solarelementen als Funktion der integralen Bestrahlungsstärke E_L bis zu einer Solarkonstanten gemessen. Hierzu werden die Prüflinge zusätzlich zu einer stationären Be-

strahlung E_L mit einer modulierten, monochromatischen Strahlungsleistung homogen ausgeleuchtet und der differentielle Kurzschlußstrom ΔI_{sc} mit lock-in Technik als Funktion der Wellenlänge analysiert.

$$\Delta s(\lambda ,E_L) = \Delta I_{sc} /(\frac{\partial \phi}{\partial \lambda} \Delta \lambda \ ) = \Delta i_{sc} / (\frac{\partial E}{\partial \lambda} \Delta \lambda)$$

Abb. 1 zeigt als Beispiel derartige $\Delta s(\lambda)$-Spektren eines Si-Solarelements bei verschiedenen Bestrahlungsstärken E_L einer gleichbleibend "sonnenähnlichen" Zusatzbestrahlung (Halogenglühlampe mit Wasserfilter). Die Faltung von $\Delta s(\lambda)$ mit der spektralen Bestrahlungsstärke des vorzugebenden AMn-Spektrums führt auf die differentielle Empfindlichkeit

$$\Delta s_{AMn}(E_L) = (E_{AMn})^{-1} \cdot \int_{(AMn)} \Delta s(\lambda ,E_L) \left(\frac{\partial E}{\partial \lambda} \right)_{AMn} d\lambda ,$$

die bezüglich der spektralen Verteilung AMn eine integrale Größe ist (in Abb. 2 das terristrische AM1.5-Spektrum nach /1/).

Bezogen auf den durch den Betrag der wirksamen Bestrahlungsstärke gegebenen "Arbeitspunkt" ist $\Delta s(E_L)$ jedoch eine differentielle Empfindlichkeit, aus der durch Integration die für den Wirkungsgrad benötigte integrale Empfindlichkeit folgt

$$s_{AMn,L} = (E_{AMn})^{-1} \int_0^{E_{AMn}} \Delta s_{AMn}(E_L) \cdot dE_L .$$

Für den Verlauf in Abb. 2 gilt beispielsweise

$s_{AM1.5,L}$ = 0,27mA/mW. Dieser Wert entspricht etwa $0,999 \cdot \Delta s_{AM1.5}(E_L{=}E_{AM1.5})$.

Der Index L weist darauf hin, daß das so gewonnene Ergebnis $s_{AMr,L}$ noch von der spektralen Zusammensetzung der stationären Zusatzbestrahlung E_L abhängen darf. Untersuchungen über den Einfluß der spektralen Verteilung von E_L auf den Verlauf $\Delta s = \Delta s(E_L)$ sind in Vorbereitung.

Die Auswertung der Kurvenschar $\Delta s(\lambda ,E_L)$ eines Solarelements gemäß Abb. 1 mit $(E_L + \frac{\partial E}{\partial \lambda}\Delta \lambda)$ als Parameter liefert Informationen über:

(i)   die Abweichung der Kurzschlußstromdichte $i_{sc}(E_L)$ von der Linearität;

(ii)   den Wert der integralen Empfindlichkeit $s_{AMn,(L)}$ bezüglich AMn (s.o);

(iii) die Beeinflussung der spektralen Empfindlichkeit durch die Gesamtheit der Bestrahlung je nach Bestrahlungsstärke und Wellenlänge.

Eine Nichtlinearität wie in Abb. 1 ist dabei gekoppelt mit der Zunahme der spektralen Empfindlichkeit $\Delta s(\lambda_i, \frac{\partial E}{\partial \lambda}\Delta \lambda)$ bei gegebener Wellenlänge λ_i durch Bestrahlung auch aus anderen Spektralbereichen $\frac{\partial E}{\partial \lambda} (\lambda{\neq}\lambda_i)\Delta \lambda$. Außerdem besteht die Möglichkeit, daß $\Delta s_{AMn}(E_L)$ konstant ist, obwohl sich das $\Delta s(\lambda ,E_L)$-Spektrum mit E_L ändert; dies ist beim Solarelement in Abb. 2 oberhalb von $10^{-4}E_{AM1.5}$, wo die Rot-

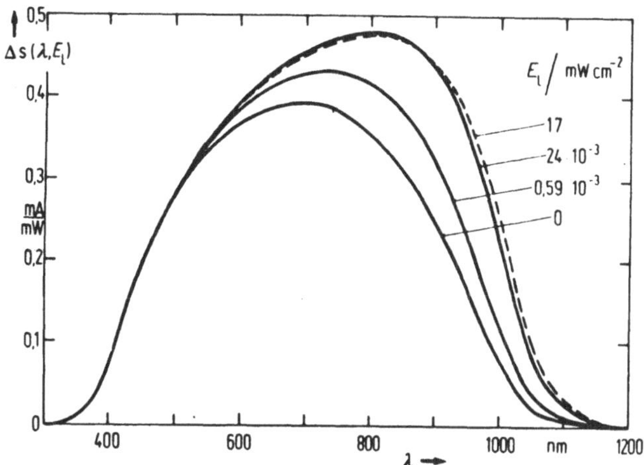

Abb. 1. Absolute differentielle spektrale Empfindlichkeit $\Delta s(\lambda, E_L)$ eines Si-Solarelementes als Funktion der Wellenlänge λ mit der integralen Bestrahlungsstärke E_L als Parameter ($t = 25\,^\circ C$).

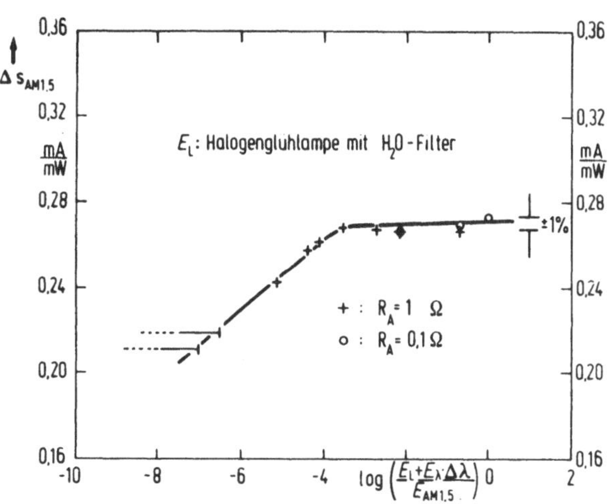

Abb. 2. Differentielle Empfindlichkeit des Si-Solarelementes von Abb. 1 bezüglich AM1.5 als Funktion der normierten Gesamtbestrahlungsstärke.

verschiebung weiter anhält (s. gestrichelte Kurve in Abb. 1), näherungsweise
erfüllt.

In diesem Zusammenhang sei nur erwähnt, daß aufgrund unserer bisherigen Untersuchungen der in Abb. 1 gezeigte Anstieg der spektralen Empfindlichkeit im langwelligen Bereich, der durch ebenfalls langwelligere Zusatzbestrahlung bevorzugt
induziert wird, typisch für Si-Solarelemente mit einer p-Typ Basis ist, wohingegen
Elemente mit einer n-Typ Basis eine wesentlich bessere Linearität aufweisen. Die
zugrundeliegende Beeinflussung der Diffusionslänge in der Basis und die dazugehörigen Rekombinationsmechanismen werden hier nicht diskutiert.

Im einfachsten Fall, d.h. wenn der Kurzschlußstrom eine lineare Funktion der Bestrahlungsstärke ist, kann alternativ so vorgegangen werden, daß die Empfindlichkeit eines Solarelements aus nur einer Messung der spektralen Empfindlichkeit bei
niedriger Bestrahlungsstärke $\frac{\partial E}{\partial \lambda} \Delta \lambda < 10^{-6} E_{AM1.5}$ (Kurve $E_L = 0$ in Abb. 1), einer
Linearitätsmessung nach der Doppel-Apertur-Methode /2/ und der Faltung mit der be-
kannten Bestrahlungsstärke E_{AMn} der Sonnenstrahlung bestimmt wird.

Bei dem in Abb. 1 und 2 gezeigten Beispiel ist die "Anfangsempfindlichkeit" mit
$s_{AM1.5,0} \lesssim 0,22 mA/mW$ um ca. 20 % kleiner als der Endwert $s_{AM1.5,L}$ und deshalb nicht
benutzbar. Eine derartige Diskrepanz sollte bei der Linearitätskontrolle sicher auf-
fallen und ein entsprechendes Solarelement als für eine herkömmliche Kalibrierung
ungeeignet ausweisen. Problematisch kann es jedoch sein, wenn die Linearität ohne
spektrale Zerlegung "nur" über weniger als 4 Zehnerpotenzen von der Solarkonstanten
abwärts überprüft wird und nicht bis herab in den Bereich der monochromatischen
Bestrahlungsstärken $\leq 10^{-6} E_{AMO}$ fortgesetzt wird, da sonst eine weitgehende Linearität vorgetäuscht werden kann (s. Abb. 2).

Desgleichen kann es auch nicht befriedigend sein, die hohen monochromatischen
Bestrahlungsstärken eines durchstimmbaren Farbstofflasers (abgesehen von dem
technischen Aufwand und dem begrenzten Spektralbereich) zu benutzen, da der Einfluß der gleichzeitigen, spektral benachbarten Bestrahlung fehlt. Denn da
$s(\lambda_i)$ nicht nur von $\frac{\partial E}{\partial \lambda}(\lambda_i)\Delta\lambda_i$ abhängig zu sein braucht (s.o.), ist das Integral
der partiellen Kurzschlußströme aufgrund monochromatischer Bestrahlung nicht immer
mit dem Kurzschlußstrom der Gesamtbestrahlung identisch.

3. Einsatz eines Sonnensimulators

Die direkte (Einzel-)Messung der integralen Empfindlichkeit setzt den Einsatz
eines Sonnensimulators voraus - die Sonnenbestrahlung selbst ist im allgemeinen
zu wenig definiert und reproduzierbar. Mit Hilfe der gemessenen relativen spektralen Empfindlichkeit ist die Abweichung vom gewünschten AMn-Spektrum durch Faltung
zu korrigieren /1,3/. Die von dem Sonnensimulator in der Meßebene erzeugte relative spektrale Bestrahlungsstärke und die Gesamtbestrahlungsstärke sind zuvor
durch Vergleich mit einem Strahlungsnormal bzw. mit einem unselektiven Empfänger
zu bestimmen.

Ohne diese noch ausstehende Korrektur der spektralen Fehlanpassung wurden Abweichungen zwischen der Empfindlichkeit $s_{AM1.5,L}$ (s.o.) und dem mit einem
kommerziellen Sonnensimulator gewonnenen Wert $s_{AM1.5,sim}$ von 5 % bis 15 % festgestellt, die von der Verteilung der spektralen Empfindlichkeit der Solarelemente
($9\ \% \leq \eta_{AM1.5} \leq 14\ \%$) abhängen. Untersuchungen über systematische Fehler und
eventuell verbleibende Abweichungen nach der Fehlanpassungs-Korrektur für verschiedene Typen von Solarelementen sind in Vorbereitung.

Diese Arbeit wurde vom Bundesministerium für Forschung und Technologie im
Rahmen des Projekts "Energieforschung und Energietechnologie" gefördert.

Literatur

/1/ Specification Nr. 101 of the CEC, Issue 2 (1980)
     BOGUS, K.: ESA-Report TEC/KB/sml/3551 (1981)
/2/ BISCHOFF, K.: Z. Instr. <u>69</u>, 143 (1961)
/3/ KAASE, H.: ESA-Report SP-147 (1980), p. 163

VDI-Technologiezentrum:

Industrielle Umsetzung der Lasertechnologie
Industrial Transfer of Laser Technology

The Commercial Laser Market in Western Europe

W.R.F. GOSLING, Consultant

1b Park Drive, Harrogate, HG2 9AY, U.K.

1. Introduction

A description of the market for any product could, if the term 'market' is
used in its narrowest sense, be confined to scrutiny of applications and
end-users alone. This would be to imply however that development of the
market is determined wholly by the demand side of the equation whereas the
supply side has an equally important influence, arguably a greater one when
the products are derived from a new technology as is the case with lasers.

In this presentation therefore the laser market is defined more broadly;
that is in terms of the products, applications, participants, industry
structure, market size and prospects.

2. Products, Participants and Applications

The products supplied by the Industry are lasers and laser systems and
these are now used in a wide range of applications.

Lasers are assembled from special components whilst in laser systems the
laser is assembled with electro-optic and mechanical ancillary items.
These ancillaries may also be special (in the sense that they have no
application other than as part of a laser system) or non-special but
nevertheless essential to the successful application of the system. In
general both types of ancillary are involved and they can account for
anything between 15% and 90% of the sales value of a system depending on
the application.

Thus the income of a laser systems company is in general derived as much
from the sale of non-laser items as lasers themselves and the market size
as a whole is strongly dependent on the added value represented by these
items.

This situation arises because end-users are interested primarily in the application of lasers and, with the notable exception of research organisations, few have the desire or capability to design and produce their own systems. Thus the supply side of the market whilst 'driven' by the laser manufacturers, includes systems houses and original equipment manufacturers (OEM's) whose presence is essential to its maintenance and growth. In order to have control over their markets many laser manufacturers therefore act as systems houses as well.

OEM's are distinguished from systems houses in that they supply and commission a complete system of which the laser system is only a part. A systems house only has a direct relationship with the end-user when sales are made direct. When he sells through an OEM that role is taken over in every respect including the important one of acting as 'prime contractor' in the legal sense. OEM's come into play when the end-users are such that special marketing and distribution methods are called for that are beyond the capability of the laser systems house. As a rule such customers also prefer to deal with one supplier during sales negotiations and when service back up is required. Perhaps the best examples of OEM systems are those supplied in large volume calling for consumer marketing and support (e.g. video disc players).

The OEM route to the market is not adopted lightly however since it involves a certain loss of control for the laser systems company and affects sales because discounts must be given. In some cases the would be OEM lacks the technical skill to back up the laser sub-system and can only refer the customer to the laser system supplier. This causes delay and customer dissatisfaction. The direct or 'retrofit' approach is therefore often preferred and can be attractive to the end-user when he has non-laser equipment options that he might like to start with having in mind the possibility of laser conversion at a later date (eg. construction machine control).

Similar arguments apply when considering the alternatives of direct selling or selling through dealers. Dealers (or at least agents) are essential when exporting since local knowledge (business procedures, language) must be available to the laser company. A good dealer whether it be for domestic or export sales, will also relieve the laser company of promotion and distribution costs which will justify the discount he receives.

USA laser companies play a very important part in meeting demand in Western Europe and in most cases act through appointed distributors having inventory holdings. Market leaders such as Spectra Physics and Coherent have set up subsidiaries for this purpose.

3. Industry Structure

The laser industry in Western Europe divdes into two main blocks; the indigenous manufacturers and the representatives (subsidiaries, distributors, dealers) of foreign companies, largely USA based. In 1980 there were about 80 indigenous manufacturers of lasers and laser systems and 55 USA companies with about 80 representative organisations here. There is a considerable amount of overlap however with many of the indigenous manufacturers employing imported lasers in their systems and many of them also acting as distributors or dealers.

With two USA companies (Spectra Physics and Coherent) accounting for nearly 50% of sales the prominent role of USA companies is likely to continue; indeed there is some evidence to suggest that it is strengthening with hitherto inactive companies beginning to establish representation here.

Outside the two market leaders the industry consists of a much larger number of companies with substantially lower sales of laser products. Some of these are large organisations such as Siemens, Thomson CSF and Ferranti; many are small concerns (20 employees or less) in the general business of electrooptics and by no means solely dedicated to lasers.

Indigenous manufacturers employed around 1300 personnel on laser related activities in 1980 and the distributors of non-European lasers and laser systems about 500. However, a multiplying factor of between 4 and 5 is thought to apply when calculating the corresponding number of jobs that have been created with OEM's and dealers. Against this the numbers employed by end-users is in some instances reduced by the introduction of lasers for industrial applications since use of lasers increases productivity.

At present in Western Europe the jobs with suppliers of specialist components and ancillaries sustained by the demand from laser companies is relatively small in number, the multiplying factor being less than unity. This is in marked contrast to the situation in the USA where the multiplying factor is around 3. The dependence of indiginous manufacturers on non-European (mainly USA) sources of components and ancillaries has led to this inbalance.

4. <u>Market Size and Projections</u>

Market size may be defined in terms of supply (sales) and end-user demand
(purchases). Supply generally exceeds demand in Western Europe because
there is some exporting (including re-exporting of imported products from
the USA), some intertrading between laser companies and a certain amount of
inventory building by OEM's. Even if this were not so there would be a
difference in monetary terms due to the value added by OEM's in the form of
non-laser items and the distribution services provided by dealers.

The territorial spread of demand appears relatively even with general
economic activity but supply has been more concentrated, the FDR, France
and the UK   accounting for 81% of sales but only 66% of purchases in
1980.

Lasers and laser systems from USA companies accounted for $95 million of
the $186 million sales made in 1980. The corresponding imports from the
USA were $54 million indicating an added value of $41 million (mainly
distribution) by their West European distributors and associated OEM's.
Japanese and Canadian imports were also involved but to a much lesser
extent.

Total sales of lasers in 1980 amounted to $110 million. Gas lasers
accounted for 50% of this figure. Research applications employed each of
the laser types in one way or another and accounted for about 47% of sales.

A near tenfold growth in demand to around $1.5 billion by 1990 was
indicated coming mainly from the emergence of 'new' applications. The
established application of lasers as research tools and in construction and
alignment accounted for 51% of demand in 1980 but was expected to fall to
20% by 1990.

The projected growth could lead to the creation of around 80,000 new jobs
by 1990, a figure that could be increased to nearer 120,000 should there be
a significant transfer or takeover of manufacturing and applications
technology from the USA. If at the same time there were to be a
corresponding decrease in the dependence of indiginous manufacturers on
non-European specialist components and ancillaries, this figure could
increase to around 170,000.

The projections assumed that certain developments in laser technology would take place notably an increase in CO laser power output in the range above 1.0 Kw and the increasing availability of long-life semiconductor lasers with improved coherence. The improvements have been or are being realised in most respects and this, with ongoing applications R and D and effective marketing, is maintaining growth. The relatively low level of semiconductor laser sales achieved in 1980 will amost certainly be subject to a marked increase as the impact of optical communications and video applications begins to be felt.

5. Conclusions

Lasers and laser systems are established as the basis of a major segment of the electro-optics industry in Western Europe. Encouraged by the increasing acceptance of lasers in a wide range of applications, its growth seems assured but is likely to remain strongly influenced by the activities of USA companies here.

State of Laser Markets and Technology in the U.S.[*]

GARY K. KLAUMINZER, President
Questek Inc.
7 De Angelo Drive
Bedford, MA 01730
U.S.A.

In 1982 the worldwide sales of lasers and laser systems reached
US$2 billion. This total was split almost equally between commercial
and non-commercial (military and government contract) markets. Table
1 below shows the sales by specific market segment.

TABLE 1: 1982 WORLDWIDE LASER SYSTEM SALES (US$M)

COMMERCIAL	
SCIENTIFIC	85
MEDICAL	85
MATERIALS PROCESSING	75
SEMICONDUCTOR FAB	55
ALIGNMENT	60
INFORMATION PROCESSING	350
GRAPHIC ARTS	260
COMMUNICATION	50
TOTAL	1020
MILITARY	800
GOVERNMENT R & D	240
TOTAL	2060

U.S. manufacturers accounted for approximately 65% of the commercial
sales.

The laser component of commercial sales in 1982 was US$290 million,
with U.S. manufacturers accounting for 80% of this total. The
purchase of U.S. lasers by several European manufacturers of graphic
arts and information processing systems accounts for the higher U.S.
contribution to laser sales compared to system sales.

---

[*] A portion of the material presented is from a marketing study
    completed by the author for Laser Focus magazine. Please consult
    Laser Focus for details on this study.

Table 2 below shows by laser type both 1982 commercial sales and number of manufacturers. Note that a company producing more than one type of laser is counted for each type. Given the approximately 70 companies with significant commercial laser sales in 1982, the average company produces two types of lasers, each with annual sales of US$2M. The large number of companies and low-average sales per company are consistent with the youthful nature of the laser industry.

TABLE 2: WORLDWIDE COMMERCIAL LASER SALES AND TOTAL MANUFACTURERS

	1982 US$M	TOTAL MFRS
GAS LASERS		
HELIUM-NEON	44	12
ION-ARGON, KRYPTON	70	8
HELIUM-CADMIUM	6	3
CO_2 - CW	50	24
- PULSED	9	6
NITROGEN	2	3
EXCIMER	8	5
OTHER	5	7
TOTAL	194	68
DYE LASERS	20	14
SOLID STATE LASERS		
RUBY	4	8
ND:YAG	37	25
ND:GLASS	4	5
ALEXANDRITE	1	1
TOTAL	46	39
SEMICONDUCTOR DIODE LASERS	31	19
TOTAL	290	140

For the next decade, the major growth markets will be communications, medical, materials processing, and semiconductor fabrication. The estimated 1992 sales for these markets and average percentage growth per year are shown in Table 3 below.

TABLE 3: MAJOR GROWTH MARKETS FOR THE 1980's

	1992 SALES	GROWTH %/YR
COMMUNICATIONS	US$1000M	35
MEDICAL	800	25
MATERIALS PROCESSING	700	25
SEMICONDUCTOR FABRICATION	350	20

656

Major R & D trends in the U.S. for commercial lasers to support these markets are described below.

1. <u>Compact high power CO_2 lasers for materials processing.</u> Fast axial flow or cross axial flow technology is reducing the cavity length in CO_2 lasers in the 1kW power range. Prices are decreasing while reliability appears to be increasing. Of note is that much of this technology has its origins in Europe.

2. <u>Small sealed-tube CO_2 lasers for medical applications.</u> Several companies are developing compact 20-40 watt CO_2 lasers for outpatient surgical treatment in clinics as opposed to major hospitals. The small size and low cost of these lasers are major factors in this market. Unfortunately, tube lifetime in most designs has not yet reached an acceptable level; thus most surgical CO_2 lasers sold today are of the older flowing-tube design.

3. <u>High-reliability small blue lasers.</u> For film-writing and high-speed printing, several manufacturers are perfecting helium-cadmium or air-cooled argon lasers. The specific applications may determine which type of laser is used.

4. <u>High power UV excimer lasers.</u> Lasers with up to 100 watts of average power are now available commercially, and such lasers are attractive for semiconductor fabrication and materials processing applications. Excimer laser technology is advancing rapidly, and in most cases reliability is not adequate for the industrial environment.

5. <u>High average power solid state "slab" lasers.</u> The slab geometry overcomes many of the thermal problems associated with the traditional rod design in Nd:YAG and Nd:glass lasers. Powers in the one kilowatt range are now possible, a three-fold increase over that achieved with the rod geometry. These lasers are attractive for many wleding and drilling applications.

6. <u>Low cost and long life semiconductor diode lasers.</u> Communications and information processing markets require both low cost and long life from high performance diode lasers. Major U.S. communications companies are developing such alsers for their own use.

The trend for laser markets in the 1980's is definitely positive.
By 1992, commercial systems sales should reach US$6 billion for an
annual growth rate of 20% per year. Laser sales should grow at a
slightly slower rate to US$1.4 billion reflecting a trend toward
systems as opposed to laser manufacturing. However, the technology
of laser production will remain with the laser companies; most
systems manufacturers will continue to purchase lasers rather than
build their own.

Laser Applications in Japanese Industry

K. KAKIZAKI
Toshiba Corporation, Manufacturing Engineering Laboratory
8 Shinsugita-cho, Isogo-ku, Yokohama, Japan 235

1. Introduction

Japanese laser industry is expanding rapidly in recent years.
Important application areas are materials processing, information
handling, audio and video equipment, optical fiber communication in
addition to measurement and laboratory science. But it is difficult
to estimate the market size. The market this year for materials
processing alone is estimated to be 10 billion yen (US \$40 million)
with YAG to CO_2 ratio of 2 to 1.
In this review laser applications in manufacturing processes,
lasers and laser-based equipment, and laser-related R&D in Japan
will be briefly described.

2. Applications in Manufacturing Processes

Measurement using low power gas lasers and materials processing
using YAG lasers are well established techniques. The former
includes precise distance measurement using laser interferometers
and alignment of various machine parts. YAG lasers have been
debugged and become a reliable tool for micromachining. Use of
optical fibers eases problems in beam delivery to otherwise
difficult-to-reach work pieces of laser powers up to 300W average.
Recently high-power CO_2 lasers with powers up to 5kW have become
available from Japanese manufacturers. Many companies are
contemplating introduction of lasers into procuction lines. Some
twenty companies are estimated to have lasers with powers of 1kW
and above. There are at least three Japanese companies offering
multi-kW models in addition to ten or so other foreign and Japanese
companies offering low power models.
Unfortunately users usually do not reveal uses of their lasers or
sometimes the fact that they have. So I will present some of our
activities at Toshiba as a manufacturer and user of lasers.

We estimate that close to 1000 YAG lasers are being used for
materials processing. Toshiba has sold about 400 YAG lasers in the
past, and expects to ship over 100 this year. Major applications
are micromachining of electronics and precision machine parts. We
offer flash-lamp-pumped normal-pulse lasers for welding, drilling,
and cutting with output waveform and pulse width optimized for each
application. Lasers of this type constitute about 80% of the total
sales volume. CW lasers with optional Q-switch are for trimming
scribing, and marking.

Fig.1 shows an example of YAG laser welding employed at one of
Toshiba factories.[1] This system features beam splitting, fiber
delivery, multipoint simultaneous spot welding of color CRT
electron gun parts. Last year a total of 4 million pieces were
laser welded in ten of the fully automated production lines. The
laser method reduced welding splash and size of the gun, and
improved performance.

Estimated 400 CO_2 lasers are in use in Japan. Low power units (500W
and below) have been mostly imported, and are used for sheet metal
cutting and other non-metal processing applications. Statistics for
high power units are not available. Fig.2 shows an example of CO_2
laser marking and cutting system.[2] Four other multi-kilowatt CO_2
lasers are operational at Toshiba factories for various processing
applications.

3. Other Applications

Several companies have released optical disk files and laser
printers for "office automation." For home entertainment, laser
video disk players using He-Ne lasers and compact audio disk
players using laser diodes are being marketed. They are a combined
result of low-cost high-quality lasers and other peripheral
components and precision assembly engineering.

Medical application is an interesting and promising area, but it is
not a widely accepted practice excepting argon laser eye
coagulators.

Optical fiber communication is another huge application area, but
this will not be discussed here.

Finally I will introduce a novel application of YAG laser for
burden profile measurement of the blast furnace. Fig.3 shows the
schematic diagram of the measurement scheme.[3]

4. Laser Devices for Applications

Most lasers for industrial applications are available from Japanese
manufacturers. Users of laser application systems insist on careful
examination before purchase and prompt servicing in case of machine
failures that lasers such as YAG and CO_2 have to be assembled in
Japan. Some foreign companies, therefore, have opened application
labs or formed joint ventures in this country. Critical components
like YAG rods, flash lamps, and high-power transmitting CO_2 optics
are still imported. Production techniques have been developed, but
the prices are not competetive yet.

In contrast the lasers for laboratory and scientific applications
call for special performance rather than low cost, and in many
cases imported lasers are used such as high-power argon, dye, and
excimers. The fusion project, however, relies on Japanese
manufacturers for development of key components like laser glass,
lamps, and high-power optical components.

5. Laser-Related R&D

A large number of personnel in government, universities, and
industry are involved in R&D. There are probably 50 small meetings
held each year in addition to large annual conferences held by
professional societies of physics, applied physics, electrical
engineers, mechanical engineers, The Laser Society, and Japan
Society of Laser Technology. The total number of the papers
presented may reach 1000.

Industrial laboratories have short-range programs such as improving
existing lasers and developing new products using lasers.
Especially active is electronics industry in the development of
audio and video disk players, office machines, and laser machining
techniques.

R&D in optical fiber communication have involved many university
labs in addition to industry, government and the Nippon Telephone
and Telegraph Public Corporation. This was partly because even a
tiny research team with modest budget could find an interesting
subject and contribute to theoretical and experimental research in
all phases of design, fabrication, and evaluation of semiconductor
lasers, optical fibers, components, and systems.

Government organizations such as MITI sponsor long range, large
scale projects that have important future potentials but too costly

and risky for the industry to invest on. Such projects include "Flexible Manufacturing System Complex Provided with Laser (FY1977-1984)" and "Optical Measurement and Control System (FY1979-1986)." Another large project is the laser fusion. Transfer of technology is somewhat limited among different companies but is good within a company in Japan, because employment is supposed to be for lifetime and large companies tend to do everything within themselves and to be independent.

References
(1) SHIMOI,Y. et al.: To be presented at 33rd CIRP Meeting, August 20-23, 1983, Harrogate, UK.
(2) FUJIWARA,S. et al.: CLEO Paper THN4, May 19, 1983, Baltimore, MD, USA.
(3) FUJIMORI,Y. et al.: CLEO Paper WF6, April 14, 1982, Phoenix, AZ, USA.

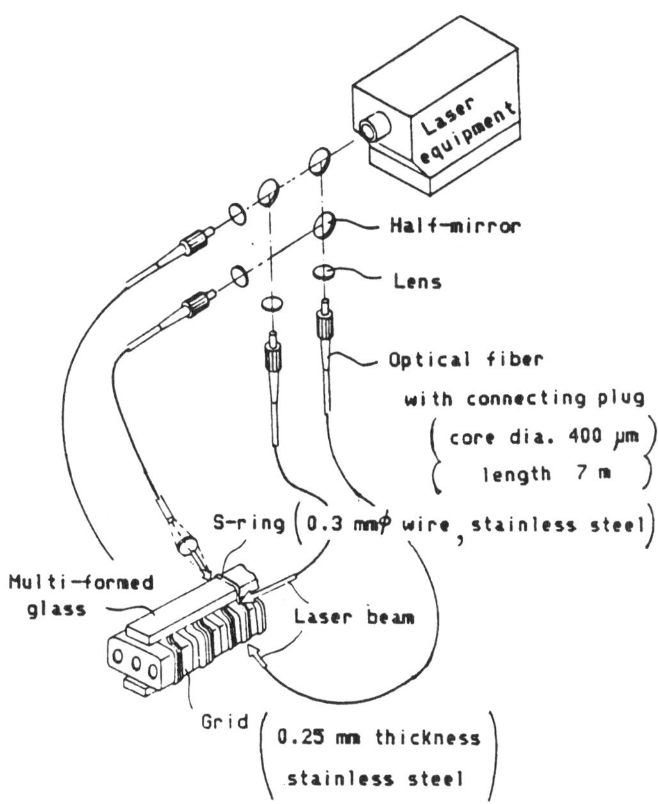

Fig. 1. Four-point simultaneous welding of color CRT electron gun parts

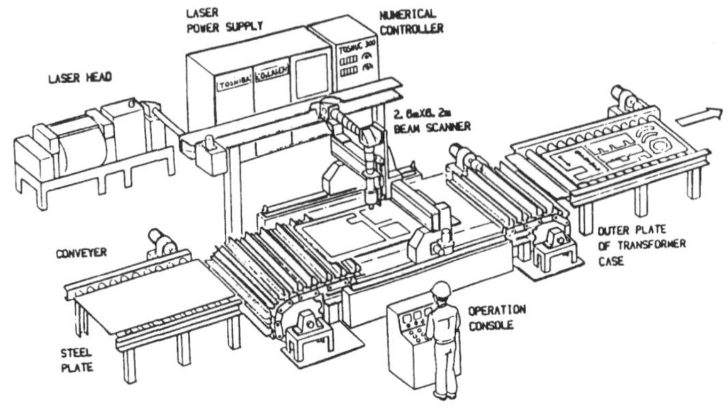

Fig. 2. CO_2 laser cutting and marking system with 2.6m x 6.2m beam scanner

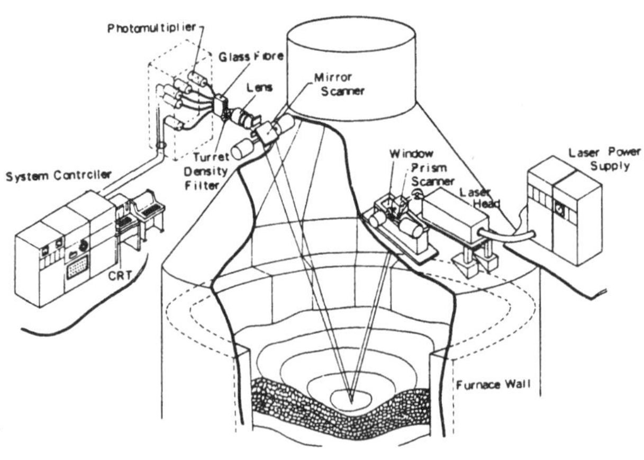

Fig. 3. Measurement of burden profile in blast furnace

R & D Programs and Industrial Application of Lasers in Italy

A. SONA

CISE S.p.A. - Segrate (Milano) Italy

At the end of 1978 a five year Project on High power Lasers was started
by the Italian National Research Council. The main objectives were:
i)   - to investigate and promote the use of lasers in material proces-
        sing for automated manufacturing systems.
ii)   - to study medical applications of lasers mainly in general and en
        doscopic surgery and phototherapy.
iii) - to study high power laser technologies and develop prototypes
        for mechanical and medical applications
The aim was also to assist the italian laser industry in entering
the extremely competitive laser market.

I - <u>Metalworking and Material Processing</u>

The three high energy density processes namely:welding,cutting,and dril
ling were first investigated both theoretically and experimentally in
various application centers. This resulted in the collection of a large set of data on optimal working parameters for various processes and
materials.
Some examples of the results are the following:

. <u>Welding procedures already assessed for</u>:
   . Austenitic stainless steels for general use
   . High Strength Low Alloyed Steels for pipelines
   . Bonding of iron plates for cars bodies
   . Longitudinal welding of stainless steel pipes
   . Filler wire assisted welding of high thickness, irregular geometry
     plates

. <u>Cutting and drilling procedures for</u>:
   . Stainless steel and carbon steel plates
   . Titanium plates even with complex geometry
   . Glass and quartz plates
   . Ceramic scribing and drilling
   . Cutting of plastic materials such ad PVC, Acrylic, polycarbonate
   . Cutting of wood leather textiles etc.

Processes requiring medium energy density were also considered. Procedures for heat treatments and surface alloying on medium and high carbon steels and cladding of stainless steel to increase wear resistance were established.

. <u>Actions for technology transfer</u>:

i) - A network of five mechanical application centers has been set up with the aim to provide preliminary informations, "on the spot" demonstrations, feasibility studies and processing cost estimates. More than two hundreds italian industries have submitted feasibility studies to the application centers.

ii)- Technical meetings and demonstrations were arranged in many italian industrial areas followed by specific contacts with the industrial operators.

iii)- A data base on material processing parameters is being completed to provide an immediate release of information.

As a result the diffusion of technology is reasonably advanced and the increasing number of italian industries already using lasers or considering laser processes is a confirm of the validity of this approach.

II - <u>MEDICAL APPLICATIONS</u>

Surgical methods and tools were developed and evaluated in the following fields:
- General surgery and endoscopic surgery mainly for gastro-intestinal treatments, for ginecology and urology
- Laser microsurgery for laringology and ginecology
Phototherapy and photochemiotherapy by lasers were also investigated with special attention to treatments with laser activated hematoporphyrin to selectively destray malignant tissues.
The results were made available for clinical use in four laser application centers to provide an efficient diffusion of the laser therapy.
As an example in laringology and in ginecology surgical treatments on a day-hospital basis are now corrently made in place of conventional treatments requiring typically three to six days stay in hospital. The Hpd treatment is also being applied on a clinical basis on selected patients.
Specialized technical meetings were arranged in many italian medical centers yo provide proper information.

III - <u>Laser Sources Development</u>

Laser prototypes for mechanical and medical applications were developed with the following technologies and power levels
- High power CW CO_2 lasers:.

- Transverse flow E-Beam sustained discharge:up to 10 kW
  - Transverse flow self sustained discharge:up to 5 kW (also with modular structure)
  - Fast axial flow:up to 2.5 kW
- High power Nd-YAG lasers:
  - Pulsed emission:300 W average power
  - CW emission:100 W power

In addition pulsed dye lasers with 200 W average power, Excimer lasers, and guided wave CO_2 lasers were developed for medical applications.

At the present time the italian industry can provide a broad line of lasers for medical applications.

Lasers for mechanical applications are presently manufactured with power levels below the 1 kW level. Multichilowatt units will be commercially available at the beginning of 1984.

Laser prototypes with output power up to 10 kW are currently been used for research and demonstrations in the application centers.

A key point for the success of the project was the contribution of interdisciplinary teams of experts both for mechanical and medical appli cations.

IV - <u>Projections of laser systems demand with particular consideration to the italian market</u>

By 1990 the commercial segment of the laser and laser-systems market is expected to reach the 6 billion $ level (at constant 1982 monetary values) compared to 0.5 billion $ in 1982.

The West-European fraction of the whole market will be about 35% while the Italian percentage of the European sales is expected to be 12%.

Projections for demand in Western Europe for industrial material processing laser systems indicate a 15% to 20% growth-rate during the 80's. This rate could suffer from a prolonged european economic recession. Anyway, laser systems sales have grown even in 1982 and, considering the drop in machine tool sales in recent months, we can say that laser systems have done well and are expected to do better.

The market for material working systems should reach 260 million $ in 1990 in West Europe. The percentage that laser sources will represent in total sales of laser systems is expected to be around 30%. For instance, the laser sources market in Italy should reach $ 4 million in 1985. The Italian laser industry will begin to produce multikilowatt sources in 1984 while laser automated manufacturing systems have already been realized by Italian industries.

Medical sales are becoming increasingly important in West Europe. This segment of the market is growing at a very high rate and should total at least 100 million $ in 1990.

The Italian laser industry can provide a complete set of medical sources and systems (gas, dye, and solid state lasers) and we expect a considerable growth of this segment of the market in our country. Projected demand (in millions $) for material processing laser systems in Italy

	1982	1985	1990
laser sources only	2.5	4	10
laser and laser systems	7.0	13	40

Laser Technology in Great Britain

I.J. SPALDING and A.C. WALKER
UKAEA Culham Laboratory
Abingdon, Oxon, OX14 3DB, UK

Abstract Over 100 firms in the UK are active in laser and related opto-electronic
technologies. About one tenth of these are involved in manufacturing complete
lasers or laser systems for industrial applications. Published work in materials-
processing has primarily involved Nd-YAG and CO_2 lasers; illustrative examples are
the trepanning of cooling holes, and the cladding of components, for the aerospace
industry. Fibre-optics and data-transmission form another important growth area.
Work at Culham has ranged from the single-pass welding of steels having thicknesses
up to 25mm to the development of on-line two-dimensional laser beam monitoring (for
quality assurance), and computer-controlled equipment. Our own experience, together
with that reported elsewhere, is used to essay a 'forward-look' on possible market
developments and trends in technology transfer.

The British Laser Industry The main specialization of firms tabulated in a
selection of trade guides as being actively involved in laser technology are listed
in Table 1. Column (i) sums the principle laser-related activity claimed by each
company, and column (ii) the number of companies claiming to work (at least to some
degree) under each heading.

Table 1. (Principle activities within UK)

	(i)	(ii)
Laser Manufacturer	9	9
Laser Systems Manufacturer	10	15
Optics and Coatings	29	31
Associated Equipment (modulators, detectors etc.)	20	27
Fibre Optics Equipment	18	19
Sales Agents	11	22
	97	–

The list includes three commercial off-shoots from University departments, some
foreign-owned subsidiaries (manufacturing in the UK), and various companies
specializing in military markets (which will not be discussed in this paper). Total

industrial activity is certainly larger than this list suggests (particularly in fibre optics communications): for example firms such as Desitech supplying turn-key laser-based XY tables and FLS (robotic systems) were not included in the tabulations used as the source material for Table 1, and a few contract R&D organisations working in the industrial laser field, (including ourselves) have also been omitted.

Lasers manufactured commercially within the UK span a wavelength range from the ultraviolet to the far infrared. They include excimer; copper vapour; He-Ne; ion; L.E.D.; Nd-YAG; tunable diode; waveguide, TEA, sealed-tube slow/fast axial-flow, and transverse-flow CO_2 systems. (Academic research is also broadly based, and includes work on free electron lasers. The SERC Central Laser Facility at the Rutherford Appleton Laboratory acts as a focal point for Universities interested in very high intensity multiterawatt laser beams, the generation of ultra-dense plasmas, and related studies on non-linear interactions with matter etc.)

<u>Industrial Applications</u> The lasers listed above are finding an increasing, and diverse range of industrial applications which include: alignment, metrology, quality assurance (e.g. detection of surface flaws on paper and similar materials), non-destructive testing, pollution-monitoring and laser Doppler aenemometry. However, in market terms perhaps the most significant areas are associated with information technology and laser materials-processing in the electronic, printing, manufacturing/engineering and related industries. The following section concentrates on the last area, and in particular on topics with which we are most familiar at Culham. (The oral presentation will be freely illustrated with slides provided by a typical cross-section of the firms mentioned in Table 1).

<u>Materials Processing at Culham</u> A range of commercially available axial-flow CO_2 lasers are normally used for machining or welding non-metals and steels up to a few mm thick. For work at higher powers continuously-rated multikilowatt transverse-flow CO_2 lasers have been developed. Four industrially-rated 5kW (CL5) units have been run routinely on a single-shift basis - with a cumulative life exceeding 15,000 hours. At full power, not less than 80% of the beam can be focused into an f/4 spot diameter of 0.3mm. On-line measurements of the (two-dimensional) radial intensity distribution at a range of axial positions are available using a beam-scanning quality assurance unit. This prototype unit has an insertion loss of a few percent and data is available from its pyroelectric detector within 2 secs; a microprocessor then deconvolves and displays this information on a VDU for on-line laser or process control. One of these CL5 lasers has been installed at Springfields Nuclear Laboratory. Two others have recently been combined at Culham to provide an industrially-rated 10kW (CL10) laser. Typical single-pass autogeneous mild-steel

welding speeds achieved with the CL5 and CL10 lasers at Culham are shown in Fig. 1. Whilst aerodynamic windows are not always essential on these systems, one has been fully developed for use on CL10. Lasers have proved to be very flexible processing tools. They have been reliably routed by mirrors over distances of tens of metres to a number of separate workstations, on demand. They can weld or cut components whilst others are loaded for an entirely different process at an adjacent workstation. When the inherently high processing speed of the laser is combined with efficient beam utilisation in this way, one achieves the high production capacity required of a fully-automated and flexible manufacturing system (FMS). If the workpiece is three-dimensional it may prove more convenient to couple the stationary laser to the work by means of a flexible beam guide, in which all the mirrors remain automatically in alignment. We have successfully operated such systems with power capacities up to 1kW, with and without robotic control of the head, (Fig. 2). Such systems are scalable to powers of many kW.

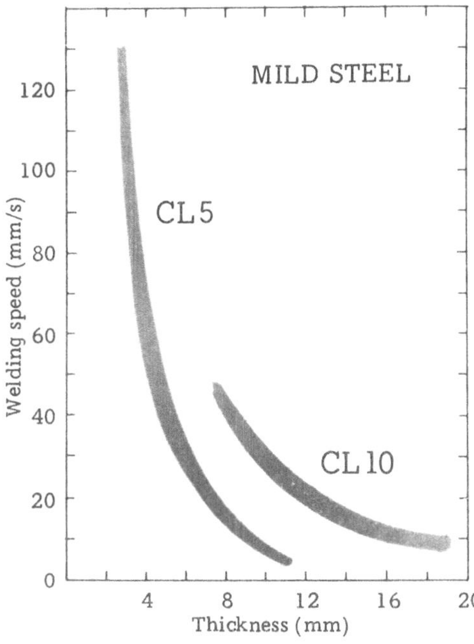

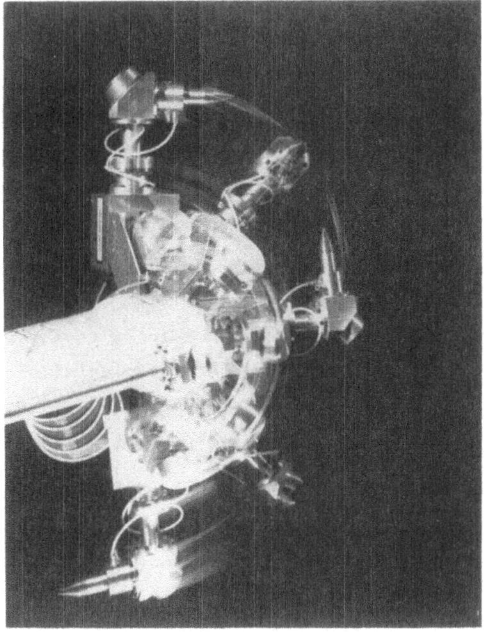

Fig. 1. Illustrative welding speeds vs. thickness for the Culham 5 & 10kW (modular) laser systems.

Fig. 2. 5-axis 1kW laser beam guide, coupled to an ASEA industrial robot. Overlapping camera shots demonstrate the flexibility of the system for 3D machining. (FLS Ltd).

Typical sub-kilowatt laser cutting applications include disassembling radio-active fast-reactor fuel subassemblies at Dounreay N.P.D.E.; and the mass production of a novel (contoured) security thread using microprocessor control techniques for the new £50 bank note recently introduced by the Bank of England. For both cutting and welding it is not simply laser power but mode quality and stability which is important. For example, single-pass, autogeneous welds in materials as thick as 25mm steel and 16mm titanium have been made over the past year with the CL10 system. Such welds have low distortion and shrinkage and the process is, of course, achieved out-of-vacuum at high speed. Adequate process control can ensure weld closures on circular components with high integrity and reproducibility. Conventional filler-wire equipment has been used successfully on components having poor fit-up. Multikilowatt systems such as CL5 can also heat-treat at rates of $\sim$ 30-500mm^2 s^{-1}, depending on the case depth and material of interest. Scanning electron micrographs reveal marked differences in wear scars for different treatments. Hard facing has been undertaken using both powder and wire feed. At an incident power of 5kW material wastage during application is low; surface roughness of dense, hard, well-bonded stellite layers on stainless steel is typically < 0.1mm about the mean height.

<u>Development of the Laser Market</u> High technology markets such as this, with a broad range of existing and potential applications, are more difficult to quantify than, for example, the market for soap! However, an indication of possible growth areas can be obtained from a look at products now being sold and developed. We will therefore summarize a few comments from various UK laser manufacturers, in order to provide a 'snap-shot' of current trends.

Ferranti plc. have been working with CO_2 lasers since the early 1960's and have sold several hundred MF400 (400W slow axial flow) lasers throughout the world. The latest additions to the range are their MFK (1kW) laser, plus the fast-transverse flow CL5 (5kW) and CL10 (10kW) lasers developed at Culham. Other possible extensions to this range are being planned, based on work at the UK Welding Institute. On the applications side they are currently engaged on the integration of lasers and robotics in collaboration with the UKAEA.

Control Lasers are manufacturing, and further developing, fast-axial-flow multikilowatt cw CO_2 lasers of a type initially marketed in the UK by British Oxygen. Control have a track record in the supply of 5-axis systems for cutting 3-D objects, based on their 2kW laser design.

Electrox Ltd have expanded their range of fast-axial-flow CO_2 lasers by the recent introduction of their M1000 1kW laser, a scaled up version of their existing 450W

device. Its operation is compatible with computer control and includes full 0-1kW
adjustability, and pulsed mode at up to 2.5kHz. The device is packaged within a
single integrated unit for easy installation and maintenance. Electrox see
applications in both high speed cutting and welding.

JK Lasers have specialised for many years in Nd-YAG-related lasers for scientific
applications (e.g. pico-second studies in chemistry and photobiology), and in the
provision of automated systems for fine-scale industrial cutting or drilling. A
turn-key CNC 5-axis system has, for example, recently been supplied for trepanning
cooling holes in aeroengine components. Their merger with Lumonics Ltd of Canada
has given both companies access to wider markets and complimentary technologies.

Scientifica-Cook Ltd manufacture a wide range of He-Ne lasers with output powers
from 1mW to 40mW. They have recently introduced their latest industrial laser-
alignment systems (LM-10 and RB-10) for both linear and plane x-y alignment and also
a unique eye-testing instrument (LASERSPEC), based on the speckle effect, suitable
for rapid eye test in schools, hospitals and industry etc.

Edinburgh Instruments, who manufacture high-stability low-power cw gas lasers, was
started by staff from Heriot-Watt University. They have recently developed a range
of radio-frequency excited waveguide CO_2 lasers in a major Scottish collaboration
with Ferranti. These compact lasers are sealed devices with guaranteed operating
lifetime per gas-fill in excess of 2,500 hours and are already finding numerous
applications in industry, medicine and research. The output power range now extends
to 15W, with higher power models in an advanced state of pre-production prototype
testing.

Laser Applications Ltd, an off-shoot from Hull University, manufacture a range of cw
and pulsed gas lasers for both research and industrial applications, including high
average power 200Hz devices and single shot systems up to 50J per pulse. Recently
developed is a continuously tunable high-pressure CO_2 laser producing single mode,
250mJ, 50ns pulses at 10Hz. They also offer a variety of RF excited waveguide CO_2
lasers and claim the highest powers per unit length available for such devices.

Oxford Lasers, another university associated company, manufactures a range of UV and
visible pulsed gas lasers. They have recently announced that, with minor
modifications, their Cu25 (5kHz, 25W) copper-vapour laser can be operated as a gold
laser giving 9 watts at 627.8nm and that a 20kHz option is also available, giving
10W with Cu-vapour. The output of their rare-gas halide laser KX1 has now been
increased to 350mJ pulse energy in KrF and a modified version of this device can now

give 30mJ per pulse on the vacuum UV (158nm) transition of F_2 - claimed to be the highest energy available at this wavelength.

<u>Conclusion</u> Within both the UK and other EEC countries there is now an impressive range of laser technology available. It seems probable that the market will be expanded more by innovative development of novel turn-key industrial systems than by the mere duplication of laser suppliers around Europe.

Lasertechnologie in der Bundesrepublik Deutschland

G. Rauscher, München

1) Einleitung

Lasertechnologie ist in der Bundesrepublik Deutschland, wie in anderen Ländern, kein einheitliches Ganzes zu dem präzise Aussagen gemacht werden können. Herstellung von Lasern und deren Anwendungen z.B. als Fertigungsmittel sind zwei wesentliche Teilbereiche, wobei der Herstellungsbereich unterteilt werden muß in:

* Herstellung von Laserstrahlungsquellen und deren Komponenten
  Diese Produkte werden z.B. als OEM-Teile in Geräte, Systeme oder Maschinen eingebaut (OEM = original equipment manufacture).

* Herstellung von Geräten, Systemen und Maschinen,
  die als Schlüsselelement einen Laser enthalten. Diese Laser werden teils selbst hergestellt, teils als OEM-Teil bezogen. In vielen Fällen, insbesondere in der Materialbearbeitung, werden die Systeme kundenspezifisch entwickelt.

* Anwendung der Lasertechnologie
  mit Geräten, Systemen und Maschinen, die einen Laser enthalten und die mit Vorteil industriell eingesetzt werden.

2) Anwendung

Es ist sicher nicht falsch zu sagen, daß Laser Schlüsselelemente werden für immer mehr Produkte und Anwendungen in Bereichen, in denen weltweit die höchsten Innovationsraten registriert werden (z.B. Elektronik, Maschinenbau, Chemie).

Die Anwendung der Laser in der Bundesrepublik Deutschland ist breit gefächert; typisch sind viele kleine Marktnischen, eine Vielzahl von

Lasertypen und in sehr großem Maße der Einsatz von aufgabenorientier-
ten Spezialgeräten.

Bild 1 gibt einen Überblick über den industriellen Lasereinsatz und
über die jeweils vorwiegend verwendeten bzw. in Entwicklung befindlichen Laser.

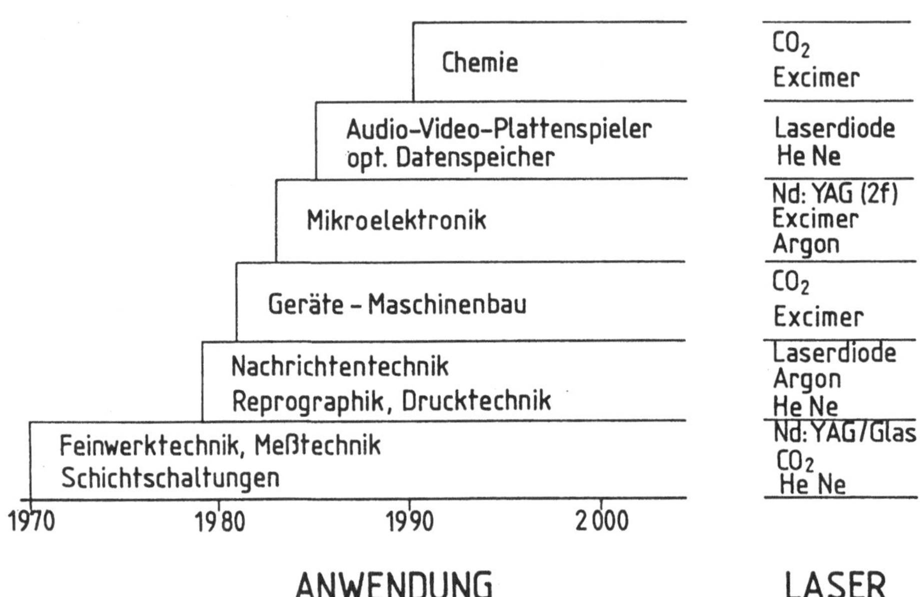

Bild 1. Laseranwendung in der Bundesrepublik Deutschland
        und verwendete (bzw. im Planungsstadium) Lasertypen

Bild 2 zeigt wie vielfältig die mit dem Laser durchzuführenden
Verfahrensschritte geworden sind und welche Industriezweige schon
heute den Laser als Werkzeug einsetzen. Für eine breite industrielle Nutzung sind allerdings heute nur wenige Felder "erschlossen".

Als Standardverfahren für die Laseranwendung können heute gelten:

* Abgleich elektronischer Schaltungen. Hierzu sind keine deutschen
  Geräte auf dem Markt, mit Ausnahme von Handarbeitsplätzen mit
  einfachen Abgleichstrategien.

* Punktschweißen in der Feinwerktechnik, insbesondere im TV-Röhren-
  bau. Auf dem Gebiet setzen europäische Firmen den Standard, Deutschland ist auf der Herstellerseite nur mit Systemen kleiner und
  mittlerer Leistung vertreten.

Verfahren	Industriezweige
messen	
justieren	
steueren	
übertragen	Elektrotechnik
	Geräte-Maschinenbau
fügen	
trennen	KFZ - Industrie
härten	
legieren	Hoch-Tiefbau
	Drucktechnik
Reaktionsprozesse einleiten	
selektive Reaktionen erzeugen	Medizintechnik
ätzen	
abscheiden	Feinwerktechn. Industrie
	Opt. Industrie
abgleichen	
abtragen	

Bild 2. Übersicht über Laserverfahren und Anwender
(Industriezweige)

* <u>Beschriften</u> von Werkstücken, Schildern, Bauteilen usw. Von den
beiden Verfahrensvarianten (Schreiben von Zeichen - Abbilden von
Masken) wurden in Deutschland Systeme vom Typ "Schreiben" entwickelt, die Spitzenprodukte darstellen. Die verwendeten Laserstrahlquellen sind jedoch ausnahmslos ausländischer Herkunft.

* <u>Schneiden von Blechen</u> und Kunststoffen. Für dieses Feld liefern
deutsche Firmen Systeme mit Weltstandardniveau, vorwiegend allerdings mit importierten Laserstrahlquellen.

* <u>Längenmessung und Justage in der Bautechnik.</u> Für diese Anwendung
stehen Geräte und Laser deutscher Fertigung zur Verfügung; der
Standard für Laserinterferometer und für Nivelliergeräte der Bau-
industrie wird allerdings in den USA eingesetzt.

* <u>Belichten in der Drucktechnik.</u> Recorder für die Herstellung von
Farbauszügen (Tiefdrucktechnik) enthalten Laserquellen mit schnellen Modulatoren. Ein deutsches Unternehmen ist Marktführer auf
diesem Gebiet, verwendet aber importierte Laser (Ionenlaser geringer Leistung).

3) Lasermarkt der Bundesrepublik Deutschland

Auf dem Lasermarkt der Bundesrepublik gibt es Firmen, die

* OEM-Laser oder OEM-Komponenten liefern
* Geräte/Systeme mit OEM-Lasern bauen
* vollständige Geräte/Systeme herstellen.

Die Marktgröße zeigen die Bilder 3 und 4. 1980 (durch die Rezession 81-82 dürfen die Marktdaten ungefähr auch für 1983 stimmen) betrug der Markt in Deutschland für Laser und Lasergeräte DM 116'.

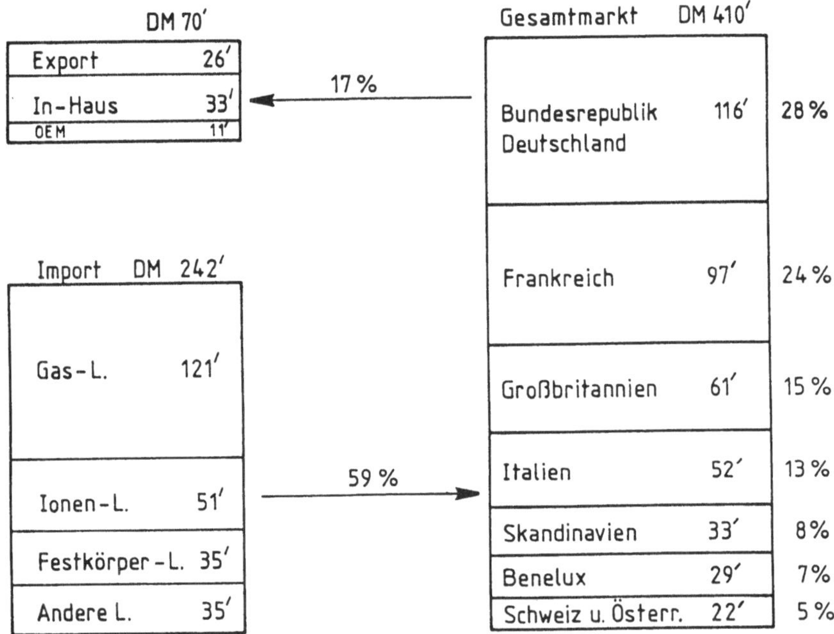

Bild 3. Laser- und Lasergerätemarkt (zivile Anwendung) in Europa (West) für 1980. /1/

Bei einem Import von nahezu 60% (die Zahl dürfte auch in erster Näherung für die Bundesrepublik zutreffen) betrug die Wertschöpfung nur knapp DM 50'. Der Anteil der Materialbearbeitung dürfte 1980 knapp die Hälfte des Gesamtmarktes betragen haben. Die Prognose des Teilmarktes "Materialbearbeitung" zeigt Bild 4. Der Markt soll bis 1990 auf ca. DM 250' anwachsen.

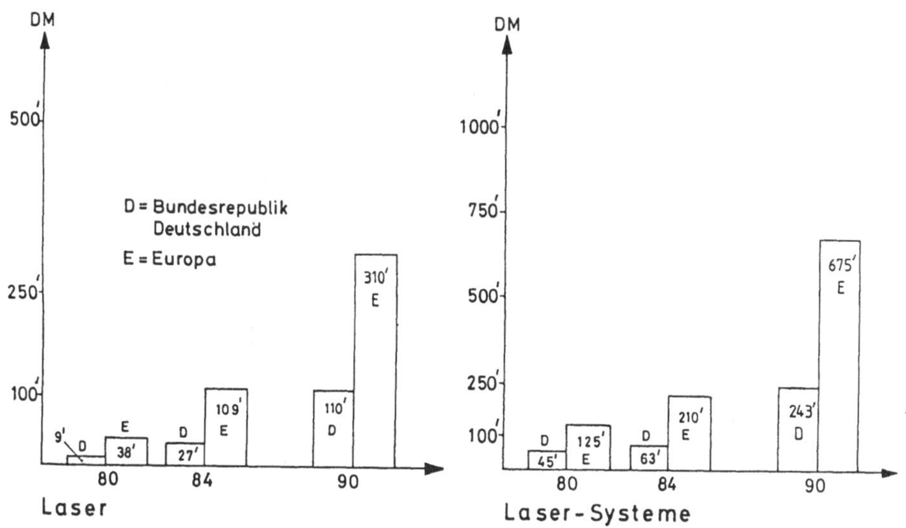

Bild 4. Marktprognose für Laser und Lasergeräte für das Gebiet der
Materialbearbeitung (Verwendung in Informationen aus /1/)

Hersteller	Produkte	Zahl der Hersteller (Deutsche Firmen)	Bedeutung Markt Europa / Welt	
Vollständige Geräte/Systeme	Blech u. Kunstoffschneide-systeme	1	—	—
	Punktschweißsysteme	2	—	—
	(Schnelldrucker)	(1)	(+)	(+)
	Med. Systeme	2	+	+
OEM - Teile (Laser, Laserkomponenten	HeNe – Laser	2	+	+
	CO_2-Laser (500w–1kW)	2	+	(+)
	Nd: Glas – Laser	1	+	—
	Opt. Komponenten	> 6	+	—
	Excimer – Laser	1	+	+
	Farbstofflaser	1	+	—
Durch Zukauf v. OEM-Laser vollständige Geräte Systeme	Blechschneidemaschinen	2	+	—
	Keramikbohrsysteme	2	—	—
	Holzbearbeitungssysteme (Bandstahlschnittwerkzeuge)	2	+	—
	Abgleichplätze	1	—	—
	Beschriftungssysteme	3	+	—

Bild 5. Deutsche Laserhersteller und deren Bedeutung auf dem
Europa- bzw. Weltmarkt

Die Zahl der Firmen auf dem Lasermarkt beträgt heute in Deutschland
ca. 50-60, in der Mehrzahl allerdings Vertretungen ausländischer
Firmen. Die beiden großen Laserfirmen haben auch in Deutschland
eine sehr starke Marktstellung.
Die Bundesrepublik ist ein sehr interessanter Markt für die Lasertechnologie, denn die Mehrzahl der deutschen Firmen sind aufge-
schlossen für neue Methoden und neue Verfahren. Allerdings ist die
Bundesrepublik als Markt zu klein, um mit dem Inlandsmarkt die be-
trächtlichen Entwicklungskosten für neue, insbesondere leistungs-
starke Laserstrahlquellen und für komplexe Lasersysteme (z.B. Be-
arbeitungssysteme für die Mikroelektronik) abdecken zu können. Dies
ist einer der Gründe weshalb deutsche Firmen auf dem europäischen
Markt noch eine gewisse, auf dem Weltmarkt aber praktisch keine
Rolle spielen, eine Ausnahme bilden Excimerlaser (Einsatzgebiet F+E)
und HeNe-Laser (Bild 5).

4) <u>Forschung und Entwicklung, Technologietransfer</u>

In der Bundesrepublik wird relativ breit Forschung an Lasern betrieben an den Instituten der Hochschulen und Universitäten, den Instituten der Max-Planck-Gesellschaft, der Fraunhofer-Gesellschaft, von
Batelle und DFVLR.
An den gleichen Instituten wird auch Forschung mit Laser betrieben,
wesentliche Forschungsfelder sind Laserphysik, Wirkung von Strahlung auf Materie, Meßtechnik, Spectroskopie, Diagnostik, Plasma-
physik, chem. Reaktionskinetik und physikalische Chemie.
Der Technologietransfer von den Forschungsinstituten zu den Industrie-
unternehmen geht in Deutschland kaum auf direktem Wege, da die Auf-
gabenstellungen zu verschieden sind (Bild 6).

Für neue Lösungen, für spezifische Anwendungen, ist eine Art Ingenieurbüro dazwischen zu schalten. Dieses kann ein Entwicklungslabor
im Anwenderunternehmen sein, es kann ein praxisnah orientiertes
Institut sein (z.B. ein Institut für Auftragsentwicklung) oder es
kann ein Ingenieurbüro mit einschlägigen Spezialisten sein. Regelmäßige Seminare und Workshops könnten den Technologietransfer von
den Forschungsinstituten zu den Anwendern unterstützen und beschleunigen. Die Seminare könnten auch umgekehrt einen Informations-
fluß von den Geräteherstellern/Anwendern zu den Instituten hin in
Gang bringen.

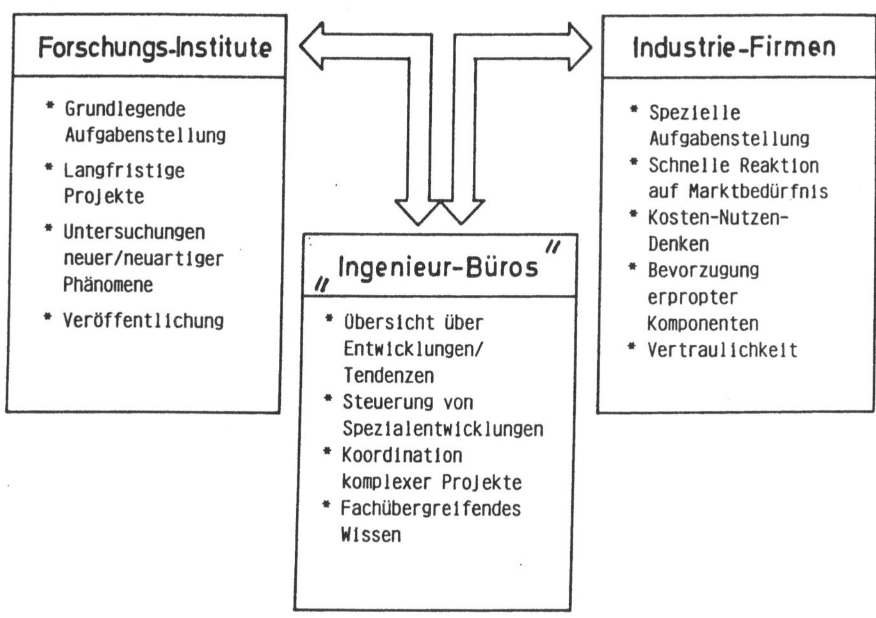

Bild 6. Technologietransfer in Deutschland

Den Standard der Laser und der notwendigen Komponenten bestimmen
heute US-Firmen. Um abzuschätzen, auf welchen Teilgebieten ein Auf-
holen am ehesten möglich sein wird, kann man die Laserstrahlquellen
in drei Lasergruppen zusammenfassen. In Tabelle 1 sind für die
3 Gruppen die notwendigen Herstellungstechnologien, der Entwick-
lungsstand und die Weiterentwicklungsmöglichkeiten angegeben.

Bei den Lasergruppen "kleine Leistung" "mittlere Leistung" ist die
Entwicklung ziemlich abgeschlossen, über Marktanteile wird die Ra-
tionalisierung bei der Serienherstellung entscheidend sein. Bei
der Gruppe "hohe Leistung" ist die Entwicklung noch voll im Fluß,
auf diesem Gebiet würde ein Einstieg am ehesten möglich sein. Allerdings darf nicht vergessen werden, daß für die Entwicklung von
Lasern hoher Leistung auch Entwicklungskosten in beträchtlicher
Höhe anfallen. Diese Überlegung gilt allerdings nur für bekannte
Lasertypen, für neue Entwicklungen kann die Entscheidung anders
ausfallen. Ein Beispiel hierfür ist der an der Markteinführungs-
schwelle befindlichen CO_2-Laser vom Typ waveguide mit Leistungen
30 Watt.

Literatur
/1/ FROST and SULLIVAN: The Commercial Laser Market In Europe (1980)

Tabelle 1. Laser-Leistungsklassen

Laser-Leistungsklasse Beispiele Laser	Technologien für Laserbau bzw. wesentliche Komponenten	Entwicklungsstand Weiterentwicklung

Laser kleiner Leistung
N : $<$ 100 mW
 	
HeNe-Laser, Ionenlaser
N_2-Laser
	* Metall-Glas-Technik
	  (Röhrenbau)
	* Stromversorgung (Hochspannung)
	  geringer Leistung
		* Entwicklung abgeschlossen
		* Serienprodukte, große Stück-
		  zahlen
		* Fertigungsrationalisierung

Laser mittlerer Leistung
N : $<$ 500 W
 	
Nd:YAG-Laser
CO_2-Laser (längs-geströmt, TEA-Typ[1])
	* Feinwerktechnik
	* Opt. Komponenten
	* Schaltnetzteile und elek-
	  tronische Komponenten für
	  Stromversorgung
		* Nur noch wenig Entwicklung
		* Serienprodukte, kleine Stückzahlen
		  teilweise Einzelgerätebau,
		* Vereinfachung,
		* Fertigungsrationalisierung
		  (bei steigenden Stückzahlen)

Laser hoher Leistung
N : $>$ 1 kW
 	
CO_2-Laser (quergeströmt, fast flow-System[2])
Excimer-Laser
Wasserstoff-Fluorid-Laser
	* Plasmaphysik
	* Gasdynamik
	* Maschinenbau
	* Starkstromtechnik für
	  Netzgeräte, einschl.
	  schnelle Schaltelemente
	* Störsichere Elektronik
	* Optische Komponenten
		* Entwicklung noch im Anfang
		* Gerätebau (Einzelfertigung)
		* Grundlagen für mechanischen
		  und optischen Aufbau, Strom-
		  versorgung mit schnellen Schalt-
		  komponenten, Strahlsteuerung
		  und Fokussierung

<u>Anmerkungen:</u>   1) TEA Abkürzung aus <u>T</u>ransversely <u>E</u>xcited <u>A</u>tmospheric-Pressure, TEA-Laseraufbau für Pulslaser (µs-ns)

2) Resonator besteht aus mehreren Entladungsstrecken   Entladung und Gasströmung sind parallel zur Resonatorachse.

Optoelektronik in der Technik 1983
Optoelectronics in Enginering 1983

Vorträge des 6. Internationalen Kongresses
Proceedings of the 6th International Congress
Laser 83

Herausgeber/Editor: **W. Waidelich**

1984. 537 Abbildungen. Etwa 710 Seiten
DM 118,-. ISBN 3-540-12779-8

Inhaltsübersicht: Sitzungsleiter. – Referenten. – Laser-Systeme für Forschung. – Optoelektronische Komponenten und Sensoren. – Optronisches und lasertechnisches Messen und Prüfen. – Laser in der Materialbearbeitung. – Optoelektronische Signalübertragung. – Laser und Optoelektronik in der Weltraumtechnik. – Laser in der Umweltmeßtechnik. – Optoelektronische Displays. – Laser-Chemie. – Laser Sicherheit. – Photovoltaische Solartechnik.– VDI-Technologiezentrum: Industrielle Umsetzung der Lasertechnologie.

In zweijährigem Rhythmus finden in München internationale Kongresse und Fachmessen über Laserphysik und Lasertechnik statt. Sie behandeln die sich besonders schnell entwickelnden optoelektronischen Einrichtungen und ihre Einsatzmöglichkeiten vor allem bei ingenieurwissenschaftlichen und medizinischen Problemen.
Im vorliegenden Band werden die 120 Vorträge zusammengefaßt, die auf dem 6. Kongreß dieser Art im Sommer 1983 zum Themenbereich „Optoelektronik in der Technik" gehalten worden sind. An ihnen waren rund 250 Autoren aus der gesamten wissenschaftlichen Fachwelt beteiligt. Die Vorträge überdecken ein Spektrum, das in 12 Gruppen gegliedert werden mußte und von grundlegenden, Laserphysik und Laserchemie betreffenden Diskussionsbeiträgen über die schon allgemein bekannteren Anwendungsgebiete wie Meßtechnik und Materialbearbeitung bis hin zur optoelektronischen Solartechnik reicht. Der parallel erscheinende Band bringt die Vorträge des Kongresses „LASER 83" zum Themenbereich „Optoelektronik in der Medizin".

Springer-Verlag
Berlin
Heidelberg
NewYork
Tokyo

Optoelectronics in Medicine

Proceedings of the 5th International Congress Laser 81

Editor: **W. Waidelich**

1982. 150 figures (11 figures in colour). XI, 239 pages
DM 62,-. ISBN 3-540-10968-4

Contents: Milestones in Laser Medicine. – Laser in Surgery. – Laser in Urology. – Laser in Dermatology. – Laser in Gynaecology. – Laser in Otorhinopharyncology. – Photobiology and Laser Photomedicine. – Laser and Optoelectronics in Medical Diagnosis. – Laser in Dental Technique.

Important new developments in the medical application of laser technology were reported at Laser 81, the fifth in a series of international congress devoted to the dissemination of recent results in laser research. The proceedings of this congress are presented in this book.
It opens with an introductory survey by L. Goldman, the founder of laser medicine. This is followed by 26 technical papers describing the specialized use of lasers and optoelectronics in various fields, including
- the thermal action of high-power laser radiation in surgery, urology, dermatology, gynaecology, otorhinopharyncology, and dentistry
- the coherence and monochromaticity of low power laser radiation in diagnosis and therapy
- the photobiologic action of laserlight in wound healing and tumor irradiation.
Clinicians and laser technologists alike will find these articles an valuable source of information and insight into the advantages of optoelectronics in medicine.

Optoelektronik in der Technik
Optoelectronics in Engineering

Vorträge des 5. Internationalen Kongresses
Proceedings of the 5th International Congress Laser 81

Herausgeber/Editor: **W. Waidelich**

1982. 504 Abbildungen. XXII, 580 Seiten (266 Seiten in Englisch). DM 88,-. ISBN 3-540-10969-2

Im vorliegenden Band werden die 94 Vorträge zusammengefaßt, die auf dem 4. Kongreß dieser Art im Sommer 1981 zum Themenbereich **Optoelektronik in der Technik** gehalten worden sind. An ihnen waren rund 200 Autoren aus der gesamten wissenschaftlichen Fachwelt beteiligt. Die Vorträge überdecken ein Spektrum, das in 12 Gruppen gegliedert werden mußte und von grundlegenden, Laserphysik und Laserchemie betreffenden Diskussionsbeiträgen über die schon allgemein bekannteren Anwendungsgebiete wie Meßtechnik und Materialbearbeitung bis hin zur optoelektronischen Solartechnik reicht.

Springer-Verlag
Berlin
Heidelberg
New York
Tokyo